The Refinery of the Future

The Refinery of the Future

Second Edition

James G. Speight
CD&W Inc., Laramie, Wyoming,
United States

ELSEVIER
Gulf Professional Publishing
An imprint of Elsevier

Gulf Professional Publishing is an imprint of Elsevier
50 Hampshire Street, 5th Floor, Cambridge, MA 02139, United States
The Boulevard, Langford Lane, Kidlington, Oxford, OX5 1GB, United Kingdom

British Library Cataloguing-in-Publication Data
A catalogue record for this book is available from the British Library

Library of Congress Cataloging-in-Publication Data
A catalog record for this book is available from the Library of Congress

ISBN: 978-0-12-816994-0

For Information on all Gulf Professional Publishing publications visit our website at https://www.elsevier.com/books-and-journals

Publisher: Joe Hayton
Acquisitions Editor: Katie Hammon
Editorial Project Manager: Chris Hockaday
Production Project Manager: Prasanna Kalyanaraman
Cover Designer: Mark Rogers

Typeset by MPS Limited, Chennai, India

Contents

Preface

The petroleum refining industry includes integrated process operations that are engaged in refining crude petroleum into refined petroleum products, especially liquid fuels such as gasoline and diesel as well as processes that produce raw materials for the petrochemical industry.

Over the past four decades, the refining industry has experienced significant changes in oil market dynamics, resource availability, and technological advancements. Advancements made in exploration, production, and refining technologies allow utilization of resources such as heavy oil and tar sand bitumen that were considered economically and technically unsuitable in the middle decades of the past century. Along with the many challenges, it is imperative for refiners to raise their operations to new levels of performance. Merely extending today's performance incrementally will fail to meet most company's performance goals.

Petroleum refining in the 21st century may continue to be shaped by the factors such as consolidation of oil companies, dramatic changes in market demand, customization of products, and a decrease in the API gravity and sulfur content of the petroleum feedstocks. In fact, in addition to a (hopeful but unlikely) plentiful supply of petroleum, the future of the refining industry will base on the following factors such as (1) increased operating costs or investments due to stringent environmental requirements for facilities and products and (2) accelerating globalization resulting in stronger international petroleum price scenarios. The effect of these factors is likely to reduce refinery profit margins further, and petroleum companies worldwide will need to make significant changes in their operation and structure to be competitive on global basis. Development and incorporation of novel technologies to cope with the future challenges are essential to remain competitive in the business.

As global petroleum consumption increases and resources are depleted, it is expected that conventional oil production will decline within the next two to three decades, and the production of oil from residua, heavy oil, and tar sand bitumen will increase significantly. In fact, over the next decade, refineries will need to adapt to receiving heavier oils as well as a range of biofeedstocks. It is conceivable that current refineries could not handle such a diverse slate of feedstocks without experiencing shutdowns and related problems.

As feedstocks to refineries change, there must be an accompanying change in refinery technology as petroleum feedstocks are becoming highly variable. At the same time, more stringent antipollution regulations are forcing greater restrictions on fuel specifications. There are fundamental limitations on how far current processes can go in achieving proper control over feedstock behavior. This means a movement from conventional means of refining heavy feedstocks by using (typically) a cooking process to more development and use of more innovative processes that will produce the maximum yields of liquid fuels (or other desired products) fuels from the feedstock.

With the changing face of refining, the use of tailor-made technologies will provide the edge to move ahead. This is more so when the refining margins are shrinking day to day due to several emerging factors such as stringent product qualities, strict environmental regulations, changing feedstock properties, and high prices for petroleum feedstocks.

Thus the need for the development of upgrading processes continues in order to fulfill the product market demand as well as to satisfy environmental regulations. One area, in particular, the need for residuum conversion, technology has emerged as result of declining residual fuel oil market and the necessity to upgrade crude oil residua beyond the capabilities of the visbreaking, cooking, and low-severity hydrodesulfurization processes.

With the arrival of the 21st century, the refining industry has entered a significant transition period, and it is not surprising that refinery operations have evolved to include a range of *next-generation processes* as the demand for transportation fuels, and fuel oil has shown a steady growth. These processes are different from one another in terms of the method and product slates and will find employment in refineries according to their respective features. The primary goal of these processes is to convert heavy feedstocks, such as residua, to lower boiling products, and such processes are noted in this book.

This book will present to the reader the evolution of refinery processes during the last century and develop as well as the means by which refinery processes will evolve during the next three-to-five decades. Chapters will contain material relevant to (1) comparisons of current feedstocks with heavy oil and biofeedstocks, (2) evolution of refineries since the 1950s, (3) properties and refinability of heavy oil and biofeedstocks, (4) thermal processes versus hydroprocesses, and (5) evolution of products to match the environmental market.

Technological advances are on the horizon for alternate sources of transportation fuels. For example, gas-to-liquids and biomass-to-liquids are just two of the concepts currently under development. However, the state of many of these technologies coupled with the associated infrastructure required to implement them leaves traditional refining of petroleum hydrocarbons for transportation fuels as the *modus operandi* for the foreseeable future, which in this context is seen to be 50 years. The near future challenge for refiners will be how to harness new technologies to remain alive in a changing global marketplace.

It is imperative for refiners to raise their operations to new levels of performance. Merely extending current process performance incrementally will fail to meet most future performance goals. To do this, it will be necessary to reshape refining technology to be more adaptive to changing feedstocks and product demand and to explore the means by which the technology and methodology of refinery operations can be translated not only into increased profitability but also into survivability.

Furthermore, there is considerable uncertainty surrounding the future of unconventional crude oil production in the United States. Environmental regulations could either preclude unconventional production or, more likely, raise the cost significantly. If future US laws limited and/or taxed greenhouse gas emissions, these laws will lead to substantial increase in the costs of production of fuels from unconventional sources. In addition to increases in the volumes of carbon dioxide, restrictions on access to water also could prove costly, especially in the arid or semiarid Western States. In addition, environmental restrictions on land use could preclude unconventional oil production in some areas of the United States. Such policies could open to way to the increased use of biomass feedstocks as a means of supplementing production of hydrocarbon fuels.

However, the *refinery of the future* will be more technology focused than today—a trend that has already commenced. It will make only high-value products, with one of those products being power. The refinery will be a *clean refinery* with a high-value, energy-efficient installation insofar as the refinery environmentally benign and the carbon footprint of the refinery will be negligible. The refinery will also be a *smart refinery* insofar as it will highly to accept a variety of feedstocks (including biomass) with operations managed around chemical principles and quantitative chemical reaction engineering concepts.

Chapter 1, Feedstock Types and Properties, introduces the book by giving an overview of the properties of various feedstocks, including biomass. Chapter 2, Introduction to Refining Processes, and Chapter 3, Refining Chemistry, focus on refining processes and refining chemistry. Chapters 4–9 deal with distillation, thermal cracking, catalytic cracking, deasphalting and dewaxing, hydrotreating and desulfurization, and hydrocracking, respectively, including the introduction of biotechnology into the refinery (Chapter 8: Desulfurization, Denitrogenation, and Demetalization). Chapters 10–15 are new to this edition and present the potential for the accommodation of nonfossil fuel feedstocks and the means by which these feedstocks can be used to provide various types of fuels.

By understanding the evolutionary changes that have occurred to date, this book will satisfy the needs of engineers and scientists at all levels from academia to the refinery and help them understand the current refining industry and prepare for process changes and evolution of the industry.

The target audience includes engineers, scientists, and students who want an update on petroleum processing and the direction of the industry over the next 50 years. Nontechnical readers, with help from the extensive glossary, will also benefit from reading the book.

James G. Speight

CD&W Inc., Laramie, Wyoming, United States

March 2020

Feedstock types and properties

1

1.1 Introduction

The modern crude oil industry began in the later years of the 1850s with the discovery and subsequent commercialization of crude oil in Pennsylvania in 1859 (Bell, 1945; Yergin, 1991; Bower, 2009). The modern refining era can be said to have commenced in 1862 with the first appearance of crude oil distillation. The story of the discovery of the character of crude oil is somewhat circuitous but worthy of mention, in the historical sense (Burke, 1996).

At a time when the carbonation of water was being investigated, Joseph Priestley became involved in attempting to produce such liquid since it was to be used a cure for scurvy during the second expedition made by Captain Cook in 1771. Priestley decided to make a contribution to the success of the expedition and set himself to invent a drink that would cure scurvy. During his experiments at a brewery near his home in Leeds, he had discovered the properties of the carbon dioxide (he called it "fixed air") given off by the fermenting beer vats. One of these properties was that when water was placed in a flat dish for a time above the vats, it acquired a pleasant, acidulous taste that reminded Priestley of seltzer mineral waters.

Experiments convinced him that the medicinal qualities of seltzer might be due to the air dissolved in it. Pouring water from one glass to another for 3 minutes in the fixed air above a beer vat achieved the same effect. By 1772 he had devised a pumping apparatus that would impregnate water with fixed air, and the system was set up on board ships Resolution and Adventure in time for the voyage by Cook. It was a great success. Meanwhile, the politics expounded by Priestley continued to dog him. His support for the French Revolution was seen as particularly traitorous, and in 1794 a mob burned down his house and laboratory. As a result, Priestley (who escaped the wrath of the mob) took ship for Pennsylvania, where he settled in Northumberland, honored by his American hosts as a major scientific figure. Then one night, while dining at Yale, he met a young professor of chemistry. The result of their meeting would change the life of the United States in the 20th Century.

It may have been because the young man at dinner that night, Benjamin Silliman, was a hypochondriac (rather than the fact that he was a chemist) that subsequent events took the course they did. Silliman imagined that he suffered from lethargy, vertigo, nervous disorders, and whatever else he could think of. In common with other invalids, he regularly visited health spas such as Saratoga Springs, New York (at the expense of his mother), and he knew that such places were only for the rich. The meeting with Priestley apparently moved Silliman to decide to make the mineral-water cure available to the common people (also at the expense of his mother).

The Refinery of the Future. DOI: https://doi.org/10.1016/B978-0-12-816994-0.00001-4

In 1809 Silliman set up in business with an apothecary named Darling, assembled apparatus to impregnate 50 bottles of water a day and opened two soda-water fountains in New York City, one at the Tontine Coffee House and one at the City Hotel. The decor was hugely expensive (a lot of gilt), and they only sold 70 glasses on opening day. But Darling was optimistic. A friend of Priestley visited and declared that drinking the waters would prevent yellow fever but in spite of Silliman's hopes that the business would make him rich, by the end of the summer the endeavor was a disastrous flop. It would be many more decades before the soda fountain became a cultural icon in America!

Silliman cast around for some other way to make money. Two years earlier, he had analyzed the contents of a meteor that had fallen on Weston, Connecticut, and this research had enhanced his scientific reputation. So he decided to offer his services (as a geologist) to mining companies. His degree had been in law: he was as qualified for geology as he was to be Yale professor of chemistry. The geology venture prospered, and by 1820 Silliman was in great demand for field trips, on which he took his son, Benjamin, Jr. When he retired in 1853, his son took up where he had left off, as professor of General and Applied chemistry at Yale (this time, with a degree in the subject). After writing a number of chemistry books and being elected to the National Academy of Sciences, Benjamin, Jr. took up lucrative consulting posts, as his father had done, with the Boston City Water Company and various mining enterprises.

In 1855 one of these asked him to research and report on some mineral samples from the new Pennsylvania Rock Oil Company. After several-month work, Benjamin, Jr. announced that approximately 50% of the black tar-like substance could be distilled into first-rate burning oils (which would eventually be called kerosene and paraffin) and that an additional 40% of what was left could be distilled for other purposes, such as lubrication and gaslight. On the basis of this single report, a company was launched to finance the drilling of the Drake Well at Oil Creek, Pennsylvania, and in 1857 it became the first well to produce crude oil. It would be another 50 years before the reference by Silliman to other fractions available from the oil through extra distillation would provide gasoline for the combustion engine of the first automobile. The report by Silliman changed the world because it made possible an entirely new form of transportation and helped turn the United States into an industrial superpower.

After completion of the first well (by Edwin Drake), the surrounding areas were immediately leased and extensive drilling took place. Crude oil output in the United States increased from approximately 2000 barrels (1 barrel, bbl = 42 US gallons = 35 imperial gallons = 5.61 ft^3 = 158.8 L) in 1859 to nearly 3,000,000 bbl in 1863 and approximately 10,000,000 bbl in 1874. In 1861 the first cargo of oil, contained in wooden barrels, was sent across the Atlantic to London, and by the 1870s, refineries, tank cars, and pipelines had become characteristic features of the industry, mostly through the leadership of Standard Oil that was founded by John D. Rockefeller. Throughout the remainder of the 19th century, the United States and Russia were the two areas in which the most striking developments took place.

At the outbreak of World War I in 1914, the two major producers were the United States and Russia, but supplies of oil were also being obtained from Indonesia, Rumania, and Mexico. During the 1920s and 1930s, attention was also focused on other areas for oil production, such as the United States, the Middle East, and Indonesia. At this time, European and African countries were not considered major oil-producing areas. In the post-1945 era, Middle Eastern countries continued to rise in importance because of new discoveries of vast reserves. The United States, although continuing to be the biggest producer, was also the major consumer and thus was not a major exporter of oil. At this time, oil companies began to roam much farther in the search for oil, and significant discoveries in Europe, Africa, and Canada thus resulted.

Crude oil refining has grown increasingly complex in the last 20 years. Lower quality crude oil (such as heavy crude oil, extra heavy oil, and tar sand bitumen), crude oil price volatility, and environmental regulations that require cleaner manufacturing processes and higher performance products present new challenges to the refining industry. Improving processes and increasing the efficiency of energy use with technology research and development are keys to meeting the challenges and maintaining the viability of the refining industry in the United States and the production of the hydrocarbon fuels upon which the modern world is dependent.

It is the purpose of this chapter to present a general description of the types of feedstocks that are currently accepted by refineries and to illustrate the evolution of the acceptance of these feedstocks from the original conventional crude oil for which the refineries were constructed.

In the simplest definition, a refinery feedstock is the crude oil produced from a reservoir (a geological formation) by means of one or more wells drilled into the formation that is destined for processing in a refinery. By this means the crude oil is transformed into one or more components and/or finished products (Parkash, 2003; Gary et al., 2007; Speight, 2014, 2017; Hsu and Robinson, 2017).

1.2 Terminology

Even though crude oil and its derivatives have been used for millennia, it is only in the last decade or so that some attempts have been made to standardize the nomenclature and terminology. But confusion may still exist. Therefore it is the purpose of this section to provide some semblance of order into the disordered state that exists in the segment of crude oil technology that is known as terminology.

1.2.1 Conventional crude oil

The term "crude oil" and the equivalent term "petroleum" cover a wide assortment of materials consisting of mixtures of hydrocarbon derivatives and other compounds containing variable amounts of sulfur, nitrogen, and oxygen, which may vary widely in volatility, specific gravity, and viscosity. Metal-containing constituents, notably those compounds that contain vanadium and nickel, usually occur in the more viscous crude oils in amounts up to several thousand parts per million and can have serious consequences during processing of these feedstocks (Speight, 2014, 2019c). Because petroleum is a mixture of widely varying constituents and proportions, its physical properties also vary widely and the color from colorless to black. Thus the definition of crude oil has been varied, unsystematic, diverse, and often archaic. Furthermore, the terminology of crude oil is a product of many years of growth. Thus the long established use of an expression, however inadequate it may be, is altered with difficulty, and a new term, however precise, is at best adopted only slowly.

Because of the need for a thorough understanding of crude oil and the associated technologies, it is essential that the definitions and the terminology of crude oil science and technology be given prime consideration. This will aid in a better understanding of crude oil, its constituents, and its various fractions. Of the many forms of terminology that have been used not all have survived, but the more commonly used are illustrated here. Particularly troublesome, and more confusing, are those terms that are applied to the more viscous materials, for example, the use of the terms bitumen and asphalt. This part of the text attempts to alleviate much of the confusion that exists, but it must be remembered that the terminology of crude oil is still open to personal choice and historical usage.

Crude oil is a mixture of gaseous, liquid, and solid hydrocarbon compounds that occur in sedimentary rock deposits throughout the world and also contains small quantities of nitrogen-containing compounds, oxygen-containing compounds, and sulfur-containing compounds as well as trace amounts of metallic constituents (Speight, 1990, 2014, 2015).

In the unrefined state, crude oil has minimal value, but when refined it provides high-value liquid fuels, solvents, lubricants, and many other products (Parkash, 2003; Gary et al., 2007; Speight, 2014, 2017; Hsu and Robinson, 2017). The fuels derived from crude oil contribute approximately one-third to one-half of the total world energy supply and are used not only for transportation fuels (i.e., gasoline, diesel fuel, and aviation fuel, among others) but also to heat buildings. Crude oil products have a wide variety of uses that vary from gaseous and liquid fuels to near-solid machinery lubricants. In addition, the residue of many refinery processes, asphalt—a once-maligned by-product—is now a premium value product for highway surfaces, roofing materials, and miscellaneous waterproofing uses.

Crude oil is a mixture of compounds boiling at different temperatures that can be separated into a variety of different generic fractions by distillation. And the terminology of these fractions has been bound by utility and often bears little relationship to composition. In fact, the molecular boundaries of crude oil cover a wide range of boiling points and carbon numbers of hydrocarbon compounds and other compounds containing nitrogen, oxygen, and sulfur, as well as metallic (porphyrin) constituents. However, the actual boundaries of such a crude oil map can only be arbitrarily defined in terms of boiling point and carbon number (Speight, 2014). In fact, crude oil is so diverse that materials from different sources exhibit different boundary limits, and for this reason alone it is not surprising that crude oil has been difficult to map in a precise manner.

Since there is a wide variation in the properties of crude oil (Speight, 2014, 2015), the proportions in which the different constituents occur vary with origin (Speight, 2014). Thus some crude oils have higher proportions of the lower boiling components and others (such as extra heavy oil and tar sand bitumen) have higher proportions of higher boiling components (asphaltic components and residuum).

1.2.2 High-acid crudes

High-acid crude oils are crude oil that contains considerable proportions of naphthenic acid derivatives. The term "naphthenic acid," as commonly used in the petroleum industry, refers collectively to all of the organic acids present in the crude oil (Shalaby, 2005).

The name was originally derived from the early discovery of monobasic carboxylic acids in petroleum but there is a variety of organic acids to be present in crude oil. These include fatty acids as low in molecular weight as formic and acetic as well as saturated and unsaturated acids based on single and multiple five- and six-membered rings. The general chemical formula of naphthenic acids is $R(CH_2)_nCOOH$, where R is one or more cyclopentane ring and n is typically greater than 12, although the naphthenic acid fraction is now known to have complex compositional heterogeneity (Zhang et al., 2004) and range of molecular weight (Baugh et al., 2005). The amounts of the naphthenic acids present in crude oils vary from one crude to another and variations of the constituents of the naphthenic acid fraction include variations in molecular weight, boiling point, and ring structure can influence both their fraction characteristics and chemical reactivity.

1.2.2.1 Properties and character of naphthenic acids

By the original definition a naphthenic acid is a monobasic carboxyl group attached to a saturated cycloaliphatic structure. However, it has been a convention accepted in the oil industry that all organic acids in crude oil are called naphthenic acids. Naphthenic acids in crude oils are now known to be mixtures of low- to high-molecular weight acids and the naphthenic acid fraction also contains other acidic species.

Naphthenic acids can be very water-soluble to oil-soluble depending on their molecular weight, process temperatures, salinity of waters, and fluid pressures. In the water phase, naphthenic acids can cause stable reverse emulsions (oil droplets in a continuous water phase). In the oil phase with residual water, these acids have the potential to react with a host of minerals, which are capable of neutralizing the acids. The main reaction product found in practice is the calcium naphthenate soap (the calcium salt of naphthenic acids).

In addition, naphthenic acids have been found to cause the formation of soaps. The alkali metals soaps/salts, sodium, and potassium naphthenates are water-soluble and water-dispersible, giving tight emulsions and poor oil-in-water qualities.

Naphthenic acid soaps of the alkaline earth metals are insoluble in normal oil-field brines, with a pH greater than 7 at normal upstream process temperatures.

Normally, the end result of formation of low-molecular weight acidic species is treated in the overheads in refineries. A combined approach to front end treating at crude inlet to heaters and preheat exchangers should be considered. It is commonly assumed that acidity in crude oils is related to carboxylic acid species, that is, components containing a $-COOH$ functional group. While it is clear that carboxylic acid functionality is an important feature (60% of the ions have two or more oxygen atoms), a major portion (40%) of the acid types are not carboxylic acids. In fact, naphthenic acids are a mixture of different compounds which may be polycyclic and may have unsaturated bonds, aromatic rings, and hydroxyl groups. Even the carboxylic acids are more diverse than expected, with approximately 85% containing more heteroatoms than the two oxygen atoms needed to account for the carboxylic acid groups. Examining the distribution of component types in the acid fraction reveals that there is a broad distribution of species.

The total acid matrix is therefore complex and it is unlikely that a simple titration, such as the traditional methods for measurement of the total acid number (TAN), can give meaningful results to use in predictions of problems. An alternative way of defining the relative organic acid fraction of crude oils is therefore a real need in the oil industry, both upstream and downstream.

1.2.2.2 Naphthenic acid chemistry

Naphthenic acids in crude oil cause corrosion which often occurs in the same places as high-temperature sulfur attack such as heater tube outlets, transfer lines, column flash zones, and pumps (Shalaby, 2005). Furthermore, naphthenic acids alone or in combination with other organic acids such as phenols can cause corrosion at temperatures as low as 65°C (150°F) up to 420°C (790°F) (Gorbaty et al., 2001; Kittrell, 2006). Crude oils with a TAN higher than 0.5 and crude oil fractions with a TAN higher than 1.5 are considered to be potentially corrosive between the temperature of 230°C and 400°C (450°F and 750°F).

Corrosion by naphthenic acids typically has a localized pattern, particularly at areas of high velocity and, in some cases, where condensation of concentrated acid vapors can occur in crude distillation units. The attack also is described as lacking corrosion products. Damage is in the form of unexpected high corrosion rates on alloys that would normally be expected to resist sulfidic corrosion (particularly steels with more than 9% w/w chromium). In some cases, even very highly alloyed materials [i.e., 12% w/w chromium, type 316 stainless steel (SS), and type 317 SS], and in severe cases even 6% w/w Mo SS have been found to exhibit sensitivity to corrosion under these conditions.

The corrosion reaction processes involve the formation of iron naphthenates:

$$\begin{aligned}
&Fe + 2RCOOH = Fe(RCOO)_2 + H_2 \\
&Fe(RCOO)_2 + H_2S = FeS + 2RCOOH
\end{aligned} \tag{1.1}$$

The iron naphthenates are soluble in oil and the surface is relatively film free. In the presence of hydrogen sulfide, a sulfide film is formed that can offer some protection depending on the acid concentration. If the sulfur-containing compounds are reduced to hydrogen sulfide, the formation of a potentially protective layer of iron sulfide occurs on the unit walls and corrosion is reduced (Kane and Cayard, 2002; Yépez, 2005). When the reduction product is water, coming from the reduction of sulfoxides, the naphthenic acid corrosion is enhanced (Yépez, 2005).

Thermal decarboxylation can occur during the distillation process during which the temperature of the crude oil in the distillation column can be as high as 400°C (750°F):

$$RCO_2H \rightarrow RH + CO_2$$

However, not all acidic species in crude oil are derivatives of carboxylic acids (RCOOH) and some of the acidic species are resistant to high temperatures. For example, acidic species appear in the vacuum residue after having been subjected to the inlet temperatures of an atmospheric distillation tower and a vacuum distillation tower (Speight and Francisco, 1990). In addition, for the acid species that are volatile, naphthenic acids are most active at their boiling point and the most severe corrosion generally occurs on condensation from the vapor phase back to the liquid phase.

1.2.2.3 Total acid number and laboratory testing

The naphthenic acids' content in crude oils is expressed as the TAN, which is measured in units of milligrams of potassium hydroxide required to neutralize a gram of oil.

Regardless of the source, the acids present in the oil cause much corrosion in the refinery equipment. The most common current measures of the corrosive potential of a crude oil are the neutralization number or TAN. These are total acidity measurements determined by base titration. Commercial experience reveals that while such tests may be sufficient for providing an indication of whether any given crude may be corrosive, the tests are poor quantitative indicators of the severity of corrosion.

Current methods for the determination of the acid content of hydrocarbon compositions are well established (ASTM D664), which includes potentiometric titration in nonaqueous conditions to clearly defined end points as detected by changes in millivolt readings versus volume of titrant used. A color indicator method (ASTM D974) is also available.

Potentiometric titration:

> In this method (ASTM D664) the sample is normally dissolved in toluene and propanol with a little water and titrated with alcoholic potassium hydroxide (if sample is acidic). A glass electrode and reference electrode are immersed in the sample and connected to a voltmeter/potentiometer. The meter reading (in millivolts) is plotted against the volume of

titrant. The end point is taken at the distinct inflection of the resulting titration curve corresponding to the basic buffer solution.

Color indicating titration:

In this test method (ASTM D974) an appropriate pH color indicator (such as phenolphthalein) is used. The titrant is added to the sample by means of a burette and the volume of titrant used to cause a permanent color change in the sample is recorded from which the TAN is calculated. It can be difficult to observe color changes in crude oil solutions. It is also possible that the results from the color indicator method may or may not be the same as the potentiometric results.
Test method ASTM D1534 is similar to ASTM D974 in that they both use a color change to indicate the end point. ASTM D1534 is designed for electric insulating oils (transformer oils), where the viscosity will not exceed 24 cSt at 40°C. The standard range of applications is for oils with an acid number between 0.05 and 0.50 mg KOH/g, which is applicable to the transformer oils.
Test method ASTM D3339 is also similar to ASTM D974 but is designed for use on smaller oil samples. ASTM D974 and ASTM D664 roughly use a 20 g sample; ASTM D3339 uses a 2.0 g sample.

In terms of repeatability (which is the difference between successive test results obtained by the same operator with the same apparatus under constant operating conditions on identical test material), data acquired using ASTM D664 were found to be within ±7% of the mean 95% of the time for fresh oils using the inflection point method or ±12% of the mean for used oils with the buffer end point method. On the other hand, when using ASTM D974, a sample that having an acid number of 0.15 could vary from 0.10 to 0.20 and, when using ASTM D664, the acid number could vary from 0.17 to 0.13.

In terms of reproducibility (the difference between two single independent results obtained by different operators working in different laboratories on identical test material), 95% of the time, the reproducibility of ASTM D664 is ±20% of the mean for fresh oils using the inflection point method or ±44% of the mean for used oil using the buffer end point method. For example, if a mean acid number was 0.10, the results could be expected to vary from an acid number of 0.14 to an acid number of 0.06, 95% of the time. When using ASTM D974, the analyses (on the same oil) from multiple laboratories could vary from 0.09 to 0.01. Furthermore, according to ASTM, the AN obtained by this standard (ASTM D664) may or may not be numerically the same as that obtained in accordance with test methods ASTM D974 and ASTM D3339.

In addition, the TAN values as conventionally analyzed in accordance with standard test methods (ASTM D664) do not correlate at all with their risk of forming naphthenates or other soaps during production in oilfields. The TAN of oil has frequently been used to quantify the presence of naphthenic acids, because the carboxylic acid components of oils are believed to be largely responsible for oil acidity. However, more recent research has begun to highlight deficiencies in relying upon this method for such a direct correlation and the TAN is no longer considered to be such a reliable indicator.

Furthermore, the ASTM D974 test method is an older method and used for distillates, while the ASTM D664 test method is more accurate but measures acid gases and hydrolyzable salts in addition to organic acids. These differences are important on crude oils but less significant on distillates, and the Nalco NAT testing is more precise for quantifying the naphthenic acid content (Haynes, 2006).

Inorganic acids, esters, phenolic compounds, sulfur compounds, lactones, resins, salts, and additives such as inhibitors and detergents interfere with both the methods. In addition, these ASTM methods do not differentiate between naphthenic acids, phenols, carbon dioxide, hydrogen sulfide, mercaptans, and other acidic compounds present in the oil.

1.2.3 Opportunity crudes

Opportunity crudes (also called challenging crude) are generally characterized by a variety of properties undesirable to a refiner, such as high TAN, high sulfur, nitrogen, and the content of aromatic derivatives, as well as high viscosity.

The main characteristics typical of opportunity crudes include (1) high TAN, TAN > 1.0 mg KOH/g sample, (2) high sulfur content, >1% w/w, and (3) low American Petroleum Institute (API) gravity or specific gravity, <26 degrees API or >0.9 g/mL. In addition to these typical properties, they may also present processing challenges due to high levels of water, salt, metals, solids, asphaltene incompatibility, high pour point or high conductivity. Depending on their characteristics, these crudes impact multiple units in the refinery.

Opportunity crudes are those that are sold at a discount (relative to benchmark crude oil). The discount is most often due to the increased processing costs or risks associated with the crude. Because of the economic advantages, many refiners are looking increasingly at processing higher levels of opportunity crude oils in their crude slates. There are many potential problems associated with processing opportunity crudes and to properly manage the potential risk of bringing in a new crude the refiner should consider the impact that crude will have on reliability and operations due to (1) corrosion from naphthenic acid derivatives in the high-temperature regions, (2) desalter upsets, (3) increased fouling, (4) diesel cetane reduction, (5) corrosion in the overhead regions from increased low-molecular weight organic acids, and (6) product stability problems.

Although these crudes can cause corrosion and fouling problems in a downstream facility, they offer attractive discounts in crude prices. Effective methods and technologies to upgrade and process opportunity crudes can resolve these problems and provide attractive margins to the refiner. Defining crude characteristics and understanding how unit operations might be impacted will help the refiner explore various control strategies and profit on the opportunity of processing these challenging crudes.

Thus processing opportunity crude can play an important role in refinery profitability, but the risks are high because these crude oils usually come laden with contaminants that cause high maintenance costs and equipment losses due to excessive corrosion. In particular, overhead corrosion is caused by the mineral salts, magnesium, calcium, and sodium chloride that are hydrolyzed to produce volatile hydrochloric acid, causing a highly corrosive condition in the overhead exchangers. Therefore these salts present a significant contamination in opportunity crude oils. Other contaminants in opportunity crude oils which are shown to accelerate the hydrolysis reactions are inorganic clays and organic acids.

As refiners target improved operating margins, one of the more likely methods to accomplish this is through the processing of opportunity crudes, sometimes called challenging crudes. Opportunity crudes may command a discounted price due to known processing issues, or they may be new crudes with unknown or poorly understood properties and processing challenges. The economics of processing opportunity crudes is often so attractive that more and more refineries are updating their strategy for purchasing these difficult crudes. The experience gained from treating over 50 high-acid crude units over the last 20 years is used to manage the risks of processing new opportunity crudes. When developing a strategy for processing opportunity crudes, it is necessary to consider the total impact on the refinery—whether the effect is positive or negative. This paper shows an effective way to analyze opportunity crudes for potential negative impacts on the process. Risk-managing techniques for corrosivity studies, desalter emulsion stability, fouling prediction, and stability issues are reported. Listed are laboratory and field evaluations utilizing on-line monitoring systems, corrosion probes, and corrosion coupons. Proper monitoring strategy is critical to successfully managing the risk of processing opportunity crudes. Finally, the use of high-temperature corrosion inhibitors was successfully evaluated as a means to mitigate naphthenic acid corrosion.

Some of these problems can be predicted, but others are too dependent on unit operations to predict but can be controlled. The solution to managing the risk against these problems is to perform a thorough risk assessment on the unit expected to run the opportunity crude. For a refiner a successful strategy for processing opportunity crudes involves identifying their processing issues and assessing the inherent risks associated with running a particular crude or crude blend. With this knowledge the refiner can anticipate the possible processing problems and implement cost-effective mitigation measures.

1.2.4 Foamy oil

Foamy oil is oil-continuous foam that contains dispersed gas bubbles produced at the well head from heavy oil reservoirs under solution-gas drive. The nature of the gas dispersions in oil distinguishes foamy oil behavior from conventional heavy oil. The gas that comes out of solution in the reservoir does not coalesce into large gas bubbles nor into a continuous flowing gas phase. Instead it remains as small bubbles entrained in the crude oil, keeping the effective oil viscosity low while providing expansive energy that helps drive the oil toward the producing. Foamy oil accounts for unusually high production in heavy oil reservoirs under solution-gas drive (Sun et al., 2013).

Foamy oil behavior is a unique phenomenon associated with production of heavy oils. It is believed that this mechanism contributes significantly to the abnormally high production rate of heavy oils observed in the Orinoco Belt. During the production of heavy oil from solution-gas drive reservoirs, the oil is pushed into the production wells by energy supplied by the dissolved gas. As fluid is withdrawn from the production wells, the pressure in the reservoir declines and the gas that was dissolved in the oil at high pressure starts to come out of solution (hence, foamy oil). As pressure declines further with continued removal of fluids from the production wells, more gas is released from solution and the gas already released expands in volume. The expanding gas, which at this point is in the form of isolated bubbles, pushes the oil out of the pores and provides energy for the flow of oil into the production well. This process is very efficient until the isolated gas bubbles link up and the gas itself starts flowing into the production well. Once the gas flow starts, the oil has to compete with the gas for available flow energy. Thus in some heavy oil reservoirs, due to the properties of the oil and the sand and also due to the production methods, the released gas forms foam with the oil and remains subdivided in the form of dispersed bubbles much longer.

1.2.5 Tight oil

Tight oil is a low-viscosity crude oil (sometimes erroneously referred to as shale oil —by way of definition, shale oil is the liquid product produced by the decomposition of the kerogen component of oil shale) that is confined in the pore spaces of these impermeable shale formations. On the other hand, oil shale is a kerogen-rich petroleum source rock that was not buried under the correct maturation conditions to experience the temperatures required to generate oil and gas (Speight, 2014). Economic production from tight oil formations requires the same hydraulic fracturing and often uses the same horizontal well technology used in the production of tight gas.

Tight formations (such as shale formations) are heterogeneous and vary widely over relatively short distances. Tight oil reservoirs subjected to fracking can be divided into four different groups: (1) Type I has little matrix porosity and permeability—leading to fractures dominating both storage capacity and fluid-flow pathways, (2) Type II has low matrix porosity and permeability, but here the matrix provides storage capacity, while fractures provide fluid-flow paths, (3) Type III are microporous reservoirs with high matrix porosity but low matrix permeability, thus giving induced fractures dominance in fluid-flow paths, and (4) Type IV is macroporous reservoirs with high matric porosity and permeability; thus the matrix provides both storage capacity and flow paths, while fractures only enhance permeability.

Even in a single horizontal drill hole, the amount recovered may vary, as may recovery within a field or even between adjacent wells. This makes evaluation of plays and decisions regarding the profitability of wells on a particular lease difficult. Production of oil from tight formations requires a gas cap representing at least 15%–20% natural gas in the reservoir pore space to drive the oil toward the borehole; tight reservoirs that contain only oil cannot be economically produced but such reserves may be limited (Wachtmeister et al., 2017).

1.2.6 Heavy crude oil

There are also other types of crude oil that are different from the conventional crude oil insofar as they are much more difficult to recover from the subsurface reservoir. These materials have a much higher viscosity (and lower API gravity) than conventional crude oil, and primary recovery of these crude oil types usually requires thermal stimulation of the reservoir. When crude oil occurs in a reservoir that allows the crude material to be recovered by pumping operations as a free-flowing dark- to light-colored liquid, it is often referred to as conventional petroleum. Heavy (high-density) crude oil is more difficult to recover from the subsurface reservoir than conventional (low-density) crude oil. The definition of heavy crude oil is usually based on the API gravity or viscosity, and the definition is quite arbitrary, although there have been attempts to rationalize the definition based upon viscosity, API gravity, and density (Fig. 1.1).

For many years, petroleum and heavy crude oil were very generally defined in terms of physical properties. For example, heavy crude oil was considered to be crude oil that had gravity somewhat less than 20 degrees API with the heavy crude oils falling into the range 10–15 degrees API. For example, Cold Lake heavy crude oil has an API gravity that is equal to 12 degrees and extra heavy oils, such as tar sand bitumen, usually have an API gravity in the range of 5–10 degrees (Athabasca bitumen = 8 degrees API). Residua would vary depending upon the

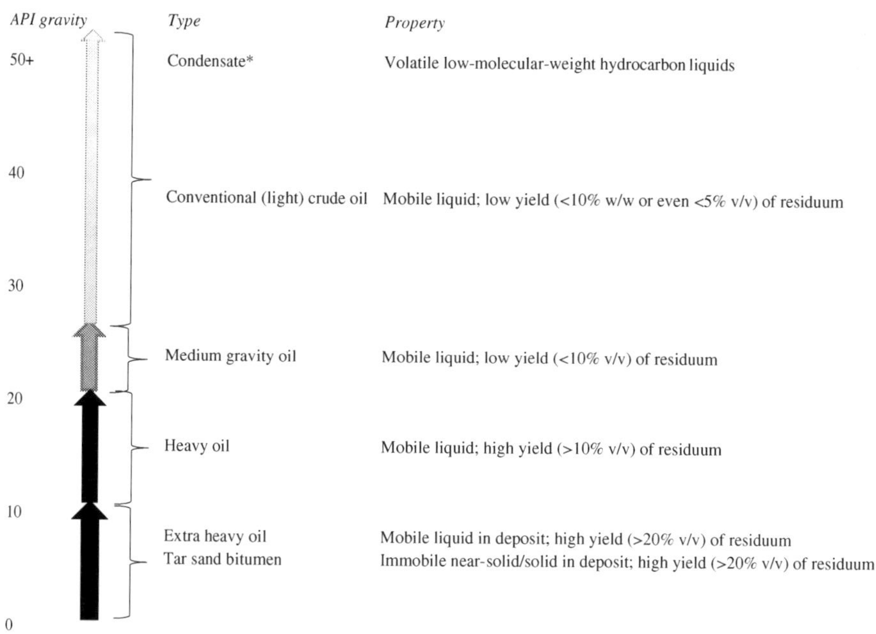

Figure 1.1 General description of various feedstocks by API gravity. *Included for comparison. *API*, American Petroleum Institute.

temperature at which distillation was terminated but usually vacuum residua are in the range of 2–8 degrees API (Speight, 2000, 2014, 2017).

Heavy crude oil has a much higher viscosity (and lower API gravity) than conventional crude oil, and primary recovery of these crude oil types usually requires thermal stimulation of the reservoir. The generic term "heavy crude oil" is often applied to a crude oil that has less than 20 degrees API and usually, but not always, a sulfur content higher than 2% w/w (Speight, 2000). Furthermore, in contrast to conventional crude oils, heavy crude oils are darker in color and may even be black. On the other hand, extra heavy oil is the material that occurs in the solid or near-solid state and is generally incapable of free flow under ambient conditions (tar sand bitumen).

Thus the generic term "heavy crude oil" is often applied inconsistently to crude oil that has an API gravity of less than 20 degrees. Other definitions classify heavy crude oil as heavy crude oil having an API gravity less than 22 degrees API, or less than 25 degrees API and usually, but not always, a sulfur content higher than 2% w/w (Ancheyta and Speight, 2007). Furthermore, in contrast to conventional crude oils, heavy crude oils are darker in color and may even be black. The term "heavy crude oil" has also been arbitrarily used to describe both the heavy crude oils that require thermal stimulation of recovery from the reservoir and the bitumen in bituminous sand (tar sand) formations from which the heavy bituminous material is recovered by a mining operation.

1.2.7 Extra heavy oil

Extra heavy oil is a nondescript term (related to viscosity) of little scientific meaning that is usually applied to tar sand bitumen–like material. The general difference is that extra heavy oil, which may have properties similar to tar sand bitumen in the laboratory but, unlike immobile tar sand bitumen in the deposit, has some degree of mobility in the reservoir or deposit (Tables 1.1 and 1.2) (Delbianco and Montanari, 2009; Speight, 2014). An example is the extra heavy oil of the Zaca-Sisquoc extra heavy oil (sometimes referred to as the Zaca-Sisquoc bitumen) that has an API gravity on the order of 4.0–6.0 degrees. The reservoir has average depth of 3500 ft, average thickness of 1700 ft, average temperature of 51°C–71°C (125°F–160°F), and sulfur in the range of 6.8%–8% w/w (Isaacs, 1992; Villarroel and Hernández, 2013). The deposit temperature is certainly equal to or above the pour point of the oil (Isaacs, 1992). This renders the oil capable of being pumped as a liquid from the deposit because of the high deposit temperature (which is higher than the pour point of the oil). The same rationale applied to the extra heavy oil found in the Orinoco deposits.

Thus extra heavy oil is a material that occurs in the solid or near-solid state and generally has mobility under reservoir conditions. While this type of oil resembles tar sand bitumen and does not flow easily, extra heavy oil is generally recognized as having mobility in the reservoir compared to tar sand bitumen, which is typically incapable of mobility (free flow) under reservoir conditions. It is likely that the mobility of extra heavy oil is due to a high reservoir temperature (that is higher

Table 1.1 Comparison of selected properties of Athabasca tar sand bitumen (Alberta, Canada) and Zuata extra heavy oil (Orinoco, Venezuela).

		Athabasca bitumen	Zuata extra heavy oil
Whole oil	API gravity	8	8
	Sulfur (% w/w)	4.8	4.2
Resid ($650^\circ F_+$)	% v/v	85	86
	Sulfur (% w/w)	5.4	4.6
	Ni + V (ppm)	420	600
	CCR (% w/w[a])	14	15

API, American Petroleum Institute.
[a]*CCR*, Conradson carbon residue.

Table 1.2 Simplified use of pour point and reservoir/deposit temperature to differentiate between heavy oil, extra heavy oil, and tar sand bitumen.

Oil	Location	Temperature	Effect on oil
Heavy oil	Reservoir or deposit	Higher than oil pour point	Fluid and/or mobile
	Surface/ambient	Higher than oil pour point	Fluid and/or mobile
Extra heavy oil	Reservoir or deposit	Higher than oil pour point	Fluid and/or mobile
	Surface/ambient	Lower than oil pour point	Solid to near-solid Immobile
Tar sand bitumen	Reservoir or deposit	Lower than oil pour point	Solid to near-solid Immobile
	Surface/ambient	Lower than oil pour point	Solid to near-solid Immobile

than the pour point of the extra heavy oil) or due to other factors are variable and subject to localized conditions in the reservoir.

1.2.8 Tar sand bitumen

The term "bitumen" (also, on occasion, referred to as native asphalt, and extra heavy crude oil) includes a wide variety of reddish brown to black materials of semisolid, viscous to brittle character that can exist in nature with no mineral impurity or with mineral matter contents that exceed 50% w/w. Bitumen is frequently found filling pores and crevices of sandstone, limestone, or argillaceous sediments, in which case the organic and associated mineral matrix is known as rock asphalt (Abraham, 1945; Hoiberg, 1964). The expression tar sand is commonly used in the crude oil industry to describe sandstone reservoirs that are impregnated with a high-boiling, high-density viscous black crude oil that cannot be retrieved through a well by conventional production techniques (FE-76-4, above). However, the term "tar sand" is actually a misnomer; more correctly, the name tar is usually applied to the high-boiling product remaining after the destructive distillation of coal or other organic matter (Speight, 2013).

Bitumen is a naturally occurring material that is found in deposits where the permeability is low and passage of fluids through the deposit can only be achieved by prior application of fracturing techniques. Tar sand bitumen is a high-boiling material with little, if any, material boiling below 350°C (660°F) and the boiling range approximates the boiling range of an atmospheric residuum.

Tar sands have been defined in the United States (FE-76-4, United States Congress, 1976) as

> *...the several rock types that contain an extremely viscous hydrocarbon which is not recoverable in its natural state by conventional oil well production methods including currently used enhanced recovery techniques. The hydrocarbon-bearing rocks are variously known as bitumen-rocks oil, impregnated rocks, oil sands, and rock asphalt.*

The recovery of the bitumen depends to a large degree on the composition and construction of the sands. Generally, the bitumen found in tar sand deposits is an extremely viscous material that is immobile under reservoir conditions and cannot be recovered through a well by the application of secondary or enhanced recovery techniques.

The bitumen in tar sand formations requires a high degree of thermal stimulation for recovery to the extent that some thermal decomposition may have to be induced. Current recovery operations of bitumen in tar sand formations involve use of a mining technique.

It is incorrect to refer to native bituminous materials as tar or pitch. Although the word tar is descriptive of the black, viscous bituminous material, it is best to avoid its use with respect to natural materials and to restrict the meaning to the volatile or near-volatile products produced in the destructive distillation of such organic substances as coal (Speight, 2013). In the simplest sense, pitch is the distillation residue of the various types of tar.

Thus alternative names, such as bituminous sand or oil sand, are gradually finding usage, with the former name (bituminous sands) more technically correct. The term "oil sand" is also used in the same way as the term "tar sand," and these terms are used interchangeably throughout this text.

However, in order to define conventional crude oil, heavy crude oil, extra heavy oil, and tar sand bitumen, the use of a single physical parameter such as viscosity is not sufficient. Physical properties such as API gravity, elemental analysis, and composition fall short of giving an adequate definition. It is the properties of the bulk deposit and, most of all, the necessary recovery methods that form the basis of the definition of these materials. Only then is it possible to classify crude oil, heavy crude oil, extra heavy oil, and tar sand bitumen.

Table 1.3 Comparison of the properties of tar sand bitumen (Athabasca) with the properties of conventional crude oil.

Property	Bitumen (Athabasca)[a]	Crude oil
Specific gravity	1.01–1.03	0.85–0.90
API gravity	5.8–8.6	25–35
Viscosity (cp)		
38°C/100°F	750,000	<200
100°C/212°F	11,300	
Pour point (°F)	>50	ca. -20
Elemental analysis (wt.% w/w)		
Carbon	83.0	86.0
Hydrogen	10.6	13.5
Nitrogen	0.5	0.2
Oxygen	0.9	<0.5
Sulfur	4.9	<2.0
Ash	0.8	0.0
Nickel (ppm)	250	<10.0
Vanadium (ppm)	100	<10.0
Composition (wt.% w/w)		
Asphaltenes (pentane)	17.0	<10.0
Resins	34.0	<20.0
Aromatics	34.0	>30.0
Saturates	15.0	>30.0
Carbon residue (% w/w)		
Conradson	14.0	<10.0

API, American Petroleum Institute.
[a]Extra heavy oil (e.g., Zuata extra heavy oil) has a similar analysis to tar sand bitumen (Table 2.11) but some mobility in the deposit because of the relatively high temperature of the deposit.

Tar sands, also variously called oil sands or bituminous sands, are loose-to-consolidated sandstone or a porous carbonate rock, impregnated with bitumen, a high-boiling high-density asphaltic material with an extremely high viscosity that is immobile under reservoir conditions and vastly different to conventional crude oil (Speight, 1990, 1997, 2008). It is therefore worth noting here the occurrence and potential supply of these materials. On an international note the bitumen in tar sand deposits represents a potentially large supply of energy. However, many of the reserves are available only with some difficulty and that optional refinery scenarios will be necessary for conversion of these materials to liquid products because of the substantial differences in character between conventional crude oil and tar sand bitumen (Table 1.3).

Because of the diversity of available information and the continuing attempts to delineate the various world tar sand deposits, it is virtually impossible to present accurate numbers that reflect the extent of the reserves in terms of the barrel unit. Indeed, investigations into the extent of many of the world tar sand deposits are continuing at such a rate that the numbers vary from 1 year to the next.

Accordingly, the data quoted here must be recognized as approximate with the potential of being quite different at the time of publication.

1.3 Occurrence and reserves

1.3.1 Conventional crude oil

Crude oil has been used as an energy resource for more than 6000 years and it is by far the most commonly used source of energy, especially as the source of liquid fuels, and use of crude oil is projected to continue at east in current amounts—in spite of the various aspects of petropolitics that can cause disruption in supply to the oil-importing countries as well as the evolving of more stringent environmental regulations—for at least five decades (Speight, 2011, 2014, 2017; BP, 2019).

In recent years the average quality of crude oil has deteriorated and offers less of the prime hydrocarbon products that were produced several decades ago. This is reflected in a progressive decrease in API gravity (i.e., increase in density) and a rise in sulfur content (Speight, 2014, 2015). However, it is now believed that there has been a recent tendency for the quality of crude oil feedstocks to stabilize. Be that as it may, refineries have had to adapt to the changing nature of crude oil feedstocks to produce the needed liquid fuels. Furthermore, the declining reserves of conventional crude oil (crude oil from tight low permeability formations notwithstanding) have resulted in an increasing need to develop options to upgrade the abundant supply of known reserves of heavy crude oil. In addition, there is considerable focus and renewed efforts on adapting recovery techniques to the production of extra heavy oil and tar sand bitumen (Speight, 2008, 2014, 2016).

Fossil fuels are those fuels, namely, natural gas, crude oil and heavy crude oil, extra heavy oil, tar sand bitumen, coal and oil shale that have been produced by the decay of plant remains over geological time (Speight, 1990, 2008, 2014, 2019b). They are carbon-based and represent a vast source of energy. Resources such as extra heavy oil and tar sand bitumen in tar sand formations are also discussed in this text. These represent an unrealized potential, with liquid fuels from crude oil being only a fraction of those that could ultimately be produced from extra heavy oil and tar sand bitumen.

Crude oil is scattered throughout the Earth's crust, which is divided into natural groups or strata, categorized in order of their antiquity (Table 1.4). These divisions are recognized by the distinctive systems of organic debris (as well as fossils, minerals, and other characteristics) that form a chronological time chart that indicates the relative ages of the Earth's strata. It is generally acknowledged that carbonaceous materials such as crude oil occur in all these geological strata from the Precambrian to the recent, and the origin of crude oil within these formations is a question that remains open to conjecture and the basis for further investigation.

Crude oil occurs underground, at various pressures depending on the depth. Because of the pressure, it contains considerable natural gas in solution. Crude oil underground is much more fluid than it is on the surface and is generally mobile

Table 1.4 The geologic timescale.

Era	Period	Epoch	Approximate duration (millions of years)	Approximate number of years ago (millions of years)
Cenozoic	Quaternary	Holocene	10,000 Years ago to the present	
		Pleistocene	2	0.01
	Tertiary	Pliocene	11	2
		Miocene	12	13
		Oligocene	11	25
		Eocene	22	36
		Paleocene	71	58
Mesozoic	Cretaceous		71	65
	Jurassic		54	136
	Triassic		35	190
Paleozoic	Permian		55	225
	Carboniferous		65	280
	Devonian		60	345
	Silurian		20	405
	Ordovician		75	425
	Cambrian		100	500
Precambrian			3380	600

under reservoir conditions because the elevated temperatures (the geothermal gradient) in subterranean formations decrease the viscosity. Although the geothermal gradient varies from place to place, it is generally on the order of 25°C−30°C/km (15°F/1000 ft or 120°C/1000 ft, i.e., 0.015°C/ft of depth or 0.012°C/ft of depth).

Crude oil is derived from aquatic plants and animals that lived and died hundreds of millions of years ago. Their remains mixed with mud and sand in layered deposits that, over the millennia, were geologically transformed into sedimentary rock. Gradually the organic matter decomposed and eventually formed crude oil (or a related precursor), which migrated from the original source beds to more porous and permeable rocks, such as sandstone and siltstone, where it finally became entrapped. Such entrapped accumulations of crude oil are called reservoirs. A series of reservoirs within a common rock structure or a series of reservoirs in separate but neighboring formations is commonly referred to as an oil field. A group of fields is often found in a single geologic environment known as a sedimentary basin or province.

The major components of crude oil are hydrocarbon derivatives, compounds of hydrogen and carbon that display great variation in their molecular structure. The simplest hydrocarbon derivatives are the paraffin derivatives. This broad series extends from methane, which forms natural gas, through liquids that are refined into gasoline, to crystalline waxes. A series of cyclic hydrocarbon derivatives (also called naphthene derivatives) ranges from volatile liquids such as naphtha to high-molecular weight substances isolated as the asphaltene fraction. Another group of cyclic hydrocarbon derivatives is known as the aromatic derivatives; the chief compound in this series is benzene, a popular raw material for making petrochemicals (Speight, 2014, 2019a).

Nonhydrocarbon constituents of crude oil include organic derivatives of nitrogen, oxygen, sulfur, and the metals nickel and vanadium. Most of these impurities are removed during refining.

Geologic techniques can determine only the existence of rock formations that are favorable for oil deposits, not whether oil is actually there. Drilling is the only sure way to ascertain the presence of oil. With modern rotary equipment, wells can be drilled to depths of more than 30,000 ft. Once oil is found, it may be recovered (brought to the surface) by the pressure created by natural gas or water within the reservoir. Crude oil can also be brought to the surface by injecting water or steam into the reservoir to raise the pressure artificially, or by injecting such substances as carbon dioxide, polymers, and solvents to reduce crude oil viscosity. Thermal recovery methods are frequently used to enhance the production of heavy crude oil when the extraction of the oil from the reservoir is impeded by the resistance to flow at the temperature of the reservoir.

Crude oil is transported to refineries by pipelines, which can often carry more than 500,000 barrels per day, or by ocean-going tankers. The basic refinery process is distillation that separates the crude oil into fractions of differing volatility. After the distillation, other physical methods are employed to separate the mixtures, including absorption, adsorption, solvent extraction, and crystallization. After physical separation into such constituents as low-boiling and high-boiling naphtha, kerosene, and low-boiling and high-boiling gas oils, selected crude oil fractions may be subjected to conversion processes, such as thermal cracking (i.e., coking; Chapter 5: Thermal Cracking) and catalytic cracking (Chapter 6: Catalytic Cracking). In the most general terms, cracking breaks the large molecules of heavier gas oils into the smaller molecules that form the low-boiling more valuable naphtha fractions.

Reforming (Chapter 2) changes the structure of straight-chain paraffin molecules into branched-chain isoparaffin derivatives and cyclic aromatic derivatives. The process is widely used to raise the octane number of gasoline obtained by distillation of paraffin-base crude oil (Speight, 2014).

In addition, and in keeping with the preferential use of conventional (low-density) crude oil as well as maturation effect in the reservoir, crude oil available currently to the refinery is somewhat different in composition and properties to those available approximately 50 years ago (Parkash, 2003; Gary et al., 2007, Speight, 2014, 2017; Hsu and Robinson, 2017). The current crude oils are somewhat more viscous insofar as they have higher proportions of nonvolatile (asphaltic) constituents. In fact, by the standards of yesteryear, many of the crude oils currently in use would have been classified as viscous high-density feedstocks, bearing in mind that they may not approach the definitions currently used for heavy crude oil. Changes in feedstock character, such as this tendency to heavier materials, require adjustments to refinery operations to handle these heavier crude oils to reduce the amount of coke formed during processing and to balance the overall product slate.

1.3.2 Heavy crude oil

The Western Hemisphere has approximately 69% v/v of the world's technically recoverable heavy crude oil and 82% of the technically recoverable natural bitumen (Meyer and Attanasi, 2003). In contrast the Eastern Hemisphere has approximately 85% v/v of the conventional crude oil reserves in the world.

1.3.3 Extra heavy oil

1.3.3.1 Tar sand bitumen

The bitumen in tar sand deposits is estimated to be at least two trillion barrels (2×10^{12} bbl) in the Canadian Athabasca tar sand deposits and two trillion barrels (2×10^{12} bbl) in the Venezuelan Orinoco tar sand deposits, compared to approximately one-to-two billion barrels ($1-2 \times 10^9$ bbl) of conventional oil worldwide, most of it in Saudi Arabia and other Middle Eastern countries (BP, 2019). Eighty-one percent of the global known recoverable tar sand bitumen is in the Alberta, Canada.

In spite of the high estimations of the reserves of bitumen, the two conditions of vital concern for the economic development of tar sand deposits are the concentration of the resource, or the percent bitumen saturation, and its accessibility, usually measured by the overburden thickness. Recovery methods are based either on mining combined with some further processing or operation on the oil sands in situ. The mining methods are applicable to shallow deposits, characterized by an overburden ratio (i.e., overburden depth to thickness of tar sand deposit). For example, indications are that for the Athabasca deposit, no more than 15% w/w of the in-place deposit is available within current concepts of the economics and technology of open-pit mining; this 10% w/w portion may be considered as the proven reserves of bitumen in the deposit.

1.4 Ultimate (elemental) composition

With few exceptions the proportions of the elements (carbon, hydrogen, nitrogen, oxygen, sulfur, and metals) in crude oil (whatever and wherever the source) vary over fairly narrow limits:

carbon, 83.0%–87.0% w/w
hydrogen, 10.0%–14.0% w/w
nitrogen, 0.1%–2.0% w/w
oxygen, 0.05%–1.5% w/w
sulfur, 0.05%–6.0% w/w
metals (Ni and V), <1000 ppm

The narrow range of variation is quite surprising when the variation of the precursors is considered (Speight, 2014) and even more surprising when one considers the wide variation in physical properties from the low-density more mobile crude oils at one extreme to the heavier asphaltic crude oils at the other extreme (Charbonnier et al., 1969; Draper et al., 1977). In addition, when the many localized or regional variations in maturation conditions are assessed, it is perhaps surprising that the ultimate compositions are so similar. Perhaps this observation, more than any other observation, is indicative of the similarity in nature of the precursors from one site to another.

The elemental analysis of tar sand bitumen has also been widely reported (Meyer and Steele, 1981; Speight, 2014), but the data suffer from the disadvantage that identification of the source is too general and is often not specific to any site within the tar sand area. In addition, the analysis is quoted for separated bitumen, which may have been obtained by any one of several procedures and may therefore not be representative of the total bitumen on the sand (Speight, 2014, 2015).

Like conventional crude oil, of the data that are available, the elemental composition of oil sand bitumen is generally constant and, like the data for crude oil, falls into a narrow range (Speight, 1990):

carbon: 83.4% ± 0.5% w/w
hydrogen: 10.4% ± 0.2% w/w
nitrogen: 0.4% ± 0.2% w/w
oxygen: 1.0% ± 0.2% w/w
sulfur: 5.0% ± 0.5% w/w
metals (Ni and V): >1000 ppm

The major exception to these narrow limits is the oxygen content of bitumen, which can vary from as little as 0.2% w/w to as high as 4.5% w/w. This is not surprising, since when oxygen is estimated by difference the analysis is subject to the accumulation of all of the errors in the other elemental data. In addition, bitumen is susceptible to aerial oxygen and the oxygen content is very dependent upon the sample history. Furthermore, the ultimate composition of the Alberta bitumen does not appear to be influenced by the proportion of bitumen in the oil sand or by the particle size of the tar sand minerals.

Bitumen from US tar sand has an ultimate composition similar to that of the Athabasca bitumen (Speight, 1990). As already noted previously, when the many localized or regional variations in maturation conditions are assessed, it is perhaps surprising that the ultimate compositions are so similar.

1.5 Chemical composition

1.5.1 Crude oil

Crude oil contains an extreme range of organic functionality and molecular size. In fact the variety is so great that it is unlikely that a complete compound-by-compound description for even a single crude oil would not be possible. As already noted, the composition of crude oil can vary with the location and age of the field in addition to any variations that occur with the depth of the individual well. Two adjacent wells are more than likely to produce crude oil with very different characteristics.

The molecular composition of crude oil can be described in terms of three classes of compounds: saturates, aromatic derivatives, and compounds bearing heteroatoms (sulfur, oxygen, or nitrogen). Within each class, there are several families of related compounds, and these are (1) saturated constituents include normal (straight-chain) alkane derivatives, branched alkane derivatives, and cycloalkane derivatives (paraffin derivatives, isoparaffin derivatives, and naphthene derivatives); (2) alkene constituents (olefin derivatives) are rare to the extent of being considered an oddity; (3) mono-aromatic constituents range from benzene to multiple fused ring analogs such as naphthalene derivatives and phenanthrene derivatives as well as higher ring systems; (4) thiol constituents (mercaptan derivatives) which contain sulfur as do thioethers and thiophenes forms; and (5) nitrogen-containing and oxygen-containing constituents are more likely to be found in polar forms such as pyridine derivatives, pyrrole derivatives, phenol derivatives, carboxylic acid derivatives, amide derivatives, as well as nonpolar forms such as ether derivatives. The distribution and characteristics of these molecular species account for the rich variety of crude oils.

The hydrocarbon content of crude oil may be as high as 97% w/w (e.g., in low-density paraffinic crude oil) or as low as 50% w/w or less as illustrated by the high-boiling high-density asphaltic crude oils. Nevertheless, crude oils with as little as 50% w/w hydrocarbon components are still assumed to retain most of the essential characteristics of the hydrocarbon derivatives. It is, nevertheless, the nonhydrocarbon (sulfur, oxygen, nitrogen, and metal) constituents that play a large part in determining the processability of the crude oil (Rossini et al., 1953). But there is more to the composition of crude oil than the hydrocarbon content. The inclusion of organic compounds of sulfur, nitrogen, and oxygen serves only to present crude oils as even more complex mixtures, and the appearance of appreciable amounts of these nonhydrocarbon compounds causes some concern in the refining of crude oils. Even though the concentration of nonhydrocarbon constituents (i.e., those organic compounds containing one or more sulfur, oxygen, or nitrogen atoms) in certain fractions may be quite small, they tend to concentrate in the higher boiling fractions of crude oil. Indeed, their influence on the processability of the crude oil is important irrespective of their molecular size and the fraction in which they occur.

The presence of traces of nonhydrocarbon compounds can impart objectionable characteristics to finished products, leading to discoloration and/or lack of stability during storage. On the other hand, catalyst poisoning and corrosion are the most noticeable effects during refining sequences when these compounds are present. It is therefore not surprising that considerable attention must be given to the nonhydrocarbon constituents of crude oil as the trend in the refining industry, of late, has been to process more heavy crude oil as well as residua that contain substantial proportions of these nonhydrocarbon materials.

1.6 Fractional composition

1.6.1 Distillation

Distillation is a means of separating chemical compounds (usually liquids) through differences in their respective vapor pressures. In the mixture the components evaporate, such that the vapor has a composition determined by the chemical properties of the mixture. Distillation of a given component is possible, if the vapor has a higher proportion of the given component than the mixture. This is caused by the given component having a higher vapor pressure and a lower boiling point than the other components.

By the nature of the process, it is theoretically impossible to completely separate and purify the individual components of crude oil when the possible number of isomers is considered for the individual carbon numbers that occur within the paraffin family. When other types of compounds are included, such as the aromatic derivatives and heteroatom derivatives, even though the maturation process might limit the possible number of isomeric permutations (Tissot and Welte, 1978), the potential number of compounds in crude oil is still (in a sense) astronomical.

However, crude oil can be separated into a variety of fractions on the basis of the boiling points of the crude oil constituents. Such fractions are primarily identified by their respective boiling ranges and, to a lesser extent, by chemical composition. However, it is often obvious that as the boiling ranges increase, the nature of the constituents remains closely similar and it is the number of the substituents that caused the increase in boiling point.

1.6.1.1 Gases and naphtha

Methane is the main hydrocarbon component of crude oil gases with lesser amounts of ethane, propane, butane, isobutane, and some C_4^+ hydrocarbon derivatives. Other gases, such as hydrogen, carbon dioxide, hydrogen sulfide, and carbonyl sulfide, are also present.

Saturated constituents with lesser amounts of mono-aromatic and di-aromatic derivatives dominate the naphtha fraction. While naphtha covers the boiling range of gasoline, most of the raw crude oil naphtha molecules have low octane number. However, most raw naphtha is processed further and combined with other process naphtha and additives to formulate commercial gasoline.

Within the saturated constituents in crude oil gases and naphtha, every possible paraffin from methane (CH_4) hydrocarbon to *n*-decane *n*-$C_{10}H_{22}$ (normal decane) is present. Depending upon the source, one of these low-boiling paraffin derivatives may be the most abundant compound in a crude oil reaching several percent. The isoparaffin derivatives begin at C4 with isobutane as the only isomer of *n*-butane. The number of isomers grows rapidly with carbon number and there may be increased difficulty in dealing with multiple isomers during analysis.

In addition to aliphatic molecules, the saturated constituents consist of cycloalkane derivatives (naphthene derivatives) with predominantly five-carbon or six-carbon rings. Methyl derivatives of cyclopentane and cyclohexane, which are commonly found at higher levels than the parent unsubstituted structures, may be present (Tissot and Welte, 1978). Fused ring dicycloalkane derivatives such as *cis*-decahydronaphthalene (*cis*-decalin) and *trans*-decahydronaphthalene (*trans*-decalin), and hexahydro-indan are also common but bicyclic naphthene derivatives separated by a single bond, such as cyclohexyl cyclohexane, are not.

The numerous aromatic constituents in crude oil naphtha begin with benzene, but its C_1-C_3 alkylated derivatives are also present (Tissot and Welte, 1978). Each of the alkyl benzene homologs through the 20 isomeric C4 alkylbenzenes has been isolated from crude oil along with various C_5-derivatives. Benzene derivatives having fused cycloparaffin rings (naphthene–aromatic derivatives) such as indane and tetralin have been isolated along with a number of their methyl derivatives. Naphthalene is included in this fraction, while the 1-methyl and 2-methyl naphthalene derivatives and higher homologs of fused two-ring aromatic derivatives appear in the mid-distillate fraction.

Sulfur-containing compounds are the only heteroatom compounds to be found in this fraction. Usually, the total amount of the sulfur in this fraction accounts for less than 1% w/w of the total sulfur in the crude oil. In naphtha from high-sulfur (sour) crude oil, 50%–70% w/w of the sulfur may be in the form of mercaptans (thiols). Over 40 individual thiols have been identified, including all the isomeric C_1-C_6 compounds plus some C_7 and C_8 isomers plus thiophenol. In naphtha from low-sulfur (sweet) crude oil, the sulfur is distributed between sulfides (thioethers) and thiophenes. In these cases the sulfides may be in the form of both linear (alkyl sulfides) and five-ring or six-ring cyclic (thiacyclane) structures. The sulfur structure distribution tends to follow the distribution hydrocarbon derivatives, that is, naphthenic oils with a high cycloalkane content tend to have a high thiacyclane content. Typical alkyl thiophene derivatives in naphtha have multiple short-side chains or exist as naphthene–thiophene derivatives. Methyl and ethyl disulfides have been confirmed to be present in some crude oils in analyses that minimized their possible formation by oxidative coupling of thiols (Aksenova and Kanayanov, 1980; Freidlina and Skorova, 1980).

1.6.1.2 Middle distillates

Saturated species are the major component in the mid-distillate fraction of crude oil but aromatic derivatives, which now include simple compounds with up to three aromatic rings, and heterocyclic compounds are present and represent a larger portion of the total. Kerosene, jet fuel, and diesel fuel are all derived from raw middle distillate which can also be obtained from cracked and hydroprocessed refinery streams.

Within the saturated constituents, the concentration of *n*-paraffin derivatives decreases regularly from C_{11} to C_{20}. Two isoprenoid species (pristane = 2,6,10,14-tetramethylpentadecane and phytane = 2,6,10,14-tetramethyl hexadecane) are generally present in crude oils in sufficient concentration to be seen as irregular peaks alongside the n-C_{17} and n-C_{18} peaks in a gas chromatogram. These isoprene derivatives, believed to arise as fragments of ancient precursors, have relevance as simple biomarkers, to the genesis of crude oil. The distribution of pristane and phytane relative to their neighboring n-C_{17} and n-C_{18} peaks has been used to aid the identification of crude oils and to detect the onset of biodegradation. The ratio of pristane to phytane has also been used for the assessment of the oxidation and reduction environment in which ancient organisms were converted into crude oil.

Mono-cycloparaffin and di-cycloparaffin derivatives with five or six carbons per ring constitute the bulk of the naphthene derivatives in the middle distillate boiling range, decreasing in concentration as the carbon number increases (Tissot and Welte, 1978) and the alkylated naphthene derivatives may have a single long–side chain as well as one or more methyl or ethyl groups.

The most abundant aromatic derivatives in the mid-distillate boiling fractions are di-methyl and tri-methyl naphthalene derivatives. Other one- and two-ring aromatic derivatives are undoubtedly present in small quantities as either naphthene or alkyl homologs in the C_{11}–C_{20} range. In addition to these homologs of alkylbenzenes, tetralin, and naphthalenes, the mid-distillate contains some fluorene derivatives and phenanthrene derivative. The phenanthrene structure appears to be favored over that of anthracene structure (Tissot and Welte, 1978) and this appears to continue through the higher boiling fractions of crude oil (Speight, 2014, 2015).

The five-membered heterocyclic constituents in the mid-distillate range are primarily the thiacyclane derivatives, benzothiophene derivatives, and dibenzothiophene derivatives with lesser amounts dialkyl-, diaryl, and aryl-alkyl sulfides (Aksenova and Kanayanov, 1980; Freidlina and Skorova, 1980). Alkylthiophenes are also present. As with the naphtha fractions, these sulfur species account for a minimal fraction of the total sulfur in the crude.

Although only trace amounts (usually ppm levels) of nitrogen are found in the middle distillate fractions, both neutral and basic nitrogen compounds have been isolated and identified in fractions boiling below 345°C (650°F) (Hirsch et al., 1974). Pyrrole and indole derivatives account for the approximately two-thirds of the nitrogen, while the remainder is found in the basic alkylated pyridine and alkylated quinoline compounds.

The saturate constituents contribute less to the vacuum gas oil (VGO) than the aromatic constituents but more than the polars that are now present at percentage rather than trace levels. VGO is occasionally used as a heating oil but most commonly it is processed by catalytic cracking to produce naphtha or extraction to yield lubricating oil.

Within the VGO, saturates, distribution of paraffin derivatives, isoparaffin derivatives, and naphthene derivatives is highly dependent upon the crude oil source. Generally, the naphthene constituents account for approximately 60% w/w of the saturate constituents but the overall range of variation is from <20% w/w to >80% w/w. In most samples the n-paraffin derivatives from C_{20} to C_{44} are still present in sufficient quantity to be detected as distinct peaks in gas chromatographic analysis. Some (but not all) crude oils show a preference for odd-numbered alkane derivatives. Both the distribution and the selectivity toward odd-numbered hydrocarbon derivatives are considered to reflect differences in the petrogenesis of the crude oil.

The bulk of the saturated constituents in VGO consist of isoparaffin derivatives and especially naphthene species, although isoprenoid compounds, such as squalane (C_{30}) and lycopane (C_{40}), have been detected. Analytical techniques show that the naphthene derivatives contain from one to more than six fused rings accompanied by alkyl substitution. For mono-aromatic and di-aromatic derivatives, the alkyl substitution typically involves several methyl and ethyl substituents. Hopanes and steranes have also been identified and are also used as internal markers for estimating biodegradation of crude oils during bioremediation processes.

The aromatic derivatives in VGO may contain one to six fused aromatic rings that may bear additional naphthene rings and alkyl substituents in keeping with their boiling range. Mono-aromatic and di-aromatic derivatives account for approximately 50% w/w of the aromatic derivatives in crude oil VGO samples. Analytical data show the presence of up to four fused naphthenic rings on some aromatic compounds. This is consistent with the suggestion that these species originate from the aromatization of steroids. Although present at lower concentration, alkyl benzenes and naphthalenes show one long−side chain and multiple short−side chains.

The fused ring aromatic compounds (having three or more rings) in crude oil include phenanthrene, chrysene, and picene as well as fluoranthene, pyrene, benzo(a)pyrene, and benzo(ghi)perylene.

The most abundant reported individual phenanthrene compounds appear to be the 3-derivatives. In addition, phenanthrene derivatives outnumber anthracene derivatives by as much as 100:1. In addition, chrysene derivatives are favored over pyrene derivative.

Heterocyclic constituents are significant contributors to the VGO fraction. In terms of sulfur compounds, thiophene and thiacyclane sulfur predominate over sulfide sulfur. Some molecules even contain more than one sulfur atom. The benzothiophenes and dibenzothiophenes are the prevalent thiophene forms of sulfur.

In the VGO range, the nitrogen-containing compounds include higher molecular weight pyridines, quinolines, benzoquinoline derivatives, amides, indoles, carbazole, and molecules with two nitrogen atoms (diaza compounds) with three and four aromatic rings are especially prevalent (Green et al., 1989). Typically, approximately one-third of the compounds are basic, that is, pyridine and its benzologs, while the remainder is present as neutral species (amides and carbazoles). Although benzoquinoline and dibenzo-quinoline derivatives found in crude oil are rich in satirically hindered structures, hindered and unhindered structures have been found to be present at equivalent concentrations in source rocks. This has been rationalized as geo-chromatography, in which the less polar (hindered) structures moved more readily to the reservoir and are less likely to adsorbed on to the rock.

Oxygen levels in the VGO parallel the nitrogen content. Thus the most commonly identified oxygen compounds are the carboxylic acids and phenols, collectively called naphthenic acids.

1.6.1.3 Vacuum residua

This fraction, the vacuum bottoms (1050°F$^+$), is the most complex of crude oil. Vacuum residua contain the majority of the heteroatoms originally in the crude oil and molecular weight of the constituents range, as near as can be determined subject to method dependence, up to several thousand. The fraction is so complex that the characterization of individual species is virtually impossible, no matter what claims have been made or will be made. Separation of vacuum residua by group type becomes difficult and confused because of the multisubstitution of aromatic and naphthenic species as well as by the presence of multiple functionalities in single molecules.

Classically, *n*-pentane or *n*-heptane precipitation is used as the initial step for the characterization of vacuum residuum (Speight, 2014; ASTM International, 2019a–2019i). The insoluble fraction, the pentane-asphaltene fraction or the heptane-asphaltene fraction, may be as much as 50% w/w of a vacuum residuum. The pentane-soluble or heptane-soluble portion (maltenes) of the residuum is then fractionated chromatographically into several solubility or adsorption classes for characterization. However, in spite of claims to the contrary, the method is not a separation by chemical type. Kit is a separation by solubility and adsorption.

The separation of the asphaltene constituents does, however, provide a simple way to remove some of the highest molecular weight and most polar components but the asphaltene fraction is so complex that compositional detail based on average parameters is of questionable value. The use of ion-exchange chromatography has offered some indications of chemical types within the complex high-molecular weight fractions (Green et al., 1989).

For the 565°C$^+$ (1050°F$^+$) fractions of crude oil, the levels of nitrogen and oxygen may begin to approach the concentration of sulfur. These elements consistently concentrate in the most polar fractions to the extent that every molecule contains more than one heteroatom. At this point, structural identification is somewhat fruitless and characterization techniques are used to confirm the presence of the functionalities found in lower boiling fractions such as acids, phenols, nonbasic (carbazole-type) nitrogen, and basic (quinoline-type) nitrogen.

Several models have been proposed based on the observed functionalities, apparent molecular weight, and elemental analysis of the fraction but whether or not these models offer insights into the nature, and behavior of the asphaltene constituents remains open to speculation and question (Speight, 2014, 2019c).

The nickel and vanadium that are concentrated into the vacuum residuum appear to occur in two forms (1) porphyrin derivatives and (2) nonporphyrin derivatives (Reynolds, 1998). Because the metalloporphyrins can provide insights into crude oil maturation processes, they have been studied extensively and several families of related structures have been identified. On the other hand, the nonporphyrin metals remain not clearly identified, although some studies suggest that some of the metals in these compounds still exist in a tetrapyrrole (porphyrin-type) environment.

It is more than likely that, in a specific residuum molecule, the heteroatoms are arranged in different functionalities, making an incredibly complex molecule. Considering how many different combinations are possible, the chances of determining every structure in a residuum are very low. Because of this seemingly insurmountable task, it may be better to determine ways of utilizing the residuum rather attempting to determine (at best questionable) molecular structures.

1.6.2 Solvent methods

The use of solvents invokes the concept of the solubility (or insolubility) of a solute in the chosen solvent. The solubility of a solute is the maximum quantity of solute that can dissolve in a certain quantity of solvent or quantity of solution at a specified temperature. The main factors that have an effect on solubility are (1) the nature of the solute and solvent, (2) the temperature, and (3) the pressure. The rate of solution is a measure of how fast a substance dissolves. Some of the factors determining the rate of solution are (1) the size of the particles, (2) whether or not the mixture is stirred, (3) the amount of solute already dissolved, and (4) the temperature.

In order for a solvent to dissolve a solute, the particles of the solvent must be able to separate the particles of the solute and occupy the intervening spaces. Polar solvent molecules can effectively separate the molecules of other polar substances. This happens when the positive end of a solvent molecule approaches the negative end of a solute molecule. A force of attraction then exists between the two molecules. The solute molecule is pulled into solution when the force overcomes the attractive force between the solute molecule and its neighboring solute molecule. Ethyl alcohol and water are examples of polar substances that readily dissolve in each other. Polar solvents can generally dissolve solutes that are ionic.

Fractionation of crude oil by distillation is an excellent means by which the volatile constituents can be isolated and studied. However, the nonvolatile residuum, which may actually constitute from 1% to 60% w/w of the crude oil, cannot be fractionated by distillation without the possibility of thermal decomposition, and as a result alternative methods of fractionation have been developed.

The distillation process separates light (lower molecular weight) and heavy (higher molecular weight) constituents by virtue of their volatility and involves the participation of a vapor phase and a liquid phase. These are, however, physical processes that involve the use of two liquid phases, usually a solvent phase and an oil phase.

Solvent methods have also been applied to crude oil fractionation on the basis of molecular weight. The major molecular weight separation process used in the laboratory as well as in the refinery is solvent precipitation. Solvent precipitation occurs in a refinery in a deasphalting unit (Chapter 7: Deasphalting and Dewaxing) and is essentially an extension of the procedure for separation by molecular weight, although some separation by polarity might also be operative. The deasphalting process is usually applied to the higher molecular weight fractions of crude oil such as atmospheric and vacuum residua for the production of asphalt or demetallized deasphalted oil.

1.6.2.1 Asphaltene separation

The systematic separation of crude oil by treatment with solvents has been practiced for several decades (Girdler, 1965). If chosen carefully, solvents effect a separation between the constituents of conventional crude oil, heavy oil, extra heavy oil, tar sand bitumen, and crude oil residua according to differences in molecular weight and aromatic character. The nature and the quantity of the components separated depend on the conditions of the experiment, namely, the degree of dilution temperature and the nature of the solvent.

On the basis of the solubility in a variety of solvents, it has become possible to distinguish among the various constituents of crude oil, heavy crude oil, and tar sand bitumen (Fig. 1.2). Highly paraffin crude oil may contain only small portions of asphaltenes. Crude oil generally does not contain carboids and carbenes that are, for the purposes of this text, considered to be the products of thermal processes. Hence, residua from cracking distillation of cracking processes may contain 2% w/w or more of the carbene and carboid fractions.

Thus the separation of crude oil into two fractions, asphaltenes and maltenes, is conveniently achieved using means of low-molecular weight paraffin hydrocarbon derivatives, which were recognized to have selective solvency for hydrocarbon derivatives, and simple relatively low-molecular weight hydrocarbon derivatives. The more complex, higher molecular weight compounds are precipitated particularly well by addition of 40 volumes of *n*-pentane or *n*-heptane in the methods generally preferred at present (Speight, 2014, 2019c), although hexane is used on occasion (Yan et al., 1997), which is not in keeping with the recommended standard methods using pentane or heptane (Speight, 2014, 2015). It is no doubt a separation of the chemical components with the most complex structures from the mixture, and this fraction, which should correctly be called *n*-pentane asphaltenes or *n*-heptane asphaltenes, is qualitatively and quantitatively reproducible (Fig. 1.2).

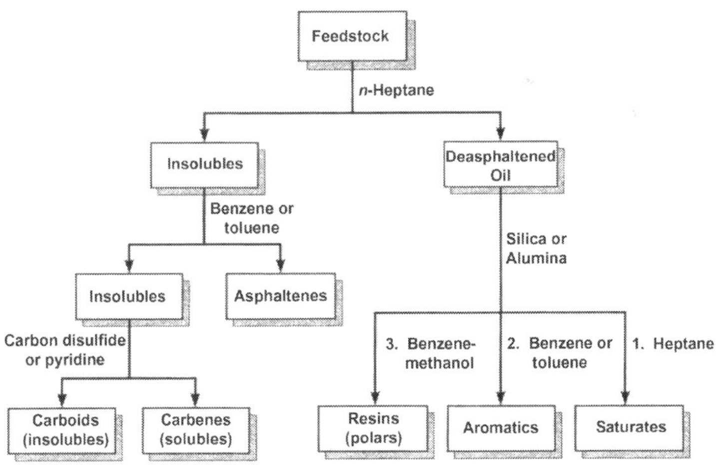

Figure 1.2 General fractionation scheme for crude oil, heavy oil, extra heavy oil, and tar sand bitumen*. *Since the procedure for the heavier feedstocks (extra heavy oil and tar sand bitumen as well as residua) is diffusion-controlled, it is customary of mix these feedstocks with an equal volume of toluene and before adding the heptane—in such cases, the ratio of heptane to the actual feedstock is 1 part feedstock to 80 parts heptane to mitigate any effects from the added the toluene (Speight, 2015).

Variation in the solvent type also causes significant changes in asphaltene yield. The solvent power of the solvents (i.e., the ability of the solvent to dissolve asphaltenes) increases in the order

$$2-\text{Methyl paraffin (iso-paraffin)} < n\text{-paraffin} < \text{terminal olefin}$$

Cycloparaffin derivatives (naphthene derivatives) have a remarkable effect on asphaltene yield and give results totally unrelated to those from any other nonaromatic solvent (Speight, 2014, 2015). For example, when cyclopentane, cyclohexane, or their methyl derivatives are employed as precipitating media, only approximately 1% w/w of the material remains insoluble.

In any method used to isolate asphaltenes as a separate fraction, standardization of the technique is essential. For many years the method of asphaltene separation was not standardized, and even now it remains subject to the preferences of the standard organizations of different countries. The use of both the *n*-pentane and *n*-heptane has been widely advocated, and although *n*-heptane is becoming the deasphalting liquid of choice, this is by no means a hard-and-fast rule. And it must be recognized that large volumes of solvent may be required to effect a reproducible separation, similar to amounts required for consistent asphaltene separation. It is also preferable that the solvents be of sufficiently low-boiling point that complete removal of the solvent from the fraction can be effected and, most important, the solvent must not react with the feedstock. Hence, there has been a preference for hydrocarbon liquids. Although the several standard methods are available, they are not unanimous in the particular hydrocarbon liquid or in a ratio of hydrocarbon liquid to feedstock.

Method	Deasphalting liquid	Volume (mL/g)
ASTM D893	*n*-Pentane	10
ASTM D2007	*n*-Pentane	10
ASTM D3279	*n*-Heptane	100
ASTM D4124	*n*-Heptane	100

However, it must be recognized that some of these methods were developed for use with feedstocks other than extra heavy oil and tar sand bitumen. Therefore adjustments in the methods may be necessary to ensure efficient separation.

Insofar as industrial solvents are very rarely one compound, it was also of interest to note that the physical characteristics of two different solvent types, in this case benzene and *n*-pentane, are additive on a mole fraction basis and also explain the variation of solubility with temperature (Speight, 2014, 2015). The data also show the effects of blending a solvent with the bitumen itself and allowing the resulting solvent-heavy oil blend to control the degree of bitumen solubility. Varying proportions of the hydrocarbon alter the physical characteristics of the oil to such an extent that the amount of precipitate (asphaltenes) can be varied accordingly within a certain range.

At constant temperature the quantity of precipitate first increases with increasing ratio of solvent to feedstock and then reaches a maximum. In fact, there are indications that when the proportion of solvent in the mix is <35% w/w, little or no asphaltenes are precipitated (Speight, 2014, 2015). When pentane and the lower molecular weight hydrocarbon solvents are used in large excess, the quantity of precipitate and the composition of the precipitate change with increasing temperature. One particular example is the separation of asphaltenes from using *n*-pentane. At ambient temperatures (21°C, 70°F), the yield of the asphaltene fractions is 17% w/w but at 35°C (95°F), the yield of the asphaltene fraction is on the order of 22.5% w/w using the same feedstock–pentane ratio. This latter precipitate is in fact asphaltenes plus resins; similar effects have been noted with other hydrocarbon solvents at temperatures up to 70°C (160°F). These results are self-explanatory when it is realized that the heat of vaporization H^v and the surface tension γ, from which the solubility parameters are derived, both decrease with increasing temperature.

Contact time between the hydrocarbon and the feedstock (especially feedstocks such as heavy crude oil, extra heavy oil, and tar sand bitumen) also plays an important role in asphaltene separation (Speight, 2014, 2019c). The yields of the asphaltene fraction reaches a maximum after approximately 8 hours, which may be ascribed to the time required for the asphaltene particles to agglomerate into particles of a filterable size as well as the diffusion-controlled nature of the process, since the heavier feedstocks also need time for the hydrocarbon to penetrate their mass.

1.6.2.2 Fractionation

After removal of the asphaltene fraction, further fractionation of crude oil is also possible by variation of the hydrocarbon solvent. For example, liquefied gases, such as propane and butane, precipitate as much as 50% w/w of the residuum or bitumen. The precipitate is a black, tacky, semisolid material, in contrast to the pentane-precipitated asphaltenes, which are usually brown, amorphous solids. Treatment of the propane precipitates with pentane then yields the insoluble brown, amorphous asphaltenes and soluble, near-black, semisolid resins, which are, as near as can be determined, equivalent to the resins isolated by adsorption techniques (Speight, 2014).

There are also claims that solvent treatment at low temperatures (−4°C to −20°C, −4°F to 25°F) assists in the fractionation of the maltenes. The hydrocarbon solvents pentane and hexane have been claimed adequate for this purpose but may not be successful with maltenes from bitumen or from residua. The author has had considerable success using acetone at −4°C (25°F) for the fractionation of maltenes from material other than cracked residua.

Other miscellaneous fractionation procedures involving the use of solvents are available (Speight, 2014).

The disadvantages of an all-solvent separation technique are that, first, in some instances, low temperatures (e.g., 0°C to −10°C and the like) are advocated as a means of effecting oil fractionation with solvents (Speight, 2014, 2015). Such requirements may cause inconvenience in a typical laboratory operation by requiring a permanently cool temperature during the separation. Second, it must be recognized that large volumes of solvent may be required to effect a reproducible separation in the same manner as the amounts required for consistent asphaltene separation (ASTM D2007, ASTM D4124, ASTM D893). Finally, it is also essential that the solvent be of sufficiently low-boiling point so that complete removal of the solvent from the product fraction can be effected. Although not specifically included in the three main disadvantages of the all-solvent approach, it should also be recognized that the solvent must not react with the feedstock constituents. In addition, caution is still required to ensure that there is no interaction between the solvent and the solute.

1.6.3 Adsorption methods

The most common industrial solvents are activated clay, carbon, silica gel, and alumina, because they present enormous surface areas per unit weight and result in fractions given operational (common or trivial) names that are subject to the separation procedure: (1) the saturates fraction, (2) the aromatics fraction, (3) the resin fraction, and (4) the asphaltene fraction (Fig. 1.2). However, each fraction is a complex mixture of a variety of chemical constituents and the common (trivial) names are not meant to indicate specific chemical types (Speight, 2014).

By definition the saturate fraction consists (or should consist) of paraffin derivatives and cycloparaffin derivatives (naphthene derivatives). The single-ring naphthene derivatives, or cycloparaffin derivatives, present in crude oil are primarily alkyl-substituted cyclopentane and cyclohexane. The aromatics fraction consists of those compounds containing an aromatic ring and vary from mono-aromatic derivatives (containing one benzene ring in a molecule) to di-aromatic derivatives (substituted naphthalene) to tri-aromatic (three-ring) derivatives (i.e., substituted phenanthrene derivatives). The resins fraction contains low-molecular weight functional (polar) species.

1.6.3.1 General methods

Separation of crude oil, heavy crude oil, extra heavy oil, tar sand bitumen, and residua by adsorption chromatography essentially commences with the preparation of a porous bed of finely divided solid adsorbent (Hoiberg, 1964). The adsorbent is usually contained in an open tube (column chromatography); the sample is introduced at one end of the adsorbent bed and induced to flow through the bed by means of a suitable solvent.

It is essential that, before application of the adsorption technique to the crude oil, the asphaltenes first be completely removed, for example, by any of the methods outlined in the previous section. The prior removal of the asphaltenes is essential insofar as they are usually difficult to remove from the earth or clay and may actually be irreversibly adsorbed on the adsorbent. The proportions of each fraction are subject to the ratio of fuller's earth to *n*-pentane or *n*-hexane soluble materials.

Other methods of fractionation by the use of adsorbents include separation of the maltene fraction by elution with *n*-heptane from silica gel into two fractions named aromatic derivatives and nonaromatic derivatives and is, in fact, a separation into the two broad groups called resins and oils in other methods. The silica gel method may also be modified to produce three fractions: (1) nonaromatic derivatives eluted with *n*-heptane, (2) aromatic derivatives eluted with benzene, and (3) compounds that contain oxygen as well as sulfur and nitrogen, eluted with pyridine. Prior separation of the asphaltenes renders the procedure especially suitable and convenient for use with extra heavy oil and tar sand bitumen. Other modifications include successive elution with *n*-pentane, benzene, carbon tetrachloride, and ethanol.

Alumina has also been used as an adsorbent and involves (1) precipitation of asphaltene fraction with normal pentane or with normal heptane, (2) elution of oils from alumina with pentane, and (3) elution of resins from alumina with a methanol–benzene mixture (Speight, 2014, 2015, 2019c). In fact, the choice of the adsorbent appears to be arbitrary, as does the choice of the various solvents or solvent blend. The use of ill-defined adsorbents, such as earths or clays, is a disadvantage in that certain components of the crude oil may undergo changes (e.g., polymerization) caused by the catalytic nature of the adsorbent and can no longer be extracted quantitatively. Furthermore, extraction of the adsorbed components may require the use of solvents of comparatively high solvent power, such as chloroform or pyridine, which may be difficult to remove from the product fractions.

1.6.3.2 US Bureau of Mines-American Petroleum Institute and Saturates, Aromatic Derivatives, Resins, and Asphaltenes methods

There are two procedures that have received considerable attention over the years and these are (1) the US Bureau of Mines-API method and (2) the Saturates, Aromatic, Resins, and Asphaltenes (SARA) method (Speight, 2014, 2015). This latter method is often also called the Saturates, Aromatics Derivatives, Polars, Asphaltenes method. These two methods are used as representing the standard methods of crude oil fractionation. Other methods are also noted, especially when the method has added further meaningful knowledge to compositional studies.

The SARA method is essentially an extension of the API method that allows more rapid separations by placing the two ion-exchange resins and the ferric chloride ($FeCl_3$)-clay-anion-exchange resin packing into a single column (Speight, 2014, 2015). The adsorption chromatography of the nonpolar part of the same is still performed in a separation operation. Since the asphaltene content of crude oil (and synthetic fuel) feedstocks is often an important aspect of processability, an important feature of the SARA method is that the asphaltenes are separated as a group. Perhaps more important is that the method is reproducible and applicable to a large variety of feedstocks.

The selection of any separation procedure depends primarily on the information desired in relation to the feedstock.

Another issue related to the use of any adsorption-based fractionation scheme is the nature of the adsorbent. In the early reports of crude oil fractionation (Pfeiffer, 1950), clay often appeared as an adsorbent to effect the separation of the feedstock into various constituent fractions. However, clay (fuller's earth, Attapulgus Clay, and the like) is often difficult to define with any degree of precision from one batch to another. Variations in the nature and properties of the clay can, and will, cause differences not only in the yields of composite fractions but also in the distribution of the compound types in those fractions. In addition, irreversible adsorption of the more polar constituent to the clay can be a serious problem when further investigations of the constituent fractions are planned.

One option for resolving this problem has been the use of more standard adsorbents, such as alumina and silica. These materials are easier to define and are often accompanied by guarantees of composition and type by various manufacturers. They also tend to irreversibly adsorb less of the feedstock than clay. Once the nature of the adsorbent is guaranteed, reproducibility becomes a reality. Without reproducibility the analytic method does not have credibility.

1.6.3.3 ASTM methods

There are three ASTM methods that provide for the separation of a feedstock into four or five constituent fractions, and it is interesting to note that as the methods have evolved there has been a change from the use of pentane to heptane to separate the asphaltene fraction (ASTM D2007, ASTM D4124) (Speight, 2014, 2015). This is, in fact, in keeping with the production of a more consistent fraction that represents these higher molecular weight, more complex constituents of crude oil (Girdler, 1965; Speight, 2014, 2015, 2019c).

In summary the terminology used for the identification of the various methods might differ. However, in general terms, group-type analysis of crude oil is often identified by the acronyms for the names: PONA (paraffin derivatives, olefin derivatives, naphthene derivatives, and aromatic derivatives), PIONA (paraffin derivatives, isoparaffin derivatives, olefin derivatives, naphthene derivatives, and aromatic derivatives), PNA (paraffin derivatives, naphthene derivatives, and aromatic derivatives), PINA (paraffin derivatives, isoparaffin derivatives, naphthene derivatives, and aromatic derivatives), or SARA (saturates, aromatic derivatives, resins, and asphaltenes). However, it must be recognized that the fractions produced by the use of different adsorbents will differ in content and will also be different from fractions produced by solvent separation techniques.

The variety of fractions isolated by these methods and the potential for the differences in composition of the fractions makes it even more essential that the method is described accurately and that it reproducible not only in any one laboratory but also between various laboratories.

1.7 Crude oil products

Crude oil products and fuels, in contrast to petrochemicals, are those bulk fractions that are derived from crude oil and have commercial value as a bulk product (Speight, 2014, 2019a). Fuels produced from crude oil are also crude oil products and are the fuels by which industry is driven. A brief mention of the properties of these fuels is worthy of brief description here as it helps one to focus on the future needs and direction of the crude oil industry. More detailed descriptions of crude oil fuels are available (Speight, 2014).

Natural gas, which is predominantly methane, occurs in underground reservoirs separately or in association with crude oil (Speight, 2014, 2019b). The principal types of gaseous fuels are oil (distillation) gas, reformed natural gas, and reformed propane or liquefied petroleum gas (LPG). LPG is the term applied to certain specific hydrocarbon derivatives and their mixtures, which exist in the gaseous state under atmospheric ambient conditions but can be converted to the liquid state under conditions of moderate pressure at ambient temperature. These are the low-boiling hydrocarbon derivatives fraction of the paraffin series, derived from refinery processes, crude oil stabilization plants, and natural gas processing plants comprising propane ($CH_3CH_2CH_3$), butane ($CH_3CH_2CH_2CH_3$), isobutane [$CH_3CH(CH_3)CH_3$], and to a lesser extent propylene ($CH_3CH=CH_2$) or butylene ($CH_3CH_2CH=CH_2$). Mixed gas is a gas prepared by adding natural gas or liquefied petroleum gas to a manufactured gas, giving a product of better utility and higher heat content or Btu value.

The compositions of natural, manufactured, and mixed gases can vary so widely, no single set of specifications could cover all situations. The requirements are usually based on performances in burners and equipment, on minimum heat content, and on maximum sulfur content.

Synthesis gas (syngas) is the name given to a gas mixture that is generated by the gasification of a carbon containing fuel (e.g., crude oil coke) to a gaseous product that contains varying amounts of carbon monoxide and hydrogen. The name synthesis gas originates from their use as intermediates in creating synthetic natural gas and for producing ammonia and/or methanol. Syngas is also used as an intermediate in producing synthetic fuels via the Fischer–Tropsch reaction.

In the strictest sense, synthesis gas consists primarily of carbon monoxide and hydrogen, although carbon dioxide and nitrogen may also be present. The chemistry of synthesis gas production is relatively simple but the reactions are often much more complex that indicated by simple chemical equations (Speight, 2008).

Gasoline is mixture of hydrocarbon derivatives that usually boil below 180°C (355°F) or, at most, below 200°C (390°F). Gasoline is manufactured to meet specifications and regulations and not to achieve a specific distribution of hydrocarbon derivatives by class and size. However, chemical composition often defines properties. For example, volatility is defined by the individual hydrocarbon constituents and the lowest boiling constituent(s) defines the volatile as determined by certain test methods.

Automotive gasoline typically contains approximately almost 200 (if not several hundred) hydrocarbon compounds. The relative concentrations of the compounds vary considerably depending on the source of crude oil, refinery process, and product specifications. Typical hydrocarbon chain lengths range from C_4 through C_{12} with a general hydrocarbon distribution consisting of alkane derivatives (4%–8% w/w), alkene derivatives (2%–5% w/w), isoalkane derivatives (25%–40% w/w), cycloalkane derivatives (3%–7% w/w), cycloalkene derivatives (1%–4% w/w), and aromatic derivatives (20%–50% w/w). However, these proportions vary greatly.

Thermal cracking and catalytic cracking, once used to supplement the gasoline supplies produced by distillation, are now the major processes used to produce gasoline. In addition, other methods used to improve the quality of gasoline and increase its supply include polymerization, alkylation, isomerization, and reforming.

Aviation gasoline is the form of motor gasoline that has been especially prepared for use for aviation piston engines and is composed of paraffin derivatives and isoparaffin derivatives (50%–60% w/w), moderate amounts of naphthene derivatives (20%–30% w/w), small amounts of aromatic derivatives (10% w/w), and usually no olefin derivatives, whereas motor gasoline may contain up to 30% w/w olefin derivatives and up to 40% w/w aromatic derivatives. It has an octane number suited to the engine, a freezing point on the order of −60°C (−76°F), and a distillation range usually within the limits of 30°C–180°C (86°F–356°F) compared to −1°C–200°C (30°F–390°F) for automobile gasoline.

Kerosene (also spelled kerosine, also called paraffin or paraffin oil) is a flammable pale yellow or colorless oily liquid with a characteristic odor. It is obtained from crude oil and used for burning in lamps and domestic heaters or furnaces, as a fuel or fuel component for jet engines, and as a solvent for greases and insecticides.

Kerosene is intermediate in volatility between gasoline and gas oil. It is a medium oil distilling between 150°C and 300°C (300°F–570°F). Kerosene has a flash point on the order of 25°C (77°F) and is suitable for use as an illuminant when burned in a wide lamp. The term "kerosene" is also too often incorrectly applied to various fuel oils, but a fuel oil is actually any liquid or liquid crude oil product that produces heat when burned in a suitable container or that produces power when burned in an engine.

Jet fuel is a low-boiling crude oil distillate that is available in several forms suitable for use in various types of jet engines. The major jet fuels used by the military are JP-4, JP-5, JP-6, JP-7, and JP-8. Briefly, JP-4 is a wide-cut fuel developed for broad availability. JP-6 is a higher cut than JP-4 and is characterized by fewer impurities. JP-5 is specially blended kerosene, and JP-7 is high flash point special kerosene used in advanced supersonic aircraft. JP-8 is kerosene modeled on Jet A-l fuel (used in civilian aircraft). From what data are available, typical hydrocarbon chain lengths characterizing JP-4 range from C_4 to C_{16}. Aviation fuels consist primarily of straight and branched alkane derivatives and cycloalkane derivatives. Aromatic hydrocarbon derivatives are limited to 20%–25% w/w of the total mixture because they produce smoke when burned. A maximum of 5% w/w alkene derivatives is specified for JP-4. The approximate distribution by chemical class is straight-chain alkane derivatives (32% w/w), branched alkane derivatives (31% w/w), cycloalkane derivatives (16% w/w), and aromatic hydrocarbon derivatives (21% w/w).

Fuel oil is classified in several ways but generally may be divided into two main types: distillate fuel oil and residual fuel oil. Distillate fuel oil is vaporized and condensed during a distillation process and thus has a definite boiling range and does not contain high-boiling constituents. A fuel oil that contains any amount of the residue from crude distillation of thermal cracking is a residual fuel oil. The terms "distillate fuel oil" and "residual fuel oil" are losing their significance, since fuel oil is now made for specific uses and may be either distillates or residuals or mixtures of the two. The terms "domestic fuel oil," "diesel fuel oil," and "heavy fuel oil" are more indicative of the uses of fuel oils.

Diesel fuel oil is also a distillate fuel oil that distills between 180°C and 380°C (356°F and 715°F). Several grades are available depending on uses such as (1) diesel oil for diesel compression ignition such as cars, trucks, and marine engines and (2) low-boiling heating oil for industrial and commercial uses.

No. 1 fuel oil is a crude oil distillate that is one of the most widely used of the fuel oil types. It is used in atomizing burners that spray fuel into a combustion chamber, where the tiny droplets bum while in suspension. It is also used as a carrier for pesticides, as a weed killer, as a mold release agent in the ceramic and pottery industry, and in the cleaning industry. It is found in asphalt coatings, enamels, paints, thinners, and varnishes. No. 1 fuel oil is a low-boiling crude oil distillate (straight-run kerosene) consisting primarily of hydrocarbon derivatives in the range of C_9-C_{16}. Fuel oil No. l is very similar in composition to diesel fuel; the primary difference is in the additives.

No. 2 fuel oil is a crude oil distillate that may be referred to as domestic or industrial. The domestic fuel oil is usually lower boiling and a straight-run product. It is used primarily for home heating. Industrial distillate is a cracked product or a blend of both. It is used in smelting furnaces, ceramic kilns, and packaged boilers. No. 2 fuel oil is characterized by hydrocarbon chain lengths in the $C_{11}-C_{20}$ range. The composition consists of aliphatic hydrocarbon derivatives (straight-chain alkane derivatives and cycloalkane derivatives) (64% w/w), 1%–2% w/w unsaturated hydrocarbon derivatives (alkene derivatives) (1%–2% w/w), and aromatic hydrocarbon derivatives (including alkyl benzenes and 2-ring, 3-ring aromatic derivatives) (35% w/w) but contains only low amounts of the polycyclic aromatic hydrocarbon derivatives (<5% w/w).

No. 6 fuel oil (also called Bunker C oil or residual fuel oil) is the residuum from crude oil after naphtha-gasoline, No. 1 fuel oil, and No. 2 fuel oil have been removed. No. 6 fuel oil can be blended directly to heavy fuel oil or made into asphalt. Residual fuel oil is more complex in composition and impurities than distillate fuels. Limited data are available on the composition of No. 6 fuel oil. Polycyclic aromatic hydrocarbon derivatives (including the alkylated derivatives) and metal-containing constituents are components of No. 6 fuel oil.

Diesel fuel oil is essentially the same as furnace fuel oil, but the proportion of cracked gas oil is usually less, since the high aromatic content of the cracked gas oil reduces the cetane value of the diesel fuel. Under the broad definition of diesel fuel, many possible combinations of characteristics (such as volatility, ignition quality, viscosity, gravity, stability, and other properties) exist. To characterize diesel fuels and thereby establish a framework of definition and reference, various classifications are used in different countries. An example is ASTM D975 in the United States in which grades No. 1-D and 2-D are distillate fuels, the types most commonly used in high-speed engines of the mobile type, in medium-speed stationary engines, and in railroad engines. Grade 4-D covers the class of more viscous distillates and, at times, blends of these distillates with residual fuel oils. No. 4-D fuels are applicable for use in low- and medium-speed engines employed in services involving sustained load and predominantly constant speed.

Coke (a solid fuel) is the residue left by the destructive distillation of crude oil residua. That formed in catalytic cracking operations is usually nonrecoverable, as it is often employed as fuel for refinery processes. The composition of crude oil coke varies with the source of the crude oil, but in general, large amounts of high-molecular weight complex hydrocarbon derivatives (rich in carbon but correspondingly poor in hydrogen) make up a high proportion. The solubility of crude oil coke in carbon disulfide has been reported to be as high as 50%–80% w/w, but this is in fact a misnomer, since the coke is the insoluble, honeycomb material that is the end product of thermal processes.

1.8 Petrochemicals

A petrochemical is any chemical manufactured from crude oil and natural gas as distinct from fuels and other products (Speight, 2014, 2019a), derived from crude oil and natural gas, and used for a variety of commercial purposes. The definition has been broadened to include the whole range of organic chemicals. In many instances a specific chemical included among the petrochemicals may also be obtained from other sources, such as coal, coke, or vegetable products. For example, materials such as benzene and naphthalene can be made from either crude oil or coal, while ethyl alcohol may be of petrochemical or vegetable origin. However, in the current context, petrochemicals can be conveniently divided into two groups: (1) primary petrochemicals and (2) intermediates and derivatives.

Primary petrochemicals include (1) olefin derivatives, such as ethylene and propylene and butadiene; (2) aromatic derivatives, such as benzene, toluene, and the xylene isomers; and (3) methanol. Petrochemical intermediates are generally produced by chemical conversion of primary petrochemicals to form more complicated derivative products. Petrochemical derivative products can be made in a variety of ways: directly from primary petrochemicals; through intermediate products which still contain only carbon and hydrogen; and through intermediates that incorporate chlorine, nitrogen, or oxygen in the finished derivative. In some cases, they are finished products; in others, more steps are needed to arrive at the desired composition.

The chemical industry is in fact the chemical process industry by which a variety of chemicals are manufactured. The chemical process industry is, in fact, subdivided into other categories that are (1) chemicals and allied products in which chemicals are manufactured from a variety of feedstocks and may then be put to further use, (2) rubber and miscellaneous products that focus on the manufacture of rubber and plastic materials, and (3) crude oil refining and related industries which, on the basis of prior chapters in this text, are now self-explanatory. Thus the petrochemical industry falls under the subcategory of crude oil and related industries.

References

Abraham, H., 1945. Asphalts and Allied Substances. Van Nostrand, New York.

Aksenova, V.S., Kanayanov, V.F., 1980. Regularities in composition and structures of native sulfur compounds from petroleum. In: Proceedings 9th International Symposium on Organic Sulfur Chemistry, Riga, USSR, June 9–14.

Ancheyta, J., Speight, J.G., 2007. Heavy oils and residua. In: Ancheyta, J., Speight, J.G. (Eds.), Hydroprocessing of Heavy Oils and Residua. CRC-Taylor & Francis Group, Boca Raton, FL, Chapter 1.

ASTM International, 2019a. Annual Book of Standards. ASTM International, West Conshohocken, PA.

ASTM International, 2019b. Standard test method for acid number of petroleum products by potentiometric titration. Annual Book of Standards. ASTM International, West Conshohocken, PA, ASTM D664.

ASTM International, 2019c. Standard test method for insolubles in used lubricating oils. Annual Book of Standards. ASTM International, West Conshohocken, PA, ASTM D893.

ASTM International, 2019d. Standard test method for acid and base number by color-indicator titration. Annual Book of Standards. ASTM International, West Conshohocken, PA, ASTM D974.

ASTM International, 2019e. Standard test method for approximate acidity in electrical insulating liquids by color-indicator titration. Annual Book of Standards. ASTM International, West Conshohocken, PA, ASTM D1534.

ASTM International, 2019f. Standard test method for characteristic groups in rubber extender and processing oils and other petroleum-derived oils by the clay-gel absorption chromatographic method. Annual Book of Standards. ASTM International, West Conshohocken, PA, ASTM D2007.

ASTM International, 2019g. Standard test method for n-heptane insolubles. Annual Book of Standards. ASTM International, West Conshohocken, PA, ASTM D3279.

ASTM International, 2019h. Standard test method for acid number of petroleum products by semi-micro color indicator titration. Annual Book of Standards. ASTM International, West Conshohocken, PA, ASTM D3339.

ASTM International, 2019i. Standard test method for separation of asphalt into four fractions. Annual Book of Standards. ASTM International, West Conshohocken, PA, ASTM D4124.

Baugh, T.D., Grande, K.V., Mediaas, H., Vindstad, J.E., Wolf, N.O., 2005. The discovery of high molecular weight naphthenic acids (ARN acid) responsible for calcium naphthenate deposits. In: SPE International Symposium on Oilfield Scale, Aberdeen, United Kingdom, May 11–12.

Bell, H.S., 1945. American Petroleum Refining. Van Nostrand, New York.

Bower, T., 2009. Oil: Money, Politics, and Power in the 21st Century. Grand Central Publishing, New York.

BP, 2019. Statistical Review of World Energy, 68th ed. BP PLC, London. Available from: https://www.bp.com/content/dam/bp/business-sites/en/global/corporate/pdfs/energy-economics/statistical-review/bp-stats-review-2019-full-report.pdf.

Burke, J., 1996. The Pinball Effect. Little, Brown and Company, New York, p. 128, et seq.

Charbonnier, R.P., Draper, R.G., Harper, W.H., Yates, A., 1969. Analyses and characteristics of oil samples from Alberta. Information Circular IC 232. Department of Energy Mines and Resources, Mines Branch, Ottawa.

Delbianco, A., Montanari, R., 2009. Encyclopedia of Hydrocarbons, Volume III/New Developments: Energy, Transport, Sustainability. Eni S.p.A., Rome.

Draper, R.G., Kowalchuk, E., Noel, G., 1977. Analyses and Characteristics of Crude Oil Samples Performed Between 1969 and 1976. Report ERP/ERL 77-59 (TR). Energy, Mines, and Resources, Ottawa.

Freidlina, I.K., Skorova, A.E. (Eds.), 1980. Organic Sulfur Chemistry. Pergamon Press, New York.

Gary, J.G., Handwerk, G.E., Kaiser, M.J., 2007. Petroleum Refining: Technology and Economics, fifth ed. CRC Press, Taylor & Francis Group, Boca Raton, FL.

Girdler, R.B., 1965. Proc. Assoc. Asphalt Paving Technologists 34. p. 45.

Gorbaty, M.L., Martella, D.J., Sartori, G., Savage, D.W., Ballinger, B.H., Blum, S.C., et al., 2001. Process for Neutralization of Petroleum Acids Using Overbased Detergents, United States Patent 6,054,042. April 25.

Green, J.A., Green, J.B., Grigsby, R.D., Pearson, C.D., Reynolds, J.W., Sbay, I.Y., et al., 1989. Analysis of Heavy Oils: Method Development and Application to Cerro Negro Heavy Petroleum, NIPER-452 (DE90000200), Vols. I and II. Research Institute, National Institute for Petroleum and Energy Research (NIPER), Bartlesville, OK.

Haynes, D. 2006. Naphthenic acid bearing refinery feedstocks and corrosion abatement. In: AIChE—Chicago Symposium.

Hirsch, D.E., Cooley, J.E., Coleman, H.J., Thompson, C.J., 1974. Qualitative characterization of aromatic concentrates of crude oils from GPC analysis. Report 7974. Bureau of Mines, U.S. Department of the Interior, Washington, DC.

Hoiberg, A.J., 1964. Bituminous Materials: Asphalts, Tars, and Pitches. John Wiley and Sons, New York.

Hsu, C.S., Robinson, P.R. (Eds.), 2017. Handbook of Petroleum Technology. Springer International Publishing AG, Cham.

Isaacs, C.M., 1992. Preliminary petroleum geology background and well data for oil samples in the Cooperative Monterey Organic Geochemistry Study, Santa Maria and Santa Barbara-Ventura Basins, California. Open-File Report No. USGS 92-539-F. United States Geological Survey, Reston, VA.

Kane, R.D., Cayard, M.S., 2002. A Comprehensive Study on Naphthenic Acid Corrosion. Paper No. 02555. Corrosion 2002. NACE International, Houston, TX.

Kittrell, N., 2006. Removing Acid From Crude Oil. Crude Oil Quality Group, New Orleans, LA.

Meyer, R.F., Steele, C.F. (Eds.), 1981. The Future of Heavy Crude and Tar Sands. McGraw-Hill, New York.

Meyer, R.F., Attanasi, E.D., 2003. Heavy Oil and Natural Bitumen—Strategic Petroleum Resources. Fact Sheet 70-03. United States Geological Survey, Washington, DC. Available from: http://pubs.usgs.gov/fs/fs070-03/fs070-03.html.

Parkash, S., 2003. Refining Processes Handbook. Gulf Professional Publishing, Elsevier, Amsterdam.

Pfeiffer, J.H., 1950. The Properties of Asphaltic Bitumen. Elsevier BV, Amsterdam.

Reynolds, J.G., 1998. Metals and heteroatoms in heavy crude oils. In: Speight, J.G. (Ed.), Petroleum Chemistry and Refining. Taylor & Francis Publishers, Washington, DC, pp. 63–102. , Chapter 3.

Rossini, F.D., Mair, B.J., Streif, A.J., 1953. Hydrocarbons From Petroleum. Reinhold, New York.

Shalaby, H.M., 2005. Refining of Kuwait's heavy crude oil: materials challenges. In: Proceedings Workshop on Corrosion and Protection of Metals, Arab School for Science and Technology, Kuwait, December 3–7.

Speight, J.G. (Ed.), 1990. Fuel Science and Technology Handbook. Marcel Dekker, New York.

Speight, J.G., 1997. fourth ed. Kirk-Othmer Encyclopedia of Chemical Technology, Vol. 23. John Wiley & Sons Inc., Hoboken, NJ, p. 717.

Speight, J.G., 2000. The Desulfurization of Heavy Oils and Residua, second ed. Marcel Dekker, New York.

Speight, J.G., 2008. Handbook of Synthetic Fuels: Properties, Processes, and Performance. McGraw-Hill, New York.

Speight, J.G., 2011. An Introduction to Petroleum Technology, Economics, and Politics. Scrivener Publishing, Beverly, MA.

Speight, J.G., 2013. The Chemistry and Technology of Coal, third ed. CRC Press, Taylor & Francis Group, Boca Raton, FL.

Speight, J.G., 2014. The Chemistry and Technology of Petroleum, fifth ed. CRC Press, Taylor & Francis Group, Boca Raton, FL.

Speight, J.G., 2015. Handbook of Petroleum Product Analysis, second ed. John Wiley & Sons Inc., Hoboken, NJ.

Speight, J.G., 2016. Introduction to Enhanced Recovery Methods for Heavy Oil and Tar Sands, second ed. Gulf Professional Publishing Company, Elsevier, Oxford.

Speight, J.G., 2017. Handbook of Petroleum Refining. CRC Press, Taylor & Francis Group, Boca Raton, FL.

Speight, J.G., 2019a. Handbook of Petrochemical Processes. CRC Press, Taylor & Francis Group, Boca Raton, FL.

Speight, J.G., 2019b. Natural gas: a basic handbook, GPC Books, second ed. Gulf Publishing Company, Houston, TX.

Speight, J.G., 2019c. The asphaltene fraction – demystified. Pet. Chem. Ind. Int. 2 (4). Available from: https://opastonline.com/wp-content/uploads/2019/10/the-asphaltene-fraction-demystified-pcii-19.pdf.

Speight, J.G., Francisco, M.A., 1990. Studies in petroleum composition IV: changes in the nature of chemical constituents during crude oil distillation. Rev. Inst. Francais Pétrole 45, 733.

Sun, X., Zhang, Y., Li, X., Cui, G., Gu, J., 2013. A case study on foamy oil characteristics of the Orinoco Belt, Venezuela. Adv. Pet. Explor. Dev. 5 (1), 37–41.

Tissot, B.P., Welte, D.H., 1978. Petroleum Formation and Occurrence. Springer-Verlag, New York.

United States Congress, 1976. Public law FE-76-4. In: United States Library of Congress, Washington, DC.

Villarroel, T., Hernández, R., 2013. Technological developments for enhancing extra heavy oil productivity in fields of the Faja Petrolifera del Orinoco (FPO), Venezuela. Proceedings AAPG Annual Convention and Exhibition, Pittsburgh, Pennsylvania, May 19–22. American Association of Petroleum Geologists, Tulsa, OK.

Wachtmeister, H., Linnea Lund, L., Aleklett, K., Höök, M.M., 2017. Production decline curves of tight oil wells in eagle ford shale. Nat. Resour. Res. 26 (3), 365–377.

Yan, J., Plancher, H., Morrow, N.R., 1997. Paper no. SPE 37232. In: SPE International Symposium on Oilfield Chemistry, Houston, TX.

Yépez, O., 2005. Influence of different sulfur compounds on corrosion due to naphthenic acid. Fuel 84 (1), 97–104.

Yergin, D., 1991. The Prize: The Epic Quest for Oil, Money, and Power. Simon & Schuster, New York.

Zhang, A., Ma, Q., Goddard, W.A., Tang, Y., 2004. Improved processes to remove naphthenic acids. Annual Technical Progress Report. DE-FC26-02NT15383. United States Department of Energy, Washington, DC, April 28.

Introduction to refining processes

2

2.1 Introduction

Crude oil refining (producing a product from crude oil that has a different character and properties to the starting crude oil) can be traced back over 5000 years to the time of the Sumerians when asphalt materials and crude oil–type products were isolated not only from areas where natural seepage occurred but also by distillation (Parkash, 2003; Gary et al., 2007; Speight, 2014, 2017; Hsu and Robinson, 2017). Crude oil refining is a recent technology and many innovations evolved during the 20th century. Furthermore, as feedstocks change in character, these innovations will need to continue until crude oil refining as is currently known becomes applied to, and adapted to, other sources of energy (Speight, 2008).

Currently, the crude oil refining industry converts crude oil into a wide slate of products, which include liquefied petroleum gas, naphtha, gasoline, kerosene, aviation fuel, diesel fuel, fuel oils, lubricating oils, wax, asphalt, and feedstocks for the petrochemical industry.

Refinery activities start with receipt of crude oil for storage at the refinery that is followed by dewatering and desalting and includes all of the subsequent handling and refining operations. Refining terminates with storage preparatory to shipping the refined products from the refinery. The industry employs a wide variety of processes and the processes selected from refining (processing flow scheme) is determined by the composition of the crude oil feedstock and the chosen slate of crude oil products. The selection and arrangement of these processes will vary among refineries, and few, if any, employ all of these processes.

A refinery is a group of integrated manufacturing plants that vary in number with the variety of products produced (Fig. 2.1) which are selected to give a balanced production of saleable products in amounts that are in accord with the demand for each. To prevent the accumulation of nonsaleable products, the refinery must be flexible and able to change operations as needed. The complexity of crude oil is emphasized insofar as the actual amounts of the products vary significantly from one crude oil to another (Parkash, 2003; Gary et al., 2007; Speight, 2014, 2017; Hsu and Robinson, 2017).

The refining industry has been the subject of the four major forces that affect most industries and have hastened the development of new crude oil refining processes: (1) the demand for products such as gasoline, diesel, fuel oil, and jet fuel; (2) feedstock supply, specifically the changing quality of crude oil and geopolitics between different countries and the emergence of alternate feed supplies such as natural gas, tar sand bitumen, coal, and oil shale; (3) environmental regulations that

The Refinery of the Future. DOI: https://doi.org/10.1016/B978-0-12-816994-0.00002-6

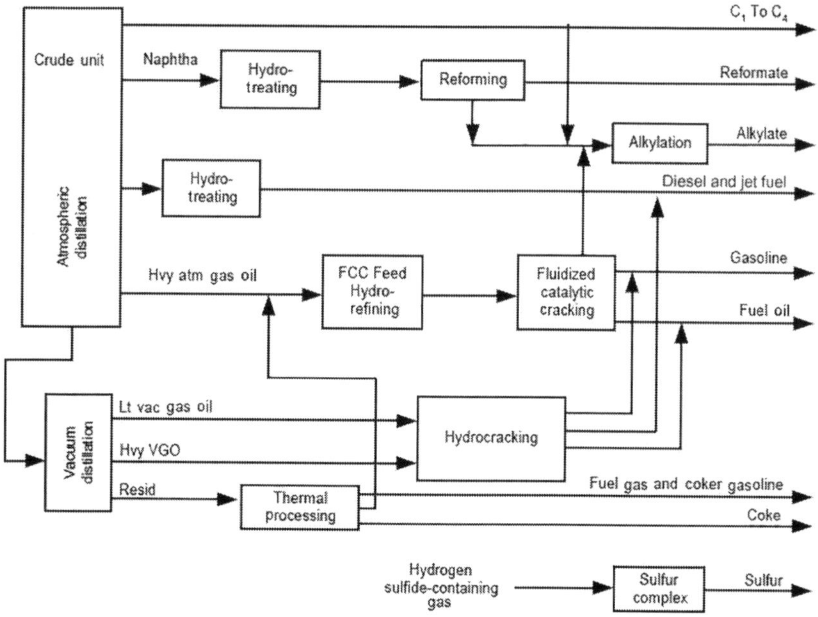

Figure 2.1 Schematic overview of a modern refinery.

include more stringent regulations in relation to sulfur in gasoline and diesel; and (4) technology development such as new catalysts and processes.

Crude oil refineries were originally designed and operated to run within a narrow range of crude oil feedstock and to produce a relatively fixed slate of crude oil products. Since the 1970s refiners had to increase their flexibility in order to adapt to a more volatile environment. Several possible paths may be used by refiners to increase their flexibility within existing refineries. Examples of these paths are change in the severity of operating rules of some process units by varying the range of inputs used, thus achieving a slight change in output. Alternatively refiners can install new processes, and this alternate scenario offers the greatest flexibility but is limited by the constraint of strict complementarily of the new units with the rest of the existing plant and involves a higher risk than the previous ones. It is not surprising that many refiners decide to modify existing processes.

The means by which a modern refinery operates depends not only on the nature of the crude oil feedstock (or, nowadays, the different crude oils that make up the blended feedstock) but also on its configuration (i.e., the number of types of the processes that are employed to produce the desired product slate), which is strongly influenced by the specific demands of a market. Therefore refineries need to be constantly adapted and upgraded to remain viable and responsive to ever-changing patterns of crude supply and product market demands. As a result, refineries have been introducing increasingly complex in order to produce higher yields of, for example, hydrocarbon fuels, from higher boiling fractions and residua.

Whatever the choice, refinery practice continues to evolve and (as will be seen in the relevant chapters following) new processes are installed in live with older modified process. The purpose of this chapter is to present to the reader a general overview of refining that, when taken into the context of the following chapters, will show some of the differences that occurring in refineries.

This chapter presents an overview of crude oil refining in order for the reader to place each process in the correct context of the refinery. From this description is it convenient to understand the evolution of the crude oil refinery and thence to speculate on the refineries of the future.

This section deals with the various processes used in refineries and the way in which these processes fit into the overall scheme of crude oil refining.

2.2 Refinery configurations

The configuration of any refinery-type may vary from refinery to refinery. Some refineries may be more oriented toward the production of gasoline (large reforming and/or catalytic cracking), whereas the configuration of other refineries may be more oriented toward the production of middle distillates such as jet fuel and gas oil.

A modern refinery is a highly complex and integrated system separating and transforming crude oil into a wide variety of products, including transportation fuels, residual fuel oils, lubricants, and many other products. The simplest refinery configuration is the topping refinery (Fig. 2.2) that is designed to prepare feedstocks for petrochemical manufacture or for the production of industrial fuels in remote oil-producing areas.

The topping refinery consists of tankage, a distillation unit, recovery facilities for gases and low-boiling hydrocarbon derivatives, and the necessary utility systems (steam, power, and water-treatment plants). Topping refineries produce large quantities of unfinished oils and are highly dependent on local markets, but the addition of hydrotreating and reforming units to this basic configuration results in a more flexible hydroskimming refinery (Fig. 2.3), which can also produce desulfurized distillate fuels and high-octane gasoline. These refineries may produce up to half of their output as residual fuel oil, and they face increasing market loss as the demand for low-sulfur (even no-sulfur), high-sulfur fuel oil increases.

The most versatile modern refinery configurations are the catalytic cracking refinery (Fig. 2.4) and the coking refinery (Fig. 2.5). These refineries incorporate all the basic units found in both the topping and hydroskimming refineries, but it also feature gas oil conversion plants such as catalytic cracking and hydrocracking units, olefin conversion plants such as alkylation or polymerization units, and coking units for residuum conversion to reduce or eliminate the production of residual fuels. Modern catalytic cracking and coking refineries produce high outputs of gasoline, with the balance distributed between liquefied petroleum gas, jet fuel, diesel fuel, and a small quantity of coke. Many such refineries also incorporate solvent

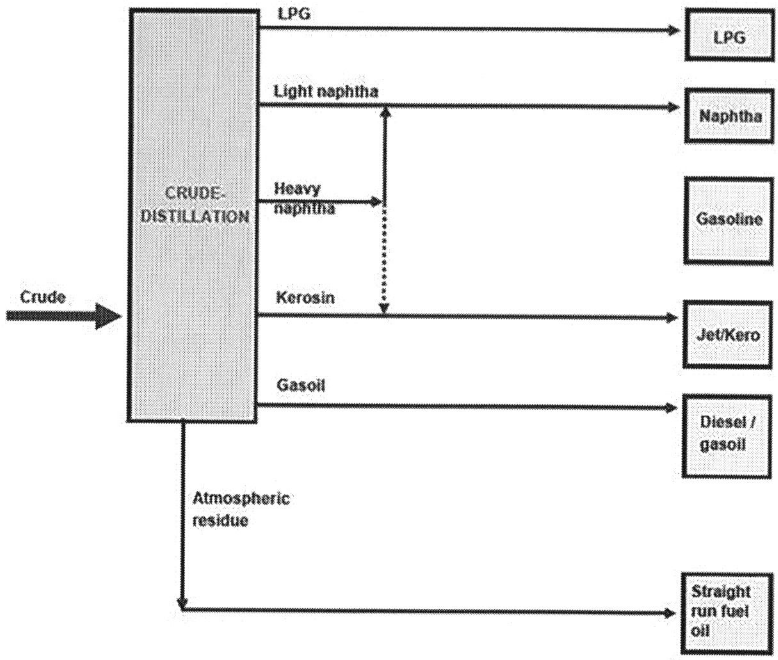

Figure 2.2 A topping refinery.

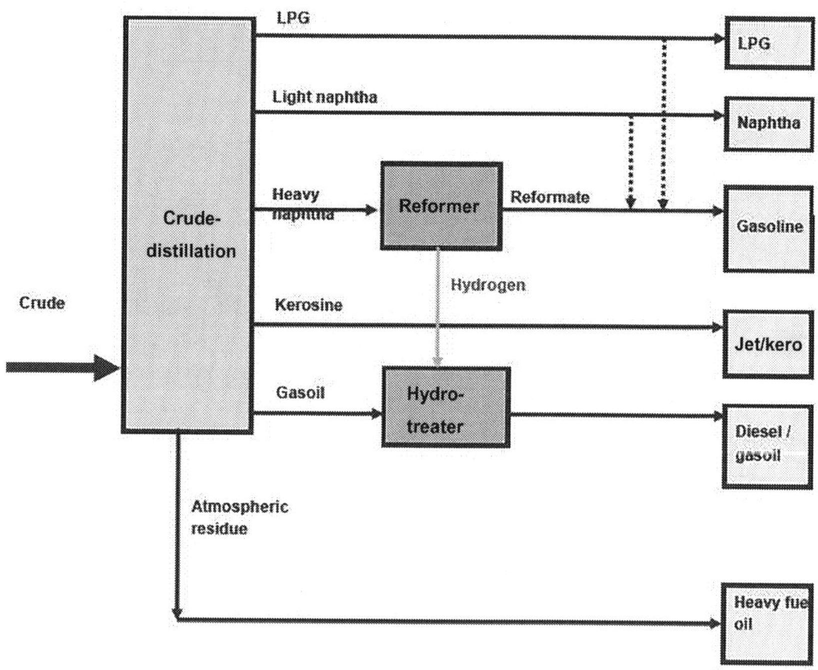

Figure 2.3 A hydroskimming refinery.

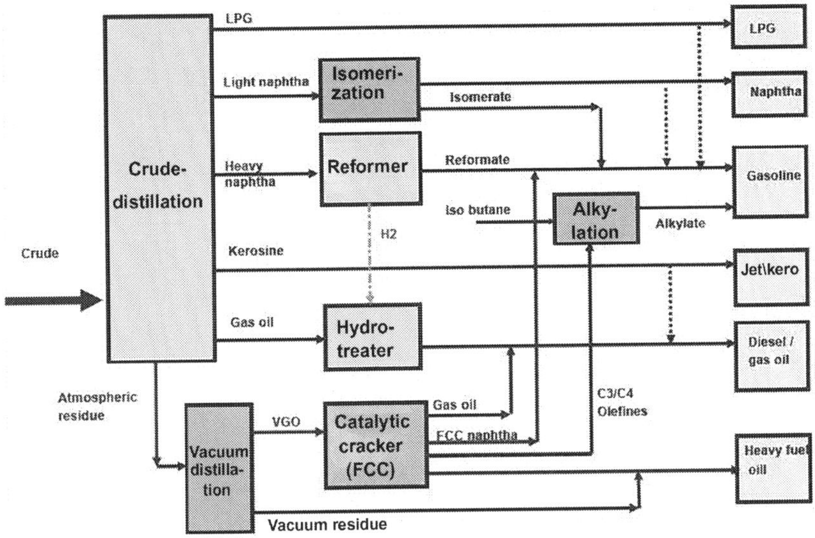

Figure 2.4 A catalytic cracking refinery.

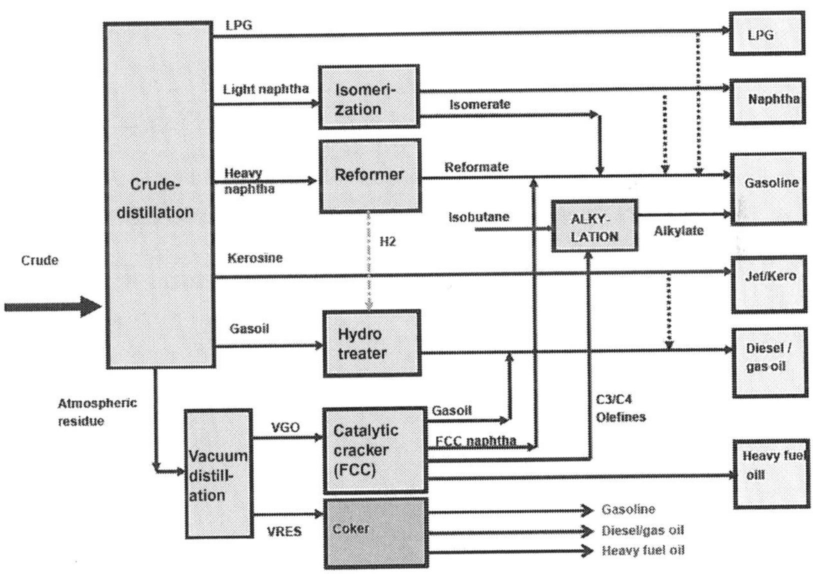

Figure 2.5 A coking refinery.

extraction processes for manufacturing lubricants and petrochemical units with which to recover propylene, benzene, toluene, and xylenes for further processing into polymers.

To convert crude oil into desired products in an economically feasible and environmentally acceptable manner. Refinery processes for crude oil are generally divided into four categories: (1) separation processes, of which distillation is the prime example; (2) conversion processes, of which coking and catalytic cracking are prime example; (3) finishing processes, of which hydrotreating to remove sulfur is a prime example; and (4) product improvement which involves product treating so that the product can meet sales specifications (Parkash, 2003; Gary et al., 2007; Speight, 2014, 2017; Hsu and Robinson, 2017).

2.3 Refinery processes

2.3.1 Dewatering and desalting

Crude oil is recovered from the reservoir mixed with a variety of substances: gases, water, and dirt (minerals). Thus refining actually commences with the production of fluids from the well or reservoir and is followed by pretreatment operations that are applied to the crude oil either at the refinery or prior to transportation. Pipeline operators, for instance, are insistent upon the quality of the fluids put into the pipelines; therefore any crude oil to be shipped by pipeline or, for that matter, by any other form of transportation must meet rigid specifications in regard to water and salt content. In some instances, sulfur content, nitrogen content, and viscosity may also be specified.

Field separation, which occurs at a field site near the recovery operation, is the first attempt to remove the gases, water, and dirt that accompany crude oil coming from the ground. The separator may be no more than a large vessel that gives a quieting zone for gravity separation into three phases: gases, crude oil, and water containing entrained dirt.

Desalting is a water-washing operation performed at the production field and at the refinery site for additional crude oil cleanup (Fig. 2.6). Desalting involves the mixing of heated crude oil with water in a large cylindrical tank. The water is separated from the crude oil in a separating vessel by adding demulsifier chemicals to assist in breaking the emulsion and/or, more commonly, by applying a high potential electric field across the settling vessel to coalesce the polar salt water droplets.

The crude oil from the separators may contain water and dirt, water washing can remove much of the water-soluble minerals and entrained solids. If these crude oil contaminants are not removed, they can cause operating problems during refinery processing, such as equipment plugging and corrosion as well as catalyst deactivation.

The usual practice is to blend crude oils of similar characteristics, although fluctuations in the properties of the individual crude oils may cause significant variations in the properties of the blend over a period of time. Blending several crude oils prior to refining can eliminate the frequent need to change the processing conditions that may be required to process each of the crude oils individually.

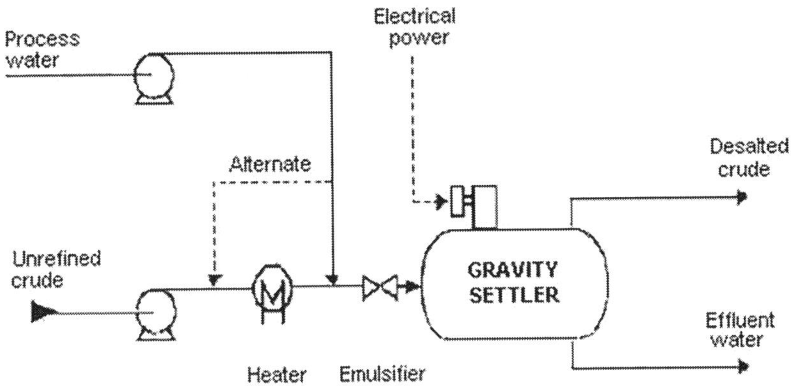

Figure 2.6 An electrostatic desalting unit.
Source: OSHA Technical Manual, Section IV, Chapter 2: Petroleum Refining Processes. http://www.osha.gov/dts/osta/otm/otm_iv/otm_iv_2.html.

However, simplification of the refining procedure is not always the end result. Incompatibility of different crude oils, which can occur if, for example, a paraffinic crude oil is blended with high-boiling asphaltic oil, can cause sediment formation in the unrefined feedstock or in the products, thereby complicating the refinery process (Speight, 2014).

2.3.2 Distillation

Distillation (Chapter 4: Distillation) was the first method by which crude oil was refined. In the early stages of refinery development, when illuminating and lubricating oils were the main products, distillation was the major and often only refinery process. At that time gasoline was a minor, but more often unwanted, product. As the demand for gasoline increased, conversion processes were developed because distillation could no longer supply the necessary quantities of this volatile product (Brooks et al., 1954).

The original distillation method involved a batch operation in which the still was a cast-iron vessel mounted on brickwork over a fire, and the volatile materials were passed through a pipe or gooseneck that led from the top of the still to a condenser. The latter was a coil of pipe (worm) immersed in a tank of running water. Heating a batch of crude oil caused the more volatile, lower boiling components to vaporize and then condense in the worm to produce naphtha. As the distillation progressed, the higher boiling components became vaporized and were condensed to produce kerosene, the major crude oil product of the time. When all of the possible kerosene had been obtained, the material remaining in the still was discarded. The still was then refilled with crude oil and the operation repeated. The capacity of the stills at that time was usually several barrels (bbl) of crude oil. (1 bbl = 42 US gallons = 34.97

imperial gallons = 158.9 liters of crude oil). It often required 3 days or even more to run (distill) a batch of crude oil.

The simple distillation as practiced in the 1860s and 1870s was notoriously inefficient. The kerosene was more often than not contaminated by naphtha, which distilled during the early stages, or by a high-boiling oil, which distilled from the residue during the final stages of the process. The naphtha generally rendered the kerosene so flammable explosions accompanied that ignition. On the other hand, the presence of higher boiling constituents adversely affected the excellent burning properties of the kerosene and created a great deal of smoke. This condition could be corrected by redistilling (rerunning) the kerosene, during which process the more volatile fraction (front-end) was recovered as additional naphtha, while the kerosene residue (tail) remaining in the still was discarded.

The capacity of the early was usually several barrels of crude oil and if often required 3 or more days to distill (run) a batch of crude oil and was notoriously inefficient. The kerosene was more often than not contaminated by naphtha, which distilled during the early stages, or by a high-boiling fraction, which distilled from the residue during the final stages of the process. The naphtha generally rendered the kerosene so flammable explosions accompanied that ignition. On the other hand, the presence of heavier oil adversely affected the excellent burning properties of the kerosene and created a great deal of smoke. This condition could be corrected by redistilling (rerunning) the kerosene, during which process the more volatile fraction (front-end) was recovered as additional naphtha, while the kerosene residue (tail) remaining in the still was discarded.

The 1880s saw the introduction of the continuous distillation of crude oil. The method employed a number of stills coupled together in a row (battery), and each still was heated separately and was hotter than the preceding one. The stills were arranged so that oil flowed by gravity from the first to the last. Crude oil in the first still was heated so that a low-boiling naphtha fraction distilled from it before the crude oil flowed into the second still, where a higher temperature caused the distillation of a heavier naphtha fraction. The residue then flowed to the third still where an even higher temperature caused kerosene to distill. The oil thus progressed through the battery to the last still, where destructive distillation (thermal decomposition; cracking) was carried out to produce more kerosene. The residue from the last still was removed continuously for processing into lubricating oils or for use as fuel oil.

In the early 1900s a method of partial (or selective) condensation was developed to allow a more exact separation of crude oil fractions. A partial condenser (van Dyke tower) was inserted between the still and the conventional water-cooled condenser. The lower section of the tower was packed with stones and insulated with brick so that the heavier less volatile material entering the tower condensed and drained back into the still. Noncondensed material passed into another section where more of the less volatile material was condensed on air-cooled tubes, and the condensate was withdrawn as a crude oil fraction. The volatile (overhead) material from the air-cooled section entered a second tower that also contained air-cooled tubes and often produced a second fraction. The volatile material remaining at this stage was then condensed in a water-cooled condenser to yield a third fraction.

2.3.2.1 Atmospheric distillation

The present-day crude oil distillation unit is a collection of distillation units and, in contrast to the early battery units of the 1800s, a tower is used in the modern-day refinery (Fig. 2.7) and achieves a controlled and efficient degree of fractionation (separation).

The feed to a distillation tower is heated by flow through pipes arranged within a large furnace. The heating unit is known as a pipe still heater or pipe still furnace, and the heating unit and the fractional distillation tower make up the essential parts of a distillation unit or pipe still. The pipe still furnace heats the feed to a predetermined temperature—usually a temperature at which a predetermined portion of the feed will change into vapor. The vapor is held under pressure in the pipe in the furnace until it discharges as a foaming stream into the fractional distillation tower. Here the nonvolatile liquid portion of the feed descends to the bottom of the tower to be pumped away as a bottom nonvolatile product, while the vapors pass up the tower to be fractionated into gas oils, kerosene, and naphtha.

Pipe still furnaces vary greatly and, in contrast to the early units where capacity was usually 200–500 bbl per day, can accommodate 25,000 bbl or more of crude oil per day. The walls and ceiling are insulated with firebrick, and the interior of

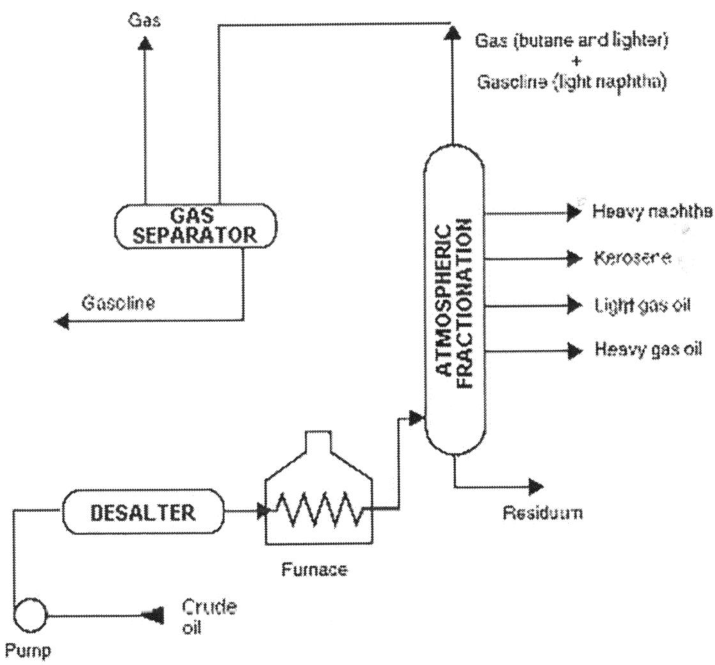

Figure 2.7 An atmospheric distillation unit.
Source: OSHA Technical Manual, Section IV, Chapter 2: Petroleum Refining Processes. http://www.osha.gov/dts/osta/otm/otm_iv/otm_iv_2.html.

the furnace is partially divided into two sections: a smaller convection section where the oil first enters the furnace and a larger section (fitted with heaters) where the oil reaches its highest temperature.

All of the primary fractions from a distillation unit are equilibrium mixtures and contain some proportion of the lower boiling constituents that are characteristic of a lower boiling fraction. The primary fractions are stripped of these constituents (stabilized) before storage or further processing.

2.3.2.2 Vacuum distillation

Vacuum distillation as applied to the crude oil refining industry is truly a technique of the 20th century and has since wide use in crude oil refining. Vacuum distillation evolved because of the need to separate the less volatile products, such as lubricating oils, from the crude oil without subjecting these high-boiling products to cracking conditions. The boiling point of the heaviest cut obtainable at atmospheric pressure is limited by the temperature (approximately 350°C; approximately 660°F) and residence time at which the residue starts to decompose (crack). When the feedstock is required for the manufacture of lubricating oils, further fractionation without cracking is desirable, and this can be achieved by distillation under vacuum conditions.

Operating conditions for vacuum distillation (Fig. 2.8) are usually 50–100 mm of mercury (atmospheric pressure: 760 mm of mercury). In order to minimize large fluctuations in pressure in the vacuum tower, the units are necessarily of a larger diameter than the atmospheric units. Some vacuum distillation units have diameters on the order of 45 ft. (14 m). By this means, a high boiling gas oil may be obtained as an overhead product at temperatures of approximately 150°C (300°F), and lubricating oil cuts may be obtained at temperatures of 250°C–350°C (480°F–660°F), feed and residue temperatures being kept at a temperature and residence time so that cracking does not occur. The partial pressure of the products is reduced still further by the injection of steam. The steam added to the column, principally for the stripping of asphalt in the base of the column, is superheated in the convection section of the heater.

The fractions obtained by vacuum distillation of the reduced crude (atmospheric residuum) from an atmospheric distillation unit depend on whether or not the unit is designed to produce lubricating or vacuum gas oils. In the former case the fractions include (1) high-boiling gas oil, which is an overhead product and is used as catalytic cracking stock or, after suitable treatment, a low-boiling lubricating oil; (2) lubricating oil (usually three fractions: low boiling, intermediate boiling, and high boiling), which is obtained as a side-stream product; and (3) residuum, which is the bottom product and may be used to produce asphalt, which may also be blended with a gas oil fraction to produce a high-boiling fuel oil.

2.3.2.3 Azeotropic distillation and extractive distillation

Distillation units as currently designed and used in refineries are incapable of producing specific product fractions. In order to accommodate this type of product

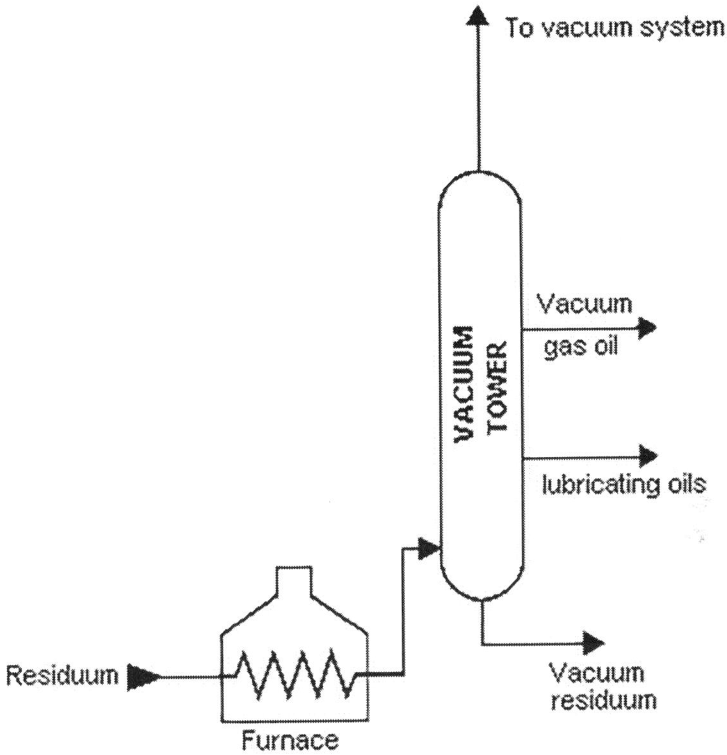

Figure 2.8 A vacuum distillation unit.
Source: OSHA Technical Manual, Section IV, Chapter 2: Petroleum Refining Processes. http://www.osha.gov/dts/osta/otm/otm_iv/otm_iv_2.html.

demand, refineries have, in the latter half of this century, incorporated azeotropic distillation and extractive distillation in their operations. When it is desired to separate close-boiling components, the addition of a nonindigenous component will form an azeotropic mixture with one of the components of the mixture thereby lowering the boiling point by the formation of an azeotrope and facilitate separation by distillation.

The separation of these components of similar volatility may become economic if an entrainer can be found that effectively changes the relative volatility. It is also desirable that the entrainer be reasonably cheap, stable, nontoxic, and readily recoverable from the components. In practice, it is probably this last-named criterion that limits severely the application of extractive and azeotropic distillation. The majority of successful processes are those in which the entrainer and one of the components separate into two liquid phases on cooling if direct recovery by distillation is not feasible. A further restriction in the selection of an azeotropic entrainer is that the boiling point of the entrainer be in the range of 10°C–40°C (18°F–72°F) below that of the components.

2.3.3 Thermal (noncatalytic) processes

Cracking distillation (thermal decomposition with simultaneous removal of distillate) was recognized as a means of producing the valuable lower boiling product (kerosene) from heavier nonvolatile materials (Chapter 5: Thermal Cracking). In the early days of the process (1870–1900), the technique was very simple—a batch of crude oil was heated until most of the kerosene had been distilled from it, and the overhead material had become dark in color. At this point, distillation was discontinued and the high-boiling constituents were held in the hot zone, during which time some of the high molecular weight components were decomposed to produce lower molecular weight products. After a suitable time, distillation was continued to yield low-boiling oil (kerosene) instead of the high-boiling oil that would otherwise have been produced.

The yields of kerosene products were usually markedly increased by means of cracking distillation but the technique was not suitable for gasoline production. As the need for gasoline arose in the early 1900s, the necessity of prolonging the cracking process became apparent and a process known as pressure cracking evolved.

Pressure cracking was a batch operation in which, for example, gas oil was heated to approximately 425°C (800°F) in stills that had been reinforced to operate at pressures as high as 95 psi (6.4 atmospheres). The gas oil was held under maximum pressure for 24 hours while the temperature was maintained. Distillation was then started during the next 48 hours to produce a lower boiling distillate that contained gasoline components. This distillate was treated with sulfuric acid to remove unstable gum-forming components and then redistilled to produce a cracked gasoline.

The large-scale production of cracked gasoline was first developed by Burton in 1912. The process employed batch distillation in horizontal shell stills and operated at approximately 400°C (approximately 750°F) and 75–95 psi. Although it was the first successful method of converting heavier oils into gasoline, heating a bulk volume of oil was soon considered cumbersome, and during the years 1914–22 a number of successful continuous cracking processes were developed.

Of these processes, the tube-and-tank cracking process is not only typical of the early (post-1900) cracking units but also is one of the first units on record in which the concept of reactors (soakers) being onstream/offstream is realized.

In the tube-and-tank process a feedstock (at that time a gas oil) was preheated by exchange with the hot products from the unit pumped into the cracking coil, which consisted of several hundred feet of very strong pipe that lined the inner walls of a furnace where oil or gas burners raised the temperature of the gas oil to 425°C (800°F). The hot gas oil passed from the cracking coil into a large reaction chamber (soaker) where the gas oil was held under the temperature and pressure conditions long enough for the cracking reactions to be completed. The cracking reactions formed coke that, in the course of several days, filled the soaker. The gas oil stream was then switched to a second soaker, and the first soaker was cleaned out by drilling operations similar to those used in drilling an oil well.

The volatile products left the onstream soaker to enter an evaporator (cracked residuum separator) maintained under a much lower pressure than the soaker where, because of the lower pressure, all of the cracked material, except the cracked residuum, became vaporized. The vapor left the top of the separator where it was distilled into separate fractions—gases, gasoline, and gas oil (which often included kerosene). The tar that was deposited in the separator was pumped out for use as asphalt or as a high-boiling fuel oil.

Such a concept departs from the true batch concept and allowed a greater degree of continuity. In fact, the tube-and-tank cracking was the forerunner of the delayed coking operation.

As refining technology evolved throughout the 20th century, feedstocks for cracking processes became the residuum or high-boiling distillate from a distillation unit. In addition, the residual oils produced as the end products of distillation processes, and even some of the heavier virgin oils often contain substantial amounts of asphaltic materials, which preclude use of the residuum as fuel oils or lubricating stocks. However, subjecting these residua directly to thermal processes has become economically advantageous, since, on the one hand, the end result is the production of lower boiling salable materials; on the other hand, the asphaltic materials in the residua are regarded as the unwanted coke-forming constituents.

With this as the goal a number of thermal processes, such as tar separation (flash distillation), vacuum flashing, visbreaking, and coking, came into wide usage by refiners and were directed at upgrading feedstocks by removal of the asphaltic fraction.

2.3.3.1 Thermal cracking

The majority of the thermal cracking processes use temperatures of 455°C–540°C (850°F–1005°F) and pressures of 100–1000 psi—the Dubbs process may be taken as a typical application of an early thermal cracking operation. In the process the feedstock (atmospheric residuum, reduced crude) is preheated by direct exchange with the cracking products in the fractionating columns. Cracked gasoline and heating oil are removed from the upper section of the column. Low-boiling and high-boiling distillate fractions are removed from the lower section and are pumped to separate heaters. Higher temperatures are used to crack the more refractory low-boiling distillate fraction. The streams from the heaters are combined and sent to a soaking chamber where additional time is provided to complete the cracking reactions. The cracked products are then separated in a low-pressure flash chamber where a high-boiling fuel oil is removed as bottoms. The remaining cracked products are sent to the fractionating columns.

Mild cracking conditions, with a low conversion per cycle, favor a high yield of gasoline components, with low gas and coke production, but the gasoline quality is not high, whereas more severe conditions give increased gas and coke production and reduced gasoline yield (but of higher quality). With limited conversion per cycle, the heavier residues must be recycled, but these recycle oils become increasingly refractory upon repeated cracking, and if they are not required as a fuel oil stock, they may be coked to increase gasoline yield or refined by means of a hydrogen process.

The antiknock requirements of modern automobile engines together with the different nature of crude oils (compared to those of 50 or more years ago) have reduced the ability of the thermal cracking process to produce gasoline constituents on an economic basis. Very few new units have been installed since the 1960s, and some refineries may still operate the older cracking units.

2.3.3.2 Visbreaking

Visbreaking (viscosity breaking) is essentially a process of the post-1940s era and was initially introduced as a mild thermal cracking operation that could be used to reduce the viscosity of residua to allow the product to meet fuel oil specifications. Alternatively, the visbroken residua could be blended with low-boiling product fractions to produce fuel oils of acceptable viscosity. By reducing the viscosity of the residuum, visbreaking reduces the amount of low-boiling heating oil that is required for blending to meet the fuel oil specifications. In addition to the major product, fuel oil, material in the gas oil and gasoline boiling range is produced. The gas oil may be used as additional feed for catalytic cracking units or as heating oil.

In a typical visbreaking operation (Fig. 2.9), residuum is passed through a furnace where it is heated to a temperature of 480°C–510°C (895°F–950°F) under an outlet pressure of approximately 100 psi. The heating coils in the furnace are arranged to provide a soaking section of low heat density, where the charge remains until the visbreaking reactions are completed, and the cracked products are then passed into a flash-distillation chamber. The overhead material from this chamber is then fractionated to produce a low-quality gasoline as an overhead product and low-boiling gas oil as bottom. The liquid products from the flash chamber are cooled with a gas oil flux and then sent to a vacuum fractionator. This yields a high-boiling gas oil distillate and a residual tar of reduced viscosity.

2.3.3.3 Coking

Coking is a thermal process for the continuous conversion of high-boiling, low-grade oils into lower boiling products. Unlike visbreaking, coking involved compete thermal conversion of the feedstock into volatile products and coke (Parkash, 2003; Gary et al., 2007; Speight, 2014, 2017; Hsu and Robinson, 2017). The feedstock is typically a residuum, and the products are gases, naphtha, fuel oil, gas oil, and coke. The gas oil may be the major product of a coking operation and serves primarily as a feedstock for catalytic cracking units. The coke obtained is usually used as fuel but specialty uses such as electrode manufacture, production of chemicals, and metallurgical coke are also possible and increase the value of the coke. For these uses, the coke may require treatment to remove sulfur and metal impurities.

2.3.3.3.1 Delayed coking

Delayed coking (Chapter 5: Thermal Cracking) is a semicontinuous process (Fig. 2.10) (reminiscent of the tube-and-tank process), in which the heated charge is

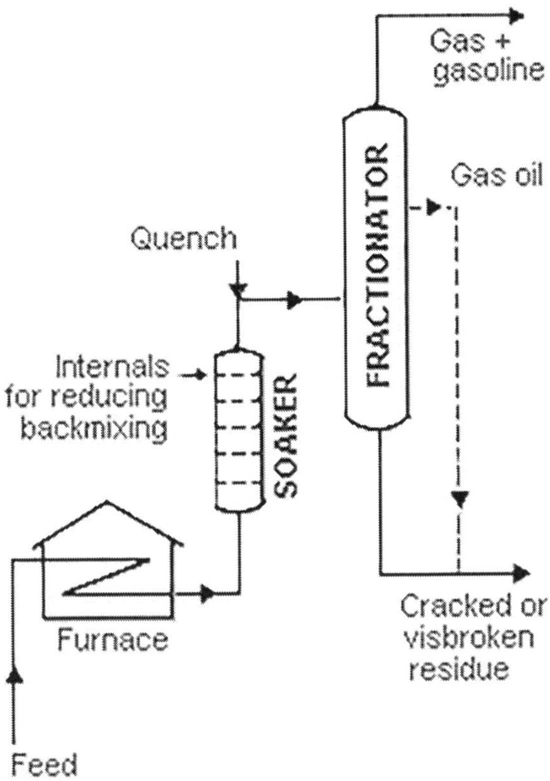

Figure 2.9 A soaker visbreaker.
Source: OSHA Technical Manual, Section IV, Chapter 2: Petroleum Refining Processes. http://www.osha.gov/dts/osta/otm/otm_iv/otm_iv_2.html.

transferred to large soaking (or coking) drums, which provide the long residence time needed to allow the cracking reactions to proceed to completion. The feed to these units is normally an atmospheric residuum although cracked residua are also used.

The feedstock is introduced into the product fractionator where it is heated, and lower boiling fractions are removed as a side stream. The fractionator bottoms, including a recycle stream of high-boiling product, are then heated in a furnace whose outlet temperature varies from 480°C to 515°C (895°F to 960°F). The heated feedstock enters one of a pair of coking drums where the cracking reactions continue. The cracked products leave as overheads, and coke deposits form on the inner surface of the drum. To give continuous operation, two drums are used; while one is onstream, the other is being cleaned. The temperature in the coke drum ranges from 415°C to 450°C (780°F to 840°F) with pressures from 15 to 90 psi.

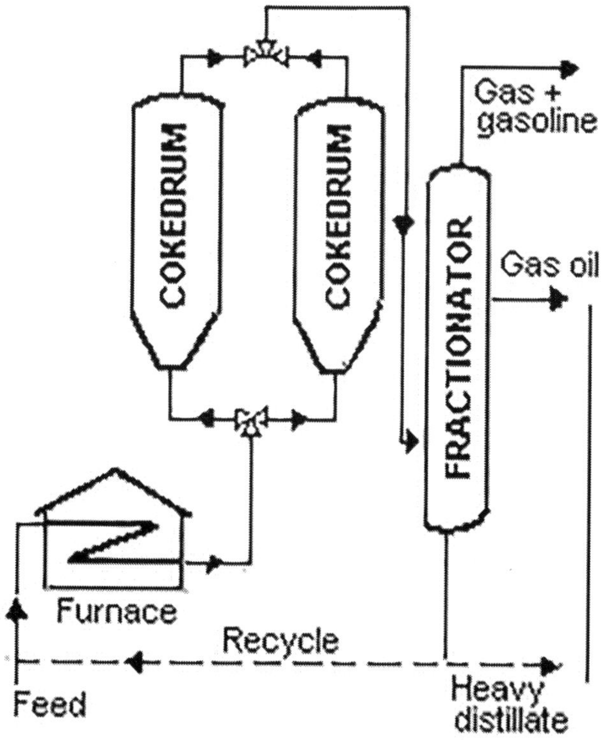

Figure 2.10 A delayed coker.
Source: OSHA Technical Manual, Section IV, Chapter 2: Petroleum Refining Processes. http://www.osha.gov/dts/osta/otm/otm_iv/otm_iv_2.html.

Overhead products go to the fractionator, where naphtha and heating oil fractions are recovered. The nonvolatile material is combined with preheated fresh feed and returned to the reactor. The coke drum is usually onstream for approximately 24 hours before becoming filled with porous coke after which the coke is removed hydraulically. Normally, 24 hours are required to complete the cleaning operation and prepare the coke drum for subsequent use onstream.

2.3.3.3.2 Fluid Coking

Fluid coking (Chapter 5: Thermal Cracking) is a continuous process (Fig. 2.11) that uses the fluidized-solids technique to convert atmospheric and vacuum residua to more valuable products. The residuum is coked by being sprayed into a fluidized bed of hot, fine coke particles, which permits the coking reactions to be conducted at higher temperatures and shorter contact times than can be employed in delayed coking. Relative to the delayed coking process, the fluid coking process results in decreased yields of coke and higher quantities of more valuable liquid products.

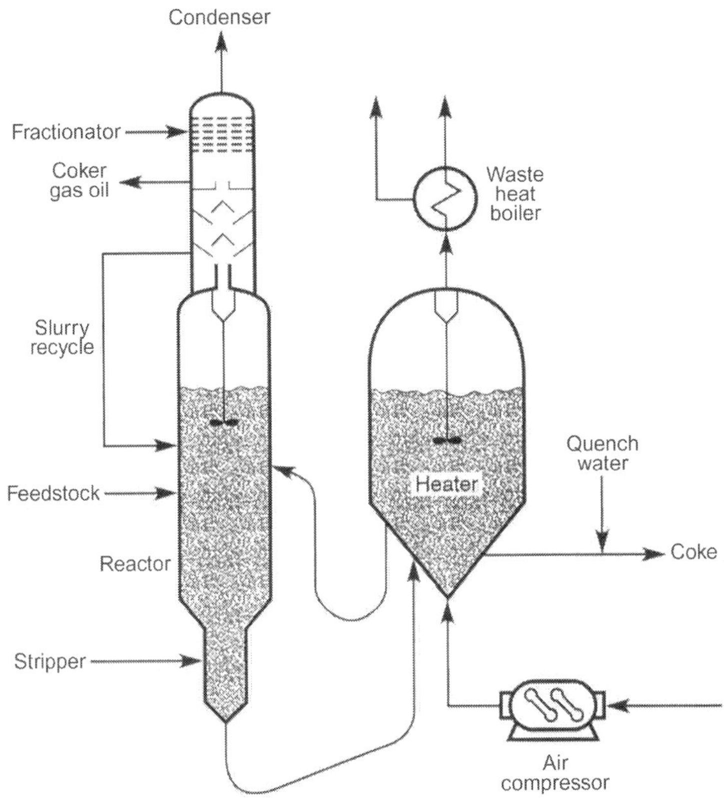

Figure 2.11 A fluid coker.

Fluid coking uses two vessels: a reactor and a burner; coke particles are circulated between these to transfer heat (generated by burning a portion of the coke) to the reactor. The reactor holds a bed of fluidized coke particles, and steam is introduced at the bottom of the reactor to fluidize the bed.

Flexicoking (Fig. 2.12) is also a continuous process that is a direct descendent of fluid coking. The unit uses the same configuration as the fluid coker but has a gasification section in which excess coke can be gasified to produce refinery fuel gas. The flexicoking process was designed during the late 1960s and the 1970s as a means by which excess coke-make could be reduced in view of the gradual incursion of the heavier feedstocks in refinery operations. Such feedstocks are notorious for producing high yields of coke ($>15\%$ by weight) in thermal and catalytic operations.

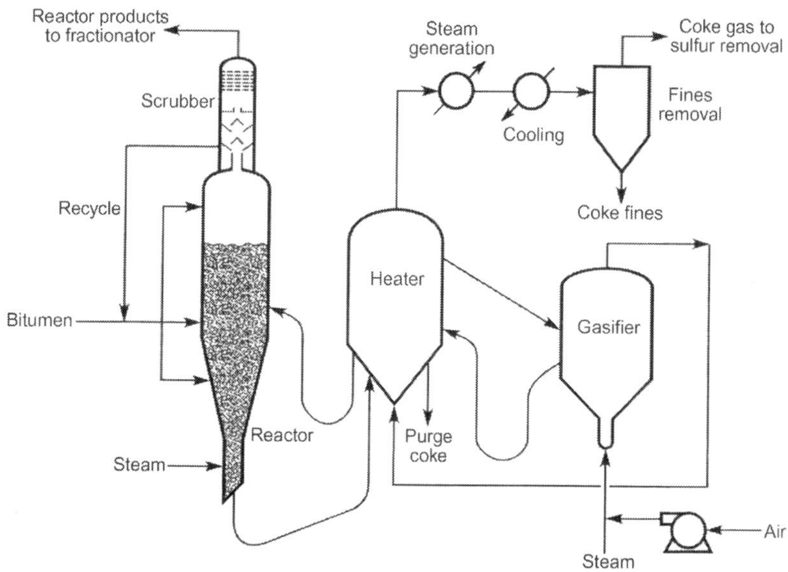

Figure 2.12 Flexicoking process.

2.3.4 Catalytic cracking processes

In the 1930s thermal cracking units produced approximately 50% of the total gasoline. The octane number of this gasoline was approximately 70 compared to 60 for straight-run (distilled) gasoline. The thermal reforming and polymerization processes that were developed during the 1930s could be expected to further increase the octane number of gasoline to some extent, but an additional innovation was needed to increase the octane number of gasoline to enhance the development of more powerful automobile engines.

In 1936 a new cracking process opened the way to higher octane gasoline—this involved the use of a catalyst and became known as catalytic cracking. The catalyst directs the course of the cracking reactions to produce more of the desired higher octane hydrocarbon products. In theory the catalyst is not consumed in the process but, in practice, because of coke and metals lay down on the catalyst, frequent catalyst renewal is necessary—the timing of the catalyst renewal is feedstock dependent.

The products of catalytic cracking have a number of advantages over thermal cracking: (1) the gasoline produced has a higher octane number, (2) the catalytically cracked gasoline consists largely of isoparaffins and aromatics that have high-octane numbers, and (3) the catalytically cracked gasoline has greater chemical stability than monoolefins and diolefins that are present in much greater quantities in thermally cracked gasoline.

Substantial quantities of olefinic gases suitable for polymer gasoline manufacture and smaller quantities of methane, ethane, and ethylene are produced by catalytic cracking. These products can also be used for manufacture of petrochemicals.

Sulfur compounds are changed in such a way that the sulfur content of catalytically cracked gasoline is lower than in thermally cracked gasoline. Catalytic cracking produces less cracked residuum and more of the useful gas oils (which can be used as hydrocracker feedstocks) than thermal cracking. However, during the cracking reaction, carbonaceous material is deposited on the catalyst, which markedly reduces its activity, and removal of the deposit is very necessary. This is usually accomplished by burning the catalyst in the presence of air until catalyst activity is reestablished.

The several processes (Parkash, 2003; Gary et al., 2007; Speight, 2014, 2017; Hsu and Robinson, 2017) currently employed in catalytic cracking differ mainly in the method of catalyst handling, although there is overlap with regard to catalyst type and the nature of the products. The catalyst, which may be an activated natural or synthetic material, is employed in bead, pellet, or microspherical form and can be used as a fixed bed, moving bed, or fluid bed. The fixed-bed process was the first process to be used commercially and uses a static bed of catalyst in several reactors, which allows a continuous flow of feedstock to be maintained. Thus the cycle of operations consists of (1) flow of feedstock through the catalyst bed, (2) discontinuance of feedstock flow and removal of coke from the catalyst by burning, and (3) insertion of the reactor onstream. The moving-bed process uses a reaction vessel (in which cracking takes place) and a kiln (in which the spent catalyst is regenerated), and catalyst movement between the vessels is provided by various means.

2.3.4.1 Fluid-bed catalytic cracking

In the fluid-bed process (Fig. 2.13), the powdered catalyst is circulated as a fluidized solid with the feedstock. The several fluid catalytic cracking processes in use differ primarily in mechanical design. Side-by-side reactor–regenerator construction along with unitary vessel construction (the reactor either above or below the regenerator) is the two main mechanical variations.

2.3.4.2 Moving-bed catalytic cracking

The moving-bed process uses a reaction vessel in which cracking takes place and a kiln in which the spent catalyst is regenerated, and catalyst movement between the vessels is provided by various means.

In an example of the moving-bed process, vapor-liquid flows downward through the reactor concurrently with catalyst. The catalyst is purged by steam at the base of the reactor and gravitates into the kiln, or regeneration is accomplished by the use of air injected into the kiln. Approximately 70% of the carbon on the catalyst is burned in the upper kiln burning zone and the remainder in the

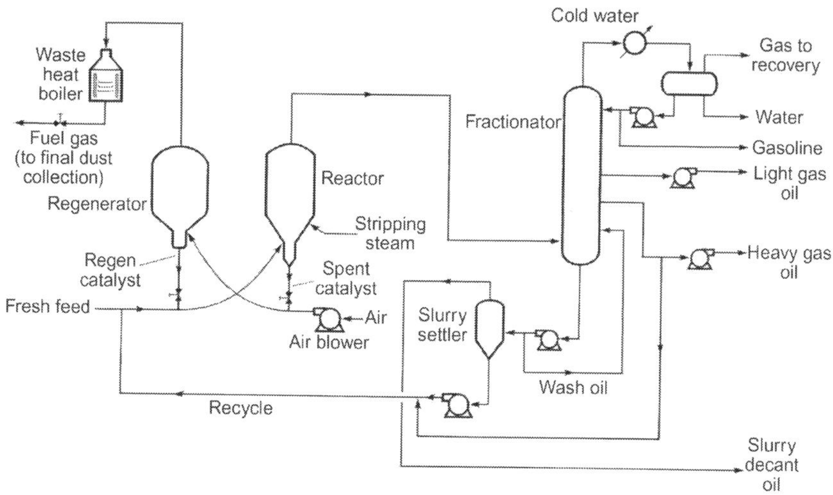

Figure 2.13 A FCC unit. *FCC*, Fluid catalytic cracking.

bottom-burning zone. Regenerated, cooled catalyst enters the lift pot, where low-pressure air transports it to the surge hopper above the reactor for reuse.

2.3.4.3 Fixed-bed catalytic cracking

The fixed-bed process was the first to be used commercially and uses a static bed of catalyst in several reactors that allows a continuous flow of feedstock to be maintained. Thus the cycle of operations consists of (1) flow of feedstock through the catalyst bed, (2) discontinuance of feedstock flow and removal of coke from the catalyst by burning, and (3) insertion of the reactor onstream.

In the fixed-bed process the catalyst in the form of small lumps or pellets was made up in layers or beds in several (four or more) catalyst-containing drums called converters. Feedstock vaporized at approximately 450°C (840°F) and less than 7–15 psi pressure passed through one of the converters where the cracking reactions took place. After a short time, deposition of coke on the catalyst rendered it ineffective, and using a synchronized valve system, the feed stream was turned into a neighboring converter while the catalyst in the first converter was regenerated by carefully burning the coke deposits with air. After approximately 10 minutes the catalyst was ready to go onstream again.

2.3.4.4 Catalysts

Natural clays have long been known to exert a catalytic influence on the cracking of oils, but it was not until 1936 that the process using silica–alumina catalysts was developed sufficiently for commercial use. The major innovations of the 20th century lie not only in reactor configuration and efficiency but also in catalyst development.

Cracking crude oil fractions occurs over many types of catalytic materials, but high yields of desirable products are obtained with hydrated aluminum silicates. These may be either activated (acid-treated) natural clays of the bentonite type of synthesized silica–alumina or silica–magnesia preparations. Their activity to yield essentially the same products may be enhanced to some extent by the incorporation of small amounts of other materials such as the oxides of zirconium, boron (which has a tendency to volatilize away on use), and thorium. Natural and synthetic catalysts can be used as pellets or beads and also in the form of powder; in either case replacements are necessary because of attrition and gradual loss of efficiency. It is essential that they are stable to withstand the physical impact of loading and thermal shocks and that they withstand the action of carbon dioxide–, air-, and nitrogen-containing compounds and steam. They also should be resistant to sulfur and nitrogen compounds, and synthetic catalysts, or certain selected clays, appear to be better in this regard than average untreated natural catalysts.

The catalysts are porous and highly adsorptive, and their performance is affected markedly by the method of preparation. Two chemically identical catalysts having pores of different sizes and distribution may have different activity, selectivity, temperature coefficients of reaction rates, and responses to poisons. The intrinsic chemistry and catalytic action of a surface may be independent of pore size, but small pores produce different effects because of the manner in which hydrocarbon vapors are transported into and out of the pore systems.

2.3.5 Hydroprocesses

Hydrogenation processes for the conversion of crude oil fractions and crude oil products may be classified as (1) nondestructive hydrogenation (2) destructive hydrogenation (Chapter 8: Desulfurization, Denitrogenation, and Demetalization, and Chapter 9: Hydrocracking).

Nondestructive hydrogenation (hydrogenation, hydrotreating) is generally used for the purpose of improving product quality without appreciable alteration of the boiling range. Mild processing conditions are employed so that only the more unstable materials are attacked. Nitrogen, sulfur, and oxygen compounds undergo reaction with the hydrogen to remove ammonia, hydrogen sulfide, and water, respectively. Unstable compounds that might lead to the formation of gums, or insoluble materials, are converted to more stable compounds.

Destructive hydrogenation (hydrogenolysis, hydrocracking) is characterized by the conversion of the higher molecular weight constituents in a feedstock to lower boiling products. Such treatment requires severe processing conditions and the use of high hydrogen pressures to minimize polymerization and condensation reactions that lead to coke formation.

2.3.5.1 Hydrotreating

Hydrotreating and hydroprocessing are similar processes used to remove impurities such as sulfur, nitrogen, oxygen, halides, and trace metal impurities that may deactivate process catalysts (Chapter 8: Desulfurization, Denitrogenation, and Demetalization). Hydrotreating also upgrades the quality of fractions by converting olefins and diolefins to paraffins for the purpose of reducing gum formation in fuels. Hydroprocessing, which typically uses residuals from the crude distillation units, also cracks these heavier molecules to lower boiling more saleable products. Both hydrotreating and hydroprocessing units are usually placed upstream of those processes in which sulfur and nitrogen could have adverse effects on the catalyst, such as catalytic reforming and hydrocracking units. The processes utilize catalysts in the presence of substantial amounts of hydrogen under high pressure and temperature to react the feedstocks and impurities with hydrogen. The reactors are nearly all fixed bed with catalyst replacement or regeneration done after months or years of operation often at an off-site facility. In addition to the treated products, the process produces a stream of fuel gases as well as hydrogen sulfide and ammonia. The treated product and hydrogen-rich gas are cooled after they leave the reactor before being separated. The hydrogen is recycled to the reactor.

The commercial processes for hydrotreating (hydrofinishing) crude oil fractions all operate in essentially the same manner. The feedstock is heated and passed with hydrogen gas through a tower or reactor filled with catalyst pellets. The reactor is maintained at a temperature of 260°C–425°C (500°F–800°F) at pressures from 100 to 1000 psi, depending on the particular process, the nature of the feedstock and the degree of hydrogenation required. After leaving the reactor, excess hydrogen is separated from the treated product and recycled through the reactor after removal of hydrogen sulfide. The liquid product is passed into a stripping tower where steam removes dissolved hydrogen and hydrogen sulfide, and after cooling the product is taken to product storage or, in the case of feedstock preparation, pumped to the next processing unit.

Hydrotreating (Fig. 2.14) is carried out by charging the feed to the reactor, together with hydrogen in the presence of catalysts such as tungsten–nickel sulfide,

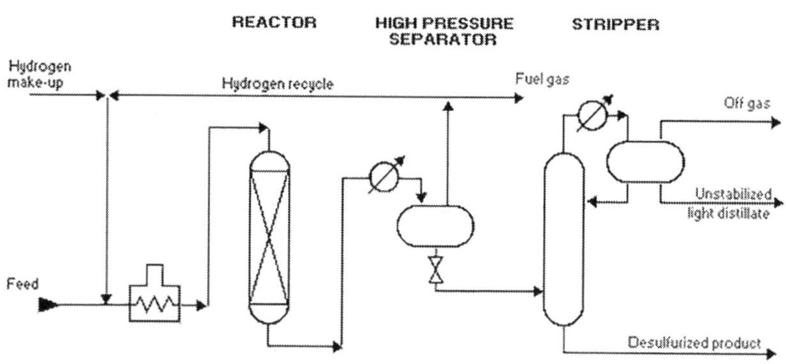

Figure 2.14 A distillate hydrotreater for hydrodesulfurization.
Source: OSHA Technical Manual, Section IV, Chapter 2: Petroleum Refining Processes. http://www.osha.gov/dts/osta/otm/otm_iv/otm_iv_2.html.

cobalt–molybdenum–alumina, nickel oxide–silica–alumina, and platinum–alumina. Most processes employ cobalt–molybdena catalysts that generally contain approximately 10% of molybdenum oxide and less than 1% of cobalt oxide supported on alumina. The temperatures employed are in the range of 260°C–345°C (500°F–655°F), while the hydrogen pressures are approximately 500–1000 psi.

The reaction generally takes place in the vapor phase but, depending on the application, may be a mixed-phase reaction. Generally, it is more economical to hydrotreat high-sulfur feedstocks prior to catalytic cracking than to hydrotreat the products from catalytic cracking. In this option, several benefits accrue as follows: (1) sulfur is removed from the catalytic cracking feedstock, and corrosion is reduced in the cracking unit; (2) carbon formation during cracking is reduced so that higher conversions result; and (3) the cracking quality of the gas oil fraction is improved.

2.3.5.2 Hydrocracking

Hydrocracking is a refining technology that such as hydrotreating falls under the general umbrella of hydroprocessing (Chapter 9: Hydrocracking). The outcome is the conversion of a variety of feedstocks to a range of products and units to accomplish this goal can be found at various points in a refinery (Parkash, 2003; Gary et al., 2007; Speight, 2014, 2017; Hsu and Robinson, 2017).

Thus hydrocracking is a more recent process development compared to the older thermal cracking, visbreaking, and coking. In fact, the use of hydrogen in thermal processes is perhaps the single most significant advance in refining technology during the 20th century, and the ability of refiners to cope with the renewed trend toward distillate production from heavier feedstocks with low atomic hydrogen/carbon ratios has created a renewed interest in hydrocracking.

Catalytic hydrocracking normally utilizes a fixed-bed catalytic cracking reactor with cracking occurring under substantial pressure (1200–2000 psi) in the presence of hydrogen. Feedstocks to hydrocracking units are often those fractions that are the most difficult to crack and cannot be cracked effectively in catalytic cracking units. These include (1) middle distillate fractions, (2) cycle oils, (3) residual fuel oils, and (4) residua. The feedstock is often first hydrotreated to remove impurities before being sent to the catalytic hydrocracker. Sometimes, hydrotreating is accomplished by using the first reactor of the hydrocracking process to remove impurities. Water has a detrimental effect on some hydrocracking catalysts and must be removed before being fed to the reactor. The water is removed by passing the feed stream through a silica gel or molecular sieve dryer. Depending on the products desired and the size of the unit, catalytic hydrocracking is conducted in either single-stage or multistage reactor processes. Most catalysts consist of a crystalline mixture of silica–alumina with small amounts of rare-earth metals.

The history of the process goes back to the late 1920s when it was realized that there was a need for gasoline of a higher quality than that obtained by catalytic cracking led to the development of the hydrocracking process. One of the first plants to use hydrocracking was commissioned for the commercial hydrogenation of brown coal was commissioned at Leuna in Germany. Tungsten sulfide was used as a catalyst in this one-stage unit, in which high reaction pressures, 2900–4350 psi, were applied. In 1939 Imperial Chemical Industries in Britain developed the second-stage catalyst for a plant for production of aviation gasoline.

During World War II, two-stage processes were applied on a limited scale in Germany, Britain, and the United States. In Britain, feedstocks were creosote from coal tar and gas oil from crude oil. In the United States, Standard Oil of New Jersey operated a plant at Baton Rouge, producing gasoline from a Venezuelan kerosene/low-boiling gas oil fraction. Operating conditions in those units were comparable: approximate reaction temperature 400°C (750°F) and reaction pressures of 2900–4350 psi. After the war, commercial hydrocracking was very expensive, but by the end of the 1950s the process had become economic. The development of improved catalyst made it possible to operate the process at considerably lower pressure, which resulted in a reduction in equipment wall thickness as well as advances in reactor design.

The older hydrogenolysis type of hydrocracking practiced in Europe during and after World War II used tungsten sulfide (WS_2) or molybdenum sulfide (MoS) as catalysts. These processes required high reaction temperatures and operating pressures, sometimes in excess of approximately 3000 psi for continuous operation. Modern hydrocracking processes were initially developed for converting refractory feedstocks to gasoline and jet fuel. Process and catalyst improvements and modifications have made it possible to yield products from gases and naphtha to furnace oils and catalytic cracking feedstocks. The zeolites most frequently used in commercial hydrocracking catalysts are partially dealuminated and low sodium, or high-silica, type Y zeolites in hydrogen or rare-earth forms.

Hydrocracking processes (Fig. 2.15) were initially developed for converting refractory feedstocks (such as gas oils) to gasoline and jet fuel, but process and catalyst improvements and modifications have made it possible to yield products from gases and naphtha to furnace oils and catalytic cracking feedstocks.

2.3.6 Reforming

Reforming processes (molecular rearrangement processes) are used for product improvement and are processes in which the molecular structure of the feedstock is reorganized but without significant effects on the molecular weight (Parkash, 2003; Gary et al., 2007; Speight, 2014, 2017; Hsu and Robinson, 2017). An example is the conversion of *n*-hexane (molecular weight: 86) to cyclohexane (molecular weight: 84) or cyclohexane to benzene (molecular weight: 78). These processes reform or rearrange one particular molecular type to another thereby changing the properties of the product relative to the feedstock.

2.3.6.1 Thermal reforming

Thermal reforming was a natural development from older thermal cracking processes; cracking converts heavier oils into gasoline, whereas reforming converts

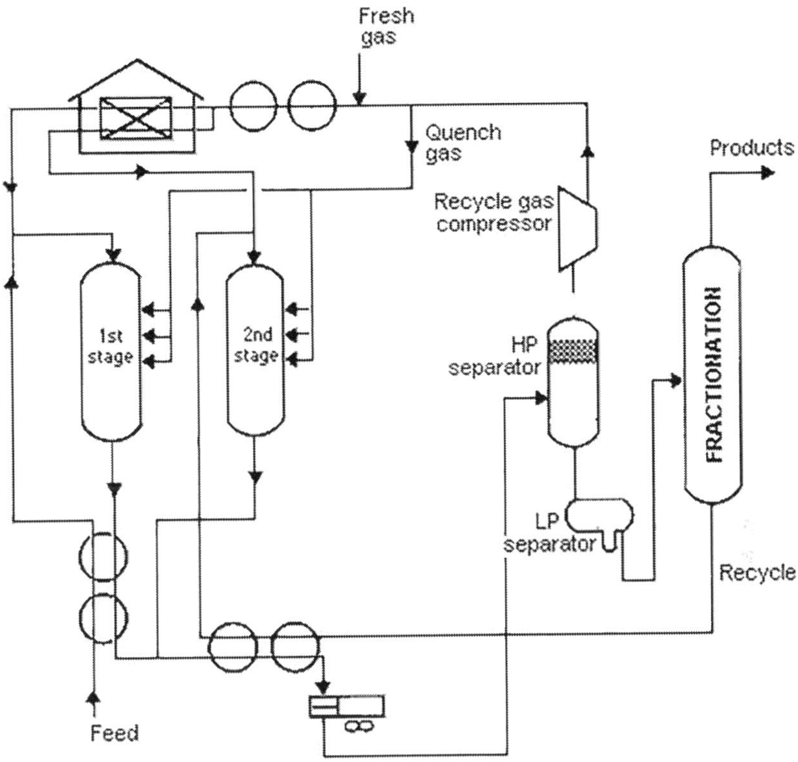

Figure 2.15 A single-stage or two-stage (optional) hydrocracking unit.
Source: OSHA Technical Manual, Section IV, Chapter 2: Petroleum Refining Processes. http://www.osha.gov/dts/osta/otm/otm_iv/otm_iv_2.html.

(reforms) gasoline into higher octane gasoline. The equipment for thermal reforming is essentially the same as for thermal cracking, but higher temperatures are used.

In carrying out thermal reforming a feedstock such as naphtha (end point: 205°C, 400°F) or straight-run gasoline is heated to 510°C–595°C (950°F–1100°F) in a furnace with pressures on the order of 400–1000 psi (27–68 atmospheres). As the heated naphtha leaves the furnace, it is cooled or quenched by the addition of cold naphtha. The material then enters a fractional distillation tower where any high-boiling products are separated. The remainder of the reformed material leaves the top of the tower to be separated into gases and higher octane reformate.

The products of thermal reforming are gases, reformate, and residual oil or nonvolatile residuum, the latter being formed in very small amounts (approximately 1%). The amount and quality of the reformate product is dependent on the temperature. Generally, the higher the reforming temperature, the higher the octane number, but the lower the yield of reformate.

More recent modifications of the thermal reforming process due to the inclusion of hydrocarbon gases with the feedstock are known as gas reversion and polyforming. Thus olefinic gases produced by cracking and reforming can be converted into liquids boiling in the gasoline range by heating them under high pressure. Since the resulting liquids (polymers) have high-octane numbers, they increase the overall quantity and quality of gasoline produced in a refinery.

2.3.6.2 Catalytic reforming

Catalytic reforming uses catalytic reactions to process primarily low-octane high-boiling straight-run (from the crude distillation unit) gasolines and naphthas into high-octane aromatics (including benzene). There are four major types of reactions that occur during reforming processes: (1) dehydrogenation of naphthenes to aromatics, (2) dehydrocyclization of paraffins to aromatics, (3) isomerization, and (4) hydrocracking. The dehydrogenation reactions are very endothermic, requiring that the hydrocarbon stream be heated between each catalyst bed. All but the hydrocracking reaction release hydrogen that can be used in the hydrotreating or hydrocracking processes. Fixed-bed or moving-bed processes are utilized in a series of three to six reactors. Feedstocks to catalytic reforming processes are usually hydrotreated first to remove sulfur, nitrogen, and metallic contaminants. In continuous reforming processes, catalysts can be regenerated one reactor at a time, once or twice per day, without disrupting the operation of the unit. In semiregenerative units, regeneration of all reactors can be carried out simultaneously after 3 to 24 months of operation by first shutting down the process. Because the recent reformulated gasoline rules have limited the allowable amount of benzene in gasoline, catalytic reforming is being used less as an octane enhancer than in past years.

The catalytic reforming process is a process of the 1950s and showed phenomenal growth in 1953–59 time period.

Like thermal reforming, catalytic reforming converts low-octane gasoline into high-octane gasoline (reformate). When thermal reforming could produce reformate with research octane numbers of 65–80 depending on the yield, catalytic reforming produces reformate with octane numbers on the order of 90–95. Catalytic reforming is conducted in the presence of hydrogen over hydrogenation–dehydrogenation catalysts, which may be supported on alumina or silica–alumina. Depending on the catalyst, a definite sequence of reactions takes place, involving structural changes in the feedstock. This more modern concept actually rendered thermal reforming somewhat obsolescent.

The commercial processes available for use can be broadly classified as (1) moving-bed processes, (2) fluid-bed processes, and (3) fixed-bed processes. The moving-bed and fluid-bed processes use mixed nonprecious metal oxide catalysts in units equipped with separate regeneration facilities. Fixed-bed processes use predominantly platinum-containing catalysts in units equipped for cycle, occasional, or no regeneration.

Catalytic reformer feeds are saturated (i.e., not olefinic) materials; in the majority of cases that feed may be a straight-run naphtha, but other by-product low-octane naphtha (e.g., coker naphtha) can be processed after treatment to remove olefins and other contaminants. Hydrocracker naphtha that contains substantial quantities of naphthenes is also a suitable feed.

Dehydrogenation is a main chemical reaction in catalytic reforming, and hydrogen gas is consequently produced in large quantities. The hydrogen is recycled though the reactors where the reforming takes place to provide the atmosphere necessary for the chemical reactions and also prevents the carbon from being deposited on the catalyst, thus extending its operating life. An excess of hydrogen above whatever is consumed in the process is produced, and, as a result, catalytic reforming processes are unique in that they are the only crude oil refinery processes to produce hydrogen as a by-product.

Catalytic reforming usually is carried out by feeding a naphtha (after pretreating with hydrogen if necessary) and hydrogen mixture to a furnace where the mixture is heated to the desired temperature, 450°C–520°C (840°F–965°F), and then passed through fixed-bed catalytic reactors at hydrogen pressures of 100–1000 psi (7–68 atmospheres) (Fig. 2.16). Normally, pairs of reactors are used in series with heaters are located between adjoining reactors in order to compensate for the endothermic reactions taking place. As many as four or five reactors may be onstream in series while one or more is being regenerated and the onstream cycle of any one reactor may vary from several hours to many days, depending on the feedstock and reaction conditions.

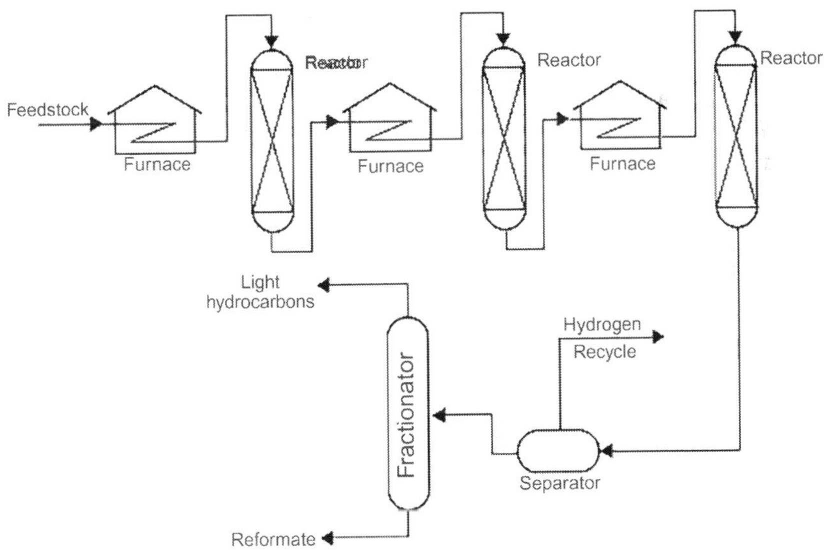

Figure 2.16 Catalytic reforming.
Source: OSHA Technical Manual, Section IV, Chapter 2: Petroleum Refining Processes. http://www.osha.gov/dts/osta/otm/otm_iv/otm_iv_2.html.

2.3.6.3 Catalysts

The catalysts used in catalytic reforming processes are principally molybdena–alumina, chromia–alumina, or platinum on a silica–alumina or alumina base. The nonplatinum catalysts are widely used in regenerative process for feeds containing, for example, sulfur, which poisons platinum catalysts, although pretreatment processes (e.g., hydrodesulfurization) may permit platinum catalysts to be employed.

The purpose of platinum on the catalyst is to promote dehydrogenation and hydrogenation reactions, that is, the production of aromatics, participation in hydrocracking, and rapid hydrogenation of carbon-forming precursors. For the catalyst to have an activity for isomerization of both paraffins and naphthenes and to participate in paraffin dehydrocyclization, it must have an acid activity. The balance between these two activities is most important in a reforming catalyst. In fact, in the production of aromatics from cyclic saturated materials (naphthenes), it is important that hydrocracking be minimized to avoid loss of the desired product, and thus the catalytic activity must be moderated relative to the case of gasoline production from a paraffinic feed, where dehydrocyclization and hydrocracking play an important part.

2.3.7 Isomerization

The conversion of *n*-paraffins to the respective branched-chain isomers (isomerization) yields gasoline components of high-octane rating in this lower boiling range (Parkash, 2003; Gary et al., 2007; Speight, 2014, 2017; Hsu and Robinson, 2017). Conversion is obtained in the presence of a catalyst (aluminum chloride activated with hydrochloric acid), and it is essential to inhibit side reactions such as cracking and olefin formation.

Isomerization found initial commercial applications during World War II for making high-octane aviation gasoline components and additional feed for alkylation units. The lowered alkylate demands in the period after World War II led to the majority of the butane isomerization units being shut down. In recent years the greater demand for high-octane lead-free gasoline has resulted in new butane isomerization units being installed.

2.3.7.1 Processes

Isomerization may take place in the vapor phase, with the activated catalyst supported on a solid phase, or in the liquid phase with a dissolved catalyst. In the process, pure butane or a mixture of isomeric butanes (Fig. 2.17) is mixed with hydrogen (to inhibit olefin formation) and passed to the reactor, at 110°C–170°C (230°F–340°F) and 200–300 psi (14–20 atmospheres). The product is cooled, the hydrogen separated and the cracked gases are then removed in a stabilizer column. The stabilizer bottom product is passed to a superfractionator where the normal butane is separated from the isobutane.

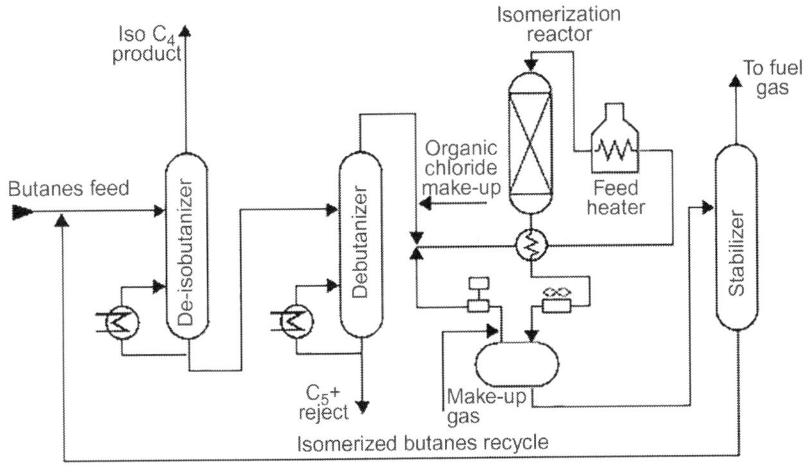

Figure 2.17 A butane isomerization unit.
Source: OSHA Technical Manual, Section IV, Chapter 2: Petroleum Refining Processes. http://www.osha.gov/dts/osta/otm/otm_iv/otm_iv_2.html.

The present isomerization processes are used with the objective of providing additional feedstock for alkylation units or high-octane fractions for gasoline blending. Straight-chain paraffins (*n*-butane, *n*-pentane, *n*-hexane) are converted to respective isocompounds by continuous catalytic (aluminum chloride, noble metals) processes. Natural gasoline or low-boiling straight-run gasoline can provide feed by first fractionating as a preparatory step. High volumetric yields (>95%) and 40%–60% conversion per pass are characteristic of the isomerization reaction.

2.3.7.2 Catalysts

During World War II aluminum chloride was the catalyst used to isomerize butane, pentane, and hexane. Since then, supported metal catalysts have been developed for use in high-temperature processes that operate in the range of 370°C–480°C (700°F–900°F) and 300–750 psi (20–51 atmospheres), while aluminum chloride plus hydrogen chloride are universally used for the low-temperature processes.

Nonregenerable aluminum chloride catalyst is employed with various carriers in a fixed-bed or liquid contactor. Platinum or other metal catalyst processes utilized fixed-bed operation and can be regenerable or nonregenerable. The reaction conditions vary widely depending on the particular process and feedstock, 40°C–480°C (100°F–900°F) and 150–1000 psi (10 68 atmospheres).

2.3.8 Alkylation processes

The combination of olefins with paraffins to form higher isoparaffins is termed alkylation (Parkash, 2003; Gary et al., 2007; Speight, 2014, 2017; Hsu and

Robinson, 2017). Since olefins are reactive (unstable) and are responsible for exhaust pollutants, their conversion to high-octane isoparaffins is desirable when possible. Although alkylation is possible without catalysts, commercial processes use aluminum chloride, sulfuric acid, or hydrogen fluoride as catalysts, when the reactions can take place at low temperatures, minimizing undesirable side reactions, such as polymerization of olefins.

Alkylation is used to produce a high-octane gasoline blending stock from the isobutane formed primarily during catalytic cracking and coking operations, but also from catalytic reforming, crude distillation, and natural gas processing (Mokhatab et al., 2006; Speight, 2016). Alkylation combines an olefin and an isoparaffin using either a sulfuric acid or hydrofluoric acid catalyst. The products are alkylates, including propane and butane liquids.

Alkylate (the product of an alkylation process) is composed of a mixture of isoparaffins that have octane numbers that vary with the olefins from which they were made. Butylenes produce the highest octane numbers, propylene the lowest, and pentylene derivatives the intermediate octane numbers. All alkylates, however, have high-octane numbers (>87).

2.3.8.1 Processes

Alkylation processes are directed toward production of high-octane liquids for aviation gasoline. The sulfuric acid process was introduced in 1938, and hydrogen fluoride alkylation was introduced in 1942. Rapid commercialization took place during the war to supply military needs, but many of these plants were shut down at the end of the war.

In the 1950s aviation gasoline demand started to decline, but motor gasoline quality requirements rose sharply. To aid in the improvement of the economics of the alkylation process and also the sensitivity of the premium gasoline pool, additional olefins were gradually added to alkylation feed. New plants were built to alkylate propylene and the butylenes (butanes) produced in the refinery rather than the butane–butylene stream formerly used.

The alkylation reaction as now practiced in crude oil refining is the union, through the agency of a catalyst, of an olefin (ethylene, propylene, butylene, and amylene) with isobutane to yield high-octane branched-chain hydrocarbon derivatives in the gasoline boiling range. Olefin feedstock is derived from the gas produced in a catalytic cracker, while isobutane is recovered by refinery gases or produced by catalytic butane isomerization.

To accomplish this, either ethylene or propylene is combined with isobutane at 50°C–280°C (125°F–450°F) and 300–1000 psi (20–68 atmospheres) in the presence of metal halide catalysts such as aluminum chloride. Conditions are less stringent in catalytic alkylation; olefins (propylene, butylene derivatives, or pentylene derivatives) are combined with isobutane in the presence of an acid catalyst (sulfuric acid or hydrofluoric acid) at low temperatures and pressures (1°C–40°C, 30°F–105°F, and 14.8–150 psi; 1–10 atmospheres) (Fig. 2.18). When the concentration of acid becomes less than 88% v/v, some of the acid must be removed and

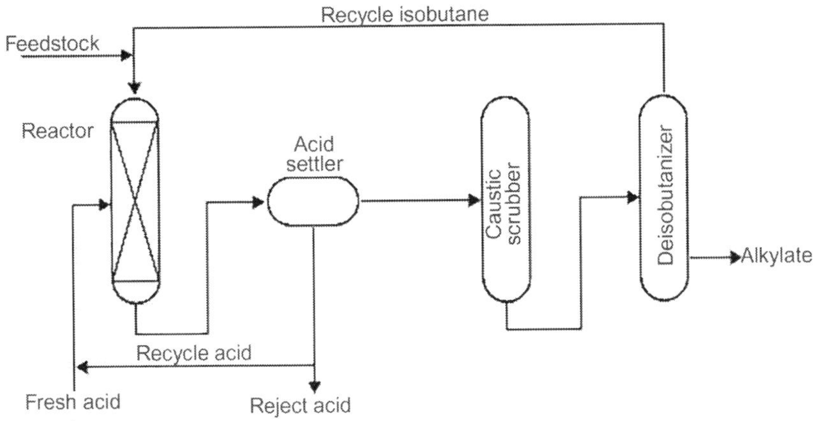

Figure 2.18 An alkylation unit (sulfuric acid catalyst).
Source: OSHA Technical Manual, Section IV, Chapter 2: Petroleum Refining Processes. http://www.osha.gov/dts/osta/otm/otm_iv/otm_iv_2.html.

replaced with stronger acid. In the hydrofluoric acid process the slipstream of acid is redistilled. Dissolved polymerization products are removed from the acid as thick dark oil.

Hydrofluoric acid alkylation units require special engineering design, operator training, and safety equipment precautions to protect operators from accidental contact with hydrofluoric acid that is an extremely hazardous substance. In the sulfuric acid process the sulfuric acid removed must be regenerated in a sulfuric acid plant that is generally not a part of the alkylation unit and may be located off-site.

2.3.8.2 Catalysts

Sulfuric acid, hydrogen fluoride, and aluminum chloride are the general catalysts used commercially. Sulfuric acid is used with propylene and higher boiling feeds, but not with ethylene, because it reacts to form ethyl hydrogen sulfate. The acid is pumped through the reactor and forms an air emulsion with reactants, and the emulsion is maintained at 50% acid.

Aluminum chloride is not widely used as an alkylation catalyst but when employed, hydrogen chloride is used as a promoter, and water is injected to activate the catalyst as an aluminum chloride/hydrocarbon complex. Hydrogen fluoride is used for alkylation of higher boiling olefins, and the advantage of hydrogen fluoride is that it is more readily separated and recovered from the resulting product.

2.3.9 Polymerization processes

In the crude oil industry, polymerization is the process by which olefin gases are converted to liquid products that may be suitable for gasoline (polymer gasoline) or other liquid fuels (Parkash, 2003; Gary et al., 2007; Speight, 2014, 2017; Hsu and Robinson, 2017). The feedstock usually consists of propylene and butylenes from cracking processes or may even be selective olefins for dimer, trimer, or tetramer production.

Polymerization is a process that can claim to be the earliest process to employ catalysts on a commercial scale. Catalytic polymerization came into use in the 1930s and was one of the first catalytic processes to be used in the crude oil industry.

Polymerization is occasionally used to convert propene and butene to high-octane gasoline blending components. The process is similar to alkylation in its feed and products but is often used as a less expensive alternative to alkylation. The reactions typically take place under high pressure in the presence of a phosphoric acid catalyst. The feed must be free of sulfur, which poisons the catalyst; basic materials, which neutralize the catalyst; and oxygen, which affects the reactions. The propene and butene feed is washed first with caustic to remove mercaptans (molecules containing sulfur), then with an amine solution to remove hydrogen sulfide, then with water to remove caustics and amines, and finally dried by passing through a silica gel or molecular sieve dryer.

2.3.9.1 Processes

Thermal polymerization is regarded as not being as effective as catalytic polymerization but has the advantage that it can be used for saturated materials that cannot be induced to react by catalysts. The process consists of vapor-phase cracking of, for example, propane and butane followed by prolonged periods at the high temperature (510°C–595°C, 950°F–1100°F) for the reactions to proceed to near completion.

Olefins can also be conveniently polymerized by means of an acid catalyst (Fig. 2.19). Thus the treated, olefin-rich feed stream is contacted with a catalyst

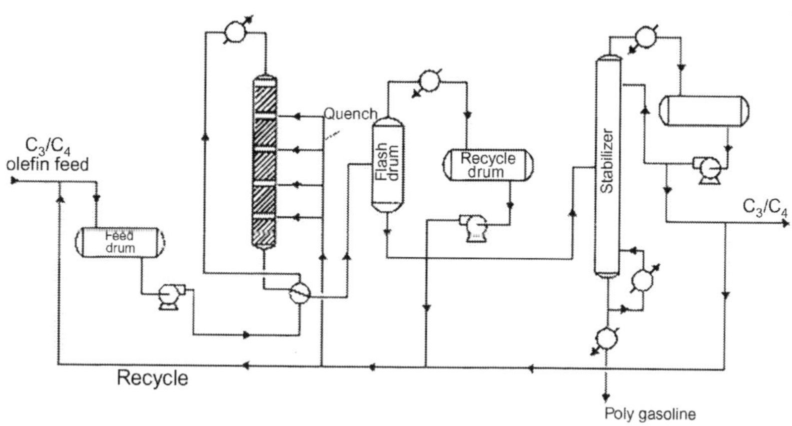

Figure 2.19 A polymerization unit.

(sulfuric acid, copper pyrophosphate, phosphoric acid) at 150°C–220°C (300°F–425°F) and 150–1200 psi (10–81 atmospheres), depending on feedstock and product requirement.

2.3.9.2 Catalysts

Phosphates are the principal catalysts used in polymerization units; the commercially used catalysts are liquid phosphoric acid, phosphoric acid on kieselguhr, copper pyrophosphate pellets, and phosphoric acid film on quartz. The latter is the least active, but the most used and easiest one to regenerate simply by washing and recoating; the serious disadvantage is that tar must occasionally be burned off the support. The process using liquid phosphoric acid catalyst is far more responsible to attempts to raise production by increasing temperature than the other processes.

Propane deasphalting was developed to produce lubricating oil base stocks by extracting asphaltenes and resins from the residuals of the vacuum distillation unit. Propane is usually used to remove asphaltenes due to its unique solvent properties. At lower temperatures, paraffins are very soluble in propane, and at higher temperatures, all hydrocarbon derivatives are almost insoluble in propane (Parkash, 2003; Gary et al., 2007; Speight, 2014, 2017; Hsu and Robinson, 2017).

Solvent deasphalting is essentially a solvent extraction process (Parkash, 2003; Gary et al., 2007; Speight, 2014, 2017; Hsu and Robinson, 2017). Low-boiling paraffins such as propane, butane, or their mixtures are used as the solvent. Low-boiling liquid paraffin solvent partially or completely dissolves the constituents of low degree of aromaticity, which causes asphaltenes and resins to precipitate. The dissolved components, known as deasphalted oil (DAO), are recovered from solvent in flash column and stripper.

The major process variables are temperature, pressure, solvent-to-oil ratio, and solvent type. Pressure and temperature are both variables because the solvent power of low-boiling hydrocarbon is approximately proportional to the density of the solvent. Higher temperature always results in decreased yield of DAO. On the other hand, increasing solvent-to-oil ratio increases the recovery of DAO with increase in viscosity. However, for the given product quality that can be maintained with change in temperature, solvent-to-oil ratio increases the yield of DAO. It has been shown that solvent power of paraffin solvent increases with increase in solvent molecular weight.

Solvent deasphalting processes are a major part of refinery operations (Parkash, 2003; Gary et al., 2007; Speight, 2014, 2017; Hsu and Robinson, 2017) and are not often appreciated for the tasks for which they are used. In the solvent deasphalting processes an alkane is injected into the feedstock to disrupt the dispersion of components and causes the polar constituents to precipitate. Propane (or sometimes propane/butane mixtures) is extensively used for deasphalting and produces a DAO and propane deasphalter bottoms (also referred to as propane deasphalter asphalt, PDA, or propane deasphalter tar, PD tar) (Dunning and Moore, 1957). Propane has unique solvent properties; at lower temperatures (38°C–60°C; 100°F–140°F),

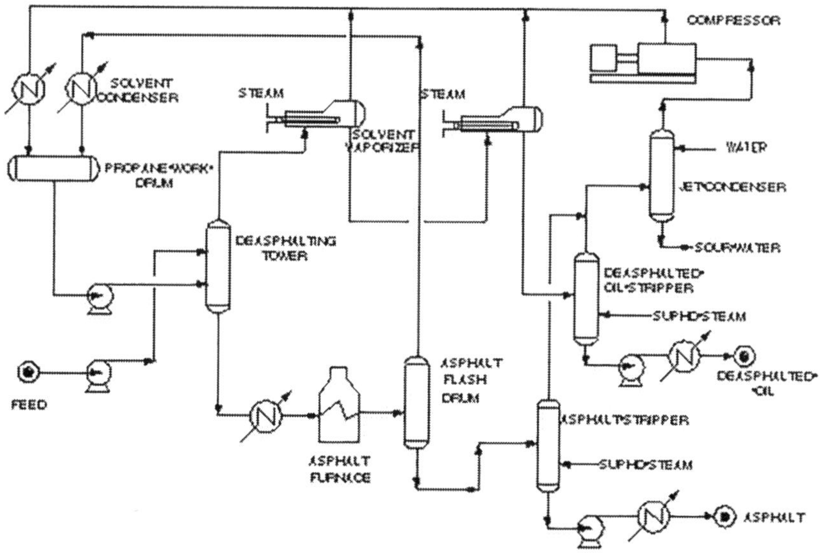

Figure 2.20 Propane deasphalting.
Source: OSHA Technical Manual, Section IV, Chapter 2: Petroleum Refining Processes. http://www.osha.gov/dts/osta/otm/otm_iv/otm_iv_2.html.

paraffins are very soluble in propane, and at higher temperatures (approximately 93°C; 200°F), all hydrocarbon derivatives are almost insoluble in propane.

A solvent deasphalting unit (Fig. 2.20) processes the residuum from the vacuum distillation unit and produces DAO, used as feedstock for a fluid catalytic cracking unit, and the asphaltic residue (deasphalter tar, deasphalter bottoms) which, as a residual fraction, can only be used to produce asphalt or as a blend stock or visbreaker feedstock for low-grade fuel oil. Solvent deasphalting processes have not realized their maximum potential. With ongoing improvements in energy efficiency, such processes would display its effects in a combination with other processes. Solvent deasphalting allows removal of sulfur and nitrogen compounds as well as metallic constituents by balancing yield with the desired feedstock properties (Parkash, 2003; Gary et al., 2007; Speight, 2014, 2017; Hsu and Robinson, 2017).

The propane deasphalting process is similar to solvent extraction in that a packed or baffled extraction tower or rotating disk contactor is used to mix the oil feedstocks with the solvent. In the tower method, 4–8 vol of propane are fed to the bottom of the tower for every volume of feed flowing down from the top of the tower. The oil, which is more soluble in the propane, dissolves and flows to the top. The asphaltene and resins flow to the bottom of the tower where they are removed in a propane mix. Propane is recovered from the two streams through two-stage flash systems followed by steam stripping in which propane is condensed and removed by cooling at high pressure in the first stage and at low pressure in the second stage.

The asphalt recovered can be blended with other asphalts or high-boiling fuels or can be used as feed to the coker.

2.3.9.3 Dewaxing

Solvent dewaxing processes are designed to remove wax from lubricating oils to give the product with good fluidity characteristics at low temperatures (e.g., low pour points) rather than from the whole crude oil (Parkash, 2003; Gary et al., 2007; Speight, 2014, 2017; Hsu and Robinson, 2017). The mechanism of solvent dewaxing involves either the separation of wax as a solid that crystallizes from the oil solution at low temperature or the separation of wax as a liquid that is extracted at temperatures above the melting point of the wax through preferential selectivity of the solvent—the former mechanism is the usual basis for commercial dewaxing processes.

Dewaxing of lubricating oil base stocks is necessary to ensure that the oil will have the proper viscosity at lower ambient temperatures. Two types of dewaxing processes are used: selective hydrocracking and solvent dewaxing. In selective hydrocracking, one or two zeolite catalysts are used to selectively crack the wax paraffins. Solvent dewaxing is more prevalent. In solvent dewaxing the oil feed is diluted with solvent to lower the viscosity, chilled until the wax is crystallized, and then filtered to remove the wax. Solvents used for the process include propane and mixtures of methyl ethyl ketone (MEK) with methyl isobutyl ketone or MEK with toluene.

In the 1930s two types of stocks, naphthenic and paraffinic, were used to make motor oils. Both types were solvent extracted to improve their quality, but in the high-temperature conditions encountered in service, the naphthenic type could not stand up as well as the paraffinic type. Nevertheless, the naphthenic type was the preferred oil, particularly in cold weather, because of its fluidity at low temperatures. Previous to 1938 the highest quality lubricating oils were of the naphthenic type and were phenol treated to pour points of −40°C to −7°C (−40°F to 20°F), depending on the viscosity of the oil. Paraffinic oils were also available and could be phenol treated to higher quality oil, but their wax content was so high that the oils were solid at room temperature.

The lowest viscosity paraffinic oils were dewaxed by the cold press method to produce oils with a pour point of 2°C (35°F). The low-boiling paraffin distillate oils contained a paraffin wax that crystallized into large crystals when chilled and could thus readily be separated from the oil by the cold press filtration method. The more viscous paraffinic oils (intermediate- and high-boiling paraffin distillates) contained amorphous or microcrystalline waxes, which formed small crystals that plugged the filter cloths in the cold press and prevented filtration. Because the wax could not be removed from intermediate- and high-boiling paraffin distillates, the high-quality, high-viscosity lubricating oils in them could not be used except as cracking stock.

Methods were therefore developed to dewax these high-viscosity paraffinic oils. The methods were essentially alike in that the waxy oil was dissolved in a solvent that would keep the oil in solution; the wax separated as crystals when the

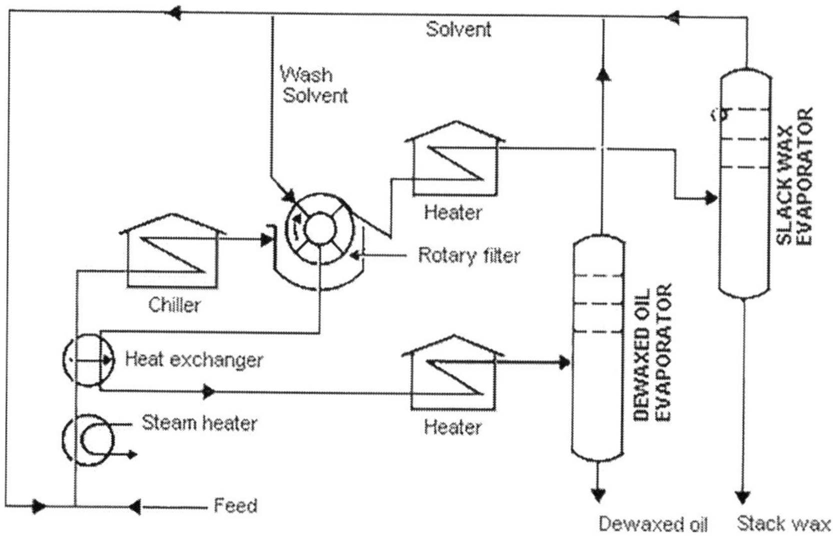

Figure 2.21 A solvent dewaxing unit.
Source: OSHA Technical Manual, Section IV, Chapter 2: Petroleum Refining Processes. http://www.osha.gov/dts/osta/otm/otm_iv/otm_iv_2.html.

temperature was lowered. The processes differed chiefly in the use of the solvent. Commercially used solvents were naphtha, propane, sulfur dioxide, acetone–benzene, trichloroethylene, ethylene dichloride–benzene (Barisol), MEK–benzene (benzol), methyl-*n*-butyl ketone, and methyl-*n*-propyl ketone.

The process as now practiced involves mixing the feedstock with one to four times its volume of the ketone (Fig. 2.21) (Parkash, 2003; Gary et al., 2007; Speight, 2014, 2017; Hsu and Robinson, 2017). The mixture is then heated until the oil is in solution and the solution is chilled at a slow, controlled rate in double-pipe, scraped-surface exchangers. Cold solvent, such as filtrate from the filters, passes through the 2-in. annular space between the inner and outer pipes and chills the waxy oil solution flowing through the inner 6-in. pipe.

2.3.10 Gas cleaning

Gas cleaning processes are an essential part of refinery operations, especially when the as stream in intended for use in synthesis gas production (Chapter 12: Synthesis Gas and the Fischer–Tropsch process) or in the petrochemical section of the refinery (Speight, 2019b). To include a description of all of the possible process variations for gas cleaning is beyond the scope of this book and is covered in detail elsewhere (Maddox, 1982; Kohl and Riesenfeld, 1985; Maddox et al., 1985; Kohl and Nielsen, 1997; Mokhatab et al., 2006; May-Britt, 2008; Speight, 2013, 2014, 2017, 2019a).

The gas cleaning systems may comprise different processes and methods depending on the specific requirements. The most common technologies utilized are venturi-type scrubbers, disintegrators, bag filters, and electrostatic precipitators. However, the selection of a cleaning technology (or technologies) is site specific, case sensitive, and dependent on the gas utilization requirements and whatever the source of the gas (Ramírez et al., 2015). But first it is important to recognize the composition of the gas so that the appropriate technologies can be selected for use to remove the specific contaminants. Thus the cleaning of the raw gas is needed strictly in order to remove contaminants and potential catalyst poisons as well as to achieve the qualitative composition required before the gas is sold for use.

A gas stream, as produced by any of the refinery processes, is typically saturated with water vapor and contains, in addition to methane (CH_4) and carbon dioxide (CO_2), various amounts of hydrogen sulfide (H_2S) as well as other contaminants that vary in amounts depending upon the composition of the feedstocks and the process parameters employed for production of the gas. The properties of the contaminants are such that failure to remove them will make the gas unusable and even poisonous to the user. For example, hydrogen sulfide is a toxic gas, with a specific, unpleasant odor, similar to rotten eggs, forming acidic products in combination with the water vapor in the gas stream which can also result in corrosion of equipment. To prevent this, and similar example of the adverse effects of other contaminants, the gas stream must be dried, and any contaminants removed before the gas stream is sent to one or more conversion units or for sales.

Typically, in the gas processing industry (also called the gas cleaning industry or the gas refining industry), the feedstock is not used directly as fuel because of the complex nature of the feedstock and the presence of one or more of the aforementioned impurities that are corrosive or poisonous. It is, therefore, essential that any gas should be contaminant free when it enters any one of the various reactors (Parkash, 2003; Gary et al., 2007; Speight, 2014, 2017; Hsu and Robinson, 2017). Typically, the primary raw gas stream has been subjected to chemical and/or physical changes (refining) after being produced. For many applications the quality of the gas stream has to be improved.

The main contaminants (but not necessarily the only contaminants) that may require removal in gas cleaning systems are hydrogen sulfide, water, carbon dioxide, and halogenated compounds. Desulfurization prevents corrosion and avoids concentrations of the toxic hydrogen sulfide for safety in use. Also, if hydrogen sulfide and other sulfur-containing species such as thiol derivatives (RSH, also called mercaptan derivatives) and carbonyl sulfide (COS) are not removed, combustion of the gas produces sulfur dioxide and sulfur trioxide, which is even more poisonous than hydrogen sulfide. At the same time, the presence of sulfur dioxide in the gas stream lowers the dew point (the temperature to which the gas must be cooled to become saturated with water vapor) of in the stack gas. The sulfurous acid formed (H_2SO_3) by reaction of the sulfur dioxide and the water vapor as well as reaction of carbon dioxide and the water vapor to form highly corrosive acidic species:

$$SO_2 + H_2O \rightarrow H_2SO_3$$

$$CO_2 + H_2O \rightarrow H_2CO_3$$

Thus the removal of water from the gas stream is also essential, not only because of the potential for the reaction of water with contaminants in the gas stream but also because of potential accumulation of water vapor condensing in the gas line, the formation of a corrosive acidic solution when hydrogen sulfide is dissolved or to achieve low dew points when the gas stream is stored under elevated pressures in order to avoid condensation and freezing. Also, if the noncombustible carbon dioxide is not removed, the energy content of the gas stream is diluted and there may also be an environmental impact due to the presence and emission of this contaminant.

Generally, the contaminants are categorized as (1) particulate matter; (2) condensable hydrocarbon derivatives, including tar products; (3) alkali metals, such as sodium and potassium; (4) nitrogen-containing derivatives, including ammonia and hydrogen cyanide; (5) sulfur-containing derivatives, such as hydrogen sulfide, COS, and carbon disulfide; (6) and halogen-containing derivatives, such as hydrogen chloride and hydrogen bromide, and hydrogen fluoride.

In fact, the constituents make up the tar products that are formed during gasification when biomass decomposes in pyrolysis and gasification reactions. Primary tar compounds are mostly oxygenated compounds that are decomposition products of biomass. These compounds react further and form secondary and tertiary tar compounds, which consist of compounds that do not exist in the source biomass. The secondary tar typically consists of alkylated one-ring and two-ring aromatic compounds (including heterocyclic compounds), whereas tertiary tar consists of aromatic hydrocarbon derivatives such as benzene, naphthalene, and various polycyclic aromatic hydrocarbon derivatives (PAHs, also called polynuclear aromatic compounds, PNAs). Tar constituents are typically classified on the basis of the number of rings in the constituents or by boiling point distribution or physical properties. More generally, tar is defined as aromatic compounds that are higher boiling than benzene, and in addition, an operational definition for tar depending on the end-use application has been used.

Process selectivity indicates the preference with which the process removes one acid gas component relative to (or in preference to) another. For example, some processes remove both hydrogen sulfide and carbon dioxide; other processes are designed to remove hydrogen sulfide only. It is very important to consider the process selectivity for, say, hydrogen sulfide removal compared to carbon dioxide removal that ensures minimal concentrations of these components in the product, thus the need for consideration of the carbon dioxide and hydrogen sulfide in the gas stream.

2.3.11 Ancillary operations

Refineries typically utilize primary and secondary wastewater treatment. Primary wastewater treatment consists of the separation of oil, water, and solids. After primary treatment the wastewater can be discharged to a publicly owned treatment work or undergo secondary treatment before being discharged directly to surface waters under a National Pollution Discharge Elimination System permit. In secondary treatment, dissolved oil and other organic pollutants may be consumed biologically by microorganisms.

Sulfur is removed from a number of refinery process off-gas streams (sour gas) in order to meet the SO_x emissions limits of the Clean Air Act (CAA) and to recover saleable elemental sulfur. Process off-gas streams, or sour gas, from the coker, catalytic cracking unit, hydrotreating units, and hydroprocessing units can contain high concentrations of hydrogen sulfide mixed with the refinery fuel gases. Before elemental sulfur can be recovered, the fuel gases (primarily methane and ethane) need to be separated from the hydrogen sulfide. This is typically accomplished by dissolving the hydrogen sulfide in a chemical solvent. Solvents most commonly used are amines, such as diethanolamine.

A number of chemicals (mostly alcohols and ethers) are added to motor fuels to either improve performance or meet federal and state environmental requirements. Since the 1970s, alcohols (methanol and ethanol) and ethers have been added to gasoline to increase octane levels and reduce carbon monoxide generation in place of the lead additives that were being phased out as required by the various environmental regulations. The more stringent regulations established minimum and maximum amounts of chemically combined oxygen in motor fuels as well as an upper limit on vapor pressure. As a result, alcohol additives have been increasingly supplemented or replaced with a number of different ethers that are better able to meet both the new oxygen requirements and the vapor pressure limits.

The most common ethers being used as additives are methyl tertiary butyl ether (MTBE) and tertiary amyl methyl ether (TAME). Many of the larger refineries manufacture their own supplies of MTBE and TAME by reacting isobutylene and/or isoamylene with methanol. Smaller refineries usually buy their supplies from chemical manufacturers or the larger refineries. Isobutylene is obtained from a number of refinery sources, including the low-boiling naphtha from the fluid catalytic cracking unit and the coking unit, the by-product from steam cracking of naphtha or low-boiling hydrocarbon derivatives during the production of ethylene and propylene, catalytic dehydrogenation of isobutane, and conversion of tertiary butyl alcohol recovered as a by-product in the manufacture of propylene oxides.

Heat exchangers are used throughout crude oil refineries to heat or cool crude oil process streams. The heat exchangers consist of bundles of pipes, tubes, plate coils, or steam coils enclosing heating or cooling water, steam, or oil to transfer heat indirectly to or from the oil process stream. The bundles are cleaned periodically to remove accumulations of scales, sludge, and any oily residues.

Blowdown systems provide for the safe handling and disposal of liquid and gases that are either automatically vented from the process units through pressure relief valves or that are manually drawn from units. Recirculated process streams and cooling water streams are often manually purged to prevent the continued buildup of contaminants in the stream. Part or all of the contents of equipment can also be purged to the blowdown system prior to shut down before normal or emergency. Blowdown systems utilize a series of flash drums and condensers to separate the blowdown into its vapor and liquid components.

Blending is the final operation in crude oil refining and consists of mixing the products in various proportions to meet specifications such as vapor pressure, specific gravity, sulfur content, viscosity, octane number, cetane index, initial boiling point, and pour point. The blending process can be carried out in-line or in batch blending tanks (Parkash, 2003; Gary et al., 2007; Speight, 2014, 2017; Hsu and Robinson, 2017).

Storage tanks are used throughout the refining process to store crude oil and intermediate process feeds for cooling and further processing. Finished crude oil products are also kept in storage tanks before transport off-site. Storage tank bottoms are mixtures of iron rust from corrosion, sand, water, and emulsified oil and wax, which accumulate at the bottom of tanks. Liquid tank bottoms (primarily water and oil emulsions) are periodically drawn off to prevent their continued buildup. Tank bottom liquids and sludge are also removed during periodic cleaning of tanks for inspection.

2.4 The future

Crude oil refining has evolved continuously in response to changing consumer demand for better and different products. The original requirement was to produce kerosene as a cheaper and better source of light than whale oil. The development of the internal combustion engine led to the production of gasoline and diesel fuels. The evolution of the airplane created a need first for high-octane aviation gasoline and then for jet fuel, a sophisticated form of the original product, kerosene. Present-day refineries produce a variety of products, including many required as feedstocks for the petrochemical industry.

Not surprisingly, crude oil refining has become more complex in the decades since the end of World War II. The infusion of high-boiling crude oil (low-quality crude oil), price volatility, and environmental regulations that require cleaner manufacturing processes and higher performance products present new challenges to the refining industry. Improving processes and increasing the efficiency of energy use with technology research and development are key to meeting the challenges and maintaining the viability of the crude oil refining industry.

Refining produces not only fuels but raw materials used as feedstock for the petrochemical industry (plastics, pharmaceuticals, agrochemicals, etc.). Crude oil is separated and converted into a variety of products. Modern refineries are able to alter processing configurations and change the product slate to meet demand, which varies seasonally and geographically.

The success of current operations is notwithstanding, the challenges facing the refining industry will focus on the diversity of the feedstocks. Even within the crude oil family of feedstocks where elemental analysis varies over a relatively narrow range, changes to refining technology are required to produce optimum yield of desired products.

Another unique foreseeable disruption coming to the industry is the anticipated inclusion of biomass feedstocks and the changes that this will bring to refining (Speight, 2011). Indeed, much of the intellectual property embodied in the current refinery operations will have to change as wide variations in feedstock composition occur, and attempts are made to produce the necessary hydrocarbon fuels from a wide variety of biomass feedstocks. Any yet, the refining industry will survive—being one of the most resilient industries to commence operations during the past 150 years.

In the 21st century the crude oil industry must prepare to address many important challenges. Major forces for change include (1) continuing concern for the environment, (2) governmental regulation and policy, (3) higher consumer expectations for fuels and fuel delivery systems, and (4) global competition. In many cases, technology research and development will be needed to meet these challenges and maintain the health and profitability of the industry.

The refining industry can be regarded as unique insofar as very few industries have to deal with a feedstock–product chain beginning at a natural resource that has to be recovered from a subterranean formation and proceed through the application of a variety of processes all the way through to the end-use consumer. Furthermore, it is imperative for refiners to raise their operations to new levels of performance. Merely extending current process performance incrementally will most likely fail to meet most future performance goals. To do this, it will be necessary to reshape refining technology to be more adaptive to changing feedstocks and product demand and to explore the means by which the technology and methodology of refinery operations can be translated not only into increased profitability but also into survivability.

Development and inclusion of novel technologies to cope up with future challenges are essential for refineries to remain competitive. Furthermore, there is considerable uncertainty surrounding the future of unconventional crude oil production in the United States. Environmental regulations could either preclude unconventional production or, more likely, raise the cost significantly. If future US laws limited and/or taxed greenhouse gas emissions, these laws will lead to substantial increase in the costs of production of fuels from unconventional sources. In addition to increases in the volumes of carbon dioxide, restrictions on access to water also could prove costly, especially in the arid or semiarid western states. In addition, environmental restrictions on land use could preclude unconventional oil production in some areas of the United States. Such policies could open to way to the increased use of biomass feedstocks as a means of supplementing production of hydrocarbon fuels.

From the abovementioned descriptions (this chapter and Chapter 1: Feedstock Types and Properties), it is possible to understand the evolution of the crude oil refinery and thence to speculate on the changes might be put in place in the refinery of the future.

References

Dunning, H.N., Moore, J.W., 1957. Propane removes asphalts from crudes. Pet. Refin. 36 (5), 247–250.

Gary, J.G., Handwerk, G.E., Kaiser, M.J., 2007. Petroleum Refining: Technology and Economics, fifth ed. CRC Press, Taylor & Francis Group, Boca Raton, FL.

Hsu, C.S., Robinson, P.R. (Eds.), 2017. Handbook of Petroleum Technology. Springer International Publishing AG, Cham.

Kohl, A.L., Riesenfeld, F.C., 1985. Gas Purification, 4th ed. Gulf Publishing Company, Houston, TX.

Kohl, A.L., Nielsen, R.B., 1997. Gas Purification. Gulf Publishing Company, Houston, TX.

Maddox, R.N., 1982. Gas conditioning and processing, Gas and Liquid Sweetening, Vol. 4. Campbell Publishing Co, Norman, OK.

Maddox, R.N., Bhairi, A., Mains, G.J., Shariat, A., 1985. In Acid and Sour Gas Treating Processes. S.A. Newman (Ed.). Gulf Publishing Company, Houston, TX, Chapter 8.

May-Britt, H., 2008. Membranes in gas separation. Handbook of Membrane Separations. CRC Press, Taylor & Francis Group, Boca Raton, FL, pp. 65–106.

Mokhatab, S., Poe, W.A., Speight, J.G., 2006. Handbook of Natural Gas Transmission and Processing. Elsevier BV, Amsterdam.

Parkash, S., 2003. Refining Processes Handbook. Gulf Professional Publishing, Elsevier, Amsterdam.

Ramírez, M., Gómez, J.M., Cantero, D., 2015. Chapter 13: Biogas: sources, purification and uses. In: Sharma, U.C., Kumar, S., Prasad, R. (Eds.), Hydrogen and Other Technologies. Studium Press LLC, New Delhi, pp. 296–323.

Speight, J.G., 2008. Synthetic fuels handbook: properties. Processes and Performance. McGraw-Hill, New York.

Speight, J.G., 2011. A biorefinery. The Biofuels Handbook. Royal Society of Chemistry, London, Part 1, Chapter 4.

Speight, J.G., 2014. The Chemistry and Technology of Petroleum, fifth ed. CRC Press, Taylor & Francis Group, Boca Raton, FL.

Speight, J.G., 2016. Introduction to Enhanced Recovery Methods for Heavy Oil and Tar Sands, second ed. Gulf Professional Publishing Company, Elsevier, Oxford.

Speight, J.G., 2017. Handbook of Petroleum Refining. CRC Press, Taylor & Francis Group, Boca Raton, FL.

Speight, J.G., 2019a. Natural Gas: A Basic Handbook, second ed. Gulf Publishing Company, Elsevier, Cambridge, MA.

Speight, J.G., 2019b. Handbook of Petrochemical Processes. CRC Press, Taylor & Francis Group, Boca Raton, FL.

Refining chemistry

3

3.1 Introduction

Crude oil is unstable for use in the unrefined state and must be processed into its various products, which have hydrogen content different from that of the original feedstock leading to the production not only of better products but also of salable materials. The major refinery products are liquefied petroleum gas (LPG), gasoline, jet fuel, solvents, kerosene, middle distillates, residual fuel oil, lubricating oil, and asphalt. In addition, the formation of coke from the higher molecular weight and polar constituents of a given feedstock is detrimental to process efficiency and to catalyst performance (Speight, 1987; Dolbear, 1998).

Refining crude oil is used thermal decomposition, catalytic decomposing, pressure, and chemicals to rearrange the structures and bonding patterns of the molecules into different hydrocarbon molecules and compounds. In many cases the chemical structure of the constituents of the products does not reflect the chemical structures of the constituents originally in the unprocessed crude oil. It is only in low-temperature solvent processes, such as deasphalting, the original structures of the constituents are retained in the deasphalted oil and asphalt.

Understanding refining chemistry not only allows an explanation of the means by which these products can be formed from crude oil but also offers a chance of predictability. This is very necessary when the different types of crude oil accepted by refineries are considered. And the major processes by which these products are produced from crude oil constituents involve thermal decomposition leading to many modern products. Furthermore, it must be recognized that refining involves a wide variety of chemical reactions but the production of liquid fuels is the focus of a refinery.

A refinery is a complex network of integrated unit processes for the purpose of producing a variety of products from crude oil (Chapter 2: Introduction to Refining Processes) (Parkash, 2003; Gary et al., 2007; Speight, 2014, 2017; Hsu and Robinson, 2017). Refined products establish the order in which the individual refining units will be introduced, and the choice from among several types of units and the size of these units is dependent upon economic factors. The trade-off among product types, quantity, and quality influences the choice of one kind of processing option over another.

Even though refining chemistry might be represented by relatively simple equations, the chemistry of conversion process may be quite complex, and an understanding of the chemistry involved in the conversion of a crude oil to a variety of products is essential to an understanding of refinery operations.

The Refinery of the Future. DOI: https://doi.org/10.1016/B978-0-12-816994-0.00003-8

In fact, the complexity of the individual reactions occurring in an extremely complex mixture and the interference of the products with those from other components of the mixture is unpredictable. Or the interference of secondary and tertiary products with the course of a reaction and, hence, with the formation of primary products may also be cause for concern. Hence, caution is advised when applying the data from model compound studies to the behavior of crude oil, especially the molecularly complex high-boiling oils. These have few, if any, parallels in organic chemistry.

It is the purpose of this chapter to serve as an introduction to the chemistry involved in these conversion processes so that the subsequent chapters dealing with refining and especially the refinery of the future are easier to visualize and understand. However, understanding refining chemistry from the behavior of model compounds under refining conditions is not as straightforward as it may appear since this type of chemical representation often ignores the interactions between the various molecular species in the feedstock.

3.2 Cracking

3.2.1 Thermal cracking

With the dramatic increases in the number of gasoline-powered vehicles, distillation processes (Chapter 2: Introduction to Refining Processes) were not able to completely fill the increased demand for gasoline. In 1913, the thermal cracking process was developed and is the phenomenon by which higher boiling (higher molecular weight) constituents in crude oil are converted into lower boiling (lower molecular weight) products' application of elevated temperatures (usually in the order of >350°C, >660°F).

Thermal cracking is the oldest and, in principle, the simplest refinery conversion process. The temperature and pressure depends on the type of feedstock and the product requirements as well as the residence time (Speight, 2003, 2014, 2017). Thermal cracking processes allow the production of lower molecular weight products such as the constituents of LPG and naphtha/gasoline constituents from higher molecular weight fraction such as gas oils and residua. The simplest thermal cracking process—visbreaking (Chapter 2: Introduction to Refining Processes)—is used to upgrade fractions such as distillation residua (Chapter 2: Introduction to Refining Processes) to produce fuel oil that meets specifications.

Two extremes of the thermal cracking in terms of product range are represented by high-temperature processes—(1) steam cracking or (2) pyrolysis.

Steam cracking is a process in which feedstock is decomposed into lower molecular weight (often unsaturated) products saturated hydrocarbon derivatives. In the process a gaseous or liquid hydrocarbon feed such as ethane or naphtha is diluted with steam and briefly heated in a furnace (at approximately 850°C, 1560°F) in the absence of oxygen at a short residence time (often in the order of milliseconds). After the cracking temperature has been reached, the products are rapidly quenched in a heat exchanger. The products produced in the reaction depend on the composition of the feedstock, the feedstock/steam ratio, the cracking temperature, and the residence time.

Pyrolysis processes require temperatures in the order of 750°C–900°C (1380°F–1650°F) to produce high yields of low-molecular-weight products, such as ethylene, for petrochemical use. Delayed coking, which uses temperature in the order of 500°C (930°F) is used to produced distillates from nonvolatile residua as well as coke for fuel and other uses—such as the production of electrodes for the steel and aluminum industries.

3.2.1.1 General chemistry

In thermal cracking processes, some of the lower molecular weight products are expelled from the system as gases, gasoline-range materials, kerosene-range materials, and the various intermediates that produce other products such as coke. Materials that have boiling ranges higher than gasoline and kerosene may (depending upon the refining options) be referred to as recycle stock, which is recycled in the cracking equipment until conversion is complete.

Thermal cracking is a free-radical chain reaction. A free radical (in which an atom or group of atoms possessing an unpaired electron) is very reactive (often difficult to control) and it is the mode of reaction of free radicals that determines the product distribution during thermal cracking (i.e., noncatalytic thermal decomposition). In addition, a significant feature of hydrocarbon free radicals is the resistance to isomerization during the existence of the radical. For example, thermal cracking does not produce any degree of branching in the products (by migration of an alkyl group) other than that already present in the feedstock. Nevertheless, the chemistry of thermal cracking typically involves the following chemical reactions:

1. Initiation reaction, where a single molecule breaks apart into two free radicals. Only a small fraction of the feedstock constituents may actually undergo initiation, which involves breaking the bond between two carbon atoms, rather than the thermodynamically stronger bond between a carbon atom and a hydrogen atom.

   $CH_3CH_3 \rightarrow 2CH_3\bullet$

2. Hydrogen abstraction reaction in which the free radical abstracts a hydrogen atom from another molecule:

   $CH_3\bullet + CH_3CH_3 \rightarrow CH_4 + CH_3CH_2\bullet$

3. Radical decomposition reaction in which a free radical decomposes into an alkene derivative:

   $CH_3CH_2\bullet \rightarrow CH_2=CH_2 + H\bullet$

4. Radical addition reaction in which a radical reacts with an alkene derivative to form a single, larger free radical:

   $CH_3CH_2\bullet + CH_2=CH_2 \rightarrow CH_3CH_2CH_2CH_2\bullet$

5. Termination reaction in which two free radicals react with each other to produce the products—two common forms of termination reactions are recombination reactions (in which two radicals combine to form one molecule) and disproportionation reactions (in which one free radical transfers a hydrogen atom to the other to produce an alkene derivative and an alkane):

$$CH_3\bullet + CH_3CH_2\bullet \rightarrow CH_3CH_2CH_3$$

$$CH_3CH_2\bullet + CH_3CH_2\bullet \rightarrow CH_2 = CH_2 + CH_3CH_3$$

The smaller free radicals, hydrogen, methyl, and ethyl are more stable than the larger radicals. They will tend to capture a hydrogen atom from another hydrocarbon, thereby forming a saturated hydrocarbon and a new radical. In addition, many thermal cracking processes and different chemical reactions occur simultaneously. Thus an accurate explanation of the mechanism of the thermal cracking reactions is difficult. The primary reaction is the decomposition of higher molecular weight species into lower molecular weight products.

As the molecular weight of the hydrocarbon feedstock increases, the reactions become much more complex leading to a wider variety of products. For example, using a more complex hydrocarbon (dodecane, $C_{12}H_{26}$) as the example, two general types of reaction occur during cracking:

1. The decomposition of high-molecular-weight constituents into lower molecular weight constituents (primary reactions):

$$CH_3(CH_2)_{10}CH_3 \rightarrow CH_3(CH_2)_8CH_3 + CH_2 = CH_2$$

$$CH_3(CH_2)_{10}CH_3 \rightarrow CH_3(CH_2)_7CH_3 + CH_2 = CHCH_3$$

$$CH_3(CH_2)_{10}CH_3 \rightarrow CH_3(CH_2)_6CH_3 + CH_2 = CHCH_2CH_3$$

$$CH_3(CH_2)_{10}CH_3 \rightarrow CH_3(CH_2)_5CH_3 + CH_2 = CH(CH_2)_2CH_3$$

$$CH_3(CH_2)_{10}CH_3 \rightarrow CH_3(CH_2)_4CH_3 + CH_2 = CH(CH_2)_3CH_3$$

$$CH_3(CH_2)_{10}CH_3 \rightarrow CH_3(CH_2)_3CH_3 + CH_2 = CH(CH_2)_4CH_3$$

$$CH_3(CH_2)_{10}CH_3 \rightarrow CH_3(CH_2)_2CH_3 + CH_2 = CH(CH_2)_5CH_3$$

$$CH_3(CH_2)_{10}CH_3 \rightarrow CH_3CH_2CH_3 + CH_2 - CH(CH_2)_6CH_3$$

$$CH_3(CH_2)_{10}CH_3 \rightarrow CH_3CH_3 + CH_2 = CH(CH_2)_7CH_3$$

$$CH_3(CH_2)_{10}CH_3 \rightarrow CH_4 + CH_2 = CH(CH_2)_8CH_3$$

2. Reactions by which some of the primary products interact to form higher molecular weight materials (secondary reactions):

$$CH_2 = CH_2 + CH_2 = CH_2 \rightarrow CH_3CH_2CH = CH_2$$

$$RCH = CH_2 + R^1CH = CH_2 \rightarrow \text{cracked residuum} + \text{coke} + \text{other products}$$

Thus from the chemistry of the thermal decomposing of pure compounds (and assuming little interference from other molecular species in the reaction mixture), predictions can be made of the product types that arise from the thermal cracking of various feedstocks. However, during thermal cracking, all of the reactions illustrated previously can and do occur simultaneously.

n-Paraffin derivatives are believed to form, initially, higher molecular weight material, which subsequently decomposes as the reaction progresses; other paraffinic materials and (terminal) olefin derivatives are produced. An increase in pressure maintains the lower molecular weight products in the liquid phase and promotes the formation of higher molecular weight materials.

Data obtained from the thermal decomposition of pure compounds indicate certain decomposition characteristics that permit predictions to be made of the product types that arise from the thermal cracking of various feedstocks. For example, normal paraffin derivatives are believed to form, initially, higher molecular weight material, which subsequently decomposes as the reaction progresses. An increase in pressure inhibits the formation of low-molecular-weight gaseous products and therefore promotes the formation of higher molecular weight materials.

Furthermore, for saturated hydrocarbon derivatives, the connecting link between gas-phase pyrolysis and liquid-phase thermal degradation is the concentration of alkyl radicals. In the gas phase, alkyl radicals are present in low concentration and undergo unimolecular radical decomposition reactions to form α-olefin derivatives and smaller alkyl radicals. In the liquid phase, alkyl radicals are in much higher concentration and prefer hydrogen abstraction reactions to radical decomposition reactions. It is this preference for hydrogen abstraction reactions that gives liquid-phase thermal degradation a broad product distribution.

Branched paraffin derivatives react somewhat differently to the normal paraffin derivatives during cracking processes and produce substantially higher yields of olefin derivatives having one fewer carbon atom than the parent hydrocarbon. Cyclohexane produces hydrogen, ethylene, butadiene, and benzene: alkyl-substituted cycloparaffin derivatives decompose by means of scission of the alkyl chain to produce an olefin and (predominantly) methylcyclohexane or (to a lesser extent) ethyl cyclohexane.

The aromatic ring is fairly resilient to decomposition at moderate cracking temperatures (350°C–500°C, 660°F–930°F). However, alkylated aromatic derivatives, like the alkylated naphthene derivatives, are more prone to dealkylation than to ring destruction yield (predominantly) the methyl aromatic derivatives. Ring destruction of the benzene derivatives occurs above 500°C (930°F), but condensed aromatic derivatives may undergo ring destruction at somewhat lower temperatures (450°C, 840°F).

Generally, the relative ease of cracking of the various types of hydrocarbon derivatives of the same molecular weight is given in the following descending order:

(1) paraffin derivatives; (2) olefin derivatives; (3) cycloparaffin derivatives, also known as naphthene derivatives; and (4) aromatic derivatives. To remove any potential confusion, paraffin derivatives are the least stable and aromatic derivatives are the most stable.

Within any type of hydrocarbon series, the higher boiling hydrocarbon derivatives tend to crack easier than the lower boiling derivatives. Paraffin derivatives are by far the easiest hydrocarbon derivatives to crack with the rupture most likely to occur between the first and second carbon bonds in the lower boiling paraffin derivatives. However, as the molecular weight of the paraffin molecule increases, rupture tends to occur nearer the middle of the molecule. The main secondary reactions that occur in thermal cracking are polymerization and condensation.

3.2.1.2 Asphaltene chemistry

Crude oil, heavy crude oil, and residua contain heptane-insoluble asphaltene constituents and resin constituents that, because of the content of polynuclear aromatic compounds and polar functionalities, provide hurdles to conversion. The high thermal stability of polynuclear aromatic systems prevents thermal decomposition to lower boiling point products and usually results in the production of substantial yields of thermal coke. Furthermore, the high concentrations of heteroatom compounds (nitrogen, oxygen, sulfur) and metals (vanadium and nickel) in heavy crude oil, extra heavy oil, tar sand bitumen, and residua have an adverse effect on catalysts. Therefore process choice often favors thermal process but catalytic processes can be used as long as catalyst replacement and catalyst regeneration is practiced.

Asphaltene constituents and, to a lesser extent, resin constituents can cause major problems in refineries through unanticipated coke formation and/or through excessive coke formation. Recognition of this is a step in the direction of mitigating the problem.

Improvement in high-boiling viscous feedstock conversion may be sought in the use of specific chemical additives. However, to improve the conversion of the high-boiling viscous feedstocks it is necessary to understand the chemistry of conversion.

The thermal decomposition of the more complex asphaltene constituents has received some attention (Magaril and Aksenova, 1968, 1970; Magaril and Ramazaeva, 1969; Magaril et al., 1970, 1971; Parkash, 2003; Gary et al., 2007; Speight, 2014, 2017; Hsu and Robinson, 2017).

The thermal reaction is believed to be first-order although there is the potential that it is, in fact, a multiorder reaction process, but because of the multiplicity of the reactions that occur, it appears as a pseudo-first-order process. However, it is definite that there is an induction period before coke begins to form that seems to be triggered by phase separation of reacted asphaltene product (Magaril et al., 1971; Speight, 1987, 2014). The organic nitrogen originally in the asphaltenes invariably undergoes thermal reaction to concentrate in the nonvolatile coke. In scheme the chemistry of asphaltene coking has been suggested to involve the thermolysis of thermally labile bonds to form reactive species that react with each other

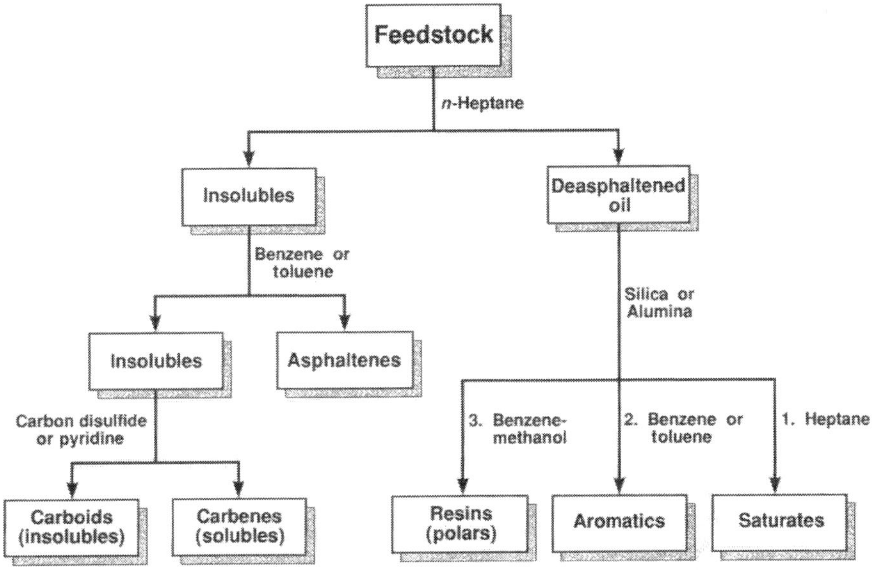

Figure 3.1 General fractionation scheme and nomenclature of petroleum fraction; carbenes and carboids are thermally generated product fractions.

(condensation) to form coke. However, not all the original aromatic carbon in the asphaltene constituents forms coke. Volatile aromatic species are eliminated during thermal decomposition, and it must be assumed that some of the original aliphatic carbon plays a role in coke formation.

It is more likely that the initial reactions of asphaltene constituents involve thermolysis of pendant alkyl chains to form lower molecular weight higher polar species (carbenes and carboids; Fig. 3.1) which then react to form coke. The reactions involve unimolecular thermolysis of aromatic-alkyl systems of the asphaltenes to produce volatile species (paraffin derivatives and olefin derivatives) and nonvolatile species (aromatic derivatives) (Fig. 3.2) (Speight, 1987, 1994; Schabron and Speight, 1997).

It is also interesting to note that although the aromaticity of the resin and asphaltene constituents is approximately equivalent to the yield of thermal coke, not all the original aromatic carbon in the asphaltenes forms coke. Volatile aromatic species are eliminated during thermal decomposition, and it must be assumed that some of the original aliphatic carbon plays a role in coke formation. Its precise nature has yet to be determined, but the process can be represented as involving a multireaction process involving series and parallel reactions (Parkash, 2003; Gary et al., 2007; Speight, 2014, 2017; Hsu and Robinson, 2017).

As examples of thermal cracking, in the delayed coking process, the feedstock is heated to high temperatures (480°C–500°C; 895°F–930°F) in a furnace and then reaction is allowed to continue in a cylindrical, insulated drum. The volatile products pass overhead into a fractionator and coke accumulates in the drum. Any

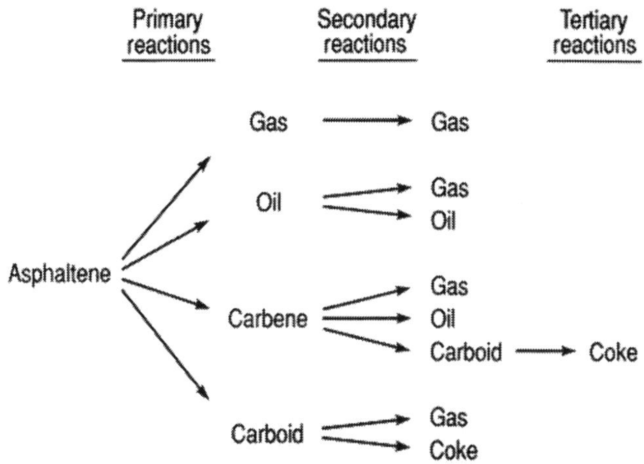

Figure 3.2 Multireaction sequence for the thermal decomposition of asphaltene constituents.

high-boiling liquid product from the fractionator is recycled to the coker furnace. When the drum fills up with coke, the reacting feedstock is directed to a second drum. The coke is removed from the first drum by hydraulic drilling and cutting after which the drum is ready for the next 16–24 hour reaction cycle. During this process the asphaltene and resin constituents in the feedstock are converted to coke in accordance with their respective carbon residue values (c. 50% w/w for asphaltene constituents and c. 35% w/w for resin constituents) (Parkash, 2003; Gary et al., 2007; Speight, 2014, 2017; Hsu and Robinson, 2017).

Nitrogen species also appear to contribute to the pattern of the thermolysis insofar as the carbon–carbon bonds adjacent to ring nitrogen undergo thermolysis quite readily (Speight, 1998, 2014). Thus the initial reactions of asphaltene decomposition involve thermolysis of aromatic-alkyl bonds that are enhanced by the presence of heterocyclic nitrogen (Speight, 1987). Thus the molecular species within the asphaltene fraction, which contain nitrogen and other heteroatoms (and have lower volatility than the pure hydrocarbon derivatives), are the prime movers in the production of coke (Speight, 1987). Such species, containing various polynuclear aromatic systems, can be denuded of the attendant hydrocarbon moieties and are undoubtedly insoluble (Bjorseth, 1983; Dias, 1987, 1988) in the surrounding hydrocarbon medium. The next step is gradual carbonization of these heteroatom-rich entities to form coke.

Thus coke formation is a complex thermal process involving both chemical reactions and thermodynamic behavior. The challenges facing process chemistry and physics are determining (1) the means by which crude oil constituents thermally decompose, (2) the nature of the products of thermal decomposition, (3) the subsequent decomposition of the primary thermal products, (4) the interaction of the products with each other, (5) the interaction of the products with the original constituents, and (6) the influence of the products on the composition of the liquids.

The challenge in mitigating the formation of thermal coke is to eliminate or modify the prime chemical reactions in the formation of incompatible products during the processing of feedstocks containing asphaltenes, particularly those reactions in which the insoluble lower molecular weight products (carbenes and carboids; Fig. 3.1) are formed (Speight, 1987, 2014, 2017; Parkash, 2003; Gary et al., 2007; Hsu and Robinson, 2017).

Thus the challenges facing process chemistry and physics are determining (1) the means by which crude oil constituents thermally decompose, (2) the nature of the products of thermal decomposition, (3) the subsequent decomposition of the primary thermal products, (4) the interaction of the products with each other, (5) the interaction of the products with the original constituents, and (6) the influence of the products on the composition of the liquids.

3.2.1.3 Process chemistry

Thermal cracking processes are commonly used to convert crude oil residua into distillable liquid products, although thermal cracking processes as used in the early refineries are no longer in use. Examples of modern thermal cracking processes are visbreaking and coking (delayed coking, fluid coking, and flexicoking). In all of these processes the simultaneous formation of sediment or coke limits the conversion to usable liquid products. However, for the purposes of this section, the focus will be on the visbreaking and hydrocracking processes.

To study the thermal chemistry of crude oil constituents, it is appropriate to select the visbreaking process (Fig. 3.3) (Chapter 5: Thermal Cracking), which is primarily a means of reducing the viscosity of high-boiling feedstocks, such as residua, by controlled thermal decomposition insofar as the products emanating

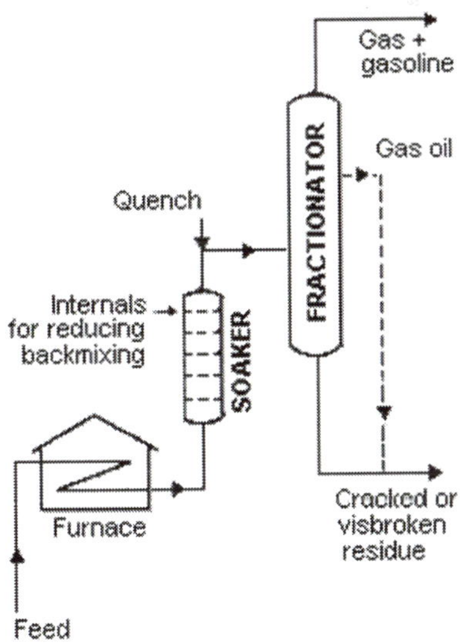

Figure 3.3 The visbreaking process using a soaker.

from the reactor are quenched before complete conversion can occur (Speight, 2014, 2017). The low residence times are required to avoid coking reactions, although additives can help to suppress coke deposits on the tubes of the furnace (Parkash, 2003; Gary et al., 2007; Speight, 2014, 2017; Hsu and Robinson, 2017).

The process consists of a reaction furnace, followed by quenching with a recycled oil, and fractionation of the product mixture. The severity of the reaction is controlled by the flow rate through the furnace and the temperature; typical conditions are 475°C–500°C (885°F–930°F) at the furnace exit with a residence time of 1–3 minutes (Parkash, 2003; Gary et al., 2007; Speight, 2014, 2017; Hsu and Robinson, 2017). The operating pressure in the furnace tubes is close to atmospheric but is dependent on the degree of vaporization of the feedstock and products the residence time in the hot zone before quenching. A lower operating pressure reduces the residence time of the liquid phase.

Conversion of residua in visbreaking is complex but generally follows first-order reaction kinetics. The stability of products from a visbreaker is an issue since the products might contain polar species that have been denuded of some of the alkyl chains, which, on the basis of solubility, might be more rightly called carbenes and carboids (Fig. 3.1), but an induction period is required for phase separation or agglomeration to occur. Such products might initially be soluble in the liquid phase but after the induction period, cooling, and/or diffusion of the products, incompatibility (phase separation, sludge formation, agglomeration) occurs.

Obviously, the temperature and residence time of the asphaltene constituents in the reactor are key to the successful operation of a visbreaker, which must operate in a temperature regime and residence time regime that do not promote the formation of insoluble sediment (often referred to as coke). However, there is a break point above which considering might be increased but the possibility of sediment deposition increases (Fig. 3.4). At the temperatures and residence times outside of the most beneficial temperature and residence time regimes, thermal changes to the asphaltene constituents cause phase separation of a solid product that then progresses to coke.

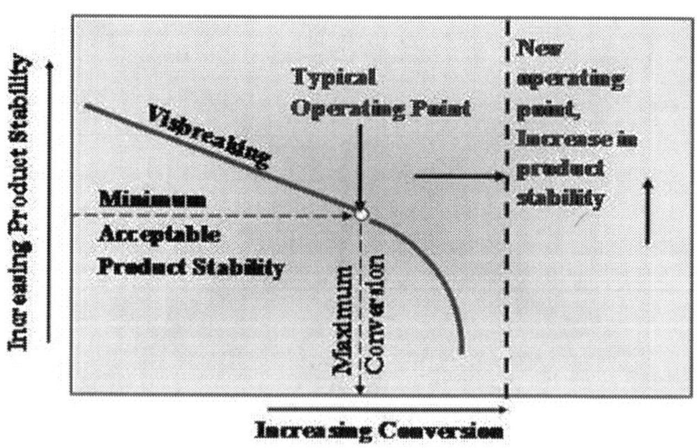

Figure 3.4 Representation of the break point above which maximum conversion is assured but product stability (i.e., inhibition of sediment formation) is less certain.

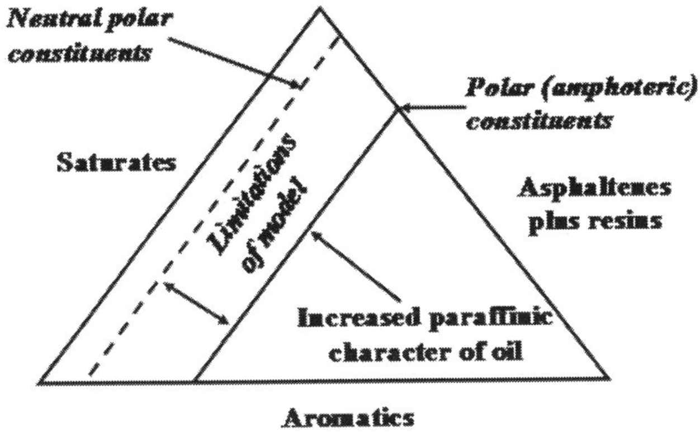

Figure 3.5 The limitations of the visbreaking process when predictions are based on average parameters for high-asphaltene feedstocks.

For example, the most polar (amphoteric) constituents of the asphaltene fraction are more reactive than the less polar constituents (Parkash, 2003; Gary et al., 2007; Speight, 2014, 2017; Hsu and Robinson, 2017). The thermal products from these constituents form first and separate out from the reaction matrix before other products (Fig. 3.5).

Knowing the actual nature of the subtypes of the asphaltene constituents is obviously beneficial and will allow steps to be taken to correct any such unpredictable occurrence. Indeed, the concept of hydrovisbreaking (visbreaking in the presence of hydrogen) could be of valuable assistance when high-asphaltene feedstocks are used.

The net overall rate of thermal cracking follows the Arrhenius relationship, that is:

$$k = Ae^{-E/RT}$$

where k is the rate, A is the frequency factor, E is the activation energy, and R is the gas constant.

Thus the rate of thermal cracking reaction(s) is, exponentially, a function of temperature. The degree of conversion is, therefore, a function of both time and temperature. This allows the use of lower temperature if the contact time (in the reactor) is increased providing there is no change in the chemistry of the reaction with increases in contact time or, in some cases, increase in temperature and decrease in contact time.

The activation energy for thermal; cracking of many hydrocarbon derivatives is generally constant at approximately 55,000 cal/mol. However, the frequency varies considerably since it is based on molecular size as well as hydrocarbon type and, to some extent, stereochemical factors. Generally, the rate of the thermal cracking reaction increases with molecular size and decreases with aromaticity.

3.2.2 Catalytic cracking

Catalytic cracking is the thermal decomposition of crude oil constituents' hydrocarbon derivatives in the presence of a catalyst (Pines, 1981). Gasoline produced by catalytic cracking is richer in branched paraffin, cycloparaffin, and aromatic derivatives, which all serve to increase the quality of the gasoline. Catalytic cracking also results in the production of the maximum amount of butenes and butanes (C_4H_8 and C_4H_{10}) rather than ethylene and ethane (C_2H_4 and C_2H_6).

Fluid catalytic cracking is a commonly used process, and a modern oil refinery will typically include a cat cracker (Chapter 2: Introduction to Refining Processes) to satisfy the demand for gasoline and diesel fuel. In the reactor, catalyst particles are suspended in a rising flow of feedstock in a fluidized bed.

3.2.2.1 General chemistry

The use of a catalyst permits alternate routes for cracking reactions, usually by lowering the free energy of activation for the reaction. The acid catalysts first used in catalytic cracking were amorphous solids composed of approximately 87% silica (SiO_2) and 13% alumina (Al_2O_3) and were designated low-alumina catalysts. However, this type of catalyst is now being replaced by crystalline aluminosilicates (zeolites) or molecular sieves.

Catalytic cracking is an ionic process involving carbonium ions, which are hydrocarbon ions having a positive charge on a carbon atom, which can occur by:

1. addition of a proton from an acid catalyst to an olefin and
2. abstraction of a hydride ion (H^-) from a hydrocarbon by the acid catalyst or by another carbonium ion.

However, carbonium ions are not formed by cleavage of a carbon–carbon bond.

Once the carbonium ions are formed, the modes of interaction constitute an important means by which product formation occurs during catalytic cracking. For example, isomerization by hydride ion shift or by methyl group shift occurs readily. The trend is stabilization of the carbonium ion by movement of the charged carbon atom toward the center of the molecule, which accounts for the isomerization of α-olefin derivatives to internal olefin derivatives when carbonium ions are produced. Cyclization occurs by internal addition of a carbonium ion to a double bond with the formation of a cyclic carbonium ion, which results in aromatization of the cyclic intermediate.

The major catalytic cracking reaction exhibited by paraffin derivatives is carbon–carbon bond scission into a lower boiling paraffin and an olefin derivative. Bond rupture occurs at certain definite locations on the paraffin molecule, rather than randomly as in thermal cracking. For example, paraffin derivatives tend to crack toward the center of the molecule, the long chains cracking in several places simultaneously. Normal paraffin derivatives usually crack at β carbon–carbon bonds or still nearer the center of the molecule. On the other hand, iso-paraffin derivatives tend to rupture between carbon atoms that are β to or further removed from a tertiary carbon. In either case, catalytic cracking tends to yield products containing three or four carbon atoms rather than the one- or two-carbon atom molecules produced in thermal cracking.

Olefin derivatives are the most reactive class of hydrocarbon derivatives in catalytic cracking and tend to crack from 1000 to 10,000 times faster than in thermal processes. Severe cracking conditions destroy olefin derivatives almost completely, except for those in the low-boiling gasoline and gaseous hydrocarbon range and, as in the catalytic cracking of paraffin derivatives, iso-olefin derivatives crack more readily than *n*-olefin derivatives.

Olefin derivatives also tend to undergo rapid isomerization and yield mixtures with an equilibrium distribution of double-bond positions. In addition, the chain-branching isomerization of olefin derivatives is fairly rapid and often reaches equilibrium. These branched-chain olefin derivatives can then undergo hydrogen transfer reactions with naphthene and other hydrocarbon derivatives. Other olefin reactions include polymerization as well as condensation to yield aromatic molecules, which in turn may be the precursors of coke formation.

Cycloparaffin derivatives (naphthene derivatives) catalytically crack more readily than paraffin derivatives but not as readily as olefin derivatives. Naphthene cracking occurs by both ring and chain rupture and yields olefin and paraffin derivatives, but the formation of methane and the C_2 hydrocarbon derivatives (ethane, CH_3CH_3, ethylene, $CH_2=CH_2$, and acetylene, $CH\equiv CH$) is relatively minor if present at all. Naphthene derivatives catalytically crack more readily than paraffin derivatives but not as readily as olefin derivatives. Naphthene cracking occurs by both ring and chain rupture and yields olefin and paraffin derivatives, but the formation of methane and the C_2 hydrocarbon derivatives (ethane, CH_3CH_3, ethylene, $CH_2=CH_2$, and acetylene, $CH\equiv CH$) is relatively minor if present at all.

Naphthene derivatives (from C_9 and higher molecular weight) produce considerable amounts of aromatic hydrocarbon derivatives during catalytic cracking. Alkylated benzenes undergo nearly quantitative dealkylation to benzene without apparent ring degradation below 500°C (930°F). However, polymethyl benzene derivatives undergo disproportionation and isomerization with very little benzene formation.

Like the paraffin derivatives, naphthene derivatives do not appear to isomerize before cracking. However, the naphthenic hydrocarbon derivatives (from C_9 upward) produce considerable amounts of aromatic hydrocarbon derivatives during catalytic cracking. Reaction schemes similar to that outlined here provide possible routes for the conversion of naphthene derivatives to aromatic derivatives.

Aromatic hydrocarbon derivatives exhibit wide variations in their susceptibility to catalytic cracking. The benzene ring is relatively inert, and condensed-ring compounds, such as naphthalene, anthracene, and phenanthrene, crack very slowly. When these aromatic derivatives crack a substantial part of their conversion is reflected in the amount of coke deposited on the catalyst. Alkylbenzenes with attached groups of C_2 or larger primarily form benzene and the corresponding olefin derivatives, and heat sensitivity increases as the size of the alkyl group increases.

Alkylated benzenes undergo nearly quantitative dealkylation to benzene without apparent ring degradation below 500°C (930°F). However, polymethyl benzene derivatives undergo disproportionation, isomerization, and cyclization with very little benzene formation.

3.2.2.2 Coke formation

Coke formation is a malignant side reaction of catalytic cracking. The carbene/carboid-type species can produce other products, but the most damaging product will be one which remains on the catalyst surface and cannot be desorbed. This results in the formation of coke, or remains in a noncoke form but effectively blocks the active sites of the catalyst.

Currently, there is no consensus as to the detailed chemistry of coke formation. There is, however, much evidence and good reason to believe that catalytic coke is formed from carbenium ions that undergo addition, dehydrogenation and cyclization, and elimination side reactions in addition to the main-line chain propagation processes.

Catalytic cracking can be represented by simple reaction schemes (see, e.g., Speight, 2014). However, questions have arisen as to how the cracking of paraffin derivatives is initiated. The Lewis site mechanism is the most obvious since it proposes that a carbenium ion is formed by the abstraction of a hydride ion from a saturated hydrocarbon by a strong Lewis acid site: a tri-coordinated aluminum species. On Brønsted sites a carbenium ion may be readily formed from an olefin by the addition of a proton to the double bond or, more rarely, via the abstraction of a hydride ion from a paraffin molecular by a Brønsted proton. This latter process requires the formation of hydrogen as an initial product. This concept, for various reasons, was often neglected.

A general reaction sequence for coke formation from paraffin derivatives involves oligomerization, cyclization, and dehydrogenation of small molecules at active sites within zeolite pores:

alkene derivatives→oligomers
oligomers→naphthene derivatives
naphthene derivatives→aromatic derivatives
aromatic derivatives→coke

Whether or not these are the true steps to coke formation can only be surmised. The issue with this reaction sequence is that it ignores sequential reactions in favor of consecutive reactions. And it must be accepted that the chemistry leading up to coke formation is a complex process, consisting of many sequential and parallel reactions.

There is a complex and little understood relationship between coke content, catalyst activity, and the chemical nature of the coke. For instance, the C-to-H ratio of coke depends on how the coke was formed; its exact value will vary from system to system and catalyst decay is not related in any simple way to the hydrogen-to-carbon atomic ratio of the coke, or to the total coke content of the catalyst, or any simple measure of coke properties. Moreover, despite many and varied attempts, there is currently no consensus as to the detailed chemistry of coke formation. There is, however, much evidence and good reason to believe that catalytic coke is formed from carbenium ions that undergo addition, dehydrogenation and cyclization, and elimination side reactions in addition to the main-line chain propagation processes.

3.3 Hydroprocesses

Hydroprocesses (hydrogenation processes) for the conversion of crude oil and crude oil products may be classified as destructive and nondestructive. The former (destructive hydrogenation: hydrogenolysis or hydrocracking) is characterized by the rupture of carbon–carbon bonds and is accompanied by hydrogen saturation of the fragments to produce lower boiling products. Such treatment requires rather high temperatures and high hydrogen pressures, the latter to minimize coke formation. Many other reactions, such as isomerization, dehydrogenation, and cyclization, can occur under these conditions (Dolbear, 1998; Speight, 2014, 2017, 2000, Parkash, 2003; Gary et al., 2007; Hsu and Robinson, 2017).

On the other hand, nondestructive hydrogenation (hydrotreating, hydrofining) is generally used for the purpose of improving product quality (or even feedstock quality prior to, say, fluid catalytic cracking) without appreciable alteration of the boiling range. Hydrotreating is essentially a means of eliminating nitrogen, oxygen, and sulfur as ammonia, water, and hydrogen sulfide, respectively.

Hydrogenation without simultaneous cracking is used for saturating olefin derivatives or for converting aromatic derivatives to naphthene derivatives. Under atmospheric pressure, olefin derivatives can be hydrogenated up to approximately 500°C (930°F), but beyond this temperature dehydrogenation commences. Application of pressure and the presence of catalysts make it possible to effect complete hydrogenation at room or even cooler temperature; the same influences are helpful in minimizing dehydrogenation at higher temperatures.

3.3.1 *Hydrocracking*

Hydrocracking is a thermal process (>350°C, >660°F) in which hydrogenation accompanies cracking. Relatively high pressure (1000–5000 psi) is employed, and the overall result is usually a change in the character or quality of the products. The reactions are catalyzed by dual-function catalysts in which the cracking function is provided by silica–alumina (or zeolite) catalysts, and platinum, tungsten oxide, or nickel provides the hydrogenation function.

3.3.1.1 *General chemistry*

Generally, all of the initial reactions of thermal and catalytic cracking occur, but some of the secondary reactions are inhibited or stopped by the presence of hydrogen. For example, the yields of olefin derivatives and the secondary reactions that result from the presence of these materials are substantially diminished, and branched-chain paraffin derivatives undergo demethanation. The methyl groups attached to secondary carbons are more easily removed than those attached to tertiary carbon atoms, whereas methyl groups attached to quaternary carbons are the most resistant to hydrocracking.

The effect of hydrogen on naphthenic hydrocarbon derivatives is mainly that of ring scission followed by immediate saturation of each end of the fragment produced. For example, methyl-cyclopentane is converted (over a platinum-carbon catalyst) to 2-methylpentane, 3-methylpentane, and *n*-hexane.

Aromatic hydrocarbon derivatives are resistant to hydrogenation under mild conditions, but under more severe conditions the main reactions are conversion of the aromatic to naphthenic rings and scissions within the alkyl side chains—the naphthene derivatives may also be converted to paraffin derivatives.

Polynuclear aromatic derivatives are more readily attacked than the single-ring compounds, the reaction proceeding by a stepwise process in which one ring at a time is saturated and then opened. For example, naphthalene is hydrocracked over molybdenum oxide-molecular-weight paraffin derivatives (C_6).

3.3.1.2 Asphaltene chemistry

There have also been many attempts to focus attention on the asphaltenes during hydrocracking studies (Dolbear, 1998; Speight, 2014). In terms of hydroprocessing, the means by which asphaltene constituents are desulfurized, as one step of a hydrocracking operation, is also suggested as part of the process. This concept can then be taken one step further to show the dealkylation of the aromatic systems as a definitive step in the hydrocracking process (Speight, 1987).

If catalytic processes are employed, complex molecules (such as those occurring in the asphaltene fraction), or those formed during the process, are not sufficiently mobile (or are too strongly adsorbed by the catalyst) to be saturated by the hydrogenation components. Such molecular species eventually degrade to coke, which deactivates the catalyst sites and eventually interferes with the process.

It may be that the chemistry of hydrocracking has to be given serious reconsideration insofar as the data show that the initial reactions of the asphaltene constituents appear to be the same as the reactions under thermal conditions where hydrogen is not present. Rethinking of the process conditions and the potential destruction of the catalyst by the deposition of carbenes and carboids require further investigation of the chemistry of asphaltene hydrocracking.

If these effects are prevalent during hydrocracking high-asphaltene feedstocks, the option may be to hydrotreat the feedstock first and then to hydrocrack the hydrotreated feedstock. There are indications that such hydrotreatment can (at some obvious cost) act beneficially in the overall conversion of the feedstocks to liquid products.

3.3.1.3 Catalysts

The reactions of hydrocracking require a dual-function catalyst with high cracking and hydrogenation activities. The catalyst base, such as acid-treated clay, usually supplies the cracking function or alumina or silica–alumina that is used to support the hydrogenation function supplied by metals, such as nickel, tungsten, platinum, and palladium. These highly acid catalysts are very sensitive to nitrogen compounds in the feed, which break down the conditions of reaction to give ammonia and neutralize the acid sites.

Hydrocracking catalysts typically contain separate hydrogenation and cracking functions. Palladium sulfide and promoted group VI sulfides (nickel–molybdenum or nickel–tungsten) provide the hydrogenation function. These active compositions saturate aromatic derivatives in the feed and olefin derivatives formed in the cracking and protect the catalysts from poisoning by coke. Zeolites or amorphous silica–alumina provide the cracking functions. The zeolites are usually type Y (faujasite), ion exchanged to replace sodium with hydrogen and make up 25%–50% of the catalysts. Pentasils (silicalite or ZSM-5) may be included in dewaxing catalysts.

Hydrocracking catalysts, such as nickel (5% by weight) on silica–alumina, work best on feedstocks that have been hydrotreated to low nitrogen and sulfur levels. The nickel catalyst then operates well at 350°C–370°C (660°F–700°F) and a pressure of approximately 1500 psi to give good conversion of feed to lower boiling liquid fractions with minimum saturation of single-ring aromatic derivatives and a high iso-paraffin to *n*-paraffin ratio in the lower molecular weight paraffin derivatives.

Catalysts containing platinum or palladium (approximately 0.5% wet) on a zeolite base appear to be somewhat less sensitive to nitrogen than are nickel catalysts, and successful operation has been achieved with feedstocks containing 40 ppm nitrogen. This catalyst is also more tolerant of sulfur in the feed, which acts as a temporary poison, the catalyst recovering its activity when the sulfur content of the feed is reduced. With catalysts of higher hydrogenation activity, such as platinum on silica–alumina, direct isomerization occurs. The product distribution is also different, and the ratio of low- to intermediate-molecular-weight paraffin derivatives in the breakdown product is reduced.

Catalyst poisoning can be minimized by mild hydrogenation to remove nitrogen, oxygen, and sulfur from feedstocks in the presence of more resistant catalysts, such as cobalt–molybdenum–alumina (Co–Mo–Al_2O_3).

3.3.2 Hydrotreating

Hydrotreating (Chapter 8: Desulfurization, Denitrogenation, and Demetalization) is the (relatively) low-temperature removal of heteroatomic species by treatment of a feedstock or product in the presence of hydrogen (Chapter 8: Desulfurization, Denitrogenation, and Demetalization). Hydrocracking (Fig. 3.6) is the thermal decomposition of a feedstock in which carbon–carbon bonds are cleaved in addition to the removal of heteroatomic species (Chapter 9: Hydrocracking).

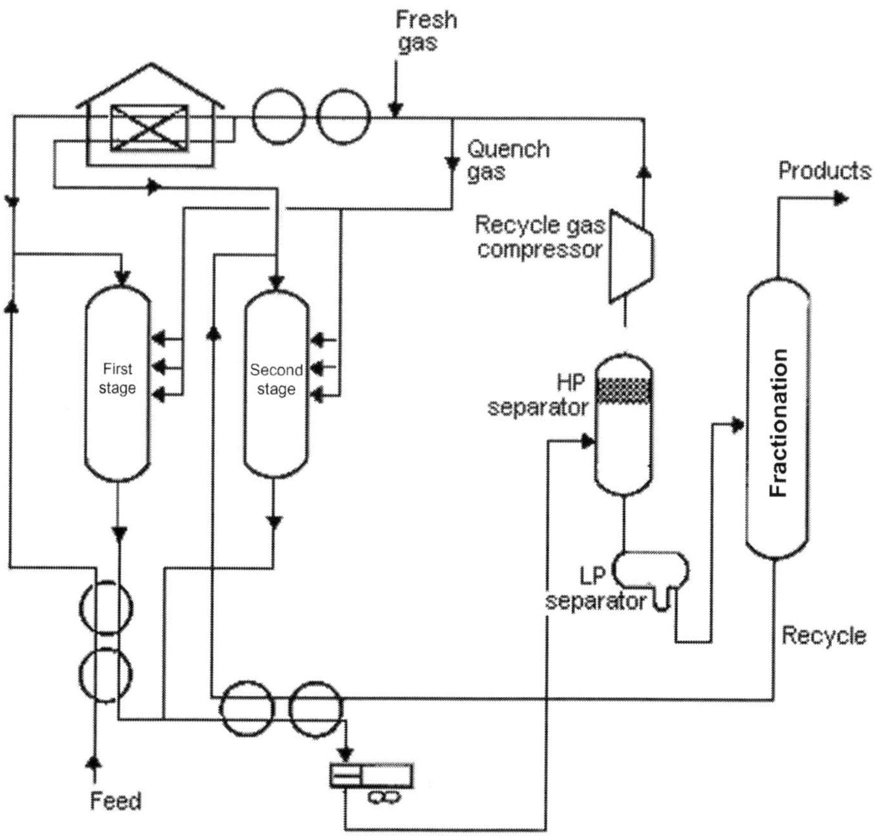

Figure 3.6 A two-stage hydrocracking unit.

3.3.2.1 General chemistry

Hydrotreating (i.e., hydrogenation without simultaneous cracking) (Chapter 8: Desulfurization, Denitrogenation, and Demetalization) is used for saturating olefin derivatives or for converting aromatic derivatives to naphthene derivatives as well as for heteroatom removal. Under atmospheric pressure, olefin derivatives can be hydrogenated up to approximately 500°C (930°F), but beyond this temperature dehydrogenation commences.

The presence of hydrogen changes the nature of the products (especially the decreasing coke yield) by preventing the buildup of precursors that are incompatible in the liquid medium and form coke (Magaril and Aksenova, 1968, 1970; Magaril and Ramazaeva, 1969; Magaril et al., 1970).

In contrast to the visbreaking process, in which the general principle is the production of products for use as fuel oil, the hydroprocessing is employed to produce a slate of products for use as liquid fuels. Nevertheless, the decomposition of asphaltenes is, again, an issue, and just as models consisting of large polynuclear aromatic systems are inadequate to explain the chemistry of visbreaking, they are also of little value for explaining the chemistry of hydrocracking.

3.3.2.2 Asphaltene chemistry

The asphaltene constituents present complex modes of thermal cracking (Parkash, 2003; Gary et al., 2007; Speight, 2014, 2017; Hsu and Robinson, 2017) and, in fact, it is at this point that the thermal chemistry of model compounds decreases in use in terms of understanding the thermal cracking of crude oil. As stated previously, the thermal behavior of model compounds may not (does not) reflect the true thermal behavior of a complex mixture such as crude oil and the thermal cracking of crude oil residua cannot be described by a single activation energy model (Schucker, 1983). The complexity of the individual reactions occurring in a residuum of similar mixture and the interference of the products with those from other components of the mixture is unpredictable. Or the interference of secondary and tertiary products with the course of a reaction and, hence, with the formation of primary products may also be cause for concern. Hence, caution is advised when applying the data from model compound studies to the behavior of crude oil, especially the molecularly complex high-boiling oils. These have few, if any, parallels in organic chemistry.

Recognition that the thermal behavior of crude oil is related to composition has led to a multiplicity of attempts to establish crude oil and its fractions as compositions of matter. As a result, various analytical techniques have been developed for the identification and quantification of every molecule in the lower boiling fractions of crude oil. However, the name crude oil does not describe a composition of matter but rather a mixture of various organic compounds that includes a wide range of molecular weights and molecular types that exist in balance with each other (Parkash, 2003; Gary et al., 2007; Speight, 2014, 2017; Hsu and Robinson, 2017).

In a mixture as complex as crude oil, the reaction processes can only be generalized because of difficulties in analyzing not only the products but also the feedstock as well as the intricate and complex nature of the molecules that make up the feedstock. The formation of coke from the higher molecular weight and polar constituents (resin and asphaltene fractions) of crude oil is detrimental to process efficiency and to catalyst performance. Although, little has been acknowledged here of the role of low-molecular-weight polar species (resins) in coke formation, the resins are presumed to be lower molecular weight analogs of the asphaltenes (Parkash et al., 1980). This being the case, similar reaction pathways may apply.

Deposition of solids or incompatibility is still possible when asphaltenes interact with catalysts, especially acidic support catalysts, through the functional groups, for example, the basic nitrogen species just as they interact with adsorbents. And there is a possibility for interaction of the asphaltene with the catalyst through the agency of a single functional group in which the remainder of the asphaltene molecule remains in the liquid phase. There is also a less desirable option in which the asphaltene reacts with the catalyst at several points of contact, causing immediate incompatibility on the catalyst surface.

During the early stages of the hydrotreating process, the chemistry of the asphaltene constituents follows the same routes as the thermal chemistry (Speight, 2014). Thus initially there is an increase in the amount of asphaltene constituents followed by a decrease indicating that, in the early stages of the process, resin constituents are being converted to asphaltene material by aromatization and by some dealkylation.

In addition, aromatization and dealkylation of the original asphaltene constituents yields asphaltene products that are of higher polarity and lower molecular weight that then original asphaltene constituents. Analogous to the thermal processes, this produces an overall asphaltene fraction that is more polar material and also of lower molecular weight. As the hydrotreating process proceeds, the amount of asphaltene constituents precipitated decreases due to conversion of the asphaltene constituents to products. At more prolonged onstream times, there is a steady increase in the yield of the asphaltene constituents. This is accompanied by a general increase in the molecular weight of the precipitated material.

As predicted from the chemistry of the thermal reactions of the asphaltene constituents, there is a steady increase in aromaticity (reflected as a decrease in the hydrogen/carbon atomic ratio) with onstream time. This is due to (1) the aromatization of naphthene ring systems that are present in asphaltene constituents, (2) the cyclodehydrogenation of alkyl chains to form other naphthene ring systems, (3) the dehydrogenation of the new naphthene ring systems to form more aromatic rings, and (4) the dealkylation of aromatic ring systems.

As the reaction progresses the aromatic carbon atoms in the asphaltene constituents show a general increase and the degree of substitution of the aromatic rings decreases. Again this is in keeping with the formation of products from the original asphaltene constituents (carbenes, carboids, and eventually coke) that have an increased aromaticity and decreased number of alkyl chains as well as a decrease in the alkyl chain length. Thus, as the reaction progresses with increased onstream time, new asphaltene constituents are formed that, relative to the original asphaltene constituents, the new species have increased aromaticity coupled with a lesser number of alkyl chains that are shorter than the original alky chains.

3.3.2.3 Catalysts

A wide variety of metals are active hydrogenation catalysts; those of most interest are nickel, palladium, platinum, cobalt, iron, nickel-promoted copper, and copper chromite. Special preparations of the first three are active at room temperature and atmospheric pressure. The metallic catalysts are easily poisoned by sulfur-containing and arsenic-containing compounds, and even by other metals. To avoid such poisoning, less effective but more resistant metal oxides or sulfides are frequently employed, generally those of tungsten, cobalt, chromium, or molybdenum.

Alternatively, catalysts poisoning can be minimized by mild hydrogenation to remove nitrogen, oxygen, and sulfur from feedstocks in the presence of more resistant catalysts, such as cobalt–molybdenum–alumina ($Co-Mo-Al_2O_3$). The reactions involved in nitrogen removal are somewhat analogous to those of the sulfur compounds and follow a stepwise mechanism to produce ammonia and the relevant substituted aromatic compound.

Hydrotreating catalysts consist of metals impregnated on a porous alumina support. Almost all of the surface area is found in the pores of the alumina (200–300 m^2/g) and the metals are dispersed in a thin layer over the entire alumina surface within the pores. This type of catalyst does display a huge catalytic surface for a small weight of catalyst. Cobalt (Co), molybdenum (Mo), and nickel (Ni) are the most commonly used metals for desulfurization catalysts. The catalysts are manufactured with the metals in an oxide state. In the active form they are in the sulfide state, which is obtained by sulfiding the catalyst either prior to use or with the feed during actual use. Any catalyst that exhibits hydrogenation activity will catalyze hydrodesulfurization to some extent. However, the group VIB metals (chromium, molybdenum, and tungsten) are particularly active for desulfurization, especially when promoted with metals from the iron group (iron, cobalt, and nickel).

Hydrodesulfurization and demetallization occur simultaneously on the active sites within the catalyst pore structure. Sulfur and nitrogen occurring in residua are converted to hydrogen sulfide and ammonia in the catalytic reactor and these gases are scrubbed out of the reactor effluent gas stream. The metals in the feedstock are deposited on the catalyst in the form of metal sulfides and cracking of the feedstock to distillate produces a laydown of carbonaceous material on the catalyst; both events poison the catalyst and activity or selectivity suffers. The deposition of carbonaceous material is a fast reaction that soon equilibrates to a particular carbon level and is controlled by hydrogen partial pressure within the reactors. On the other hand, metal deposition is a slow reaction that is directly proportional to the amount of feedstock passed over the catalyst.

Removal of sulfur from the feedstock results in a gradual increase in catalyst activity, returning almost to the original activity level. As with ammonia, the concentration of the hydrogen sulfide can be used to control precisely the activity of the catalyst. Nonnoble metal-loaded zeolite catalysts have an inherently different response to sulfur impurities since a minimum level of hydrogen sulfide is required to maintain the nickel–molybdenum and nickel–tungsten in the sulfide state.

3.4 Other reactions

In addition to the chemistry described previously, there are several other reactions that occur during crude oil refining. Each of these reactions can occur during (1) thermal conversion processes, (2) catalytic conversion processes, and (3) during product improvement processes such as reforming, isomerization, alkylation, and polymerization. Each differs from the chemical reactions described earlier and is worthy of separate treatment here.

3.4.1 Dehydrogenation

Dehydrogenation is the removal of hydrogen from the parent molecule. For example, at 550°C (1025°F) *n*-butane (C_4H_{10}) loses hydrogen to produce butene-1 ($CH_3CH_2CH=CH_2$) and butene-2 ($CH_3CH=CHCH_3$). The development of selective catalysts, such as chromic oxide (chromia, Cr_2O_3) on alumina (Al_2O_3) has rendered the dehydrogenation of paraffin derivatives to olefin derivatives particularly effective, and the formation of higher molecular weight material is minimized.

The extent of dehydrogenation (vis-à-vis carbon–carbon bond scission) during the thermal cracking of crude oil varies with the starting material and operating conditions, but because of its practical importance, methods have been found to increase the extent of dehydrogenation and, in some cases, to render it almost the only reaction.

Naphthene derivatives are somewhat more difficult to dehydrogenate than paraffin derivatives, and cyclopentane derivatives form only aromatic derivatives if a preliminary step to form the cyclohexane structure can occur. Alkyl derivatives of cyclohexane usually dehydrogenate at 480°C–500°C (895°F–930°F), and polycyclic naphthene derivatives are also quite easy to dehydrogenate thermally. In the presence of catalysts, cyclohexane and its derivatives are readily converted into aromatic derivatives; reactions of this type are prevalent in catalytic cracking and reforming. Benzene and toluene are prepared by the catalytic dehydrogenation of cyclohexane and methylcyclohexane, respectively.

Polycyclic naphthene derivatives can also be converted to the corresponding aromatic derivatives by heating at 450°C (840°F) in the presence of a chromia–alumina (Cr_2O_3–Al_2O_3) catalyst. Alkyl aromatic derivatives also dehydrogenate to various products. For example, styrene is prepared by the catalytic dehydrogenation of ethylbenzene. Other alkylbenzenes can be dehydrogenated similarly; iso-propyl benzene yields α-methyl styrene.

3.4.2 Dehydrocyclization

Catalytic aromatization involving the loss of 1 mol of hydrogen followed by ring formation and further loss of hydrogen has been demonstrated for a variety of paraffin derivatives (typically *n*-hexane and *n*-heptane). Thus *n*-hexane can be converted to benzene, heptane is converted to toluene, and octane is converted to ethylbenzene and *o*-xylene. Conversion takes place at low pressures, even atmospheric, and at temperatures above 300°C (570°F), although 450°C–550°C (840°F–1020°F) is the preferred temperature range.

The catalysts are metals (or their oxides) of the titanium, vanadium, and tungsten groups and are generally supported on alumina; the mechanism is believed to be dehydrogenation of the paraffin to an olefin, which in turn is cyclized and dehydrogenated to the aromatic hydrocarbon. In support of this, olefin derivatives can be converted to aromatic derivatives much more easily than the corresponding paraffin derivatives.

3.4.3 Isomerization

The importance of isomerization in crude oil-refining operations is twofold. First, the process is valuable in converting *n*-butane into iso-butane, which can be alkylated to liquid hydrocarbon derivatives in the gasoline boiling range. Second, the process can be used to increase the octane number of the paraffin derivatives boiling in the gasoline boiling range by converting some of the *n*-paraffin derivatives present into iso-paraffin derivatives.

The process involves contact of the hydrocarbon and a catalyst under conditions favorable to good product recovery. The catalyst may be aluminum chloride promoted with hydrochloric acid or a platinum-containing catalyst. Both are very reactive and can lead to undesirable side reactions along with isomerization. These side reactions include disproportionation and cracking, which decrease the yield and produce olefinic fragments that may combine with the catalyst and shorten its life. These undesired reactions are controlled by such techniques as the addition of inhibitors to the hydrocarbon feed or by carrying out the reaction in the presence of hydrogen.

Paraffin derivatives are readily isomerized at room temperature, and the reaction is believed to occur by means of the formation and rearrangement of carbonium ions. The chain-initiating ion R^+ is formed by the addition of a proton from the acid catalyst to an olefin molecule, which may be added, presented as an impurity, or formed by dehydrogenation of the paraffin.

Except for butane, the isomerization of paraffin derivatives is generally accompanied by side reactions involving carbon–carbon bond scissions when catalysts of the aluminum halide type are used. Products boiling both higher and lower than the starting material are formed, and the disproportionation reactions occur with the pentanes and higher paraffin derivatives ($>C_5$) are caused by unpromoted aluminum halide. A substantial pressure of hydrogen tends to minimize these side reactions.

The ease of paraffin isomerization increases with molecular weight, but the extent of disproportionation reactions also increases. Conditions can be established under which isomerization takes place only with the butanes, but this is difficult for the pentanes and higher hydrocarbon derivatives. At 27°C (81°F) over aluminum bromide ($AlBr_3$), the equilibrium mixture of *n*-pentane and iso-pentane, contains over 70% of the branched isomer; at 0°C (32°F) approximately 90% of the branched isomer is present. Higher and lower boiling hydrocarbon products, hexanes, heptanes, and iso-butane are also formed in side reactions even at 0°C (32°F) and in increased amounts when the temperature is raised. Although the thermodynamic conditions are favorable, neo-pentane [$C(CH_3)_4$] does not appear to isomerize under these conditions.

Olefin derivatives are readily isomerized—the reaction involves either movement of the position of the double bond (hydrogen-atom shift) or skeletal alteration (methyl group shift). The double-bond shift may also include a reorientation of the groups around the double bond to cause a *cis–trans* isomerization. Thus 1-butene is isomerized to a mixture of *cis*- and *trans*-2-butene. *Cis* (same side) and *trans* (opposite side) refer to the spatial arrangement of the methyl groups with respect to the double bond.

Olefin derivatives having a terminal double bond are the least stable. They isomerize more rapidly than those in which the double bond carries the maximum number of alkyl groups.

Naphthene derivatives can isomerize in various ways; for example, in the case of cyclopropane (C_3H_6) and cyclobutane (C_4H_8), ring scission can occur to produce an olefin. Carbon–carbon rupture may also occur in any side chains to produce polymethyl derivatives, whereas cyclopentane (C_5H_{10}) and cyclohexane (C_6H_{12}) rings may expand and contract, respectively.

The isomerization of alkyl aromatic derivatives may involve changes in the side-chain configuration, disproportionation of the substituent groups, or their migration within the molecule. The conditions needed for isomerization within attached long side chains of alkylbenzenes and alkyl naphthalene derivatives are also those for the scission of such groups from the ring. Such isomerization, therefore, does not take place unless the side chains are relatively short. The isomerization of ethylbenzene to xylenes, and the reverse reaction, occurs readily.

Disproportionation of attached side chains is also a common occurrence; higher and lower alkyl substitution products are formed. For example, xylenes disproportionate in the presence of hydrogen fluoride–boron trifluoride or aluminum chloride to form benzene, toluene, and higher alkylated products; ethylbenzene in the presence of boron trifluoride forms a mixture of benzene and 1,3-diethyl benzene.

3.4.4 Alkylation

Alkylation in the crude oil industry refers to a process for the production of high-octane motor fuel components by the combination of olefin derivatives and paraffin derivatives. The reaction of iso-butane with olefin derivatives, using an aluminum chloride catalyst, is a typical alkylation reaction.

In acid-catalyzed alkylation reactions, only paraffin derivatives with tertiary carbon atoms, such as iso-butane and iso-pentane react with the olefin. Ethylene is slower to react that the higher olefin derivatives. Olefin derivatives higher than propene may complicate the products by engaging in hydrogen-exchange reactions.

Cycloparaffin derivatives, especially those containing tertiary carbon atoms, are alkylated with olefin derivatives in a manner similar to the iso-paraffin derivatives; the reaction is not as clean, and the yields are low because of the several side reactions that take place.

Aromatic hydrocarbon derivatives are more easily alkylated than the iso-paraffin derivatives by olefin derivatives. Cumene (iso-propylbenzene) is prepared by alkylating benzene with propene over an acid catalyst. The alkylating agent is usually an olefin, although cyclopropane, alkyl halides, aliphatic alcohols, ethers, and esters may also be used. The alkylation of aromatic hydrocarbon derivatives is presumed to occur through the agency of the carbonium ion.

Thermal alkylation is also used in some plants, but like thermal cracking, it is presumed to involve the transient formation of neutral free radicals and therefore tends to be less specific in production distribution.

3.4.5 Polymerization

Polymerization is a process in which a substance of low molecular weight is transformed into one of the same composition but of higher molecular weight while maintaining the atomic arrangement present in the basic molecules. It has also been described as the successive addition of one molecule to another by means of a functional group, such as that present in an aliphatic olefin.

In the crude oil industry, polymerization is used to indicate the production of, say, gasoline components that fall into a specific (and controlled) molecular weight range, hence the term polymer gasoline. Furthermore, it is not essential that only one type of monomer be involved:

$$CH_3CH = CH_2 + CH_2 = CH_2 \rightarrow CH_3CH_2CH_2CH = CH_2$$

This type of reaction is correctly called copolymerization, but polymerization in the true sense of the word is usually prevented, and all attempts are made to terminate the reaction at the dimer or trimer (three monomers joined together) stage. It is the 4- to 12-carbon compounds that are required as the constituents of liquid fuels. However, in the petrochemical section of the refinery, polymerization, which results in the production of, say, polyethylene, is allowed to proceed until materials of the required high molecular weight have been produced.

3.5 Instability and incompatibility

The study of the chemical reactivity of crude oil would not be complete without some attention to the phenomena of instability and incompatibility. Both are a result of chemical reactions that form degradation products or other undesirable changes in the original chemical properties of crude oil and crude oil products. Furthermore, it is the various analytical methods (also discussed here) that provide the data that point to the reason for problems in the refinery or for the failure of products to meet specifications and to perform as desired.

The focus of incompatibility studies has usually been on the whole crude oil and specifically on the asphaltene fraction. The asphaltene constituents are the highest molecular weight and most polar fractions found in crude oil (Parkash, 2003; Gary et al., 2007; Speight, 2014, 2017; Hsu and Robinson, 2017). The characteristics of the asphaltene constituents and the amount in crude oil depend to a greater or lesser extent on the crude oil source. During refining of crude oil, asphaltene constituents will end up in a high percentage in the residual fuels as the low-boiling fractions (gasoline, jet fuel, etc.) are removed from the oil through cracking and visbreaking.

The problems with the asphaltene constituents have increased due to the need to extract even the heaviest crude oils as well as the trend to extract large amounts of low-boiling fractions out of crude oil by visbreaking and cracking processes.

Deposition of asphaltene constituents is the consequence of instability of the crude oil or the crude oil product (e.g., fuel oil). The asphaltene constituents are stabilized by resin constituents (Parkash, 2003; Gary et al., 2007; Speight, 2014, 2017; Hsu and Robinson, 2017) and maintained in the crude oil due to this stabilization. Asphaltene dispersants are substitutes for the natural resin constituents and are able to keep the asphaltene constituents dispersed to prevent flocculation/aggregation and phase separation. Dispersants will also clean up sludge in the fuel system and they have the ability to adhere to surface of materials that are insoluble in the oil and convert them into stable colloidal suspensions.

At the other end of the molecular weight scale and refining, the heteroatoms (particularly nitrogen, sulfur, and trace metals) (Chapter 1: Feedstock Types and Properties) that are present in crude oil and might be also expected to be present in liquid fuels and other products from crude oil. And, indeed, this is often the case, although there may have been some skeletal changes induced by the refining processes. Oxygen is much more difficult to define in crude oil and liquid fuels. However, it must be stressed that instability/incompatibility is not directly related to the total nitrogen, oxygen, or sulfur content. The formation of color/sludge/sediment is a result of several factors. Perhaps the main factor is the location and nature of the heteroatom which, in turn, determines reactivity (Mushrush and Speight, 1995).

Incompatibility in crude oil products can be linked to the presence of several different deleterious heteroatomic compound classes and their potential reactions (Mushrush and Speight, 1995).

In general, the chemical reaction sequence for sediment formation can be envisaged as being dependent upon the most reactive of the various heteroatomic species that are present in fuels. The worst case scenario would consist of a high olefin fuel with both high indole concentration and a catalytic trace of sulfonic acid species. This reaction matrix would lead to rapid degradation. However, just as there is no one specific distillate product, there is also no one mechanism of degradation. In fact, the mechanism and the functional groups involved will give a general but not specific mode of incompatibility (Mushrush and Speight, 1995; Speight, 2014).

When various feedstocks are blended at the refinery, incompatibility can be explained by the onset of acid—base-catalyzed condensation reactions of the various organo-nitrogen compounds in the individual blending stocks. These are usually very rapid reactions with practically no observed induction time period. In addition, when the product is transferred to a storage tank or some other holding tank, incompatibility can occur by the free-radical hydroperoxide-induced polymerization of active olefin derivatives. This is a relatively slow reaction, because the observed increase in hydroperoxide concentration is dependent on the dissolved oxygen content.

Incompatibility during refining can occur in a variety of processes, either by intent (such as in the deasphalting process) or inadvertently when the separation is detrimental to the process. Thus separation of solids occurs whenever the solvent characteristics of the liquid phase are no longer adequate to maintain polar and/or high-molecular-weight material in solution.

Examples of such occurrences are (1) the separation of asphaltene constituents separation, which occurs when the paraffinic nature of the liquid medium increases; (2) the separation of wax constituents, which occurs when there is a drop in temperature or the aromaticity of the liquid medium increases; and (3) the formation of sludge/sediment in a reactor, which occurs when the solvent characteristics of the liquid medium change, so that asphaltic or wax materials separate, coke formation that occurs at high temperatures and commences when the solvent power of the liquid phase is not sufficient to maintain the coke precursors in solution, and sludge/sediment formation in fuel products, which occurs because of the interplay of several chemical and physical factors (Mushrush and Speight, 1995).

Product complexity and the means by which the product is evaluated have made the industry unique among the process industries (Speight, 2015; ASTM, 2019). But product complexity has also brought to the fore issues such as instability and incompatibility. Product complexity becomes even more disadvantageous when various fractions from different types of crude oil are blended or are allowed to remain under conditions of storage (prior to use) and a distinct phase separates from the bulk product. The adverse implications of this for refining the fractions to salable products increase (Mushrush and Speight, 1995; Parkash, 2003; Gary et al., 2007; Speight, 2014, 2017; Hsu and Robinson, 2017).

3.5.1 Definitions and terminology

Therefore it is appropriate here to define some of the terms that are used in the liquid fuels field so that their use later in the text will be more apparent and will also alleviate some potential for misunderstanding. The general scientific areas of instability and incompatibility are complex and have been considered to be nothing better than a black art because not all of the reactions that contribute to instability and incompatibility have been defined (Mushrush and Speight, 1995; Speight, 2014).

Briefly, the term incompatibility refers to the formation of a precipitate (or sediment) or separate phase when two liquids are mixed. The term instability is often used in reference to the formation of color, sediment, or gum in the liquid over a period of time and is usually due to chemical reactions, such as oxidation, and is chemical rather than physical. This term may be used to contrast the formation of a precipitate in the near term (almost immediately).

The phenomenon of instability is often referred to as incompatibility, and more commonly known as sludge formation, and sediment formation, or deposit formation. In crude oil and its products, instability often manifests itself in various ways (Stavinoha and Henry, 1981; Mushrush and Speight, 1995; Speight, 2014). Hence, there are different ways of defining each of these terms but the terms are often used interchangeably.

Gum formation (ASTM D525, 2019) alludes to the formation of soluble organic material, whereas sediment is the insoluble organic material. Storage stability (or storage instability) (ASTM D381, 2019; ASTM D4625, 2019) is a term used to describe the ability of the liquid to remain in storage over extended periods of time without appreciable deterioration as measured by gum formation and/or the formation sediment. Thermal stability is also defined as the ability of the liquid to withstand relatively high temperatures for short periods of time without the formation of sediment (i.e., carbonaceous deposits and/or coke). Thermal oxidative stability is the ability of the liquid to withstand relatively high temperatures for short periods of time in the presence of oxidation and without the formation of sediment or deterioration of properties (ASTM D3241, 2019) and there is standard equipment for various oxidation tests (ASTM D4871, 2019). Stability is also as the ability of the liquid to withstand long periods at temperatures up to 100°C (212°F) without degradation.

Existent gum is the name given to the nonvolatile residue present in the fuel as received for test (ASTM D381, 2019). In this test the sample is evaporated from a beaker maintained at a temperature of 160°C–165°C (320°F–330°F) with the aid of a similarly heated jet of air. This material is distinguished from the potential gum that is obtained by aging the sample at an elevated temperature.

Thus potential gum is determined by the accelerated gum test (ASTM D873, 2019) that is used as a safeguard of storage stability and can be used to predict the potential for gum formation during prolonged storage. In this test the fuel is heated for 16 hours with oxygen under pressure in a bomb at 100°C (212°F). After this time, both the gum content and the solid precipitates are measured. A similar test, using an accelerated oxidation procedure, is also in use for determining the oxidative stability of diesel fuel (ASTM D2274, 2019), steam turbine oil, distillate fuel oil (ASTM D2274, 2019), and lubricating grease (ASTM D942, 2019).

Dry sludge is defined as the material separated from crude oil and crude oil products by filtration, which is insoluble in heptane. Existent dry sludge is the dry sludge in the original sample as received and is distinguished from the accelerated dry sludge obtained after aging the sample by chemical addition or heat. The existent dry sludge is distinguished from the potential dry sludge that is obtained by aging the sample at an elevated temperature.

The existent dry sludge is operationally defined as the material separated from the bulk of a crude oil or crude oil product by filtration, which is insoluble in heptane. The test is used as an indicator of process operability and as a measure of potential downstream fouling.

An analogous test, the thin-film oven test (ASTM D1754, 2019) is used to indicate the rate of change of various physical properties such as penetration (ASTM D5, 2019), viscosity (ASTM D2170, 2019), and ductility (ASTM D113, 2019) after a film of asphalt or bitumen has been heated in an oven for five hours at 163°C (325°F) on a rotating plate. A similar test is available for the stability of engine oil by the thin-film oxygen uptake test (ASTM D4424, 2019).

This test establishes the effects of heat and air based on changes incurred in the abovementioned physical properties measured before and after the oven test. The allowed rate of changes in the relevant bitumen properties after the exposure of the tested sample to the oven test are specified in the relevant specifications (ASTM D3381, 2019).

Attractive as they may seem, any tests that involve accelerated oxidation of the sample must be used with caution and consideration of the chemistry. Depending on the constituents of the sample, it is quite possible that the higher temperature and extreme conditions (oxygen under pressure) may not be truly representative of the deterioration of the sample under storage conditions. The higher temperature and the oxygen under pressure might change the chemistry of the system and produce products that would not be produced under ambient storage conditions. An assessment of the composition of the fuel prior to storage and application of the test will assist in this determination.

Thus crude oil constituents are incompatible when sludge, semisolid, or solid particles (for convenience here, these are termed secondary products to distinguish them from the actual crude oil product) are formed during and after blending. This phenomenon usually occurs prior to use. If the secondary products are marginally soluble in the blended crude oil product, use might detract from solubility of the secondary products and they will appear as sludge or sediment that can be separated by filtration or by extraction (ASTM D4310, 2019). When the secondary products are truly insoluble, they separate and settle out as a semisolid or solid phase floating in the fuel or are deposited on the walls and floors of containers. In addition, secondary products usually increase the viscosity of the crude oil product. Standing at low temperatures will also cause a viscosity change in certain fuels and lubricants (ASTM D2532, 2019). Usually the viscosity change might be due to separation of paraffin derivatives as might occur when diesel fuel and similar engines are allowed to cool and stand unused overnight in low-temperature climates.

There is also the suggestion that sediments and deposits originate from the inorganic constituents of crude oil. They may be formed from the inherent components of the crude oil (i.e., the metalloporphyrin constituents) or from the ingestion of contaminants by the crude oil during the initial processing operations. For example, crude oil is known to pick up iron and other metal contaminants from contact with pipelines and pumps.

3.5.2 General chemistry

The chemistry and physics of incompatibility can, to some extent, be elucidated (Mushrush and Speight, 1995) but many unknowns remain. In addition to the chemical aspects, there are also aspects such as the attractive force differences, such as (1) specific interactions between like/unlike molecules, which include hydrogen bonding and electron donor–acceptor phenomena; (2) field interactions such as dispersion forces and dipole–dipole interactions; and (3) any effects imposed on the system by the size and shape of the interacting molecular species.

Such interactions are not always easy to define and, thus, the measurement of instability and incompatibility has involved visual observations, solubility tests of hot filtration sediment, and gum formation. However, such methods are often considered to be after-the-fact methods insofar as they did not offer much in the way of predictability. In refinery processes (Chapter 2: Introduction to Refining Processes), predictability is not just a luxury, it is a necessity. The same principle must be applied to the measurement of instability and incompatibility. Therefore methods are continually being sought to aid in achieving this goal.

In addition to the gravimetric methods, there have also been many attempts to use crude oil and/or product characteristics and their relation to the sludge and deposit formation tendencies. In some cases a modicum of predictability is the outcome but, in many cases, the data appear as preferred ranges and are subject to personal interpretation. Therefore caution is advised.

3.5.3 Test methods

The instability/incompatibility of crude oil and of crude oil products is manifested in the formation of sludge, sediment, and general darkening in color of the liquid (ASTM D1500, 2019).

Sludge (or sediment) formation takes one of the following forms: (1) material dissolved in the liquid, (2) precipitated material, and (3) material emulsified in the liquid. Under favorable conditions, sludge or sediment will dissolve in the crude oil or product with the potential of increasing the viscosity. Sludge or sediment, which is not soluble in the crude oil (ASTM D473, 2019; ASTM D1796, 2019; ASTM D2273, 2019; ASTM D4007, 2019; ASTM D4807, 2019; ASTM D4870, 2019), may either settle at the bottom of the storage tanks or remain in the crude oil as an emulsion. In most of the cases the smaller part of the sludge/sediment will settle satisfactorily, the larger part will stay in the crude oil as emulsions. In any case there is a need of breaking the emulsion, whether it is a water-in-oil emulsion or whether it is the sludge itself, which has to be separated into the oily phase and the aqueous phase. The oily phase can be then processed with the crude oil and the aqueous phase can be drained out of the system.

Phase separation can be accomplished by either the use of suitable surface-active agents allowing for sufficient settling time or the use of a high-voltage electric field for breaking such emulsions after admixing water at a rate of approximately 5% and at a temperature of approximately 100°C (212°F).

Chemical and electrical methods for sludge and water removal, often combined with chemical additives, have to be used for breaking such emulsions. Each emulsion has its own structure and characteristics: water-in-oil emulsions where the oil is the major component or oil-in-water emulsions where the water is the major component. The chemical and physical nature of the components of the emulsion plays a major role in their susceptibility to the various surface-active agents used for breaking them.

Therefore appropriate emulsion breaking agents have to be chosen very carefully, usually with help of previous laboratory evaluations. Water- or oil-soluble demulsifiers, the latter being often nonionic surface-active alkylene oxide adducts, are used for this purpose. But, as had been said in the foregoing, the most suitable demulsifier has to be chosen for each case from a large number of such substances in the market, by a prior laboratory evaluation.

3.5.4 Determination of instability and incompatibility

A number of experimental methods are available for estimation of the factors that influence instability/incompatibility. These factors have been explored and attempts made to estimate the character of the fuel or product with varied results.

3.5.4.1 Elemental analysis

The ultimate analysis (elemental composition) of crude oil and its products is not reported to the same extent as for coal (Speight, 1994, 2014, 2015). Nevertheless, there are ASTM (2019) procedures for the ultimate analysis of crude oil and crude oil products but many such methods may have been designed for other materials.

Of the data that are available, the proportions of the elements in crude oil vary only slightly over narrow limits: carbon: 83.0%–87.0%, hydrogen: 10.0%–14.0%, nitrogen: 0.10%–2.0%, oxygen: 0.05%–1.5%, and sulfur: 0.05%–6.0% (Chapter 1: Feedstock Types and Properties). And yet, there is a wide variation in physical properties from the lower boiling more mobile crude oils at one extreme to the extra heavy crude oil at the other extreme (Chapter 1: Feedstock Types and Properties). In terms of the instability and incompatibility of crude oil and crude oil products, the heteroatom content appears to represent the greatest influence. In fact, it is not only the sulfur and nitrogen content of crude oil are important parameters in respect of the processing methods that have to be used in order to produce fuels of specification sulfur concentrations but also the type of sulfur and nitrogen species in the oil. There could well be a relation between nitrogen and sulfur content and crude oil (or product) stability; higher nitrogen and sulfur crude oils are suspect of higher sludge forming tendencies.

3.5.4.2 Density/specific gravity

In the earlier years of the crude oil industry, density and specific gravity [with the American Petroleum Institute (API) gravity] were the principal specifications for feedstocks and refinery products. They were used to give an estimate of the most desirable product, that is, kerosene, in crude oil. At the present time a series of standard tests exists for determining density and specific gravity (Speight, 2014, 2015).

There is no direct relation between the density and specific gravity of crude oils to their sludge forming tendencies, but crude oil having a higher density (thus a lower API gravity) is generally more susceptible to sludge formation, presumably because of the higher content of the polar/asphaltic constituents.

3.5.4.3 Volatility

Crude oil can be subdivided by distillation into a variety of fractions of different boiling ranges or cut points. In fact distillation was, and still is, the method for feedstock evaluation for various refinery options. Indeed, volatility is one of the major tests for crude oil products and it is inevitable that the majority of all products will, at some stage of their history, be tested for volatility characteristics.

As an early part of characterization studies, a correlation was observed between the quality of crude oil fractions and their heteroatom content (Fig. 3.7) since gasoline, kerosene, diesel fuel, and lubricating oil are made up of hydrocarbon constituents containing high proportions of hydrogen. Thus it is not surprising that test to determine the volatility of crude oil and crude oil products was among the first to be defined.

The very nature of the distillation process by which residua are produced (Chapter 2: Introduction to Refining Processes), that is, removal of distillate without thermal decomposition, dictates that the majority of the heteroatoms that are predominantly in the higher molecular weight fractions will be concentrated in the

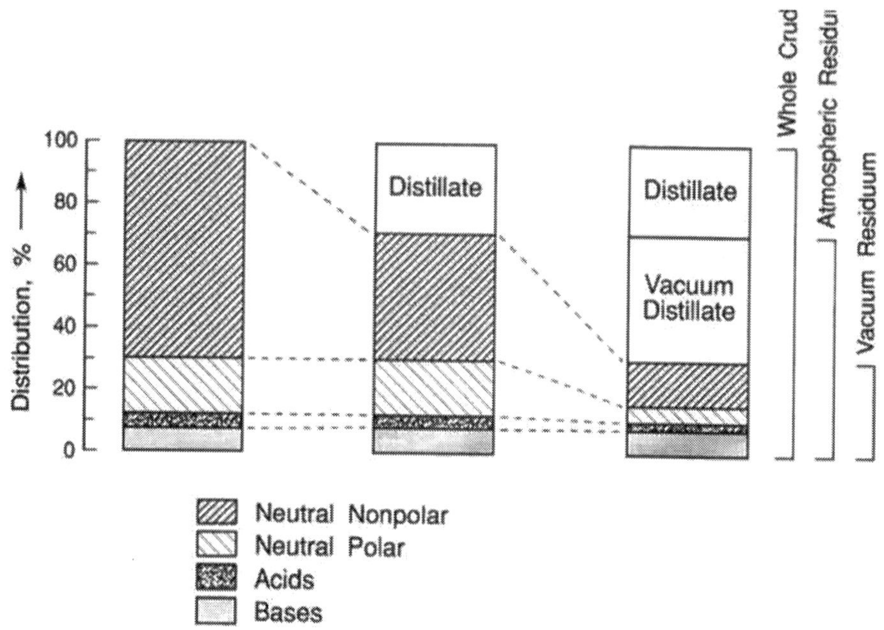

Figure 3.7 Schematic representation of the distribution of polar (heteroatomic) groups after distillation.

higher boiling products and the residuum (Speight, 2000, 2014, 2015, 2017). Thus the inherent nature of the crude oil and the means by which it is refined can seriously influence the stability and incompatibility of the products.

Heavier crude oils, yielding higher amounts of residua, tend to form more sludge during storage compared to conventional low-density crude oils.

3.5.4.4 Viscosity

The viscosity of a feedstock varies with the origin and type of the crude oil and also with the nature of the chemical constituents, particularly the polar functions where intermolecular interactions can occur. For example, there is a gradation of viscosity between conventional crude oil, heavy crude oil, extra heavy oil, and tar sand bitumen (Speight, 2008, 2014, 2017).

Viscosity is a measure of fluidity properties and consistencies at given temperatures. Heavier crude oils, that is, crude oils of lower API gravity, have usually higher viscosity. Increases of viscosity during storage indicate either an evaporation of volatile components or formation of degradation products dissolving in the crude oil.

3.5.4.5 Asphaltene content

Asphaltenes are dark brown to black friable solids that have no definite melting point and usually intumesce on heating with decomposition to leave a carbonaceous residue. They are obtained from crude oil by the addition of a nonpolar solvent (such as a liquid hydrocarbon). Liquids used for this purpose are *n*-pentane and *n*-heptane (Parkash, 2003; Gary et al., 2007; Speight, 2014, 2017; Hsu and Robinson, 2017). Usually, the asphaltene fraction is removed by filtration through paper but more recently a membrane method has come into use. Liquid propane is used commercially in processing crude oil residues for asphaltenes and resins. Asphaltenes are soluble in liquids such as benzene, toluene pyridine, carbon disulfide, and carbon tetrachloride.

The asphaltene fraction of feedstocks (Fig. 3.1) is particularly important because as the proportion of this fraction increases, there is (a) concomitant increase in thermal coke yield and an increase in hydrogen demand as well as catalyst deactivation. The constituents of the asphaltenes form coke quite readily, which is of particular interest in terms of the compatibility/incompatibility of the coke precursors (Speight, 1994).

The effect of asphaltenes and the micelle structure and the state of dispersion also merit some attention. The degree of dispersion of asphaltenes is higher in the more naphthenic/aromatic crude oils because of higher solvency of naphthene derivatives and aromatic derivatives over paraffinic constituents. This phenomenon also acts in favor of the dissolution of any sludge that may form thereby tending to decrease sludge deposition. However, an increase in crude oil often accompanies sludge dissolution.

The higher the asphaltene content, the greater the tendency of the crude oil is to form sludge, especially when blended with other noncompatible stocks.

3.5.4.6 Pour point

The pour point defines the cold properties of crude oils and crude oil products, that is, the minimal temperature at which they still retain their fluidity (ASTM D97, 2019). Therefore pour point also indicates the characteristics of crude oils: the higher the pour point, the more paraffinic is the oil and vice versa. Higher pour point crude oils are waxy and therefore they tend to form wax-like materials that enhance sludge formation.

To determine the pour point (ASTM D97, 2019; ASTM D5853, 2019; ASTM D5949, 2019; ASTM D5950, 2019; ASTM D5985, 2019) the sample is contained in a glass test tube fitted with a thermometer and immersed in one of three baths containing coolants. The sample is dehydrated and filtered at a temperature 25°C (45°F) higher than the above anticipated cloud point. It is then placed in a test tube and cooled progressively in coolants held at −1°C to +2°C (30°F–35°F), −18°C to −20°C (−4°F to 0°F), and −32°C to −35°C (−26°F to −31°F), respectively. The sample is inspected for cloudiness at temperature intervals of 1°C (2°F). If conditions or oil properties are such that reduced temperatures are required to determine the pour point, alternate tests are available that accommodate the various types of samples.

3.5.4.7 Acidity

The acidity of crude oil or crude oil products is usually measured in terms of the acid number that is defined as the number of milli-equivalents per gram of alkali required to neutralize the acidity of the crude oil sample.

Acidity due to the presence of inorganic constituents is not expected to be present in crude oils, but organic acidity might be found. Acidic character is composed of contributions from strong organic acids and other organic acids. Values above 0.15 mg potassium hydroxide per gram are considered to be significantly high. Crude oils of higher acidities may exhibit a tendency of instability.

The acid imparting agents in crude oils are naphthenic acids and hydrosulfides (thiols, mercaptans, R-SH). These are sometimes present in the crude oil originally in small and varying concentrations. Normally, the total acidity of crude oils is in the range of 0.1–0.5 mg potassium hydroxide per gram, although higher values are not exceptional.

Free hydrogen sulfide is often present in crude oils, a concentration of up to 10 ppm being acceptable in spite of its toxic nature. However, higher hydrogen sulfide concentrations are sometimes present, 20 ppm posing serious safety hazards. Additional amounts of hydrogen sulfide can form during the crude oil processing, when hydrogen reacts with some organic sulfur compounds converting them to hydrogen sulfide. In this case it is referred to as potential hydrogen sulfide, contrary to free hydrogen sulfide.

Acidity can also form by bacterial action insofar as some species of aerobic bacteria can produce organic acids from organic nutrients. On the other hand, anaerobic sulfate-reducing bacteria can generate hydrogen sulfide, which, in turn, can be converted to sulfuric acid (by bacterial action).

3.5.4.8 Metal content

The majority of crude oils contain metallic constituents that are often determined as combustion ash (ASTM D482). This is particularly so for the heavier feedstocks. These constituents, of which nickel and vanadium are the principal metals, are very influential in regard to feedstock behavior in processing operations.

The metal (inorganic) constituents of crude oil or a liquid fuel arise from either those present in the crude oil originally or those picked up by the crude oil during storage and handling. The former are mostly metallic substances such as vanadium, nickel, sodium, iron, and silica, and the latter may be contaminants such as sand, dust, and corrosion products.

Incompatibility, leading to deposition of the metals (in any form) on to the catalyst, leads to catalyst deactivation whether it is by physical blockage of the pores or destruction of reactive sites. In the present context the metals must first be removed if erroneously high carbon residue data are to be avoided. Alternatively, they can be estimated as ash by complete burning of the coke after carbon residue determination.

Metal contents above 200 ppm are considered to be significant, but the variations are very large. The higher the ash content the higher is the tendency of the crude oil to form sludge or sediment.

3.5.4.9 Water content, salt content, and bottom sediment and water

Water content (ASTM D4007, 2019), salt content, and bottom sediment/water (BS&W) (ASTM D1796, 2019; ASTM D4007, 2019) indicate the concentrations of aqueous contaminants, present in the crude either originally or picked up by the crude during handling and storage. Water and salt content of crude oils produced at the field can be very high, forming sometimes its major part. The salty water is usually separated at the field, usually by settling and draining, surface-active agents electrical emulsion breakers (desalters) are sometimes employed. The water and salt contents of crude oil supplied to the buyers are function of the production field. Water content below 0.5%, salt content up to 20 lb per 1000 bl, and BS&W up to 0.5% are considered to be satisfactory.

Although the centrifuge methods are still employed (ASTM D1796, 2019; ASTM D4007, 2019), many laboratories prefer the Dean and Stark adaptor (ASTM D95). The apparatus consists of a round-bottom flask of capacity 50 mL connected to a Liebig condenser by a receiving tube of capacity 25 mL, graduated in 0.1 mL. A weighed amount, corresponding to approximately 100 mL of oil, is placed in the flask with 25 mL of dry toluene. The flask is heated gently until the 25 mL of toluene has distilled into the graduated tube. The water distilled with the toluene separates to the bottom of the tube where the volume is recorded as milliliter, or the weight as milligram, or percent.

To determine the sediment in crude oil or in a crude oil product, the method involves solvent extraction using a Soxhlet extractor.

The Karl Fischer titration method, the Karl Fischer titration method, and the colorimetric Karl Fischer titration method still find wide application in many laboratories for the determination of water in liquid fuels, specifically the water content of aviation fuels.

The higher the BS&W content, the higher sludge and deposit formation rates can be expected in the stored crude oil.

3.6 The future

The initial stages of the thermal decomposition of asphaltenes have been least well understood of the thermal decomposition process. It is known that the overall thermal decomposition involves scission of carbon–carbon bonds, aromatization of naphthenic ring systems, formation of lower molecular weight material, and formation of coke. The presence of hydrogen during the thermal decomposition of asphaltenes can alter the product slate. However, the chemistry remains complex and representation of the chemistry of asphaltene thermolysis is speculative and has been done simply by the use of nonchemical formulas.

If no side reactions occur, very residence long times at low temperature should be equivalent to very short times at high temperature. Thermal reactions, however, can give rise to a variety of different reactions, so that selectivity for a given product changes with temperature and pressure. In fact, the nature of the thermal processes varies considerably.

The formation of solid sediments, or coke, during thermal processes is a major limitation on processing. Furthermore, the presence of different types of solids shows that solubility controls the formation of solids. And the tendency for solid formation changes in response to the relative amounts of the low-boiling products, middle distillates, and residues and to changing chemical composition during the process.

One of the postulates of coke formation involves the production of coke by a sequence of polymerization and condensation steps from the lowest boiling (lower molecular weight) to the heaviest (higher molecular weight) fractions. While this concept is acceptable for pure aromatic hydrocarbon derivatives, the formation of coke by thermal cracking the constituents of crude oil asphaltenes requires a modification of this concept. Indeed, present evidence (Speight, 1991, 1994, 2014) shows that the reaction pathway of the asphaltenes to coke is much more complex and involves at least several reaction stages.

Various findings offer some indication of the proposed chemistry of coke or sediment formation during crude oil processing. However, in each case the aromatization of naphthene rings and the potential or reality of cyclodehydrogenation reactions are ignored. In addition, the thermal reactions of the resin constituents are not addressed.

The asphaltene constituents rely upon the constituents of resins for stability within the oil medium. Removal or thermal alteration of the resin constituents will also cause the asphaltene constituents to separate as a distinct phase. Such processes would also lead to increased aromaticity of the polynuclear aromatic–naphthenic (often referenced as redistribution of hydrogen within constituents) core and further enhance the insolubility (phase separation) of the dealkylated polynuclear product.

The available data suggest that the thermal cracking of the high-molecular-weight constituents of crude oil (leading to the formation of volatile products and coke) is a complex process involving both chemical reactions and thermodynamic behavior. Reactions that contribute to this process are (1) cracking of side chains from aromatic groups, (2) dehydrogenation of naphthene derivatives to form aromatic derivatives, (3) condensation of aliphatic structures to form aromatic derivatives, (4) condensation of aromatic derivatives to form higher fused-ring aromatic derivatives, and (5) dimerization or oligomerization reactions. Loss of side chains always accompanies thermal cracking, and dehydrogenation and condensation reactions are favored by hydrogen-deficient conditions.

The importance of the use of solvents to mitigate coke formation during the thermal decomposition of crude oil has been recognized for many years but the effects have been correctly ascribed to hydrogen donor reactions that have an inhibiting effect on the formation of the molecular species that are prone to phase separation. The separation of the phases depends on the solvent characteristics of the liquid. Addition of aromatic solvents suppresses phase separation, whereas paraffin derivatives enhance separation. Microscopic examination of coke particles often shows evidence for mesophase, spherical domains that exhibit the anisotropic optical characteristics of liquid crystals.

The phase separation phenomenon that is the prelude to coke formation can also be explained by use of the solubility parameter (δ) for crude oil fractions and for the solvents (Speight, 1994, 2014, 2015). Although little is known related to the solubility parameter of crude oil fractions, there has been a noteworthy attempt to define the solubility parameter ranges for different fossil fuel liquids.

When catalytic processes are employed, complex molecules (such as those that may be found in the original resin fraction and in the original asphaltene fraction or those formed during the process) are not sufficiently mobile (or are too strongly adsorbed by the catalyst) to be saturated by the hydrogenation components. Hence, these molecular species continue to condense and eventually degrade to coke. These deposits deactivate the catalyst sites and eventually interfere with the process.

As these concepts show, characterization data can be used as an integral part of understanding the thermal chemistry of asphaltenes as a start in predicting product yield and product distribution; predictions of product quality will follow.

References

ASTM, 2019. Annual Book of Standards. ASTM International, West Conshohocken, PA.

ASTM D5, 2019. Standard test method for penetration of bituminous materials. Annual Book of Standards. ASTM International, West Conshohocken, PA.

ASTM D97, 2019. Standard test method for pour point of petroleum products. Annual Book of Standards. ASTM International, West Conshohocken, PA.

ASTM D113, 2019. Standard test method for ductility of asphalt materials. Annual Book of Standards. ASTM International, West Conshohocken, PA.

ASTM D381, 2019. Standard test method for gum content in fuels by jet evaporation. Annual Book of Standards. ASTM International, West Conshohocken, PA.

ASTM D473, 2019. Standard test method for sediment in crude oils and fuel oils by the extraction method. Annual Book of Standards. ASTM International, West Conshohocken, PA.

ASTM D525, 2019. Standard test method for oxidation stability of gasoline (induction period method). Annual Book of Standards. ASTM International, West Conshohocken, PA.

ASTM D873, 2019. Standard Test Method for Oxidation Stability of Aviation Fuels (Potential Residue Method).

ASTM D942, 2019. Standard test method for oxidation stability of lubricating greases by the oxygen pressure vessel method. Annual Book of Standards. ASTM International, West Conshohocken, PA.

ASTM D1500, 2019. Standard test method for ASTM color of petroleum products (ASTM Color Scale). Annual Book of Standards. ASTM International, West Conshohocken, PA.

ASTM D1754, 2019. Standard test method for effects of heat and air on asphaltic materials (thin-film oven test). Annual Book of Standards. ASTM International, West Conshohocken, PA.

ASTM D1796, 2019. Standard test method for water and sediment in fuel oils by the centrifuge method (laboratory procedure). Annual Book of Standards. ASTM International, West Conshohocken, PA.

ASTM D2170, 2019. Standard test method for kinematic viscosity of asphalts. Annual Book of Standards. ASTM International, West Conshohocken, PA.

ASTM D2273, 2019. Standard test method for trace sediment in lubricating oils. Annual Book of Standards. ASTM International, West Conshohocken, PA.

ASTM D2274, 2019. Standard test method for oxidation stability of distillate fuel oil (accelerated method). Annual Book of Standards. ASTM International, West Conshohocken, PA.

ASTM D2532, 2019. Standard test method for viscosity and viscosity change after standing at low temperature of aircraft turbine lubricants. Annual Book of Standards. ASTM International, West Conshohocken, PA.

ASTM D3241, 2019. Standard test method for thermal oxidation stability of aviation turbine fuels. Annual Book of Standards. ASTM International, West Conshohocken, PA.

ASTM D3381, 2019. Standard specification for viscosity-graded asphalt binder for use in pavement construction. Annual Book of Standards. ASTM International, West Conshohocken, PA.

ASTM D4007, 2019. 1 Standard test method for water and sediment in crude oil by the centrifuge method (laboratory procedure). Annual Book of Standards. ASTM International, West Conshohocken, PA.

ASTM D4310, 2019. Standard test method for determination of sludging and corrosion tendencies of inhibited mineral oils. Annual Book of Standards. ASTM International, West Conshohocken, PA.

ASTM D4424, 2019. Standard test method for butylene analysis by gas chromatography. Annual Book of Standards. ASTM International, West Conshohocken, PA.

ASTM D4625, 2019. Standard test method for middle distillate fuel storage stability at 43°C (110°F). Annual Book of Standards. ASTM International, West Conshohocken, PA.

ASTM D4807, 2019. Standard test method for sediment in crude oil by membrane filtration. Annual Book of Standards. ASTM International, West Conshohocken, PA.

ASTM D4870, 2019. Standard test method for determination of total sediment in residual fuels. Annual Book of Standards. ASTM International, West Conshohocken, PA.

ASTM D4871, 2019. Standard guide for universal oxidation/thermal stability test apparatus. Annual Book of Standards. ASTM International, West Conshohocken, PA.

ASTM D5853, 2019. Standard test method for pour point of crude oils. Annual Book of Standards. ASTM International, West Conshohocken, PA.

ASTM D5949, 2019. Standard test method for pour point of petroleum products (automatic pressure pulsing method). Annual Book of Standards. ASTM International, West Conshohocken, PA.

ASTM D5950, 2019. Standard test method for pour point of petroleum products (automatic tilt method). Annual Book of Standards. ASTM International, West Conshohocken, PA.

ASTM D5985, 2019. Standard test method for pour point of petroleum products (rotational method). Annual Book of Standards. ASTM International, West Conshohocken, PA.

Bjorseth, A., 1983. Handbook of Polycyclic Aromatic Hydrocarbons. Marcel Dekker Inc, New York.

Dias, J.R., 1987. Handbook of Polycyclic Hydrocarbons. Part A. Benzenoid Hydrocarbons. Elsevier, New York.

Dias, J.R., 1988. Handbook of Polycyclic Hydrocarbons. Part B. Polycyclic Isomers and Heteroatom Analogs of Benzenoid Hydrocarbons. Elsevier, New York.

Dolbear, G.E., 1998. Chapter 7: Hydrocracking: reactions, catalysis, and processes. In: Speight, J.G. (Ed.), Petroleum Chemistry and Refining. Taylor & Francis Publishers, Washington, DC, pp. 175–198.

Gary, J.G., Handwerk, G.E., Kaiser, M.J., 2007. Petroleum Refining: Technology and Economics, fifth edition CRC Press, Taylor & Francis Group, Boca Raton, Florida.

Hsu, C.S., Robinson, P.R. (Eds.), 2017. Handbook of Petroleum Technology. Springer International Publishing AG, Cham, Switzerland.

Magaril, R.Z., Aksenova, E.L., 1968. Study of the mechanism of coke formation in the cracking of petroleum resins. Int. Chem. Eng. 8, 727–729.

Magaril, R.Z., Aksenova, E.I., 1970. Mechanism of coke formation in the thermal decomposition of asphaltenes. Khim. Tekhnol. Topl. Masel. (7), 22–24.

Magaril, R.Z., Ramazaeva, L.F., 1969. Study of coke formation in the thermal decomposition of asphaltenes in solution. Izv. Vyssh. Ucheb. Zaved. Neft Gaz 12 (1), 61–64.

Magaril, R.Z., Ramazaeva, L.F., Aksenova, E.I., 1970. Kinetics of coke formation in the thermal processing of petroleum. Khim. Tekhnol. Topliv Masel 15 (3), 15–16.

Magaril, R.Z., Ramazeava, L.F., Aksenova, E.I., 1971. Kinetics of formation of coke in the thermal processing of crude oil. Int. Chem. Eng. 11, 250–251.

Mushrush, G.W., Speight, J.G., 1995. Petroleum Products: Instability and Incompatibility. Taylor & Francis Publishers, Washington, DC.

Parkash, S., Moschopedis, S.E., Speight, J.G., 1980. The thermal decomposition of resins. Fuel 59, 64.

Pines, H., 1981. The Chemistry of Catalytic Hydrocarbon Conversions. Academic Press, New York.

Schabron, J.F., Speight, J.G., 1997. An evaluation of the delayed coking product yield of heavy feedstocks using asphaltene content and carbon residue. Rev. de l'Institut Français de Pétrole 52, 73–85.

Schucker, R.C., 1983. Thermogravimetric determination of the coking kinetics of Arab heavy vacuum residuum. Ind. Eng. Chem. Process Design Dev. 22, 615–619.

Speight, J.G., 1987. Initial reactions in the coking of residua. Preprints Am. Chem. Soc. Div. Petrol. Chem 32 (2), 413.

Speight, J.G., 1994. Chapter 2: Chemical and physical studies of petroleum asphaltenes. In: Yen, T.F., Chilingarian, G.V. (Eds.), Asphaltenes and Asphalts, I. Developments in Petroleum Science, vol. 40. Elsevier BV, Amsterdam, The Netherlands.

Speight, J.G., 2000. The Desulfurization of Heavy Oils and Residua, Second ed. Marcel Dekker Inc, New York.

Speight, J.G., 2003. Chapter 2: Thermal cracking of petroleum. In: Ikan, R. (Ed.), Natural and Laboratory-Simulated Thermal Geochemical Processes. Kluwer Academic Publishers Inc, Dordrecht, The Netherlands.

Speight, J.G., 2008. Handbook of Synthetic Fuels: Properties, Processes, and Performance. McGraw-Hill, New York.

Speight, J.G., 2014. The Chemistry and Technology of Petroleum, fifth ed. CRC Press, Taylor & Francis Group, Baca Raton, FL.

Speight, J.G., 2015. Handbook of Petroleum Product Analysis, second ed. John Wiley & Sons Inc, Hoboken, NJ.

Speight, J.G., 2017. Handbook of Petroleum Refining. CRC Press, Taylor & Francis Group, Baca Raton, FL.

Distillate fuel stability and cleanliness. In: Stavinoha, L.L., Henry, C.P. (Eds.), Special Technical Publication No. 751. American Society for Testing and Materials, Philadelphia.

Distillation

4

4.1 Introduction

Crude oil in the unrefined state is of limited value and of limited use and refining, which is a series of steps by which the crude oil is converted into salable products, is required to produce the products in the amounts dictated by the market (Parkash, 2003; Gary et al., 2007; Speight, 2014a, 2017). In fact, a refinery is essentially a group of manufacturing plants that vary in number with the variety of products produced and the processes are and products manufactured to give a balanced operation.

For example, the manufacture of products from the lower boiling portion of crude oil automatically produces a certain amount of higher boiling components. If the latter cannot be sold as, say, high-boiling fuel oil, they accumulate until refinery storage facilities are full. To prevent the occurrence of such a situation the refinery must be flexible and able to change operations as needed. Typically, this means the inclusion of a wider variety of processes to accommodate the ever-changing demands of the market—such as the inclusion of a cracking process to change an excess of high-boiling fuel oil to more gasoline with coke as the residual product or inclusion of a vacuum distillation process to separate the high-boiling oil to lubricating oil stocks and asphalt.

Crude oil products may be partly or totally blended fractions insofar as they have compositions that are impossible to achieve by direct separation of these materials from crude oil. They result from chemical processes that change the molecular nature of selected portions of crude oil; in other words, they are the products of refining or they are refined products (Parkash, 2003; Gary et al., 2007; Speight, 2014a, 2017; Hsu and Robinson, 2017).

The crude oil refinery of the 21st century is a much more complex operation than those refineries of 100–120 years ago. Early refineries were predominantly distillation units, perhaps with ancillary units to remove objectionable odors from the various product streams. The refinery of the 1930s was somewhat more complex but was essentially a distillation unit, but at this time cracking and coking units were starting to appear in the scheme of refinery operations. These units were not what is used in current refineries as cracking and coking units but were the forerunners of modern units. Also at this time, asphalt was becoming a recognized crude oil product. Finally, current refineries are a result of major evolutionary trends and are highly complex operations. Most of the evolutionary adjustments to refineries have occurred during the decades since the commencement of World War II. In the crude oil industry, as in many other industries, *supply and demand* is the key factor in efficient and economic operation, innovation being also a key factor.

The Refinery of the Future. DOI: https://doi.org/10.1016/B978-0-12-816994-0.00004-X

In simple refineries, cut points can be changed slightly to vary yields and balance products, but the more common practice is to produce relatively narrow fractions and then process (or blend) to meet product demand. Since all these primary fractions are equilibrium mixtures, they all contain some proportion of the lower boiling constituents characteristic of a lower boiling fraction and so are stripped of these constituents, or stabilized, before further processing or storage. Thus gasoline is stabilized to a controlled butanes–pentanes content, and the overhead may be passed to superfractionators, towers with a large number of plates that can produce nearly pure C_1–C_4 hydrocarbon derivatives (methane to butanes, CH_4 to C_4H_{10}).

Kerosene and gas oil fractions are obtained as sidestream products from the atmospheric tower (*primary tower*), and these are treated in stripping columns (i.e., vessels of a few bubble trays) into which steam is injected—and the volatile overhead from the stripper is returned to the primary tower. Steam is usually introduced by the stripping section of the primary column to lower the temperature at which fractionation of the heavier ends of the crude can occur.

Crude oil refining, as it is currently practiced, is a very recent science and for the purposes of this chapter will be acknowledged as such. Many innovations have evolved during the 20th century and it is the purpose of the present chapter to illustrate the evolution of crude oil distillation from the early processes to those in use at the present day.

4.2 Current processes and equipment

In early refineries, distillation was the primary means by which products were separated from crude oil. As the technologies for refining evolved into the 20th century, refineries became much more complex but distillation remained the prime means by which crude oil is refined. Indeed, the distillation section of a modern refinery is the most flexible unit in the refinery since conditions can be adjusted to process a wide range of refinery feedstocks from the lighter (low density) crude oils to the heavier crude oils, more viscous crude oils. Generally, the maximum permissible temperature (in the vaporizing furnace or heater) to which the feedstock can be subjected is 350°C (660°F). The rate of thermal decomposition increases markedly above this temperature, although higher temperatures (up to approximately 395°C, 745°F) are part of the specifications for some distillation units—serious cracking does not occur at these higher temperatures but is subject to the properties of the crude oil feedstock and the residence time of the feedstock in the hot zone. If unplanned cracking occurs within a distillation unit, coke deposition can occur in the heater pipes or in the tower itself, resulting in failure of the distillation unit.

Distillation has remained a major refinery process and a process to which crude oil that enters the refinery is subjected. A multitude of separations are accomplished by distillation, but its most important and primary function in the refinery is its use for the separation of crude oil into component fractions (Chapter 2: Introduction to Refining Processes). Thus it is possible to obtain products ranging from gaseous materials taken off the top of the distillation column to a nonvolatile atmospheric residuum (*bottoms, reduced crude*) with correspondingly lower boiling materials (gas, gasoline, naphtha, kerosene, and gas oil) taken off at intermediate points (Parkash, 2003; Gary et al., 2007; Speight, 2014a, 2017; Hsu and Robinson, 2017).

The reduced crude may then be processed by vacuum or steam distillation to separate the high-boiling lubricating oil fractions without the danger of decomposition, which occurs at high (>350°C, 660°F) temperatures, and the amount of decomposition is subject to the residence time of the feedstock in the distillation unit. Indeed, atmospheric distillation may be terminated with a lower boiling fraction (*boiling cut*) if it is thought that vacuum or steam distillation will yield a better quality product or if the process appears to be economically more favorable (Parkash, 2003; Gary et al., 2007; Speight, 2014a, 2017; Hsu and Robinson, 2017).

Of all the units in a refinery the distillation section, comprising the atmospheric unit (Fig. 4.1) and the vacuum unit (Fig. 4.2), is required to have the greatest flexibility in terms of variable quality of feedstock and range of product yields (Parkash,

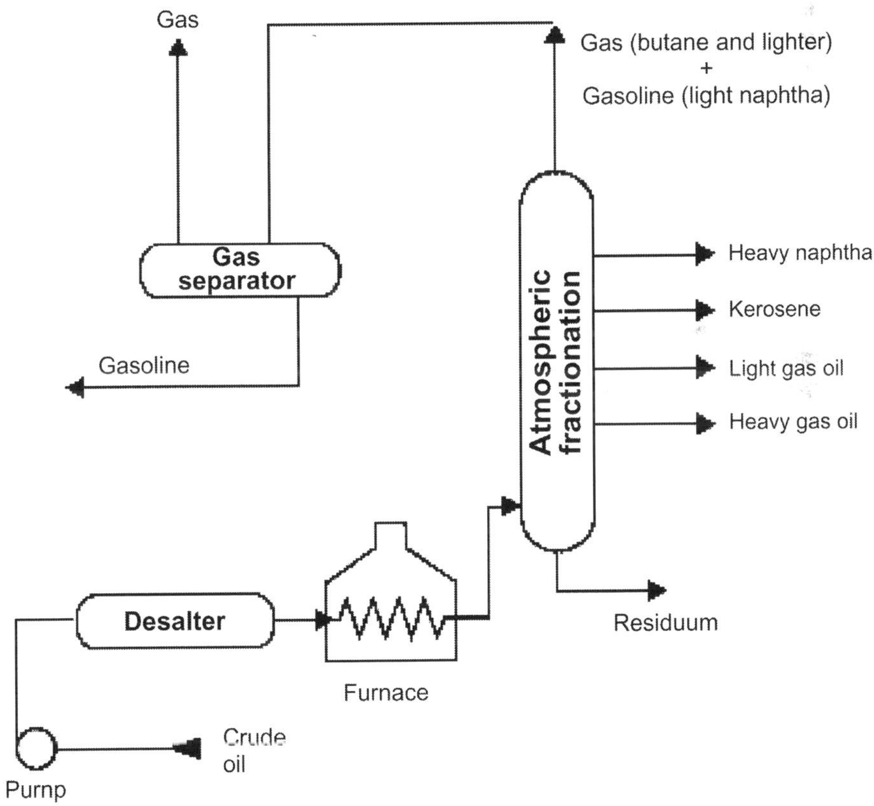

Figure 4.1 An atmospheric distillation unit.
Source: Modified from OSHA Technical Manual, Section IV, Chapter 2: Petroleum Refining Processes. http://www.osha.gov/dts/osta/otm/otm_iv/otm_iv_2.html.

2003; Gary et al., 2007; Speight, 2014a, 2017; Hsu and Robinson, 2017). The maximum permissible temperature of the feedstock in the vaporizing furnace is the factor limiting the range of products in a single-stage (atmospheric) column. Thermal decomposition or *cracking* of the constituents begins as the temperature of the oil approaches 350°C (660°F) and the rate increases markedly above this temperature. This thermal decomposition is generally regarded as being undesirable because the coke-like material produced tends to be deposited on the tubes with consequent formation of hot spots and eventual failure of the affected tubes. In the processing of lubricating oil stocks, an equally important consideration in the avoidance of these high temperatures is the deleterious effect on the lubricating properties. However, there are occasions when cracking distillation might be regarded as beneficial and the still temperature will be adjusted accordingly. In such a case the products will be named accordingly using the prefix *cracked*, for example, *cracked residuum* in which case the term *pitch* (Chapter 1: Feedstock Types and Properties) is applied.

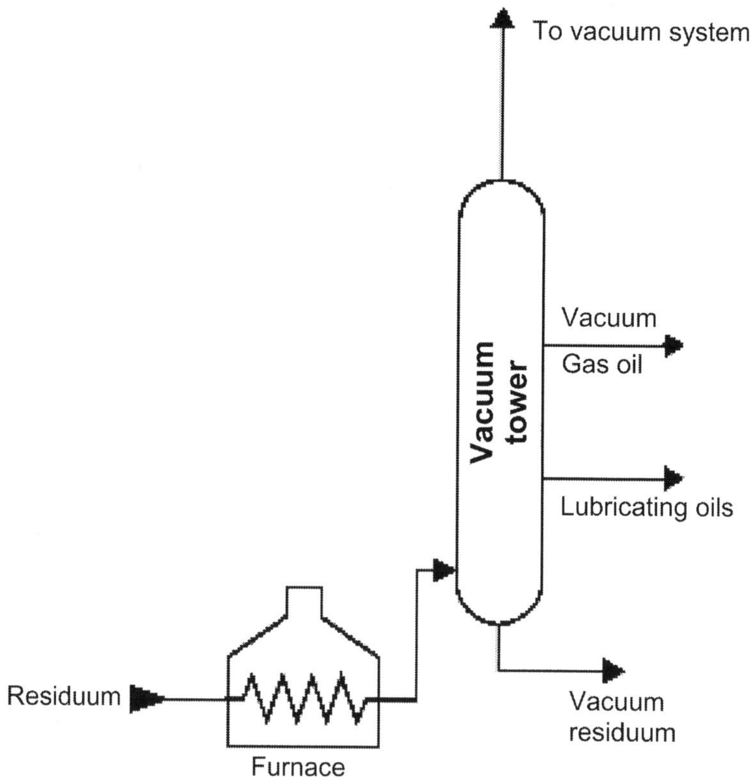

Figure 4.2 A vacuum distillation unit.
Source: Modified from OSHA Technical Manual, Section IV, Chapter 2: Petroleum Refining Processes. http://www.osha.gov/dts/osta/otm/otm_iv/otm_iv_2.html.

4.2.1 Atmospheric distillation

Distillation columns are the most commonly used separation units in a refinery. Operation is based on the difference in boiling temperatures of the liquid mixture components, and on recycling countercurrent gas–liquid flow. The properly organized temperature distribution up the column results in different mixture compositions at different heights. While multicomponent interphase mass transfer is a common phenomenon for all column types, the flow regimes are very different depending on the internal elements used. The two main types are a tray column and a packed column, the latter equipped with either random or structured packing. Different types of distillation columns are used for different processes, depending on the desired liquid holdup, capacity (flow rates), and pressure drop but each column is a complex unit, combining many structural elements.

It is a common practice to use furnaces to heat the feedstock only when distillation temperatures above 205°C (400°F) are required. Lower temperature (such as that used in the redistillation of naphtha and similar low-boiling products) are provided by heat exchangers and/or steam reboilers.

The feed to a fractional distillation tower is heated by flow through pipe arranged within a large furnace. The heating unit is known as a *pipestill heater* or *pipestill furnace*, and the heating unit and the fractional distillation tower make up the essential parts of a distillation unit or pipestill. The pipestill furnace heats the feed to a predetermined temperature, usually a temperature at which a calculated portion of the feed changes into vapor. The vapor is held under pressure in the pipestill furnace until it discharges as a foaming stream into the fractional distillation tower. Here the vapors pass up the tower to be fractionated into gas oil, kerosene, and naphtha, while the nonvolatile or liquid portion of the feed descends to the bottom of the tower to be pumped away as a bottom product and.

Pipestill furnaces vary greatly in size, shape, and interior arrangement and can accommodate 25,000 bbl or more of crude oil per day. The walls and ceiling are insulated with firebrick, and gas or oil burners are inserted through one or more walls. The interior of the furnace is partially divided into two sections: a smaller convection section where the oil first enters the furnace and a larger section into which the burners discharge and where the oil reaches its highest temperature.

Heat exchangers are also used to preheat the feedstock before it enters the furnace. These exchangers are bundles of tubes arranged within a shell so that a stream passes through the tubes in the opposite direction of a stream passing through the shell. Thus cold crude oil, by passing through a series of heat exchangers where hot products from the distillation tower are cooled, before entering the furnace and saving of heat in this manner, may be a major factor in the economical operation of refineries.

Steam reboilers may take the form of a steam coil in the bottom of the fractional distillation tower or in a separate vessel. In the latter case the bottom product from the tower enters the reboiler where part is vaporized by heat from the steam coil. The hot vapor is directed back to the bottom of the tower and provides part of the heat needed to operate the tower. The nonvolatile product leaves the reboiler and passes through a heat exchanger, where its heat is transferred to the feed to the tower. Steam may also be injected into a fractional distillation tower not only to provide heat but also to induce boiling to take place at lower temperatures. Reboilers generally increase the efficiency of fractionation, but a satisfactory degree of separation can usually be achieved more conveniently by the use of a *stripping* section.

The *stripping operation* (see Section 5.1) occurs in that part of the tower below the point at which the feed is introduced. The more volatile components are stripped from the descending liquid. Above the feed point (the rectifying section) the concentration of the less volatile component in the vapor is reduced.

The tower is divided into a number of horizontal sections by metal *trays* or *plates*, and each is the equivalent of a still. These force a rising vapor to pas though a pool of descending liquid. Therefore the more trays, the more redistillation, and hence the better is the fractionation or separation of the mixture fed into the tower. A tower for fractionating crude oil may be 13 ft. in diameter and 85 ft. high, but a tower stripping unwanted volatile material from gas oil may be only 3 or 4 ft. in diameter and 10 ft. high. Towers concerned with the distillation of liquefied gases are only a few feet in diameter but may be up to 200 ft. in height. A tower used in the fractionation of crude oil may have from 16 to 28 trays, but one used in the fractionation of liquefied gases may have 30–100 trays. The feed to a typical tower enters the vaporizing or flash zone, an area without trays. The majority of the trays are usually located above this area. The feed to a bubble tower, however, may be at any point from top to bottom with trays above and below the entry point, depending on the kind of feedstock and the characteristics desired in the products.

Liquid collects on each tray to a depth of, say, several inches and the depth controlled by a *dam* or *weir*. As the liquid level rises, excess liquid spills over the weir into a channel (downspout), which carries the liquid to the tray below.

The temperature of the trays is progressively cooler from bottom to top (Fig. 4.3). The bottom tray is heated by the incoming heated feedstock, although in some instances a steam coil (*reboiler*) is used to supply additional heat. As the hot vapors pass upward in the tower, condensation occurs onto the trays until refluxing (simultaneous boiling of a liquid and condensing of the vapor) occurs on the trays. Vapors continue to pass upward through the tower, whereas the liquid on any particular trays spills onto the tray below, and so on until the heat at a particular point is too intense for the material to remain liquid. It then becomes vapor and joins the other vapors passing upward through the tower. The whole tower thus simulates a collection of several (or many) stills, with the composition of the liquid at any one point or on any one tray remaining fairly consistent. This allows part of the refluxing liquid to be tapped off at various points as *sidestream* products. Thus in the distillation of crude oil, low-boiling naphtha and gases are removed as vapor from the top of the tower, high-boiling naphtha, kerosene, and gas oil are removed as sidestream products, and reduce crude is taken from the bottom of the tower.

The efficient operation of the distillation, or fractionating, tower requires the rising vapors to mix with the liquid on each tray. This is usually achieved by installing

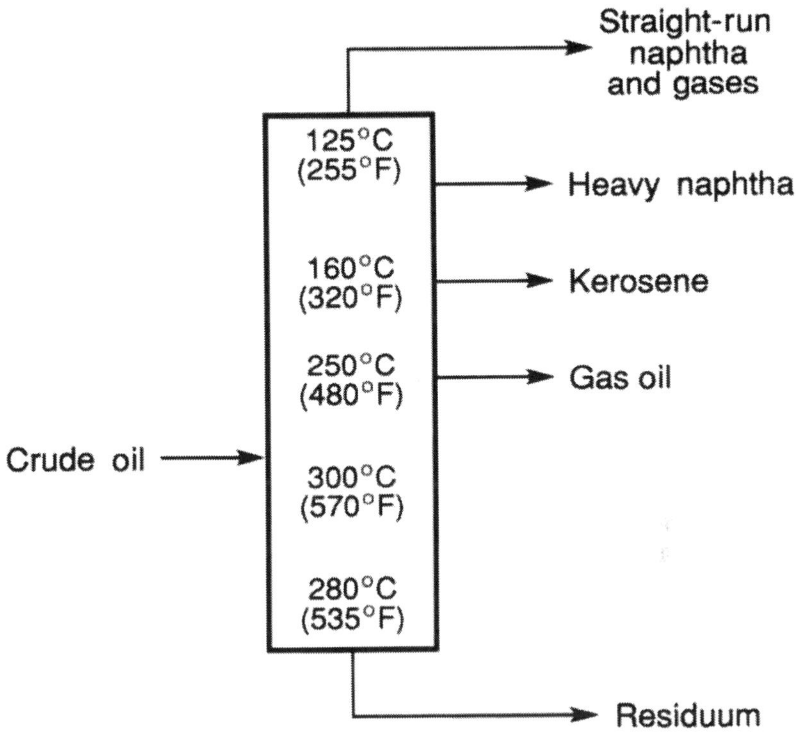

Figure 4.3 Representation of temperature profiles within an atmospheric distillation tower.

a short chimney on each hole in the plate and a cap with a serrated edge (*bubble cap*, hence *bubble-cap tower*) over each chimney (Fig. 4.4). The cap forces the vapors to go below the surface of the liquid and to bubble up through it. Since the vapors may pass up the tower at substantial velocities, the caps are held in place by bolted steel bars.

Perforated trays are also used in fractionating towers. This tray is similar to the bubble-cap tray but has smaller holes (approximately 3 in., 6 mm, vs 2 in., 50 mm). The liquid spills back to the tray below through weirs and is actually prevented from returning to the tray below through the holes by the velocity of the rising vapors. Needless to say, a minimum vapor velocity is required to prevent return of the liquid through the perforations.

The specifications for most crude oil products make it extremely difficult to obtain marketable material by distillation only. In fact, the purpose of atmospheric distillation is considered the provision of fractions that serve as feedstocks for intermediate refining operations and for blending. Generally this is carried out at atmospheric pressure, although conventional (low-density) crude oils may be *topped* at an elevated pressure and the residue then distilled at atmospheric pressure.

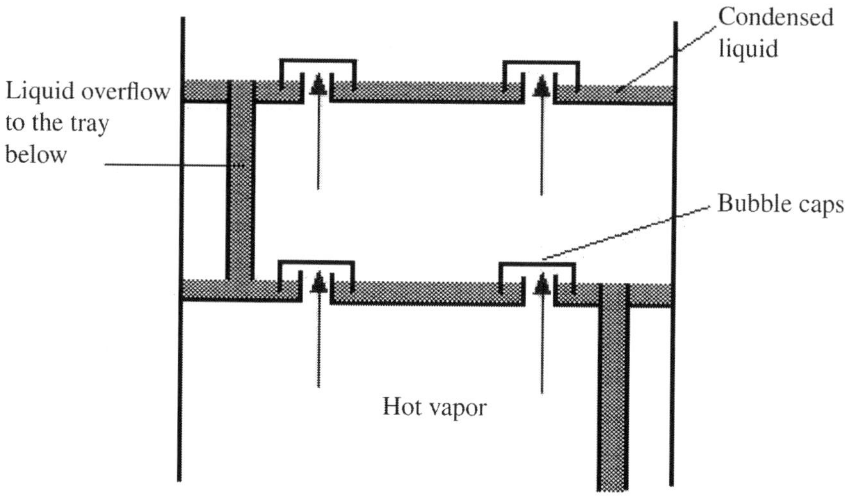

Figure 4.4 A bubble cap tray.

The *topping* operation differs from normal distillation procedures insofar as the majority of the heat is directed to the feed stream rather than by reboiling the material in the base of the tower. In addition, products of volatility intermediate between that of the overhead fractions and bottoms (residua) are withdrawn as sidestream products. Furthermore, steam is injected into the base of the column and the sidestream strippers to adjust and control the initial boiling range (or point) of the fractions.

Topped crude oil must always be *stripped* with steam to elevate the flash point or to recover the final portions of gas oil. The composition of the topped crude oil is a function of the temperature of the vaporizer (or *flasher*). In addition, the properties of the residuum are very dependent upon the extent of volatiles removal either by atmospheric distillation or by vacuum distillation (Table 4.1).

4.2.2 Vacuum distillation

The boiling range of the highest boiling fraction that can be produced at atmospheric pressure is limited by the temperature at which the residue starts to decompose or *crack*. If the stock is required for the manufacture of lubricating oils, further fractionation without cracking may be desirable, and this may be achieved by distillation under vacuum.

Vacuum distillation evolved because of the need to separate the less volatile products, such as lubricating oils, from the crude oil without subjecting these high boiling products to cracking conditions. The boiling range of the highest boiling fraction obtainable at atmospheric pressure is limited by the temperature

Table 4.1 Properties of various residua.

Feedstock	Gravity	Sulfur	Nitrogen	Nickel	Vanadium	Asphaltenes	Carbon residue
	API					(heptane)	(Conradson)
		wt.%	wt.%	ppm	ppm	wt.%	wt.%
Arabian Light, >650°F	17.7	3.0	0.2	10.0	26.0	1.8	7.5
Arabian Light, >1050°F	8.5	4.4	0.5	24.0	66.0	4.3	14.2
Arabian Heavy, > 650°F	11.9	4.4	0.3	27.0	103.0	8.0	14.0
Arabian Heavy, >1050°F	7.3	5.1	0.3	40.0	174.0	10.0	19.0
Alaska, North Slope, >650°F	15.2	1.6	0.4	18.0	30.0	2.0	8.5
Alaska, North Slope, >1050°F	8.2	2.2	0.6	47.0	82.0	4.0	18.0
Lloydminster (Canada), >650°F	10.3	4.1	0.3	65.0	141.0	14.0	12.1
Lloydminster (Canada), >1050°F	8.5	4.4	0.6	115.0	252.0	18.0	21.4
Kuwait, >650°F	13.9	4.4	0.3	14.0	50.0	2.4	12.2
Kuwait, >1050°F	5.5	5.5	0.4	32.0	102.0	7.1	23.1
Tia Juana, >650°F	17.3	1.8	0.3	25.0	185.0		9.3
Tia Juana, >1050°F	7.1	2.6	0.6	64.0	450.0		21.6
Taching, >650°F	27.3	0.2	0.2	5.0	1.0	4.4	3.8
Taching, >1050°F	21.5	0.3	0.4	9.0	2.0	7.6	7.9
Maya, >650°F	10.5	4.4	0.5	70.0	370.0	16.0	15.0

(c. 350°C; c. 660°F) at which the residue starts to decompose or *crack*, unless *cracking distillation* is preferred. When the feedstock is required for the manufacture of lubricating oils, further fractionation without cracking is desirable and this can be achieved by distillation under vacuum (reduced pressure) conditions.

The distillation of high-boiling lubricating oil stocks may require pressures as low as 15–30 mmHg (0.29–0.58 psi), but operating conditions are more usually 50–100 mmHg (0.97–1.93 psi). Volumes of vapor at these pressures are large and pressure drops must be small to maintain control, so vacuum columns are necessarily of large diameter. Differences in vapor pressure of different fractions are relatively larger than for lower boiling fractions, and relatively few plates are required. Under these conditions, high-boiling gas oil may be obtained as an overhead product at temperatures of approximately 150°C (300°F). Lubricating oil fractions may be obtained as sidestream products at temperatures of 250°C–350°C (480°F–660°F). The feedstock and residue temperatures being kept below the temperature of 350°C (660°F), above which the rate of thermal decomposition increases and cracking occurs (Chapter 3: Refining Chemistry). The partial pressure of the hydrocarbon derivatives is effectively reduced yet further by the injection of steam. The steam added to the column, principally for the stripping of asphaltic constituents in the base of the column, is superheated in the convection section of the heater.

At the point where the heated feedstock is introduced in the vacuum column (the *flash zone*) the temperature should be high and the pressure as low as possible to obtain maximum distillate yield. The flash temperature is restricted to approximately 420°C (790°F), however, in view of the cracking tendency of the feedstock constituents. Vacuum is maintained with vacuum ejectors and lately also with liquid ring pumps. In the older type high-vacuum units the required low hydrocarbon partial pressure in the flash zone could not be achieved without the use of *lifting* steam that acts in a similar manner as the stripping steam of atmospheric distillation units. This type of units is called *wet* units. One of the latest developments in vacuum distillation has been the deep vacuum flashers, in which no steam is required. These *dry* units operate at very low flash zone pressures and low pressure drops over the column internals. For that reason the conventional reflux sections with fractionation trays have been replaced by low pressure–drop spray sections. Cooled reflux is sprayed via a number of specially designed spray nozzles in the column countercurrent to the up-flowing vapor. This spray of small droplets comes into close contact with the hot vapor, resulting in good heat and mass transfer between the liquid and vapors phase.

When trays similar to those used in the atmospheric column are used in vacuum distillation, the column diameter may be extremely high, up to 45 ft. To maintain low pressure drops across the trays the liquid seal must be minimal. The low holdup and the relatively high viscosity of the liquid limit the tray efficiency, which tends to be much lower than in the atmospheric column. The vacuum is maintained in the column by removing the noncondensable gas that enters the column by way of the feed to the column or by leakage of air.

The fractions obtained by vacuum distillation of reduced crude depend on whether the run is designed to produce lubricating or vacuum gas oils. In the former case the fractions include the following:

- *high-boiling gas oil*, an overhead product and is used as catalytic cracking stock or, after suitable treatment, a low-boiling lubricating oil;
- *lubricating oil* (usually three fractions: low-boiling, intermediate-boiling, and high-boiling), obtained as a sidestream product; and
- *residuum*, the nonvolatile product that may be used directly as asphalt or to asphalt.

The residuum may also be used as a feedstock for a coking operation or blended with gas oils to produce a high-boiling fuel oil. However, if the *reduced* crude is not required as a source of lubricating oils, the lubricating and high-boiling gas oil fractions are combined or, more likely, removed from the *residuum* as one fraction and used as a catalytic cracking feedstock.

The continued use of atmospheric and vacuum distillation has been a major part of refinery operations during this century and no doubt will continue to be employed, at least into the beginning decades of the 21st century, as the primary refining operation.

Three types of high-vacuum units for long residue upgrading have been developed for commercial application: (1) feedstock preparation units, (2) lube oil high-vacuum units, and (3) high-vacuum units for asphalt production.

The feedstock preparation units make a major contribution to deep conversion upgrading and produce distillate feedstocks for further upgrading in catalytic crackers, hydrocracking units, and coking units. To obtain an optimum waxy distillate quality a wash oil section is installed between feed flash zone and waxy distillate draw off. The wash oil produced is used as fuel component or recycled to feed. The flashed residue (short residue) is cooled by heat exchange against long residue feed. A slipstream of this cooled short residue is returned to the bottom of the high-vacuum column as quench to minimize cracking (maintain low bottom temperature).

Lube oil high-vacuum units are specifically designed to produce high-quality distillate fractions for lube oil manufacturing. Special precautions are, therefore, taken to prevent thermal degradation of the distillates produced. The units are of the *wet* type. Normally, three sharply fractionated distillates are produced (spindle oil, low density machine oil, and medium density machine oil). Cut points between those fractions are typically controlled on their viscosity quality. Spindle oil and low density light machine oil are subsequently steam-stripped of any steam-volatile constituents in dedicated strippers. The distillates are further processed to produce lubricating base oil. The short residue is normally used as feedstock for the solvent deasphalting process to produce deasphalted oil, an intermediate for bright stock manufacturing. High-vacuum units for asphalt production are designed to produce straight-run asphalt and/or feedstocks for residuum blowing to produce *blown asphalt* that meets specifications. In principle, these units are designed on the same basis as feed preparation units, which may also be used to provide feedstocks for asphalt manufacturing.

Deep cut vacuum distillation involves a revamp of the vacuum distillation unit to cut deeper into the residue is one of the first options available to the refiner. In addition to the limits of the major equipment, other constraints include (1) the VGO quality specification required by downstream conversion units, (2) the minimum flash zone pressure achievable, and (3) the maximum heater outlet temperature achievable without excessive cracking. These constraints typically limit the cut point (true boiling point) to 560°C–590°C (1040°F–1100°F) although units are designed for cut points (true boiling point) as high as 625°C (1160°F).

4.2.3 Columns

Distillation columns (distillation towers) are made up of several components, each of which is used either to transfer heat energy or enhance material transfer. A typical distillation column consists of several major parts:

1. A vertical shell where separation of the components is carried out.
2. Column internals such as trays, or plates, or packings that are used to enhance component separation.
3. A reboiler to provide the necessary vaporization for the distillation process.
4. A condenser to cool and condense the vapor leaving the top of the column.
5. A reflux drum to hold the condensed vapor from the top of the column so that liquid (reflux) can be recycled back to the column.

The vertical shell houses the column internals and together with the condenser and reboiler constitutes a distillation column (Fig. 4.5).

In a crude oil distillation unit the feedstock liquid mixture is introduced usually near the middle of the column to a tray known as the feed tray. The feed tray divides the column into a top (enriching, rectification) section and a bottom (stripping) section. The feed flows down the column where it is collected at the bottom in the reboiler. Heat is supplied to the reboiler to generate vapor. The source of heat input can be any suitable fluid, although in most chemical plants this is normally steam. In refineries the heating source may be the output streams of other columns. The vapor raised in the reboiler is reintroduced into the unit at the bottom of the column. The liquid removed from the reboiler is known as the bottoms.

The vapor moves up the column, and as it exits the top of the unit, it is cooled by a condenser. The condensed liquid is stored in a holding vessel known as the reflux drum. Some of this liquid is recycled back to the top of the column and this is called the reflux. The condensed liquid that is removed from the system is known as the distillate or top product. Thus there are internal flows of vapor and liquid within the column as well as external flows of feeds and product streams, into and out of the column.

The column is divided into a number of horizontal sections by metal trays or plates, and each is the equivalent of a still. The more trays, the more redistillation, and hence the better is the fractionation or separation of the mixture fed into the tower. A tower for fractionating crude oil may be 13 ft. in diameter and 85 ft. high according to a general formula:

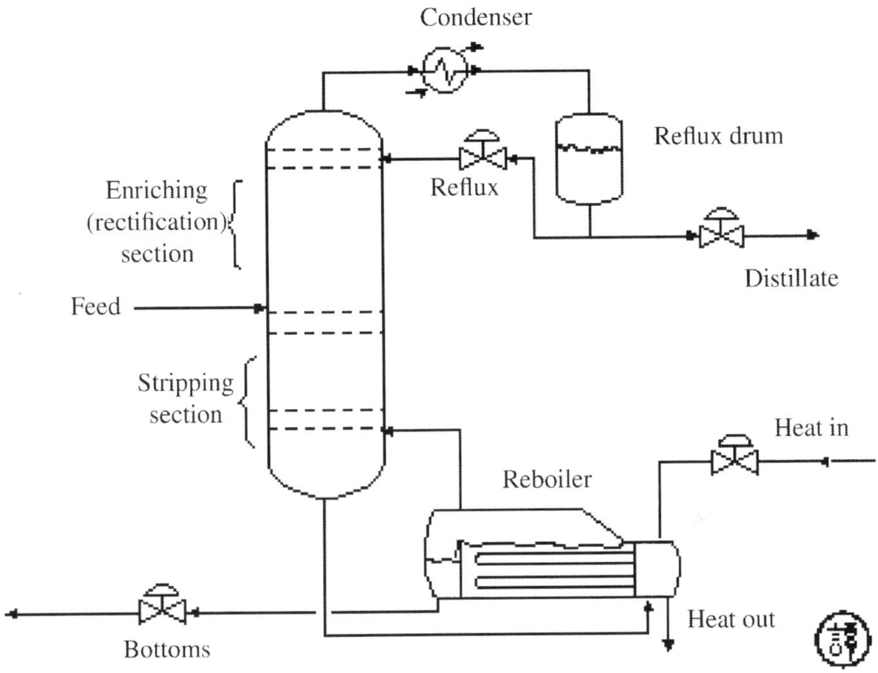

Figure 4.5 Individual parts of an atmospheric distillation column.

$$c = 220d^2r$$

where c is the capacity in bbl/day, d is the diameter in feet, and r is the amount of residuum expressed as a fraction of the feedstock (Parkash, 2003; Gary et al., 2007; Speight, 2014a, 2017; Hsu and Robinson, 2017).

A tower stripping unwanted volatile material from gas oil may be only 3 or 4 ft. in diameter and 10 ft. high with less than 20 trays. Towers concerned with the distillation of liquefied gases are only a few feet in diameter but may be up to 200 ft. in height. A tower used in the fractionation of crude oil may have from 16 to 28 trays, but one used in the fractionation (superfractionation) of liquefied gases may have 30–100 trays. The feed to a typical tower enters the vaporizing or flash zone, an area without trays. The majority of the trays are usually located above this area. The feed to a bubble tower, however, may be at any point from top to bottom with trays above and below the entry point, depending on the kind of feedstock and the characteristics desired in the products.

4.2.4 Tray columns

The tray column typically combines the open flow channel with weirs, downcomers, and heat exchangers. Free surface flow over the tray is disturbed by gas bubbles coming through the perforated tray, and possible leakage of liquid dropping through the upper tray.

Usually, trays are horizontal, flat, specially prefabricated metal sheets, which are placed at a regular distance in a vertical cylindrical column. Trays have two main parts: (1) the part where vapor (gas) and liquid are being contacted—the *contacting area* and (2) the part where vapor and liquid are separated, after having been contacted—the *downcomer area*.

Classification of trays is based on (1) the type of plate used in the contacting area, (2) the type and number of downcomers making up the downcomer area, (3) the direction and path of the liquid flowing across the contacting area of the tray, (4) the vapor (gas) flow direction through the (orifices in) the plate, and (5) the presence of baffles, packing or other additions to the contacting area to improve the separation performance of the tray.

Common plate types, for use in the *contacting area* are (1) *bubble cap* tray in which caps are mounted over risers fixed on the plate (Fig. 4.4)—the caps come in a wide variety of sizes and shapes, round, square, and rectangular (tunnel) and (2) *sieve* trays come with different hole shapes [round, square, triangular, rectangular (slots), and star], various hole sizes (from approximately 2 mm to approximately 25 mm) and several punch patterns (triangular, square, and rectangular), and (3) the *valve* tray that also is available in a variety of valve shapes (round, square, rectangular, triangular), valve sizes, valve weights (light and heavy), orifice sizes and either as fixed or floating valves.

Trays usually have one or more downcomers. The type and number of downcomers used mainly depends on the amount of *downcomer area* required to handle the liquid flow. Single pass trays are trays with one downcomer delivering the liquid from the next higher tray, a single bubbling area across which the liquid passes to contact the vapor and one downcomer for the liquid to the next lower tray.

Trays with multiple downcomers and hence multiple liquid passes can have a number of layout geometries. The downcomers may extend, in parallel, from wall to wall, as in. The downcomers may be rotated 90 (or 180) degrees on successive trays. The downcomer layout pattern determines the liquid *flow path* arrangement and liquid *flow direction* in the *contacting area* of the trays.

Giving a preferential direction to the vapor flowing through the orifices in the plate will induce the liquid to flow in the same direction. In this way, liquid flow rate and flow direction, as well as liquid height, can be manipulated. The presence of *baffles*, *screen mesh* or *demister mats*, loose or restrained dumped *packing* and/or the addition of other devices in the *contacting area* can be beneficial for improving the contacting performance of the tray, namely, its separation efficiency.

The most important parameter of a tray is its separation performance and four parameters are of importance in the design and operation of a tray-column: (1) the level of the tray efficiency, in the normal operating range; (2) the vapor rate at the "upper limit," that is the maximum vapor load, (3) the vapor rate at the "lower limit," that is the minimum vapor load, and (4) the tray pressure drop.

The separation performance of a tray is the basis of the performance of the column as a whole. The primary function of, for instance, a distillation column is the separation of a feed stream in (at least) one top product stream and one bottom product stream. The quality of the separation performed by a column can be judged from the purity of the top and bottom product streams. The specification of the impurity levels in the top and bottom streams and the degree of recovery of pure products set the targets for a successful operation of a distillation column. It is evident that tray efficiency is influenced by (1) the specific component under consideration (this holds specially for multicomponent systems in which the efficiency can be different for each component), because of different diffusivities, diffusion interactions, and different stripping factors and (2) the vapor flow rate; usually increasing the flow rate increases the effective mass transfer rate, while it decreases the contact time, at the same time. These counteracting effects lead to a roughly constant efficiency value, for a tray in its normal operating range. Upon approaching the lower operating limit a tray starts weeping and efficiency is reduced.

4.2.5 Packed columns

A packed column is similar to a trickle-bed reactor, where liquid film flows down over the packing surface in contact with the upward gas flow. A small fragment of packing geometry can be accurately analyzed assuming the periodic boundary conditions, which allows calibration of the porous media model for a big packing segment.

The packing in a distillation column creates a surface for the liquid to spread on, thereby providing a high surface area for mass transfer between the liquid and the vapor.

4.3 Other processes

Atmospheric distillation and vacuum distillation provide the primary fractions from crude oil to use as feedstocks for other refinery processes for conversion convert into products. Many of these subsequent processes involve fractional distillation and some of the procedures are so specialized and used with such frequency that they are identified by name.

4.3.1 Stripping

Stripping is a fractional distillation operation carried out on each sidestream product immediately after it leaves the main distillation tower. Since perfect separation is not accomplished in the main tower, unwanted components are mixed with those of the sidestream product. The purpose of stripping is to remove the more volatile components and thus reduce the flash point of the sidestream product. Thus a sidestream product enters at the top tray of a stripper, and as it spills down the four to six trays, steam injected into the bottom of the stripper removes the volatile components. The steam and volatile components leave the top of the stripper to return to the main tower. The stripped sidestream product leaves at the bottom and, after being cooled in a heat exchanger, goes to storage. Since strippers are short, they are arranged one above another in a single tower; each stripper, however, operates as a separate unit.

4.3.2 Rerunning

Rerunning is a general term covering the *redistillation* of any material and indicating, usually, that a large part of the material is distilled overhead. Stripping, in contrast, removes only a relatively small amount of material as an overhead product. A rerun tower may be associated with a crude distillation unit that produces wide boiling range naphtha as an overhead product. By separating the wide-cut fraction into low-boiling naphtha and high-boiling naphtha, the rerun tower acts in effect as an extension of the crude distillation tower.

The product from chemical treating process of various fractions may be rerun to remove the treating chemical or its reaction products. If the volume of material being processed is small, a *shell still* may be used instead of a continuous fractional distillation unit. The same applies to gas oils and other fractions from which the *front end* or *tail* must be removed for special purposes.

4.3.3 Stabilization and light end removal

The gaseous and more volatile liquid hydrocarbon derivatives produced in a refinery are collectively known as *light hydrocarbons* or *light ends* (more correctly referred to as low-boiling hydrocarbon derivatives). Light ends are produced in relatively small quantities from crude oil and in large quantities when gasoline is manufactured by cracking and reforming. When a naphtha or gasoline component at the time of its manufacture is passed through a condenser, most of the light ends do not condense and are withdrawn and handled as a gas. A considerable part of the light ends, however, can remain dissolved in the condensate, thus forming a liquid with a high vapor pressure.

Liquids with high vapor pressures may be stored in refrigerated tanks or in tanks capable of withstanding the pressures developed by the gases dissolved in the liquid. The more usual procedure, however, is to separate the light ends from the liquid by a distillation process generally known as *stabilization*. Enough of the light ends are removed to make a stabilized liquid, that is, a liquid with a low enough vapor pressure to permit its storage in ordinary tanks without loss of vapor. The simplest stabilization process is a *stripping* process. Low-boiling naphtha from a crude tower, for example, may be pumped into the top of a tall, small-diameter fractional distillation tower operated under a pressure of 50–80 psi. Heat is introduced at the bottom of the tower by a steam reboiler. As the naphtha cascades down the tower, the light ends separate and pass up the tower to leave as an overhead product. Since reflux is not used, considerable amounts of liquid hydrocarbon derivatives pass overhead with the light ends.

Stabilization is usually a more precise operation than that just described. An example of more precise stabilization can be seen in the handling of the mixture of hydrocarbon derivatives produced by cracking. The overhead from the atmospheric distillation tower that fractionates the cracked mixture consists of light ends and cracked gasoline with light ends dissolved in it. If the latter is pumped to the usual type of tank storage, the dissolved gases cause the gasoline to boil, with consequent loss of the gases and some of the liquid components. To prevent this the gasoline and the gases dissolved in it are pumped to a stabilizer maintained under a pressure of approximately 100 psi and operated with reflux. This fractionating tower makes a cut between the highest boiling gaseous component (butane) and the lowest boiling liquid component (pentane). The bottom product is thus a liquid free of all gaseous components, including butane; hence the fractionating tower is known as a debutanizer. The debutanizer bottoms (gasoline constituents) can be safely stored, whereas the overhead from the debutanizer contains the butane, propane ethane, and methane fractions. The butane fraction, which consists of all the hydrocarbon derivatives containing four carbon atoms, is particularly needed to give easy starting characteristics to motor gasoline. It must be separated from the other gases and blended with motor gasoline in amounts that vary with the season: more in the winter and less in the summer. Separation of the butane fraction is effected by another distillation in a fractional distillation tower called a depropanizer, since its purpose is to separate propane and the lower boiling gases from the butane fraction.

The depropanizer is very similar to the debutanizer, except that it is smaller in diameter because of the smaller volume being distilled and is taller because of the larger number of trays required to make a sharp cut between the butane and propane fractions. Since the normally gaseous propane must exist as a liquid in the tower, a pressure of 200 psi is maintained. The bottom product, known as the butane fraction, stabilizer bottoms, or refinery casinghead, is a high-vapor-pressure material that must be stored in refrigerated tanks or pressure tanks. The depropanizer overhead, consisting of propane and lower boiling gases, is used as a petrochemical feedstock or as a refinery fuel gas, depending on the composition.

A *depentanizer unit* is a fractional distillation tower that removes the pentane fraction from a debutanized (butane-free) fraction. The *depentanizer unit* is similar to debutanizers and has been introduced recently to segregate the pentane fractions from cracked gasoline and reformate. The pentane fraction when added to a premium gasoline makes this gasoline extraordinarily responsive to the demands of an engine accelerator.

The gases produced as overhead products from crude distillation, stabilization, and depropanizer units may be delivered to a gas absorption plant for the recovery of small amounts of butane and higher boiling hydrocarbon derivatives. The gas absorption plant consists essentially of two towers. One tower is the absorber where the butane and higher boiling hydrocarbon derivatives are removed from the lower boiling gases.

This is done by spilling a light (low-density) oil (*lean oil*) down the absorber over trays similar to those in a fractional distillation tower. The gas mixture enters at the bottom of the tower and rises to the top. As it does this, it contacts the lean oil, which absorbs the butane and higher boiling hydrocarbon derivatives but not the lower boiling hydrocarbon derivatives. The latter leave the top of the absorber as dry gas. The lean oil that has become enriched with butane and higher boiling hydrocarbon derivatives is now termed *fat oil*. This is pumped from the bottom of the absorber into the second tower, where fractional distillation separates the butane and higher boiling hydrocarbon derivatives as an overhead fraction and the oil, once again lean oil, as the bottom product.

The condensed butane and higher boiling hydrocarbon derivatives are included with the refinery *casinghead bottoms* or *stabilizer bottoms*. The dry gas is frequently used as fuel gas for refinery furnaces. It contains propane and propylene, however, which may be required for liquefied crude oil gas for the manufacture of polymer gasoline or petrochemicals. Separation of the propane fraction (propane and propylene) from the lower boiling gases is accomplished by further distillation in a fractional distillation tower similar to those previously described and particularly designed to handle liquefied gases. Further separation of hydrocarbon gases is required for petrochemical manufacture.

4.3.4 Superfractionation

The term *superfractionation* is sometimes applied to a highly efficient fractionating tower used to separate ordinary crude oil products. For example, to increase the yield of furnace fuel oil, high-boiling naphtha may be redistilled in a tower that is capable of making a better separation of the naphtha and the fuel oil components. The latter, obtained as a bottom product, is diverted to furnace fuel oil.

Fractional distillation as normally carried out in a refinery does not completely separate one crude oil fraction from another. One product overlaps another, depending on the efficiency of the fractionation, which in turn depends on the number of trays in the tower, the amount of reflux used, and the rate of distillation. Kerosene, for example, normally contains a small percentage of hydrocarbon derivatives that (according to their boiling points) belong in the naphtha fraction and a small percentage that should be in the gas oil fraction. Complete separation is not required for the ordinary uses of these materials, but certain materials, such as solvents for particular purposes (hexane, heptane, and aromatics), are required as essentially pure compounds. Since they occur in mixtures of hydrocarbon derivatives they must be separated by distillation and with no overlap of one hydrocarbon with another. This requires highly efficient fractional distillation towers specially designed for the purpose and referred to as superfractionators. Several towers with 50–100 trays operated with a *high reflux ratio* may be required to separate a single compound with the necessary purity.

4.3.5 Azeotropic distillation

Azeotropic distillation is the use of a third component to separate two close-boiling components by means of the formation of an azeotropic mixture between one of the original components and the third component to increase the difference in the boiling points and facilitates separation by distillation.

All compounds have definite boiling temperatures, but a mixture of chemically dissimilar compounds sometimes causes one or both of the components to boil at a temperature other than that expected. For example, benzene boils at 80°C (176°F), but if it is mixed with hexane, it distills at 69°C (156°F). A mixture that boils at a temperature lower than the boiling point of either of the components is called an azeotropic mixture.

Two main types of azeotropes exist, that is, the *homogeneous azeotrope*, where a single liquid phase is in the equilibrium with a vapor phase, and the *heterogeneous azeotropes*, where the overall liquid composition which form two liquid phases, is identical to the vapor composition. Most methods of distilling azeotropes and low relative volatility mixtures rely on the addition of specially chosen chemicals to facilitate the separation.

The five methods for separating azeotropic mixtures are as follows:

1. *Extractive distillation* and *homogeneous azeotropic distillation* where the liquid separating agent is completely miscible.
2. *Heterogeneous azeotropic distillation*, or more commonly, *azeotropic distillation* where the liquid separating agent (the *entrainer*) forms one or more azeotropes with the other components in the mixture and causes two liquid phases to exist over a wide range of compositions. This immiscibility is the key to making the distillation sequence work.
3. *Distillation using ionic salts*. The salts dissociate in the liquid mixture and alters the relative volatilities sufficiently that the separation become possible.
4. *Pressure-swing distillation* where a series of column operating at different pressures are used to separate binary azeotropes which change appreciably in composition over a moderate pressure range or where a separating agent which forms a pressure-sensitive azeotrope is added to separate a pressure-insensitive azeotrope.
5. *Reactive distillation* where the separating agent reacts preferentially and reversibly with one of the azeotropic constitutes. The reaction product is then distilled from the nonreacting components and the reaction is reversed to recover the initial component. In fact, the chemical process industries have shown increasing interest in the development of reactive separation processes that combine reaction and separation mechanisms into a single, integrated unit. Such processes bring several important advantages among which are increase of reaction yield and selectivity, overcoming thermodynamic restrictions, for example, azeotropes, and considerable reduction in energy, water, and solvent consumption. Important examples of reactive separations are reactive distillation and reactive absorption. Due to strong interactions of chemical reaction and heat and mass transfer, the process behavior of reactive separation processes may be adaptable not only to crude oil refineries but also for the separation of alternate fuels (Noeres et al., 2003; Sundmacher and Kienle, 2003; Harmsen, 2007; Kiss, 2009, 2011; Cardenas-Guerra et al., 2010).

In *simple distillation* a multicomponent liquid mixture is slowly boiled in a heated zone and the vapors are continuously removed as they form and, at any instant in time, the vapor is in equilibrium with the liquid remaining on the still. Because the vapor is always richer in the more volatile components than the liquid, the liquid composition changes continuously with time, becoming more and more concentrated in the least volatile species. A simple *distillation residue curve* is a means by which the changes in the composition of the liquid residue curves on the pot changes over time changes with time. A *residue curve map* is a collection of the liquid residue curves originating from different initial compositions. Residue curve maps contain the same information as phase diagrams but represent this information in a way that is more useful for understanding how to synthesize a distillation sequence to separate a mixture.

All of the residue curves originate at the light (lowest boiling) pure component in a region, move toward the intermediate boiling component, and end at the highest boiling pure component in the same region. The lowest temperature nodes are termed as *unstable nodes*, as all trajectories leave from them, while the highest temperature points in the region are termed *stable nodes*, as all trajectories ultimately reach them. The point that the trajectories approach from one direction and end in a different direction (as always is the point of intermediate boiling component) are termed *saddle point*. Residue curve that divide the composition space into different distillation regions are called distillation boundaries.

Many different residue curve maps are possible when azeotropes are present. Ternary mixtures containing only one azeotrope may exhibit six possible residue curve maps that differ by the binary pair forming the azeotrope and by whether the azeotrope is minimum or maximum boiling. By identifying the limiting separation achievable by distillation, residue curve maps are also useful in synthesizing separation sequences combining distillation with other methods.

However, the separation of components of similar volatility may become economical if an *entrainer* can be found that effectively changes the relative volatility. It is also desirable that the entrainer be reasonably cheap, stable, nontoxic, and readily recoverable from the components. In practice it is probably this last criterion that severely limits the application of extractive and azeotropic distillation. The majority of successful processes, in fact, are those in which the entrainer and one of the components separate into two liquid phases on cooling if direct recovery by distillation is not feasible.

A further restriction in the selection of an azeotropic entrainer is that the boiling point of the entrainer be in the range 10°C–40°C (18°F–72°F) below that of the components. Thus although the entrainer is more volatile than the components and distills off in the overhead product, it is present in a sufficiently high concentration in the rectification section of the column.

4.3.6 Extractive distillation

Extractive distillation is the use of a third component to separate two close-boiling components in which one of the original components in the mixture which is extracted by the third component and retained in the liquid phase to facilitate separation by distillation.

Using acetone–water as an extractive solvent for butanes and butenes, butane is removed as overhead from the extractive distillation column with acetone–water charged at a point close to the top of the column. The bottoms product of butenes and the extractive solvent is fed to a second column where the butenes are removed as overhead. The acetone–water solvent from the base of this column is recycled to the first column. Extractive distillation may also be used for the continuous recovery of individual aromatics, such as benzene, toluene, or xylene(s), from the appropriate crude oil fractions. *Prefractionation* concentrates a single aromatic cut into a close-boiling cut, after which the aromatic concentrate is distilled with a solvent (usually phenol) for benzene or toluene recovery. Mixed cresylic acids (cresol derivatives and methyl phenol derivatives) are used as the solvent for xylene recovery.

Extractive distillation is successful because the solvent is specially chosen to interact differently with the components of the original mixture, thereby altering their relative volatilities. Because these interactions occur predominantly in the liquid phase, the solvent is continuously added near the top of the extractive distillation column so that an appreciable amount is present in the liquid phase on all of the trays below. The mixture to be separated is added through second feed point further down the column. In the extractive column the component having the greater volatility, not necessarily the component having the lowest boiling point, is taken overhead as a relatively pure distillate. The other component leaves with the solvent via the column bottoms. The solvent is separated from the remaining components in a second distillation column and then recycled back to the first column.

One of the most important steps in developing a successful (economical) extractive distillation sequence is selecting a good solvent. In general, selection criteria for the solvent include the following:

- Should enhance significantly the natural relative volatility of the key component.
- Should not require an excessive ratio of solvent to nonsolvent (because of cost of handling in the column and auxiliary equipment).
- Should remain soluble in the feed components and should not lead to the formation of two phases.
- Should be easily separable from the bottom product.
- Should be inexpensive and readily available.
- Should be stable at the temperature of the distillation and solvent separation.
- Should be nonreactive with the components in the feed mixture.
- Should have a low latent heat.
- Should be noncorrosive and nontoxic.

No single solvent or solvent mixture satisfies all of the criteria for use in extractive distillation. However, the following solvent selection criteria assist in choosing the best possible solvent:

1. Screen by functional group or chemical family.
   a. Select candidate solvent from the high boiling homologous series of both low-boiling and high-boiling key components.
   b. Select candidate solvents from groups that tend to give positive (or no) deviations from Raoult's law for the key component desire in the distillate and negative (or no) deviations for the other key.
   c. Select solvents that are likely to cause the formation of hydrogen bonds with the key component to be removed in the bottoms, or disruption of hydrogen bonds with the key to be removed in the distillate. Formation and disruption of hydrogen bonds are

often associated with strong negative and positive deviations, respectively, from Raoult's Law.

   d. Select candidate solvents from chemical groups that tend to show higher polarity than one key component or lower polarity than the other.

2. Identify the individual candidate solvents.

   a. Select only candidate solvents that boil at least 30°C–40°C above the key components to ensure that the solvent is relatively nonvolatile and remains largely in the liquid phase. With this boiling point difference the solvent should also not form azeotropes with the other components.

   b. Rank the candidate solvents according to their selectivity at infinite dilution.

   c. Rank the candidate solvents by the increase in relative volatility caused by the addition of the solvent.

Residue curve maps are of limited use at the preliminary screening stage because there is usually insufficient information available to sketch them, but they are valuable and should be sketched or calculated as part of the second stage of the solvent selection.

In general, none of the fractions or combinations of fractions separated from crude oil is suitable for immediate use as crude oil products. Each fraction must be separately refined by processes that vary with the impurities in the fraction and the properties required in the finished product. The simplest treatment is the washing of a fraction with a lye solution to remove sulfur compounds. The most complex is the series of treatments—solvent treating, dewaxing, clay treating or hydrorefining, and blending—required to produce lubricating oils. On rare occasions no treatment of any kind is required. Some crude oils yield a low-boiling gas oil fraction that is suitable as furnace fuel oil or as a diesel fuel.

4.3.7 Process options for viscous feedstocks

In order to further distill heavy crude oil, extra heavy oil tar sand bitumen, or residuum, or topped crude, reduced pressure is required to prevent thermal cracking and the process takes place in one or more vacuum distillation towers.

The principles of vacuum distillation resemble those of fractional distillation, and except that larger diameter columns are used to maintain comparable vapor velocities at the reduced pressures, the equipment is also similar. The internal designs of some vacuum towers are different from atmospheric towers in which random packing and demister pads are used instead of trays. A typical first-phase vacuum tower may produce gas oil, lubricating-oil base stock, and a high-boiling residuum for propane deasphalting. A second-phase tower operating at lower vacuum may distill surplus residuum from the atmospheric tower, which is not used for lube-stock processing, and surplus residuum from the first vacuum tower not used for deasphalting.

Vacuum towers are typically used to separate catalytic cracking feedstock from surplus residuum and heavy crude oil, extra heavy oil, and tar sand bitumen have fewer components distilling at atmospheric pressure and under vacuum than conventional crude oil. Nevertheless, some heavy crude oils still pass through the distillation stage of a refinery before further processing is undertaken. In addition, a vacuum tower has recently been installed at the Syncrude Canada plant to offer an additional process option for upgrading tar sand bitumen (Speight, 2005, 2014a, 2017). The installation of such a tower as a means of refining viscous feedstocks (with the possible exception of the residua that are usually produced through a vacuum tower) is a question of economics and the ultimate goal of the refinery in terms of product slate. After distillation the residuum from the heavy crude oil might pass to a cracking unit such as visbreaking or coking to produce salable products. Catalytic cracking of the residuum or the whole heavy crude oil is also an option but is very dependent on the constituents of the feedstock and their interaction with the catalyst.

The development of the *catalytic distillation* or *reactive distillation* which unite in the same equipment catalyst and distillation devices finds its main applications for reversible reactions, such as methyl *tert*-butyl ether ethyl tributyl ether synthesis, so as to shift an unfavorable equilibrium by continuous reaction product withdrawal (DeCroocq, 1997). But catalytic distillation can provide also several advantages in selective hydrogenation of C_3, C_4, and C_5 cuts for the manufacture of petrochemicals (Speight, 2019). Inserting the catalyst in the fractionation column improves mercaptans removal, catalyst fouling resistance, and selective hydrogenation performances by modifying the reaction mixture composition along the column.

Thus there is the potential for applying a related concept to the deep distillation of heavy crude oil.

4.4 The future

The distillation unit (along with the catalytic cracking unit) is the core unit of the modern crude oil refinery (Parkash, 2003; Gary et al., 2007; Speight, 2014a, 2017; Hsu and Robinson, 2017).

Heavy crude oil and/or sour crude oil, which require more energy-intensive processing than conventional sweet (low-density and low-sulfur) crude oils, are expected to contribute a growing fraction of hydrocarbon fuels production. As existing reserves of oil are depleted and there is greater worldwide competition for the light, low-sulfur crude oils, refiners will increasingly utilize heavy crude oil and/or sour crude oil to meet demand.

The influx of such crude oils is causing noticeable change to the operation of distillation units in terms of the actual unit internals, unit operation, and prevention of corrosion (Speight, 2014b). Innovations to the distillation units will most likely be more subtle than a complete restructuring of the distillation section of the refinery and will focus on (1) energy savings, (2) changes to the internal packing to prevent fouling within the distillation system, and (3) the use of metal alloy systems to mitigate corrosion.

Some of the process options and improvements mentioned below may already be under consideration, but where they have not been considered, the potential for adoption will be high.

4.4.1 Distillation units

Distillation is one of the most energy intensive operations in the crude oil refinery and is used throughout the refinery to separate process products. The incoming flow is heated, after which the products are separated on the basis of boiling points (Chapter 2: Introduction to Refining Processes).

A crude oil distillation unit is one of a critically important unit in the refinery—the main function of the unit is to separate the crude oil into the constituents fractions many kinds of crude oil products. The General design of the unit includes the crude oil preheater, which are either cabin or vertical cylindrical furnace, for adjusting the crude feed properties and increasing the temperature of the feedstock (Parkash, 2003; Gary et al., 2007; Speight, 2014a, 2017; Hsu and Robinson, 2017).

The crude oil distillation process is a large energy consumer and a large waste generator and, as a result, there is the need to balance the economics of processing crude oil to viable fractions. Various advanced integrated and distributed distillation methods can be employed to minimize waste and thus optimize performance oil. The scale of crude oil fractionation/distillation is generally large, involving atmospheric distillation of crude oil, vacuum distillation of bottoms residue, main fractionation of gaseous effluent from catalytic cracking, and main distillation of effluent from thermal coking of various crude oil fractions. The equipment, which consumes large quantities of energy, and there is a move—which will continue—to use the more advanced thermodynamics of multicomponent distillation.

In the near-term, retrofitting will be used to increase the efficiency of the process by maximizing the use of existing equipment. Examples of retrofit objectives are to (1) increase the throughput, (2) change the feedstock, (3) increase the production or the quality of the products, (4) reduce the energy demand, and (5) reduce the atmospheric emissions. The retrofit methodology for crude oil distillation systems will require rigorous simulation and optimizations procedures to optimize the process conditions and to explore structural modifications to increase the capacity and the energy efficiency of the system.

Another issue that has received, and will continue to receive, attention is the formation of a carbonaceous deposit (coke) which tends to occur at the inner side of the crude oil tube in the furnace during the preheating process. The coking and coating of the inner tube leads to the coke accumulating, the product quality deteriorating, the pressure drop increasing across the furnace, the energy consumption increasing, and the tube wall carbonization corroding.

Generally, decoking is executed on demand based on the measured heat loss, or performed every fixed period of time. However, due to the different rate of coke formation of the various crude oils, as well as the difference in cost of product, process operation, and maintenance operation, both on-demand and fixed maintenance practices are not optimal method for decoking. Modeling of the process by which the coke is formed and accumulation rate of the coke at the inner tube of the furnace will be needed for the optimization, prediction, and scheduling of the decoking operation.

This will require a novel approach that allows the distillation and heat integration arrangements to be designed and optimized simultaneously. In a retrofit operation, there will be the need to improve the design for increased energy efficiency, or for increased throughput, brings special challenges. Changes to operating conditions throughout the system, column internals, and heat exchange arrangements must be optimized simultaneously as well as unit design (Table 4.2).

Also, there are numerous heaters and exchangers in the atmospheric and vacuum distillation units that are the sources of ignition, spills, leaks, and fugitive emissions. Variations in pressure, temperature, or liquid levels may occur if automatic controls fail. Control of temperature, pressure, and reflux within operating parameters is needed to prevent thermal cracking within the towers. Relief systems must be used for overpressure and operations monitored to prevent the crude from entering the reformer charge.

Energy efficiency opportunities exist in the heating side and by optimizing the distillation column—such opportunities will no doubt be put into practice. For example, optimization of the reflux ratio of the distillation column can produce significant energy savings. The efficiency of a distillation column is determined by the characteristics of the feed. If the characteristics of the feed have changed over time or compared to the design conditions, operational efficiency can be improved.

Damaged or worn internals result in increased operation costs. As the internals become damaged, efficiency decreases and pressure drops rise. This causes the column to run at a higher reflux rate over time. With an increased reflux rate, energy costs will increase accordingly. Replacing the trays with new ones or adding a high performance packing can have the column operating like the day it was brought online. If operating conditions have seriously deviated from designed operating conditions, the investment may have a relative short payback. When replacing the trays, it will be worthwhile to consider new efficient tray designs. New tray designs can result in enhanced separation efficiency and decrease pressure drop, which will result in reduced energy consumption. When considering new tray designs, the number of trays should be optimized.

An energy efficient design for a crude distillation unit has recently been developed by Elf in which the crude preheater and the distillation column were redesigned. The crude preheat train was separated in several steps to recover fractions

Table 4.2 Future actions for the distillation unit.

Action	Outcome
Optimize the operating conditions	Reduction in the energy consumption
Modify the design	Reduction in the energy consumption
Different distillation configurations	Increase unit efficiency
Optimize the configuration of new units	Reduction in the energy consumption and increase unit efficiency

at different temperatures. The distillation tower was redesigned to work at low pressure and the outputs were changed to link to the other processes in the refinery and product mix of the refinery. The design resulted in reduced fuel consumption and better heat integration. Such options will be more obvious in future refineries.

In general, refineries in the future will optimize energy use through more efficient heat exchange and heat integration, better controls, and adopting energy-saving approaches to very energy-intensive process units (e.g., furnaces and distillation towers).

Effective integration of controls and practices to increase energy efficiency (e.g., pipe insulation) would result in higher levels of energy optimization. Refineries will maximize their ability to produce energy on-site by increasing the use of cogeneration to generate both heat and power, and in some cases would be producing electricity for sale back to the local grid. In many cases, high-efficiency turbines and steam generators would be used to achieve a high thermal efficiency in cogeneration and power-generation systems.

As feedstocks change in composition, the distillation unit will be required to achieve higher degrees of efficiency to produce the precursors to hydrocarbon fuels as well as feedstocks for other units that will eventually produce hydrocarbon fuels through cracking. This will more likely be achieved by changes in the internals of the distillation units as well as changes to the overlay use of the units. The overall effects will be for refineries to create the option to take deeper cuts into the crude oil feedstock leaving a harder redid to be used as feedstocks for the cracking units.

Reboilers consume a large part of total refinery energy use as part of the distillation process. By using chilled water the reboiler duty can in principal be lowered by reducing the overhead condenser temperature.

Fouling and *foaming* are frequently at the top of a list of well-known but little understood problems in distillation towers. The difficulty is the unpredictable characteristics that are hard to repeat or correlate in lab experiments or industrial operations. Chemical reactions and surface phenomena in fouling and foaming systems can further complicate their predictability. Several techniques for dealing with such unpredictable problems include monitoring of tower conditions, selection of tower internals, and pretreatment of recycle streams. These methods will be improved and developed to the point where they are operative in all distillation units.

New control technologies that provide a more accurate estimation of distillation-column flooding will be required. A distillation column flooding predictor which is a pattern-recognition system that identifies patterns of transient tower instabilities, which precede tray flooding, will be very advantageous. This would allow for more stable long-term operation, resulting in greater efficiency and throughput increase.

There is a tendency to focus on the debottlenecking of the column internals, which is basically employing the traditional approach to improve the hydraulic performance of the column internals, that is, to get more hydraulic capacity through an existing distillation column.

One method to avoid column debottlenecking is to retrofit the column with the internals that allow more vapor or liquid traffic to pass through without the loss of efficiency. In this kind of project, major changes to the existing column are required to replace the existing internals with the new one that debottleneck the column. For example, advanced trays (Nye trays and multiple downcomer trays) and structured packings require the removal of the existing tray support rings and the existing internals. The use of better internals obviously promotes debottlenecking, but this is normally achieved at the expense of a huge cost. Of concern to us are the costs of new types of internals, costs of removing the existing internals, and the opportunity cost of extensive plant downtime.

In the near term the useful capacity of column internals, either with trays, structured, or random packing will be maximized and optimized with customized feed distributors, downcomers, and other fabricated components.

Refinery distillation efficiency is limited by the undesired coke deposition, resulting in a significant loss of distillation efficiency. When a residuum is heated to pyrolysis temperatures (> 350°C, 650°F), there is typically an induction period before coke formation begins. To avoid fouling, refiners often stop heating well before coke forms, using arbitrary criteria, but cessation of the heating can result in less than maximum distillate yield.

Over the past three decades a better understanding of the chemistry and physics of coking has evolved (Chapter 3: Refining Chemistry), and improved designs based on primary internals have allowed an increase in the amount of gas oil produced with increases in cut point from approximately 520°C (970°F) to 590°C (1.095°F). As continuing inroads are made into the chemistry of coking, future distillation units will show improvements in the design of the internals leading to process equivalents of the laboratory spinning band distillation units.

The concept of a nonconventional distillation column is not new. This type of column has found its application in the crude oil refining, such as the use of side strippers, and the cryogenic air separation, such as the use of side rectifiers. Other forms of these columns such as the prefractionator arrangement, the fully thermally coupled column and the dividing wall column, however, have started to attract the interests of both process designers and operating companies. The justifications for their use have been prompted by the less energy consumption and greater efficiency when compared with the conventional fractionation system.

With the potential for an increase in the influx of heavy crude oil, extra heavy oil, and tar sand bitumen to refineries, there may be a resurgence of interest in the application of *reactive distillation* in refineries. Although invented in 1921, the industrial application of reactive distillation did not take place before the 1980s.

Reactive distillation is a process where the still is also a chemical reactor. Separation of the product from the reaction mixture does not need a separate distillation step, which saves energy (for heating) and materials. This technique is especially useful for equilibrium-limited reactions and conversion can be increased far beyond what is expected by the equilibrium due to the continuous removal of reaction products from the reactive zone. This helps reduce capital and investment costs and may be important for sustainable development due to a lower consumption of resources.

The suitability of reactive distillation for a particular reaction depends on various factors such as volatilities of reactants and products along with the feasible reaction and distillation temperature. Hence, the use of reactive distillation for every reaction may not be feasible. Exploring the candidate reactions for reactive distillation is an area that needs considerable attention to expand the domain of reactive distillation processes.

The benefits of applying reactive distillation include (1) increased speed and improved efficiency; (2) lower costs—reduced equipment use, energy use, and handling; (3) less waste and fewer byproducts; and (4) improved product quality–reducing opportunity for degradation because of less heat.

However, the conditions in the reactive column are suboptimal both as a distillation column and a chemical reactor, since the reactive column combines these. In addition, the introduction of an in situ separation process in the reaction zone or vice versa leads to complex interactions between vapor–liquid equilibrium, mass transfer rates, diffusion and chemical kinetics, which poses a great challenge for design and synthesis of these systems. Side reactors, where a separate column feeds a reactor and vice versa, are better for some reactions if the optimal conditions of distillation and reaction differ too much.

In a similar vein, there is also a greater potential for the application of extractive distillation.

Extractive distillation is distillation in the presence of a miscible, high boiling, relatively nonvolatile component, and the solvent, which forms no azeotrope with the other components in the mixture. The method is used for mixtures having a low value of relative volatility, nearing unity. Such mixtures cannot be separated by simple distillation, because the volatility of the two components in the mixture is nearly the same, causing them to evaporate at nearly the same temperature at a similar rate, making normal distillation impractical.

The method of extractive distillation uses a separation solvent, which is generally nonvolatile, has a high boiling point and is miscible with the mixture, but does not form an azeotropic mixture. The solvent interacts differently with the components of the mixture, thereby causing their relative volatilities to change. This enables the new three-part mixture to be separated by normal distillation. The original component with the greatest volatility separates out as the top product. The bottom product consists of a mixture of the solvent and the other component, which can again be separated easily because the solvent does not form an azeotrope with it. The bottom product can be separated by any of the methods available.

It is important to select a suitable separation solvent for this type of distillation. The solvent must alter the relative volatility by a wide enough margin for a successful result. The quantity, cost, and availability of the solvent should be considered. The solvent should be easily separable from the bottom product and should not react chemically with the components or the mixture or cause corrosion in the equipment (Speight, 2014b).

Membranes may offer future alternatives to distillation. Membranes have started to enter the refinery for hydrogen recovery (see above) but are also being developed for other separations.

Current membrane systems will probably be most effective in hybrid distillation processes to perform a first, crude, low-energy, low-cost separation, leaving the polishing operation for distillation. If high selectivity could be achieved with membranes, there is the potential to replace distillation in many separation processes. For example, the separation of olefins (ethylene and propylene) from paraffins (ethane and propane) on a commercial scale is accomplished almost exclusively via cryogenic distillation. This separation processes is very energy intensive and a less energy-intensive method for separating olefins from paraffins would, therefore, be extremely beneficial. The separation of olefins from paraffins could be accomplished with less energy if selective facilitated transport membranes were to be used. A complete olefin/paraffin separation with membrane processes alone would require membranes with high selectivity.

Improved isomer separations could significantly reduce the energy consumption required for the manufacture of certain chemical products. The isomer separations that could be achieved using membranes include the separation of p-xylene from other xylenes. Molecular sieve membranes could also be used in the membrane reactor configuration for the manufacture of *p*-xylene and linear olefins and paraffins if their corresponding isomerization reactions were enhanced. In this case a chemical synthesis would be performed in conjunction with a closely coupled, but separate, membrane separation device. For example, *p*-xylene could be selectively produced by an equilibrium redistribution of mixed isomeric xylenes coupled with a selective transport of the product through a membrane (Parkash, 2003; Gary et al., 2007; Speight, 2014a, 2017; Hsu and Robinson, 2017).

An extensive study funded by DOE focused on membranes for different separations (gas/gas and fluid/fluid) and studied the current state-of-the-art and the potential applications in crude oil refining. The report concludes that membrane technology will definitely enter the refinery, although further research is needed to develop appropriate membrane materials that can withstand the environment found in crude oil refining processes. Also, membrane technology should be evaluated as an integrated part of the specific process for which it is being implemented to warrant the full energy savings potential.

4.4.2 Combating corrosion

Refinery distillation units run as efficiently as possible to reduce costs. One of the major issues that occurs in distillation units and decreases efficiency is corrosion of the metal components found throughout the process line of the hydrocarbon refining process (Speight, 2014b). Corrosion causes the failure of parts in addition to dictating the shutdown schedule of the unit, which can cause shutdown of the refinery. Attempts to block such corrosive influences will be a major issue of future refineries.

Furthermore, in addition to the corrosive properties of high-acid crude oils, sulfur may be present in crude oil as hydrogen sulfide (H_2S), as compounds (such as mercaptan derivatives, sulfide derivatives, disulfide derivatives, and thiophene derivatives), or as elemental sulfur. Each crude oil has different amounts and types of sulfur compounds but, generally, the proportion, stability, and complexity of the compounds are greater in the heavier more viscous feedstock (Speight, 2014a, 2017). Hydrogen sulfide is a primary contributor to corrosion in refinery processing units. Other corrosive substances are elemental sulfur and mercaptans.

Crude oils often contain inorganic salts such as sodium chloride, magnesium chloride, and calcium chloride in suspension or dissolved in entrained water (brine). These salts must be removed or neutralized before processing to prevent catalyst poisoning, equipment corrosion, and fouling. Salt corrosion is caused by the hydrolysis of some metal chlorides to hydrogen chloride (HCl) and the subsequent formation of hydrochloric acid when crude is heated. Hydrogen chloride may also combine with ammonia to form ammonium chloride (NH_4Cl), which causes fouling and corrosion.

The sections of the process susceptible to corrosion include (but may not be limited to) preheat exchanger (hydrogen chloride and hydrogen sulfide), preheat furnace and bottoms exchanger (hydrogen sulfide and sulfur compounds), atmospheric tower and vacuum furnace (hydrogen sulfide, sulfur compounds, and organic acids), vacuum tower (hydrogen sulfide and organic acids), and overhead (hydrogen sulfide, hydrogen chloride, and water). Where sour crudes are processed, severe corrosion can occur in furnace tubing and in both atmospheric and vacuum towers where metal temperatures exceed 230°C (450°F). Wet hydrogen sulfide also will cause cracks in steel. When processing high-nitrogen crudes, nitrogen oxides can form in the flue gases of furnaces and these oxides are corrosive to steel when cooled to low temperatures (nitric and nitrous acids are formed) in the presence of water.

As a first step in the refining process, to reduce corrosion, plugging, and fouling of equipment and to prevent poisoning the catalysts in processing units, these contaminants must be removed by desalting (dehydration). However, the desalting operation does not always remove all of the corrosive elements, and hydrogen chloride may be a product of the thermal treatment that occurs as part of the distillation process. Inadequate desalting can cause fouling of heater tubes and heat exchangers throughout the refinery. Fouling restricts product flow and heat transfer and leads to failures due to increased pressures and temperatures. Corrosion, which occurs due to the presence of hydrogen sulfide, hydrogen chloride, naphthenic (organic) acids, and other contaminants in the crude oil, also causes equipment failure. Neutralized salts (ammonium chlorides and sulfides), when moistened by condensed water, can cause corrosion.

Corrosion occurs in various forms in the distillation section of the refinery and is manifested by events such as pitting corrosion from water droplets, embrittlement from chemical attack if the dewatering and desalting unit has not operated efficiently, and stress corrosion cracking from sulfide attack.

High temperature crude corrosivity of distillation units will continue to be a major concern to the refining industry. The presence of naphthenic acid and sulfur compounds considerably increases corrosion in the high temperature parts of the distillation units, and equipment failures have become a critical safety and reliability issue. The difference in process conditions, materials of construction, and blend processed in each refinery and especially the frequent variation in crude diet increases the problem of correlating corrosion of a unit to a certain type of crude oil. In addition, a large number of interdependent parameters influence the high temperature crude corrosion process.

Damage is typically in the form of unexpected high corrosion rates on alloys that would normally be expected to resist sulfidic corrosion. In many cases, even very highly alloyed materials have been found to exhibit sensitivity to corrosion under these conditions.

Naphthenic acid corrosion is differentiated from sulfidic corrosion by the nature of the corrosion (pitting and impingement) and by its severe attack at high velocities in crude distillation units. Crude feedstock heaters, furnaces, transfer lines, feed and reflux sections of columns, atmospheric and vacuum columns, heat exchangers, and condensers are among the type of equipment subject to this type of corrosion.

From a materials standpoint, carbon steel can be used for refinery components. Carbon steel is resistant to the most common forms of corrosion, particularly from hydrocarbon impurities at temperatures below 205°C (400°F), but other corrosive chemicals and high temperature environments prevent its use everywhere. Common replacement materials are low alloy steel containing chromium and molybdenum, with stainless steel containing more chromium dealing with more corrosive environments. More expensive materials commonly used are nickel titanium and copper alloys. These are primarily saved for the most problematic areas where extremely high temperatures or very corrosive chemicals are present.

In fact, the increasing move in the crude oil slate (crude oil from tight-low-permeability formations notwithstanding), there is the need to design units that will be (as near as possible) not susceptible to corrosion. The sections most susceptible to corrosion from hydrogen chloride (HCl) and hydrogen sulfide (H_2S) include the preheat exchanger; preheat furnace and bottoms exchanger from hydrogen sulfide and sulfur compounds; atmospheric tower and vacuum furnace from hydrogen sulfide, sulfur compounds, and organic acids; vacuum tower from H_2S and organic acids; and overhead from hydrogen sulfide, hydrogen chloride, and water. Where sour crudes are processed, severe corrosion can occur in furnace tubing and in both atmospheric and vacuum towers where metal temperatures exceed 230°C (450°F). Wet hydrogen sulfide also will cause cracks in steel leading to spills, leaks, and fugitive emissions.

When processing high-nitrogen crudes, nitrogen oxides can form in the flue gases of furnaces. Nitrogen oxides are corrosive to steel when cooled to low temperatures in the presence of water. Aging refineries have fugitive emissions from these sources that the authors do not believe are adequately accounted for. Chemicals are used to control corrosion by hydrogen chloride produced in distillation units. Ammonia (NH_3) may be injected into the overhead stream prior to initial condensation and/or an alkaline solution may be injected into the hot crude-oil feed. If sufficient wash water is not injected, deposits of ammonium chloride can form causing serious corrosion. Crude feedstock may contain appreciable amounts of water in suspension. This water phase can separate during start-up and, along with water remaining in the tower from steam purging, settle in the bottom of the tower. When the water is heated to its boiling point, an instantaneous vaporization explosion can occur upon contact with the oil in the unit. This can lead to devastating damage to the columns and catastrophic release of vapors and liquid.

Attempts to mitigate corrosion will continue to use a complex system of monitoring, preventative repairs and careful use of materials. *Monitoring methods* include both off-line checks taken during maintenance and online monitoring. Off-line checks measure corrosion after it has occurred, telling the engineer when equipment must be replaced based on the historical information he has collected.

Online systems are a more modern development and are revolutionizing the way corrosion is, and will be, approached. There are several types of online corrosion monitoring technologies such as linear polarization resistance, electrochemical noise, and electrical resistance. Online monitoring generally had slow reporting rates in the past (minutes or hours) and been limited by process conditions and sources of error but newer technologies can report rates up to twice per minute with much higher accuracy (referred to as real-time monitoring). This allows process engineers to treat corrosion as another process variable that can be optimized in the system. Immediate responses to process changes allow the control of corrosion mechanisms, so they can be minimized while also maximizing production output.

Materials methods will include selecting the proper material for the application. In areas of minimal corrosion, cheap materials are preferable, but when bad corrosion can occur, more expensive but longer lasting materials should be used. Other materials methods will be in the form of protective barriers between corrosive substances and the equipment metals. These can be either a lining of refractory material such as standard Portland cement or other special acid-resistant cements that are shot onto the inner surface of the vessel. Also available are thin overlays of more expensive metals that protect cheaper metal against corrosion without requiring lots of material.

Blending of crude feedstocks will continue to be used to mitigate the effects of corrosion as will blending, inhibition, materials upgrading, and process control. Blending will be used to reduce the naphthenic acid content of the feed, thereby reducing corrosion to an acceptable level. However, while blending of viscous crude oils heavy and light crude oils can change shear stress parameters and might also help reduce corrosion, there is also the potential for incompatibility of the constituents of the viscous crude oil and light crude oil.

Chemicals are used to control corrosion by hydrochloric acid produced in distillation units. For example, caustic is added to desalting water to neutralize acids and reduce corrosion. They are also added to desalted crude in order to reduce the amount of corrosive chlorides in the tower overheads. They are used in some refinery treating processes to remove contaminants from hydrocarbon streams.

Ammonia may be injected into the overhead stream prior to initial condensation and/or an alkaline solution may be carefully injected into the hot crude-oil feed. If sufficient wash water is not injected, deposits of ammonium chloride can form and cause serious corrosion. The crude feedstock may contain appreciable amounts of water in suspension that can separate during start-up and, along with water remaining in the tower from steam purging, settle in the bottom of the tower. This water can be heated to the boiling point and create an instantaneous vaporization explosion upon contact with the oil in the unit.

More extreme treatment, in terms of the use of chemicals, would be the addition of inorganic such as calcium oxide or magnesium oxide to trap the sulfur compounds in the distillation unit. This would be especially helpful in the case of cracking distillation where additional volatile and corrosive sulfur species are generated. In either case, any excess oxides and the calcium sulfide or magnesium sulfide would collect in the residuum. There may be a benefit to have such inorganic materials in any asphalt produced from the residue.

Injection of corrosion inhibitors will provide some protection for specific fractions that are known to be particularly severe. However, monitoring will need to be adequate in this case to check on the effectiveness of the treatment. Process control changes may provide adequate corrosion control if there is the possibility of reducing charge rate and temperature.

Process changes will include any action to remove or at least reduce the amount of acid gas present and to prevent accumulation of water on the tower trays. Material upgrading will include the lining of distillation tower tops with alloys resistant to hydrochloric acid. Design changes will be used to prevent the accumulation of water and will include coalescers and water draws. The application of chemicals will include the injection of a neutralizer to increase the pH and a corrosion inhibitor. The presence of many acids, such as naphthenic acids and carbon dioxide (in water), can buffer the environment and require greater use of neutralizers. However, the use of excessive amounts of neutralizers will cause plugging of trays and corrosion under the salt deposits.

4.4.3 Refinery feedstocks

Finally, a comment on the expected changes in the refinery feedstock slate that will arise because of the depletion of crude oil resources that brings serious challenges to the crude oil economy is facing serious challenges. Biorefining has gradually become a new method used to produce energy and chemical products, which parallels crude oil refining. Biomass is used as a raw material to produce some middle platform compounds by thermochemical, chemical, or biological methods, which are then processed into biofuels or chemicals. Biorefining offers the possibility of achieving sustainable production of bioenergy and biomaterials.

Biorefining (Chapter 11: Production of Fuels From Nonfossil Fuel Feedstocks) is a new concept in industrial manufacturing, which will become a new manufacturing technology paradigm, becoming more and more important because of its use of alternative resources and its environmental benefits.

Compared with crude oil refining, biorefining is unique for the following reasons: (1) biorefinery feedstocks are carbohydrates while crude oil refining involves hydrocarbons without oxygen molecules and lacking functional groups, thus biorefining is more economical from the point of view of atom utilization and, therefore, it is cost-effective due to the avoidance of the high-cost oxidation process, (2) compared with the crude oil-derived fuels, biodiesel obtained from biorefining has better oxidation stability and higher energy density, which can reduce carbon deposits in the engine, so that operating costs are only half those of the conventional diesel apparatus, and (3) in the crude oil refining process, the resource-rich region is separated from the end market, which means long-distance transport is required, while biomass resources are usually found near to the market.

With the advances of technology a number of bio-based products that are able to replace or partially replace crude oil-based products—such as ethanol, lactic acid, citric acid, biodiesel, and 1,3-propanediol—have arisen. These products can be divided into many groups according to carbohydrates platform, oil platform, and thermochemical platform. Although there are a variety of products from biorefining, many of them are not needed in large quantities. Therefore it would be not be economical to prepare all of these products directly from the basic raw materials. Hence, it is necessary to develop many platform products, such as succinic acid, fumaric acid, or malic acid, which can then be economically converted into various chemicals (Speight, 2019).

References

Cardenas-Guerra, J.C., Lopez-Arenas, T., Lobo-Oehmichen, R., Perez-Cisneros, E.S., 2010. A reactive distillation process for deep hydrodesulfurization of diesel: multiplicity and operation aspects. Comput. Chem. Eng. 34, 196–209.

DeCroocq, D., 1997. Major scientific and technical challenges about development of new processes in refining and petrochem. Rev. Inst. Français de Pétrole 52 (5), 469–489.

Gary, J.G., Handwerk, G.E., Kaiser, M.J., 2007. Petroleum Refining: Technology and Economics, fifth ed. CRC Press, Taylor & Francis Group, Boca Raton, FL.

Harmsen, G.J., 2007. Reactive distillation: the frontrunner of industrial process intensification: a full review of commercial applications, research, scale-up, design and operation. Chem. Eng. Process. 46 (9), 774–780.

Kiss, A.A., 2009. Novel process for biodiesel by reactive absorption. Sep. Purif. Technol. 69, 280–287.

Kiss, A.A., 2011. Heat-integrated reactive distillation process for synthesis of fatty esters. Fuel Process. Technol. 92, 1288–1296.

Noeres, C., Kenig, E.Y., Gorak, A., 2003. Modelling of reactive separation processes: reactive absorption and reactive distillation. Chem. Eng. Process. 42, 157–178.

Parkash, S., 2003. Refining Processes Handbook. Gulf Professional Publishing, Elsevier, Amsterdam.

Speight, J.G., 2005. Natural bitumen (tar sands) and heavy oil, Coal, Oil Shale, Natural Bitumen, Heavy Oil and Peat, from Encyclopedia of Life Support Systems (EOLSS),

2005. Developed under the Auspices of the UNESCO, EOLSS Publishers, Oxford, <http://www.eolss.net>.

Speight, J.G., 2014a. The Chemistry and Technology of Petroleum, fifth ed. CRC Press, Taylor & Francis Group, Boca Raton, FL.

Speight, J.G., 2014b. Oil and Gas Corrosion Prevention. Gulf Professional Publishing, Elsevier, Oxford.

Speight, J.G., 2017. Handbook of Petroleum Refining. CRC Press, Taylor & Francis Group, Boca Raton, FL.

Speight, J.G., 2019. Handbook of Petrochemical Processes. CRC Press, Taylor & Francis Group, Boca Raton, FL.

Sundmacher, K., Kienle, A., 2003. Reactive Distillation – Status and Future Directions. Wiley-VCH Verlag GmbH & Co. KGaA, Weinheim.

Thermal cracking

5

5.1 Introduction

Balancing product yield and market demand, without the manufacture of large quantities of fractions having low commercial value, has long required processes for the conversion of hydrocarbon derivatives of one molecular weight range and/or structure into some other molecular weight range and/or structure. Basic processes for this are cracking processes in which high-boiling constituents of crude oil are cracked (thermally decomposed) into lower molecular weight, lower boiling molecules, although reforming, alkylation, polymerization, and hydrogen-refining processes have wide applications in producing premium-quality products (Parkash, 2003; Gary et al., 2007; Speight, 2014, 2017; Hsu and Robinson, 2017).

Thermal cracking is a high-temperature process (in some cases pressure is also applied) in which the constituents of crude oil are decomposed to break molecular bonds and yield products having a lower molecular weight than the feedstock. The process is one of several cracking methods used in the crude oil industry to process crude oil and other crude oil products for commercial use (Parkash, 2003; Gary et al., 2007; Speight, 2014, 2017; Hsu and Robinson, 2017).

Cracking processes allow the production of lower boiling products such as methane, ethane, propane and butane [liquefied petroleum gas (LPG)], and naphtha from higher molecular weight fraction or constituents such as gas oil and residua. Thermal cracking is currently used to produce low-boiling fractions or distillates, burner fuel, and/or crude oil coke. The major process variables are (1) feedstock type, (2) temperature, and (3) pressure and residence time, which need to be considered to maximize the yield of distillates and minimize the yield of coke.

It is generally recognized that the most important part of any refinery is the gasoline (and liquid fuels) manufacturing facilities. Distillation (Chapter 4: Distillation) has remained a major refinery process and a process to which almost all refinery crude oils are subject as the first refining process (not counting the dewatering and desalting process). However, not all crude oils yield the same distillation products. In fact, the nature of the crude oil dictates the processes that may be required for refining. However, with the increased use of heavier crude oils (which contain significant amounts of residua), distillation alone is not sufficient to produce the necessary amounts of gasoline. Hence, the need for cracking facilities to increase the production of gasoline components.

Among the processes that have been used for gasoline and liquid fuels production are (1) thermal cracking, (2) catalytic cracking, (3) thermal reforming, (4) catalytic reforming, (5) polymerization, (6) alkylation, and (7) coking. Each of these

The Refinery of the Future. DOI: https://doi.org/10.1016/B978-0-12-816994-0.00005-1

processes may be carried out in a number of ways, which differ in details of operation, or essential equipment, or both (Parkash, 2003; Gary et al., 2007; Speight, 2014, 2017; Hsu and Robinson, 2017).

When crude oil fractions are heated to temperatures over 350°C (660°F) in the absence of a catalyst, the rates of thermal decomposition to proceed at significant rates (Chapter 3: Refining Chemistry). The severity of thermal processing (the combination of reaction time and temperature to achieve a given conversion) determines the conversion and the product characteristics. For example, sufficiently high temperatures and long-residence-times convert crude oil constituents entirely to gases and coke and so cracking conditions are controlled to produce as much as possible of the desired product, which is usually gasoline but may be cracked gases for petrochemicals or a lower viscosity oil for use as a fuel oil. The feedstock (cracking stock) may be almost any fraction obtained from crude oil, but the greatest amount of cracking is carried out on gas oil. Residua are also feedstocks for cracking cracked but the processes are somewhat different from those used for gas oils.

In addition, thermal cracking processes may be classified by the types of products obtainable from crude oil. These processes may also be characterized by the physical state (liquid and/or vapor phase) in which the decomposition occurs. The state depends on the nature of the feedstock as well as conditions of pressure and temperature (Parkash, 2003; Gary et al., 2007; Speight, 2014, 2017; Hsu and Robinson, 2017).

From the chemical viewpoint the products of cracking are very different from those obtained directly from crude oil. When a 12-carbon atom hydrocarbon typical of straight-run gas oil is cracked, there are several potential reactions that can occur that lead to a variety of products (Chapter 3: Refining Chemistry), for example:

$$CH_3(CH_2)_{10}CH_3 \rightarrow CH_3(CH_2)_8CH_3 + CH_2 = CH_2$$

$$CH_3(CH_2)_{10}CH_3 \rightarrow CH_3(CH_2)_7CH_3 + CH_2 = CHCH_3$$

$$CH_3(CH_2)_{10}CH_3 \rightarrow CH_3(CH_2)_6CH_3 + CH_2 = CHCH_2CH_3$$

$$CH_3(CH_2)_{10}CH_3 \rightarrow CH_3(CH_2)_5CH_3 + CH_2 = CH(CH_2)_2CH_3$$

$$CH_3(CH_2)_{10}CH_3 \rightarrow CH_3(CH_2)_4CH_3 + CH_2 = CH(CH_2)_3CH_3$$

$$CH_3(CH_2)_{10}CH_3 \rightarrow CH_3(CH_2)_3CH_3 + CH_2 = CH(CH_2)_4CH_3$$

$$CH_3(CH_2)_{10}CH_3 \rightarrow CH_3(CH_2)_2CH_3 + CH_2 = CH(CH_2)_5CH_3$$

$$CH_3(CH_2)_{10}CH_3 \rightarrow CH_3CH_2CH_3 + CH_2 = CH(CH_2)_6CH_3$$

$$CH_3(CH_2)_{10}CH_3 \rightarrow CH_3CH_3 + CH_2 = CH(CH_2)_7CH_3$$

$$CH_3(CH_2)_{10}CH_3 \rightarrow CH_4 + CH_2 = CH(CH_2)_8CH_3$$

The products are dependent on temperature and residence time, and the simple reactions shown previously do not take into account the potential for isomerization of the products or secondary and even tertiary reactions that can (and do) occur.

The hydrocarbon derivatives with the least thermal stability are the paraffin derivatives, and the olefin derivatives produced by the cracking of paraffin derivatives are also reactive. Cycloparaffin derivatives (naphthenes) are less easily cracked, their stability depending mainly on any side chains present, but ring splitting may occur, and dehydrogenation can lead to the formation of unsaturated naphthenes and aromatics. Aromatics are the most stable (refractory) hydrocarbon derivatives, the stability depending on the length and stability of side chains. Very severe thermal cracking of high-boiling feedstocks can result in condensation reactions of ring compounds, yielding a high proportion of coke (Speight, 1986, 2014, 2017; Parkash, 2003; Gary et al., 2007; Hsu and Robinson, 2017).

The gases formed by cracking are particularly important because of their chemical properties and their quantity. Only relatively small amounts of paraffinic gases are obtained from crude oil, and these are chemically inactive. Cracking produces both paraffinic gases (e.g., propane, C_3H_8) and olefinic gases (e.g., propylene, C_3H_6); the latter are used in the refinery as the feed for polymerization plants where high-octane polymer gasoline is made. In some refineries the gases are used to make alkylate, a high-octane component for aviation and motor gasoline. In particular, the cracked gases are the starting points for many petrochemicals (Speight, 2014, 2019).

It is the purpose of this chapter to present thermal cracking processes in the light of their use and development in modern refineries as well as potential innovations that will be used in the refineries of the future.

5.2 Early processes

As the demand for gasoline increased with the onset of automobile sales, the issue of how to produce more gasoline from less crude oil was solved in 1913 by the incorporation of cracking units into refinery operations in which fractions higher boiling than gasoline were converted into gasoline by thermal decomposition.

The origins of cracking are unknown. There are records that illustrate the use of naphtha in Greek fire almost 2000 years ago (Speight, 2014), but whether the naphtha was produced naturally by distillation or by cracking distillation is not clear. Cracking was used commercially in the production of oils from coal and oil shale before the beginning of the modern crude oil industry. From this the discovery that the higher boiling materials from crude oil could be decomposed to lower molecular weight products (craking distillation) was used to increase the production of kerosene (Parkash, 2003; Gary et al., 2007; Speight, 2014, 2017; Hsu and Robinson, 2017).

The precise origins of the modern version of cracking distillation, as applied in the modern crude oil industry, are unknown. However, it is essential to recognize that the production of volatile product by the destructive distillation of wood and coal was known for many years, if not decades or centuries, before the birth of the modern crude oil industry. Indeed, the production of spirits of fire (i.e., naphtha, the flammable constituent of Greek fire) was known from early times. The occurrence of bitumen at Hit (Mesopotamia) that was used as mastic by the Assyrians was further developed for use in warfare through the production of naphtha by destructive distillation.

At the beginning of the 20th century, the earliest process (thermal cracking) that was used to satisfy the rising demand for automotive fuel consisted of heating higher boiling crude oil distillates (for which there was a low market requirement) in pressurized reactors and thereby producing more valuable lower boiling fractions such as gasoline, kerosene, and distillate fuels.

The yields of gasoline and kerosene fractions were usually markedly increased by means of cracking distillation, but the technique was not entirely suitable for gasoline production. As the need for gasoline arose, the necessity of prolonging the cracking process became apparent and led to a process known as pressure cracking.

Pressure cracking was a batch operation in which feedstock was heated to approximately 425C (800F) in stills (shell stills) especially reinforced to operate at pressures as high as 95 psi for 24 hours. Distillation was then started and during the next 48 hours 70, a low-boiling distillate was obtained, which was treated with sulfuric acid to remove unstable gum-forming components (olefin derivatives and diolefin derivatives) and then redistilled to produce a naphtha (cracked gasoline, boiling range <205°C, <400°F) and residual fuel oil (Parkash, 2003; Gary et al., 2007; Speight, 2014, 2017; Hsu and Robinson, 2017).

The Burton cracking process for the large-scale production of naphtha (cracked gasoline) was first used in 1912. The process employed batch distillation in horizontal shell stills and operated at approximately 400°C (c. 750°F) and 75–95 psi and was the first successful method of converting higher boiling feedstocks to gasoline. However, batch heating gas oil was considered inefficient, and during the years 1914–22 a number of successful continuous cracking processes were developed. In these processes, gas oil was continuously pumped through a unit that heated the gas oil to the required temperature, held it for a time under pressure, and then the cracked product was discharged into a distillation unit for separation into gases, gasoline, gas oil, and cracked residuum (often called tar).

The tube-and-tank cracking process is typical of the early continuous cracking processes. Gas oil, preheated by exchange with the hot products of cracking, was pumped into the cracking coil (up to several 100 ft. long) that lined the inner walls of a furnace where oil or gas burners raised the temperature of the gas oil to 425°C (800°F). The hot gas oil passed from the cracking coil into a reaction chamber (soaker) where the gas oil was held under these temperature and pressure conditions until the cracking reactions to be completed. The cracking reactions formed coke, which over the course of several days filled the soaker. The gas oil stream was then switched to a second soaker, and drilling operations similar to those used in drilling an oil well cleaned out the first soaker. The cracked material (other than coke) left the on-stream soaker to enter an evaporator (tar separator) maintained under a much lower pressure than the soaker, where, because of the lower pressure, all the cracked material except the tar became vaporized. The vapor left the top of the separator, where it was distilled into separate fractions: gases, gasoline, and gas oil. The tar that was deposited in the separator was pumped out for use as asphalt or as a heavy fuel oil.

With the exception of the lower temperature, the tube-and-tank process is the forerunner to the delayed coking process. The Burton process, like the tube-and-tank cracking process, was one of the earliest thermal cracking processes in 1912 which operated at 370°C–400°C (700°F–750°F) and 90 psi. Shortly thereafter, in 1921, a more advanced thermal cracking process that operated at 750°F–860°F (400°C–460°C) was developed (Dubbs process).

The majority of regular thermal cracking processes use temperatures on the order of 455°C–540°C (850°F–1005°F) and pressures in the range of 100–1000 psi. The Dubbs process (which was used extensively by many refineries until the early 1940s until catalytic cracking came into use) may be taken as a typical application of conventional thermal cracking. The feedstock (reduced crude) is preheated by direct exchange with the cracked products in the fractionating columns. Cracked gasoline and middle-distillate fractions are removed from the upper section of the column. Lower boiling and heavy distillate fractions are removed from the lower section and are pumped to separate heaters. Higher temperatures are used to crack the more refractory fraction. The streams from the heaters are combined and sent to a soaking chamber, where additional time is provided to complete the cracking reactions. The cracked products are then separated in a low-pressure flash chamber where a heavy fuel oil is removed as bottoms. The remaining cracked products are sent to the fractionating columns.

During World War II, improved refining processes involving the use of catalysts led to further improvements in the quality of transportation fuels and further increased their supply. These improved processes, including catalytic cracking (Chapter 6: Catalytic Cracking), enabled the crude oil industry to meet the fuel demands of high-performance combat aircraft and, after the war, to supply increasing quantities of transportation fuels.

As refining technology evolved throughout the 20th century, the feedstocks for cracking processes became the residuum or gas oil from a distillation unit. In addition, the residual oil produced as the end product of distillation processes, and even some of the higher boiling crude oil constituents, often contain substantial amounts of asphaltic materials, which preclude use of the residuum as fuel oils or lubricating stocks (Parkash, 2003; Gary et al., 2007; Speight, 2014, 2017; Hsu and Robinson, 2017).

However, subjecting these residua directly to thermal processes has become economically advantageous since, on the one hand, the end result is the production of lower boiling products but, on the other hand, the asphaltene constituents and the resin constituents that are concentrated in residua are precursors to high yields of thermal (i.e., noncatalytic) coke (Chapter 3: Refining Chemistry).

Although new thermal cracking units are now under development for heavy crude oil, extra heavy oil, and tar sand bitumen (Speight, 2008, 2014), processes

Table 5.1 Comparison of visbreaking with delayed coking and fluid coking.

Thermal cracking
Prime purpose: conversion: prime purpose
Semicontinuous process
High conversion
Process configuration: various
Visbreaking
Prime purpose: viscosity reduction
Mild (470°C–495°C; 880°F–920°F) heating at pressures of 50–200 psi
Reactions quenched before going to completion
Low conversion (10%) to products boiling less than 220°C (430°F)
Heated coil or drum (soaker)
Delayed coking
Prime purpose: conversion
Moderate (480°C–515°C; 900°F–960°F) heating at pressures of 90 psi
Reactions allowed to proceed to completion
Complete conversion of the feedstock
Soak drums (845°F–900°F) used in pairs (one on stream and one off stream being decoked)
Coke yield: 20%–40% by weight (dependent upon feedstock)
Fluid coking
Prime purpose: conversion
Severe (480°C–565°C; 900°F–1050°F) heating at pressures of 10 psi
Reactions allowed to proceed to completion
Complete conversion of the feedstock
Coke bed fluidized with steam; heat dissipated throughout the fluid bed
Higher yields of light ends (<C_5) than delayed coking
Less coke make than delayed coking (for one particular feedstock)

that can be regarded as having evolved from the original concept of thermal cracking are visbreaking and the various coking processes (Table 5.1).

5.3 Commercial processes

As crude oil slates into refineries continue to become heavier (lower American Petroleum Institute gravity and higher sulfur content), environmental restrictions are expected to significantly reduce the demand for high-sulfur residual fuel oil. Light sweet crude oils will continue to be available and in even greater demand than in the past, and refiners will be faced with the choice of purchasing light sweet crude oils at a premium price, or adding bottom of the barrel upgrading capability, through additional new investment, to reduce the production of high-sulfur residual fuel oil and increase the production of low-sulfur distillate transportation fuels.

These issues will push refiners toward more extensive bottom of the barrel upgrading by (1) increasing low-sulfur distillate fuel demand that will require increased cracking capacity, (2) heavier crude oil slates that will result in a greater high-sulfur residual fuel oil production, if conversion is not added, and (3) environmental restrictions that will result in reduced demand for high-sulfur residual fuel oil.

Low pressures (<100 psi) and temperatures in excess of 500°C (930°F) tend to produce lower molecular weight hydrocarbon derivatives than those produced at higher pressures (400–1000 psi) and at temperatures up to 500°C (930°F). The reaction time is also important; light feeds (gas oils) and recycle oils require longer reaction times than the readily cracked heavy residues.

Thermal cracking processes offer viable methods of conversion of heavy feedstocks because they enable low operating pressure while involving high operating temperature without requiring expensive catalysts. Currently, the widest operated residuum conversion processes are visbreaking and the various delayed coking processes (Parkash, 2003; Gary et al., 2007; Sawarkar et al., 2007; Joshi et al., 2008; Speight, 2014, 2017; Hsu and Robinson, 2017).

5.3.1 Visbreaking

Visbreaking (viscosity reduction, viscosity breaking), a mild form of thermal cracking, was developed in the late 1930s to produce more desirable and valuable products (Parkash, 2003; Gary et al., 2007; Joshi et al., 2008; Speight, 2014, 2017; Hsu and Robinson, 2017).

Mild cracking conditions (low feedstock conversion per cycle) favor a high yield of naphtha with low gas production and decreased coke production. With limited conversion per cycle, the higher boiling residues must be recycled. However, the recycled oils become increasingly refractory with each pass through the thermal zone, and if such oils are not required as a fuel oil stock they may be subjected to a coking operation to increase gasoline yield or refined by means of a hydrogen process.

Visbreaking is a relatively mild thermal cracking operation used to reduce the viscosity of residua to produce fuel oil that meets specifications (Parkash, 2003; Gary et al., 2007; Speight, 2014, 2017; Hsu and Robinson, 2017). The process uses the approach of mild thermal cracking insofar as the thermal reactions are not allowed to proceed to completion and are interrupted by quenching.

The process is a relatively low-cost and low-severity approach to improving the viscosity characteristics of the residue without attempting significant conversion to distillates. Low-residence-times are required to avoid coke formation, although additives can help to suppress coke deposits on the tubes of the furnace. By reducing the viscosity of the nonvolatile fraction, visbreaking reduces the amount of the more valuable distillate heating oil that is required for blending to meet the fuel oil specifications. The process is also used to reduce the pour point of a waxy residue.

Visbreaking conditions range from 455°C to 510°C (850°F to 950°F) at a short-residence-time and from 50 to 300 psi at the heating coil outlet. It is the short-residence-time that brings to visbreaking the concept of being a mild thermal reaction in contrast to, for example, the delayed coking process where residence times are much longer and the thermal reactions are allowed to proceed to completion. The visbreaking process uses a quench operation to terminate the thermal reactions. Liquid-phase cracking takes place under these low-severity conditions to produce some naphtha, as well as material in the kerosene and gas oil boiling range. The gas oil may be used as additional feed for catalytic cracking units, or as heating oil.

In the process the feedstock (usually residuum or tar sand bitumen) is passed through a furnace where it is heated to a temperature of 480°C (895°F) under an outlet pressure of approximately 100 psi (Fig. 5.1). The cracked products are then passed into a flash-distillation chamber. The overhead material from this chamber is then fractionated to produce naphtha and light gas oil. The liquid products from the flash chamber are cooled with a gas oil flux and then sent to a vacuum fractionator. This yields a heavy gas oil (HGO) distillate and a residuum of reduced viscosity (Table 5.2). A 5%–10% conversion of residuum to naphtha is usually sufficient

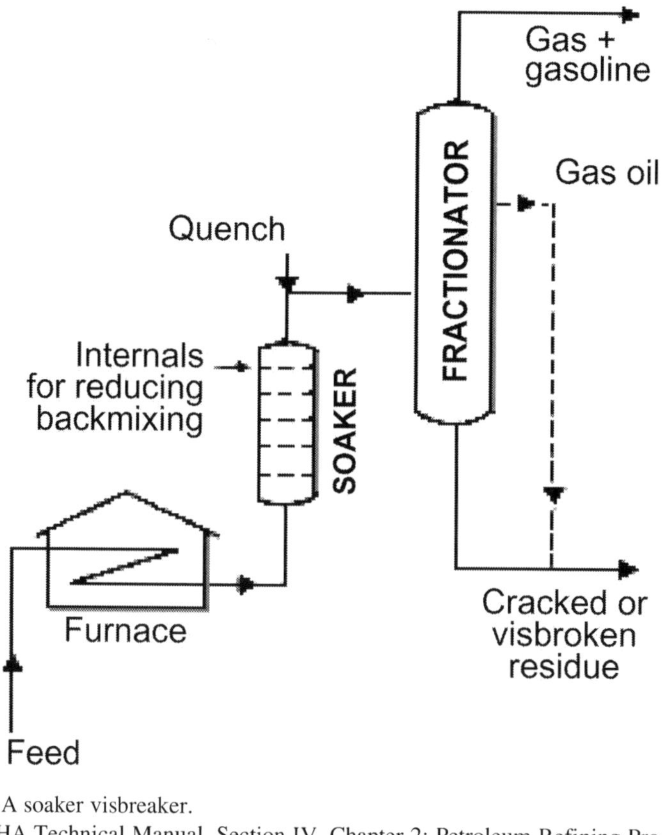

Figure 5.1 A soaker visbreaker.
Source: OSHA Technical Manual, Section IV, Chapter 2: Petroleum Refining Processes. http://www.osha.gov/dts/osta/otm/otm_iv/otm_iv_2.html.

Table 5.2 Examples of product yields and properties from visbreaking Athabasca tar sand bitumen and feedstocks having a similar American Petroleum Institute (API) gravity.

	Arabian light	**Arabian light**	**Iranian light**	
Feedstock	Vacuum Residuum	Vacuum Residuum	Vacuum Residue	Athabasca Bitumen
API gravity	7.1	6.9	8.2	8.6
Carbon residue[a]	20.3		22.0	13.5
Sulfur, wt.%	4.0	4.0	3.5	4.8
Product yields,[b] vol.%				
Naphtha (<425°F, <220°C)	6.0	8.1	4.8	7.0
Light gas oil (425°F–645°F, 220°C–340°C)	16.0	10.5	13.1	21.0
Heavy gas oil (645°F–1000°F, 340°C–540°C)		20.8	b	35.0
Residuum	76.0	60.5	79.9	34.0
API gravity	3.5	0.8	5.5	
Carbon residue[a]				
Sulfur, wt.%	4.7	4.6	3.8	

[a]Conradson.
[b]A blank product yield line indicates that the yield of the lower boiling product has been included in the yield of the higher boiling product.

to afford at least an approximate fivefold reduction in viscosity. Reduction in viscosity is also accompanied by a reduction in the pour point. An alternative option is to use lower furnace temperatures and longer times, achieved by installing a soaking drum between the furnace and the fractionator. The disadvantage of this approach is the need to remove coke from the soaking drum.

Two visbreaking processes are commercially available: the soaker visbreaker and the coil visbreaker.

The soaker–visbreaking process (a relatively low-temperature high-residence-time process) (Fig. 5.1) (Parkash, 2003; Gary et al., 2007; Speight, 2014, 2017; Hsu and Robinson, 2017) achieves a minor degree of conversion within the heater, but the majority of the conversion occurs in a reaction vessel (soaker) that holds the two-phase effluent at an elevated temperature for a predetermined length of time to allow cracking to occur before being quenched. The oil then passes to a fractionator. Lower temperatures are used in soaker visbreaker than in coil visbreaking. The comparatively long duration of the cracking reaction is used instead to achieve the desired results.

Product quality and yields from the coil and soaker drum design are essentially the same at a specified severity being independent of visbreaker configuration. By providing the residence time required for achieving the desired reaction, the soaker drum design allows the heater to operate at a lower outlet temperature (thereby saving fuel) but there are disadvantages.

The main disadvantage of the soaker–visbreaking process is the decoking operation of the heater and soaker drum, although decoking requirements of the soaker drum design is not as frequent as those of the coil-type visbreaker; the soaker design requires more equipment for coke removal and handling. The customary practice of removing coke from a drum is to cut it out with high-pressure water thereby producing a significant amount of coke-laden water that needs to be handled, filtered, and then recycled for use again.

The coil visbreaking process (high-temperature-short-residence-time process) (Parkash, 2003; Gary et al., 2007; Speight, 2014, 2017; Hsu and Robinson, 2017) differs from soaker–visbreaking insofar as the coil process achieves conversion by high-temperature cracking within a dedicated soaking coil in the furnace. Products exiting the furnace are quenched to halt the cracking reactions—this is frequently achieved by heat exchange with the virgin material being fed to the furnace or a stream of cold oil (usually gas oil) is used to the same effect and the gas oil is recovered and reused. The extent of the cracking reaction is controlled by regulation of the speed of flow of the feedstock through the furnace tubes. The quenched oil then passes to a fractionator where the products of the cracking (gas, LPG, gasoline, gas oil, and tar) are separated and recovered.

The main advantage of the coil-type design is the two-zone fired heater that provides better control of the material being heated and, with the coil-type design, decoking of the heater tubes is accomplished more easily by the use of steam-air decoking.

The two different types of visbreaking (the soaker visbreaker and the coil visbreaker) are very similar. The only difference is that the soaker visbreaker works with a lower temperature with a higher residence time and the coil visbreaker operates at a higher temperature with a lower residence time. The end result is that the soaker visbreaker has lower energy consumption for the same visbreaking severity.

The higher heater outlet temperature specified for a coil visbreaker is an important advantage of coil visbreaking. The higher heater outlet temperature is used to recover significantly higher quantities of heavy visbroken gas oil. This capability cannot be achieved with a soaker visbreaker without the addition of a vacuum flasher.

In terms of product yield, there is little difference between the two options (soaker visbreaker compared to coil visbreaker). However, each offers significant advantages in particular situations. For example, the cracking reaction forms coke as a by-product. In coil visbreaking, this lays down in the tubes of the furnace and will eventually lead to fouling or blocking of the tubes.

Briefly, fouling (a deposit buildup in refinery processes that impedes heat transfer and/or reduces throughput) is the leading cause of diminished efficiency and productivity in refineries. The energy lost due to this inefficiency must be supplied by burning additional fuel or reducing feed. While most fouling is caused by the deposition of heavier hydrocarbon species coming directly from the crude oil, a small undetermined percentage is related to corrosion and scale deposits, either actively participating as loose corrosion products or by scale acting as a substrate for hydrocarbon deposition.

Fouling will also occur in the drum of a soaker visbreaker, though the lower temperatures used in the soaker drum lead to fouling at a much slower rate. Coil visbreakers therefore require frequent decoking. Soaker drums require far less frequent attention but their being taken out of service normally requires a complete halt to the operation.

The lower temperatures used in the soaker approach mean that these units use less fuel. In cases where a refinery buys fuel to support process operations, any savings in fuel consumption could be extremely valuable. In such cases, soaker visbreaking may be advantageous.

The Shell soaker visbreaking process is suitable for the production of fuel oil by residuum (atmospheric residuum, vacuum residuum, or solvent deasphalter bottoms) viscosity reduction with maximum production of distillates. The basic configuration of the process includes the heater, soaker, and fractionator and more recently a vacuum flasher to recover more distillate products (Parkash, 2003; Gary et al., 2007; Speight, 2014, 2017; Hsu and Robinson, 2017). The cut point of the heavy gas–oil stream taken from the vacuum flasher is approximately 520°C (970°F).

In the process the feedstock is preheated before entering the visbreaker heater, where the residue is heated to the required cracking temperature. Heater effluent is sent to the soaker drum where most of the thermal cracking and viscosity reduction take place under controlled conditions. Soaker drum effluent is flashed and then quenched in the fractionator, and the flashed vapors are fractionated into gas, naphtha, gas oil, and visbreaker residue. The visbreaker residue is steam-stripped in the bottom of the fractionator and pumped through the cooling circuit for further processing. Visbreaker gas oil, which is recovered as a side stream, is steam-stripped, cooled, and sent for further processing.

As expected, product yields (Table 5.3) are dependent on feed type and product specifications. The heavy gas–oil stream for the visbreaker can be used as feedstock for a thermal distillate cracking unit or for a catalytic cracker for the production of lower boiling distillate products.

However, a recurring issue with the soaker visbreaker is the need to periodically decoke the soaker drum and the inability of the soaker process to easily adjust to

Table 5.3 Typical product yields for Middle East vacuum residuum for the Shell soaker–visbreaking process.

Product	Yield (% by weight)	
Gas	2	
Naphtha	4	End point 165°C
Light gas oil	12	End point 350°C
Heavy gas oil	18	End point 520°C
Vacuum-flashed residuum	64	

changes in feedstock quality because of the need to fine tune two process variables, temperature, and residence time. Recent combination of the visbreaking technology and the addition of new coil visbreaker design features have provided the coil process with a competitive advantage over the traditional soaker visbreaker process. Limitations in heater run length are no longer a problem for the coil visbreaker. Advances in visbreaker coil heater design now allow for the isolation of one or more passes through the heater for decoking, eliminating the need to shut the entire visbreaker down for furnace decoking.

Most of the existing visbreakers are the soaker type, which utilize a soaker drum in conjunction with a fired heater to achieve conversion and reduce the temperature required to achieve conversion while producing a stable residue product, thereby increasing the heater run length and reducing the frequency of unit shut down for heater decoking.

Overall, the main limitation of the visbreaking process, and for that matter all thermal processes, is that the products can be unstable due to the presence of unsaturated products. For example, thermal cracking at low pressure produces olefin derivatives (and diolefin derivatives) particularly in the naphtha fraction. These olefin derivatives give a very unstable product that tends to undergo secondary reactions to form gum and intractable nonvolatile tar.

The reduction in viscosity of distillation residua tends to reach a limiting value with conversion, although the total product viscosity can continue to decrease. The minimum viscosity of the unconverted residue can lie outside the range of allowable conversion if sediment begins to form. When shipment of the visbreaker product by pipeline is the process objective, addition of a diluent such as gas condensate can be used to achieve a further reduction in viscosity.

In spite of the abovementioned various limitation outlines, visbreaking has much potential and, in fact, remains an important, relatively inexpensive bottom-of-the-barrel upgrading process in many areas of the world.

Other variations of visbreaking technology include the Tervahl T and Tervahl H processes (Peries et al., 1988). The Tervahl T alternative includes only the thermal section to produce a synthetic crude oil with better transportability by having reduced viscosity and greater stability. The Tervahl H alternative adds hydrogen that also increases the extent of the desulfurization and decreases the carbon residua.

5.3.2 Coking processes

Coking is a thermal process for the continuous conversion of residua into lower boiling products. The feedstock can be atmospheric, vacuum, or cracked residuum and the products are gases, naphtha, fuel oil, gas oil, and coke. The gas oil may be the major product of a coking operation and serves primarily as a feedstock for catalytic cracking units. The coke can be used as fuel but processing for specialty uses, such as electrode manufacture, production of chemicals, and metallurgical coke, is also possible. For these latter uses the coke may require treatment to remove sulfur and metal impurities—calcined crude oil coke can be used for making anodes for aluminum manufacture, and a variety of carbon or graphite products such as brushes for electrical equipment.

Coking processes utilize longer reaction times than visbreaking, and the reactions are allowed to proceed to completion. To accomplish this, drums or chambers (reaction vessels) are employed, but it is necessary to use such vessels in pairs so that coke removal can be accomplished in those vessels not on-stream without interrupting the semicontinuous nature of the process.

5.3.2.1 Delayed coking

Delayed coking is a semicontinuous (semibatch) process in which the heated charge is transferred to large soaking (or coking) drums, which provide the long-residence-time needed to allow the cracking reactions to proceed to completion. The process (Fig. 5.2) is used for treating residua (Table 5.4) and uses long reaction times in the liquid phase to convert the residue fraction of the feed to gases, distillates, and coke. The coke is highly aromatic product that also retains sulfur, nitrogen, and metals and is enriched in these elements relative to the feedstock. In the process the feedstock is introduced into the product fractionator where it is heated, and lighter fractions are removed as side-stream products. The fractionator bottoms, including a recycle stream of heavy product, are then heated in a furnace (outlet temperature: 480°C–515°C, 895°F–960°F). The heated feedstock enters one of a pair of coking drums where the cracking reactions continue. The cracked products leave as an overhead stream, and coke deposits form on the inner surface of the drum. To allow continuous operation, two drums are used—while one is on stream, the other is being cleaned. The temperature in the coke drum ranges from 415°C to 450°C (780°F to 840°F) at pressures from 15 to 90 psi. The drum operation cycle is typically 48 hours—a coke drum is usually on-stream for approximately 24 hours before becoming filled with porous coke, and the following procedure is used to remove the coke: (1) the coke deposit is cooled with water, (2) one of the heads of the coking drum is removed to permit the drilling of a hole through the center of the deposit, and (3) a hydraulic cutting device, which uses multiple high-pressure water jets, is inserted into the hole and the wet coke is removed from the drum.

Fractionators separate the overhead products into fuel gas (low molecular weight gases up to and including ethane), propane and propylene ($CH_3CH_2CH_3$ and $CH_3CH=CH_2$), butane and butylene ($CH_3CH_2CH_2CH_3$ and $CH_3CH_2CH=CH_2$), naphtha, light gas oil, and HGO. Yields and product quality vary widely due to the broad range of feedstock types available for delayed coking units, and there is a decrease in overhead yield with increase in asphaltene content of the feedstock (Schabron and Speight, 1997; Parkash, 2003; Gary et al., 2007; Speight, 2014, 2017; Hsu and Robinson, 2017).

In the past, many delayed cokers were designed to provide complete conversion of atmospheric residue to naphtha, kerosene, and other low-boiling products. However, some cokers have been designed to minimize coke and produce heavy coker gas oil (HCGO) that is catalytically upgraded. The yield slate for a delayed

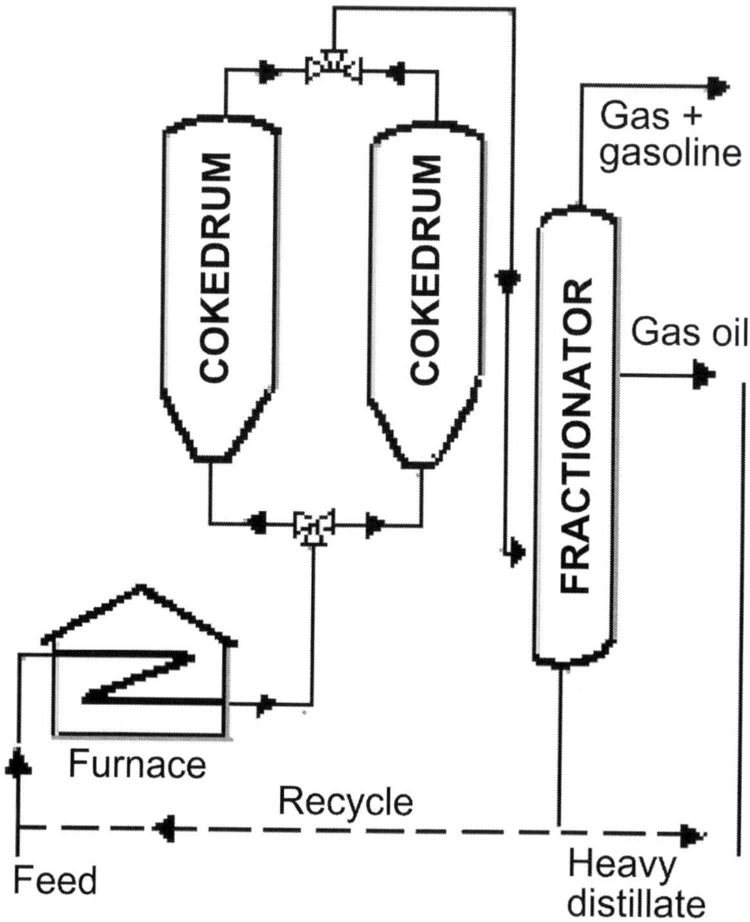

Figure 5.2 A delayed coker.
Source: OSHA Technical Manual, Section IV, Chapter 2: Petroleum Refining Processes. http://www.osha.gov/dts/osta/otm/otm_iv/otm_iv_2.html.

coker can be varied to meet a refiner's objectives through the selection of operating parameters.

Delayed coking has an increasingly important role to play in the integration of modem crude oil refineries because of the inherent flexibility of the process to handle even the heaviest of residues. The flexibility of operation inherent in delayed coking permits refiners to process a wide variety of crude oils, including those containing heavy, high-sulfur residua.

Low-pressure coking is a process designed for a once-through, low-pressure operation. The process is similar to delayed coking except that recycling is not usually practiced and the coke chamber operating conditions are 435°C (815°F) with a pressure on the order of 25 psi. Excessive coking is inhibited by the addition of

Table 5.4 Product yields and product properties for delayed coking of Athabasca tar sand bitumen and similar low American Petroleum Institute (API) gravity feedstocks.

	Kuwait	West Texas	Tia Juana	Alaska NS	Arabian Light	Athabasca
Feedstock	Residuum	Residuum	Residuum	Residuum	Residuum	Bitumen
API gravity	6.7	8.9	8.5	7.4	6.9	7.3
Carbon residue[a]	19.8	17.8	22.0	18.1		17.9
Sulfur, wt.%	5.2	3.0	2.9	2.0	4.0	5.3
Product yields, vol.%						
Naphtha (95°F–425°F, 35°C–220°C)	26.7	28.9	25.6	12.5	19.1	20.3
Light gas oil (425°F–645°F, 220°C–340°C)	28.0	16.5	26.4	b	b	b
Heavy gas oil (645°F–1000°F, 340°C–540°C)	18.4	26.4	13.8	51.2	48.4	58.8
Coke	30.2	28.4	33.0	27.2	32.8	21.0
Sulfur, wt.%	7.5	4.5		2.6	5.6	8.0

[a]Conradson.
[b]A blank product line indicates that the yield of the lower boiling product has been included in the yield of the higher boiling product.

water to the feedstock in order to quench and restrict further reactions of the reactive intermediates.

High-temperature coking is a semicontinuous process designed to convert asphaltic residua to gas oil and coke as the primary products. In the process the feedstock is transported to the heater (370°C, 700°F, 30 psi) and finally to the coking unit, where temperatures may be as high as 980°C–1095°C (1800°F–2000°F). Volatile materials are fractionated, and after the cycle is complete, coke is collected for sulfur removal before storage.

5.3.2.2 Fluid coking

Fluid coking is a continuous process developed in the mid-1950s that uses the fluidized solids technique to convert atmospheric and vacuum residua as well as cracked residua to more valuable products (Table 5.5) (Parkash, 2003; Gary et al., 2007; Speight, 2014, 2017; Hsu and Robinson, 2017). The yield of distillates from fluid coking can be improved by reducing the residence time of the cracked vapors.

In the process the hot feedstock is sprayed on to a fluidized bed of hot, fine coke particles, which permits the coking reactions to be conducted at higher temperatures and shorter contact times than can be employed in delayed coking. These conditions result in decreased yields of coke; greater quantities of more valuable liquid product are recovered in the fluid-coking process.

The process uses two vessels—a reactor and a burner (Fig. 5.3) (Parkash, 2003; Gary et al., 2007; Speight, 2014, 2017; Hsu and Robinson, 2017). The reactor holds a bed of fluidized coke particles, and steam is introduced at the bottom of the reactor to fluidize the bed. The feedstock coming from the bottom of a vacuum tower at 260°C–370°C (500°F–700°F) is injected directly into the reactor. The temperature in the coking vessel ranges from 480°C to 565°C (900°F to 1050°F), and the pressure is close to atmospheric pressure so the incoming feed is partly vaporized and partly deposited on the fluidized coke particles. The feedstock then cracks and vaporizes, leaving a residue that dries to form coke. The vapor products pass through cyclones that remove most of the entrained coke particles.

The vapor is discharged into the bottom of a scrubber, where the products are cooled to condense a heavy tar containing the remaining coke dust, which is recycled to the coking reactor. The upper part of the scrubber tower is a fractionating zone from which coker gas oil is withdrawn and then fed to a catalytic cracking unit; naphtha and gas are taken overhead to condensers.

In the reactor the coke particles flow down through the vessel into a stripping zone at the bottom. Steam displaces the product vapors between the particles, and the coke then flows into a riser that leads to the burner. Steam is added to the riser to reduce the solids loading and to induce upward flow.

The typical bed temperature in the burner is on the order of 590°C–650°C (1095°F–1200°F), and air is added as needed to maintain the temperature by burning part of the product coke. The pressure in the burner may range from 5 to

Table 5.5 Product yields and product properties for fluid coking low American Petroleum Institute (API) gravity feedstocks (vacuum residua).

	LA Basin	Kuwait	Hawkins	Tia Juana	Arabian Heavy	Iranian Heavy	Bachaquero	Zaca
API gravity	6.7	5.6	4.2	8.5	4.4	5.1	2.6	4.7
Carbon residue[a]	17.0	21.8	24.5	22.0	24.4	21.4	21.4	19.0
Sulfur, wt.%	2.1	5.5	4.3	2.9	5.3	3.4	3.7	7.8
Product yields, v/v%								
Naphtha (95°F–425°F, 35°C–220°C)	17.0	21.0	19.5	20.7	15.0	15.4	14.7	20.5
Gas oil (425°F–1000°F, 220°C–540°C)	62.0	48.0	53.0	48.3	47.7	55.1	48.3	61.0
Coke, wt.%	21.0	28.0	27.5	20.0	30.4	26.4	32.9	17.5

[a]Conradson.

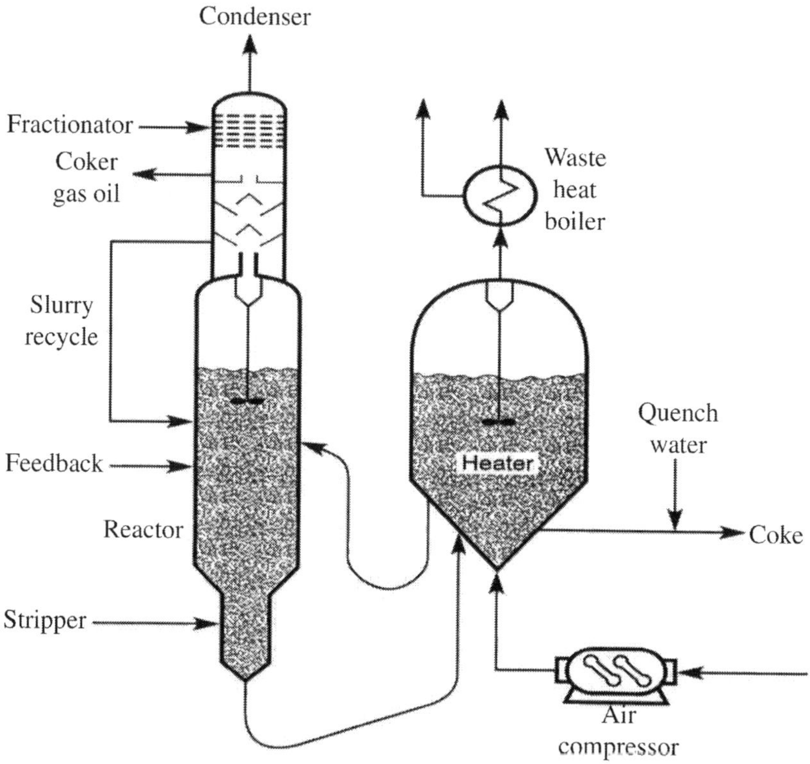

Figure 5.3 A fluid coker.

25 psi. Flue gases from the burner bed pass through cyclones and discharge to the stack. Hot coke from the bed is returned to the reactor through a second riser assembly.

Coke is one of the products of the process, and it must be withdrawn from the system to keep the solids inventory from increasing. The net coke produced is removed from the burner bed through a water-quench elutriator drum and cooled coke is withdrawn and sent to storage.

The yields of products are determined by (1) the properties of the feedstock, (2) the temperature of the fluid bed, and (3) the residence time of the feedstock in the bed. The use of a fluidized bed reduces the residence time of the vapor-phase products in comparison to delayed coking, which, in turn, reduces cracking reactions leading to reduced yields of coke and increased yields of gas oil and olefin derivatives. An increase of 5°C (9°F) in the operating temperature of the fluid-bed reactor typically increases gas yield by 1% w/w and naphtha by approximately 1% w/w.

The disadvantage of burning the coke to generate process heat is that sulfur from the coke is liberated as sulfur dioxide. The gas stream from the coke burner also contains carbon monoxide (CO), carbon dioxide (CO_2), and nitrogen (N_2).

An alternate approach is to use a coke gasifier to convert the carbonaceous solids to a mixture of carbon monoxide (CO), carbon dioxide (CO_2), and hydrogen (H_2).

5.3.2.3 Flexicoking

Flexicoking is a direct offshoot or descendent of fluid coking and uses the same configuration as the fluid coker but includes a gasification section in which excess coke can be immediately gasified to produce fuel gas (Fig. 5.4) (Parkash, 2003; Gary et al., 2007; Speight, 2014, 2017; Hsu and Robinson, 2017). The flexicoking process was designed during the late 1960s and the 1970s as a means by which excess coke could be reduced in view of the gradual incursion into refinery operations of the heavy crude oil, extra heavy oil, and tar sand bitumen—such feedstocks are notorious for producing high yields of coke (> 15% by weight) in both thermal and catalytic processes.

In the process, all else being equal, excess coke is converted to a low-heating value gas in a fluid bed gasifier with steam and air. The air is supplied to the gasifier to maintain temperatures of 830°C–1000°C (1525°F–1830°F) but is

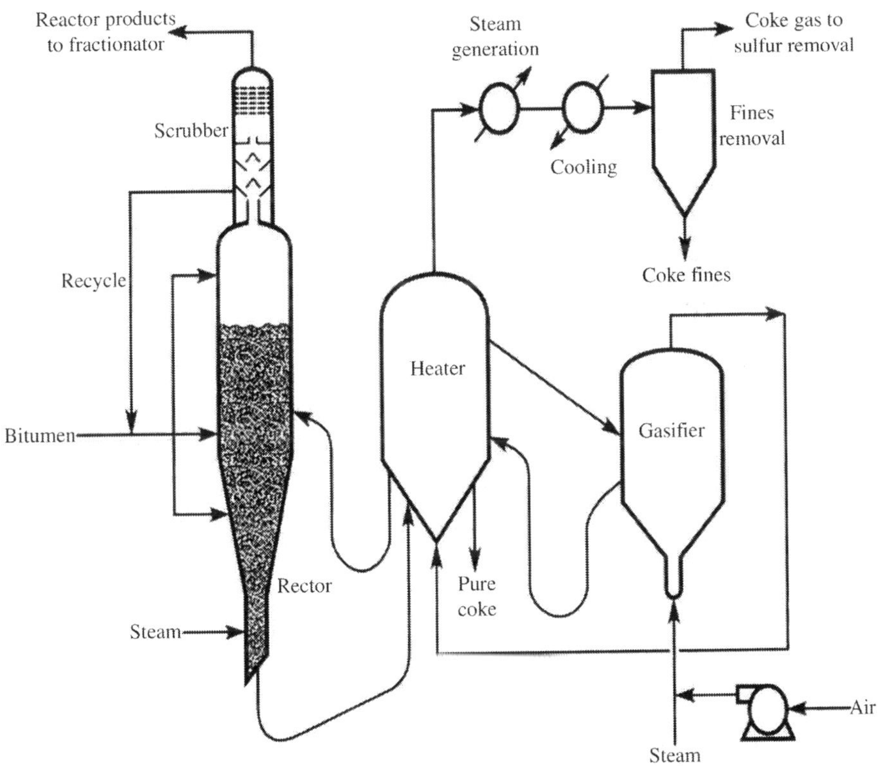

Figure 5.4 The flexicoking process.

insufficient to burn all of the coke. Under these reducing conditions, the sulfur in the coke is converted to hydrogen sulfide, which can be scrubbed from the gas prior to combustion. A typical gas product from the gasifier, after removal of hydrogen sulfide, contains carbon monoxide (CO, 18%), carbon dioxide (CO_2, 10%), hydrogen (H_2, 15%), nitrogen (N_2, 51%), water (H_2O, 5%), and methane (CH_4, 1%).

Yields of liquid products from flexicoking are the same as from fluid coking, because the coking reactor is unaltered. The main drawback of gasification is the requirement for a large additional reactor, especially if high conversion of the coke is required. Units are designed to gasify 60%−97% of the coke from the reactor. Even with the gasifier, the product coke will contain more sulfur than the feed, which limits the attractiveness of even the most advance of coking processes.

5.4 Process options for heavy feedstocks

The limitations of processing the more complex difficult-to-convert heavy crude oil, residua, extra heavy oil, tar sand bitumen, and residua depend to a large extent on the amount of nonvolatile higher molecular weight constituents, which also contain the majority of the heteroatoms (i.e., nitrogen, oxygen, sulfur, and metals such as nickel and vanadium) (Chapter 1: Feedstock Types and Properties). The chemistry of the thermal reactions of some of these constituents dictates that certain reactions, once initiated, cannot be reversed and proceed to completion and coke is the eventual product (Chapter 3: Refining Chemistry) (Parkash, 2003; Gary et al., 2007; Speight, 2014, 2017; Hsu and Robinson, 2017).

However, one answer to the need for conversion of the heavier feedstocks is the installation and use of one or more thermal cracking process. It is in the area of processing heavy feedstocks such as tar sand bitumen that process innovations have been initiated over the past two decades.

Thermal cracking processes offer attractive methods of conversion of heavy crude oil, extra heavy oil, tar sand bitumen, and residua because they enable low operating pressure while involving high operating temperature, without requiring expensive catalysts. Currently, the widest operated conversion processes are (1) visbreaking, (2) delayed coking, and (3) fluid coking, which also include flexicoking—these are attractive processes for refineries from an economic point of view.

Upgrading residua, which are similar in character to some heavy crude oil, extra heavy oil, and tar sand bitumen, began with the introduction of desulfurization processes that were designed to reduce the sulfur content of residua as well as some heavy crude oils and products therefrom. In the early days the goal was desulfurization but, in later years, the processes were adapted to a 10%−30% partial conversion operation, as intended to achieve desulfurization and obtain low-boiling fractions simultaneously, by increasing severity in operating conditions.

Refinery evolution has seen the introduction of a variety of heavy feedstock cracking processes (some use catalysts and are, of necessity, included here). These processes are different from one another in cracking method, cracked product patterns, and product properties and will be employed in refineries according to their respective features.

5.4.1 Aquaconversion process

The Aquaconversion process (Fig. 5.5) is a hydrovisbreaking technology that uses catalyst-activated transfer of hydrogen from water added to the feedstock. Reactions that lead to coke formation are suppressed, and there is no separation of asphaltene-type material (Parkash, 2003; Gary et al., 2007; Speight, 2014, 2017; Hsu and Robinson, 2017). The important aspect of the Aquaconversion technology is that it does not produce any solid by-product such as coke nor requires any hydrogen source or high-pressure equipment. In addition, the Aquaconversion process can be implanted in the production area, and thus the need for external diluent and its transport over large distances is eliminated.

5.4.2 Asphalt coking technology process

The asphalt coking technology (ASCOT) process is a residual oil upgrading process that integrates the delayed coking process and the deep solvent deasphalting process (low energy deasphalting) (Speight, 2014, 2017). Removing the deasphalted oil fraction prior to application of the delayed coking process has two benefits: (1) in the coking process, this fraction is thermally cracked to extinction, degrading this material as an fluid catalytic cracking (FCC) feedstock, and (2) thermally cracking this material to extinction results in conversion of a significant portion to coke.

In the process the vacuum residuum is brought to the desired extraction temperature and then sent to the extractor where the solvent (straight run naphtha, coker

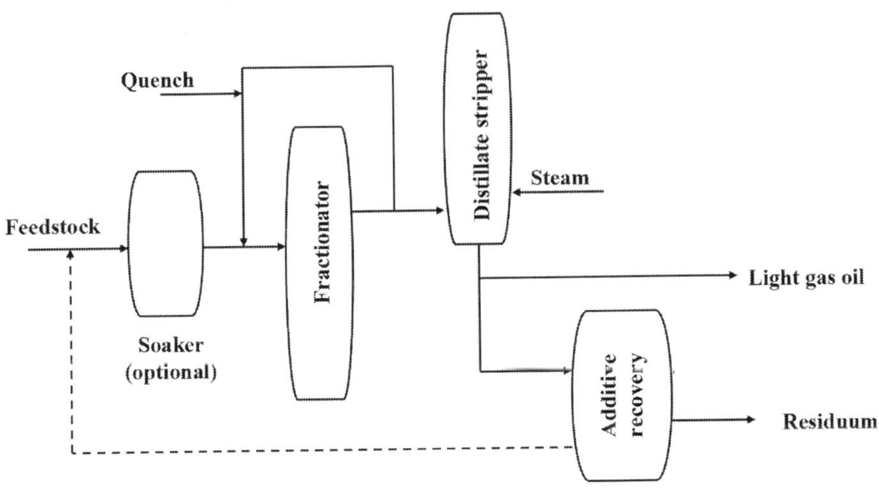

Figure 5.5 The Aquaconversion process.

naphtha) flows upward and extracts soluble material from the down-flowing feedstock. The solvent-deasphalted phase leaves the top of the extractor and flows to the solvent-recovery system where the solvent is separated from the deasphalted oil and recycled to the extractor. The deasphalted oil is sent to the delayed coker where it is combined with the HCGO from the coker fractionator and sent to the HCGO stripper where low-boiling hydrocarbon derivatives are stripped off and returned to the fractionator. The stripped deasphalted oil/heavy coker gas–oil mixture is removed from the bottom of the stripper and used to provide heat to the naphtha stabilizer–reboiler before being sent to battery limits as a cracking stock. The raffinate phase containing the asphalt and some solvent flows at a controlled rate from the bottom of the extractor and is charged directly to the coking section.

The solvent contained in the asphalt and deasphalted oil is condensed in the fractionator overhead condensers, where it can be recovered and used as lean oil for a propane/butane recovery in the absorber, eliminating the need to recirculate lean oil from the naphtha stabilizer. The solvent introduced in the coker heater and coke drums results in a significant reduction in the partial pressure of asphalt feed, compared with a regular delayed coking unit. The low asphalt partial pressure results in low coke and high liquid yields in the coking reaction.

With the ASCOT process, there is a significant reduction in by-product fuel as compared to either solvent deasphalting or delayed coking and the process can be tailored to process a specific quantity or process to a specific quality of cracking stock (Parkash, 2003; Gary et al., 2007; Speight, 2014, 2017; Hsu and Robinson, 2017).

5.4.3 Cherry-P (comprehensive heavy ends reforming refinery) process

The Cherry-P process is a process for the conversion of heavy crude oil or residuum into distillate and a cracked residuum (Speight, 2014, 2017). In this process the principal aim is to upgrade heavy crude oil residues at conditions between those of conventional visbreaking and delayed coking. Although coal is added to the feedstock, it is not intended to be a coprocessing feedstock but the coal is intended to act as a scavenger to prevent the buildup of coke on the reactor wall.

In the process the feedstock is mixed with coal powder in a slurry mixing vessel (without a catalyst or hydrogen), heated in the furnace and fed to the reactor where the feedstock undergoes thermal cracking reactions for several hours at a temperature higher than 400°C–450°C (750°F–840°F) and under pressure (70–290 psi) with a residence time on the order of 1–5 hours. Gas and distillate from the reactor are sent to a fractionator and the cracked residuum residue is extracted out of the system after distilling low-boiling fractions by the flash drum and vacuum flasher to adjust its softening point. Distillable product yields of 44% by weight on total feed are reported (Parkash, 2003; Gary et al., 2007; Speight, 2014, 2017; Hsu and Robinson, 2017).

5.4.4 Decarbonizing process

The thermal decarbonizing process (not to be confused with the propane decarbonizing process, which a deasphalting process) is designed not only to minimize coke and gasoline yields but also, at the same time, to produce maximum yields of gas oil. Decarbonizing in this sense of the term should not be confused with propane decarbonizing, which is essentially a solvent deasphalting process (Parkash, 2003; Gary et al., 2007; Speight, 2014, 2017; Hsu and Robinson, 2017).

Thermal decarbonizing is essentially the same as the delayed coking process, but lower temperatures and pressures are employed. For example, heater outlet temperatures may be 485°C (905°F) and coke drum temperatures may be of the order of 415°C (780°F), while pressures range from 10 to 25 psi.

5.4.5 ET-II process

The ET-II process is a thermal cracking process for the production of distillates and cracked residuum for use as metallurgical coke and is designed to accommodate feedstocks such as viscous feedstocks, atmospheric residua, and vacuum residua (Speight, 2014, 2017). The distillate (cracked oil) is suitable as a feedstock to hydrocracker and FCC. The basic technology of the ET-II process is derived from that of the original Eureka process.

In the process the feedstock is heated up to 350°C (660°F) by passage through the preheater and fed into the bottom of the fractionator, where it is mixed with recycle oil, and the high-boiling fraction of the cracked oil. The ratio of recycle oil to feedstock is within the range of 0.1%–0.3% by weight. The feedstock mixed with recycle oil is then pumped out and fed into the cracking heater, where the temperature is raised to approximately 490°C–495°C (915°F–925°F) and the outflow is fed to the stirred-tank reactor where it is subjected to further thermal cracking. Both cracking and condensation reactions take place in the reactor.

The cracked oil and gas products, together with steam from the top of the reactor, are introduced into the fractionator where the oil is separated into two fractions, cracked light oil and vacuum gas oil (VGO), and pitch (Parkash, 2003; Gary et al., 2007; Speight, 2014, 2017; Hsu and Robinson, 2017).

5.4.6 Eureka process

The Eureka process is a thermal cracking process to produce a cracked oil and aromatic residuum from heavy residual materials (Aiba et al., 1981; Parkash, 2003; Gary et al., 2007; Speight, 2014, 2017; Hsu and Robinson, 2017). In this process (Fig. 5.6) the feedstock, usually a vacuum residuum, is fed to the preheater and then enters the bottom of the fractionator, where it is mixed with the recycle oil. The mixture is then fed to the reactor system that consists of a pair of reactors operating alternately. In the reactor, thermal cracking reaction occurs in the presence of superheated steam that is injected to strip the cracked products out of the reactor and supply a part of heat required for cracking reaction. At the end of the reaction

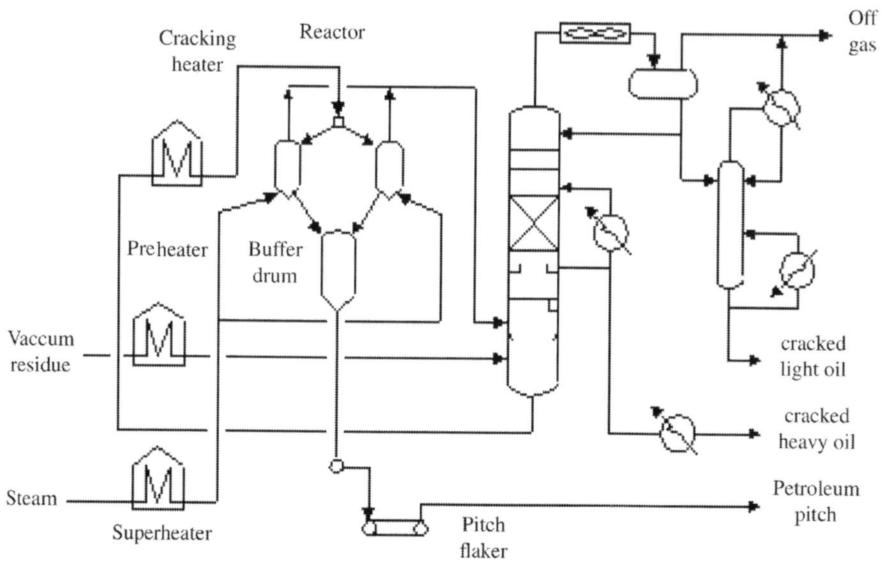

Figure 5.6 The Eureka process.

the bottom product is quenched. The oil and gas products (and steam) pass from the top of the reactor to the lower section of the fractionator, where a small amount of entrained material is removed by a wash operation. The upper section is an ordinary fractionator, where the heavier fraction of cracked oil is drawn as a side stream. The process bottoms (pitch) can be used as boiler fuel, as partial oxidation feedstock for producing hydrogen and carbon monoxide, and as binder pitch for manufacturing metallurgical coke (Parkash, 2003; Gary et al., 2007; Speight, 2014, 2017; Hsu and Robinson, 2017).

The process reactions proceed at lower cracked oil partial pressure by injecting steam into the reactor, keeping crude oil pitch in a homogeneous liquid state and, unlike a conventional delayed coker, a higher cracked oil yield can be obtained. A wide range of residua can be used as feedstock, such as atmospheric and vacuum residue of crude oil, various cracked residues, asphalt products from solvent deasphalting, and native asphalt. After hydrotreating the cracked oil is used as feedstock for a fluid catalytic cracker or hydrocracker.

The original Eureka process uses two batch reactors, while the newer ET-II and the high conversion soaker cracking (HSC) processes both employ continuous reactors.

5.4.7 Fluid thermal cracking process

The FTC process is a heavy crude oil and residuum upgrading process in which the feedstock is thermally cracked to produce distillate and coke (Miyauchi et al.,

1987; Parkash, 2003; Gary et al., 2007; Speight, 2014, 2017; Hsu and Robinson, 2017).

The feedstock, mixed with recycle stock from the fractionator, is injected into the cracker and is immediately absorbed into the pores of the particles by capillary force and is subjected to thermal cracking. In consequence the surface of the noncatalytic particles is kept dry and good fluidity is maintained allowing a good yield of, and selectivity for, middle-distillate products. Hydrogen-containing gas from the fractionator is used for the fluidization in the cracker. Excessive coke caused by the metals accumulated on the particle is suppressed under the presence of hydrogen. The particles with deposited coke from the cracker are sent to the gasifier, where the coke is gasified and converted into carbon monoxide (CO), hydrogen (H_2), carbon dioxide (CO_2), and hydrogen sulfide (H_2S) with steam and air. Regenerated hot particles are returned to the cracker.

5.4.8 High conversion soaker cracking process

The HSC process is a cracking process designed for moderate conversion, higher than visbreaking but lower than coking (Parkash, 2003; Gary et al., 2007; Speight, 2014, 2017; Hsu and Robinson, 2017). The process features less gas-make and a higher yield of distillate compared to other thermal cracking processes. The process can be used to convert a wide range of feedstocks with high-sulfur and metals content, including heavy crude oil, extra heavy oil, tar sand bitumen, residua, and visbroken residua. As a note of interest, the HSC process employs continuous reactors, whereas the original Eureka process uses two batch reactors.

In the process the preheated feedstock enters the bottom of the fractionator, where it is mixed with the recycle oil. The mixture is pumped up to the charge heater and fed to the soaking drum (c. atmospheric pressure, steam injection at the top and bottom), where sufficient residence time is provided to complete the thermal cracking. In the soaking drum the feedstock and some product flow downward passing through a number of perforated plates, while steam with cracked gas and distillate vapors flow through the perforated plates countercurrently.

The volatile products from the soaking drum enter the fractionator where the distillates are fractionated into desired product oil streams, including a HGO fraction. The cracked gas product is compressed and used as refinery fuel gas after sweetening. The cracked oil product after hydrotreating is used as FCC or hydrocracker feedstock. The residuum is suitable for used as boiler fuel, road asphalt, binder for the coking industry, and as a feedstock for partial oxidation.

5.4.9 Mixed-phase cracking

Mixed-phase cracking (also called liquid-phase cracking) is a continuous thermal decomposition process for the conversion of heavy feedstocks to products boiling in the gasoline range. The process generally employs rapid heating of the feedstock (kerosene, gas oil, reduced crude, or even whole crude), after which it is passed to a reaction chamber and then to a separator where the vapors are cooled. Overhead products from the flash chamber are fractionated to gasoline components and recycle stock, and flash chamber bottoms are withdrawn as a heavy fuel oil. Coke formation, which may be considerable at the process temperatures (400°C–480°C, 750°F–900°F), is minimized by the use of pressures in excess of 350 psi.

5.4.10 Selective cracking

Selective cracking is a thermal conversion process that utilizes different conditions depending on feedstock composition. For example, a heavy crude oil may be cracked at 494°C–515°C (920°F–960°F) and 300–500 psi; a lighter gas oil may be cracked at 510°C–530°C (950°F–990°F) and 500–700 psi. (Moschopedis et al., 1998; Parkash, 2003; Gary et al., 2007; Speight, 2014, 2017; Hsu and Robinson, 2017).

Each feedstock has its own particular characteristics that dictate the optimum conditions of temperature and pressure for maximum yields of the products. These factors are utilized in selective combination of cracking units in which the more refractory feedstocks are cracked for longer periods of time or at higher temperatures than the less stable feedstocks, which are cracked at lower temperatures.

The process eliminates the accumulation of stable low-boiling material in the recycle stock and also minimizes coke formation from high-temperature cracking of the higher boiling material. The end result is the production of fairly high yields of gasoline, middle distillates, and olefin gases.

5.4.11 Shell thermal cracking

The Shell thermal distillate cracking unit is based on the principle of converting high-boiling feedstocks to lower boiling products (Parkash, 2003; Gary et al., 2007; Speight, 2014, 2017; Hsu and Robinson, 2017). Thermal cracking of heavy crude oil takes place in the liquid phase in a furnace, at elevated pressure and temperature, and the products are residuum and distillate products. In the process the feedstock (such as the HGO from the atmospheric distillation unit or the VGO from the vacuum distillation unit) is sent to a surge drum. Liquid from this drum is pumped to the distillate heater, which typically operates at a pressure of approximately 490°C (915°F) and 290 psi. Under these conditions the cracking reactions take place in the liquid phase.

Fluid from the distillate heater is then routed to the combi tower where separation is achieved between residue, gas oil, and lighter products. In addition, an HGO fraction is taken from the combi tower, returned to the surge drum, and then recycled through the distillate heater. The bottom product of the combi tower is routed to a vacuum flasher where HGO is recovered from the residuum stream and routed back to the distillate heater. The vacuum-flashed residuum from the vacuum flasher can be routed to fuel oil blending or can be used internally as refinery fuel. The recycling of HGO from both the vacuum flasher and the combi tower to the distillate heater means that all of the HGO is converted. Light gas oil from the combi tower is first stripped and is then routed to a hydrotreater. Alternatively, the light gas oil can be used as cutter stock.

The feedstock and product requirements of the thermal distillate cracking process are flexible, and the process has the capability to optimize conversion through adjustment of the HGO recycle rate.

The Shell deep thermal gas oil process is a combination of the Shell deep thermal conversion process and the Shell thermal gas oil process. In this alternative high conversion scheme, the HGO from the atmospheric distillation unit and the VGO from the vacuum flasher or vacuum distillation unit are cracked in a distillate thermal cracking heater into lower-boiling range gas oil.

5.4.12 Tervahl-T process

The Tervahl-T process (Fig. 5.7) offers options that allow the process to accommodate differences in the feedstock as well as the desired sale of products. In the process (Parkash, 2003; Gary et al., 2007; Speight, 2014, 2017; Hsu and Robinson, 2017) the feedstock is heated to the desired temperature using the coil heater and heat recovered in the stabilization section and held for a specified residence time in the soaking drum. The soaking drum effluent is quenched and sent to a conventional stabilizer or fractionator where the products are separated into the desired streams (Parkash, 2003; Gary et al., 2007; Speight, 2014, 2017; Hsu and Robinson, 2017). The gas produced from the process is used for fuel.

In the related Tervahl H process (a hydrogenation process but covered here for convenient comparison with the Tervahl-T process), the feedstock and hydrogen-rich stream are heated using heat-recovery techniques and fired heater and held in

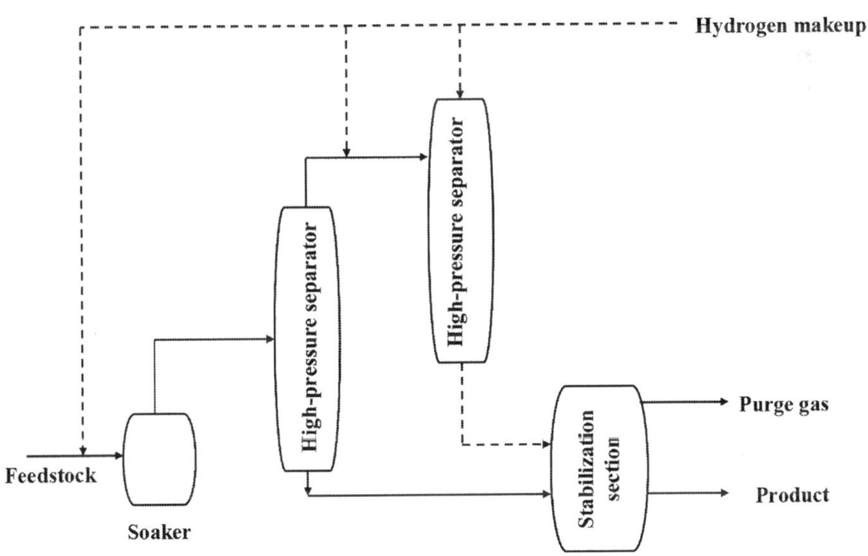

Figure 5.7 The Tervahl process.

the soak drum as in the Tervahl T process. The gas and oil from the soaking drum effluent are mixed with recycle hydrogen and separated in the hot separator where the gas is cooled passed through a separator and recycled to the heater and soaking drum effluent. The liquids from the hot and cold separator are sent to the stabilizer section where purge gas and synthetic crude are separated. The gas is used as fuel and the synthetic crude can now be transported or stored.

5.5 The future

Crude oil refining is growing increasingly complex due to lower quality crude oil (Parkash, 2003; Gary et al., 2007; Speight, 2014, 2017; Hsu and Robinson, 2017), crude oil price volatility, and environmental regulations that require cleaner manufacturing processes and higher performance products. In many cases, technology research and development have led to increasingly complex processing sequences and process units as the key to meeting these challenges and maintaining the health and profitability of the industry.

The conventional thermal upgrading methods are known for the high energy demand and low products yield (Parkash, 2003; Gary et al., 2007; Speight, 2014, 2017; Hsu and Robinson, 2017). Reducing the energy requirements and/or the process yields of the various technologies will increase the feasibility of the thermal cracking processes. Part of the energy requirements is the need to reduce coke formation by converting the constituents of the asphaltene fraction and the resin fraction to distillable products. Thus controlling the process conditions during the thermal cracking is crucial due to their impact of the process parameters on product yield and quality. This also requires a knowledge of the behavior of the constituents of these fractions, and studies related to the erroneous postulate of an average structure for each fraction are not only misleading but incorrect also.

While process innovations have been introduced in the form of varying process options, several using piggy-back techniques (where one process works in close conjunction with another process; see abovementioned text), there are other options that have not yet been introduced or even invented but may well fit into the refinery of the future. However, one answer to the need for conversion of the heavier feedstocks is the installation and use of one or more thermal cracking process. It is in the area of processing heavy feedstocks such as tar sand bitumen that process innovations have been initiated over the past two decades.

In addition to separation by distillation (Chapter 4: Distillation), alternative, less energy-intensive methods for converting crude fractions to the desired products could have a large energy impact. Visbreaking and/or coking processes are potential candidates. Existing processes could also be improved through redesign, or incorporation of practices that improve heat transfer or reduce process heating requirements (e.g., heat integration, waste heat recovery, better monitoring, and maintenance practices).

For example, a new generation of thermal processes has added additional flexibility to the refinery by converting the heavy bottom feed to lighter feedstocks and coke. Modern coking processes are mainly used to prepare a feed for the FCC unit. The most important processes are delayed coking and flexicoking and used at various refineries around the world. The further development of such processes will continue in the belief, with some justification, that cojoining two or more different process units will satisfy refining needs. However, while new processes are developed by many of the suppliers to the oil industry, most of these represent slight changes to previous designs, improving productivity, energy efficiency, and lowering production costs. However, there are also important new approaches to reactor design.

Nevertheless, because of their relative simplicity of design and straightforward thermal approach (some might say "brute force approach"), visbreaking and delayed coking processes will not be ignored or absent from the refinery of the future. These processes should not be ignored because they have the ability to adapt, by virtue of their relative simplicity, to the changing markets of heavy feedstock processing, tar sand bitumen processing, and biofeedstock processing. However, new and improved approaches are important for production of crude oil products. These will include advances in current methods, the minimization of process energy losses, and improved conversion efficiency—in particular (1) mitigation of fouling in heat exchangers and (2) improved conversion efficiency.

Visbreaking may the most underestimated and/or undervalued process in a refinery, although the process is not seen as making major comeback in US refineries (Marano, 2003) but this opinion may require some reevaluation.

Originally developed to produce fuel oil that met specifications, the process may find rejuvenated use not only for heavy feedstock (including tar sand bitumen) but also for biofeedstocks. These visbreaking process possesses sufficient hardware flexibility to accommodate feedstock blending (crude oil feedstocks and biofeedstocks) and of the unit as well as a high measure of reliability and predictive operations/maintenance thereby minimizing unplanned shutdowns.

Severity of visbreaker operation is generally limited by the stability requirement of the blended fuel oil and the extent of fouling and coke lay down in the visbreaker heater. The former requirement means that the stability of the residue must be sufficient to ensure that the finished fuel resulting from blending with diluents (that are less aromatic than the residue) is stable and that asphaltene flocculation does not occur. Where the residue is converted to an emulsion, blend stability is improved and severity/conversion can be increased, subject to acceptable levels of heater fouling and coke deposition (Miles, 2009). Operational modifications, such as increasing steam injection or recycling heavy distillates from the visbreaker fractionator, may help one to mitigate coking tendency and enhance yield, while some relatively low-cost options to increase heater capacity might be implemented in certain instances.

Foster Wheeler/Universal Oil Products (UOP) now offers commercial, proprietary technology that allows coil visbreaker operators to recover incremental HGO with an end point of approximately 450°C (840°F) for FCC/hydrocracker feed, without having to resort to vacuum flashing. Higher conversion versions of visbreaking are available from companies, including Shell Global Solutions and Foster Wheeler/UOP (Aquaconversion process) while still achieving a stable fuel oil product.

In terms of processing biofeedstocks, many biofeedstocks have a high-oxygen content and high-minerals content which could (even when blended) disqualify the use of the biomaterial as a feedstock to a hydroprocessing unit. Refiners are very wary of high-oxygen and high-mineral feedstocks because of the increased hydrogen requirements (hydrogen is an expensive refinery commodity) to remove the oxygen from the hydrocarbon products with the appearance of the additional hydrogen as water.

However, blending a biofeedstock with a resid as feedstock to a visbreaking unit to produce additional fuel products is a concept that could pay dividends and provide refineries with a source of fuels to supplement crude oil feedstocks. In the visbreaker the feedstock is converted to overhead (volatile products) and coke (if the unit is operated beyond the typical operating point or coke-forming threshold) (parka). The majority of the nitrogen, sulfur, and minerals appear in the coke. Oxygen often appears in the volatile product as water and carbon dioxide, unfortunately removing valuable hydrogen from the internal hydrogen management system.

Alternatively, another option is the preparation of a feedstock that is acceptable to a refinery. In particular, any process that reduces the mineral matter in the biofeedstock and reduces the oxygen content in the biofeed would be a benefit.

This can be accomplished by one or two preliminary treatment steps (such as the visbreaking process) in which the feedstock is demineralized, and the oxygen constituents are removed as overhead (volatile) material giving the potential for the production of a fraction rich in oxygen functions that may be of some use to the chemical industry. Such a process might have to be established at a biofeedstock production site, unless the refinery has the means by which to accommodate the feedstock in an already existing unit.

In a manner similar to the visbreaking process where the biofeedstock is blended with a residuum, the biofeedstock alone would be heated in a visbreaker-type reactor (at a lower temperature than the conventional visbreaking temperature) to the point where the volatile constituents are evolved and coke starts to form. As the coke forms, the mineral matter is deposited with the coke and the oxygen constituents are deoxygenized leaving a (predominantly) hydrocarbon product, which as a liquid will ensure easy separation from the coke and mineral matter. On the other hand, delayed coking is likely to remain the workhorse of thermal cracking processes for the foreseeable future. Online spalling, decoking techniques have been developed, based on successful, similar application on delayed coker heaters.

Coker heater operation (in delayed cokers and in visbreakers) is improved by online spalling of the coker heater pipes by the addition of off-line pipe is added to the online coker heater pipes. When an online pipe is to be spalled, flow is diverted to the off-line pipe allowing for full operation of the coker heater. In another embodiment a thermal transfer resistant zone plate is movably mounted in the radiant section of the coker heater. By moving the zone plate from an operating position to a spalling position and adjusting the temperature of the plurality of burners, the temperature of the pipes in the zone of the heater radiant section to be spalled can be lowered, while the temperature in the remaining zones of the heater radiant section is fully operational.

The delayed coking process will remain a preferred residue-upgrading option because of its ability to handle the heaviest, contaminated crudes. Globally, approximately one-third of installed residue-upgrading plant is by delayed coking. Although a mature process, in recent years many developments have taken place, including (1) the development of automated coke drum unheeding devices, allowing the operator to carry out the decoking procedure safely from a remote location, (2) understanding of process parameters affecting yields, coker product qualities, and coke qualities (e.g., shot coke), and (3) design and operation of major equipment items, in particular coke drums (allowing shorter coking cycles) and the delayed coker heater (online spalling/decoking and minimization of coking in furnace tubes).

Considering the need for expanding heavy feedstock processing (Parkash, 2003; Gary et al., 2007; Speight, 2014, 2017; Hsu and Robinson, 2017), there will be a need to incorporate operational flexibility into the original design of the delayed coker and make a few key equipment choices with long-term goals in mind. In recent projects and licensing proposals, refineries are also incorporating unique long-term rationales into their designs (Wodnik and Hughes, 2005). For example, some locations intentionally leave plot space and specify design criteria around the coker to allow for easy placement of an additional pair of coke drums with minimal debottlenecking of existing assets. This decision is based upon future plans to construct either another sour-crude train or other projects to make more coker feedstock available from existing refinery units. These units utilize the benefit of being online and generating earnings to help pay for the future expansion projects. Incorporating distillate recycle in the processing scheme or at least designing the coker so that recycle technology can be added later at minimal cost is another design strategy possibility.

Distillate recycle technology increases the flexibility of operating the unit. In addition to extending furnace run length and increasing liquid product yield, distillate recycle can be used to selectively increase the desired liquid products from the coker by varying the boiling range of the material circulated. This selective product-yield maximization allows the coker to easily adapt to fluctuations in product demands. In addition, as upstream units are expanded and additional feedstock for the coker becomes available, it is nearly always economically desirable to reduce distillate recycle and increase the fresh feed rate to the unit.

Recently, cokers have also been successfully designed and constructed to process high percentages (40%–60%) of solvent deasphalter pitch in the coker feed (Wodnik and Hughes, 2005). Utilizing asphaltic bottoms from an existing solvent deasphalting facility can increase overall refinery economics by converting most of the pitch to more valuable transportation fuels. Some refiners are operating and/or designing their cokers for a blocked-out type of operation. With this strategy, plants are able to process both fuel and anode-grade coke feedstocks to maximize earnings. Based on crude availability and market conditions, these coker units will process sour resids for fuel coke production for a specified time frame and then switch to run sweet resids to make anode coke.

One of the most promising pathways to simultaneously reduce energy use and capital costs is process intensification. Process intensification is a new area of reactor development aiming at more compact reactors to dramatically reduce the size of chemical plants, reduce capital costs, and intensify the chemical reactions (Stankiewicz and Moulijn, 2000).

Process intensification aims at the design of new compact reactors, and on the combination and integration of different processes (e.g., conversion and separation). The former has given rise to the development of compact heat exchangers that work under more extreme conditions (Haslego, 2001; Moritz and Gorak, 2002). An example of process intensification in the refinery industry is reactive distillation (Chapter 4: Distillation).

The major new development area for conversion processes will be the combination of conversion and separation, that is, reactive distillation (Chapter 4: Distillation). By combining the chemical reaction and separation in one reactor, capital costs are reduced and energy efficiency is improved through better integration of these process steps.

Reactive distillation (Chapter 4: Distillation) offers a promising alternative to conventional reaction-distillation schemes (Sundmacher and Kienle, 2003). Furthermore, active removal of reaction products can help one to shift the equilibrium of the reaction and improve the conversion efficiency. Reactive distillation has mainly be used in acetate technology (e.g., methyl tertiary butyl ether production) (Moritz and Gorak, 2002). Various research institutes and technology developers aim at developing new applications of reactive distillation. In the United States, researchers developed a reactive distillation process for isomerization to produce clean high-octane isomerate. Other new developments include the use of monolithic structures that contain the catalysts (Babbich and Moulijn, 2003), reducing catalyst loss (Goetze and Bailer, 1999).

On the other hand, steam cracking (in which a hydrocarbon stream is thermally cracking in the presence of steam) to yield a complex product mixture may also find use in biofeedstock conversion and allow a higher conversion of the biofeedstock to products usable in a refinery.

Fluid-coking units will be the subject of changes over the next decades, although the changes may be minimal and more focused on internal operations of the units as well as coupling of the units with other process units. The process is valuable as it generally proves to be equal or better net realizations from delayed coking, with the spread increasing for the heavier crudes (economics being sensitive to liquid yield, coker distillate quality, and product values).

Thermal upgrading processes for the conversion of high viscosity feedstocks will remain an important issue, especially when referring to the mixed feedstock concept of blending a viscous feedstocks with nonfossil fuel feedstocks, such as biomass

(Chapter 10: Nonfossil Fuel Feedstocks). Reduction of coke formation is a necessary outcome and might be achieved (fully or partially) by processes in which the asphaltene and the resin constituents are dispersed and association, which is a lead-in to coke formation, is mitigated. The use of a solvent that is inherent within the feedstock or is inherently produced from the feedstock during the process is worthy of consideration. An example is the use of a scavenger additive (i.e., not consumed in the process) merits attention. This type of additive will act as a scavenger to associate with the initially formed coke precursor which will remain as a separate phase in the reaction mix and play no further role in the process after the coke precursors have been removed. If a further role is preferred, a scavenger such as, for example, calcium oxide that has the capability of removing sulfur (as calcium sulfide) from the reaction mix might be used. In such a case a pretreatment step such as a modified visbreaking process might be given consideration (Speight, 2012, 2014, 2017).

In summary the need for more and heavier feedstock processing units has been a trend since the start of the industry. Year by year, the density and sulfur content of available crudes have slowly been rising (Parkash, 2003; Gary et al., 2007; Speight, 2014, 2017; Hsu and Robinson, 2017). The current global crude oil market will continue to offer opportunities for bottom-of-the-barrel technology to play an important role in the refiner's continuous efforts to balance available crude qualities with the market demands. However, a pessimistic future market and a refining outlook maybe, it still seems to include ample amounts of heavier and more sour crudes in addition to avid demand for transportation fuels.

Significant amounts of proven thermal technologies are, however, available to meet this objective. The main processes for such upgrading are (1) delayed coking and (2) fluid coking with flexicoking. The decision of which specific technology to apply in a particular area will depend on the choice of the most abundant crude available, the performance and environmental specifications of the marketplace, and the economic returns necessary to justify the technology.

From an environmental standpoint, with the trend of producing high-sulfur coke (from high-sulfur resids) and the decreasing demand for the same, gasification of coke to form fuel and syngas to be used within the refinery will be the only economic and environmental option for refineries in the future, and flexicoking will become a more popular choice. In addition, it is predictable that flexicoking will see an increase in use when coking residues (poor-quality coke) and biofeedstock becomes prime feedstocks for a gasification unit.

Finally, gasification process in which the feedstock is converted (by partial oxidation) into a carbon containing gas and eventually into synthesis gas (in which the major components are hydrogen and carbon monoxide) is a thermal process that is not dealt with here but elsewhere (Chapter 10: Nonfossil Fuel Feedstocks). Synthesis gas produced from a wide spectrum of fossil fuel feedstocks and biofeedstocks can not only be used to produce hydrogen and raise steam and/or power, but it can also be a building block in the manufacture of transportation fuels (via Fischer–Tropsch) or a wide range of petrochemicals (Parkash, 2003; Gary et al., 2007; Speight, 2008, 2014, 2017, 2019; Hsu and Robinson, 2017).

In summary, there is a need to improve resid conversion processes such as visbreaking and the coking technologies. Part of the future growth will be at or near heavy crude and bitumen production sites to decrease heavy crude viscosity and improve the quality to ease transportation and open markets for crudes of otherwise marginal value. Visbreaking may be considered to be a conversion process rather than a process to produce fuel oil that meets specifications. Coking can be improved by reducing hydrocarbon gas formation and by inhibiting the formation of polynuclear aromatic products that are produced by the process and are not inherent to the feedstock. Both of these processes would benefit if a higher valued product could be produced.

References

Aiba, T., Kaji, H., Suzuki, T., Wakamatsu, T., 1981. Chem. Eng. Progr. 37.

Babbich, I.V., Moulijn, J.A., 2003. Science and technology of novel processes for deep desulfurization of oil refinery streams: a review. Fuel 82, 607–631.

Gary, J.H., Handwerk, G.E., Kaiser, M.J., 2007. Petroleum Refining: Technology and Economics, fifth ed. CRC Press, Taylor & Francis Group, Boca Raton, FL.

Goetze, L., Bailer, O., 1999. Reactive distillation with Katapak-S. Sulzer Tech. Rev. 4, 29–31.

Haslego, C., 2001. Compact condensing: new technology improves on traditional approach. Hydrocarbon Process. 80, 55–60.

Hsu, C.S., Robinson, P.R. (Eds.), 2017. Handbook of Petroleum Technology. Springer International Publishing AG, Cham, Switzerland.

Joshi, J.B., Pandit, A.B., Kataria, K.L., Kulkarni, R.P., Sawarkar, A.N., Tandon, D., et al., 2008. Petroleum residue upgradation via visbreaking: a review. Ind. Eng. Chem. Res. 47, 8960–8988.

Marano, J.J., 2003. Refinery technology profiles: gasification and supporting technologies. Report prepared for the United States Department of Energy. National Energy Technology Laboratory, Energy Information Administration, Washington, DC.

Miles, J., 2009. Maximizing distillate yields and refinery economics – an alternative solution to conventional fuel oil production or residue conversion. In: Proceedings. Session A. 14th Annual Meeting – European Refining Technology Conference. November 11.

Miyauchi, T., Tsutsui, T., Nozaki, Y., 1987. A new fluid thermal cracking process for upgrading resid. In: Paper 65B. Proceedings. Spring National Meeting. American Institute of Chemical Engineers, Houston, March 29.

Moritz, P., Gorak, A., 2002. Two in one: cost reduction thanks to reactive separation. Sulzer Tech. Rev. 1, 14–16.

Moschopedis, S.E., Ozum, B., Speight, J.G., 1998. Upgrading heavy oils. Rev. Process. Chem. Eng. 1 (3), 201–259.

Parkash, S., 2003. Refining Processes Handbook. Gulf Professional Publishing, Elsevier, Amsterdam, The Netherlands.

Peries, J.P., Quignard, A., Farjon, C., Laborde, M., 1988. Thermal and catalytic ASVAHL processes under hydrogen pressure for converting heavy crudes and conventional residues. Rev. Inst. Fr. Pétrol 43 (6), 847–853.

Sawarkar, A.N., Pandit, A.B., Samant, S.D., Joshi, J.B., 2007. Petroleum residue upgrading via delayed coking: a review. Can. J. Chem. Eng. 85, 1–24.

Schabron, J.F., Speight, J.G., 1997. An evaluation of the delayed coking product yield of heavy feedstocks using asphaltene content and carbon residue. Rev. Inst. Fr. Pétrol 52 (1), 73–85.

Speight, J.G., 1986. Upgrading heavy feedstocks. Annu. Rev. Energy 11, 253.

Speight, J.G., 2008. Synthetic Fuels Handbook: Properties, Processes and Performance. McGraw-Hill, New York.

Speight, J.G., 2012. Visbreaking: a technology of the past and the future. Scientica Iran. 19 (3), 569–573.

Speight, J.G., 2014. The Chemistry and Technology of Petroleum, fifth ed. CRC Press, Taylor & Francis Group, Boca Raton, FL.

Speight, J.G., 2017. Handbook of Petroleum Refining. CRC Press, Taylor & Francis Group, Boca Raton, FL.

Speight, J.G., 2019. Handbook of Petrochemical Processes. CRC Press, Taylor & Francis Group, Boca Raton, FL.

Stankiewicz, A.I., Moulijn, J.A., 2000. Process intensification: transforming chemical engineering. Chem. Eng. Prog. 96 (1), 22–34.

Sundmacher, K., Kienle, A. (Eds.), 2003. Reactive Distillation: Status and Future Directions. John Wiley & Sons Inc, New York.

Wodnik, R., Hughes, G.C., 2005. Delayed coking advances. Pet. Technol. Q. Q4, 1–6.

6 Catalytic cracking

6.1 Introduction

Catalytic cracking is a conversion process that can be applied to a variety of feedstocks ranging from gas oil to heavy crude oil and residuum (Parkash, 2003; Gary et al., 2007; Speight, 2014, 2017; Hsu and Robinson, 2017). The concept of catalytic cracking is basically the same as thermal cracking, but it differs by the use of a catalyst that is not (in theory) consumed in the process, and it is one of several practical applications used in a refinery that employ a catalyst to improve process efficiency and product slate.

This history of catalytic cracking started in the early-to-mid decades of the 20th century (Table 6.1). In the 1930s thermal cracking units produced approximately half of the total gasoline manufactured, the octane number of which was approximately 70 compared to 60 for straight-run gasoline. These were usually blended together with lower boiling fractions and sometimes with polymer gasoline and reformate to produce gasoline base stock with an octane number of approximately 65. The addition of tetraethyl lead (ethyl fluid) increased the octane number to approximately 70 for the regular-grade gasoline and 80 for premium-grade gasoline. The thermal reforming and polymerization processes that were developed during the 1930s could be expected to further increase the octane number of gasoline to some extent, but something new was needed to assist with octane improvement for the more powerful automobile engines.

In 1936 a new cracking process opened the way to higher octane gasoline; this process was catalytic cracking that allowed major improvements in process efficiency and product yields, and the last 80 years have seen substantial advances in the development of catalytic processes (Bradley et al., 1989; Sadeghbeigi, 1995; Parkash, 2003; Gary et al., 2007; Speight, 2014, 2017; Hsu and Robinson, 2017). This has involved not only advances in the chemistry and physics of the catalysts themselves as well as an understanding of the chemistry of catalytic cracking (Chapter 3: Refining Chemistry) but also major engineering advances in reactor design, such as the evolution of catalyst bed design from fixed beds to moving beds to fluidized beds which has led to different designs of the fluidized bed reactors currently in use (Parkash, 2003; Gary et al., 2007; Speight, 2014, 2017; Hsu and Robinson, 2017).

Catalytic cracking has a number of advantages over thermal cracking (Table 6.2) (Parkash, 2003; Gary et al., 2007; Speight, 2014, 2017; Hsu and Robinson, 2017). Gasoline produced by catalytic cracking has a higher octane number and consists largely of isoparaffins and aromatics. The isoparaffins and aromatic hydrocarbon

The Refinery of the Future. DOI: https://doi.org/10.1016/B978-0-12-816994-0.00006-3

Table 6.1 A brief history of catalytic cracking.

1915: Batch reactor catalytic cracking
- Catalyst: aluminum chloride ($AlCl_3$)—a Lewis acid, electron acceptor
- Alkane—electron (abstracted by $AlCl_3$) to produce a carbocation (C^+)
- Ionic chain reactions to crack long chains
- Product: low-boiling distillate

1936: Houdry process
- Continuous feedstock flow with multiple fixed-bed reactors
- Cracking/catalyst regeneration cycles
- Catalyst: clay, natural alumina/silica particles

1942: Thermoform catalytic cracking
- Continuous feedstock flow with moving-bed catalysts
- Catalyst: synthetic alumina/silica particles

1942: Fluid catalytic cracking
- Continuous feedstock flow with fluidized-bed catalysts

1965: New catalysts
- Synthetic alumina/silica and zeolite catalysts

Table 6.2 Comparison of thermal cracking and catalytic cracking.

Thermal cracking
- No catalyst
- Free radical reaction mechanisms
- Moderate yields of gasoline and other distillates
- Gas yields feedstock dependent
- Low-to-moderate product selectivity
- Low octane number gasoline
- Low-to-moderate yield of C_4 olefins
- Low-to-moderate yields of aromatics

Catalytic cracking
- Uses a catalyst
- More flexible in terms of product slate
- Ionic reaction mechanisms
- High yields of gasoline and other distillates
- Low gas yields
- High product selectivity
- Low *n*-alkane yields
- High octane number
- Chain-branching and high yield of C_4 olefins
- High yields of aromatics

derivatives have high octane numbers and greater chemical stability than monoolefins ($R{-}CH{=}CH_2$ or $R{-}CH{=}CH{-}R^1$) and diolefins [$R{-}CH{=}CH{-}CH{=}CH_2$ or $R{-}CH{=}CH(CH_2)_nCH{=}CH{=}R^1$]. The olefins and diolefins are present in much greater quantities in gasoline produced by thermal cracking processes. Furthermore, substantial quantities of olefin gases, suitable for polymer gasoline manufacture, and smaller quantities of methane (CH_4), ethane (CH_3CH_3), and ethylene ($CH_2=CH_2$) are produced by catalytic cracking (Chapter 3: Refining Chemistry) (Parkash, 2003; Gary et al., 2007; Speight, 2014, 2017; Hsu and Robinson, 2017). Catalytic cracking produces less residuum and more of the useful gas oil constituents than thermal cracking. Finally, the process has considerable flexibility, permitting the manufacture of both motor gasoline and aviation gasoline and a variation in the gas oil production to meet changes in the fuel oil market (Speight, 1986).

The feedstocks for catalytic cracking can be any one (or blends) of the flowing: (1) straight-run gas oil, (2) vacuum gas oil, (3) atmospheric residuum, and (4) vacuum residuum. If blends of the abovemention feedstocks are employed, compatibility of the constituents of the blends must be assured under read to conditions or excessive coke will be laid down on to the catalyst. In addition, there are several pretreatment options for the feedstocks for catalytic cracking units, and these are (1) deasphalting to prevent excessive coking on catalyst surfaces; (2) demetallization, that is, removal of nickel, vanadium, and iron to prevent catalyst deactivation; (3) use of a short residence time as a means of preparing the feedstock; and (4) hydrotreating or mild hydrocracking to prevent excessive coking in the fluid catalytic cracking unit (FCCU) (Bartholic, 1981a,b; Speight, 2000, 2004, 2014, 2017; Parkash, 2003; Gary et al., 2007; Hsu and Robinson, 2017).

Hydrotreating the feedstock to the fluid catalytic cracker (FCC) improves the yield and quality of naphtha (Table 6.3) and reduces the sulfur oxide (SO_x) emissions from the catalytic cracker unit, but it is typically a high-pressure process, and furthermore, manipulation of feedstock sulfur alone may not be sufficient to meet future gasoline performance standards. Refineries wishing to process the viscous feedstocks may only have the option to desulfurize the resulting high-sulfur naphtha produced in the process.

Table 6.3 Feedstock and product data for the fluid catalytic process with feedstock hydrotreating and without feedstock hydrotreating.

Feedstock (>370°C, >700°F)	No pretreatment	With hydrotreatment
API	15.1	20.1
Sulfur (wt.%)	3.3	0.5
Nitrogen (wt.%)	0.2	0.1
Carbon residue (wt.%)	8.9	4.9
Nickel + vanadium (ppm)	51.0	7.0
Products		
Naphtha (C5—221°C, C5—430°F) (vol.%)	50.6	58.0
Light cycle oil (221°C–360°C, 430°F–680°F) (vol.%)	21.4	18.2
Residuum (>360°C, >680°F) (wt.%)	9.7	7.2
Coke (wt.%)	10.3	7.0

Catalytic cracking in the usual commercial process involves contacting a feedstock (usually a gas oil fraction) with a catalyst under suitable conditions of temperature, pressure, and residence time. By this means, a substantial part (>50%) of the feedstock is converted into gasoline and lower boiling products, usually in a single-pass operation (Parkash, 2003; Gary et al., 2007; Speight, 2014, 2017; Hsu and Robinson, 2017). However, during the cracking reaction, carbonaceous material is deposited on the catalyst, which markedly reduces its activity, and removal of the deposit is very necessary. The carbonaceous deposit arises from the thermal decomposition of high-molecular-weight polar species (asphaltene constituents and resin constituents) (Chapter 3: Refining Chemistry) in the feedstock. Removal of the deposit from the catalyst is usually accomplished by burning in the presence of air until catalyst activity is reestablished.

6.2 Early processes

Catalytic cracking is the modern method for converting high-boiling crude oil fractions, such as gas oil, into high-quality gasoline, diesel fuel, and other value-added products.

In the 1930s thermal cracking units produced approximately 50% of the total gasoline. The octane number of this gasoline was approximately 70 compared to 60 for straight-run (distilled) gasoline. The thermal reforming and polymerization processes that were developed during the 1930s could be expected to further increase the octane number of gasoline to some extent, but an additional innovation was needed to increase the octane number of gasoline to enhance the development of more powerful automobile engines.

In 1936 a new cracking process opened the way to higher octane gasoline; this process was catalytic cracking. The process is basically the same as thermal cracking but differs by the use of a catalyst, which is not (in theory) consumed in the process and directs the course of the cracking reactions to produce more of the desired higher octane hydrocarbon products.

6.3 Commercial processes

Catalytic cracking is another innovation that truly belongs to the 20th century. It is the modern method for converting high-boiling crude oil fractions, such as gas oil, into gasoline and other low-boiling fractions. The several processes currently employed in catalytic cracking differ mainly in the method of catalyst handling, although there is an overlap with regard to catalyst type and the nature of the products. The catalyst, which may be an activated natural or synthetic material, is employed in bead, pellet, or microspherical form and can be used as a fixed-bed, moving-bed, or fluid-bed configurations.

The fixed-bed process was the first to be used commercially and uses a static bed of catalyst in several reactors, which allows a continuous flow of feedstock to be maintained. Thus the cycle of operations consists of (1) flow of feedstock through the catalyst bed, (2) discontinuance of feedstock flow and removal of coke from the catalyst by burning, and (3) insertion of the reactor on stream.

In the fixed-bed process the catalyst in the form of small lumps or pellets was made up in layers or beds in several (four or more) catalyst-containing drums (converters). The feedstock is vaporized at approximately 450°C (840°F) and less than 7–15 psi pressure passed through one of the converters where the cracking reactions took place. Deposition of coke on the catalyst renders it ineffective, and using a synchronized valve system, the feed stream was directed to a neighboring converter while the catalyst in the first converter was regenerated by carefully burning the coke deposits with air. Fixed-bed processes have now generally been replaced by moving-bed or fluid-bed processes (Parkash, 2003; Gary et al., 2007; Speight, 2014, 2017; Hsu and Robinson, 2017).

In the moving-bed process the catalyst is typically in the form of pellets that are moved continuously to the top of the unit by conveyor or pneumatic lift tubes to a storage hopper, then flow downward by gravity through the reactor, and finally to a regenerator. The regenerator and hopper are isolated from the reactor by steam seals and the cracked product is separated into recycle gas, distillate, naphtha, and wet gas.

Two advances in the technology of moving-bed catalytic cracking have greatly changed the operation over the early processes. The more important of these advances was the introduction of crystalline aluminosilicate cracking catalysts and the second advance was the development of techniques to reduce catalyst attrition and metal erosion in moving-bed cracking units.

The fluid-bed process differs from the fixed-bed and moving-bed processes insofar as the powdered catalyst is circulated essentially as a fluid with the feedstock (Sadeghbeigi, 1995). The several fluid catalytic cracking processes in use differ primarily in mechanical design (Parkash, 2003; Gary et al., 2007; Speight, 2014, 2017; Hsu and Robinson, 2017). Side-by-side, reactor-regenerator configuration or the reactor either above or below the regenerator is the main mechanical variations. From a flow standpoint, all fluid catalytic cracking processes contact the feedstock and any and recycle streams with the finely divided catalyst in the reactor.

A typical fluid catalytic cracking process involves mixing a preheated hydrocarbon charge with hot, regenerated catalyst as it enters the riser pipe leading to the reactor. The charge is combined with a recycle stream within the riser pipe, vaporized, and raised to reactor temperature (900°F–1000°F) by the hot catalyst. As the mixture travels up the riser pipe, the charge is cracked at 10–30 psi. In the more modern fluid catalytic cracking process units, all cracking takes place in the riser pipe and continues until the oil vapors are separated from the catalyst in the reactor cyclones. The resultant product stream (cracked product) is then charged to a fractionating column where it is separated into fractions, and some of the unreacted (or low-reacted) feedstock is recycled to the riser pipe. Spent catalyst is regenerated to get rid of coke that collects on the catalyst during the process. Spent catalyst flows through the catalyst stripper to the regenerator, where most of the coke deposits burn off and fresh catalyst is added as spent catalyst is removed to optimize the cracking process.

Feedstocks may range from naphtha to atmospheric residuum (reduced crude). Feed preparation (to remove metallic constituents and high-molecular-weight non-volatile materials) is usually carried out through any one of the following ways: coking, propane deasphalting, furfural extraction, vacuum distillation, viscosity breaking, thermal cracking, and hydrodesulfurization (HDS) (see, e.g., Speight, 2000).

The major process variables are temperature, pressure, catalyst–feedstock ratio (ratio of the weight of catalyst entering the reactor per hour to the weight of feedstock charged per hour), and space velocity (weight or volume of the feedstock charged per hour per weight or volume of catalyst in the reaction zone). Wide flexibility in product distribution and quality is possible through control of these variables along with the extent of internal recycling is necessary. Increased conversion can be obtained by applying higher temperature or higher pressure. Alternatively, lower space velocity and higher catalyst–feedstock ratio will also contribute to an increased conversion.

When cracking is conducted in a single stage, the more reactive hydrocarbon derivatives may be cracked, with a high conversion to gas and coke, in the reaction time necessary for reasonable conversion of the more refractory hydrocarbon derivatives. However, in a two-stage process, gas and gasoline from a short-reaction-time, high-temperature cracking operation is separated before the main cracking reactions take place in a second-stage reactor.

6.3.1 Fixed-bed processes

Historically, the Houdry fixed-bed process, which went on stream in June 1936, was the first of the modern catalytic cracking processes. It was preceded only by the McAfee batch process, which employed a metal halide (aluminum chloride) catalyst but has long since lost any commercial significance.

In the fixed-bed process, the catalyst in the form of small lumps or pellets was made up in layers or beds in several (four or more) catalyst-containing drums called converters. Feedstock vaporized at approximately 450°C (840°F) and less than 7–15 psi pressure passed through one of the converters where the cracking reactions took place. After a short time, deposition of coke on the catalyst rendered it ineffective, and using a synchronized valve system, the feed stream was turned into a neighboring converter while the catalyst in the first converter was regenerated by carefully burning the coke deposits with air. After approximately 10 minutes, the catalyst was ready to go on stream again.

Fixed-bed processes have now generally been replaced by moving-bed or fluid-bed processes.

6.3.2 Fluid-bed processes

6.3.2.1 Fluid-bed catalytic cracking

The fluid catalytic cracking process (Fig. 6.1) is the most widely used process and is characterized by the use of a finely powdered catalyst that is moved through the reactor (Fig. 6.2) and flow patterns may vary depending upon the precise configuration of the reactor (Parkash, 2003; Gary et al., 2007; Speight, 2014, 2017; Hsu and Robinson, 2017). The catalyst particles are of such a size that when aerated with air or hydrocarbon vapor, the catalyst behaves like a liquid and can be moved through pipes. Thus vaporized feedstock and fluidized catalyst flow together into a reaction chamber where the catalyst, still dispersed in the hydrocarbon vapors, forms beds in the reaction chamber and the cracking reactions take place. The cracked vapors pass through cyclones located in the top of the reaction chamber, and the catalyst powder is thrown out of the vapors by centrifugal force. The cracked vapors then enter the bubble towers where fractionation into cracked low-boiling and high-boiling gas oils, cracked gasoline, and cracked gases takes place.

Since the catalyst in the reactor becomes contaminated with coke, the catalyst is continuously withdrawn from the bottom of the reactor and lifted by means of a stream of air into a regenerator where the coke is removed by controlled burning. The regenerated catalyst then flows to the fresh feed line, where the heat in the

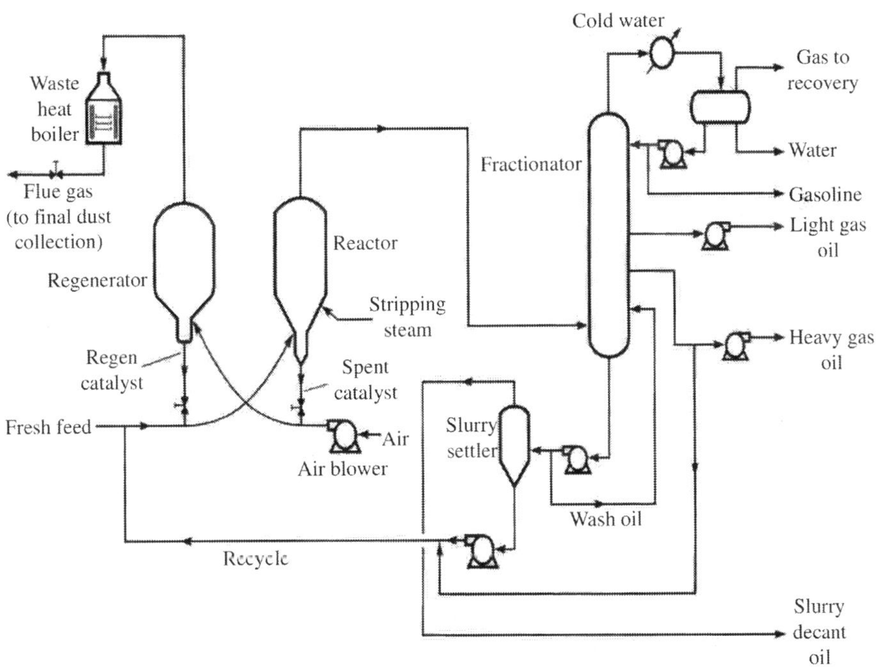

Figure 6.1 The fluid catalytic cracking process.

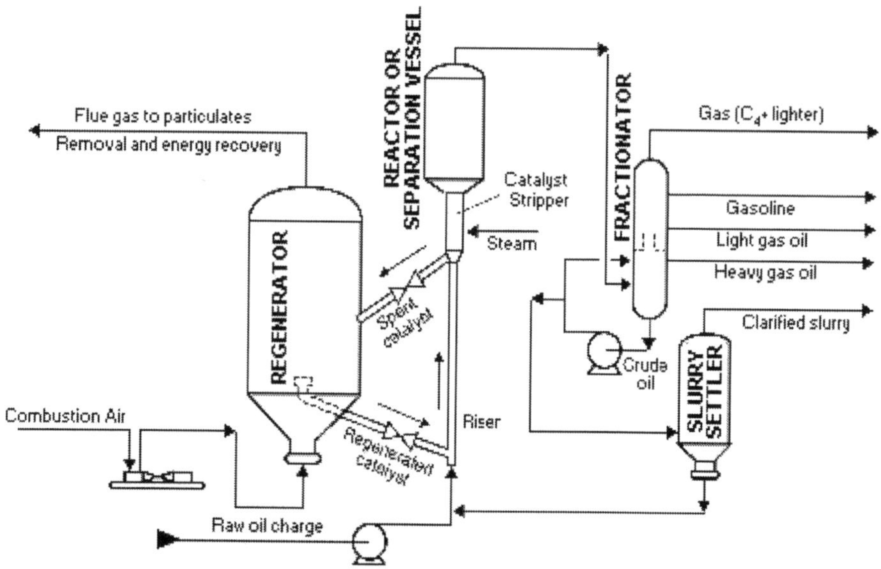

Figure 6.2 Detailed schematic a fluid catalytic cracking reactor.

catalyst is sufficient to vaporize the fresh feed before it reaches the reactor, where the temperature is about 510°C (950°F).

6.3.2.2 Model IV fluid-bed catalytic cracking unit

This unit involves a procedure in which the catalyst is transferred between the reactor and regenerator by means of U bends and the catalyst flow rate can be varied in relation to the amount of air injected into the spent-catalyst U bend. Regeneration air, other than that used to control circulation, enters the regenerator through a grid, and the reactor and regenerator are mounted side by side.

The model IV low-elevation design was preceded by the model III (1947) balanced-pressure design, the model II (1944) downflow design, and the original model I (1941) upflow design. The first commercial model IV installation in the United States was placed on stream in 1952.

6.3.2.3 Orthoflow fluid-bed catalytic cracking

This process uses the unitary vessel design that provides a straight-line flow of catalyst and thereby minimizes the erosion encountered in pipe bends. Commercial Orthoflow designs are of three types: models A and C, with the regenerator beneath the reactor, and model B, with the regenerator above the reactor. In all cases, the catalyst stripping section is located between the reactor and the regenerator. All designs employ the heat-balanced principle incorporating fresh feed–recycle feed cracking.

6.3.2.4 Shell two-stage fluid-bed catalytic cracking

Two-stage fluid catalytic cracking was devised to permit greater flexibility in shifting product distribution when dictated by demand. Thus feedstock is first contacted with cracking catalyst in a riser reactor, that is, a pipe in which fluidized catalyst and vaporized feedstock flow concurrently upward, and the total contact time in this first stage is on the order of seconds. High temperatures 470°C–565°C (875°F–1050°F) are employed to reduce undesirable coke deposits on catalyst without destruction of gasoline by secondary cracking. Other operating conditions in the first stage are a pressure of 16 psi and a catalyst–feedstock ratio of 3:1 to 50:1, and volume conversion ranges between 20% and 70% have been recorded.

All or part of the unconverted or partially converted gas oil product from the first stage is then cracked further in the second-stage fluid-bed reactor. Operating conditions are 480°C–540°C (900°F–1000°F) and 16 psi with a catalyst–oil ratio of 2:1 to 12:1. Conversion in the second stage varies between 15% and 70%, with an overall conversion range of 50%–80%.

6.3.2.5 Universal oil products fluid-bed catalytic cracking

This process is adaptable to the needs of both large and small refineries. The major distinguishing features of the process are (1) the elimination of the air riser with its attendant large expansion joints, (2) the elimination of considerable structural steel supports, and (3) a reduction in regenerator and in air-line size through use of 15–18 psi pressure operation.

6.3.3 Moving-bed processes

6.3.3.1 Airlift thermofor catalytic cracking (Socony airlift TCC process)

The airlift thermofor catalytic cracking process (Socony airlift TCC process) is a moving bed, reactor-over-generator continuous process for conversion of high-boiling gas oils into lower boiling high-quality gasoline and middle distillate fuel oils. Feed preparation may consist of flashing in a separator to obtain vapor feed, and the separator bottoms may be sent to a vacuum tower from which the liquid feed is produced.

The vapor-liquid flows downward through the reactor concurrently with the regenerated synthetic bead catalyst. The catalyst is purged by steam at the base of the reactor and gravitates into the kiln, or regeneration is accomplished by the use of air injected into the kiln. Approximately 70% of the carbon on the catalyst is burned in the upper kiln burning zone and the remainder in the bottom-burning zone. Regenerated, cooled catalyst enters the lift pot, where low-pressure air transports it to the surge hopper above the reactor for reuse.

6.3.3.2 Houdresid catalytic cracking

Houdresid catalytic cracking is a process that uses a variation of the continuously moving catalyst bed designed to obtain high yields of high-octane gasoline and low-boiling distillate from reduced crude charge.

Residua, ranging atmospheric residue to vacuum residua, including residua high in sulfur or nitrogen, can be used as the feedstock and the catalyst is synthetic or natural. Although the equipment employed is similar in many respects to that used in Houdriflow units, novel process features modify or eliminate the adverse effects and catalyst and product selectivity usually resulting when metals (such as nickel, vanadium copper and iron) are present in the fuel. The Houdresid catalytic reactor and catalyst-regenerating kiln are contained in a single vessel. Fresh feed plus recycled gas oil are charged to the top of the unit in a partially vaporized state and mixed with steam.

6.3.3.3 Houdriflow catalytic cracking

This is a continuous, moving-bed process employing an integrated single vessel for the reactor and regenerator kiln. The sweet or sour feedstock can be any fraction of the crude boiling between naphtha and atmospheric residua. The catalyst is transported from the bottom of the unit to the top in a gas lift employing compressed flue gas and steam.

The reactor feed and catalyst pass concurrently through the reactor zone to a disengager section, in which vapors are separated and directed to a conventional fractionation system. The spent catalyst, which has been steam purged of residual oil, flows to the kiln for regeneration, after which steam and flue gas are used to transport the catalyst to the reactor.

6.3.3.4 Suspensoid catalytic cracking

Suspensoid cracking was developed from the thermal cracking process carried out in tube-and-tank units (Chapter 14: A Biorefinery). Small amounts of powdered catalyst or a mixture with the feedstock are pumped through a cracking coil furnace. Cracking temperatures are 550°C–610°C (1025°F–1130°F), with pressures of 200–500 psi. After leaving the furnace the cracked material enters a separator where the catalyst and high-boiling entrained material (tar) are left behind. The cracked vapors enter a bubble tower where they are separated into two parts, gas oil and pressure distillate. The latter is separated into gasoline and gases. The spent catalyst is filtered from the organic carryover, which is used as a high-boiling industrial fuel oil.

The process is actually a compromise between catalytic and thermal cracking. The main effect of the catalyst is to allow a higher cracking temperature and to assist mechanically in keeping coke from accumulating on the walls of the tubes. The normal catalyst employed is spent clay (2–10 lb/barrel of feed) that obtained from the contact filtration of lubricating oils.

6.3.4 Process variables

Catalytic cracking is endothermic and, that being the case, heat is absorbed by the reactions, and the temperature of reaction mixture declines as the reactions proceed and a source of heat for the process is required. This heat comes from combustion of coke formed in the process. Coke is one of the important, though undesirable, products of cracking since it forms on the surface and in the pores of the catalyst during the cracking process, covering active sites and deactivating the catalysts. During regeneration, this coke is burned off the catalyst to restore catalytic activity, and like all combustion processes, the process is exothermic, liberating heat.

Most FCCUs are operated to maximize conversion to gasoline and liquefied petroleum gas (LPG). This is particularly true when building gasoline inventory for peak season demand or reducing clarified oil yield due to low market demand. Maximum conversion of a specific feedstock is usually limited by both FCCU design constraints (i.e., regenerator temperature and wet gas capacity) and the processing objectives. However, within these limitations, the FCCU operator has many operating and catalyst property variables to select from to achieve maximum conversion.

Conversion usually refers to the mount of fresh feedstock cracked to gasoline and lower boiling products and coke:

Conversion = 100 − (LCO + HCO + CO)

LCO is low-boiling cycle oil, HCO is high-boiling cycle oil, and CO is clarified oil; all are expressed as a % of the fresh feedstock.

A low conversion operation for maximum production of low-boiling cycle oil is typically 40%–60%, while a high conversion operation for maximum gasoline production is 70%–85%. Again, the range is dependent on the character of the feedstock. Each FCCU that is operated for maximum conversion at constant fresh feed quality has an optimum conversion point beyond which a further increase in conversion reduces gasoline yield and increases the yield of LPG, and the optimum conversion point is referred to as the overcracking point.

6.3.4.1 The reactor

The three main components of a FCCU are (1) the reactor, (2) the stripper, and (3) the regenerator. In the unit the catalyst and the feed and product hydrocarbon derivatives are lifted up the riser pipe to the reactor where the predominately endothermic cracking processes take place. Since the reactions are endothermic, reaction temperature declines from bottom to top. At the top the mixture enters a solid–gas separator, and the product vapors are led away. Cracked gases are separated and fractionated, the catalyst and residue, together with recycle oil from a second stage fractionator, pass to the main reactor for further cracking. The products of this second-stage reaction are gas, gasoline and gas oil streams, and recycle oil.

The coked catalyst enters the stripper where steam is added, and unreacted-reacted hydrocarbon derivatives adsorbed on the catalyst are released. The stripped catalyst is then directed into the regenerator where air is added, and the combustion of coke on the catalyst (and any hydrocarbon derivatives still adsorbed that were not stripped) occurs with the liberation of heat. Regenerator temperatures are typically 705°C–760°C (1300°F–1400°F). Heat exchangers and the circulating catalyst capture the heat evolved during regeneration to be used in preheating the reactor feed to appropriate cracking temperatures, usually in the range of 495°C–550°C (925°F–1020°F).

During operations the entire catalyst inventory is continually circulated through the unit. Catalyst residence time in the riser reactor section is typically 1–3 seconds (with current trends to even shorter residence times), and the entire reactor–stripper–regenerator cycle is less than 10 minutes. To achieve cycle times of this order, catalyst circulation rates as high as 1 ton/s in large units are required. To withstand such movement the catalyst must be sufficiently robust to withstand the operational stress.

Process temperatures are high, coke is repeatedly deposited and burned off, and the catalyst particles are moving at high speed through steel reactors and pipes. Contact between the catalyst particles and the metal walls and interparticle contact are unavoidable. Thus catalyst loss from the unit caused by poor attrition resistance can be a serious problem, since the quantities lost must be replaced by fresh catalyst additions to maintain constant unit performance. Catalyst manufacturers work hard to prevent inordinate losses due to attrition, and refineries keep a close watch on catalyst quality to be sure the produce conforms to their specifications. Therefore the robustness of the catalyst is carefully monitored and controlled to a high attrition resistance that is determined by rigorous test methods that place a semiquantitative evaluation on attrition resistance, which is generally related to breakdown with time in commercial units.

As intimated earlier, in some units, cracking does not always take place in the reactor and reaction often occurs in the vertical or upward sloped pipe called the riser (giving credence to the name riser reactor and riser pipe cracking) forming products, including coke (e.g., Bartholic, 1989). Preheated feedstock is sprayed into the base of the riser via feed nozzles where it contacts extremely hot fluidized catalyst at 1230°F–1400°F (665°C–760°C). The hot catalyst vaporizes the feed and catalyzes the cracking reactions that break down the high-molecular-weight oil into lower boiling components, including LPG, constituents, gasoline, and diesel. The catalyst–hydrocarbon mixture flows upward through the riser for just a few seconds and then the mixture is separated via cyclones. The catalyst-free hydrocarbon derivatives are routed to a main fractionator for separation into fuel gas, propane and butanes, gasoline, low-boiling cycle oils used in diesel and jet fuel, and high-boiling fuel oil.

6.3.4.2 Process parameters

As already noted, there are primary variables available to the operation of FCCUs for maximum unit conversion, for a given feedstock quality can be divided into two groups: catalytic variables and process variables. In addition to the catalyst variables (*q.v.*), there are also process variables that include (1) pressure, (2) reaction time or residence time in the reaction zone, and (3) reactor temperature. Higher conversion and coke yield are thermodynamically favored by higher pressure. However, pressure is usually varied over a very narrow range due to limited air blower horsepower. Conversion is not significantly affected by unit pressure since a substantial increase in pressure is required to significantly increase conversion.

An increase in reaction time available for cracking also increases conversion. Fresh feed rate, riser steam rate, recycle rate, and pressure are the primary operating variables that affect reaction time for a given unit configuration. Conversion varies inversely with these stream rates due to limited reactor size available for cracking. Conversion has been increased by a decrease in rate in injection of fresh feedstock. Under these circumstances, overcracking of gasoline to LPG and to dry gas may occur due to the increase in reactor residence time. One approach to offset any potential gasoline overcracking is to add additional riser steam to lower hydrocarbon partial pressure for more selective cracking. Alternatively, an operator may choose to lower reactor pressure or increase the recycle rate to decrease residence time. Gasoline overcracking may be controlled by reducing the availability of catalytic cracking sites by lowering cat/oil ratio.

Increased reactor temperature increases feedstock conversion, primarily through a higher rate of reaction for the endothermic cracking reaction and also through increased cat/oil ratio. A 5.6°C (10°F) increase in reactor temperature can increase conversion by 1%–2% absolute, but, again, this is feedstock dependent. Higher reactor temperature also increases the amount of olefins in gasoline and in the gases. This is due to the higher rate of primary cracking reactions relative to secondary hydrogen transfer reactions that saturate olefins in the gasoline boiling range and lowers gasoline octane.

However, these variables are not always available for maximizing conversion since most FCCUs operate at an optimum conversion level corresponding to a given feed rate, feed quality, set of processing objectives, and catalyst at one or more unit constraints (e.g., wet gas compressor capacity, fractionation capacity, air blower capacity, reactor temperature, regenerator temperature, and catalyst circulation). Once the optimum conversion level is found, there are very few additional degrees of freedom for changing the operating variables.

6.3.4.3 Additives

In addition to what to cracking catalyst described earlier, a series of additives has been developed that catalyze or otherwise alter the primary activity/selectivity of the catalyst or act as pollution control agents. Additives are most often prepared in microspherical form to be compatible with the primary catalysts and are available separately in compositions that (1) enhance gasoline octane and low boiling olefin formation, (2) selectively crack high-boiling cycle oil, (3) passivate vanadium and nickel present in many high-boiling feedstocks, (4) oxidize coke to carbon dioxide, and (5) reduce sulfur dioxide emissions.

Both vanadium and nickel deposit on the cracking catalyst and are extremely deleterious when present in excess of 3000 ppm on the catalyst. Formulation changes to the catalyst can improve tolerance to vanadium and nickel, but the use of additives that specifically passivate either metal is often preferred.

6.3.4.4 Coking

The formation of coke deposits has been observed in virtually every unit in operation, and the deposits can be very thick with thicknesses up to 4 ft. that have been reported (McPherson, 1984). Coke has been observed to form where condensation of hydrocarbon vapors occurs. The reactor walls and plenum offer a colder surface where hydrocarbon derivatives can condense. Higher boiling constituents in the feedstock may be very close to their dew point, and they will readily condense and form coke nucleation sites on even slightly cooler surfaces.

Unvaporized feed droplets readily collect to form coke precursors on any available surface since the high-boiling feedstock constituents do not vaporize at the mixing zone of the riser. Thus it is not surprising that residuum processing makes this problem even worse. Low residence time cracking also contributes to coke deposits since there is less time for heat to transfer to feed droplets and vaporize them. This is an observation in line with the increase in coking when short contact time riser crackers (*q.v.*) were replacing the longer residence time fluid-bed reactors.

Higher boiling feedstocks that have high aromaticity result in higher yields of coke. Furthermore, polynuclear aromatics and aromatics containing heteroatoms (i.e., nitrogen, oxygen, and sulfur) are more facile coke makers than simpler aromatics (Speight, 1987). However, feed quality alone is not a foolproof method of predicting where coking will occur. However, it is known that feedstock hydrotreaters rarely have coking problems. The hydrotreating step mitigates the effect of the coke formers, and coke formation is diminished.

The recognition that significant postriser cracking occurs in commercial catalytic cracking units resulting in substantial production of dry gas and other low-valued products (Avidan and Krambeck, 1990). There are two mechanisms by which this postriser cracking occurs, thermal and dilute-phase catalytic cracking.

Thermal cracking results from extended residence times of hydrocarbon vapors in the reactor disengaging area and leads to high dry gas yields via nonselective free radical cracking mechanisms. On the other hand, dilute-phase catalytic cracking results from extended contact between catalyst and hydrocarbon vapors downstream of the riser. While much of this undesirable cracking was eliminated in the transition from bed to riser cracking, there is still a substantial amount of nonselective cracking occurring in the dilute phase due to the significant catalyst holdup.

Many catalytic cracking units are equipped with advanced riser termination systems to minimize postriser cracking (Long et al., 1993). However, due to the complexity and diversity of catalytic cracking units, there are many variations of these systems and many such as closed cyclones and many designs are specific to the unit configuration but all serve the same fundamental purpose of reducing the undesirable postriser reactions. Furthermore, there are many options for taking advantage of reduced postriser cracking to improve yields. A combination of higher reactor temperature, higher cat/oil ratio, higher feed rate, and/or poorer quality feed is typically employed. Catalyst modification is also appropriate, and typical catalyst objectives such as low coke and dry gas selectivity are reduced in importance due to the process changes, while other features such as activity stability and bottoms cracking selectivity become more important for the new unit constraints.

Certain catalyst types seem to increase the formation of coke deposit. For example, these catalysts (some rare earth zeolites) tend to form aromatics from naphthenes as a result of secondary hydrogen transfer reactions, and the catalysts contribute to coke formation indirectly because the products that they produce have a greater tendency to be coke precursors. In addition, high zeolite content, low surface areas cracking catalysts are less efficient at heavy crude oil cracking than many amorphous catalysts because the nonzeolite catalysts contained a matrix that was better able to crack heavy crude oils and convert the coke precursors. The active matrix of some modern catalysts serves the same function.

Once coke is formed, it is matter of where it will appear. Coke deposits are most often found in the reactor (or disengager), transfer line, and slurry circuit and cause major problems in some units such as increased pressure drops, when a layer of coke reduces the flow through a pipe, or plugging, when chunks of coke spall off and block the flow completely. Deposited coke is commonly observed in the reactor as a black deposit on the surface of the cyclone barrels, reactor dome, and walls. Coke is also often deposited on the cyclone barrels 180 degrees away from the inlet. Coking within the cyclones can be potentially very troublesome since any coke spalls going down into the dip leg could restrict catalyst flow or jam the flapper valve. Either situation reduces cyclone efficiency that leads to an increase catalyst losses from the reactor. Coke formation also occurs at nozzles that can increase the nozzle pressure drop. It is possible for steam or instrument nozzles to be plugged completely, a serious problem in the case of unit instrumentation.

Coking in the transfer line between the reactor and main fractionator is also common, especially at the elbow where it enters the fractionator. Transfer line coking causes pressure drop and spalling and can lead to reduced throughput. Furthermore, any coke in the transfer line that spalls off can pass through the fractionator into the circulating slurry system where it is likely to plug up exchangers, resulting in lower slurry circulation rates and reduced heat removal. Pressure balance is obviously affected if the reactor has to be run at higher pressures to compensate for transfer line coking. On units where circulation is limited by low slide valve differentials, coke laydown may then indirectly reduce catalyst circulation. The risk of a flow reversal is also increased. In units with reactor grids, coking increases grid pressure drop, which can directly affect the catalyst circulation rate.

Shutdowns and start-ups can aggravate problems due to coking. The thermal cycling leads to differential expansion and contraction between the coke and the metal wall that will often cause the coke to spall in large pieces. Another hazard during shutdowns is the possibility of an internal fire when the unit is opened up to the atmosphere. Proper shutdown procedures that ensure the internals have sufficiently cooled before air enters the reactor will eliminate this problem. In fact, the only defense against having coke plugging problems during start-up is to thoroughly clean the unit during the turnaround and remove all the coke. If strainers on the line(s), they will have to be cleaned frequently.

The two basic principles to minimize coking are to avoid dead spots and prevent heat losses. An example of minimizing dead spots is using purge steam to sweep out stagnant areas in the disengager system. The steam prevents collection of high-boiling condensable products in the cooler regions. Steam also provides a reduced partial pressure or steam distillation effect on the high-boiling constituents and causes enhanced vaporization at lower temperatures. Steam for purging should preferably be superheated since medium-pressure low-velocity steam in small pipes with high heat losses is likely to be very wet at the point of injection and will cause more problems. Cold spots are often caused by heat loss through the walls in which case increased thermal resistance might help reduce coking. The transfer line, being a common source of coke deposits, should be as heavily insulated as possible, provided that stress related problems have been taken into consideration.

In some cases changing catalyst type or the use of an additive (*q.v.*) can alleviate coking problems. The catalyst types that appear to result in the least coke formation (not delta coke or catalytic coke) contain low or zero-earth zeolites with moderate matrix activities. Eliminating high-boiling recycle streams can lead to reduced coke formation. Since clarified oil is a desirable feedstock to make needle coke in a coker, then it must also be a potential coke maker in the disengager.

One of the trends in recent years has been to improve product yields by means of better feed atomization. The ultimate objective is to produce an oil droplet small enough so that a single particle of catalyst will have sufficient energy to vaporize it. This has the double benefit of improving cracking selectivity and reducing the number of liquid droplets that can collect to form coke nucleation sites.

6.4 Catalysts

Cracking crude oil fractions occurs over many types of catalytic materials, but high yields of desirable products are obtained with hydrated aluminum silicates. These may be either activated (acid-treated natural clays of the bentonite type) or synthesized silica–alumina or silica–magnesia preparations. Their activity to yield essentially the same products may be enhanced to some extent by the incorporation of small amounts of other materials, such as the oxides of zirconium (zirconia, ZrO_2), boron (boria, B_2O_3, which has a tendency to volatilize away on use), and thorium (thoria, ThO_2). Both the natural and the synthetic catalysts can be used as pellets or beads and also in the form of powder; in either case, replacements are necessary because of attrition and gradual loss of efficiency (DeCroocq, 1984; Parkash, 2003; Gary et al., 2007; Speight, 2014, 2017; Hsu and Robinson, 2017).

6.4.1 Catalyst properties

The catalysts must be stable to physical impact loading and thermal shocks and must withstand the action of carbon dioxide, air, nitrogen compounds, and steam. They should also be resistant to sulfur compounds; the synthetic catalysts and certain selected clays appear to be better in this regard than average untreated natural catalysts. The silica–alumina catalysts are reported to give the highest octane gasoline and silica–magnesia the largest yields, with the natural clays falling between them.

Neither silica (SiO_2) nor alumina (Al_2O_3) alone is effective in promoting catalytic cracking reactions. In fact, they (and also activated carbon) promote decomposition of hydrocarbon constituents that match the thermal decomposition patterns. Mixtures of anhydrous silica and alumina (SiO_2 and Al_2O_3) or anhydrous silica with hydrated alumina ($2SiO_2 \cdot 2Al_2O_3 \cdot 6H_2O$) are also essentially not effective. A catalyst having appreciable cracking activity is obtained only when prepared from hydrous oxides followed by partial dehydration (calcining). The small amount of water remaining is necessary for proper functioning.

The catalysts are porous and highly adsorptive, and their performance is affected markedly by the method of preparation. Two catalysts that are chemically identical but have pores of different size and distribution may have different activity, selectivity, temperature coefficient of reaction rate, and response to poisons. The intrinsic chemistry and catalytic action of a surface may be independent of pore size, but small pores appear to produce different effects because of the manner and time in which hydrocarbon vapors are transported into and out of the interstices.

Commercial synthetic catalysts are amorphous and contain more silica than is called for by the preceding formulae; they are generally composed of 10%–15% alumina (Al_2O_3) and 85%–90% silica (SiO_2). The natural materials, montmorillonite, a nonswelling bentonite, and halloysite, are hydrosilicates of aluminum, with a well-defined crystal structure and approximate composition of Al_2O_3 $4Si_2O \cdot xH_2O$. Some of the newer catalysts contain up to 25% alumina and are reputed to have a longer active life.

Commercially used cracking catalysts are insulator catalysts possessing strong acidic properties. They function as catalysts by altering the cracking process mechanisms through an alternative mechanism involving chemisorption by proton donation and desorption, resulting in cracked oil and theoretically restored catalyst. Thus it is not surprising that all cracking catalysts are poisoned by proton-accepting vanadium.

The catalyst–oil volume ratios range from 5:1 to 30:1 for the different processes, although most processes are operated to 10:1. However, for moving-bed processes the catalyst–oil volume ratios may be substantially lower than 10:1.

6.4.2 Catalyst variables

The primary variables available to the operation of FCCUs for maximum unit conversion for a given feedstock quality include catalytic variables such as (1) catalyst activity and (2) catalyst design that includes availability of cracking sites and the presence of carbon on the regenerated catalyst.

The equilibrium catalyst activity, as measured by a microactivity test (MAT), is a measure of the availability of zeolite and active matrix cracking sites for conversion. Therefore an increase in the unit activity can effect an increase in conversion and activity is increased by one or a combination of (1) increased fresh catalyst addition rate; (2) increased fresh catalyst zeolite activity; (3) increased fresh catalyst matrix activity; (4) addition of catalyst additives to trap or passivate the deleterious effects of feed nitrogen, alkalis (i.e., calcium and sodium), vanadium, and other feed metal contaminants; and (5) increased fresh catalyst matrix surface area to trap or remove feedstock contaminants.

In general, a two-digit increase in the activity as determined by the MAT activity appears to coincide with a 1% absolute increase in conversion. The increased matrix surface area improves conversion by providing more amorphous sites for cracking high-boiling range compounds in the feedstock that cannot be cracked by the zeolite. Increased zeolite, on the other hand, provides the necessary acid cracking sites for selectively cracking the amorphous cracked high-boiling compounds and lower boiling compounds.

In addition to zeolite and matrix activity, many of the physical and chemical properties of the catalyst (catalyst design) contribute to increased conversion through selectivity differences. These include zeolite type, pore size distribution, relative matrix to total surface area, and chemical composition.

Increasing the concentration of catalyst in the reactor, often referred to as cat/oil ratio, will increase the availability of cracking for maximum conversion, assuming the unit is not already operating at a catalyst circulation limit. This can be achieved by increasing reactor heat load or switching to a lower coke selective (i.e., lower delta coke) catalyst. Reactor heat load can be raised by increased reactor temperature or lower feed preheat temperature. This, in turn, increases the cat/oil ratio to maintain the unit in heat balance.

The lower the carbon on regenerated catalyst, the higher the availability of cracking sites since less coke is blocking acid cracking sites. The carbon on the regenerated catalyst is reduced by increasing regeneration efficiency through the use of carbon monoxide oxidation promoters. Carbon on the regenerated catalyst can also be reduced by more efficient air and spent catalyst contact. Increased regenerator bed levels also reduce the amount of carbon on the regenerated catalyst through increased residence time, but this must be traded off with reduced dilute-phase disengager residence time and the possibility for increased catalyst losses.

6.4.3 Catalyst treatment

The latest technique developed by the refining industry to increase gasoline yield and quality to treat the catalysts from the cracking units to remove metal poisons that accumulate on the catalyst (Gerber et al., 1999). Nickel, vanadium, iron, and copper compounds contained in catalytic cracking feedstocks are deposited on the catalyst during the cracking operation, thereby adversely affecting both catalyst activity and selectivity. Increased catalyst metal contents affect catalytic cracking yields by increasing coke formation, decreasing gasoline and butane and butylene production, and increasing hydrogen production.

The recent commercial development and adoption of cracking catalyst–treating processes definitely improve the overall catalytic cracking process economics.

6.4.3.1 Demet process

A cracking catalyst is subjected to two pretreatment steps. The first step effects vanadium removal and the second, nickel removal, to prepare the metals on the catalyst for chemical conversion to compounds (chemical treatment step) that can be readily removed through water washing (catalyst wash step). The treatment steps include the use of a sulfurous compound followed by chlorination with an anhydrous chlorinating agent (e.g., chlorine gas) and washing with an aqueous solution of a chelating agent [e.g., citric acid, $HO_2CCH_2C(OH)(CO_2H)CH_2CO_2H$, 2-hydroxy-1,2,3-propane tricarboxylic acid]. The catalyst is then dried and further treated before returning to the cracking unit.

6.4.3.2 Met-X process

This process consists of cooling, mixing, and ion-exchange separation, filtration, and resin regeneration. Moist catalyst from the filter is dispersed in oil and returned to the cracking reactor in a slurry. On a continuous basis the catalyst from a cracking unit is cooled and then transported to a stirred reactor and mixed with an ion-exchange resin (introduced as slurry). The catalyst–resin slurry then flows to an elutriator for separation. The catalyst slurry is taken overhead to a filter, and the wet filter cake is slurried with oil and pumped into the catalytically cracked feedstock system. The resin leaves the bottom of the elutriator and is regenerated before returning to the reactor.

6.5 Process options for heavy feedstocks

The processes described next are the evolutionary offspring of the fluid catalytic cracking and the residuum catalytic cracking processes. Some of these newer processes use catalysts with different silica/alumina ratios as acid support of metals such as Mo, Co, Ni, and W. In general, the first catalyst used to remove metals from oils was the conventional HDS catalyst. Diverse natural minerals are also used as raw material for elaborating catalysts addressed to the upgrading of high-boiling fractions. Among these minerals are (1) clay minerals; (2) manganese nodules; (3) bauxite activated with vanadium, nickel, chromium, iron, and cobalt, as well as and iron laterites, sepiolite minerals; and (4) mineral nickel and transition metal sulfides supported on silica and alumina. Other kinds of catalysts, such as vanadium sulfide, are generated in situ, possibly in colloidal states.

In the past decades, in spite of the difficulty of handling the viscous heavy feedstocks (such as heavy crude oil, extra heavy oil, and tar sand bitumen), residuum fluidized catalytic cracking (RFCC) has evolved to become a well-established approach for converting a significant portion of the higher boiling fractions of the crude barrel into a high-octane gasoline blending component. RFCC, which is an extension of conventional fluid catalytic cracking technology for applications involving the conversion of highly contaminated residua, has been commercially proven on feedstocks ranging from gas oil–residuum blends to atmospheric residua, as well as blends of atmospheric and vacuum residua blends. In addition to high gasoline yields, the RFCC unit also produces gaseous, distillate, and fuel oil-range products.

The product quality from the residuum fluidized catalytic cracker is directly affected by its feedstock quality. In particular, and unlike hydrotreating, the RFCC redistributes sulfur among the various products but does not remove sulfur from the products unless, of course, one discount the sulfur that is retained by ay coke formed on the catalyst. Consequently, tightening product specifications have forced refiners to hydrotreat some, or all, of the products from the resid cat cracking unit. Similarly, in the future, the emissions of sulfur oxides (SO_x) from a resid cat cracker may become more of an obstacle for residue conversion projects. For these reasons a point can be reached where the economic operability of the unit can be sufficient to justify hydrotreating the feedstock to the cat cracker.

As an integrated conversion block, residue hydrotreating and RFCC complement each other and can offset many of the inherent deficiencies related to residue conversion.

6.5.1 Asphalt residual treating process

The asphalt residual treating (ART) process is a process for increasing the production of transportation fuels and reduces high-boiling fuel oil production, without hydrocracking (Parkash, 2003; Gary et al., 2007; Speight, 2014, 2017; Hsu and Robinson, 2017).

In the process the preheated feedstock (which may be whole crude, atmospheric residuum, vacuum residuum, or tar sand bitumen) is injected into a stream of fluidized, hot catalyst (trade name: ArtCat). Complete mixing of the feedstock with the catalyst is achieved in the contactor, which is operated within a pressure–temperature envelope to ensure selective vaporization. The vapor and the contactor effluent are quickly and efficiently separated from each other, and entrained hydrocarbon derivatives are stripped from the contaminant (containing spent solid) in the stripping section. The contactor vapor effluent and vapor from the stripping section are combined and rapidly quenched in a quench drum to minimize product degradation. The cooled products are then transported to a conventional fractionator that is similar to that found in a FCCU. Spent solid from the stripping section is transported to the combustor bottom zone for carbon burn-off.

In the combustor, coke is burned from the spent solid that is then separated from combustion gas in the surge vessel. The surge vessel circulates regenerated catalyst streams to the contactor inlet for feed vaporization and to the combustor bottom zone for premixing.

The components of the combustion gases include carbon dioxide (CO_2), nitrogen (N_2), oxygen (O_2), sulfur oxides (SO_x), and nitrogen oxides (NO_x) that are released from the catalyst with the combustion of the coke in the combustor. The concentration of sulfur oxides in the combustion gas requires treatment for their removal.

6.5.2 Residue fluid catalytic cracking process

The residue fluid catalytic cracking process (HOC process) is a version of the fluid catalytic cracking process that has been adapted to conversion of residua that contain high amounts of metal and asphaltene constituents (Parkash, 2003; Gary et al., 2007; Speight, 2014, 2017; Hsu and Robinson, 2017).

In the process a residuum is desulfurized and the nonvolatile fraction from the HDS unit is charged to the residuum FCCU. The reaction system is an external vertical riser terminating in a closed cyclone system. Dispersion steam in amounts higher than that used for gas oils is used to assist in the vaporization of any volatile constituents of viscous feedstocks (such as heavy crude oil, extra heavy oil, and tar sand bitumen).

A two-stage stripper is utilized to remove hydrocarbon derivatives from the catalyst. Hot catalyst flows at low velocity in dense phase through the catalyst cooler and returns to the regenerator. Regenerated catalyst flows to the bottom of the riser to meet the feed.

The coke deposited on the catalyst is burned off in the regenerator along with the coke formed during the cracking of the gas oil fraction. If the feedstock contains high proportions of metals, control of the metals on the catalyst requires excessive amounts of catalyst withdrawal and fresh catalyst addition. This problem can be addressed by feedstock pretreatment.

The feedstocks for the process are rated on the basis of carbon residue and content of metals. Thus good-quality feedstocks have less than 5% by weight carbon residue and less than 10 ppm metals. Medium-quality feedstocks have greater than 5% but less than 10% by weight carbon residue and greater than 10 but less than 30 ppm metals. Poor-quality feedstocks have greater than 10 but less than 20% by weight carbon residue and greater than 30 but less than 150 ppm metals. Finally, bad-quality feedstocks have greater than 20% by weight carbon residue and greater than 150 ppm metals. One might question the value of this rating of the feedstocks for the HOC process since these feedstock ratings can apply to virtually many fluid catalytic cracking processes.

6.5.3 Heavy oil treating process

The heavy oil treating (HOT) process is a catalytic cracking process for upgrading viscous feedstocks such as topped crude oil, heavy oil, extra heavy oil, tar sand bitumen, vacuum residua, and solvent deasphalted bottoms using a fluidized bed of iron ore particles (Parkash, 2003; Gary et al., 2007; Speight, 2014, 2017; Hsu and Robinson, 2017).

The main section of the process consists of three fluidized reactors, and separate reactions take place in each reactor (cracker, regenerator, and desulfurizer):

Fe_3O_4 + asphaltene constituents → coke/Fe_3O_4 + oil + gas(in the cracker)

$3FeO + H_2O \rightarrow Fe_3O_4 + H_2$(in the cracker)

Coke/$Fe_3O_4 + O_2 \rightarrow 3FeO + CO + CO_2$(in the regenerator)

$FeO + SO_2 + 3CO \rightarrow FeS + 3CO_2$(in the regenerator)

$3FeS + 5O_2 \rightarrow Fe_3O_4 + 3SO_2$(in the desulfurizer)

In the reactor cracking and the steam–iron reaction takes place simultaneously under the conditions similar usual to thermal cracking. Any unconverted feedstock is recycled to the cracker from the bottom of the scrubber. The scrubber effluent is separated into hydrogen gas, LPG, and liquid products that can be upgraded by conventional technologies to priority products.

In the regenerator, coke deposited on the catalyst is partially burned to form carbon monoxide in order to reduce iron tetroxide and to act as a heat supply. In the desulfurizer, sulfur in the solid catalyst is removed and recovered as molten sulfur in the final recovery stage.

6.5.4 R2R process

The R2R process is a fluid catalytic cracking process for conversion of viscous feedstocks (Heinrich and Mauleon, 1994; Inai, 1994; Parkash, 2003; Gary et al., 2007; Speight, 2014, 2017; Hsu and Robinson, 2017). In the process, the feedstock is vaporized upon contacting hot regenerated catalyst at the base of the riser and lifts the catalyst into the reactor vessel separation chamber where rapid disengagement of the hydrocarbon vapors from the catalyst is accomplished by both a special solids separator and cyclones. The bulk of the cracking reactions takes place at the moment of contact and continues as the catalyst and hydrocarbon derivatives travel up the riser. The reaction products, along with a minute amount of entrained catalyst, then flow to the fractionation column. The stripped spent catalyst, deactivated with coke, flows into the Number 1 regenerator.

Partially regenerated catalyst is pneumatically transferred via an air riser to the Number 2 regenerator, where the remaining carbon is completely burned in a dryer atmosphere. This regenerator is designed to minimize catalyst inventory and residence time at high temperature while optimizing the coke-burning rate. Flue gases pass through external cyclones to a waste heat recovery system. Regenerated catalyst flows into a withdrawal well and after stabilization is charged back to the oil riser.

6.5.5 Reduced crude oil conversion process

In recent years, because of a trend for low-boiling products, most refineries perform the operation by partially blending residua into vacuum gas oil. However, conventional fluid catalytic cracking processes have limits in residue processing, so residue fluid catalytic cracking processes have lately been employed one after another. Because the residue fluid catalytic cracking process enables efficient gasoline production directly from residues, it will play the most important role as a residue cracking process, along with the residue hydroconversion process. Another role of the residuum fluid catalytic cracking process is to generate high-quality gasoline blending stock and petrochemical feedstock. Olefins (propene, butenes, and pentenes) serve as feed for alkylating processes, for polymer gasoline, as well as for additives for reformulated gasoline.

In the reduced crude conversion process (RCC process), the clean regenerated catalyst enters the bottom of the reactor riser where it contacts low-boiling hydrocarbon lift gas that accelerates the catalyst up the riser prior to feed injection (Hydrocarbon Processing, 1996; Parkash, 2003; Gary et al., 2007; Speight, 2014, 2017; Hsu and Robinson, 2017). At the top of the lift gas zone, the feed is injected through a series of nozzles located around the circumference of the reactor riser.

The catalyst/oil disengaging system is designed to separate the catalyst from the reaction products and then rapidly remove the reaction products from the reactor vessel. Spent catalyst from reaction zone is first steam stripped, to remove adsorbed hydrocarbon, and then routed to the regenerator. In the regenerator, all of the carbonaceous deposits are removed from the catalyst by combustion, restoring the catalyst to an active state with a very low carbon content. The catalyst is then returned to the bottom of the reactor riser at a controlled rate to achieve the desired conversion and selectivity to the primary products.

6.5.6 Shell fluid catalytic cracker process

The shell FCC process is designed to maximize the production of distillates from residua (Table 6.4) (Parkash, 2003; Gary et al., 2007; Speight, 2014, 2017; Hsu and Robinson, 2017). In the process the preheated feedstock (vacuum gas oil, atmospheric residuum) is mixed with the hot regenerated catalyst. After reaction in a

Table 6.4 Feedstock and product data for the shell fluid catalytic cracker process.

Feedstock	Residuum[a]	Residuum[a]
API	18.2	13.4
Sulfur (wt.%)	1.1	1.3
Carbon residue (wt.%)	1.2	4.7
Products		
Gasoline (C5—221°C, C5—430°F) (wt.%)	49.5	46.2
Light cycle oil (221°C–370°C, 430°F–700°F) (wt.%)	20.1	19.1
Heavy cycle oil (> 370°C, >700°F) (wt.%)	5.9	10.8
Coke (wt.%)	5.9	7.6

[a]Unspecified.

riser, volatile materials and catalyst are separated after which the spent catalyst is immediately stripped of entrained and adsorbed hydrocarbon derivatives in a very effective multistage stripper. The stripped catalyst gravitates through a short standpipe into a single vessel, simple, reliable, and yet efficient catalyst regenerator. Regenerative flue gas passes via a cyclone/swirl tube combination to a power recovery turbine. From the expander turbine the heat in the flue gas is further recovered in a waste heat boiler. Depending on the environmental conservation requirements, a reduction in the nitrogen oxide (NO_x) emissions, the sulfur oxide (SO_x), and emission of particulate matter can be achieved by inclusion of an emission control device in the flue gas train.

There is a claim that feedstock pretreatment of bitumen (by hydrogenation) prior to fluid catalytic cracking (or for that matter any catalytic cracking process) can result in enhanced yield of naphtha. It is suggested that mild hydrotreating be carried out upstream of a FCCU to provide an increase in yield and quality of distillate products. This is in keeping with earlier work (Speight and Moschopedis, 1979) where mild hydrotreating of bitumen was reported to produce low-sulfur liquids that would be amenable to further catalytic processing.

6.5.7 S&W fluid catalytic cracking process

The S&W FCC process is also designed to maximize the production of distillates from residua (Table 6.5) (Parkash, 2003; Gary et al., 2007; Speight, 2014, 2017; Hsu and Robinson, 2017). In the process the viscous feedstock is injected into a stabilized, upward flowing catalyst stream whereupon the feedstock–steam–catalyst mixture travels up the riser and is separated by a high-efficiency inertial separator. The product vapor goes overhead to the main fractionator.

The spent catalyst is immediately stripped in a staged, baffled stripper to minimize hydrocarbon carryover to the regenerator system. The first regenerator

Table 6.5 Feedstock and product data for the S&W fluid catalytic cracker process.

Feedstock	Residuum[a]	Residuum[a]
API	24.1	22.3
Sulfur (wt.%)	0.8	1.0
Carbon residue (wt.%)	4.4	6.5
Products		
Naphtha (vol.%)	61.5	60.2
Light cycle oil (vol.%)	16.6	17.5
Heavy cycle oil (vol.%)	5.6	6.6
Coke (wt.%)	7.1	7.8
Conversion (vol.%)	77.7	75.9

[a]Unspecified.

(650°C–700°C, 1200°F–1290°F) burns 50%–70% of the coke in an incomplete carbon monoxide combustion mode running countercurrently. This relatively mild, partial regeneration step minimizes the significant contribution of hydrothermal catalyst deactivation. The remaining coke is burned in the second regenerator (approximately 775°C, 1425°F) with an extremely low steam content. Hot, clean catalyst enters a withdrawal well that stabilizes its fluid qualities prior to being returned to the reaction system.

6.6 The future

Catalytic cracking is widely used in the crude oil refining industry to convert heavy crude oils into more valuable gasoline and lower boiling products. As the demand for higher octane gasoline has increased, catalytic cracking has replaced thermal cracking. Two of the most intensive and commonly used catalytic cracking processes in crude oil refining are fluid catalytic cracking and hydrocracking. In the fluid catalytic cracking process, the fine, powdery catalyst (typically zeolites, which have an average particle size of approximately 70 μm) takes on the properties of a fluid when it is mixed with the vaporized feed. Fluidized catalyst circulates continuously between the reaction zone and the regeneration zone.

Several process innovations have been introduced in the form of varying process options, some using piggyback techniques (where one process works in close conjunction with another process; see previously), there are other options that have not yet been introduced or even invented but may well fit into the refinery of the future.

In the FCC the major developments are in integration with sulfur removal to produce low-sulfur gasoline without octane loss. This development will build on the development of new catalysts (see previously).

Furthermore, recent enhancements made to RFCC units to permit feeding significant amounts of heavy crude oils, while simultaneously improving yields and service factors have focused on improved feed injection and dispersion, reduced contact time of products and catalyst, improved separation of products and catalyst, and regenerator heat removal. Traditional technology has been modified in key areas, including (1) catalyst design to accommodate higher metals feed and to minimize the amount of coke formed on the catalyst, (2) feed injection, (3) riser pipe design and catalyst/oil product separation to minimize overcracking, (4) regenerator design improvements to handle high coke output and avoid damage to catalyst structure, and (5) overall reactor/regenerator design concepts.

These developments have allowed FCCUs to substantially increase residue processing, and substantial portions of refinery residua are processes (as blends with gas oils) in fluidized bed units thereby increasing gasoline and diesel production. This will not only continue but also increase in the future. Furthermore, power recovery turbines can be installed on catalytic cracking units to produce power from the pressure of the off-gases of the catalytic cracking. The technology is commercially proven but is not in current general use.

Hydrotreating of the feedstock to the catalytic cracking unit can increase conversion by 8%–12% v/v and with most feeds, it will be possible to reduce the sulfur content of the gasoline/naphtha product to levels low enough to meet the future low-sulfur gasoline pool specifications.

With the increasing focus to reduce sulfur content in fuels, the role of desulfurization in the refinery becomes more and more important. Currently, the process of choice is the hydrotreater, in which hydrogen is added to the fuel to remove the sulfur from the fuel. Some hydrogen may be lost to reduce the octane number of the fuel, which is undesirable. Because of the increased attention for fuel desulfurization, various new process concepts are being developed with various claims of efficiency and effectiveness.

The major developments in desulfurization three main routes are advanced hydrotreating (new catalysts, catalytic distillation) and reactive adsorption (such as the use of metal oxides that will chemically abstract sulfur). Such concepts, already on stream in some refineries, the number of units performing chemical desulfurization will increase in the future.

There are also proposals (some of which have been put into practice) to bypass the atmospheric and vacuum distillation units by feeding crude oil directly into a thermal cracking process, which would provide sufficient flexibility to supply a varying need of products with a net energy savings.

Low-boiling cycle oil from catalytic cracking units will be increased by modifying feedstock composition, introducing improved catalysts and additives, and modifying operating conditions (such as recycle ratio, temperature, catalyst/oil ratio). The addition of an active alumina matrix is a common feature to help refiners increase low-boiling cycle oil production when cracking viscous feedstocks. There is some overlap in catalysts tailored for production of low-boiling cycle oil and catalysts tailored for residuum feedstocks.

Refineries equipped to process viscous feedstocks have, so far, reported high refining margins because they can take advantage of less-expensive heavy crude oils. Resid fluid catalytic cracking is an important component in the upgrading of such crudes, with unit profitability depending upon the extent to which high-boiling hydrocarbon derivatives in the feed are cracked into valuable products. The product slate, in turn, depends upon the feed characteristics, the catalyst, the hardware, and the operating conditions. Exemplifying a trend toward higher boiling feedstocks, the majority of the fluid-bed catalytic cracking units scheduled to start up before 2015 are expected to process high-boiling vacuum gas oil and/or or residuum feedstocks.

Complete process technologies for residuum catalytic cracking units are the most comprehensive approach to improve resid processing operations, but present, as might be anticipated, the most expensive. Product recycle and multiple reaction sections will be the prevalent technology trends. Improving feed injectors, riser termination and catalyst separation devices, strippers, and regenerator components are revamp options for existing units. As feedstocks get higher boiling, the trend toward a higher stripper residence time and, consequently, increased mass transfer between entrained hydrocarbon derivatives and steam will continue. Moreover, the role of the regenerator continues to evolve because of reduction requirements in the carbon dioxide emissions. The need for development of flue gas treating systems points to the need for continued innovation in that area.

Process/hardware technologies to improve low-boiling cycle oil yield from the FCCU will include improved feed injection systems and riser pipe and reaction zone designs. Finally, the use of biofeedstocks (such as animal fats and vegetable oils) in the fluid catalytic cracking will be used to increase the yield of low-boiling cycle oil and will also provide high-quality products in terms of cetane number.

In fact, over the next two decades, FCCUs will take on two additional roles—a user of biomass feedstocks and to reduce carbon dioxide emissions—to alleviate growing concerns over energy security and global warming.

The implementation of biofeedstock processing techniques in crude oil refineries can result in a competitive advantage for both refiners and society at large. First, the processes provide refineries with alternative feeds that are renewable and could be lower in cost than crude oil. Second, they can reduce the costs of producing fuels and chemicals from biofeedstocks by utilizing the existing production and distribution systems for crude oil–based products and avoiding the establishment of parallel systems. Finally, the use of biofeedstocks provide a production base for fuels and chemicals that is less threatened by changes in government policies toward fossil fuel feeds and renewable energies.

Biofeedstocks that are able to be processed in the FCCU can be categorized as biomass-derived oils (both lignocellulosic materials and free carbohydrates) or triglycerides and their free fatty acids. The operating conditions and catalysts used for each type of feed to achieve a desired product slate vary, and each feed comes with inherent advantages and disadvantages. Most of the research work completed to date has been performed on relatively pure biofeedstocks as opposed to blends of bio-based materials with traditional catalytic cracker feedstocks. Practical implementation in a refinery will, more than likely, be accompanied by blending with vacuum gas oil or residuum. In fact, biomass constituents can be blended with the feedstock and fed to a FCCU (United States Patent 6,066,751). The acidity of the oil (caused by the presence of oxygen functions) acidity can be reduced by means of a moderate thermal treatment at temperatures in the range of 320°C–420°C (610°F–790°F).

Oils derived from biofeedstocks oils are generally best upgraded by HZSM-5 or ZSM-5, as these zeolite catalysts promote high yields of liquid products and propylene. Unfortunately, these feeds tend to coke easily, and high acid numbers and undesirable by-products such as water and carbon are, and will continue to be, addi tional challenges. Waxy feeds obtained from biomass gasification followed by Fischer–Tropsch synthesis to hydrocarbon liquids and waxes (biomass-to-liquids) are especially suited for increasing low-boiling cycle oil production in the catalytic cracking unit, due to the high paraffinic character, low-sulfur content, and low-aromatics yield of the feed. A major disadvantage for biomass-to-liquid products is the intrinsically low coke yield that can disrupt the heat balance of the catalytic cracking unit. In terms of processability, triglycerides are the best suited biofeedstock for the catalytic cracker. These materials generally produce high-quality diesel, high-octane gasoline and are low in sulfur.

In general, biofeedstocks, catalytic cracker catalysts, or blends thereof achieve acceptable product yields and selectivity. As biofeedstock use in the catalytic cracking units continues toward commercialization, the market potential will begin to allow catalyst manufacturers to allocate resources for more tailored catalyst development, greatly improving the yield and quality of desired products.

In summary, the use of low-quality feedstocks (such the viscous crude oil feedstocks and nonfossil fuel feedstocks) are recognized as alternate sources for the production of high-value transportation fuels. Several process technologies have been developed to upgrade these feeds through fixed bed, moving bed, ebullated bed, slurry-phase reactor, or a combination. Hydrocracking in slurry-phase type reactor is a prominent technology to convert low value feeds into high-value transportation fuels and petrochemical products. The choice of a conversion technology is implemented by considering the feed properties, product demand, and economic benefit as well as environmental concerns.

Thus recent and forthcoming changes in the feedstocks accepted by refineries (Chapter 1: Feedstock Types and Properties, Chapter 10: Nonfossil Fuel Feedstocks), including renewables and crude oil from tight (low-permeability) formations, and longer term trends in the demand for propylene, gasoline, and middle distillates, require further developments in both catalyst and process. Detailed analyses of the catalyst pore structure, metal deposition, and zeolite deactivation are required, and the evolving analytical techniques will allow a more detailed analysis of the interaction between the matrix and the catalyst as well as a complete analysis of the pore system in the micropore and mesopore range. In situ characterization techniques will yield fundamental insights in the chemistry and dynamics of the process during individual cracking and regeneration cycles and will increase the understanding of the deactivation of the catalyst as a function of metal deposition, steaming, and coke laydown (Vogt and Weckhuysen, 2015).

It must be recognized that it is extremely important to create a clear connection between the macroscopic world of catalyst testing and real-unit performance and the microscopic world. The present microscopic and spectroscopic tools rely on the analysis of a limited number, often not more than a handful, of FCC catalyst particles, whereas the industrial performance takes place at many orders of magnitude larger. This gap needs to be bridged and proper tools will have to be developed to make this possible. These insights will allow us to fine-tune the catalyst performance in the directions required by the large-scale trends in raw material availability and product demands.

Thus even though the fluid catalytic cracking process has been onsteam for more than 70 years now, the process is still being developed further. Changes in the demand for products, and changes in the feedstock drive constant development. The following challenges need to be met for the future success of the process (1) light cycle oil maximization; (2) petrochemical feedstock maximization, such as the production of propylene; (3) flue gas emissions control; and (4) enhanced tolerance of the catalyst for metals and coke.

References

Avidan, A.A., Krambeck, F.J., 1990. FCC closed cyclone system eliminates post riser cracking. In: Proceedings. Annual Meeting. National Petrochemical and Refiners Association.

Bartholic, D.B., 1981a. United States Patent 4,243,514. Preparation of FCC Charge from Residual Fractions. January 6.

Bartholic, D.B., 1981b. United States Patent 4,263,128. Upgrading Petroleum and Residual Fractions Thereof. April 21.

Bartholic, D.B., 1989. United States Patent 4,804,459. Process for Upgrading Tar Sand Bitumen. February 14.

Bradley, S.A., Gattuso, M.J., Bertolacini, R.J., 1989. Characterization and Catalyst Development. Symposium Series No. 411. American Chemical Society, Washington, DC.

DeCroocq, D., 1984. Catalytic Cracking of Heavy Petroleum Hydrocarbons. Editions Technip, Paris.

Gary, J.H., Handwerk, G.E., Kaiser, M.J., 2007. Petroleum Refining: Technology and Economics, fifth ed. CRC Press, Taylor & Francis Group, Boca Raton, FL.

Gerber, M.A., Fulton, J.L., Frye, J.G., Silva, L.J., Bowman, L.E., Wai, C.M., 1999. Regeneration of Hydrotreating and FCC catalysts. In: Report No. PNNL-13025. US Department of Energy Contract No. DE-AC06-76RLO 1830. Pacific Northwest National Laboratory, Richland, Washington, DC.

Heinrich, G., Mauleon, J.-L., 1994. The R2R process: 21st century FCC technology. Revue Institut Français du Pétrole 49 (5), 509–520.

Hsu, C.S., Robinson, P.R. (Eds.), 2017. Handbook of Petroleum Technology. Springer International Publishing AG, Cham.

Inai, K., 1994. Operation results of the R2R process. Revue Institut Français du Pétrole 49 (5), 521–527.

Long, S.L., Johnson, A.R., Dharia, D., 1993. Advances in residual oil FCC. In: Proceedings. Annual Meeting. National Petrochemical and Refiners Association. Paper AM-93-50.

McPherson, L.J., 1984. Causes of FCC Reactor Coke Deposits Identified. Oil & Gas Journal. September 10, p. 139.

Parkash, S., 2003. Refining processes handbook. Gulf Professional Publishing. Elsevier, Amsterdam.

Sadeghbeigi, R., 1995. Fluid Catalytic Cracking: Design, Operation, and Troubleshooting of FCC Facilities. Gulf Publishing Company, Houston, TX.

Speight, J.G., 1986. Upgrading heavy feedstocks. Annu. Rev. Energy 11, 253.

Speight, J.G., 1987. Initial reactions in the coking of residua. Preprints. Am. Chem. Soc., Div. Petrol. Chem 32 (2), 413.

Speight, J.G., 2000. The Desulfurization of Heavy Oils and Residua, second ed. Marcel Dekker Inc, New York.

Speight, J.G., 2004. New approaches to hydroprocessing. Catal. Today 98 (1–2), 55–60.

Speight, J.G., 2014. The Chemistry and Technology of Petroleum, fifth ed. CRC Press, Taylor & Francis Group, Boca Raton, FL.

Speight, J.G., 2017. Handbook of Petroleum Refining. CRC Press, Taylor & Francis Group, Boca Raton, FL.

Speight, J.G., Moschopedis, S.E., 1979. The production of low-sulfur liquids and coke from Athabasca bitumen. Fuel Process. Technol. 2, 295.

Vogt, E.T.C., Weckhuysen, B.M., 2015. Fluid catalytic cracking: recent developments on the grand old lady of zeolite catalysis. Chem. Soc. Rev. 44, 7342–7370.

Deasphalting and dewaxing

7

7.1 Introduction

Solvent deasphalting is essentially a solvent-based extraction process and the required solvent is usually available within the refinery. The process separates oil from carbon-rich components, resins, and asphaltenes and makes it available to convert it to lube stock or as feedstock for other secondary processing facilities. Feedstock impurities such as sulfur and metals are concentrated in the insoluble phase. The flexibility, inherent in the process, allows wide variation in product quality to meet specific downstream process needs. The process continues to play a significant role, as one of the key process units of the modern refinery complex. Installed capacity for solvent deasphalting units worldwide is now in excess of over 140 million barrels per annum (Dickenson et al., 1997).

Solvent-based processes have been a major part of refinery operations for several decades and continue to offer opportunities for the future but are not often appreciated for the tasks for which they are used (Speight and Ozum, 2002; Sattarin et al., 2006).

In the solvent *deasphalting processes* an alkane is injected into the feedstock (such as a residuum) to disrupt the dispersion of components and causes the polar constituents to precipitate. Propane (or sometimes propane/butane mixtures) is extensively used for deasphalting and produces a deasphalted oil (DAO) and propane deasphalter asphalt (PDA or PD tar) (Dunning and Moore, 1957). The deasphalter asphalt contains most of the contaminants originally in the feedstock, for example, nickel, vanadium, and coke precursors. The DAO is normally used as feedstocks for a fluid catalytic cracking unit or for a hydrocracking unit. In recent years, attention has focused on maximizing the yield of the DAO and the quality of the asphalt for use as fuel to a circulating fluidized bed boiler.

Propane has unique solvent properties; at lower temperatures (38°C–60°C; 100°F–140°F), paraffins are very soluble in propane and at higher temperatures (about 93°C; 200°F) all hydrocarbons are almost insoluble in propane.

Solvent deasphalting processes have not realized their maximum potential. With on-going improvements in energy efficiency, such processes would display its effects in a combination with other processes. Solvent deasphalting allows removal of sulfur and nitrogen compounds as well as metallic constituents by balancing yield with the desired feedstock properties (Ditman, 1973).

In *dewaxing processes* the process is designed to remove wax from lubricating oils to give the product good fluidity characteristics at low temperatures (e.g., low pour points) rather than from the whole crude oil, as is the case with deasphalting

The Refinery of the Future. DOI: https://doi.org/10.1016/B978-0-12-816994-0.00007-5

processes. The feedstock is treated with a solvent such as methyl-ethyl-ketone (MEK) to remove this wax before further processing. The mechanism of solvent dewaxing involves either the separation of wax as a solid that crystallizes from the oil solution at low temperature or the separation of wax as a liquid that is extracted at temperatures above the melting point of the wax through preferential selectivity of the solvent. However, the former mechanism is the usual basis for commercial dewaxing processes.

7.2 Commercial deasphalting processes

Petroleum processing normally involves separation into various fractions that require further processing in order to produce marketable products. The initial separation process is distillation (Chapter 15: The Refinery of the Future and Technology Integration) in which crude oil is separated into fractions of increasingly higher boiling range fractions. Since petroleum fractions are subject to thermal degradation, there is a limit to the temperatures that can be used in simple separation processes. The crude cannot be subjected to temperatures much above 395°C (740°F), irrespective of the residence time, without encountering some thermal cracking. Therefore to separate the higher molecular weight and higher boiling fractions from crude, special processing steps must be used.

Because petroleum fractions are subject to thermal degradation, there is a limit to the temperature which can be effectively used in a separation process; that is, crude cannot be subjected to temperatures above 345°C (650°F) without the occurrence of thermal cracking. The physical separation of higher molecular weight and higher boiling fractions from the crude is first accomplished in the vacuum distillation process, which is followed by the solvent deasphalting process.

Although a crude oil might be subjected to atmospheric distillation and vacuum distillation, there may still be some valuable oils left in the vacuum residuum. These valuable oils are recovered by solvent extraction and the first application of solvent extraction in refining was the recovery of heavy lube oil base stocks by propane (C_3H_8) deasphalting. In order to recover more oil from vacuum-reduced crude, mainly for catalytic cracking feedstocks, higher molecular weight solvents, such as butane (C_4H_{10}), and even pentane (C_5H_{12}), have been employed.

7.2.1 The deasphalting process

The deasphalting process is a mature process but, as refinery operations evolve, it is necessary to include a description of the process here so that the new processes might be compared with new options that also provide for deasphalting various feedstocks. Indeed, several of these options, such as the residuum oil supercritical extraction (ROSE) process have been on-stream for several years and are included here for this same reason. Thus this section provides a one-stop discussion of solvent recovery processes and their integration into refinery operations.

The separation of residua into oil and asphalt fractions was first performed on a production scale by mixing the vacuum residuum with propane (or mixtures of *normally gaseous* hydrocarbons) and continuously decanting the resulting phases in a suitable vessel. Temperature was maintained within about 55°C (100°F) of the critical temperature of the solvent, at a level that would regulate the yield and properties of the DAO in solution and that would reject the heavier undesirable components as asphalt.

Currently, deasphalting/delayed coking or deasphalting/fluid coking processes are used frequently for residuum conversion. The high demand for petroleum coke, mainly for use in the aluminum industry, has made delayed coking a major residuum conversion process. However, many crude oils will not produce coke meeting the sulfur and metals specifications for aluminum electrodes, and coke gas oils are less desirable feedstocks for fluid catalytic cracking than virgin gas oils. In comparison the solvent deasphalting process can apply to most vacuum residua. The DAO is an acceptable feedstock for both fluid catalytic cracking and, in some cases, hydrocracking. Since it is relatively less expensive to desulfurize the DAO than the heavy vacuum residuum, the solvent deasphalting process offers a more economical route for disposing of vacuum residuum from high-sulfur crude. However, the question of disposal of the asphalt remains. Use as a road asphalt is common and as a refinery fuel is less common since expensive stack gas clean-up facilities may be required when used as fuel.

In the process the feedstock is mixed with dilution solvent from the solvent accumulator and then cooled to the desired temperature before entering the extraction tower. Because of its high viscosity, the charge oil can neither be cooled easily to the required temperature nor will it mix readily with solvent in the extraction tower. By adding a relatively small portion of solvent upstream of the charge cooler (insufficient to cause phase separation), the viscosity problem is avoided.

The feedstock, with a small amount of solvent, enters the extraction tower at a point about two-thirds up the column. The solvent is pumped from the accumulator, cooled, and enters near the bottom of the tower. The extraction tower is a multistage contactor, normally equipped with baffle trays and the heavy oil flows downward while the light solvent flows upward. As the extraction progresses, the desired oil goes to the solvent and the asphalt separate and moves toward the bottom. As the extracted oil and solvent rise in the tower, the temperature is increased in order to control the quality of the product by providing adequate reflux for optimum separation.

Separation of oil from asphalt is controlled by maintaining a temperature gradient across the extraction tower and by varying the solvent/oil ratio. The tower top temperature is regulated by adjusting the feed inlet temperature and the steam flow to the heating coils in the top of the tower. The temperature at the bottom of the tower is maintained at the desired level by the temperature of the entering solvent. The DAO-solvent mixture flows from the top of the tower under pressure control to a kettle-type evaporator heated by low-pressure steam. The vaporized solvent flows through the condenser into the solvent accumulator.

The liquid phase flows from the bottom of the evaporator, under level control, to the DAO flash tower where it is reboiled by means of a fired heater. In the flash tower, most of the remaining solvent is vaporized and flows overhead, joining the solvent from the low-pressure steam evaporator. The DAO, with relatively minor solvent, flows from the bottom of the flash tower under level control to a steam stripper operating at essentially atmospheric pressure. Superheated steam is introduced into the lower portion of the tower. The remaining solvent is stripped out and flows overhead with the steam through a condenser into the compressor suction drum where the water drops out. The water flows from the bottom of the drum under level control to appropriate disposal.

The asphalt-solvent mixture is pressured from the extraction tower bottom on flow control to the asphalt heater and on to the asphalt flash drum, where the vaporized solvent is separated from the asphalt. The drum operates essentially at the solvent condensing pressure so that the overhead vapors flow directly through the condenser into the solvent accumulator. Hot asphalt with a small quantity of solvent flows from the asphalt flash drum bottom, under level control, to the asphalt stripper that is operated at near atmospheric pressure. Superheated steam is introduced into the bottom of the stripper. The steam and solvent vapors pass overhead, join the DAO stripper overhead, and flow through the condenser into the compressor suction drum. The asphalt is pumped from the bottom of the stripper under level control, to storage.

The propane deasphalting process is similar to solvent extraction in which a packed or baffled extraction tower or rotating disk contactor is used to mix the oil feed stocks with the solvent. In the tower method, four to eight volumes of propane are fed to the bottom of the tower for every volume of feed flowing down from the top of the tower. The oil, which is more soluble in the propane dissolves and flows to the top. The asphaltene and resins flow to the bottom of the tower where they are removed in a propane mix. Propane is recovered from the two streams through two-stage flash systems followed by steam stripping in which propane is condensed and removed by cooling at high pressure in the first stage and at low pressure in the second stage. The asphalt recovered can be blended with other asphalts or heavy fuels or can be used as feed to the coker.

The yield of DAO varies with the feedstock (Table 7.1) but the DAO does make less coke and more distillate than the feedstock. Therefore the process parameters for a deasphalting unit must be selected with care according to the nature of the feedstock and the desired final products. The metals content of the DAO is relatively low and the nitrogen and sulfur contents in the DAO are also related to the yield of DAO yield (Speight and Ozum, 2002). The character of the deasphalting process is a molecular weight separation and the solvent takes a crosscut across the feedstock effecting separation by molecular weight and by polarity (Fig. 7.1).

Furthermore to the selection of the process parameters, the *choice of solvent* is vital to the flexibility and performance of the unit. The solvent must be suitable, not only for the extraction of the desired oil fraction but also for control of the yield and/or quality of the DAO at temperatures which are within the operating limits. If the temperature is too high (i.e., close to the critical temperature of the solvent) the

Table 7.1 Feedstock and product data for the deasphalting process.

Crude source	Arab	West Texas	California	Canada	Kuwait	Kuwait
Feedstock						
Crude (vol.%)	23.0	29.2	20.0	16.0	22.2	32.3
Gravity (°API)	6.8	12.0	6.3	9.6	5.6	8.1
Conradson carbon (wt.%)	15.0	12.1	22.2	18.9	24.0	19.7
SUS at 210°F	75,000	526	9600	1740	14,200	3270
Metals (wppm)						
Ni	73.6	16.0	139	46.6	29.9	29.7
V	365.0	27.6	136	30.9	110.0	89
Cu + Fe	15.5	14.8	94	40.7	13.7	7.5
Deasphalted oil						
Vol.% feed	49.8	66.0	52.8	67.8	45.6	54.8
Gravity (°API)	18.1	19.6	18.3	17.8	16.2	17.2
Conradson carbon (wt.%)	5.9	2.2	5.3	5.4	4.5	5.4
SUS at 210°F	615	113	251	250	490	656
Metals (wppm)						
Ni	3.5	1.0	8.1	3.9	0.9	0.6
V	12.4	1.3	2.3	1.4	0.7	4.0
Cu + Fe	0.2	0.8	3.5	0.2	0.8	0.8
Asphalt						
Vol.% feed	50.2	34.0	47.2	32.2	54.4	45.2
Gravity (°API)	− 1.3	− 0.9	− 5.1	− 5.1	− 1.3	− 2.0

operation becomes unreliable in terms of product yields and character. If the temperature is too low, the feedstock may be too viscous and have an adverse effect on the contact with the solvent in the tower.

Liquid propane is by far the most selective solvent among the light hydrocarbons used for deasphalting. At temperatures ranging from 38°C to 65°C (100°F to 150°F), most hydrocarbons are soluble in propane while asphaltic and resinous compounds are not thereby allowing rejection of these compounds resulting in a drastic reduction (relative to the feedstock) of the nitrogen content and the metals in the DAO. Although the DAO from propane deasphalting has the best quality, the yield is usually less than the yield of DAO produced using a higher molecular weight (higher boiling) solvent.

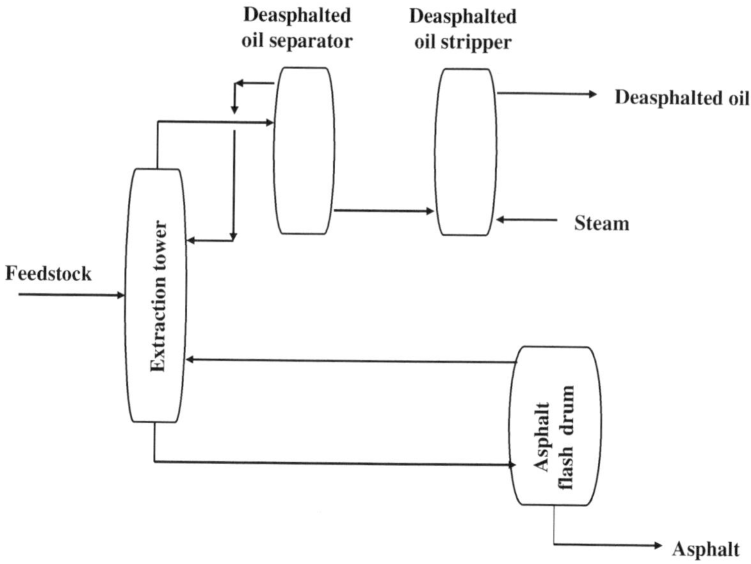

Figure 7.1 The LEDA process. *LEDA*, Low energy deasphalting.

The ratios of propane-oil required vary from 6 to 1 to 10 to 1 by volume, with the ratio occasionally being as high as 13 to 1. Since the critical temperature of propane is 97°C (206°F), this limits the extraction temperature to about 82°C (180°F). Therefore propane alone may not be suitable for high-viscosity feedstocks because of the relatively low operating temperature.

iso-Butane and *n*-butane are more suitable for deasphalting high-viscosity feedstocks since their critical temperatures are higher (134°C, 273°F and 152°C, 306°F, respectively) than that of the critical temperature of propane. Higher extraction temperatures can be used to reduce the viscosity of the heavy feed and to increase the transfer rate of oil to solvent. Although *n*-pentane is less selective for metals and carbon residue removal, it can increase the yield of DAO from a heavy feed by a factor of 2–3 over propane (Speight, 2000, 2002). However, if the content of the metals and carbon residue of the pentane-DAO is too high (defined by the ensuing process), the DAO may be unsuitable as a cracking feedstock. In certain cases the nature of the cracking catalyst may dictate that the pentane-DAO be blended with vacuum gas oil that, after further treatment such as hydrodesulfurization, produces a good cracking feedstock.

Solvent composition is an important process variable for deasphalting units. The use of a single solvent may (depending on the nature of the solvent) limit the range of feedstocks that can be processed in a deasphalting unit. When a deasphalting unit is required to handle a variety of feedstocks and/or produce various yields of DAO (as is the case in these days of variable feedstock quality), a dual solvent may be the only option to provide the desired flexibility. For example, a mixture of propane and *n*-butane might be suitable for feedstocks that vary from vacuum residua to both the heavy residue to heavy gas oils that contain asphaltic materials. Adjusting the solvent composition allows the most desirable product quantity and quality within the range of temperature control.

Besides the solvent composition, the *solvent/oil ratio* also plays an important role in a deasphalting operation. Solvent/oil ratios vary considerably and are governed by feedstock characteristics and desired product qualities and, for each individual feedstock, there is a minimum operable solvent/oil ratio. Generally, increasing the solvent-to-oil ratio almost invariably results in improving the DAO quality at a given yield but other factors must also be taken into consideration and (generalities aside) each plant and feedstock will have an optimum ratio.

The main consideration in the selection of the *operating temperature* is its effect on the yield of DAO. For practical applications the lower limits of operable temperature are set by the viscosity of the oil-rich phase. When the operating temperature is near the critical temperature of the solvent, control of the extraction tower becomes difficult since the rate of change of solubility with temperature becomes very large at conditions close to the critical point of the solvent. Such changes in solubility cause large amounts of oil to transfer between the solvent-rich and the oil-rich phases, which, in turn, causes *flooding* and/or uncontrollable changes in product quality. To mitigate such effects the upper limits of operable temperatures must lie below the critical temperature of the solvent in order to insure good control of the product quality and to maintain a stable condition in the extraction tower.

The *temperature gradient* across the extraction tower influences the sharpness of separation of the DAO and the asphalt because of internal reflux that occurs when the cooler oil/solvent solution in the lower section of the tower attempts to carry a large portion of oil to the top of the tower. When the oil/solvent solution reaches the steam-heated, higher temperature area near the top of the tower, some oil of higher molecular weight in the solvent solution is rejected because the oil is less soluble in solvent at the higher temperature. The heavier oil (rejected from the solution at the top of the tower) attempts to flow downward and causes the internal reflux. In fact, generally, the greater the temperature difference between the top and the bottom of the tower, the greater will be the internal reflux, and the better will be the quality of the DAO. However, too much internal reflux can cause tower flooding and jeopardize the process.

The *process pressure* is usually not considered to be an operating variable since it must be higher than the vapor pressure of the solvent mixture at the tower operating temperature to maintain the solvent in the liquid phase. The extractor's operating pressure is determined by the solvent, and the need to ensure the solvent/residue mixture is maintained in the liquid state. The tower pressure is usually only subject to change when there is a need to change the solvent composition or the process temperature.

Proper *contact and distribution of the oil and solvent* in the *tower* are essential to the efficient operation of any deasphalting unit. In early units, days, mixer-settlers were used as contactors but proved to be less efficient than the countercurrent contacting devices. Packed towers are difficult to operate in this process because of the large differences in viscosity and density between the asphalt phase and the solvent-rich phase.

The *extraction tower* for solvent deasphalting consists of two contacting zones: (1) a rectifying zone above the oil feed and (2) a stripping zone below the oil feed. The rectifying zone contains some elements designed to promote contacting and to avoid *channeling*. Steam-heated coils are provided to raise the temperature sufficiently to induce an oil-rich reflux in the top section of the tower. The stripping zone has disengaging spaces at the top and bottom and consists of contacting elements between the oil inlet and the solvent inlet.

A *countercurrent tower* with static baffles is widely used in solvent deasphalting service. The baffles consist of fixed elements formed of expanded metal gratings in groups of two or more to provide maximum change of direction without limiting capacity. The *rotating disk contactor* has also been employed and consists of disks connected to a rotating shaft that are used in place of the static baffles in the tower. The rotating element is driven by a variable speed drive at either the top or the bottom of the column, and operating flexibility is provided by controlling the speed of the rotating element and, thus, the amount of mixing in the contactor.

In the deasphalting process the solvent is recovered for circulation and the efficient operability of a deasphalting unit is dependent on the design of the *solvent recovery system.*

The solvent may be separated from the DAO in several ways such as conventional evaporation or the use of a flash tower. Irrespective of the method of solvent recovery from the DAO, it is usually the most efficient to recover the solvent at a temperature close to the extraction temperature. If a higher temperature for solvent recovery is used, heat is wasted in the form of high vapor temperature and, conversely, if a lower temperature is used, the solvent must be reheated, thereby requiring additional energy input. The solvent recovery pressure should be low enough to maintain a smooth flow under pressure from the extraction tower.

The asphalt solution from the bottom of the extraction tower usually contains less than an equal volume of solvent. A fired heater is used to maintain the temperature of the asphalt solution well above the foaming level and to keep the asphalt phase in a fluid state. A flash drum is used to separate the solvent vapor from asphalt with the design being such to prevent carryover of asphalt into the solvent outlet line and to avoid fouling the downstream solvent condenser. The solvent recovery system from asphalt is not usually subject to the same degree of variations as the solvent recovery system for the DAO and operation at constant temperature and pressure with a separate solvent condenser and accumulator is possible.

Asphalt from different crude oils varies considerably but the viscosity is often too high for fuel oil although, in some cases they can be blended with refinery cutter stocks to make No. 6 fuel oil. When the sulfur content of the original residuum is high, even the blend fuel oil will not be able to meet the sulfur specification of fuel oil unless stack gas clean-up is available.

The DAO and solvent asphalt are not finished products and require further processing or blending, depending on the final use. *Manufacture of lubricating oil* is on possibility, and the DAO may also be used as a *catalytic cracking feedstock* or it may be desulfurized. It is perhaps these last two options that are more pertinent to the present text and future refinery operations.

Briefly, catalytic cracking or hydrodesulfurization of atmospheric and vacuum residua from high-sulfur/high-metal crude oil is, theoretically, the best way to enhance their value. However, the concentrations of sulfur (in the asphaltene fraction) in the residua can severely limit the performance of cracking catalysts and hydrodesulfurization catalysts (Speight, 2000). Both processes generally require tolerant catalysts as well as (in the case of hydrodesulfurization) high hydrogen pressure, low space velocity, and high hydrogen recycle ratio.

For both processes the advantage of using the deasphalting process to remove the troublesome compounds becomes obvious. The DAO, with no asphaltene constituents and low metal content, is easier to process than the residua. Indeed, in the hydrodesulfurization process, the DAO may consume only 65% of the hydrogen required for direct hydrodesulfurization of topped crude oil.

As always, the use of the material rejected by the deasphalting unit remains an issue. It can be used (apart from its use for various types of asphalt) as feed to a partial oxidation unit to make a hydrogen-rich gas for use in hydrodesulfurization and hydrocracking processes. Alternatively, the asphalt may be treated in a visbreaker to reduce its viscosity, thereby minimizing the need for cutter stock to be blended with the solvent asphalt for making fuel oil, or hydrovisbreaking offers an option of converting the asphalt to feedstocks for other conversion processes.

7.2.2 Process options for heavy feedstocks

Solvent deasphalting is a separation process that represents a further step in the minimization of residual by-product fuel. However, solvent deasphalting processes, far from realizing their maximum potential for heavy feedstocks, are now under further investigation and, with on-going improvements in energy efficiency, such processes are starting to display maximum benefits when used in combination with other processes. The process takes advantage of the fact that maltenes are more soluble in light paraffinic solvents than asphaltene constituents. This solubility increases with solvent molecular weight and decreases with temperature (Girdler, 1965). As with vacuum distillation there are constraints with respect to how deep a solvent deasphalting unit can cut into the residue or how much DAO can be produced. In the case of solvent deasphalting the constraint is usually related to DAO quality specifications required by downstream conversion units.

However, solvent deasphalting has the flexibility to produce a wide range of DAO that matches the desired properties. The process has very good selectivity for asphaltene constituents (and, to a lesser extent, resin constituents) as well as metals rejection. There is also some selectivity for the rejection of carbon residue precursors, but there is less selectivity for sulfur-containing and nitrogen-containing constituents. The process is best suited for the more paraffinic vacuum residua with a somewhat lower efficiency when applied to high-asphaltene residua that contain high proportions of metals and coke-forming constituents. The advantages—disadvantages of the process are that it performs no conversion and produces a very high-viscosity byproduct deasphalter bottoms and, where high-quality DAO is required, the process is limited in the quality of feedstock that can be economically processed. In those situations where there is a ready outlet for use for the bottoms, solvent deasphalting is an attractive option for treating heavy feedstocks. One such situation is the cogeneration of steam and power, both to supply the refiner's needs and for export to nearby users.

7.2.2.1 Deep solvent deasphalting process

The deep solvent deasphalting process is an application of the LEDA (low energy deasphalting) process (Table 7.2) (RAROP, 1991, p. 91; Hydrocarbon Processing, 1998, p. 67) that is used to extract high-quality lubricating oil bright stock or prepare catalytic cracking feeds, hydrocracking feeds, hydrodesulfurization unit feeds, and asphalt from vacuum residue materials. The LEDA process uses a low-boiling hydrocarbon solvent specifically formulated to insure the most economical deasphalting design for each operation. For example, a propane solvent may be specified for a low DAO yield operation, while a higher boiling solvent, such as pentane or hexane, may be used to obtain a high DAO yield from a vacuum residuum. The deep deasphalting process can be integrated with a delayed coking operation (ASCOT process; *q.v.*). In this case the solvent can be low-boiling naphtha (Table 7.3).

LEDA operations are usually carried out in a rotating disk contractor that provides more extraction stages than a mixer-settler or baffle type column. Although not essential to the process, the rotating disk contactor provides higher quality DAO at the same yield, or higher yields at the same quality. The low energy solvent deasphalting process selectively extracts the more paraffinic components from

Table 7.2 Feedstock and product data for the low energy deasphalting process.

	Residuum[a]	Residuum[a]
Feedstock		
API	6.5	6.5
Sulfur (wt.%)	3.0	3.0
Carbon residue (wt.%)	21.8	21.8
Nickel (ppm)	46.0	46.0
Vanadium (ppm)	125.0	125.0
Products		
Deasphalted oil (vol.%)	53.0	65.0
API	17.6	15.1
Sulfur (wt.%)	1.9	2.2
Carbon residue (wt.%)	3.5	6.2
Nickel (ppm)	1.8	4.5
Vanadium (ppm)	3.4	10.3
		1250.0

[a]Unspecified.

Table 7.3 Feedstock and product data for the Demex process.

	Vacuum residuum[a]	Vacuum residuum[a]	Arabian light vacuum residuum	Arabian light vacuum residuum	Arabian light vacuum residuum	Arabian heavy vacuum residuum	Arabian heavy vacuum residuum
Feedstock							
API	7.2	7.2	6.9	6.9	6.9	3.0	3.0
Sulfur (wt.%)	4.0	4.0	4.0	4.0	4.0	6.0	6.0
Nitrogen (wt.%)	0.3	0.3	0.3	0.3	0.3	0.5	0.5
Carbon residue (wt.%)	20.8	20.8	20.8	20.8	20.8	27.7	27.7
Nickel (ppm)			23.0	23.0	23.0	64.0	64.0
Vanadium (ppm)			75.0	75.0	75.0	205.0	205.0
Nickel + vanadium (ppm)	98.0	98.0					
C6-asphaltenes (wt.%)	10.0	10.0					
C7-asphaltenes (wt.%)			10.0	10.0	10.0	15.0	15.0
Products							
Demetallized oil (vol.%)	56.0	78.0	40.0	60.0	78.0	30.0	55.0
API	16.0	12.0	18.9	15.3	12.0	16.3	12.0
Sulfur (wt.%)	2.7	3.3	2.3	2.8	3.3	3.5	4.3
Nitrogen (wt.%)	0.1	0.2	0.1	0.2	0.2	0.1	0.2
Carbon residue (wt.%)	5.6	10.7	2.9	6.4	10.7	4.8	10.1
Nickel + vanadium (ppm)	6.0	19.0	2.5	7.2	19.0	16.0	38.0
C6-asphaltenes (wt.%)	<0.1	<0.1					
Pitch (vol.%)	44.0	22.0					
API	<0.0	<0.0	<0.0	<0.0	<0.0	<0.0	<0.0
Sulfur (wt.%)	5.4	6.3	5.0	5.5	6.3	6.9	7.8
Nickel + vanadium (ppm)	201.0	341.0	154.0	216.0	341.0	364.0	515.0

[a]Unspecified.

vacuum residua while rejecting the condensed ring aromatics. As expected, DAO yields vary as a function of solvent type and quantity, and feedstock properties (Speight and Ozum, 2002).

In the process, vacuum residue feed is combined with a small quantity of solvent to reduce its viscosity and cooled to a specific extraction temperature before entering the rotating disk contactor. Recovered solvent from the high-pressure and low-pressure solvent receivers are combined, adjusted to a specific temperature by the solvent heater–cooler, and injected into the bottom section of the rotating disk contactor. Solvent flows upward, extracting the paraffinic hydrocarbons from the vacuum residuum, which is flowing downward through the rotating disk contactor.

Steam coils at the top of the tower maintain the specified temperature gradient across the rotating disk contactor. The higher temperature in the top section of the rotating disk contactor results in separation of the less soluble heavier material from the DAO mix and provides internal reflux, which improves the separation. The DAO mix leaves the top of the rotating disk contactor tower. It flows to an evaporator where it is heated to vaporize a portion of the solvent. It then flows into the high-pressure flash tower where high-pressure solvent vapors are taken overhead.

The DAO mix from the bottom of this tower flows to the pressure vapor heat exchanger where additional solvent is vaporized from the DAO mix by condensing high-pressure flash. The high-pressure solvent, totally condensed, flows to the high-pressure solvent receiver. Partially vaporized, the DAO mix flows from the pressure vapor heat exchanger to the low-pressure flash tower where low-pressure solvent vapor is taken overhead, condensed and collected in the low-pressure solvent receiver. The DAO mix flows down the low-pressure flash tower to the reboiler, where it is heated, and then to the DAO stripper, where the remaining solvent is stripped overhead with superheated steam. The DAO product is pumped from the stripper bottom and is cooled, if required, before flowing to battery limits.

The raffinate phase containing asphalt and small amount of solvent flows from the bottom of the rotating disk contactor to the asphalt mix heater. The hot, two phase asphalt mix from the heater is flashed in the asphalt mix flash tower where solvent vapor is taken overhead, condensed, and collected in the low-pressure solvent receiver. The remaining asphalt mix flows to the asphalt stripper where the remaining solvent is stripped overhead with superheated steam. The asphalt stripper overhead vapors are combined with the overhead from the DAO stripper, condensed, and collected in the stripper drum. The asphalt product is pumped from the stripper and is cooled by generating low-pressure steam.

7.2.2.2 Demex process

The Demex process is a solvent extraction demetallizing process that separates high metal vacuum residuum into demetallized oil of relatively low metal content and asphaltene of high metal content (Table 7.3) (Salazar, 1986; RAROP, 1991, p. 93; Houde, 1997) . The asphaltene and condensed aromatic contents of the demetallized oil are very low. The demetallized oil is a desirable feedstock for fixed-bed hydrodesulfurization and, in cases where the metals content and carbon residue are sufficiently low, is a desirable feedstock for fluid catalytic cracking and hydrocracking units. Overall, the Demex process is an extension of the propane deasphalting process and employs a less selective solvent to recover not only the high-quality oils but also higher molecular weight aromatics and other constituents present in the feedstock. Furthermore, the Demex process requires a much less solvent circulation in achieving its objectives, thus, reducing the utility costs and unit size significantly. The process selectively rejects asphaltenes, metals, and high molecular weight aromatics from vacuum residua. The resulting demetallized oil can then be combined with vacuum gas oil to give a greater availability of acceptable feed to subsequent conversion units.

In the process the vacuum residuum feedstock, mixed with Demex solvent recycling from the second stage, is fed to the first stage extractor. The pressure is kept high enough to maintain the solvent in liquid phase. The temperature is controlled by the degree of cooling of the recycle solvent. The solvent rate is set near the minimum required to ensure the desired separation to occur. Asphaltene constituents are rejected in the first stage. Some resins are also rejected to maintain sufficient fluidity of the asphaltene for efficient solvent recovery. The asphaltene is heated and steam stripped to remove solvent. The first stage overhead is heated by an exchange with hot solvent. The increase in temperature decreases the solubility of resins and high molecular weight aromatics (Mitchell and Speight, 1973). These precipitate in the second stage extractor. The bottom stream of this second stage extractor is recycled to the first stage. A portion of this stream can also be drawn as a separate product.

The overhead from the second stage is heated by an exchange with hot solvent. The fired heater further raises the temperature of the solvent/demetallized oil mixture to a point above the critical temperature of the solvent. This causes the demetallized oil to separate. It is then flashed and steam-stripped to remove all traces of solvent. The vapor streams from the demetallized oil and asphalt strippers are condensed, dewatered, and pumped up to process pressure for recycle. The bulk of the solvent goes overhead in the supercritical separator. This hot solvent stream is then effectively used for process heat exchange. The subcritical solvent recovery techniques, including multiple effect systems, allow much less heat recovery. Most of the low grade heat in the solvent vapors from the subcritical flash vaporization must be released to the atmosphere requiring additional heat input to the process.

7.2.2.3 MDS process

The MDS process is a technical improvement of the solvent deasphalting process, particularly effective for upgrading heavy crude oils (Table 7.4) (Kashiwara, 1980; RAROP, 1991, p. 95). Combined with hydrodesulfurization, the process is fully applicable to the feed preparation for fluid catalytic cracking and hydrocracking. The process is capable of using a variety of feedstocks, including atmospheric and

Table 7.4 Feedstock and product data for the MDS process.

	Iranian heavy atmospheric residuum	Kuwait atmospheric residuum	Khafji vacuum residuum
Feedstock			
API	17.0	16.4	5.2
Sulfur (wt.%)	2.7	3.7	5.2
Carbon residue (wt.%)	9.1	9.4	21.9
Nickel (ppm)	40.0	14.0	49.0
Vanadium (ppm)	130.0	48.0	140.0
Products			
Deasphalted oil (vol.%)	93.4	93.8	72.4
API	19.0	16.4	11.3
Sulfur (wt.%)	2.4	3.7	4.3
Carbon residue (wt.%)	5.9	9.4	10.9
Nickel (ppm)	18.0	14.0	6.0
Vanadium (ppm)	53.0	48.0	28.0
Asphalt (vol.%)	6.6	6.2	27.6
API	<0.0	<0.0	<0.0
Sulfur (wt.%)	5.4	7.2	7.3
Carbon residue (wt.%)			49.3
Nickel (ppm)	320.0	113.0	150.0
Vanadium (ppm)	1010.0	425.0	400.0

vacuum residua derived from various crude oils, tar sand bitumen, and nonvolatile products from a visbreaker.

In the process the feed and the solvent are mixed and fed to the deasphalting tower. Deasphalting extraction proceeds in the upper half of the tower. After the removal of the asphalt the mixture of DAO and solvent flows out of the tower through the tower top. Asphalt flows downward to come in contact with a countercurrent of rising solvent. The contact eliminates oil from the asphalt, the asphalt then accumulates on the bottom. DAO containing solvent is heated through a heating furnace, and fed to the DAO flash tower where most of the solvent is separated under pressure. DAO still containing a small amount of solvent is again heated and fed to the stripper, where the remaining solvent is completely removed.

Asphalt is withdrawn from the bottom of the extractor. Since this asphalt contains a small amount of solvent, it is heated through a furnace and fed to the flash tower to remove most of the solvent. Asphalt is then sent to the asphalt stripper, where the remaining portion of solvent is completely removed.

Solvent recovered from the DAO and asphalt flash towers is cooled and condensed into liquid and sent to a solvent tank. The solvent vapor leaving both strippers is cooled to remove water and compressed for condensation. The condensed solvent is then sent to the solvent tank for further recycling.

7.2.2.4 Residuum oil supercritical extraction process

The *ROSE process* is a solvent deasphalting process with minimum energy consumption using supercritical solvent recovery system and the process is of value in obtaining oils for further processing (Gearhart and Gatwin, 1976; Gearhart, 1980; RAROP, 1991, p. 97; Low et al., 1995; Hydrocarbon Processing, 1996; Northrup and Sloan, 1996).

The process used supercritical solvents and is a natural progression from propane deasphalting and allows the separation of residua into their base components (asphaltene constituents, resin constituents, and oil constituents) for recombination to optimum properties. Propane, butane, and pentane may be used as the solvent depending on the feedstock and the desired compositions. A mixer is used to blend residue with liquefied solvent at elevated temperature and pressure. The blend is pumped into the first stage separator where, through countercurrent flow of solvent, the asphaltene constituents are precipitated, separated, and stripped of solvent by steam. The overhead solution from the first tower is taken to a second stage where it is heated to a higher temperature. This causes the resin constituents to separate. The final material is taken to a third stage and heated to a supercritical temperature. This makes the oils insoluble and separation occurs. This process is very flexible and allows precise blending to required compositions.

In the process the residuum is mixed with several-fold volume of a low-boiling hydrocarbon solvent and passed into the asphaltene separator vessel. Asphaltenes rejected by the solvent are separated from the bottom of the vessel and are further processed by heating and steam stripping to remove a small quantity of dissolved solvent. The solvent free asphaltenes are sent to a section of the refinery for further processing. The main flow, solvent, and extracted oil passes overhead from the asphaltene separator through a heat exchanger and heater into the oil separator where the extracted oil is separated without solvent vaporization. The solvent, after heat exchange, is recycled to the process. The small amount of solvent contained in the oil is removed by steam stripping and the resulting vaporized solvent from the strippers is condensed and returned to the process. Product oil is cooled by heat exchange before being pumped to storage or further processing.

The deasphalting efficiency in processes using propane is of the order of 75%–83%, with an overall DAO recovery yield of the order of 50%.

Table 7.5 Feedstock and product data for the Solvahl process.

	Arabian light vacuum residuum
Feedstock	
API	9.6
Sulfur (wt.%)	4.1
Nitrogen (wt.%)	0.3
C7-asphaltenes (wt.%)	4.2
Carbon residue (wt.%)	16.4
Nickel (ppm)	19.0
Vanadium (ppm)	61.0
Products	
C4-deasphalted oil (wt.%)	70.1
API	16.0
Sulfur (wt.%)	3.3
Nitrogen (wt.%)	0.2
C7-asphaltenes (wt.%)	<0.1
Carbon residue (wt.%)	5.3
Nickel (ppm)	2.0
Vanadium (ppm)	3.0
C5-deasphalted oil (wt.%)	85.5
API	13.8
Sulfur (wt.%)	3.7
Nitrogen (wt.%)	0.2
C7-asphaltenes (wt.%)	<0.1
Carbon residue (wt.%)	7.9
Nickel (ppm)	7.0
Vanadium (ppm)	16.0

7.2.2.5 Solvahl process

The Solvahl process is a solvent deasphalting process for application to vacuum residua (Table 7.5) (RAROP, 1991, p. 9; Billon et al., 1994).

The process was developed to give maximum yields of DAO while eliminating asphaltenes and reducing metals content to a level compatible with the reliable operation of downstream units (Peries et al., 1995; Hydrocarbon Processing, 1996).

7.2.2.6 Lube deasphalting

Other facilities incorporate lube deasphalting to process vacuum residuum into lube oil base stocks. Propane deasphalting is most commonly used to remove asphaltene constituents and resins, which contribute an undesirable dark color to the lube base stocks. This process typically uses baffle towers or rotating disk contactors to mix the propane with the feed. Solvent recovery is accomplished with evaporators, and supercritical solvent recovery processes are also used in some deasphalting units.

The Duo-Sol process is used to deasphalt and extract lubricating oil feedstocks. Propane is used as the deasphalting solvent and a mixture of phenol and cresylic acids (cresols and hydroxytoluenes) are used as the extraction solvent. The extraction is conducted in a series of batch extractors followed by solvent recovery in multistage flash distillation and stripping towers.

7.3 Commercial dewaxing processes

In the 1930s two types of stocks, naphthenic and paraffinic, were used to make motor oils. Both types were solvent extracted to improve their quality, but in the high-temperature conditions encountered in service, the naphthenic type could not stand up as well as the paraffinic type. Nevertheless, the naphthenic type was the preferred oil, particularly in cold weather, because of its fluidity at low temperatures. Previous to 1938 the highest quality lubricating oils were of the naphthenic type and were phenol treated to pour points of −40°C to −7°C (−40°F to 20°F), depending on the viscosity of the oil. Paraffinic oils were also available and could be phenol treated to higher quality oil, but their wax content was so high that the oils were solid at room temperature.

7.3.1 Cold press process

The lowest viscosity paraffinic oils were dewaxed by the cold press method to produce oils with a pour point of 2°C (35°F). The light paraffin distillate oils contained a paraffin wax that crystallized into large crystals when chilled and could thus readily be separated from the oil by the cold press filtration method. The more viscous paraffinic oils (intermediate and heavy paraffin distillates) contained amorphous or microcrystalline waxes, which formed small crystals that plugged the filter cloths in the cold press and prevented filtration. Because the wax could not be removed from intermediate and heavy paraffin distillates, the high-quality, high-viscosity lubricating oils in them could not be used except as cracking stock.

Methods were, therefore, developed to dewax these high-viscosity paraffinic oils. The methods were essentially alike in which the waxy oil was dissolved in a solvent that would keep the oil in solution; the wax separated as crystals when the temperature was lowered. The processes differed chiefly in the use of the solvent. Commercially used solvents were naphtha, propane, sulfur dioxide, acetone–benzene, trichloroethylene, ethylene dichloride-benzene (*Barisol*), MEK-benzene (*benzol*), methyl-*n*-butyl ketone, and methyl-*n*-propyl ketone.

7.3.2 Solvent dewaxing process

Solvent dewaxing consists of the following steps: crystallization, filtration, and solvent recovery. In the crystallization step the feedstock is diluted with the solvent and chilled, solidifying the wax components. The filtration step removes the wax from the solution of dewaxed oil and solvent. Solvent recovery removes the solvent from the wax cake and filtrate for recycle by flash distillation and stripping.

In the first *solvent dewaxing process* (developed in 1924) the waxy oil was mixed with naphtha and filter aid (fuller's earth or diatomaceous earth). The mixture was chilled and filtered, and the filter aid assisted in building a wax cake on the filter cloth. This process is now obsolete, and most of the modern dewaxing processes use a mixture of MEK and benzene. Other ketones may be substituted for dewaxing, but regardless of what ketone is used, the process is generally known as ketone dewaxing.

The process as now practiced involves mixing the feedstock with one to four times its volume of the ketone (Scholten, 1992). The mixture is then heated until the oil is in solution and the solution is chilled at a slow, controlled rate in double-pipe, scraped-surface exchangers. Cold solvent, such as filtrate from the filters, passes through the 2-in. annular space between the inner and outer pipes and chills the waxy oil solution flowing through the inner 6-in. pipe.

To prevent wax from depositing on the walls of the inner pipe, blades, or scrapers extending the length of the pipe and fastened to a central rotating shaft, scrape off the wax. Slow chilling reduces the temperature of the waxy oil solution to 2°C (35°F), and then faster chilling reduces the temperature to the approximate pour point required in the dewaxed oil. The waxy mixture is pumped to a filter case into which the bottom half of the drum of a rotary vacuum filter dips. The drum (8 ft in diameter, 14 ft long), covered with filter cloth, rotates continuously in the filter case. Vacuum within the drum sucks the solvent and the oil dissolved in the solvent through the filter cloth and into the drum. Wax crystals collect on the outside of the drum to form a wax cake, and as the drum rotates, the cake is brought above the surface of the liquid in the filter case and under sprays of ketone that wash oil out of the cake and into the drum. A knife-edge scrapes off the wax, and the cake falls into the conveyor and is moved from the filter by the rotating scroll.

The recovered wax is actually a mixture of wax crystals with a little ketone and oil, and the filtrate consists of the dewaxed oil dissolved in a large amount of ketone. Ketone is removed from both by distillation, but before the wax is distilled, it is deoiled, mixed with more cold ketone, and pumped to a pair of rotary filters in series, where further washing with cold ketone produces a wax cake that contains very little oil. The deoiled wax is melted in heat exchangers and pumped to a distillation tower operated under vacuum, where a large part of the ketone is evaporated or flashed from the wax. The rest of the ketone is removed by heating the wax and passing it into a fractional distillation tower operated at atmospheric pressure and then into a stripper where steam removes the last traces of ketone.

An almost identical system of distillation is used to separate the filtrate into dewaxed oil and ketone. The ketone from both the filtrate and wax slurry is reused. Clay treatment or hydrotreating finishes the dewaxed oil as previously described. The wax (*slack wax*), even though it contains essentially no oil as compared to 50% in the slack wax obtained by cold pressing, is the raw material for either sweating or wax recrystallization, which subdivides the wax into a number of wax fractions with different melting points (Chapter 21).

Solvent dewaxing can be applied to light, intermediate, and heavy lubricating oil distillates, but each distillate produces a different kind of wax. Each of these waxes is actually a mixture of a number of waxes. For example, the wax obtained from light paraffin distillate consists of a series of paraffin waxes that have melting points in the range of 30°C–70°C (90°F–160°F), which are characterized by a tendency to harden into large crystals. However, heavy paraffin distillate yields a wax composed of a series of waxes with melting points in the range of 60°C–90°C (140°F–200°F), which harden into small crystals from which they derive the name of *microcrystalline wax* or *microwax*. On the other hand, intermediate paraffin distillates contain paraffin waxes and waxes intermediate in properties between paraffin and microwax.

Thus the solvent dewaxing process produces three different *slack waxes* (also known as *crude waxes* or *raw waxes*) depending on whether light, intermediate, or heavy paraffin distillate is processed. The slack wax from heavy paraffin distillate may be sold as dark raw wax, the wax from intermediate paraffin distillate as pale raw wax. The latter is treated with lye and clay to remove odor and improve color.

There are several processes in use for solvent dewaxing, but all have the same general steps, which are (1) contacting the feedstock with the solvent, (2) precipitating the wax from the mixture by chilling, and (3) recovering the solvent from the wax and dewaxed oil for recycling. The processes use benzene–acetone (*solvent dewaxing*), propane (*propane dewaxing*), trichloroethylene (*separator-Nobel dewaxing*), ethylene dichloride-benzene (*Barisol dewaxing*), and urea (*urea dewaxing*), as well as liquid sulfur dioxide-benzene mixtures.

The major processes currently in use today are the ketone dewaxing processes. Other processes that are used to a lesser degree include the Di/Me process and the propane dewaxing process. The most widely used ketone processes are the Texaco solvent dewaxing process and the Exxon Dilchill process. Both processes consist of diluting the waxy feedstock with solvent, while chilling at a controlled rate to produce a slurry. The slurry is filtered using rotary vacuum filters and the wax cake is washed with cold solvent. The filtrate is used to prechill the feedstock and solvent mixture. The primary wax cake is diluted with additional solvent and filtered again to reduce the oil content in the wax. The solvent is recovered from the dewaxed oil and wax cake by flash vaporization and recycled back into the process.

The Texaco solvent dewaxing process (also called the MEK process) uses a mixture of MEK and toluene as the dewaxing solvent, and sometimes uses mixtures of other ketones and aromatic solvents. The Exxon Dilchill dewaxing process uses a direct cold solvent dilution–chilling process in a special crystallizer in place of the scraped-surface exchangers used in the Texaco process.

The Di/Me Dewaxing process uses a mixture of dichloroethane and methylene dichloride as the dewaxing solvent. This process is used by a few refineries in Europe. The Propane Dewaxing Process is essentially the same as the ketone process except for the following: (1) propane is used as the dewaxing solvent, (2) higher pressure equipment is required, and (3) chilling is done in evaporative chillers by vaporizing a portion of the dewaxing solvent. Although this process generates a better product and does not require crystallizers, the temperature differential between the dewaxed oil and the filtration temperature is higher than for the ketone processes (higher energy costs), and dewaxing aids are required to get good filtration rates.

7.3.3 Urea dewaxing process

Urea dewaxing (Chapter 21) (Scholten, 1992) is worthy of further mention insofar as the process is highly selective and, in contrast to the other dewaxing techniques, can be achieved without the use of refrigeration. However, the process cannot compete economically with the solvent dewaxing processes for treatment of the heavier lubricating oils. But when it is applied to the lighter materials that already may have been subjected to a solvent dewaxing operation, products are obtained that may be particularly useful as refrigerator oils, transformer oils, and the like.

The process description is essentially the same as that used for solvent dewaxing with the omission of the chilling stage and the insertion of a contactor where the feedstock and the urea (with a solvent) are thoroughly mixed before filtration. The solvents are recovered from the dewaxed oil by evaporation, and the urea complex is decomposed in a urea recovery system.

Residual lubricating oils, such as cylinder oils and bright stocks, are made from paraffinic or mixed-base reduced crude oils and contain waxes of the microcrystalline type. Removal of these waxes from reduced crude produces *petrolatum*, a grease-like material that is known in a refined form as *petroleum jelly*. This material can be separated from reduced crude in several ways. The original method was cold settling, whereby reduced crude was dissolved in a suitable amount of naphtha and allowed to stand over winter until the microwax settled out. This method is still used, but the reduced crude naphtha solution is held in refrigerated tanks until the petrolatum settles out. The supernatant naphtha-oil layer is pumped to a still where the naphtha is removed, leaving cylinder stock that can be further treated to produce bright stock. The petrolatum layer is also distilled to remove naphtha and may be clay treated or acid and clay treated to improve the color.

7.3.4 Centrifuge dewaxing process

Another method of separating petrolatum from reduced crude is *centrifuge dewaxing*. In this process the reduced crude is dissolved in naphtha and chilled to −18°C (0°F) or lower, which causes the wax to separate. The mixture is then fed to a battery of centrifuges where the wax is separated from the liquid. However, the centrifuge method has now been largely displaced by solvent dewaxing methods and by more modern methods of wax removal.

7.3.5 Catalytic dewaxing process

There are also later-generation dewaxing processes that are being brought onstream in various refineries (Hargrove, 1992; Genis, 1997). Solvent dewaxing is relatively expensive for the production of low pour point oils, various catalytic dewaxing (selective hydrocracking) processes have been developed for the manufacture of lube oil base stocks. The basic process consists of a reactor containing a proprietary dewaxing catalyst followed by a second reactor containing a hydrogen finishing catalyst to saturate olefins created by the dewaxing reaction and to improve stability, color, and demulsibility of the finished lube oil.

BP has developed a *hydrocatalytic dewaxing process* that is reputed to overcome some of the disadvantages of the solvent dewaxing processes. In the process of dewaxing, waxy molecules are removed from heavy distillate fuel cuts or lube distillates.

Catalytic dewaxing is a hydrocracking process and is, therefore, operated at elevated temperatures (280°C–400°C, 550°F–750°F) and pressures (300/1500 psi). However, the conditions for a particular dewaxing operation depend upon the nature of the feedstock and the product pour point required. The catalyst employed for the process is a mordenite-type catalyst that has the correct pore structure to be selective for *n*-paraffin cracking. Platinum on the catalyst serves to hydrogenate the reactive intermediates so that further paraffin degradation is limited to the initial thermal reactions. The process has been employed to successfully dewax a wide range of naphthenic feedstocks (Hargrove et al., 1979), but it may not be suitable to replace solvent dewaxing in all cases. The process has the flexibility to fit into normal refinery operations and can be adapted for prolonged periods on-stream.

Other processes include the ExxonMobil distillate dewaxing (MDDW) process (Smith et al., 1980; Safre, 2003) by the dewaxing of which is achieved by selective cracking in which the long paraffin chains are cracked to form shorter chains using a shape-selective zeolite that rejects ring compounds and isoparaffins. In a related process (MIDW process) the paraffins are selectively isomerized using low-pressure conditions (Smith et al., 1980; Safre, 2003). This process also uses a zeolite catalyst to convert low quality gas oil into diesel fuel.

In the process the proprietary catalyst can be reactivated to fresh activity by relatively mild nonoxidative treatment. Of course, the time allowed between reactivation is a function of the feedstock, but after numerous reactivation, it is possible that there will be coke buildup on the catalyst. The process can be used to dewax a full range of lubricating base stocks and, as such, has the potential to completely replace solvent dewaxing or can even be used in combination with solvent dewaxing. This latter option, of course, serves to debottleneck existing solvent dewaxing facilities.

Both the catalytic dewaxing processes have the potential to change the conventional thoughts about dewaxing insofar as they are not solvent processes and may be looked upon (more correctly) as thermal processes rather than treatment processes. However, both provide viable alternatives to the solvent processes and offer a further advance in the science and technology of refinery operations.

Catalytic dewaxing yields various grades of lube oils and fuel components suitable for extreme winter conditions. Paraffinic (waxy) components that precipitate out at low temperatures are removed. In the universal oil product (UOP) catalytic dewaxing process the first stage saturates olefins and desulfurizes and denitrifies the feed via hydrotreating (Genis, 1997). In the second stage a dual-function, nonnoble-metal zeolite catalyst selectively adsorbs and then selectively hydrocracks the normal and near-normal long-chain paraffins to form shorter chain (nonwaxy) molecules.

Alternatively, in the Chevron Isodewaxing process, the dewaxing results from isomerizing the linear paraffins to branched paraffins by using a molecular sieve catalyst containing platinum (Miller, 1994a,b). In the Isodewaxing process, which is followed by a hydrofinishing step, waxy feedstock from a hydrocracker or hydrotreater, together with the hydrogen-containing gas, is heated and fed to the Isodewaxing reactor. The conditions in the reactor cause isomerization of *n*-paraffins to isoparaffins and. Other paraffins are cracked to highly saturated low-boiling products such as jet fuel and diesel fuel. The effluent from the Isodewaxing reactor is then sent to the hydrofinisher where hydrofinishing, including aromatics saturation provide the product. The catalysts used in the Isodewaxing and hydrofinishing units are selective for dewaxing and hydrogenation. The catalysts are at their maximum efficiency with low-sulfur and low-nitrogen feedstocks. The process generally uses a high-pressure recycle loop. Because of the conversion of wax constituents to other usable products, the process has obvious benefits over solvent dewaxing insofar as quality of the product is increased.

7.4 The future

The current global petroleum market will continue to offer opportunities for solvent-based bottom-of-the-barrel technologies to play an important role in the refiner's continuous efforts to balance available crude qualities with the market demands. A successful outcome to this scenario can only be possible with notable amounts of investments in conversion capacity.

In addition, process innovations have been introduced, several using piggy-back techniques (where one process works in close conjunction with another process) and, there are other options that have not yet been introduced or even invented but may well fit into the refinery of the future.

Significant processes utilizing proven technology are, however, available to meet this objective. The decision of which specific technology to apply in a particular area will depend on the choice of the most abundant crude available, the performance and environmental specifications of the marketplace, and the economic returns necessary to justify the technology.

7.4.1 Deasphalting

The solvent deasphalting process has been operated in refineries for several decade, usually as a piggy-back to other process or more accurately, as a preprocessing step process before application of fluid coking or fluid catalytic cracking. Nevertheless, the process has a place in future refineries that will change as the decades of the 21st century progress.

The solvent deasphalting process is a unique separation process in which residue is separated by molecular type, instead of by boiling point as in the vacuum distillation process. The UOP/FW solvent deasphalting process produces a low-contaminant DAO rich in paraffinic type molecules. These fractions can then be further processed in conventional conversion units such as a fluid catalytic cracker (FCC) unit or hydrocracking unit.

Propane deasphalting is most commonly used to remove asphaltene constituents and resin constituents which contribute undesirable properties to the lube base stocks and to catalytic cracking feedstocks. The process typically uses baffle towers or rotating disk contactors to mix the propane with the feed. Solvent recovery is accomplished with evaporators, and supercritical solvent recovery processes are also used in some deasphalting units.

Another deasphalting process is the Duo-Sol Process that is used to both deasphalt and extract lubricating oil feedstocks. Propane is used as the deasphalting solvent and a mixture of phenol and cresylic acids are used as the extraction solvent. The extraction is conducted in a series of batch extractors followed by solvent recovery in multistage flash distillation and stripping towers.

Propane remains the solvent of choice for the process (either for lube oil production or for asphalt production) due to its ability to preferentially extract paraffinic hydrocarbons and to reject most of the coke-forming precursors. In addition, a mixture of propane and butane is valuable for preparing feedstocks for catalytic cracking processes due to its ability to remove metal-bearing components. More recently, pentane deasphalting, plus hydrodesulfurization, can produce more feedstock for catalytic cracking or low-sulfur fuel oil and it is anticipated the higher boiling solvent (pentane) will see more use as the 21st century evolves.

The need for a more selective deasphalting process development has led to a process where separation of a portion of the total asphaltene fraction from feedstocks is required, a stepwise deasphalting process can selectively remove inorganic solids and heteroatom constituents from heavy feedstocks, such as tar sand bitumen (US Patent 3,779,902; Canadian Patent 969,497).

Direct hydrodesulfurization followed by hydrocracking of heavy feedstocks is one of the best methods of heavy residue upgrading in refining industry but problems emerge when metal and asphaltene contents of residue are high. In fact, the presence of these compounds adversely influences the activities of the hydrodesulfurization and hydrocracking catalysts. The solvent deasphalting process will continue to be used to improve feedstocks to hydrocracking and hydrotreating processes as it is one of the most efficient approaches to reduce metal and asphaltene contents of heavy oil cuts before sending them to hydrodesulfurization and hydrocracking units. In order to maximize the yield of DAO, important factors such as solvent composition and ratio of the solvent to the feed, temperature, pressure, and the type of extractor equipment will continue to be of future importance.

Propane has been used extensively in deasphalting asphalt-containing hydrocarbon feedstocks, especially in the preparation of high-quality lubricating oils. The use of propane has necessitated elaborate solvent cooling systems utilizing cold water, which is a relatively expensive cooling agent. In order to circumvent such technology, future units will use solvent systems that will allow operation at elevated temperatures relative to conventional propane deasphalting temperatures, thereby permitting easy heat exchange. In addition, it may be found desirable to integrate dewaxing operations with deasphalting operations by having a common solvent recovery system. This will require changes to the solvent composition and the inclusion of solvents not usually considered to be deasphalting solvents.

The introduction of the ROSE process in the 1970s made it possible to design energy-efficient, cost-effective deasphalting units of larger capacity. This breakthrough allowed solvent deasphalting technology to be used effectively as the primary bottom-of-the-barrel upgrader in a deep conversion refinery. Currently, this supercritical fluid technology is being used around the world to produce FCC feeds, lubricating oil bright stocks, deasphalted gas oil feeds for hydrotreating and hydrocracking units, specialty resins, and heavy fuel and asphalt blending components. Recent advances in supercritical separator design have led and will continue to lead to the development of new improvements in solvent deasphalting technology that have been incorporated into ROSE units. The new separator technology provides a significant increase in capacity for existing ROSE units and lower capital and operating costs for future ROSE plants. Improved computerized methods for predicting physical and transport properties for solvent, DAO, and asphaltene components at supercritical conditions are now utilized for design. These methods allow the process designer to perform meaningful process-simulation studies to ensure optimum utilization of the equipment.

Also, the UOP Demex process recovers processable oil from vacuum residues while rejecting the metal-containing constituents and the asphaltene-containing constituents. The demetallized oil product can be processed in downstream catalytic units at less severe conditions and with much lower catalyst consumption than required for residues. The asphaltenes and metals present in the vacuum residues are contained in the by-product pitch, for which there are several uses. The Demex process has its technical roots in propane deasphalting, but it uses higher molecular weight solvents, which recover higher yields of good-quality demetallized oil. Pilot plant studies and engineering innovations have made the Demex process a significant advancement over conventional deasphalting in terms of yield, product utilization, and, most importantly, energy efficiency.

In addition, as a means of energy reduction for the process, in future deasphalting units, the conventional solvent recovery scheme can be retrofitted with supercritical solvent recovery scheme to reap benefits of higher energy efficiency. Other improvements include variations in the extraction column internals.

For example, the three major properties, which influence the design of extraction column, are interfacial tension, viscosity, and density of phases. Solvent deasphalting extraction column is characterized by low interfacial tension, high viscosities of asphaltene phase, and a density difference between the phases. Extension of these property correlations for solvent deasphalting applications will be suitably validated and corrections made where necessary to improve extraction performance and yields of the products.

Other areas of future process modification will be in the extractor tower internals, studies with higher molecular weight solvent, accurate estimation of physical properties of mix stream, studies in combination with other processes and firming up design tools for supercritical solvent recovery configuration.

As noted above, the separation of asphaltic constituents from petroleum using a lower energy method for separating asphaltenes from petroleum liquid would be useful to the industry and would significantly decrease the energy required for deasphalting with solvents (e.g., propane). To make the overall process energy efficient, it is necessary that energy conservation measures are taken. Some of these measures are multistage solvent recovery and recovering solvent from DAO from extraction column above the critical temperature of the solvent; such measures reduce energy consumption in a propane deasphalting unit by approximately 30%.

Whilst the majority of the deasphalting units will continue with the tried-and-true technology alterations such as (1) variations in solvent composition, (2) temperature variations, (3) pressure versions, and (4) change to the extractor internals, there are three concepts (at least that will bring about a change to deasphalting as it is currently practiced) and these are (i) the use of membrane technology, (ii) electrical technology, and (iii) thermal technology.

The separation of asphaltene constituents and resins constituent from petroleum liquid could be achieved via high-temperature ultrafiltration through the use of a membrane. The removal of waxes from petroleum liquid could be accomplished either through ultrafiltration or microfiltration. The industry will benefit from lower cost fabrication of high-temperature membranes and modules, as well as fouling-resistant membranes and/or effective fouling control techniques, such as compositional doping, hydrodynamics, shear rate, and pressure.

It is likely that no single membrane material will have all of the desired characteristics for membrane separation technologies as applied to the deasphalting process. Important properties for emerging membrane materials include impact strength, flexibility, and thermal stability, as well as the properties required for a particular separation (i.e., flux and selectivity). The development of materials for specific applications will only be economical if the application is widely used in the refining industry.

Electrochemistry also has the potential to influence the efficiency of the deasphalting process. The asphaltic constituents of heavy feedstocks are known to be polar entities and may even contain charge within the molecules. Electric fields could be additive to selectively enhance or retard the transport of a species across a membrane. A combined-effect membrane could be constructed with both sufficient strength to permit the application of a significant pressure field and ionic transport capability. Electrode material could be applied to either side of the membrane. The promise of electrochemical separation methods is that they will be able to selectively remove the most polar constituents under controlled conditions with high efficiency.

On the other hand, thermal methods of separation will also assume a prominence in petroleum processing. The actual process may appear more as a visbreaker in which the operating parameter take the unit to a point beyond the typical operating point (Chapter 5: Thermal Cracking) so that precoke and coke are separated from the body of the oil. The amount of material separated by such a process will be feedstock and temperature dependent and solvent may or may not be required.

The process, which is a modified visbreaking operation, can be compared to be a thermal deasphalter with integrated asphaltene coking. There are a wide range of possible refining application scenarios, including resid pretreatment prior to downstream catalytic processes that are sensitive to contaminants and upgrading of low value residues such. The process also lends itself to the additive of scavenger additives (such as calcium oxide, similar metal oxides, and clay) to encourage the precoke and even asphaltene constituents to separate with minimal energy input.

The efficiency of a typical solvent deasphalting process can be successfully increased at laboratory-scale conditions by the inclusion of nanoparticles in systems with different solvent-to-oil ratios, temperatures, types of solvent, and dosage. In fact, the use of nanoparticles reduces the yield of the DAO in comparison with the traditional solvent deasphalting process and the end result is (as expected) a higher quality product. This process may be a more suitable alternative regarding the asphaltene content and the sulfur content as well as an improved API gravity and other rheological properties (Guzmán et al., 2017).

In some cases the scavenger and any adhering organic (or inorganic) material and any scavenged material will be sent to the gasifier that will become a standard installation in most refineries, if not as part of a flexicoking unit (Chapter 5: Thermal Cracking) then as a gasifier for cracked resids and biofeedstock (Chapter 10: Nonfossil Fuel Feedstocks) for hydrogen production. Indeed, the use of calcium oxide (and similar metal oxides) as scavengers in a thermal deasphalter carries an additional benefit insofar a calcium oxide (and related metal oxides) are also thermal desulfurizers—the products being calcium sulfide (or the corresponding metal sulfide) and water as well as hydrocarbons.

7.4.2 Dewaxing

Waxes of both crystalline and microcrystalline types occurring naturally in lubricating oil fractions derived from crude petroleum are undesirable components of the finished oils. Advances during the past several decades have resulted in the extensive use of solvents for dewaxing. Solvent dewaxing is used to remove wax from either distillate or residual bases stock at any stage in the refining process. There are several processes in use for solvent dewaxing, but all have the same general steps, which are (1) mixing the feedstock with a solvent, (2) precipitating the wax from the mixture by chilling, and (3) recovering the solvent from the wax and dewaxed oil for recycling by distillation and steam stripping. Usually two solvents are used: toluene, which dissolves the oil and maintains fluidity at low temperatures, and MEK, which dissolves little wax at low temperatures and acts as a wax precipitating agent. Other solvents that are sometimes used include benzene, methyl isobutyl ketone, propane, petroleum naphtha, ethylene dichloride, methylene chloride, and sulfur dioxide. In addition, there is a catalytic process used as an alternate to solvent dewaxing.

While further developments will occur in the types of catalysts designed for dewaxing processing, the dewaxing processes in future refineries will make greater use of membrane separators.

For example, an asymmetric polyimide membrane has been developed for recovering cold solvents from lube oil in the solvent dewaxing process. A membrane process can be used to prevent hold-ups in the refrigeration and recovery sections of a solvent lube plant. Recovering cold solvent directly from the filtrate and reducing the amount of solvent subjected to the heating and cooling in the solvent recovery section will result in energy savings. However, the membrane will require a high flux and must be robust and able to withstand continuous service.

In addition, with the use of microwave technology steadily growing in industrial processes and the wax interacting with the electromagnetic energy of the microwaves, it will not be surprising to see such an installation is the refinery of the future where wax may arise from additional sources (including biosources) and have different properties that are not conducive to separation by the typical dewaxing process.

Improving the cold flow properties of paraffinic feedstocks in a selective way has become an important aspect of refining during the last two decade as refiners search for more effective and cost-efficient ways to achieve improvements in cold flow properties. The growing trend is to use catalytic dewaxing to limit the use of cold flow additives, reduce kerosene blending requirements, upgrade heavier feedstocks with higher cloud and/or pour points, and, consequently, to create more room in the blending pool for heavier feedstocks. Also, increasingly stringent specifications, the rise in new types of crude from different origins or process routes, and the desire to sell products that meet cold flow property specifications result in the need to process more challenging types of feedstocks—some of them being heavier, some lighter but of different compositions, and some being significantly more paraffinic.

Flow improvers modify the wax crystallization process, by reducing the crystal size and/or the lattice formation of the solid phases, and reduce both the cold filter plugging point and the pour point. However, the cloud point—a property related to individual component characteristics and driven by the heaviest molecules within the feedstock boiling range—is also the most thermodynamically driven property. Consequently it is also the most difficult to effectively reduce by additives or by cost-effective dilution with kerosene. This becomes a greater challenge when feeds are becoming more paraffinic in nature, with the presence of longer and consequently higher cloud, linear alkanes.

Cold flow improvers can significantly reduce cold filter plugging point and pour point. Cloud point improvement using additives is typically within a couple of degrees, up to a maximum of 3°C–4°C (5°F 7°F). With hydrotreated kerosene blending, a cloud point reduction of approximately 1°C (1.8°F) is typically achieved for every 10% of kerosene added. If a cloud point improvement of more than 6°C–8°C (11°F–14°F) is desired, catalytic dewaxing is usually a more economical long-term solution than any alternative method (additives and/or kerosene blending and/or feedstock boiling range adjustment).

Improving the cold flow properties of any feedstock requires mainly modifying or removing linear alkanes (usually referred to as paraffins). This can be achieved either by a physical separation method (extraction) or by different selective chemical reactions (catalytic dewaxing). This article focuses on the latter.

References

Billon, A., Morel, F., Morrisson, M.E., Peries, J.P. 1994. Revue De Institut Français du Pétrole. 1994, vol. 49, No. 05, pp. 495–507.

Dickenson, R.L., Biasca, F.E., Schulman, B.L., Johnson, H.E., 1997. Refiner options for converting and utilizing heavy fuel oil. Hydrocarbon Process.

Ditman, J.G., 1973. Hydrocarbon Process. 52 (5), 110.

Dunning, H.N., Moore, J.W., 1957. Propane removes asphalts from crudes. Pet. Refin. 36 (5), 247–250.

Gearhart, J.A., 1980. Hydrocarbon Process. 59 (5), 150.

Gearhart, J.A., Gatwin, L., 1976. ROSE process improves resid feed. Hydrocarbon Process. 125–128.

Genis, O., 1997. In: Meyers, R.A. (Ed.), Handbook of Petroleum Refining Processes. McGraw-Hill, New York, Chapter 8.5.

Girdler, R.B., 1965. Proc. Assoc. Asphalt Paving Technologists. vol. 34, p. 45.

Guzmán, J.G., Camilo, A., Franco, C.A., Cortés, F.B., 2017. An enhanced-solvent deasphalting process: effect of inclusion of SiO_2 nanoparticles in the quality of deasphalted oil. J. Nanomater. Available from: https://doi.org/10.1155/2017/9470230Article ID 9470230.

Hargrove, J.D., 1992. In: McKetta, J.J. (Ed.), Petroleum Processing Handbook. Marcel Dekker Inc, New York, p. 558.

Hargrove, J.D., Elkes, G.J., Richardson, A.H., 1979. Oil Gas J. 77 (3), 103.

Houde, E.J., 1997. In: Meyers, R.A. (Ed.), Handbook of Petroleum Refining Processes, second ed. McGraw-Hill, New York, Chapter 10.4.

Hydrocarbon Processing, 1996. 75(11): 89 et seq.

Hydrocarbon Processing, 1998. 77(11): 53 et seq.

Kashiwara, H., 1980. Kagaku Kogaku. (7): 44.

Low, J.Y., Hood, R.L., Lynch, K.Z., 1995. Prepr. Div. Petrol. Chem. Am. Chem. Soc. 40, 780.

Miller, S.J., 1994a. In: Weitkamp, J. (Ed.), Studies in Surface Science and Catalysis, 84C. Elsevier, Amsterdam., pp. 2319–2326.

Miller, S.J., 1994b. Microporous Mater. 2, 439–450.

Mitchell, D.L., Speight, J.G., 1973. Fuel 52, 149.

Northrup, A.H., Sloan, H.D., 1996. Annual Meeting. National Petroleum Refiners Association. Houston, TX. Paper AM-96-55.

Peries, J.P., Billon, A., Hennico, A., Morrison, E., Morel, F., 1995. Proceedings Sixth UNITAR International Conference on Heavy Crude and Tar Sand, vol. 2, p. 229.

RAROP. 1991. RAROP Heavy Oil Processing Handbook. Research Association for Residual Oil Processing. T. Noguchi (Chairman). Ministry of Trade and International Industry (MITI), Tokyo.

Safre, A., 2003. Proceedings Fourth International Conference on Oil Refining and Petrochemicals in the Middle East. Abu Dhabi. January 28.

Salazar, J.R., 1986. In: Meyers, R.A. (Ed.), Handbook of Petroleum Refining Processes. McGraw-Hill, New York, Chapter 8.5.

Sattarin, M., Modarresi, M., Talachi, H., Teymori, M., 2006. Solvent deasphalting of vacuum residue in a bench-scale unit. Pet. Coal 48 (3), 14–19.

Scholten, G.G., 1992. In: McKetta, J.J. (Ed.), Petroleum Processing Handbook. Marcel Dekker Inc, New York, p. 565.

Smith, K.W., Starr, W.C., Chen, N.Y., 1980. Oil Gas J. 78 (21), 75.

Speight, J.G., 2000. The Desulfurization of Heavy Oils and Residua, second ed. Marcel Dekker Inc, New York.

Speight, J.G., Ozum, B., 2002. Petroleum Refining Processes. Marcel Dekker Inc, New York.

Desulfurization, denitrogenation, and demetalization

8

8.1 Introduction

Catalytic hydrotreating, also referred to as hydroprocessing or hydrodesulfurization (HDS), commonly appears in multiple locations in a refinery (Meyers, 1997; Speight, 2000, 2014, 2017; Parkash, 2003; Gary et al., 2007; Hsu and Robinson, 2017). Furthermore, hydrotreating, a term often used synonymously with HDS, is a catalytic refining process widely used to remove sulfur from crude oil products such as naphtha, gasoline, diesel fuel, kerosene, and fuel oil (Meyers, 1997; Speight, 2000, 2014, 2017; Parkash, 2003; Gary et al., 2007; Hsu and Robinson, 2017). The objective of the hydrotreating process is to remove sulfur as well as other unwanted compounds, for example, unsaturated hydrocarbon derivatives and nitrogen from refinery process streams. HDS processes typically include facilities for the capture and removal of the resulting hydrogen sulfide (H_2S) gas, which is subsequently converted into byproduct elemental sulfur or sulfuric acid.

Hydrotreating and desulfurization are processes for treatment of crude oil products to ensure that they meet performance specifications. During the hydrotreating and desulfurization processes several chemical conversions occur: (1) olefins are converted to alkanes, (2) aromatics are converted to cycloalkanes and alkanes, and (3) sulfur is removed (Dolbear, 1998):

$$-CH=CH- \; + H_2 \rightarrow \; -CH_2CH_2-$$

$$\text{Benzene} + H_2 \rightarrow \text{cyclohexane} + \text{hexane}$$

$$-CH_2SH + H_2 \rightarrow \; -CH_3 + H_2S$$

$$-CH_2SCH_2- \; + H_2 \rightarrow \; -CH_3 + H_2S + \; -CH_3$$

Reactions involving catalytic hydrogenation of organic substances have been known for more than 110 years, that is, prior to 1897 and the property of finely divided nickel to catalyze the fixation of hydrogen on hydrocarbon (ethylene, benzene) double bonds. In the mid-1950s the first noble metal catalytic reforming process commercialized and, at the same time, the catalytic HDS of the naphtha feed to such reformers was also commercialized. In the following decades, various proprietary catalytic HDS processes have been commercialized. Currently, virtually all of the crude oil refineries worldwide have one or more hydrotreating and/or HDS units.

The Refinery of the Future. DOI: https://doi.org/10.1016/B978-0-12-816994-0.00008-7

With the influx of higher boiling viscous feedstocks into refineries, hydroprocessing will assume a greater role in the refinery of the future. In addition to being recognized as a chemical hydrogenation process, the process uses the principle that the presence of hydrogen during a thermal reaction of a crude oil feedstock terminates many of the coke-forming reactions and enhances the yields of the lower boiling components, such as gasoline, kerosene, and jet fuel.

Hydrogenation processes for the conversion of crude oil fractions and crude oil products may be classified as destructive and nondestructive. Destructive hydrogenation (hydrogenolysis or hydrocracking) is characterized by the cleavage of carbon–carbon linkages accompanied by hydrogen saturation of the fragments to produce lower boiling products. Such treatment requires severe processing conditions and the use of high hydrogen pressures to minimize polymerization and condensation that lead to coke formation. Many other reactions, such as isomerization, dehydrogenation, and cyclization, occur under the drastic conditions employed.

On the other hand, hydrotreating is a process in which thermal decomposition is usually minimized (Figs. 8.1–8.3) (Meyers, 1997; Parkash, 2003; Gary et al., 2007; Speight, 2014, 2017; Hsu and Robinson, 2017). Hydrotreating (nondestructive hydrogenation) is generally used for the purpose of improving product quality without appreciable alteration of the boiling range. Mild processing conditions are employed so that only the more unstable materials are attacked. Thus nitrogen, sulfur, and oxygen compounds undergo hydrogenolysis to split out ammonia, hydrogen sulfide, and water, respectively. Olefins are saturated, and unstable compounds, such as diolefins, which might lead to the formation of gums or insoluble materials, are converted to more stable compounds. Heavy metals present in the feedstock are also usually removed during hydrogen processing (Parkash, 2003; Gary et al., 2007; Speight, 2014, 2017; Hsu and Robinson, 2017).

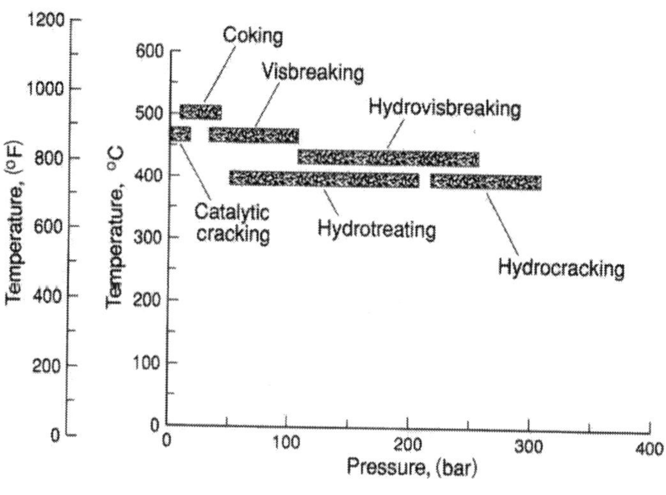

Figure 8.1 Process conditions (temperature and pressure) for refinery processes.

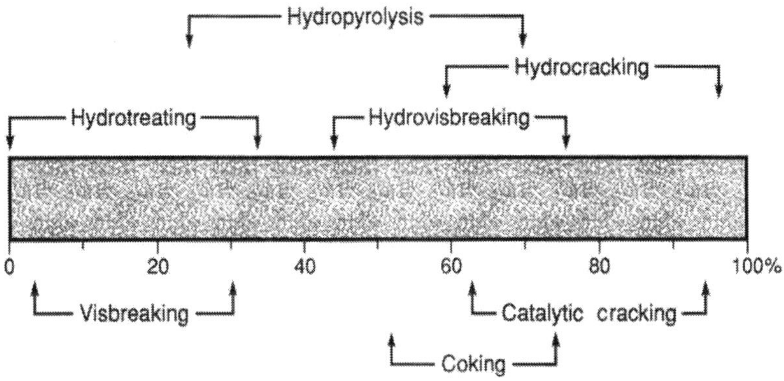

Figure 8.2 Conversion (based on feedstock) to liquids for refinery processes.

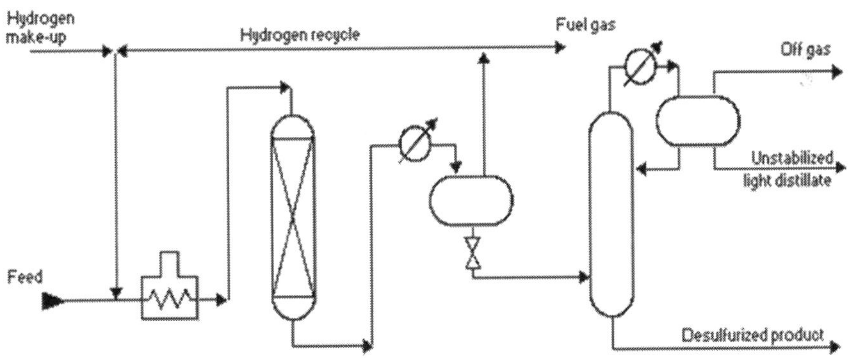

Figure 8.3 Distillate hydrotreating.
Source: Modified from OSHA Technical Manual, Section IV, Chapter 2: Petroleum Refining Processes. http://www.osha.gov/dts/osta/otm/otm_iv/otm_iv_2.html.

The major differences between hydrotreating and hydrocracking are the time at which the feedstock remains at reaction temperature and the extent of the decomposition of the nonheteroatom constituents and products. The upper limits of hydrotreating conditions may overlap with the lower limits of hydrocracking conditions. And where the reaction conditions overlap, feedstocks to be hydrotreated will generally be exposed to the reactor temperature for shorter periods; hence the reason why hydrotreating conditions may be referred to as (relatively) mild.

A growing dependence on high-heteroatom high-boiling oils and residua has emerged as a result of continuing decreasing availability of conventional crude oil through the worldwide depletion of crude oil reserves. Thus the ever growing tendency to convert as much as possible of lower grade feedstocks to liquid products is causing an increase in the total sulfur content in penultimate products. Refiners must, therefore, continue to remove substantial portions of sulfur from the lower boiling products so that the final products meet specifications. However, residua, heavy crude oil, and tar sand bitumen pose a particularly difficult problem in terms of the propensity of these feedstocks to form coke and to shorten catalyst life (Speight, 2000; Ancheyta and Speight, 2007).

8.2 Rationale for hydroprocesses

There is an approximate correlation between the quality of crude oil products and their hydrogen content (Dolbear, 1998). It so happens that desirable aviation gasoline, kerosene, diesel fuel, and lubricating oil are made up of hydrocarbon derivatives containing high proportions of hydrogen. In addition, it is usually possible to convert olefins and higher molecular weight constituents to paraffins and monocyclic hydrocarbon derivatives by hydrogen-addition processes. These facts have for many years encouraged attempts to employ hydrogenation for refining operations; despite considerable technical success, such processes were not economically possible until hydrogen became available as a result of the rise of reforming, which converts alkanes to naphthenes and thence to aromatics with the release of hydrogen.

In addition to the hydrogen content of crude oil products, there are several valid reasons for removing heteroatoms from crude oil fractions. These include the following:

1. reduction, or elimination, of corrosion during refining, handling, or use of the various products;
2. production of products having an acceptable odor and specification;
3. increasing the performance (and stability) of gasoline;
4. decreasing smoke formation in kerosene; and
5. reduction of heteroatom content in fuel oil to a level that improves burning characteristics and is environmentally acceptable.

Catalytic hydrotreating is a hydrogenation process used to remove approximately 90% of contaminants such as nitrogen, sulfur, oxygen, and metals from liquid crude oil fractions. These contaminants, if not removed from the crude oil fractions as they travel through the refinery processing units, can have detrimental effects on the equipment, the catalysts, and the quality of the finished product. Typically, hydrotreating is done prior to processes such as catalytic reforming so that the catalyst is not contaminated by untreated feedstock. Hydrotreating is also seeing increasing use prior to catalytic cracking to reduce sulfur and improve product yields and to upgrade middle-distillate crude oil fractions into finished kerosene, diesel fuel, and heating fuel oils. In addition, hydrotreating converts olefins and aromatics to saturated compounds. At one time, hydrotreating a high-boiling feedstock was not even considered because of the hydrogen demands and the detrimental effect of the feedstock on the catalyst.

In addition to HDS and hydrodenitrogenation (HDN), the removal of aromatic constituents from some product streams has also become essential. The high aromatic content in diesel fuel has been recognized both to lower the fuel quality and to contribute significantly to the formation of undesired emissions in exhaust gases.

Indeed, as a result of the stringent environmental regulations, processes for aromatic reduction in middle distillates continue to receive considerable attention.

Unsaturated compounds, such as olefins, are not indigenous to crude oil and produced during cracking processes and need to be removed from product streams because of the tendency of unsaturated compounds and heteroatomic polar compounds to form gum and sediment. On the other hand, aromatic compounds are indigenous to crude oil and some may be formed during cracking reactions. The most likely explanation is that the aromatic compounds present in product streams are related to the aromatic compounds originally present in crude oil but now having shorter alkyl side-chains. Thus in addition to olefins, product streams will contain a range of aromatic compounds that been to be removed to enable many of the product streams to meet product specifications.

Of the aromatic constituents the polycyclic aromatics are first partially hydrogenated before cracking of the aromatic nucleus takes place. The sulfur and nitrogen atoms are converted to hydrogen sulfide and ammonia, but a more important role of the hydrogenation is probably to hydrogenate the coke precursors rapidly and prevent their conversion to coke.

8.3 Process options

Although many different hydrotreater designs are marketed, they all work along the same principle—all processes use the reaction of hydrogen with the hydrocarbon feedstock to produce hydrogen sulfide and a desulfurized hydrocarbon product (Fig. 8.3).

Hydrotreating is carried out by charging the feed to the reactor, together with a portion of all the hydrogen produced in the catalytic reformer. Suitable catalysts are tungsten–nickel sulfide, cobalt–molybdenum–alumina, nickel oxide–silica–alumina, and platinum–alumina. Most processes employ cobalt–molybdenum catalysts, which generally contain approximately 10% by weight molybdenum oxide and less than 1% by weight cobalt oxide supported on alumina. The temperatures employed are in the range of 300°C–345°C (570°F–850°F), and the hydrogen pressures are on the order of 500–1000 psi.

The feedstock is preheated and mixed with hot recycle gas containing hydrogen and the mixture is passed over the catalyst in the reactor section at temperatures between 290°C and 445°C (550°F and 850°F) and pressures between 150 and 3000 psig (1 and 20.7 MPa gauge) (Table 8.1). The reactor effluent is then cooked by heat exchange, and desulfurized liquid hydrocarbon product and recycle gas are separated at essentially the same pressure as used in the reactor. The recycle gas is then scrubbed and/or purged of the hydrogen sulfide and low-boiling hydrocarbon gases, mixed with fresh hydrogen makeup, and preheated prior to mixing with hot hydrocarbon feedstock.

In some designs the feedstock is heated and then mixed with the hydrogen. The reaction temperature should not exceed 800°F (430°C) to minimize cracking. The

Table 8.1 Process parameters for hydrodesulfurization.

Parameter	Naphtha	Residuum
Temperature (°C)	300–400	340–425
Pressure (atm.)	35–70	55–170
LHSV	4.0–10.0	0.2–1.0
H_2 recycle rate (scf/bbl)	400–1000	3000–5000
Catalysts life (years)	3.0–10.0	0.5–1.0
Sulfur removal (%)	99.9	85.0
Nitrogen removal (%)	99.5	40.0

LHSV, Liquid hourly space velocity.
Source: Modified from Speight, J.G., 2007. The Chemistry and Technology of Petroleum, fourth ed. CRC Press, Taylor & Francis Group, Boca Raton, FL.

gas mixture is led over a catalyst bed of metal oxides (most often cobalt or molybdenum oxides on different metal carriers). The catalysts help the hydrogen to react with sulfur and nitrogen to form hydrogen sulfides (H_2S) and ammonia. The reactor effluent is then cooled, and the oil feed and gas mixture is then separated in a stripper column. Part of the stripped gas may be recycled to the reactor.

The recycle gas scheme is used in the HDS process to minimize physical losses of expensive hydrogen. HDS reactions require a high hydrogen partial pressure in the gas phase to maintain high desulfurization reaction rates and to suppress carbon laydown (catalyst deactivation). The high hydrogen partial pressure is maintained by supplying hydrogen to the reactors at several times the chemical hydrogen consumption rate. The majority of the unreacted hydrogen is cooled to remove hydrocarbon derivatives, recovered in the separator, and recycled for further utilization. Hydrogen is physically lost in the process by solubility in the desulfurized liquid hydrocarbon product and from losses during the scrubbing or purging of hydrogen sulfide and low-boiling hydrocarbon gases from the recycle gas.

The reaction generally takes place in the vapor phase but, depending on the application, may be a mixed-phase reaction. The reaction products are cooled in a heat exchanger and led to a high-pressure separator where hydrogen gas is separated for recycling. Liquid products from the high-pressure separator flow to a low-pressure separator (stabilizer) where dissolved low-boiling gases are removed. The product may then be fed to a reforming or cracking unit if desired.

The operating conditions in distillate HDS are dependent upon the feedstock as well as the desired degree of desulfurization or quality improvement. Kerosene and low-boiling gas oils are generally processed at mild severity and high throughput whereas low-boiling catalytic cycle oils and thermal distillates require slightly more severe conditions. Higher boiling distillates, such as vacuum gas oils and lube oil extracts, require the most severe conditions.

One of the issue-related problems in processing of high-sulfur and high-nitrogen feeds is the large quantity of hydrogen sulfide (H_2S) and ammonia (NH_3) that are produced. Substantial removal of both compounds from the recycle gas can be achieved by the injection of water in which, under the high-pressure conditions employed, both hydrogen sulfide and ammonia are very soluble compared with hydrogen and hydrocarbon gases. The solution is processed in a separate unit for the recovery of anhydrous ammonia and hydrogen sulfide.

Generally, it is more economical to hydrotreat high-sulfur feedstocks before catalytic cracking than to hydrotreat the products from catalytic cracking. The advantages are as follows:

1. The products require less finishing.
2. Sulfur is removed from the catalytic cracking feedstock, and corrosion is reduced in the cracking unit.
3. Coke formation during cracking is reduced and higher conversions result.
4. The catalytic cracking quality of the gas oil fraction is improved.

Although hydrocracking will occur during hydrotreating, attempts are made to minimize such effects but the degree of cracking is dependent on the nature of the feedstock. For example, decalin (decahydronaphthalene) cracks more readily than the corresponding paraffin analog, n-decane, $[CH_3(CH_2)_8CH_3]$ to give higher iso-paraffin to *n*-paraffin product ratios than those obtained from the paraffin. A large yield of single-ring naphthenes is also produced, and these are resistant to further hydrocracking and contain a higher than equilibrium ratio of methylcyclopentane to cyclohexane.

One of the issues that arises when hydroprocessing residua is the deposition of metals, in particular vanadium, on the catalyst. It is not possible to remove vanadium from the catalyst, which must, therefore, be replaced when deactivated, and the time taken for catalyst replacement can significantly reduce the unit time efficiency. Fixed-bed catalysts tend to plug owing to solids in the feed or carbon deposits when processing residual feeds. In addition, the exothermic reaction at high conversion gives difficult reactor design problems in heat removal and temperature control.

The problems encountered in hydrotreating high-boiling feedstocks can be directly equated to the amount of complex, higher boiling constituents that may require pretreatment (Speight and Moschopedis, 1979; Parkash, 2003; Gary et al., 2007; Speight, 2014, 2017; Hsu and Robinson, 2017). Processing these feedstocks is not merely a matter of applying know-how derived from refining conventional crude oils but requires knowledge of the composition (Chapter 1: Feedstock Types and Properties). The materials are not only complex in terms of the carbon number and boiling point ranges but also because a large part of this envelope falls into a range of model compounds and very little is known relating to the properties. It is also established that the majority of the higher molecular weight materials produce coke (with some liquids) but the majority of the lower molecular weight constituents produce liquids (with some coke).

It is the physical and chemical composition of a feedstock that plays a large part not only in determining the nature of the products that arise from refining operations but also in determining the precise manner by which a particular feedstock should be processed (Speight, 1986; Parkash, 2003; Gary et al., 2007; Speight, 2014, 2017; Hsu and Robinson, 2017). Furthermore, it is apparent that the conversion of heavy crude oils, extra heavy crude oil, tar sand bitumen, and residua requires new lines of thought to develop suitable processing scenarios (Parkash, 2003; Gary et al., 2007; Speight, 2014, 2017; Hsu and Robinson, 2017). Indeed, the use of thermal (carbon rejection) and hydrothermal (hydrogen addition) processes, which were inherent in the refineries designed to process lower boiling feedstocks, has been a particular cause for concern. This has caused the evolution of processing schemes that accommodate the higher boiling feedstocks (Parkash, 2003; Gary et al., 2007; Speight, 2014, 2017; Hsu and Robinson, 2017).

The choice of processing schemes for a given hydrotreating application depends upon the nature of the feedstock as well as the product requirements (Parkash, 2003; Gary et al., 2007; Speight, 2014, 2017; Hsu and Robinson, 2017). For higher boiling feedstocks the process is usually hydrocracking and can be simply illustrated as a single-stage or as a two-stage operation (Fig. 8.4). Variations to the process are feedstocks dependent.

For example, the single-stage process can be used to produce gasoline but is more often used to produce middle distillate from high-boiling vacuum gas oils. The two-stage process was developed primarily to produce high yields of gasoline

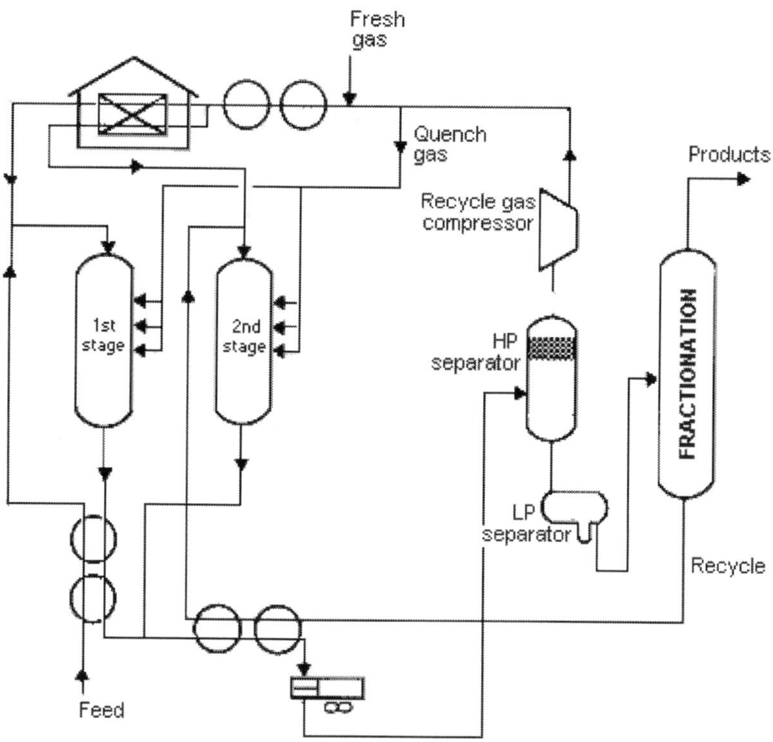

Figure 8.4 A single-stage or two-stage (optional) hydrocracking unit.
Source: Modified from OSHA Technical Manual, Section IV, Chapter 2: Petroleum Refining Processes. http://www.osha.gov/dts/osta/otm/otm_iv/otm_iv_2.html.

from straight-run gas oil, and the first stage may actually be a purification step to remove sulfur-containing (as well as nitrogen-containing) organic materials. Both processes use an extinction-recycling technique to maximize the yields of the desired product. Significant conversion of high-boiling feedstocks can be accomplished by hydrocracking at high severity (Dolbear, 1998). For some applications the products boiling up to 340°C (650°F) can be blended to give the desired final product (Parkash, 2003; Gary et al., 2007; Speight, 2014, 2017; Hsu and Robinson, 2017).

For lower boiling feedstocks the commercial processes for treating or finishing crude oil fractions with hydrogen all operate in essentially the same manner. The feedstock is heated and passed with hydrogen gas through a tower or reactor filled with catalyst pellets (Fig. 8.3). The reactor is maintained at a temperature of 260°C–425°C (500°F–800°F) at pressures from 100 to 1000 psi, depending on the particular process, the nature of the feedstock, and the degree of hydrogenation required. After leaving the reactor, excess hydrogen is separated from the treated product and recycled through the reactor after the removal of hydrogen sulfide. The liquid product is passed into a stripping tower, where steam removes dissolved hydrogen and hydrogen sulfide, and after cooling the product is run to finished product storage or, in the case of feedstock preparation, pumped to the next processing unit.

Excessive contact time and/or temperature will create coking. Precautions need to be taken when unloading coked catalyst from the unit to prevent iron sulfide fires. The coked catalyst should be cooled to below 49°C (<120°F) before removal or dumped into nitrogen-blanketed bins where it can be cooled before further handling. Antifoam additives may be used to prevent catalyst poisoning from silicone carryover in the coker feedstock. There is a potential for exposure to hydrogen sulfide or hydrogen gas in the event of a release, or to ammonia should a sour-water leak or spill occur. Phenol also may be present if high boiling-point feedstocks are processed.

Attention must also be given to the coke mitigation aspects of hydrotreating as a preliminary treatment option of feedstocks for other processes, especially higher boiling feedstocks. Although the visbreaking process (Chapter 2: Introduction to Refining Processes, and Chapter 5: Thermal cracking) reduces the viscosity of residua and partially converts the residue to lower boiling hydrocarbon derivatives and coke, the process can also be used to remove the undesirable higher molecular weight polar constituents before sending the visbroken feedstock to a catalytic cracking unit. The solvent deasphalting process (Chapter 7: Deasphalting and Dewaxing) separates the higher value liquid product [deasphalted oil (DAO)] using a low-boiling paraffinic solvent from low value asphaltene rich pitch stream. Various residuum hydrotreating (in fact, hydrocracking) processes (Chapter 9: Hydrocracking) in which the feedstock is processed under high temperature and pressure, using a robust catalyst to remove sulfur, metals, condensed aromatic, or nitrogen, and increase the hydrogen content of the residuum to a desired degree are also available. However, and increased number of options are becoming available in which the residuum is first hydrotreated (under milder conditions to remove heteroatoms and mitigate the effects of the asphaltene constituents and resin constituents) before sending the hydrotreated product to, for example, a fluid catalytic cracking unit.

In such cases, it is even more important that particular attention must be given to hydrogen management and promoting HDS and HDN (even fragmentation) of asphaltene and resin constituents, thereby producing a product that may be suitable as a feedstock for catalytic cracking with reduced catalyst destruction. The presence of a material with good solvating power to assist in the hydrotreating process is preferred. In this respect it is worth noting the reappearance of donor solvent processing of high-boiling feedstocks that has its roots in the older hydrogen donor diluent visbreaking process (Carlson et al., 1958; Langer et al., 1962; Parkash, 2003; Gary et al., 2007; Speight, 2013a, 2014, 2017; Hsu and Robinson, 2017).

Hydrogen consumption is also a parameter that varies with feedstock composition thereby indicating the need for a thorough understanding of the feedstock constituents if the process is to be employed to maximum efficiency. A convenient means of understanding the influence of feedstock on the hydrotreating process is through a study of the hydrogen content (H/C atomic ratio) and molecular weight (carbon number) of the various feedstocks or products. It is also possible to use data for hydrogen usage in residuum processing where the relative amount of hydrogen consumed in the process can be shown to be dependent upon the sulfur content of the feedstock (Parkash, 2003; Gary et al., 2007; Speight, 2014, 2017; Hsu and Robinson, 2017).

Hydrotreating processes differ depending upon the feedstock available and catalysts used. Hydrotreating can be used to improve the burning characteristics of distillates such as kerosene. Hydrotreatment of a kerosene fraction can convert aromatics into naphthenes, which are cleaner burning compounds.

Lube-oil hydrotreating uses catalytic treatment of the oil with hydrogen to improve product quality. The objectives in mild lube hydrotreating include saturation of olefins and improvements in color, odor, and acid nature of the oil. Mild lube hydrotreating also may be used following solvent processing. Operating temperatures are usually below 315°C (600°F) and operating pressures below 800 psi. Severe lube hydrotreating, at temperatures in the 315°C–400°C (600°F–750 °F) range and hydrogen pressures up to 3000 psi, is capable of saturating aromatic rings, along with sulfur and nitrogen removal, to impart specific properties not achieved at mild conditions.

Hydrotreating also can be employed to improve the quality of pyrolysis gasoline, a by-product from the manufacture of ethylene. Traditionally, the outlet for pyrolysis-produced gasoline has been motor gasoline blending, a suitable route in view of its high octane number. However, only small portions can be blended untreated owing to the unacceptable odor, color, and gum-forming tendencies of this material. The quality of pyrolysis gasoline, which is high in olefin content, can be satisfactorily improved by hydrotreating, whereby conversion of olefins into mono-olefins provides an acceptable product for motor gas blending.

In summary, hydrotreating processes differ depending upon the feedstock available and catalysts used. Hydrotreating can be used to improve the burning characteristics of distillates such as kerosene when the process converts aromatics into naphthenes, which are cleaner burning compounds. Lube-oil hydrotreating uses catalytic treatment of the oil with hydrogen to improve product quality. The objectives in mild lube hydrotreating include saturation of olefins and improvements in color, odor, and acid nature of the oil. Mild lube hydrotreating also may be used following solvent processing.

Hydrotreating also can be employed to improve the quality of pyrolysis gasoline (sometime referred to as py-gas or pygas), a by-product from the manufacture of ethylene. Traditionally, the outlet for pyrolysis gasoline has been as a gasoline blend stock, a suitable route in view of its high octane number. However, only small portions can be blended untreated owing to the unacceptable odor, color, and gum-forming tendencies of this material. The quality of pyrolysis gasoline, which is high in diolefin content, can be satisfactorily improved by hydrotreating, whereby conversion of diolefins into mono-olefins provides an acceptable product for motor gas blending. This, of course opens up the suggestion of hydrotreating of pyrolysis oil from bio-feedstocks.

8.4 Process parameters

The principal variables affecting the required severity in distillate desulfurization are (1) hydrogen partial pressure, (2) space velocity, (3) reaction temperature, and (4) feedstock properties.

8.4.1 *Hydrogen partial pressure*

The important effect of hydrogen partial pressure is the minimization of coking reactions. If the hydrogen pressure is too low for the required duty at any position within the reaction system, premature aging of the remaining portion of catalyst will be encountered. In addition, the effect of hydrogen pressure on desulfurization varies with feed boiling range. For a given feed there exists a threshold level above which hydrogen pressure is beneficial to the desired desulfurization reaction. Below this level, desulfurization drops off rapidly as hydrogen pressure is reduced.

8.4.2 *Space velocity*

As the space velocity is increased, desulfurization is decreased but increasing the hydrogen partial pressure and/or the reactor temperature can offset the detrimental effect of increasing space velocity.

8.4.3 *Reaction temperature*

A higher reaction temperature increases the rate of desulfurization at constant feed rate, and the start-of-run temperature is set by the design desulfurization level, space velocity, and hydrogen partial pressure. The capability to increase temperature as the catalyst deactivates is built into the most process or unit designs. Temperatures of 415°C (780°F) and above result in excessive coking reactions and higher than normal catalyst aging rates. Therefore units are designed to avoid the use of such temperatures for any significant part of the cycle life.

8.4.4 Feedstock effects

The character of the feedstock properties, especially the feed boiling range, has a definite effect on the ultimate design of the desulfurization unit and process flow. In agreement, there is a definite relationship between the percent by weight sulfur in the feedstock and the hydrogen requirements.

In addition, the reaction rate constant in the kinetic relationships decreases rapidly with increasing average boiling point in the kerosene and low-boiling gas oil range but much more slowly in the high-boiling gas oil range. This is attributed to the difficulty in removing sulfur from ring structures present in the entire high-boiling gas oil boding range.

The HDS of low-boiling (low-boiling) distillate (naphtha or kerosene) is one of the more common catalytic HDS processes since it is usually used as a pretreatment of such feedstocks prior to deep HDS or prior to catalytic reforming (Parkash, 2003; Gary et al., 2007; Speight, 2014, 2017; Hsu and Robinson, 2017). This is similar to the concept of pretreating residua prior to hydrocracking to improve the quality of the products (Chapter 9: Hydrocracking). HDS of such feedstocks is required because sulfur compounds poison the precious-metal catalysts used in reforming and desulfurization can be achieved under relatively mild conditions and is near quantitative (Table 8.2). If the feedstock arises from a cracking operation, HDS will be accompanied by some degree of saturation resulting in increased hydrogen consumption.

The HDS of low boiling (naphtha) feedstocks is usually a gas-phase reaction and may employ the catalyst in fixed beds and (with all of the reactants in the gaseous

Table 8.2 Hydrodesulfurization of various naphtha fractions.

Feedstock	Boiling range		Sulfur	Desulfurization[a]
	°C	°F	wt.%	%
Visbreaker naphtha	65–220	150–430	1.00	90
Visbreaker-coker naphtha	65–220	150–430	1.03	85
Straight-run naphtha	85–170	185–340	0.04	99
Catalytic naphtha (light)	95–175	200–350	0.18	89
Catalytic naphtha (heavy)	120–225	250–440	0.24	71
Thermal naphtha (heavy)	150–230	300–450	0.28	57

[a]Process conditions: Co–Mo on alumina, 260°C–370°C/500°F–700°F, 200–500 psi (1380–3440 kPa) hydrogen.
Source: Modified from Speight, J.G., 2007. The Chemistry and Technology of Petroleum, fourth ed. CRC Press, Taylor & Francis Group, Boca Raton, FL.

phase) only minimal diffusion problems are encountered within the catalyst pore system. It is, however, important that the feedstock be completely volatile before entering the reactor as there may be the possibility of pressure variations (leading to less satisfactory results) if some of the feedstock enters the reactor in the liquid phase and is vaporized within the reactor.

In applications of this type the sulfur content of the feedstock may vary from 100 ppm to 1% and the necessary degree of desulfurization to be effected by the treatment may vary from as little as 50% to more than 99%. If the sulfur content of the feedstock is particularly low, it will be necessary to presulfide the catalyst. For example, if the feedstock only has 100 to 200 ppm sulfur, several days may be required to sulfide the catalyst as an integral part of the desulfurization process even with complete reaction of all of the feedstock sulfur to, say, cobalt and molybdenum (catalyst) sulfides. In such a case, presulfiding can be conveniently achieved by addition of sulfur compounds to the feedstock or by addition of hydrogen sulfide to the hydrogen.

Generally, HDS of naphtha feedstocks to produce catalytic reforming feedstocks is carried to the point where the desulfurized feedstock contains less than 20 ppm sulfur. The net hydrogen produced by the reforming operation may actually be sufficient to provide the hydrogen consumed in the desulfurization process.

The HDS of middle distillates is also an efficient process and applications include predominantly the desulfurization of kerosene, diesel fuel, jet fuel, and heating oil that boil over the general range 250°C–400°C (480°F–750°F). However, with this type of feedstock, hydrogenation of the higher boiling catalytic cracking feedstocks has become increasingly important where HDS is accomplished alongside the saturation of condensed-ring aromatic compounds as an aid to subsequent processing.

Under the relatively mild processing conditions used for the HDS of these particular feedstocks, it is difficult to achieve complete vaporization of the feed. Process conditions may dictate that only part of the feedstock is actually in the vapor phase and that sufficient liquid phase is maintained in the catalyst bed to carry the larger molecular constituents of the feedstock through the bed. If the amount of liquid phase is insufficient for this purpose, molecular stagnation (leading to carbon deposition on the catalyst) will occur.

HDS of middle distillates causes a more marked change in the specific gravity of the feedstock, and the amount of low-boiling material is much more significant when compared with the naphtha-type feedstock. In addition, the somewhat more severe reaction conditions (leading to a designated degree of hydrocracking) also cause an overall increase in hydrogen consumption when middle distillates are employed as feedstocks in place of the naphtha.

High boiling distillates, such as the atmospheric and vacuum gas oils, are not usually produced as a refinery product but merely serve as feedstocks to other processes for conversion to lower boiling materials. For example, gas oils can be desulfurized to remove more than 80% of the sulfur originally in the gas oil with some conversion of the gas oil to lower boiling materials. The treated gas oil (which has a reduced carbon residue as well as lower sulfur and nitrogen contents relative to the untreated material) can then be converted to lower boiling products in, say, a catalytic cracker where an improved catalyst life and volumetric yield may be noted.

The conditions used for the HDS of gas oil may be somewhat more severe than the conditions employed for the HDS of middle distillates with, of course, the feedstock in the liquid phase.

In summary the HDS of the low-, middle-, and high-boiling distillates can be achieved quite conveniently using a variety of processes. One major advantage of this type of feedstock is that the catalyst does not become poisoned by metal contaminants in the feedstock since only negligible amounts of these contaminants will be present. Thus the catalyst may be regenerated several times and onstream times between catalyst regeneration (while varying with the process conditions and application) may be of the order of 3–4 years.

8.5 Reactors

8.5.1 Downflow fixed-bed reactor

The reactor design commonly used in HDS of distillates is the fixed-bed reactor design in which the feedstock enters at the top of the reactor and the product leaves at the bottom of the reactor (Fig. 8.3). The catalyst remains in a stationary position (fixed-bed) with hydrogen and crude oil feedstock passing in a downflow direction through the catalyst bed. The HDS reaction is exothermic and the temperature rises from the inlet to the outlet of each catalyst bed. With a high hydrogen consumption and subsequent large temperature rise, the reaction mixture can be quenched with cold recycled gas at intermediate points in the reactor system. This is achieved by dividing the catalyst charge into a series of catalyst beds and the effluent from each catalyst bed is quenched to the inlet temperature of the next catalyst bed.

The extent of desulfurization is controlled by raising the inlet temperature to each catalyst bed to maintain constant catalyst activity over the course of the process. Fixed-bed reactors are mathematically modeled as plug-flow reactors with very little back mixing in the catalyst beds. The first catalyst bed is poisoned with vanadium and nickel at the inlet to the bed and may be a cheaper catalyst (guard bed). As the catalyst is poisoned in the front of the bed, the temperature exotherm moves down the bed and the activity of the entire catalyst charge declines, thus requiring a raise in the reactor temperature over the course of the process sequence. After catalyst regeneration the reactors are opened and inspected, and the high metal content catalyst layer at the inlet to the first bed may be discarded and replaced with fresh catalyst. The catalyst loses activity after a series of regenerations and, consequently, after a series of regenerations, it is necessary to replace the complete catalyst charge. In the case of very high metal content feedstocks (such as residua), it is often necessary to replace the entire catalyst charge rather than to regenerate it. This is due to the fact that the metal contaminants cannot be removed by economical means during rapid regeneration, and the metals have been reported to interfere with the combustion of carbon and sulfur, catalyzing the conversion of sulfur dioxide (SO_2) to sulfate (SO_4^{2-}) that has a permanent poisoning effect on the catalyst.

Fixed-bed HDS units are generally used for distillate HDS and may also be used for residuum HDS but require special precautions in processing. The residuum must undergo two-stage electrostatic desalting so that salt deposits do not plug the inlet to the first catalyst bed and the residuum must be low in vanadium and nickel content to avoid plugging the beds with metal deposits. Hence the need for a guard bed in residuum HDS reactors.

During the operation of a fixed-bed reactor, contaminants entering with fresh feed are filtered out and fill the voids between catalyst particles in the bed. The buildup of contaminants in the bed can result in the channeling of reactants through the bed and reducing the HDS efficiency. As the flow pattern becomes distorted or restricted the pressure drop throughout the catalyst bed increases. If the pressure drop becomes high enough, physical damage to the reactor internals can result. When high-pressure drops are observed throughout any portion of the reactor, the unit is shut down and the catalyst bed is skimmed and refilled.

With fixed-bed reactors a balance must be reached between reaction rate and pressure drop across the catalyst bed. As catalyst particle size is decreased, the desulfurization reaction rate increases but so does the pressure drop across the catalyst bed. Expanded-bed reactors do not have this limitation and small 1/32 in. (0.8 mm) extrudate catalysts or fine catalysts may be used without increasing the pressure drop.

8.5.2 Upflow expanded-bed reactor

Expanded-bed reactors are applicable to distillates but are commercially used for very high-boiling, high metals, and/or dirty feedstocks having extraneous fine solids material. They operate in such a way that the catalyst is in an expanded state so that the extraneous solids pass through the catalyst bed without plugging. They are isothermal, which conveniently handles the high-temperature exotherms associated with high hydrogen consumptions. Since the catalyst is in an expanded state of motion, it is possible to treat the catalyst as a fluid and to withdraw and add catalyst during operation.

Expanded beds of catalyst are referred to as particulate fluidized insofar as the feedstock and hydrogen flow upward through an expanded bed of catalyst with each catalyst particle in independent motion. Thus the catalyst migrates throughout the entire reactor bed. Expanded-bed reactors are mathematically modeled as backmix reactors with the entire catalyst bed at one uniform temperature. Spent catalyst may be withdrawn and replaced with fresh catalyst on a daily basis. Daily catalyst addition and withdrawal eliminate the need for costly shutdowns to change out catalyst and also result in a constant equilibrium catalyst activity and product quality. The catalyst is withdrawn daily and has a vanadium, nickel, and carbon content that is representative on a macroscale of what is found throughout the entire reactor. On a microscale, individual catalyst particles have ages from that of fresh catalyst to as old as the initial catalyst charge to the unit but the catalyst particles of each age group are so well dispersed in the reactor that the reactor contents appear uniform.

In the unit the feedstock and hydrogen recycle gas enter the bottom of the reactor, pass up through the expanded catalyst bed, and leave from the top of the reactor. Commercial expanded-bed reactors normally operate with 1/32 in. (0.8 mm) extrudate catalysts that provide a higher rate of desulfurization than the larger catalyst particles used in fixed-bed reactors. With extrudate catalysts of this size the upward liquid velocity based on fresh feedstock is not sufficient to keep the catalyst particles in an expanded state. Therefore for each part of the fresh feed, several parts of product oil are taken from the top of the reactor, recycled internally through a large vertical pipe to the bottom of the reactor, and pumped back up through the expanded catalyst bed. The amount of catalyst bed expansion is controlled by the recycle of product oil back up through the catalyst bed.

The expansion and turbulence of gas and oil passing upward through the expanded catalyst bed are sufficient to cause almost complete random motion in the bed (particulate fluidized). This effect produces the isothermal operation. It also causes almost complete back-mixing. Consequently, in order to effect near complete sulfur removal (over 75%), it is necessary to operate with two or more reactors in series. The ability to operate at a single temperature throughout the reactor or reactors and to operate at a selected optimum temperature rather than an increasing temperature from the start to the end of the run result in more effective use of the reactor and catalyst contents. When all these factors are put together, that is, use of a smaller catalyst particle size, isothermal, fixed temperature throughout run, back-mixing, daily catalyst addition, and constant product quality, the reactor size required for an expanded bed is often smaller than that required for a fixed bed to achieve the same product goals. This is generally true when the feeds have high initial boiling points and/or the hydrogen consumption is very high.

8.5.3 Demetallization reactor (guard bed reactor)

Feedstocks that have relatively high metals contents (>300 ppm) substantially increase catalyst consumption because the metals poison the catalyst, thereby requiring frequent catalyst replacement. The usual desulfurization catalysts are relatively expensive for these consumption rates, but there are catalysts that are relatively inexpensive and can be used in the first reactor to remove a large percentage of the metals. Subsequent reactors downstream of the first reactor would use normal HDS catalysts. Since the catalyst materials are proprietary, it is not possible to identify them here. However, it is understood that such catalysts contain little or no metal promoters, that is, nickel, cobalt, molybdenum. Metals removal on the order of 90% has been observed with these materials.

Thus one method of controlling demetallization is to employ separate smaller guard reactors just ahead of the fixed-bed HDS reactor section. The preheated feed and hydrogen pass through the guard reactors that are filled with an appropriate catalyst for demetallization that is often the same as the catalyst used in the HDS section. The advantage of this system is that it enables replacement of the most contaminated catalyst (guard bed), where pressure drop is highest, without having to replace the entire inventory or shut down the unit. The feedstock is alternated between guard reactors while catalyst in the idle guard reactor is being replaced.

When the expanded-bed design is used, the first reactor could employ a low-cost catalyst (5% of the cost of Co/Mo catalyst) to remove the metals and subsequent reactors can use the more selective HDS catalyst. The demetallization catalyst can be added continuously without taking the reactor out of service and the spent demetallization catalyst can be loaded to more than 30% vanadium, which makes it a valuable source of vanadium.

8.6 Commercial processes

Hydrotreating technology is one of the most commonly used refinery processes, designed to remove contaminants such as sulfur, nitrogen, condensed ring aromatics, or metals. The feedstocks used in the process range from naphtha to vacuum residue, and the products in most applications are used as environmentally acceptable clean fuels. Hydrotreating technology has a long history of commercial application since 1950s, with more than 500 licensed units placed in operation worldwide. Furthermore, recent regulatory requirements to produce ultra-low-sulfur diesel (ULSD) and gasoline have created a very dynamic market as refiners must build new or revamp their existing assets to produce the low-sulfur, even no-sulfur fuels.

The crude oil industry often employs two-stage processes in which the feedstock undergoes both hydrotreating and hydrocracking. In the first, or pretreating, stage the main purpose is conversion of nitrogen compounds in the feed to hydrocarbon derivatives and to ammonia by hydrogenation and mild hydrocracking. Typical conditions are 340°C–390°C (650°F–740°F), 150–2500 psi (1–17 MPa), and a catalyst contact time of 0.5–1.5 hour; up to 1.5% w/w hydrogen is absorbed, partly by conversion of the nitrogen compounds but chiefly by aromatic compounds that are hydrogenated.

It is most important to reduce the nitrogen content of the product oil to less than 0.001% w/w (10 ppm). This stage is usually carried out with a bifunctional catalyst containing hydrogenation promoters, for example, nickel and tungsten or molybdenum sulfides, on an acid support, such as silica–alumina. The metal sulfides hydrogenate aromatics and nitrogen compounds and prevent the deposition of carbonaceous deposits; the acid support accelerates nitrogen removal as ammonia by breaking carbon–nitrogen bonds. The catalyst is generally used as 1/8 in. × 1/8 in. (0.32 cm × 0.32 cm) or 1/16 in. × 1/8 in. (0.16 cm × 0.32 cm) pellets, formed by extrusion.

Most of the hydrotreating cracking is accomplished in the first stage. Ammonia and some gasoline are usually removed from the first-stage product, and then the remaining first-stage product, which is low in nitrogen compounds, is passed over the second-stage catalyst. Again, typical conditions are 300°C–370°C

(600°F–700°F), 1500–2500 psi (10–17 MPa) hydrogen pressure, and 0.5–1.5 hours contact time; 1%–1.5% w/w hydrogen may be absorbed. Conversion to gasoline or jet fuel is seldom complete in one contact with the catalyst, so the lower boiling oils are removed by distillation of the products and the higher boiling, higher boiling product combined with fresh feed and recycled over the catalyst until it is completely converted.

8.6.1 Autofining process

The autofining process differs from other hydrorefining processes in which an external source of hydrogen is not required (Parkash, 2003; Gary et al., 2007; Speight, 2014, 2017; Hsu and Robinson, 2017). Sufficient hydrogen to convert sulfur to hydrogen sulfide is obtained by dehydrogenation of naphthenes in the feedstock.

The processing equipment is similar to that used in hydrofining (as given later). The catalyst is cobalt oxide and molybdenum oxide on alumina, and operating conditions are usually 340°C–425°C (650°F–800°F) at pressures of 100–200 psi. Hydrogen formed by dehydrogenation of naphthenes in the reactor is separated from the treated oil and is then recycled through the reactor. The catalyst is regenerated with steam and air at 200–1000 hours intervals, depending on whether conventional or viscous feedstocks have been processed. The process is used for the same purpose as hydrofining but is limited to fractions with end points no higher than 370°C (700°F).

8.6.2 Ferrofining process

This mild hydrogen-treating process was developed to treat distilled and solvent-refined lubricating oils. The process eliminates the need for acid and clay treatment. The catalyst is a three-component material on an alumina base with low hydrogen consumption and a life expectancy of 2 years or more. Process operations include heating the hydrogen–oil mixture and charging to a downflow catalyst-filled reactor. Separation of oil and gas is a two-stage operation whereby gas is removed to the fuel system. The oil is then steam stripped to control the flash point and dried in vacuum, and a final filtering step removes catalyst fines.

The catalyst in the Ferrofining process consists of the oxides of molybdenum and iron, and possibly cobalt, supported on alumina (UK Patents 913047, 913048, and 917345).

8.6.3 Gulf hydrodesulfurization process

This is a regenerative fixed-bed process to upgrade crude oil residues by catalytic hydrogenation to refined high-boiling fuel oils or to high-quality catalytic charge stocks. Desulfurization and quality improvement are the primary purposes of the process, but if the operating conditions and catalysts are varied, low-boiling distillates can be produced and the viscosity of high-boiling material can be lowered.

Long onstream cycles are maintained by reducing random hydrocracking reactions to a minimum, and whole crude oils, virgin, or cracked residua may serve as feedstock.

The catalyst is a metallic compound supported on pelletized alumina and may be regenerated in situ with air and steam or flue gas through a temperature cycle of 400°C–650°C (750°F–1200°F). Onstream cycles of 4–5 months can be obtained at desulfurization levels of 65%–75%, and catalyst life may be as long as 2 years.

The Gulf HDS process (Gulf HDS) is a commercially proven process for the removal of sulfur from crude oil residues to produce low sulfur fuel by direct HDS. The distinguishing features of the processes are (1) flexibility—the sulfur removal capacity of the plants can be modified to meet the most stringent fuel oil sulfur content specifications; (2) versatility—the process can be cojoined with riser fluid catalytic cracking to produce distillates and gasoline and this combination is more economical than conventional schemes, including delayed coking, and; (3) reliability—the process has been operating without any interruptions since the start-up date.

8.6.4 Hydrofining process

Hydrofining is a process used for reducing the sulfur content of feedstocks by treating the feedstock in the presence of a catalyst. This process can be applied to lubricating oil, naphtha, and gas oil.

The feedstock is heated in a furnace and passed with hydrogen through a reactor containing a suitable metal oxide catalyst, such as cobalt and molybdenum oxides or alumina. Hydrogen is obtained from catalytic reforming units. Reactor operating conditions range from 205°C to 425°C (400°F to 800°F) and from 50 to 800 psi, depending on the kind of feedstock and the degree of treatment required. Higher boiling feedstocks, high sulfur content, and maximum sulfur removal require higher temperatures and pressures (Parkash, 2003; Gary et al., 2007; Speight, 2014, 2017; Hsu and Robinson, 2017).

After passing through the reactor the treated oil is cooled and separated from the excess hydrogen, which is recycled through the reactor. The treated oil is pumped to a stripper tower, where hydrogen sulfide formed by the hydrogenation reaction is removed by steam, vacuum, or flue gas and the finished product leaves the bottom of the stripper tower. The catalyst is not usually regenerated; it is replaced after approximately a 1 year of use.

Lube oil hydrofining is a catalytic technology to prepare lube base stocks for further processing or it may be used as a base-stock finishing step. The process is usually integrated Exol N technology, into an Exolfining configuration, to treat the waxy raffinate from extraction upstream from the lube dewaxing unit. It saturates multiring aromatics; destroys acids; and removes sulfur, nitrogen, and color bodies to improve the color, and the color and oxidation stability of base stocks. As a finishing step, the lube hydrofining process may be used to treat a dewaxed oil product with similar color and oxidation stability benefits. Lube hydrofining, when be used on lube distillate feedstocks, can enhance the ensuing solvent extraction performance.

Lube hydrofining technology generally produced a product with (1) improved color, (2) improved oxidation stability, (3) improved antioxidant response, (4) better odor, (5) better surface properties, (6) optimal sulfur reduction, (7) significant viscosity index increase at high desulfurization levels, and (8) reduction in toxicity.

Generally, the lube hydrofining process is suitable for feedstocks from all crude sources, and the process conditions are moderate (Parkash, 2003; Gary et al., 2007; Speight, 2014, 2017; Hsu and Robinson, 2017). In an Exolfining N configuration the lube hydrofining step is fully heat integrated with the solvent extraction step. The feedstock to the lube hydrofining reactor is the hot bottoms stream from the raffinate recovery tower. In a stand-alone configuration the reactor feedstock is typically brought to reaction temperature by a combination of heat exchange with the hot hydrofinished product and furnace preheater. The fixed-bed reactor contains a hydrofinishing catalyst, and once-through or recycle hydrogen treat gas may be used. The reactor effluent is flashed to recover the unreacted hydrogen treat gas at high pressure as well as to separate the hydrogen sulfide and ammonia resulting from the hydrofinishing reactions. The hydrofinished oil is then steam stripped and dried under vacuum. As an additional option, the steam stripper tower may be designed to correct the volatility of the hydrofinished oil product.

Wax hydrofining is used to wax from a lube oil plant to produce a food-grade wax that can be used for wax paper, fruit coatings, paper cups, as a seal for medicines, and for similar applications where high purity is required. Wax hydrofining is the final critical steps for processing low-oil-content waxes from dewaxing–deoiling to produce wax with low odor, improved color, and stability. This technology is applicable to either crystalline wax from lube distillates or microcrystalline wax from DAOs. The wax hydrofining option processes low oil content (<1% by weight) wax from conventional solvent lube plant dewaxing–deoiling units under moderate hydroprocessing conditions.

Thus liquefied wax and hydrogen are preheated to reaction temperature in a conventional coil furnace and fed to a fixed-bed reactor containing a supported hydrotreating catalyst. Sulfur and nitrogen compounds are converted to hydrogen sulfide and ammonia, multiring aromatics are saturated, and trace acids and solvent residues are removed. The reactor effluent is flashed to separate hydrogen, hydrogen sulfide, and ammonia, and the wax is steam stripped and dried to improve volatility and color. Both crystalline and microcrystalline waxes can be processed in the same unit, although the types of oil molecules in microcrystalline waxes typically require more severe operating conditions. SCANfining is the next step up from hydrofining and is used to meet environmental requirements for sulfur in gasoline using a proprietary RT-225 catalyst to treat gasoline from a fluid catalytic cracking unit. The process is specially designed to achieve high selectivity for sulfur removal without excessive olefin saturation and octane loss.

Finally, as an additional option, most feedstocks to a fluid catalytic cracking unit are high sulfur with a high ratio of residue blended. This type of feedstocks results in high sulfur high-olefin gasoline. This kind of gasoline can be treated in a hydrofiner, and the hydrofined naphtha is of good quality containing more saturated hydrocarbon derivatives, through which high sulfur content problem of gasoline is resolved. Typical operating conditions are 280°C–320°C (535°F–610°F), 290 to 1160 psi with a liquid hourly space velocity (LHSV) in the range 1.5 to $-3.0\ h^{-1}$ and a hydrogen–oil ratio (v/v) of 200 to 400.

The catalysts for hydrofining typically contain tungsten oxide and/or molybdenum oxide, nickel oxide, and cobalt oxide supported on an alumina carrier. Catalysts with a lower metal content may exhibit higher activity at low temperature (Tanaka, 2004; Kumagai et al., 2005). Along similar paths the diesel oil deep desulfurization (DODD) process for removing sulfur to very low levels, and the GO-fining and Residfining processes are used for upgrading gas oils and residua, respectively, for feedstocks to a catalytic cracking unit and also to use as low sulfur fuel oil (Speight, 2013b)

8.6.5 Isomax process

The Isomax process is a two-stage, fixed-bed catalyst system that operates under hydrogen pressures from 500 to 1500 psi in a temperature range of 205°C–370°C (400°F–700°F), for example, with middle distillate feedstocks. The Isomax process takes place at high temperatures and pressures; hence, the sulfur content of the Isomax product is low.

Exact conditions depend on the feedstock and product requirements, and hydrogen consumption is of the order of 1000–1600 ft^3/bbl of feed processed. Each stage has a separate hydrogen recycling system. Conversion may be balanced to provide products for variable requirements, and recycling can be taken to extinction if necessary. Fractionation can also be handled in a number of different ways to yield desired products (Parkash, 2003; Gary et al., 2007; Speight, 2014, 2017; Hsu and Robinson, 2017).

In the process the first step serves to remove nitrogen, oxygen, sulfur, and metal contaminants from the feedstock, while the second step effects the hydrocracking into the desired hydrocarbon fractions. The first step can, therefore, be looked upon as a cleanup step, while the second step can be considered as a true hydrocracking part of the process. The first step will produce some hydrocracking; however, its main function is that of feed preparation for the second step.

8.6.6 Ultrafining process

Ultrafining is a regenerative, fixed-bed, catalytic process to desulfurize and hydrogenate refinery stocks from naphtha up to and including lubricating oil. The catalyst is cobalt–molybdenum on alumina and may be regenerated in situ using an air-stream mixture. Regeneration requires 10–20 hours and may be repeated 50 to 100 times for a given batch of catalyst; catalyst life is 2–5 years depending on feed stock (Parkash, 2003; Gary et al., 2007; Speight, 2014, 2017; Hsu and Robinson, 2017).

In the process, HDS of a cracked naphtha is effected in the presence of a transition alumina support bearing a Group VIII nonnoble metal oxide and a Group VI-B metal oxide, the atom ratio of Group VIII metal to Group VI-B metal being 1–8.

The process gives high levels of HDS compared to magnesia-containing catalysts and lower levels of olefin saturation, and less octane reduction in the desulfurized gasoline.

8.6.7 Unifining process

This is a regenerative, fixed-bed, catalytic process to desulfurize and hydrogenate refinery distillates of any boiling range (Parkash, 2003; Gary et al., 2007; Speight, 2014, 2017; Hsu and Robinson, 2017). Contaminating metals, nitrogen compounds, and oxygen compounds are eliminated, along with sulfur. The catalyst is a cobalt–molybdenum–alumina type that may be regenerated in situ with steam and air.

8.6.8 Unionfining process

Unionfining process is a fixed-bed catalytic hydrotreating technology applied to produce low-sulfur, color stable diesel fuel. The process is also used to reduce aromatics in diesel and to improve the diesel cetane number. In addition, the process can be used to prepare naphtha for catalytic reformers and to refine coker-naphtha (Parkash, 2003; Gary et al., 2007; Speight, 2014, 2017; Hsu and Robinson, 2017).

In the process (Fig. 8.5) the reactor feed is heat exchanged with the reactor effluent and reactor inlet temperature is controlled by charge heater firing. As hydrotreating is an exothermic reaction—especially when feedstocks are unsaturated—quench sections may be used to cool the reaction fluids and the same time redistribute vapors and liquids between catalyst beds. Wash water is added to the cooled reactor effluent upstream of the final cooler to minimize corrosion and prevent deposits of ammonium salts. The cooled stream enters a cold high pressure separator in order to separate out the hydrogen recycle that is recycled back to the reactor. Desulfurized product is recovered by stripping off the light (low-boiling products) ends with steam.

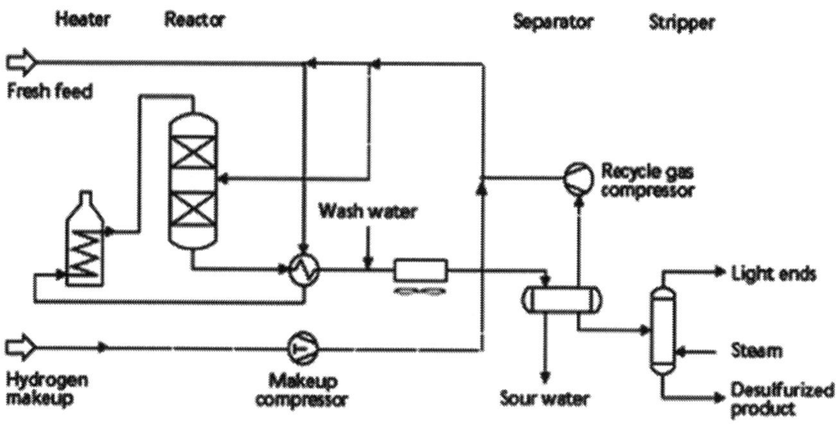

Figure 8.5 The unionfining process.

The UOP RCD Unionfining process is a fixed-bed catalytic residue hydrotreating technology that reduces the contaminants in high-boiling feedstocks, such as atmospheric residue, vacuum residue, and DAO.

The RCD Unionfining process is a fixed-bed hydrogenation process that uses base metal catalysts to remove contaminants, such as sulfur, nitrogen, organometallics, and asphaltenes, and to increase the hydrogen content of the products.

The feedstock is first filtered before combining with hydrogen and passing through the catalyst beds at elevated temperatures and low space velocities. Because of the typical high contaminant levels, a guard bed reactor is used to control pressure drop associated with large particles and solids that may be in the residue feed. Often several reactors in series are required because of the relatively low reactivity of the feedstock. The reactor effluent, which contains the products, recycle gas, and by-products, is then separated into the desired products through a combination of separators, flash stages, and eventually, fractionation equipment. The hydrogen recycle gas requires scrubbing to remove hydrogen sulfide. Then it is combined with makeup gas before being recycled to the reactors.

The ease of processing a feedstock depends on the nature of the asphaltenic molecule and the distribution of contaminants throughout the resin and asphaltene fractions. Relative processing severity is dependent on feedstock type and processing objectives. Consequently, the process operates over a large range of operating conditions: 1500–3000 psi and 0.10–1.0 LHSV. Feedstocks with high contaminants, such as vacuum residues, typically have higher pressures and lower space velocities.

8.7 Process options for heavy feedstocks

The major goal of residuum hydroconversion is cracking of residua with desulfurization, metal removal, denitrogenation, and asphaltene conversion (Speight, 2000; Ancheyta and Speight, 2007). Residuum hydroconversion process offers the production of kerosene and gas oil and of feedstocks for hydrocracking, fluid catalytic cracking, and petrochemical applications. The processes that follow are listed in alphabetical order with no other preference in mind.

Residue hydrotreating is another method for reducing high-sulfur residual fuel oil yields. This technology was originally developed to reduce the sulfur content of atmospheric residues to produce specification low sulfur residual fuel oil. Changes in crude oil quality and product demand, however, have shifted the commercial importance of this technology to include pretreating conversion unit feedstocks to minimize catalyst replacement costs, and coker feedstocks to reduce the yield and increase the quality of the by-product coke fraction. Although residue hydrotreaters are capable of processing feedstocks having a wide range of contaminants, the organometallic and asphaltene component of a feedstock typically determine its processability. Economics generally tend to limit residue hydrotreating applications to feedstocks with limitations (dictated by the process catalyst) on the content of nickel plus vanadium.

In many cases, the application of hydrotreating technology to high-boiling viscous feedstocks may also cause some cracking and, by inference, the application of hydrocracking to high-boiling feedstocks will also cause desulfurization and denitrogenation. Rather than promote unnecessary duplication of the process description, two hydrotreating processes (the RDS and VRDS process options are described together as subcategories of one process) are described here with the note that they are also amenable to hydrocracking operations we well as hydrotreating.

8.7.1 Residuum desulfurization (RDS) and vacuum residuum desulfurization (VRDS) process

Residuum hydrotreating processes have two definite roles: (1) desulfurization to supply low-sulfur fuel oils and (2) pretreatment of feed residua for residuum fluid catalytic cracking processes (Parkash, 2003; Gary et al., 2007; Speight, 2014, 2017; Hsu and Robinson, 2017). The main goal is to remove sulfur, metal, and asphaltene contents from residua and other high-boiling feedstocks to a desired level. On the other hand, the major goal of residuum hydroconversion is the cracking of residua with desulfurization, metal removal, denitrogenation, and asphaltene conversion. Residuum hydroconversion process offers the production of kerosene and gas oil and of feedstocks for hydrocracking, fluid catalytic cracking, and petrochemical applications.

The RDS/VRDS process is (like the Residfining process, *q.v.*) a hydrotreating process that is designed to hydrotreat vacuum gas oil, atmospheric residuum, or vacuum residuum to remove sulfur metallic constituents, while part of the feedstock is converted to lower boiling products. In the case of residua the asphaltene content is reduced (Parkash, 2003; Gary et al., 2007; Speight, 2014, 2017; Hsu and Robinson, 2017).

The process consists of a once-through operation of hydrocarbon feed contacting graded catalyst systems designed to maintain activity and selectivity in the presence of deposited metals. Process conditions are designed for a 6-month to 1-year operating cycle between catalyst replacements. The process is ideally suited to produce feedstocks for residuum fluid catalytic crackers or delayed coking units to achieve minimal production of residual products in a refinery.

The major product of the processes is a low-sulfur fuel oil, and the amount of gasoline and middle distillates is maintained at a minimum to conserve hydrogen. The basic elements of each process are similar and consist of a once-through operation of the feedstock coming into contact with hydrogen and the catalyst in a downflow reactor that is designed to maintain activity and selectivity in the presence of deposited metals. Moderate temperatures and pressures are employed to reduce the incidence of hydrocracking and, hence, minimize production of low-boiling distillates (Parkash, 2003; Gary et al., 2007; Speight, 2014, 2017; Hsu and Robinson, 2017). The combination of a desulfurization step and a vacuum residuum desulfurizer (VRDS) is often seen as an attractive alternate to the atmospheric residuum desulfurizer (RDS). In addition, either RDS option or the VRDS option can be coupled with other processes (such as delayed coking, fluid catalytic cracking, and solvent deasphalting) to achieve the most optimum refining performance.

8.7.2 Residfining process

The residfining process is a catalytic fixed-bed process for the desulfurization and demetallization of residua (Parkash, 2003; Gary et al., 2007; Speight, 2014, 2017; Hsu and Robinson, 2017). The process can also be used to pretreat residua to suitably low contaminant levels prior to catalytic cracking. In the process, liquid feed to the unit is filtered, pumped to pressure, preheated and combined with treat gas prior to entering the reactors. A small guard reactor would typically be employed to prevent plugging/fouling of the main reactors. Provisions are employed to periodically remove the guard while keeping the main reactors on-line. The temperature rise associated with the exothermic reactions is controlled utilizing either a gas or liquid quench. A train of separators is employed to separate the gas and liquid products. The recycle gas is scrubbed to remove ammonia and H_2S. It is then combined with fresh makeup hydrogen before being reheated and recombined with fresh feed. The liquid product is sent to a fractionator where the product is fractionated.

The different catalysts allow other minor differences in operating conditions and peripheral equipment. Primary differences include the use of higher purity hydrogen makeup gas (usually 95% or greater), inclusion of filtration equipment in most cases, and facilities to upgrade the off-gases to maintain higher concentration of hydrogen in the recycle gas. Most of the processes utilize downflow operation over fixed-bed catalyst systems, but exceptions to this are the H-Oil and LC-Fining processes (which are predominantly conversion processes) that employ upflow designs and ebullating catalyst systems with continuous catalyst removal capability, and the Shell Process (a conversion process) that may involve the use of a bunker flow reactor ahead of the main reactors to allow periodic changeover of catalyst.

The primary objective in most of the residue desulfurization processes is to remove sulfur with minimum consumption of hydrogen. Substantial percentages of nitrogen, oxygen, and metals are also removed from the feedstock. However, complete elimination of other reactions is not feasible and, in addition, hydrocracking, thermal cracking, and aromatic saturation reactions occur to some extent. Certain processes, that is, H-Oil (Chapter 9: Hydrocracking) using a single-stage or a two-stage reactor and LC-Fining (Chapter 9: Hydrocracking) using an expanded bed reactor can be designed to accomplish greater amounts of hydrocracking to yield larger quantities of lower boiling distillates at the expense of desulfurization.

Removal of nitrogen is much more difficult than the removal of sulfur. For example, nitrogen removal may be only approximately 25%–30% when sulfur removal is at a 75%–80% level. Metals are removed from the feedstock in substantial quantities and are mainly deposited on the catalyst surface and exist as metal sulfides at processing conditions. As these deposits accumulate, the catalyst pores eventually become blocked and inaccessible, thus catalyst activity is lost.

Desulfurization of residua is considerably more difficult than the desulfurization of distillates (including vacuum gas oil) because many more contaminants are present and very large, complex molecules are involved. The most difficult portion of feed in residue desulfurization is the asphaltene fraction that forms coke readily, and it is essential that these large molecules be prevented from condensing with each other to form coke, which deactivates the catalyst. This is accomplished by the selection of proper catalysts, use of adequate hydrogen partial pressure, and assuring intimate contact of the hydrogen-rich gases and oil molecules in the process design.

8.8 Catalysts

Typically, hydrotreating catalysts are usually cobalt plus molybdenum or nickel plus molybdenum (in the sulfide) forms, impregnated on an alumina base. The hydrotreating operating conditions (1000–2000 psi hydrogen and approximately 370°C, 700°F) are such that appreciable hydrogenation of aromatics does not occur. The desulfurization reactions are invariably accompanied by small amounts of hydrogenation and hydrocracking, the extent of which depends on the nature of the feedstock and the severity of desulfurization (Parkash, 2003; Gary et al., 2007; Speight, 2014, 2017; Hsu and Robinson, 2017).

Catalyst life depends on the charge stock properties and the degree of desulfurization desired. The only permanent poisons to the catalyst are metals in the feedstock that deposit on the catalyst, usually quantitatively, causing permanent deactivation as they accumulate. However, this is usually of little concern except when DAOs are used as feedstocks since most distillate feedstocks contain low amounts of metals. Nitrogen compounds are a temporary poison to the catalyst but there is essentially no effect on catalyst aging except that caused by a higher temperature requirement to achieve the desired desulfurization. Hydrogen sulfide can be a temporary poison in the reactor gas, and recycle gas scrubbing is employed to counteract this condition.

Providing that pressure drop buildup is avoided, cycles of 1 year or more and ultimate catalyst life of 3 years or more can be expected. The catalyst employed can be regenerated by normal steam–air or recycle combustion gas–air procedures. The catalyst is restored to near fresh activity by regeneration during the early part of its ultimate life. However, permanent deactivation of the catalyst occurs slowly during usage and repeated regenerations, so replacement becomes necessary.

More specifically, HDS catalysts consist of metals impregnated on a porous alumina support. Almost all of the surface area is found in the pores of the alumina (200–300 m^2/g) and the metals are dispersed in a thin layer over the entire alumina surface within the pores (Table 8.3). This type of catalyst does display a huge catalytic surface for a small weight of catalyst. Cobalt (Co), molybdenum (Mo), and nickel (Ni) are the most commonly used metals for desulfurization catalysts. The catalysts are manufactured with the metals in an oxide state. In the active form they

Table 8.3 Composition and properties of hydrotreating catalysts.

Composition	Properties		Range
Active phases (wt.%)			
MoO_3			13–20
CoO			2.5–3.5
NiO			2.5–3.5
Promoters (wt.%)			
SiO			1.0–10.0
	Surface area	m^2/g	150–500
	Pore volume	cm^3/g	0.2–0.8
	Pore diameter		
	Mesopores	nm	3.0–50.0
	Macropores	nm	100–5000
	Extrudable diameter	mm	0.8–4.0
	Extrudable length/diameter	mm	2.0–4.0
	Bulk density	kg/m^3	500–1000
Catalyst is typically composed of active phases, promoters, and a gamma-alumina carrier.			

Source: Modified from Speight, J.G., 2007. The Chemistry and Technology of Petroleum, fourth ed. CRC Press, Taylor & Francis Group, Boca Raton, FL.

are iii the sulfide state, which is obtained by sulfiding the catalyst either prior to use or with the feed during actual use. Any catalyst that exhibits hydrogenation activity will catalyze HDS to some extent. However, the Group VIB metals (chromium, molybdenum, and tungsten) are particularly active for desulfurization, especially when promoted with metals from the iron group (iron, cobalt, and nickel).

The increasing importance of HDS and HDN in crude oil processing in order to produce clean-burning fuels has led to a surge of research on the chemistry and engineering of heteroatom removal, with sulfur removal being the most prominent area of research. Most of the earlier works are focused on (1) catalyst characterization by physical methods, (2) on low-pressure reaction studies of model compounds having relatively high reactivity, (3) on process development, or (4) on cobalt–molybdenum (Co–Mo) catalysts, nickel–molybdenum catalysts (Ni–Mo), or nickel–tungsten (Ni–W) catalysts supported on alumina, often doped by fluorine or phosphorus.

HDS and demetallization occur simultaneously on the active sites within the catalyst pore structure. Sulfur and nitrogen occurring in residua are converted to hydrogen sulfide and ammonia in the catalytic reactor and these gases are scrubbed out of the reactor effluent gas stream. The metals in the feedstock are deposited on the catalyst in the form of metal sulfides and cracking of the feedstock to distillate produces a laydown of carbonaceous material on the catalyst; both events poison the catalyst and activity or selectivity suffers. The deposition of carbonaceous material is a fast reaction that soon equilibrates to a particular carbon level and is controlled by hydrogen partial pressure within the reactors. On the other hand, metal deposition is a slow reaction that is directly proportional to the amount of feedstock passed over the catalyst.

The need to develop catalysts that can carry out deep HDS and deep HDN has become even more pressing in view of recent environmental regulations limiting the amount of sulfur and nitrogen emissions. The development of a new generation of catalysts to achieve this objective of low nitrogen and sulfur levels in the processing of different feedstocks presents an interesting challenge for catalyst development.

Basic nitrogen-containing compounds in a feed diminish the cracking activity of hydrocracking catalysts. However, zeolite catalysts can operate in the presence of substantial concentrations of ammonia, in marked contrast to silica–alumina catalysts, which are strongly poisoned by ammonia. Similarly, sulfur-containing compounds in a feedstock adversely affect the noble metal hydrogenation component of hydrocracking catalysts. These compounds are hydrocracked to hydrogen sulfide, which converts the noble metal to the sulfide form. The extent of this conversion is a function of the hydrogen and hydrogen sulfide partial pressures.

Removal of sulfur from the feedstock results in a gradual increase in catalyst activity, returning almost to the original activity level. As with ammonia, the concentration of the hydrogen sulfide can be used to control precisely the chemical activity of the catalyst. Nonnoble metal-loaded zeolite catalysts have an inherently different response to sulfur impurities since a minimum level of hydrogen sulfide is required to maintain the nickel–molybdenum and nickel–tungsten in the sulfide state.

HDN is more difficult to accomplish than HDS, but the relatively smaller amounts of nitrogen-containing compounds in conventional crude oil (Chapter 1: Feedstock Types and Properties) made this of little concern to refiners. However, the trend to higher boiling feedstocks in refinery operations, which are richer in nitrogen than the conventional feedstocks, has increased the awareness of refiners to the presence of nitrogen compounds in crude feedstocks. For the most part, however, HDS catalyst technology has been used to accomplish HDN although such catalysts are not ideally suited to nitrogen removal). However, in recent years, the limitations of HDS catalysts when applied to HDN have been recognized, and there are reports of attempts to manufacture catalysts more specific to nitrogen removal (Parkash, 2003; Gary et al., 2007; Speight, 2014, 2017; Hsu and Robinson, 2017).

The character of the hydrotreating processes is chemically very simple since they essentially involve the removal of sulfur and nitrogen as hydrogen sulfide and ammonia, respectively:

$$RSR^1 + H_2 \rightarrow RH + R^1H + H_2S$$

$$R - N(R^1) - R^{11} + 3H_2 \rightarrow RH + R^1H + R^{11}H + 2NH_3$$

However, nitrogen is the most difficult contaminant to remove from feedstocks, and processing conditions are usually dictated by the requirements for nitrogen removal.

In general, any catalyst capable of participating in hydrogenation reactions may be used for HDS. The sulfides of hydrogenating metals are particularly used for HDS, and catalysts containing cobalt, molybdenum, nickel, and tungsten are widely used on a commercial basis (Parkash, 2003; Gary et al., 2007; Speight, 2014, 2017; Hsu and Robinson, 2017).

Hydrotreating catalysts are usually cobalt–molybdenum catalysts and under the conditions whereby nitrogen removal is accomplished, desulfurization usually occurs as well as oxygen removal. Indeed, it is generally recognized that fullest activity of the hydrotreating catalyst is not reached until some interaction with the sulfur (from the feedstock) has occurred, with part of the catalyst metals converted to the sulfides. Too much interaction may of course lead to catalyst deactivation.

The poisoning effect of nitrogen can be offset to a certain degree by operation at a higher temperature. However, the higher temperature tends to increase the production of material in the methane (CH_4) to butane (C_4H_{10}) range and decrease the operating stability of the catalyst so that it requires more frequent regeneration. Catalysts containing platinum or palladium (approximately 0.5 wt.%) on a zeolite base appear to be somewhat less sensitive to nitrogen than are nickel catalysts, and successful operation has been achieved with feedstocks containing 40 ppm nitrogen. This catalyst is also more tolerant of sulfur in the feed, which acts as a temporary poison, the catalyst recovering its activity when the sulfur content of the feed is reduced.

On such catalysts as nickel or tungsten sulfide on silica–alumina, isomerization does not appear to play any part in the reaction, as uncracked normal paraffins from the feedstock tend to retain their normal structure. Extensive splitting produces large amounts of low-molecular-weight (C_3–C_6) paraffins, and it appears that a primary reaction of paraffins is catalytic cracking followed by hydrogenation to form iso-paraffins. With catalysts of higher hydrogenation activity, such as platinum on silica–alumina, direct isomerization occurs. The product distribution is also different, and the ratio of low molecular weight to intermediate molecular weight paraffins in the breakdown product is reduced.

In addition to the chemical nature of the catalyst, the physical structure of the catalyst is also important in determining the hydrogenation and cracking capabilities, particularly for high-boiling feedstocks (Parkash, 2003; Gary et al., 2007; Speight, 2014, 2017; Hsu and Robinson, 2017). When gas oils and residua are used, the feedstock is present as liquids under the conditions of the reaction. Additional feedstock and the hydrogen must diffuse through this liquid before reaction can take place at the interior surfaces of the catalyst particle.

At high temperatures, reaction rates can be much higher than diffusion rates and concentration gradients can develop within the catalyst particle. Therefore the choice of catalyst porosity is an important parameter. When feedstocks are to be hydrocracked to liquefied petroleum gas and gasoline, pore diffusion effects are usually absent. High surface area (approximately 300 m^2/g) and low-to-moderate porosity (from 12 pore diameter with crystalline acidic components to 50 or more with amorphous materials) catalysts are used. With reactions involving high-molecular-weight feedstocks, pore diffusion can exert a large influence, and catalysts with a high pore diameter are necessary for more efficient conversion (Parkash, 2003; Gary et al., 2007; Speight, 2014, 2017; Hsu and Robinson, 2017).

Catalyst operating temperature can influence reaction selectivity since the activation energy for hydrotreating reactions is much lower than for hydrocracking reaction. Therefore raising the temperature in a residuum hydrotreater increases the extent of hydrocracking relative to hydrotreating, which also increases the hydrogen consumption (Speight, 2000, 2014, 2017; Parkash, 2003; Ancheyta and Speight, 2007; Gary et al., 2007; Hsu and Robinson, 2017).

Aromatic hydrogenation in crude oil refining may be carried out over supported metal or metal sulfide catalysts depending on the sulfur and nitrogen levels in the feedstock. For hydrorefining of feedstocks that contain appreciable concentrations of sulfur and nitrogen, sulfided nickel–molybdenum (Ni–Mo), nickel–tungsten (Ni–W), or cobalt–molybdenum (Co–Mo) on alumina (-Al_2O_3) catalysts are generally used, whereas supported noble metal catalysts have been used for sulfur- and nitrogen-free feedstocks. Catalysts containing noble metals on Y-zeolites have been reported to be more sulfur tolerant than those on other supports. Within the series of cobalt-promoted or nickel-promoted group VI metal (Mo or W) sulfides supported on -Al_2O_3, the ranking for hydrogenation is

$$\text{Ni–W} > \text{Ni–Mo} > \text{Co–Mo} > \text{Co–W}$$

Nickel–tungsten (Ni–W) and nickel–molybdenum (Ni–Mo) on Al_2O_3 catalysts are widely used to reduce sulfur, nitrogen, and aromatics levels in crude oil fractions by hydrotreating.

Molybdenum sulfide (MoS_2), usually supported on alumina, is widely used in crude oil processes for hydrogenation reactions. It is a layered structure that can be made much more active by addition of cobalt or nickel. When promoted with cobalt sulfide (CoS), making what is called cobalt–moly catalysts, it is widely used in HDS processes. The nickel sulfide (NiS)–promoted version is used for HDN as well as HDS. The closely related tungsten compound (WS_2) is used in commercial hydrocracking catalysts. Other sulfides (iron sulfide, FeS; chromium sulfide, Cr_2S_3; and vanadium sulfide, V_2S_5) are also effective and used in some catalysts. A valuable alternative to the base metal sulfides is palladium sulfide (PdS). Although it is expensive, palladium sulfide forms the basis for several very active catalysts.

The life of a catalyst used to hydrotreat crude oil residua is dependent on the rate of carbon deposition and the rate at which organometallic compounds decompose and form metal sulfides on the surface. Several different metal complexes exist in the asphaltene fraction of the residuum, and an explicit reaction mechanism of decomposition that would be a perfect fit for all of the compounds is not possible. However, in general terms, the reaction can be described as hydrogen (A) dissolved in the feedstock contacting an organometallic compound (B) at the surface of the hydrotreating catalyst and producing a metal sulfide (C) and a hydrocarbon (D):

$$A + B \rightarrow C + D$$

Different rates of reaction may occur with various types and concentrations of metallic compounds. For example, a medium metal-content feedstock will generally have a lower rate of demetallization compared to high-metal content feedstock. And, although individual organometallic compounds decompose according to both first- and second-order rate expressions, for reactor design, a second-order rate expression is applicable to the decomposition of residuum as a whole.

Obviously, choice of hydrogenation catalyst depends on what the catalyst designer wishes to accomplish. In catalysts to make gasoline, for instance, vigorous cracking is needed to convert a large fraction of the feed to the kinds of molecules that will make a good gasoline blending stock. For this vigorous cracking a vigorous hydrogenation component is needed. Since palladium is the most active catalyst for this, the extra expense is warranted. On the other hand, many refiners wish only to make acceptable diesel, a less demanding application. For this the less expensive molybdenum sulfides are adequate.

8.9 Bioprocesses

Petroleum biorefining, as a branch of petroleum biotechnology, has been related to the production of single cell protein (SCP) from waxy *n*-alkanes (Hamer and Al-Awadhi, 2000). Nowadays, there is an aim to be applied in upgrading of heavy crude oils. Petroleum biotechnology involves the use of wide range of conditions, milder temperature and pressure, cleaner and selective processes, lower emissions, and no-generations of undesirable by-products. Moreover, microbial and enzymatic catalysts can be manipulated and used for more specific applications. Heavy crude oils can be subjected to biorefining to get rid of most of the sulfur, nitrogen, toxic metals, and asphaltenes (Le Borgne and Quintero, 2003). Thus biotechnology has become of great interest in petroleum refining (i.e., biorefining) to accomplish (1) biodesulfurization, BDS; (2) biodenitrogenation, BDN; and (3) biodemetallization, BDM.

8.9.1 Biodesulfurization

Sulfur is the major concern for producers and refiners and has long been a key determinant of the value of crude oils. It is the third most abundant element in crude oil after carbon and hydrogen. Heavy oil, extra heavy oil, and tar sand bitumen contain 3%–6% w/w sulfur that must be removed within the refinery system. The combustion of sulfur-containing fuels would lead to the increased emissions of sulfur oxides (SO_x), the main cause of acid rains and particulate matter (PM) the main cause of black smoke associated with and gasoline-fueled and diesel-fuel vehicles. It has been reported that the emissions of PM from diesel engines are proportional to the sulfur content of the diesel (Stanislaus et al., 2010; Mohebali and Ball, 2016). If the raw refinery feedstock is subject to a desulfurization technology before entering the refinery system, the downstream desulfurization costs are minimized. Refinery feedstocks with high viscosity and high density usually contain higher amounts of the more complex sulfur-containing constituents. The aliphatic acyclic sulfides (thioether derivatives) and cyclic sulfide derivatives (also known as thiolane derivatives) are easy to remove during a HDS process or by thermal treatment. On the other hand, the sulfur that is contained in ring structures, such as thiophene derivatives and its benzologs (such as benzothiophene, dibenzothiophene, and benzonaphthothiophene) are more resistant to sulfur removal by a HDS process (Gray et al., 1996).

BDS is a biological method in which microbes or enzymes are used as a catalyst to remove organosulfur compounds, especially the recalcitrant ones, for example, dibenzothiophene and its derivatives. It can be performed aerobically or anaerobically. Its main disadvantage is its conversion rate that is much slower than HDS, since all the biological reactions are generally slower than the chemical reactions. There are three main routes for aerobic BDS: (1) complete mineralization where the end products are carbon dioxide and water; (2) the Kodama pathway, where carbon–carbon bonds are cleaved and some water soluble by-products are produced, which would significantly inhibit microbial growth dibenzothiophene oxidation; and (3) the 4S-pathway, in which the carbon skeleton is not destroyed and only sulfur is removed (El-Gendy and Speight, 2015). The first two pathways are not typically recommended for the desulfurization of liquid fuels since the efficiency of the BDS process depends on the capability of the biocatalyst to remove sulfur without altering the carbon skeleton or which can result in a reduction of the value of the fuel. But the first two process options are recommended for application to the bioremediation of oil spills and to soil or sediments polluted with crude oil and/or crude oil products (Gupta et al., 2005; Speight and Arjoon, 2012). The key techno-economic challenge to use of BDS processes is to establish a means of implementing the two-phase bioreactor system and deemulsification steps as well as the product recovery step (Ohshiro et al., 1996; Pacheco et al., 1999; McFarland, 1999).

The desulfurization of petroleum under anaerobic conditions would be attractive because it avoids costs associated with aeration, it has the advantage of liberating sulfur as a gas—hydrogen sulfide be treated with existing refinery desulfurization plants (e.g., Claus process) and does not liberate sulfate as a by-product that must be disposed by some appropriate treatment (Setti et al., 1997; Speight, 2014). Under anaerobic conditions, oxidation of hydrocarbons to undesired compounds such as colored and gum forming products is minimal (McFarland, 1999). Moreover, anaerobic microorganisms use approximately 10% of total produced energy, while aerobic ones use 50% of the produced energy. These advantages can be counted as incentives to continue research on reductive BDS. However, maintaining an anaerobic process is extremely difficult and the specific activity of most of the isolated strains have been reported to be insignificant for dibenzothiophene derivatives (Armstrong et al., 1995). Due to low reaction rates, safety and cost concerns and the lack of identification of specific enzymes and genes responsible for anaerobic desulfurization, anaerobic microorganisms effective enough for practical petroleum desulfurization have not been found yet, and anaerobic BDS process has not been developed. Due to the milder and safer process conditions of BDS, the carbon dioxide emissions and energy requirements of a BDS-based processes is estimated to be lower than that of HDS processes (Linguist and Pacheco, 1999; Singh, 2012). Moreover, the capital costs to setup a BDS process is reported to be 50% lower than that of a HDS process (Pacheco et al., 1999; Linguist and Pacheco, 1999).

The difference in desulfurization efficiency within the different microorganisms might be related to differences in cell-surface hydrophobicity, mass transport through the cell walls and membranes, as well as metabolic regulations inside the cell (Bustos-Jaimes et al., 2003). Thus the efficiency of BDS in crude oil depends mainly on the applied microorganisms and its broad versatility on different feedstocks and there is no general trend for the efficiency of microorganisms in model or real feed oils. It is well known, that, most of the aerobic microorganisms adhere to the *n*-alkanes (below n-C_{16}) which are in a liquid form at room temperature.

Crude oil refineries usually separate crude oil into several fractions and then desulfurize them separately. Moreover, a refinery can make substantial cost savings if most of the sulfur is removed from the crude oil before it is fractionated. Also, it has been suggested that due to the high content of water in crude oil, BDS of crude oil is more practicable compared to that of diesel oil and gasoline (Zhou and Zhang, 2004). Currently, the crude oil industry is increasingly dependent on heavy crude oil to meet the domestic demand for gasoline and distillate fuels. The high viscosity of many refinery feedstocks is a factor contributing to the underutilization of these valuable natural resources. Viscosity greatly complicates, and may even defeat, the extraction of many types of crude oil from the earth. It remains a concern following extraction, as high viscosity significantly hampers the pumping, transportation, refining, and handling of crude oil. Because of this, the petroleum industry has long recognized the need for a safe, economical, and effective method for reducing the viscosity of valuable fossil fuel resources. Under certain circumstances, standard-refining processes such as hydrotreating or HDS can favorably affect the viscosity of crude oil during refining. Some reduction in viscosity is also achieved through the breakdown of complex hydrocarbons (e.g., aromatic hydrocarbons) into simpler hydrocarbons of low molecular weight.

Certain technical issues hamper the applicability of BDS. One of them is the biphasic nature of the process. An aqueous phase is needed to maintain a viable biocatalyst in the BDS process and, thus, significant quantities of water must be added to the fuel. The values found in the literature generally indicate low oil-to-water volumetric ratios in the desulfurization of real fractions, ranging from 1% to 25% of oil in water.

From a cost perspective the BDS process has favorable features: (1) operation at low temperature and pressure; (2) BDS is estimated to have 70%–80% lower carbon dioxide emissions and energy consumption; (3) in the case of reaching adequate BDS efficiency level, the capital cost required for an industrial BDS process is predicted to be two-thirds of the one for a HDS process; (4) cost-effective, for BDS, the capital and operating costs are 50% and 10%–15% less than HDS, respectively (Pacheco et al., 1999).

8.9.2 Biodenitrogenation

The presence of nitrogen in petroleum is of much greater significance in refinery operations than might be expected from the small amounts present. Nitrogen compounds can be responsible for the poisoning of cracking catalysts, and they also contribute to gum formation in such products as domestic fuel oil. The trend in recent years toward cutting deeper into the crude to obtain stocks for catalytic cracking has accentuated the harmful effects of the nitrogen compounds, which are concentrated largely in the higher boiling portions.

In general, the nitrogen content of crude oil is low and generally falls within the range 0.1%–0.9%, although some crude oil may contain up to 2% nitrogen (Speight, 2012, 2014). However, crude oils with no detectable nitrogen or even trace amounts are not uncommon, but in general the more asphaltic the oil, the higher its nitrogen content. Insofar as an approximate correlation exists between the sulfur content and API gravity of crude oils (Ancheyta and Speight, 2007), there also exists a correlation between nitrogen content and the API gravity of crude oil. It also follows that there is an approximate correlation between the nitrogen content and the carbon residue: the higher the nitrogen content, the higher the carbon residue.

Although, specific BDS of crude oil and its distillates has been reasonably investigated, there is a little information about BDN of oil feed without affecting its calorific value. It has been estimated that BDN of petroleum would be beneficial for deep denitrogenation, where the classical hydroprocessing methods are costly and nonselective (Vazquez-Duhalt et al., 2002). It will also eliminate the contribution of fuel nitrogen to NO_x emissions. However, the economics of nitrogen-removal processes are affected by the amount of associated hydrocarbon lost from the fuel, during the denitrogenation process.

8.9.3 Biodemetallization

Crude oil contains metals in the form of salts (zinc, titanium, calcium, and magnesium), petroporphyrins (vanadium, copper, nickel, and iron), and other complexes that exits predominantly within the asphaltene fraction (Ali and Abbas, 2006; Speight, 2014). Thus the higher the asphaltene content of crude oil, the higher the heavy metal content (Duyck et al., 2007; Speight, 2014). In fact, the accumulation of metallic constituents in the higher molecular weight and polar fractions of crude oils plays a significant role in establishing the refining procedure (Panariti et al., 2000; Duyck et al., 2008; Speight, 2014, 2017).

Depending on the origin of crude oil, the concentration of the vanadium varies from as low as 0.1 ppm to as high as 1200 ppm, while that of nickel commonly varies from trace amounts to 150 ppm (Ali and Abbas, 2006; Duyck et al., 2007). The V/Ni ratio is constant in crude oils of common source rocks and dependent on the geological age of the rocks, where oils from Triassic or older age show a value higher than unity (Ball et al., 1960), and this ratio is also used for tracing source effects (Al-Shahristani and Al-Atyia, 1972; El-Gayar et al., 2002). However, the biodegradation of the asphaltene constituents and the resin constituents is reported to influence the Ni/V ratio in these fractions (Duyck et al., 2007).

Crude oil with a particularly high content of organometallics include Boscán crude oil, Cerro Negro crude oil, Maya crude oil, Wilmington crude oil, and Prudhoe Bay crude oil (Fish et al., 1984). The metallic constituents often occur as inorganic water-soluble forms and are easily removed during the crude desalting process, in which they are concentrated in the aqueous phase. However, the metalloporphyrins are embedded in the extremely complex structure of the asphaltene constituents and, thus, metal removal from petroporphyrins and complexes that is necessary to be addressed.

Currently, demetallization occurs during the hydrocracking process or during hydrotreating process (Ancheyta and Speight, 2007; Speight, 2014, 2017). However, feedstocks that have relatively high metals contents (> 300 ppm) substantially increase catalyst consumption because the metals poison the catalyst, thereby requiring frequent catalyst replacement. The usual desulfurization catalysts are relatively expensive for these consumption rates, but there are catalysts that are relatively inexpensive and can be used in the first reactor to remove a large percentage of the metals. Subsequent reactors downstream of the first reactor would use normal HDS catalysts. Since the catalyst materials are proprietary, it is not possible to identify them here. However, it is understood that such catalysts contain little or no metal promoters, that is, nickel, cobalt, and molybdenum.

Thus one method of controlling demetallization is to employ separate smaller *guard reactors* just ahead of the fixed-bed HDS reactor section (Speight, 2014, 2017). The preheated feedstock and hydrogen pass through the guard reactors that are filled with an appropriate catalyst for demetallization that is often the same as the catalyst used in the HDS section. The advantage of this system is that it enables replacement of the most contaminated catalyst (*guard bed*), where pressure drop is highest, without having to replace the entire inventory or shut down the unit. The feedstock is alternated between guard reactors while catalyst in the idle guard reactor is being replaced.

Unless a guard reactor is employed, a major problem during the refining of oil and other petroleum products is the continuous contamination of solid, porous catalysts by various porphyrins, metalloporphyrins, chlorins, and natural degradation products of these compounds, such as petroporphyrins, containing metals such as vanadium and nickel. The porphyrins deposits, however, consists of several different metals on the catalytic surface, including vanadium (V), nickel (Ni), titanium (Ti), iron (Fe), copper (Cu) or a combination thereof with the concentrations of V and Ni varying from a few to several hundred parts per million, depending on the type of crude oil supply. Vanadium is usually present in a concentration greater than other metals with much more than half of all V being deposited on the catalyst arising from the porphyry complex. Heavy metals poison refining catalyst. During petroleum refining, heavy metals are concentrated in the residual fraction, which is usually subjected to catalytic cracking. During the catalytic cracking, metals in the oil, deposit on the cracking catalysts, and decrease their selectivity and activity. Catalyst deactivation in cracking, hydrogenation, and HDS processes can be also occurred by pore clogging, metal deposition, deformation, and destruction of reactors (Salehizadeh et al., 2007), which, in turn, demands the catalytic process be interrupted to either replace or clean the catalyst at a high cost.

Moreover, since, both nickel and vanadium exhibit dehydrogenation activity, their presence on the catalyst particles tends to promote dehydrogenation reactions during the cracking sequence and this result in increased amounts of coke and light gases at the expense of gasoline production (Ali and Abbas, 2006). Heavy metals can be liberated into the environment, during fuel combustion in the form of ash with high concentrations of toxic metal oxides. That leads to undesirable by-products and the need for disposal (Xu et al., 1997).

Moreover, heavy metals (mostly Ni and V) are furthermore corrosive (Montiel et al., 2009). When crude oil is vacuum-distilled, the metalloporphyrins tend to be entrained in the vapors and are carried over the vacuum gas oil fraction. The typical demetallization process in petroleum industry is the solvent deasphalting process, where the lighter oils are physically separated from heavier asphaltene constituents by mixing the heavy oil/residue with a very low boiling solvent such as propane and/or butane (Speight, 1981; Farag et al., 1989; Ancheyta and Speight, 2007; Speight, 2014, 2017). Other options for demetallization include a low-temperature coking process in which the metals collect in the initially formed coke (Speight, 2014, 2017).

The distillation separates crude oils into fractions according to boiling point, so that each of the processing units following will have the feedstock that meets the required specifications. The metallic constituents concentrate in the residues (Reynolds et al., 1987). The thermal processes, such as the visbreaking and coking (Speight, 1981, 2014, 2017), typically involve redistribution of the hydrogen in the residue to produce lower boiling products containing more hydrogen while the asphaltene constituents and metals are removed in the form of coke or visbreaking residue. Filtration, using a porous membrane, is reported to be effective for the removal of nitrogen, sulfur, nickel, and vanadium as well as any other metallic constituents that have accumulated during production and transportation. There is also a method for upgrading heavy oils by solvent dissolution and ultra-filtration at high pressure (Osterhuber, 1989). The process is especially suitable for removing trace metals (mainly Ni and V), thereby reducing the Conradson carbon residue of the resulting oil.

There are also processes for the selective removal of metals from the organic moieties with a minimal conversion of the remaining petroleum. For example, metal removal using a solvent (Savastano, 1991), oxidative demetallization of petroleum asphaltene constituents and residua (Gould, 1980) and the hydro-demetallization process (Adarme et al., 1990; Bartholdy and Hannerup, 1990; Piskorz et al., 1996) are also available for demetallization of crude oils. However, these methods increase the cost of refining and usually produce secondary pollution in the environment (Hernandez et al., 1998).

8.10 The future

Environmental concerns and newly enacted rules and regulations mean the feedstock that crude oil products are expected to meet lower and lower limits on contaminates, such as sulfur and nitrogen. New regulations require the removal of sulfur compositions from liquid hydrocarbon derivatives, such as those used in gasoline, diesel fuel, and other transportation fuels.

In the HDS process, high temperatures and pressures may be required to obtain the desired low levels of sulfur. High temperature processing of olefinic naphtha, however, may result in a lower grade fuel due to the saturation of olefins leading to an octane loss. Low octane gasoline may require additional refining, isomerization, blending, and the like to produce higher quality fuels suitable for use in gasoline products. Such extra processing adds additional cost, expense, and complexity to the process and may result in other undesirable changes in the products. As a result, the future will see processes that are more chemically precise in hydrotreating and will offer higher efficiency and conversion to selected products on the base that is not always practiced currently.

Five of the most common approaches to upgrading hydrotreaters for clean-fuels production (in order of increasing capital cost) are currently and will continue (at least for the next two decades) to be (1) developing higher activity and more resilient catalysts, (2) replacing reactor internals for increased efficiency, (3) adding reactor capacity to accommodate high-boiling feedstocks and increase gasoline–diesel production, (4) increasing hydrogen partial pressure, (5) process design and hardware that are more specialized, and (6) the focus on process schemes that effectively reduce hydrogen consumption.

Finally, hydrotreating of residua requires considerably different catalysts and process flows, depending on the specific operation so that efficient hydroconversion through uniform distribution of liquid, hydrogen-rich gas, and catalyst across the reactor is assured. There will also be automated demetallization of fixed-bed systems as well as more units that operate as ebullating-bed hydrocrackers (Chapter 9: Hydrocracking).

Finally, severe hydrotreating requires high-purity (99% v/v) hydrogen, while less severe hydrotreating can employ lower purity (90% v/v) hydrogen. Refiners will continue to optimize hydrogen use by cascading hydrogen through the refinery. High purity hydrogen will continue to be used only where required and the lower purity hydrogen purged from these applications is used for services that do not require high purity hydrogen.

8.10.1 Catalyst technology

Conventional hydroprocessing catalysts are generally in the form of a carrier of a refractory oxide material on which hydrogenation metals are deposited, the choice and amount of each component being determined by the end use. Refractory oxide materials usual in the art are amorphous or crystalline forms of alumina, silica, and combinations thereof. These oxide materials can have some intrinsic catalytic activity but often only provide the support on which active metals compounds are held. Generally, the thermal stability, low surface area, and poor mechanical strength have all hindered the commercial exploitation of certain metal oxide–supported catalyst systems. The intrinsic activity of hydrogenation metals-on-catalyst is superior to alumina-based catalysts. Catalyst synthesis will attempt to harness the intrinsic activity of various metals and remedy the deficiencies that currently plague low metal loading and thermal instability by using mixed oxides.

Use of ultra-DODD of liquid hydrocarbon fuels such as gasoline, diesel, and jet fuel to satisfy new environmental regulations and fuel cell applications is receiving increased attention and will continue to receive attention. Conventional HDS technology is difficult and costly to use to remove sulfur compounds from liquid hydrocarbon fuels to levels suitable to match environmental regulations. Several non-HDS-based desulfurization technologies for use with liquid fuels have been initiated. These technologies include (1) BDS and (2) adsorptive desulfurization. Both of these technologies will continue to receive attention and have a high potential for incorporated into future refineries. In fact, the BDS technology is likely to see application during microbial enhanced oil-recovery processes.

The current trend in hydroprocessing is the treatment of high-boiling sour feeds that contain compounds such as sulfur, nitrogen, aromatics, iron, and other undesirable components. These compounds pose significant problems with catalyst poisoning; however, developments are keeping pace with increased demand. In respect of the growing demand for ULSD, low-boiling cycle oil hydrotreating is receiving much attention. Feeds such as these are typically high in heavy metals, which will require additional unit modifications and/or the installation of guard beds/reactors.

In addition, hydrotreating feedstocks prior to sending the feedstock to the fluid catalytic cracking units is another important focus and will continue to be important or even increase in importance. Many fluid catalytic cracking unit incorporate pretreaters (in the form of hydrotreating the feedstocks or guard beds/reactors) to meet their naphtha–gasoline sulfur requirements. Installation of such reactors units will necessarily increase as heavy feedstocks are incorporated into gas oils (fed to the fluid catalytic cracking unit) or become the sole feedstock for the catalytic cracking unit. Proven technology is available to remove sulfur, metals, and asphaltene content while converting an important part of the feed to lower boiling quality products. This technology will improve and operations will be varied to upgrade the following typical feedstocks: atmospheric and vacuum residua, tar sand bitumen, deasphalter bottoms, and biofeedstocks.

Thus with the increasing focus to reduce sulfur content in fuels, the role of desulfurization in the refinery becomes more and more important. Currently, the process of choice is the hydrotreater, in which hydrogen is added to the fuel to remove the sulfur from the fuel. Some hydrogen may be lost to reduce the octane number of the fuel, which is undesirable. Because of the increased attention for fuel desulfurization, various new process-concepts are being developed with various claims of efficiency and effectiveness.

The major developments in desulfurization three main routes are advanced hydrotreating (new catalysts, catalytic distillation, and processing at mild conditions), reactive adsorption (type of adsorbent used, process design), and oxidative desulfurization (catalyst and process design).

The demand for low sulfur transportation fuels requires that refiners evaluate many different options for reaching the target. Selection of catalyst types is one of the important decisions and depends on (1) feedstock, (2) operation conditions,

(3) hydrogen availability, and product properties play a role (Parkash, 2003; Gary et al., 2007; Speight, 2014, 2017; Hsu and Robinson, 2017).

Besides the issues related to the legislative drive for removing sulfur, the refiners will be faced with a growing demand for diesel fuel, which may be met by producing less low-value products such as heating oil. This can be done by converting high-boiling fractions by hydrocracking or mild hydrocracking (hydrotreating) processes or one may adopt upgrading processes, for example, for low-boiling cycle oil. The latter alternative will require innovative technology that not only removes refractory sulfur species in the presence of high amounts of nitrogen but also performs a certain degree of ring opening to reach a reasonable product cetane number. The innovative might well involve (in part) the use of guard beds or reactor or the use of scavengers (such as calcium oxide, alone or supported on the catalyst) that will remove sulfur and coke formers during the hydrotreating process.

Catalyst development will accelerate, including catalysts used for pretreating feedstocks to the fluid catalytic cracking unit that will eliminate much of the need for naphtha–gasoline posttreating. The catalysts will have multiple functions, such as (1) optimized HDS and (2) minimized hydrogen consumption. The addition of metals such as iron, tungsten, niobium, boron, and phosphorus to catalyst compositions and the use of unsupported nanoparticles will increase and begin (if not already beginning) commercial utilization.

The challenges that the refining industry is facing in regard to HDS processes in the next two decades requite for major developments within hydroprocessing catalyst technology (Speight, 2000; Ancheyta and Speight, 2007). Areas such as (1) catalyst supports, (2) catalyst morphology, and (3) reaction pathways will continue to provide new opportunities for the development of improved commercial hydrotreating catalysts.

Breakthrough technology is essential if catalysts are to be developed that are able to exhibit high activity to produce the desire products.

8.10.2 Gasoline and diesel fuel polishing

Briefly, the two desulfurization processes used for fuels purification (desulfurization) are (1) sweetening and (2) hydrotreating. Sweetening is effective only against mercaptans, which are the predominant species in low-boiling gasoline. Hydrotreating is effective against all sulfur species and is more widely used.

In the sweetening process a low-boiling naphtha stream is washed with amine to remove hydrogen sulfide and then reacted with caustic, which promotes the conversion of mercaptans to disulfides.

R-SH→RSSR

The disulfides can subsequently be extracted and removed in what is referred to as extractive sweetening.

In the hydrotreating process the feed is reacted with hydrogen, in the presence of a solid catalyst. The hydrogen removes sulfur by conversion to hydrogen sulfide, which is subsequently separated and removed from the reacted stream. As the reaction is favored by both temperature and pressure, hydrotreaters are typically designed and operated at approximately 370°C (700°F) and 1000–2000 psi hydrogen. The lower ends of the ranges typically apply to gasoline desulfurization, while gas oil desulfurization requires a more severe operation.

Hydrogen is provided in the form of treating gas at a purity that is typically around on the order of 90% by volume although gas with as little as 60% by volume hydrogen is reputed to be used. Hydrogen is produced by catalytic reformers or hydrogen generation units and distributed to the hydrotreaters through a refinery-wide network (Parkash, 2003; Gary et al., 2007; Speight, 2014, 2017; Hsu and Robinson, 2017).

In a hydrotreating unit feed and treating gas are combined and brought to the reaction temperature and pressure, prior to entering the reactor. The reactor is a vessel preloaded with solid catalyst, which promotes the reaction. The catalyst is slowly deactivated by the continuous exposure to high temperatures and by the formation of a coke layer on its surface. Refineries have to shut down the units periodically and regenerate or replace the catalyst.

The severity of operation of an existing unit can be increased by increasing the reaction temperature but there is a negative impact on the catalyst life. The severity of operation can also be increased by increasing the catalyst volume of the unit. In this case the typical solution is to add a second reactor identical to the existing one, doubling the reactor volume. The pressure of an existing unit cannot be changed to increase its severity, because the pressure is related to material of construction and thickness of metal surfaces. If higher pressure is required, the typical solution is to install a new unit and use the existing one for a less severe service.

In contrast there are also processes that do not require either of the previously stated technologies.

BDS (*q.v.*) is only one of several concepts by which gasoline and diesel fuel might be polished, that is, sulfur removed to an extremely low level, if not to a zero. At this point, it is pertinent that a brief review of the potential methods for fuel polishing shoed be introduced.

One new technology is the use of adsorption by metal oxides in which the oxides either react by physical adsorption of by chemical adsorption insofar as adsorption followed by chemical reaction is promoted. The major distinction of this type of process from conventional hydrotreating is that the sulfur in the sulfur-containing compounds adsorbs to the catalyst after the feedstock–hydrogen mixture interacts with the catalyst. The catalyst does need to be regenerated constantly.

Another option involves sulfur oxidization in which a crude oil and water emulsion is reacted with hydrogen peroxide (or another oxidizer) to convert the sulfur in sulfur-containing compounds to sulfones. The sulfones are separated from the hydrocarbon derivatives for postprocessing. The major advantages of this new technology include low reactor temperatures and pressures, short residence time, no emissions, and no hydrogen requirement. The technology preferentially treats dibenzothiophene derivatives, one of streams that are most difficult to desulfurize.

One way to add to the supply of ultra-low sulfur fuels is to turn to a nonoil-based diesel. The Fischer–Tropsch process, for example, can be used to convert natural gas to a synthetic, sulfur-free diesel fuel. Commercial viability of gas-to-liquids projects depends (in addition to capital costs) on the market for crude oil products and possible price premiums for gas-to-liquids fuels as well as the value of any byproducts.

A second way to avoid desulfurization is with biodiesel made from vegetable oil or animal fats. Although other processes are available, most biodiesel is made with a base-catalyzed reaction. In the process, fat or oil is reacted with an alcohol, such as methanol, in the presence of a catalyst to produce glycerin and methyl esters or biodiesel. The methanol is charged in excess to assist in quick conversion and recovered for reuse. The catalyst, usually sodium or potassium hydroxide, is mixed with the methanol. Biodiesel is a strong solvent and can dissolve paint as well as deposits left in fuel lines by crude oil–based diesel, sometimes leading to engine problems. Biodiesel also freezes at a higher temperature than crude oil–based diesel.

8.10.3 Biofeedstocks

There are also biofeedstock issues that will become relevant in the refinery of the future, not the least of which will be the incorporation of bio-feedstocks into refinery hydrotreaters. For example, there are already reports of (1) refining extracted bio-oils being combined with refinery streams for hydroprocessing, (2) hydrogenation of animal fats to produce a high-cetane diesel-range product, and (3) hydrogenation of palm oil. However, there are issues related to the quality of hydrotreated vegetable oils in terms of high paraffin content, low filter plugging points, and low density that must and will be resolved. Resolution of such issues will lead to recommendations on the means by which biofeedstocks can be (or will be) incorporated into existing hydrotreating units based on process efficiency as well as economic considerations.

In summary, changes in the demand for the various oil products combined with tighter product specifications has in the past and will continue to have in the future a strong impact on the role of hydroprocessing in refining. New technologies and, even more important, new catalysts will need to be developed to meet these changes. The changes in demands and specifications will need to be met by applying and improving existing technologies. Catalyst development is crucial in this respect.

In addition, hydrotreating, as applied to the production of fuels and other products from nonfossil fuel feedstocks (Chapter 10: Nonfossil Fuel Feedstocks) into renewable diesel, utilizes specialty designed catalysts compatible with the type of catalysts used for hydrotreating of fossil diesel streams to meet environmental specifications. Thus a coprocessing scheme where fossil diesel and renewable feedstocks are mixed and coprocessed is possible, producing a clean and green diesel that meets all the desired specifications. The hydrotreating may also take place in a dedicated standalone unit that produces 100% renewable drop in diesel. In either case the new feedstock components give rise to new reactions and to the formation of new byproducts, resulting in a series of challenges. These challenges can be addressed through the use of optimized catalysts and process design.

The main issue may not be to achieve full conversion of the renewable feedstock, because most naturally occurring oxygen containing species are much more reactive than the sulfur compounds in fossil fuels. The challenge in industrial operation is mainly to control the very exothermic reactions and the large amounts of hydrogen consumed, which require higher make-up hydrogen rates and larger quench gas flows, even when coprocessing smaller amounts of renewable feedstocks. Thus the refinery hydrogen balance must be checked, and the unit capacity may be lower than when only processing fossil diesel. The depletion of hydrogen, combined with high temperatures, may lead to accelerated catalyst deactivation and pressure drop buildup. Before these factors can be controlled, tailor-made catalysts and a careful selection of unit layout and reaction conditions must be employed.

References

Adarme, R., Sughrue, E.L., Johnson, M.M., Kidd, D.R., Phillips, M.D., Shaw, J.E., 1990. Demetallization of asphaltenes: thermal and catalytic effects with small pore catalysts. J. Am. Chem. Soc. 35, 614–618.

Al-Shahristani, H., Al-Atyia, M.J., 1972. Vertical migration of oil in Iraqi oil fields: evidence based on vanadium and nickel concentrations. Geochim. Cosmochim. Acta 36, 929–938.

Ali, M.A., Abbas, S., 2006. A review of methods for the demetallization of residual fuel oils. Fuel Process. Technol. 87, 573–584.

Ancheyta, J., Speight, J.G., 2007. Hydroprocessing Heavy Oils and Residua. CRC Press, Taylor & Francis Group, Boca Raton, FL.

Armstrong, S.M., Sankey, B.M., Voodoo, G., 1995. Conversion of dibenzothiophene to biphenyl by sulfate reducing bacteria isolated from oil field production facilities. Biotechnol. Lett. 17, 1133–1136.

Ball, J.S., Wenger, W.J., Hyden, H.J., Horr, C.A., Myers, A.T., 1960. Metal content of twenty-four petroleums. J. Chem. Eng. Data 5, 553–557.

Bartholdy, J., Hannerup, P.N., 1990. Hydrodemetallation in residue hydroprocessing. J. Am. Chem. Soc. 35, 619–625.

Bustos-Jaimes, I., Amador, G., Castorena, G., Le Borgne, S., 2003. Genotypic characterization of sulfur-oxidative desulfurizing bacterial strains isolated from Mexican refineries. Oil Gas Sci. Technol. 58 (4), 521–526.

Carlson, C.S., Langer, A.W., Stewart, J., Hill, R.M., 1958. Thermal hydrogenation. Transfer of hydrogen from tetralin to cracked residua. Ind. Eng. Chem. 50, 1067–1070.

Dolbear, G.E., 1998. Chapter 7: Hydrocracking reactions, catalysis, and processes. In: Speight, J.G. (Ed.), Petroleum Chemistry and Refining. Taylor & Francis, Washington, DC.

Duyck, C., Miekeley, N., Fonseca, T.C.O., Szatmari, P., Santos Neto, E.V.S., 2008. Trace element distributions in biodegraded crude oils and fractions from the Potiguar Basin, Brazil. J. Braz. Chem. Soc. 19 (5), 978–986.

Duyck, C., Miekeley, N., Porto da Silveira, C.L., Aucélio, R.Q., Campos, R.C., Grinberg, P., et al., 2007. Trace element distributions in biodegraded crude oils and fractions from the Potiguar Basin, Brazil. Spectrochim. Acta, B 62, 939–951.

El-Gendy, N.Sh, Speight, J.G., 2015. Handbook of Refinery Desulfurization, 2015. CRC Press, Taylor & Francis Group, Boca Raton, FL.

El-Gayar, M., Mostafa, M.S., Abdelfattah, A.E., Barakat, A.O., 2002. Application of geochemical parameters for classification of crude oils from egypt into source-related types. Fuel Process. Technol. 79, 13–28.

Farag, A.S., Sif El-Din, O.I., Youssef, M.H., Hassan, S.I., Farmawy, S., 1989. Solvent demetallization of heavy oil residue. Hung. J. Ind. Chem. 17 (3), 289–294.

Fish, R.H., Komlenic, J.J., Wines, B.K., 1984. Characterization and comparison of vanadyl and nickel compounds in heavy crude petroleums and asphaltenes by reverse-phase and size-exclusion liquid chromatography/graphite furnace atomic absorption spectrometry. Anal. Chem. 56, 2452–2460.

Gary, J.H., Handwerk, G.E., Kaiser, M.J., 2007. Petroleum Refining: Technology and Economics, fifth ed. CRC Press, Taylor & Francis Group, Boca Raton, FL.

Gray, K.A., Pogrebinsky, O.S., Mrachko, G.T., Xi, L., Monticello, D.J., Squires, C.H., 1996. Molecular mechanisms of biocatalytic desulfurization of fossil fuels. Nat. Biotechnol. 14 (13), 1705.

Gould, K.A., 1980. Oxidative demetallization of petroleum asphaltenes and residua. Fuel 59 (10), 733–736.

Gupta, N., Roychoudhury, P.K., Dep, J.K., 2005. Biotechnology of desulfurization of diesel: prospects and challenges. Appl. Microbiol. Biotechnol. 66, 356–366.

Hamer, G., Al-Awadhi, N., 2000. Biotechnological applications in the oil industry. Acta Biotechnol. 20, 335–350.

Hernandez, A., Mellado, R., Martinez, J., 1998. Metal accumulation and vanadium induced multidrug resistance by environmental isolates of *Escherichia hermannii* and *Enterobacter cloacae*. Appl. Environ. Microbiol. 64, 4317–4320.

Hsu, C.S., Robinson, P.R. (Eds.), 2017. Handbook of Petroleum Technology. Springer International Publishing AG, Cham.

Kumagai, H., Koyama, H., Nakamura, K., Igarashi, N., Mori, M., Tsukada, T., 2005. Hydrofining Catalyst and Hydrofining Process. United States Patent. 6,858,132. February 22.

Langer, A.W., Stewart, J., Thompson, C.E., White, H.Y., Hill, R.M., 1962. Ind. Eng. Chem. Proc. Des. Dev. 1, 309.

Le Borgne, S., Quintero, R., 2003. Biotechnological processes for refining of petroleum. Fuel Process. Technol. 81, 155–169.

Linguist, L., Pacheco, M., 1999. Enzyme-based diesel desulfurization process offers energy, CO_2 advantages. Oil Gas J. 22, 45–48.

McFarland, B.L., 1999. Biodesulfurization. Curr. Opin. Microbiol. 2, 257–264.

Meyers, R.A. (Ed.), 1997. Handbook of Petroleum Refining Processes, second ed. New York.

Mohebali, G., Ball, A.S., 2016. Biodesulfurization of diesel fuels – past, present and future perspectives. Int. Biodeter. Biodegrad. 110, 163–180.

Montiel, C., Quintero, R., Aburto, J., 2009. Petroleum biotechnology: technology trends for the future. Afr. J. Biotechnol. 8 (12), 2653–2666.

Ohshiro, T., Suzuki, K., Izumi, Y., 1996. Regulation of dibenzothiophene degrading enzyme activity of *Rhodococcus erythropolis* D-1. Ferm. Bioeng. 82, 121–124.

Osterhuber, E., 1989. Upgrading Heavy Oils by Solvent Dissolution and Ultrafiltration. US Patent No. 4,797, 200.

Pacheco, M.A., Lange, E.A., Pienkos, P.T., Yu, L.Q., Rouse, M.P., Lin, Q., Linquist, L.K. 1999. Recent advances in desulfurisation of diesel fuel. In: Proceedings Annual Meeting, NPRA. Paper No. NPRA AM-99-27, San Antonio, pp. 1–26.

Panariti, N., Del Bianco, N.A., Del Piero, G., Marchionna, M., 2000. Petroleum residue upgrading with dispersed catalysts. Part 1. Catalysts activity and selectivity. Appl. Catal. A: Gen. 204 (2), 203–213.

Parkash, S., 2003. Refining processes handbook. Gulf Professional Publishing. Elsevier, Amsterdam.

Piskorz, J., Radlein, D., Majerski, P., Scott, D., 1996. Hydrotreating of Heavy Hydrocarbon Oils in Supercritical Fluids. US Patent No. 5,496,464.

Reynolds, G.J., Biggs, W.R., Bezman, S.A., 1987. Removal of Heavy Metals from Residual Oils. ACS Symp. Ser. 344, 205–219.

Salehizadeh, H., Mousavi, M., Hatamipour, S., Kermanshahi, K., 2007. Microbial demetallization of crude oil using *Aspergillus* sp.: vanadium oxide octaethyl porphyrin (VOOEP) as a model of metallic petroporphyrins. Iran. J. Biotechnol. 5 (4), 226–231.

Savastano, C.A., 1991. Solvent extraction approach to petroleum demetallation. Fuel Sci. Technol. Int. 9 (7), 833–871.

Setti, L., Lanzarini, G., Pifferi, P.G., 1997. Whole cell biocatalysis for an oil desulfurization process. Fuel Process. Technol. 52, 145–153.

Singh, A., 2012. How specific microbial communities benefit the oil industry: biorefining and bioprocessing for upgrading petroleum oil. In: Whitby, C., Skovhus, T.L. (Eds.), Applied Microbiology and Molecular Biology in Oilfield Systems, pp. 121–178.

Speight, J.G., 1981. The Desulfurization of Heavy Oils and Residua. Marcel Dekker Inc, New York.

Speight, J.G., 1986. Upgrading heavy feedstocks. Annu. Rev. Energy 11, 253.

Speight, J.G., 2000. The Desulfurization of Heavy Oils and Residua, second ed. Marcel Dekker Inc, New York.

Speight, J.G., 2012. Crude oil assay database. Knovel. Elsevier, New York. Available from: http://www.knovel.com/web/portal/browse/display?_EXT_KNOVEL_DISPLAY_bookid = 5485&VerticalID = 0.

Speight, J.G., 2013a. The Chemistry and Technology of Coal, third ed. CRC Press, Taylor & Francis Group, Boca Raton, FL.

Speight, J.G., 2013b. Heavy and Extra Heavy Oil Upgrading Technologies. Gulf Professional Publishing, Elsevier, Oxford.

Speight, J.G., 2014. The Chemistry and Technology of Petroleum, fifth ed. CRC Press, Taylor & Francis Group, Boca Raton, FL.

Speight, J.G., 2017. Handbook of Petroleum Refining. CRC Press, Taylor & Francis Group, Boca Raton, FL.

Speight, J.G., Moschopedis, S.E., 1979. The production of low-sulfur liquids and coke from athabasca bitumen. Fuel Process. Technol. 2, 295.

Speight, J.G., Arjoon, K.K., 2012. Bioremediation of Petroleum and Petroleum Products. Scrivener Publishing, Beverly, MA.

Stanislaus, A., Marafi, A., Rana, M.S., 2010. Recent advances in the science and technology of ultra-low sulfur diesel (ULSD) production. Catal. Today 153, 1–6.

Tanaka, H., 2004. Process for Producing Hydrofining Catalyst. United States Patent 6,689,712. February 10.

Vazquez-Duhalt, R., Torres, E., Valderrama, B., Le Borgne, S., 2002. Will biochemical catalysis impact the petroleum refining industry? Energy Fuels 1 (16), 1239–1250.

Xu, G.-W., Mitchell, K.W., Moticello, D.J., 1997. Process for Demetallizing a Fossil Fuel. United States Patent 5,624,844.

Zhou, Z.Y., Zhang, K., 2004. Development situation and prospect of oil fields in China. Petrol. Explor. Dev. 31, 84–87.

Hydrocracking

9

9.1 Introduction

Hydrocracking is a refining technology that like hydrotreating (Chapter 20) also falls under the general umbrella of hydroprocessing. The outcome is the conversion of a variety of feedstocks to a range of products, and units to accomplish this goal can be found at various points in a refinery. The other technology, hydrotreating (Chapter 8: Desulfurization, Denitrogenation, and Demetalization), is a term often used synonymously with hydrodesulfurization (HDS) and is a catalytic refining process widely used to remove sulfur from crude oil products such as naphtha, gasoline, diesel fuel, kerosene, and fuel oil (Meyers, 1997; Speight, 2000, 2014, 2017; Parkash, 2003; Gary et al., 2007; Hsu and Robinson, 2017). The objective of the hydrotreating process is to remove sulfur as well as other unwanted compounds, for example, unsaturated hydrocarbon derivatives, nitrogen from refinery process streams. HDS processes typically include facilities for the capture and removal of the resulting hydrogen sulfide (H_2S) gas that is subsequently converted into by-product elemental sulfur or sulfuric acid.

The major differences between hydrocracking and hydrotreating are the time at which the feedstock remains at reaction temperature and the extent of the decomposition of the nonheteroatom constituents and products. The lower limits of hydrocracking conditions may overlap with the upper limits of hydrotreating conditions. Where the reaction conditions overlap, feedstocks to be hydrocracked will generally be exposed to the reactor temperature for longer periods; hence the reason why hydrocracking conditions may be referred to as (relatively) severe.

The history of the hydrocracking process goes back to the late 1920s when it was realized that there was a need for gasoline of a higher quality than that obtained by catalytic cracking led to the development of the hydrocracking process. One of the first plants to use hydrocracking was commissioned for the commercial hydrogenation of brown coal at Leuna in Germany. Tungsten sulfide was used as a catalyst in this one-stage unit, in which high reaction pressures, 2900–4350 psi, were applied. The catalyst displayed a very high hydrogenation activity: the aromatic feedstock, coal, and high-boiling fractions of oil, containing sulfur, nitrogen, and oxygen, were virtually completely converted into paraffins and isoparaffins. In 1939 Imperial Chemical Industries in Britain developed the second-stage catalyst for a plant that contributed largely to the supply of aviation gasoline in the subsequent years.

The older hydrogenolysis type of hydrocracking practiced in Europe during and after World War II used tungsten sulfide (WS_2) or molybdenum sulfide (MoS) as

The Refinery of the Future. DOI: https://doi.org/10.1016/B978-0-12-816994-0.00009-9

catalysts. These processes required high reaction temperatures and operating pressures, sometimes in excess of approximately 3000 psi (20,684 kPa) for continuous operation. Two-stage processes were applied on a limited scale in Germany, Britain, and United States. In Britain, feedstocks were creosote from coal tar and gas oil from crude oil. In the United States, Standard Oil of New Jersey operated a plant at Baton Rouge, producing gasoline from a Venezuelan kerosene/low-boiling gas oil fraction. Operating conditions in those units were comparable: approximate reaction temperature 400°C (750°F) and reaction pressures of 2900–4350 psi.

By the 1950s the process had become economic and the development of improved catalyst made it possible to operate the process at considerably lower pressure, viz. 1000–2200 psi. This in turn resulted in a reduction in equipment wall thickness, whereas simultaneously, advances were made in mechanical engineering, especially in the field of reactor design and heat transfer.

Modern hydrocracking processes were initially developed for converting refractory feedstocks to gasoline and jet fuel; process and catalyst improvements and modifications have made it possible to yield products from gases and naphtha to furnace oils and catalytic cracking feedstocks. The zeolites most frequently used in commercial hydrocracking catalysts are partially dealuminated and low-sodium, or high-silica, type Y zeolites in hydrogen or rare-earth forms. Other zeolites and mixtures of zeolites also are used. The zeolites often are imbedded in a high-surface-area amorphous matrix that serves as a binder. The metals can reside inside the zeolite and on the amorphous matrix (Scherzer and Gruia, 1996; Dolbear, 1998).

Furthermore, in order to maintain current gasoline and middle distillate production levels, additional conversion capacity is required because of the differential in the amount of distillates produced from conventional (low-density) light crude oil and the distillate products produced from viscous feedstocks. Process parameters (Figs. 9.1 and 9.2) emphasize the relatively severe nature of the hydrocracking process (Meyers, 1997; Speight, 2000, 2014, 2017; Parkash, 2003; Gary et al., 2007; Hsu and Robinson, 2017).

The mechanism of hydrocracking is basically similar to that of catalytic cracking, but with concurrent hydrogenation (Chapter 3: Refining Chemistry). The catalyst assists in the production of carbonium ions via olefin intermediates, and these intermediates are quickly hydrogenated under the high hydrogen partial pressures employed in hydrocracking. The rapid hydrogenation prevents adsorption of olefins on the catalyst and, hence, prevents their subsequent dehydrogenation, which ultimately leads to coke formation so that long onstream times can be obtained without the necessity of catalyst regeneration.

One of the most important reactions in hydrocracking is the partial hydrogenation of polycyclic aromatics followed by rupture of the saturated rings to form substituted monocyclic aromatics. The side chains may then be split-off to give isoparaffins. Side chains of three or four carbon atoms are easily removed from an aromatic ring during catalytic cracking, but the reaction of aromatic rings with shorter side chains appears to be quite different. For example, hydrocracking single-ring aromatics containing four or more methyl groups produces largely isobutane and

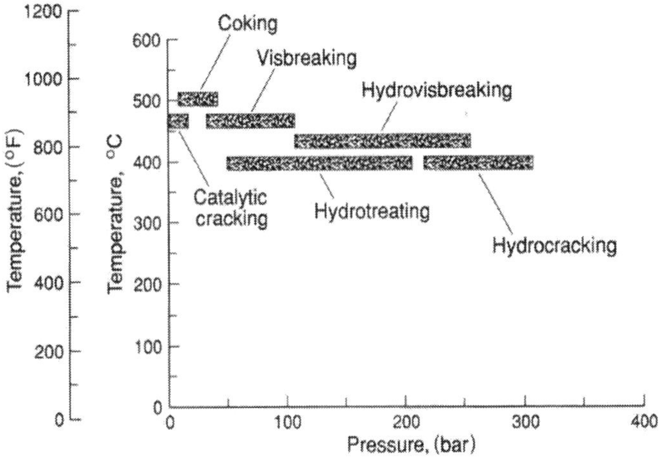

Figure 9.1 Temperature and pressure parameters for various processes.

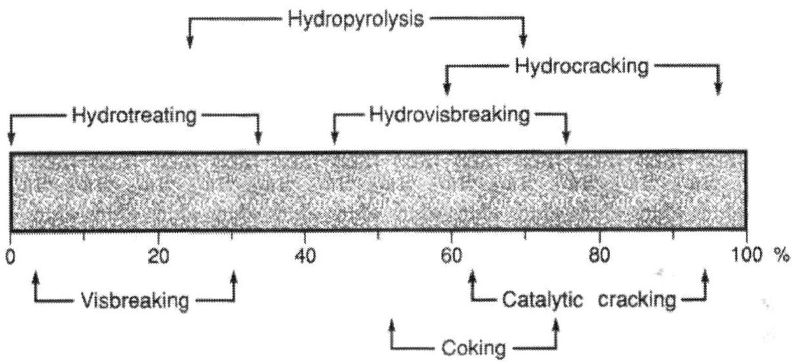

Figure 9.2 Feedstock conversion in various processes.

benzene. It may be that successive isomerization of the feed molecule adsorbed on the catalyst occurs until a four-carbon side chain is formed, which then breaks off to yield isobutane and benzene. Overall, coke formation is very low in hydrocracking since the secondary reactions, and the reactions that produce precursors to coke are suppressed as the hydrogen pressure is increased.

Hydrocracking, like any upgrading process, is evaluated on the basis of liquid yield (i.e., naphtha, distillate, and gas oil), heteroatom removal efficiency, feedstock conversion (FC), carbon mobilization (CM), and hydrogen utilization (HU), along with other process characteristics. Definition of FC, CM, and HU are as follows:

$FC = (Feedstock_{IN} - Feedstock_{OUT})/Feedstock_{IN} \times 100$
$CM = Carbon_{LIQUIDS}/Carbon_{FEEDSTOCK} \times 100$
$HU = Hydrogen_{LIQUIDS}/Hydrogen_{FEEDSTOCK} \times 100$

High CM (CM < 100%) and high HU correspond to high FC processes involving hydrogen addition such as hydrocracking. Since hydrogen is added, HU can be greater than 100%. Low CM and low HU correspond to low FC such as coking processes (Chapter 5: Thermal Cracking). Maximum efficiency from an upgrading process can be obtained by maximizing the liquid yield and its quality by minimizing the gas (C_1–C_4) yield, simultaneously. Under these operating conditions the hydrogen consumption would be the most efficient, that is, hydrogen is consumed to increase the liquid yield and its quality (Speight, 2000, 2014, 2017; Parkash, 2003; Gary et al., 2007; Hsu and Robinson, 2017).

9.2 Commercial processes

The hydrocracking process in which hydrogen is used in an attempt to stabilize the reactive fragments produced during the cracking, thereby decreasing their potential for recombination to higher boiling products and ultimately to coke. The choice of processing schemes for a given hydrocracking application depends upon the nature of the feedstock as well as the product requirements (Speight, 2000, 2014, 2017; Parkash, 2003; Ancheyta and Speight, 2007; Gary et al., 2007; Hsu and Robinson, 2017). The process can be simply illustrated as a single-stage or as a two-stage operation (Fig. 9.3).

Generally, the refinery utilizes one of three options for the process. Thus depending on the feedstock being processed and the type of plant design employed (single-stage or two-stage), flexibility can be provided to vary product distribution among the following principal end products:

The single-stage process can be used to produce gasoline but is more often used to produce middle distillate from high-boiling vacuum gas oils (VGOs). The two-stage process was developed primarily to produce high yields of gasoline from straight-run gas oil, and the first stage may actually be a purification step to remove sulfur-containing (as well as nitrogen-containing) organic materials. In terms of sulfur removal, it appears that nonasphaltene sulfur in nonasphaltene constituents may be removed before the more refractory sulfur in asphaltene constituents (Speight, 2007) thereby requiring through desulfurization. This is a good reason for processes to use an extinction-recycling technique to maximize desulfurization and the yields of the desired product. Significant conversion of high-boiling feedstocks can be accomplished by hydrocracking at high severity (Dolbear, 1998; Speight, 2000; Ancheyta and Speight, 2007). For some applications the products boiling up to 340°C (650°F) can be blended to give the desired final product.

On the whole, hydrocracking can handle a wider range of feedstock than catalytic cracking, although the latter process has seen some recent catalyst developments that narrowed the gap. There are also examples where hydrocracking is complementary rather than alternative to the other conversion process; for example,

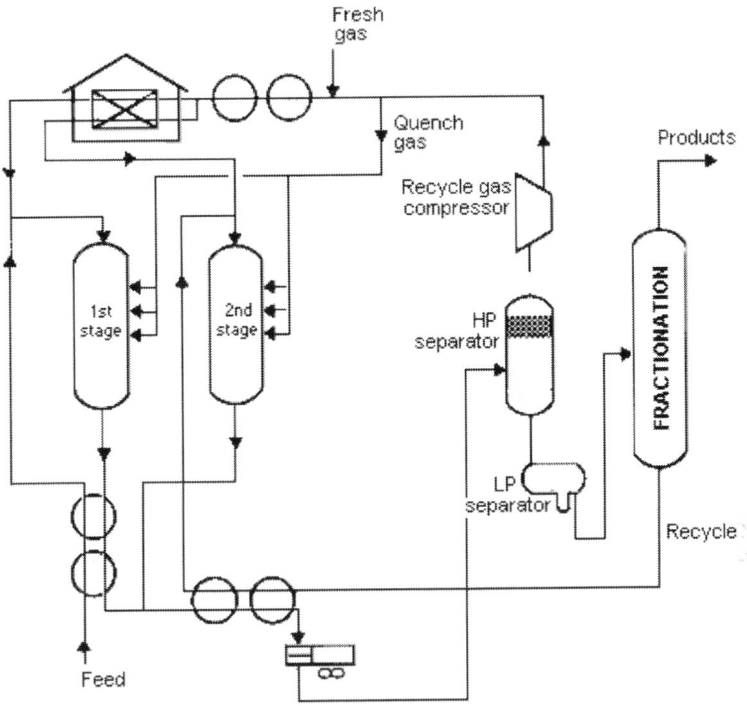

Figure 9.3 A single-stage or two-stage (optional) hydrocracking unit.
Source: Modified from OSHA Technical Manual, Section IV, Chapter 2: Petroleum Refining Processes. http://www.osha.gov/dts/osta/otm/otm_iv/otm_iv_2.html.

cycle oils, which cannot be recycled to extinction in the catalytic cracker, can be processed in the hydrocracker.

9.2.1 Process design

All hydrocracking processes are characterized by the fact that in a catalytic operation under relatively high hydrogen pressure a high-boiling oil fraction is treated to give products of lower molecular weight.

Many hydrocracking units use fixed beds of catalyst with downflow of reactants. However for higher boiling feedstocks, the H-Oil process and the LC-Fining processes employ an ebullient bed reactor in which the beds of particulate catalyst are maintained in an ebullient or fluidized condition with upflowing reactants.

When the processing severity in a hydrocracker is increased, the first reaction occurring leads to saturation of olefins present in feedstock. Next, desulfurization, denitrogenation, and deoxygenation reactions occur. These reactions constitute treating steps during which, in most cases, only limited cracking takes place. When the severity is increased further, hydrocracking reaction is initiated. They proceed at various rates, with the formation of intermediate products (e.g., saturation of aromatics) that are subsequently cracked into lower boiling products.

9.2.2 Single-stage and two-stage options

The most common form of hydrocracking process is a two-stage operation (Fig. 9.3). This flow scheme has been very popular since it can be used to maximize the yield of transportation fuels and has the flexibility to produce gasoline, naphtha, jet fuel, or diesel fuel to meet seasonal swings in product demand.

When the treating step is combined with the cracking reaction to occur in one reactor, the process is called a single-stage process. In this simplest of the hydrocracker configuration, the layout of the reactor section generally resembles that of hydrotreating unit. This configuration will find application in cases where only moderate degree of conversion (say 60% or less) is required. It may also be considered if full conversion, but with a limited reduction in molecular weight, is aimed at. An example is the production of middle distillates from high-boiling distillate oils.

The catalyst used in a single-stage process comprises a hydrogenation function in combination with a strong cracking function. Sulfide metals such as cobalt, molybdenum, and nickel provide the hydrogenation function. An acidic support, usually alumina, attends to the cracking function. Nitrogen compounds and ammonia produced by hydrogenation interfere with acidic activity of the catalyst. In the cases of high/full conversion is required, the reaction temperatures and run lengths of interest in commercial operation can no longer be adhered to. And moreover, conversion asymptotes out with increasing hydrogen pressure (Speight, 2007) so more hydrogen in the reactor is not the answer. In fact, it becomes necessary to switch to a different reactor bed system or to a multistage process, in which the cracking reaction mainly takes place in an added reactor.

In the two-stage configuration, fresh feed is preheated by heat exchange with effluent from the first reactor. It is combined with part of a not fresh gas/recycle gas mixture and passes through a first reactor for denitrogenation/denitrogenation step. These reactions, as well as those of hydrocracking, which occurs to a limited extent in the first reactor, are exothermic. The catalyst inventory is therefore divided among a number of fixed beds. Reaction temperatures are controlled by introducing part of the recycle gas as a quench medium between beds. The ensuing liquid is fractionated to remove the product made in the first reactor. Unconverted, material, with low-nitrogen content and free of ammonia, is taken as a bottom stream from the fractionation section. After heat exchange with reactor effluent and mixing with heated recycle gas, it is sent to the second reactor. Here most of the hydrocracking reactions occur. Strongly acidic catalyst with a relatively low hydrogenation activity (metal sulfides on, e.g., amorphous silica–alumina) is usually applied.

As in the first reactor, the exothermic nature of the process is controlled by recycle gas as the quench medium the catalyst beds. Effluent from the second reactor is cooled and joins first-stage effluent for separation from recycle gas and fractionation. The part of the second reactor feed that has remained unconverted is recycled to the reactor. Feedstock is thereby totally converted to the product boiling range.

In the series-flow configuration the principal difference is the elimination of first-stage cooling and gas/liquid separation and the ammonia removal step. The effluent from the first stage is mixed with more recycle gas and routed direct to the inlet of the second reactor. In contrast with the amorphous catalyst of the two-stage process, the second reactor in series flow generally has a zeolite catalyst, based on crystalline silica–alumina. AS in the two-stage process, material not converted to the product boiling range is recycled from the fractionation section.

Both two stage and series-flow hydrocracking are flexible process: they may yield, in one mode of operation, only naphtha and lower boiling products, and in a different mode, only gas oil and lower boiling products. In the naphtha mode, both configurations have comparable yield patterns. In modes for higher boiling products, kerosene and gas oil, the two-stage process is more selective because product made in the first reactor is removed from the second reactor feed. In series-flow operation, this product is partly cracked into lower boiling products in the second reactor.

Hydrocracking reactor stages usually have similar process flow schemes. The oil feed is combined with a preheated mixture of makeup hydrogen and hydrogen-rich recycle gas and heated to reactor inlet temperature via a feed-effluent exchanger and a reactor charge heater. The reactor charge heater design philosophy is based on many years of safe operation with such two-phase furnaces. The feed-effluent exchangers take advantage of special high-pressure exchanger design features developed by Chevron engineers to give leak-free end closures. From the charge heater the partially vaporized feed enters to the top of the reactor. The catalyst is loaded in separate beds in the reactor with facilities between the beds for quenching the reaction mix and ensuring good flow distribution through the catalyst.

The reactor effluent is cooled through a variety of heat exchangers, including the feed-effluent exchanger and one or more air coolers. Deaerated condensate is injected into the first-stage reactor effluent before the final air cooler in order to remove ammonia and some of the H_2S. This prevents solid ammonium disulfide from depositing in the system. A body of expertise in the field of materials selection for hydrocracker cooling trains is quite important for proper design.

The reactor effluent leaving the air cooler is separated into hydrogen-rich recycle gas, a sour water stream, and a hydrocarbon liquid stream in the high-pressure separator. The sour water–effluent stream is often then sent to a plant for ammonia recovery and for purification so water can be recycled back to the hydrocracker. The hydrocarbon-rich stream is pressure reduced and fed to the distillation section after low-boiling products are flashed off in a low-pressure separator. The hydrogen-rich gas stream from the high-pressure separator is recycled back to the reactor feed by using a recycle compressor. Sometimes with sour feeds, the first-stage recycle gas is scrubbed with an amine system to remove hydrogen. If the feed sulfur level is high, this option can improve the performance of the catalyst and result in less costly materials of construction.

The distillation section consists of a H_2S stripper and a recycle splitter. This latter column separates the product into the desired cuts. The column bottoms stream is recycled back to the second-stage feed. The recycle cut point is changed depending on the low-boiling products needed. It can be as low as 160°C (320°F) if naphtha production is maximized (for aromatics) or as high as 380°C (720°F) if a low pour point diesel is needed. Between these two extremes a recycle cut point of 260°C–285°C (500°F–550°F) results in high yields of high smoke point low-freeze-point jet fuel.

9.2.3 Process variants

A single-stage once-through unit resembles the first stage of the two-stage plant. This type of hydrocracker usually requires the least capital investment. The feedstock is not completely converted to lower boiling products. For this application the refiner must have a demand for highly refined high-boiling oil. In many refining situations, such an oil product can he used as lube oil plant feed or as FCC plant feed or in low-sulfur oil blends or as ethylene plant feed. It also lends itself to stepwise construction of a future two-stage hydrocracker for full feed conversion.

A single-stage recycle (SSREC) unit converts high-boiling oil completely into low-boiling products with a flow scheme resembling the second stage of the two-stage plant. Such a unit maximizes the yield of naphtha, jet fuel, or diesel depending on the recycle cut point used in the distillation section. This type of unit is more economical than the more complex two-stage unit when plant design capacity is less than approximately 10,000–15,000 bbl/day. Commercial SSRFC plants have operated to produce low pour point diesel fuel from waxy Middle Fast VGOs. Recent emphasis has been placed on the upgrading of lower boiling gas oils into jet fuels.

Building on the theme of one- or two-stage hydrocracking, the once-through partial conversion (OTPC) concept evolved. This concept offers the means to convert high-boiling VGO feed into high-quality gasoline, jet fuel, and diesel products by a partial conversion operation. The advantage is lower initial capital investment and also lower utilities consumption than a plant designed for total conversion. Because total conversion of the higher molecular weight compounds in the feedstock is not required, once-through hydrocracking can be carried out at lower temperatures and in most cases at lower hydrogen partial pressures than in recycle hydrocracking, where total conversion of the feedstock is normally an objective.

Recycle hydrocracking plants are designed to operate at hydrogen partial pressures from approximately 1200 to 2300 psi (8274 to 15,858 kPa) depending on the type of feed being processed. Hydrogen partial pressure is set in the design in part depending on required catalyst cycle length but also to enable the catalyst to convert high-molecular-weight polynuclear aromatic and naphthene compounds that must be hydrogenated before they can be cracked. Hydrogen partial pressure also affects properties of the hydrocracked products that depend on hydrogen uptake, such as jet fuel aromatics content and smoke point and diesel cetane number. In general, the higher the feed end point, the higher the required hydrogen partial pressure necessary to achieve satisfactory performance of the plant.

OTPC hydrocracking of a given feedstock may be carried out at hydrogen partial pressures significantly lower than required for recycle total conversion hydrocracking. The potential higher catalyst deactivation rates experienced at lower hydrogen partial pressures can be offset by using higher activity catalysts and designing the plant for lower catalyst space velocities. Catalyst deactivation is also reduced by the elimination of the recycle stream. The lower capital cost resulting from the reduction in plant operating pressure is much more significant than the increase resulting from the possible additional catalyst requirement and larger volume reactors.

A disadvantage of once-through hydrocracking compared to a recycle operation is a somewhat reduced flexibility for varying the ratio of gasoline to middle distillate that is produced. A greater quantity of naphtha can be produced by increasing conversion, and the production of jet fuel plus diesel can also be increased. But selectivity for higher boiling products is also a function of conversion. Selectivity decreases as once-through conversion increases. If conversion is increased too much, the yield of desired product will decrease, accompanied by an increase in volatile products (typically low-boiling hydrocarbon derivatives, often referred to as light ends) and gas production. Higher yields of gasoline or jet fuel plus diesel are possible from a recycle than from a once-through operation.

Middle distillate products made by once-through hydrocracking are generally higher in aromatics content of poorer burning quality than those produced by recycle hydrocracking. However, the quality is generally better than produced by catalytic cracking or from straight run. Middle distillate product quality improves as the degree of conversion increases and as hydrogen partial pressure is increased.

An important consideration, however, is that commercial hydrocracking units are often limited by design constraints of existing VGO hydrotreating units. Thus the proper choice of catalyst(s) is critical when searching for optimum performance. Typical commercial distillate hydrocracking catalysts contain both the hydrogenation (metal) and cracking (acid sites) functions required for service in existing desulfurization units.

9.3 Catalysts

The early 1960s saw increasing demand for high-octane gasoline for the high-compression-ratio engines in new high-performance cars. Demand also grew for diesel fuel for diesel—electric locomotives and low-freeze-point jet fuel. These needs were met by rapid growth in hydrocracking of the more refractory crude fractions that were not converted to gasoline and lower boiling products in the catalytic cracking units. This growth in demand was accompanied by the pioneering development of new zeolite-based hydrocracking catalysts that had improved activity and selectivity (Sherman, 1998). The need to develop catalysts that can carry out cracking and hydrogenation has become even more pressing in view of recent environmental regulations limiting the amount of sulfur and nitrogen emissions. The development of new-generation catalysts to achieve this objective of low-nitrogen and -sulfur levels in the processing of different feedstocks presents an interesting challenge.

The hydrocracking process employs high-activity catalysts that produce a significant yield of low-boiling products. Catalyst selectivity to middle distillate is a function of both the conversion level and operating temperature, with values in excess of 90% being reported in commercial operation. In addition to the increased hydrocracking activity of the catalyst, percentage desulfurization and denitrogenation at start-of-run conditions are also substantially increased. End-of-cycle is reached when product sulfur has risen to the level achieved in conventional VGO HDS process.

The deposition of coke and metals on to the catalyst diminishes the cracking activity of hydrocracking catalysts. Basic nitrogen plays a major role because of the susceptibility of such compounds for the catalyst and their predisposition to form coke (Speight, 2000, 2014, 2017; Parkash, 2003; Ancheyta and Speight, 2007; Gary et al., 2007; Hsu and Robinson, 2017). The reactions of hydrocracking require a dual-function catalyst with high cracking and hydrogenation activities (Dolbear, 1998; Speight, 2000; Ancheyta and Speight, 2007). The catalyst base, such as acid-treated clay, usually supplies the cracking function or alumina or silica–alumina that is used to support the hydrogenation function supplied by metals, such as nickel, tungsten, platinum, and palladium. These highly acid catalysts are very sensitive to nitrogen compounds in the feed, which break down the conditions of reaction to give ammonia and neutralize the acid sites. As many high-boiling gas oils contain substantial amounts of nitrogen (up to approximately 2500 ppm), a purification stage is frequently required. Denitrogenation and desulfurization can be carried out using cobalt–molybdenum or nickel–cobalt–molybdenum on alumina or silica–alumina.

Hydrocracking catalysts typically contain separate hydrogenation and cracking functions. Palladium sulfide and promoted group VI sulfides (nickel molybdenum or nickel tungsten) provide the hydrogenation function. These active compositions saturate aromatics in the feed, saturate olefins formed in the cracking, and protect the catalysts from poisoning by coke. Zeolites or amorphous silica–alumina provide the cracking functions. The zeolites are usually type Y (faujasite) ion exchanged to replace sodium with hydrogen and make up 25%–50% of the catalysts. Pentasils (silicalite or ZSM-5) may be included in dewaxing catalysts.

Hydrocracking catalysts, such as nickel (5% by weight) on silica–alumina, work best on feedstocks that have been hydrofined to low-nitrogen and -sulfur levels. The nickel catalyst then operates well at 350°C–370°C (660°F–700°F) and a pressure of approximately 1500 psi to give good conversion of feed to lower boiling liquid fractions with minimum saturation of single-ring aromatics and a high isoparaffin to n-paraffin ratio in the lower molecular weight paraffins.

The poisoning effect of nitrogen can be offset to a certain degree by operation at a higher temperature. However, the higher temperature tends to increase the production of material in the methane (CH_4) to butane (C_4H_{10}) range and decrease the operating stability of the catalyst so that it requires more frequent regeneration. Catalysts containing platinum or palladium (approximately 0.5 wt.%) on a zeolite base appear to be somewhat less sensitive to nitrogen than are nickel catalysts, and successful operation has been achieved with feedstocks containing 40 ppm nitrogen. This catalyst is also more tolerant of sulfur in the feed, which acts as a temporary poison, the catalyst recovering its activity when the sulfur content of the feed is reduced.

On such catalysts as nickel or tungsten sulfide on silica–alumina, isomerization does not appear to play any part in the reaction, as uncracked normal paraffins from the feedstock tend to retain their normal structure. Extensive splitting produces large amounts of low-molecular-weight (C_3–C_6) paraffins, and it appears that a primary reaction of paraffins is catalytic cracking followed by hydrogenation to form isoparaffins. With catalysts of higher hydrogenation activity, such as platinum on silica–alumina, direct isomerization occurs. The product distribution is also different, and the ratio of low- to intermediate-molecular-weight paraffins in the breakdown product is reduced.

In addition to the chemical nature of the catalyst, the physical structure of the catalyst is also important in determining the hydrogenation and cracking capabilities, particularly for viscous feedstocks (Speight, 2000, 2014, 2017; Parkash, 2003; Ancheyta and Speight, 2007; Gary et al., 2007; Hsu and Robinson, 2017). When gas oils and residua are used, the feedstock is present as liquids under the conditions of the reaction. Additional feedstock and the hydrogen must diffuse through this liquid before reaction can take place at the interior surfaces of the catalyst particle.

At high temperatures, reaction rates can be much higher than diffusion rates, and concentration gradients can develop within the catalyst particle. Therefore the choice of catalyst porosity is an important parameter. When feedstocks are to be hydrocracked to liquefied petroleum gas and gasoline, pore diffusion effects are usually absent. High surface area (approximately 300 m^2/g) and low-to-moderate porosity (from 12 pore diameter with crystalline acidic components to 50 or more with amorphous materials) catalysts are used. With reactions involving high-molecular-weight feedstocks, pore diffusion can exert a large influence, and catalysts with high pore diameters are necessary for more efficient conversion (Parkash, 2003; Gary et al., 2007; Speight, 2014, 2017; Hsu and Robinson, 2017).

Catalyst operating temperature can influence reaction selectivity since the activation energy for hydrotreating reactions is much lower than for hydrocracking reaction. Therefore raising the temperature in a residuum hydrotreater increases the extent of hydrocracking relative to hydrotreating, which also increases the hydrogen consumption (Speight, 2000, 2014, 2017; Parkash, 2003; Ancheyta and Speight, 2007; Gary et al., 2007; Hsu and Robinson, 2017).

Molybdenum sulfide (MoS_2), usually supported on alumina, is widely used in crude oil processes for hydrogenation reactions. It is a layered structure that can be made much more active by addition of cobalt or nickel. When promoted with cobalt sulfide (CoS), making what is called cobalt–moly catalysts, it is widely used in

HDS processes. The nickel sulfide (NiS)–promoted version is used for hydrodenitrogenation (HDN) as well as HDS. The closely related tungsten compound (WS_2) is used in commercial hydrocracking catalysts. Other sulfides (iron sulfide, FeS, chromium sulfide, Cr_2S_3, and vanadium sulfide, V_2S_5) are also effective and used in some catalysts. A valuable alternative to the base metal sulfides is palladium sulfide (PdS). Although it is expensive, palladium sulfide forms the basis for several very active catalysts.

Clay minerals have been used as cracking catalysts particularly for viscous feedstocks (Chapter 23) and have also been explored in the demetallization and upgrading of heavy crude oil (Parkash, 2003; Gary et al., 2007; Speight, 2014, 2017; Hsu and Robinson, 2017). The results indicated that the catalyst prepared was mainly active toward demetallization and conversion of the heaviest fractions of crude oils.

The cracking reaction results from attack of a strong acid on a paraffinic chain to form a carbonium ion (a carbon cation, e.g., R^+) (Dolbear, 1998). Strong acids come in two fundamental types, Brønsted and Lewis acids. Brønsted acids are the familiar proton-containing acids; Lewis acids are a broader class, including inorganic and organic species formed by positively charged centers. Both kinds have been identified on the surfaces of catalysts; sometimes both kinds of sites occur on the same catalyst. The mixture of Brønsted and Lewis acids sometimes depends on the level of water in the system.

Examples of Brønsted acids are the familiar proton-containing species such as sulfuric acid (H_2SO_4). Acidity is provided by the very active hydrogen ion (H^+) that has a very high positive charge density. It seeks out centers of negative charge such as the pi electrons in aromatic centers. Such reactions are familiar to organic chemistry students who are taught that bromination of aromatics takes place by attack of the bromonium ion (Br^+) on such a ring system. The proton in strong acid systems behaves in much the same way, adding to the pi electrons and then migrating to a site of high electron density on one of the carbon atoms.

In reactions with hydrocarbon derivatives, both Lewis and Brønsted acids can catalyze cracking reactions. For example, the proton in Brønsted acids can add to an olefinic double bond to form a carbon cation. Similarly, a Lewis acid can abstract a hydride from the corresponding paraffin to generate the same intermediate (Dolbear, 1998). Although these reactions are written to show identical intermediates in the two reactions, in real catalytic systems the intermediates would be different. This is because the carbon cations would probably be adsorbed on surface sites that would be different in the two kinds of catalysts.

Zeolites and amorphous silica–alumina provide the cracking function in hydrocracking catalysts (Sherman, 1998). Both of these have similar chemistry at the molecular level, but the crystalline structure of the zeolites provides higher activities and controlled selectivity not found in the amorphous materials.

Zeolites consist primarily of silicon, aluminum, and oxygen and host an assortment of other elements. In addition, zeolites are highly porous crystals veined with submicroscopic channels. The channels contain water (hence the bubbling at high temperatures), which can be eliminated by heating (combined with other treatments) without altering the crystal structure.

Typical naturally occurring zeolites include analcite (also called analcime) Na ($AlSi_2O_6$) and faujasite $Na_2Ca(AlO_2)_2(SiO_2)_4 \cdot H_2O$ that is the structural analog of the synthetic zeolite X and zeolite Y. Sodalite ($Na_8[(Al_2O_2)_6(SiO_2)_6]Cl_2$) contains the truncated octahedral structural unit known as the sodalite cage that is found in several zeolites. The corners of the faces of the cage are defined by either four or six Al/Si atoms that are joined together through oxygen atoms. The zeolite structure is generated by joining sodalite cages through the four Si/Al rings, so enclosing a cavity, or supercage bounded by a cube of eight sodalite cages and readily accessible through the faces of that cube (channels or pores). Joining sodalite cages together through the six Si/Al faces generates the structural frameworks of faujasite, zeolite X, and zeolite Y. In zeolites the effective width of the pores is usually controlled by the nature of the cation (M^+ or M^{2+}).

Natural zeolites form hydrothermally (e.g., by the action of hot water on volcanic ash or lava), and synthetic zeolites can be made by mixing solutions of aluminates and silicates and maintaining the resulting gel at temperatures of 100°C (212°F) or higher for appropriate periods (Swaddle, 1997). Zeolite A can form at temperatures below 100°C (212°F), but most zeolite syntheses require hydrothermal conditions (typically 150°C/300°F at the appropriate pressure). The reaction mechanism appears to involve dissolution of the gel and precipitation as the crystalline zeolite, and the identity of the zeolite produced depends on the composition of the solution. Aqueous alkali metal hydroxide solutions favor zeolites with relatively high aluminum contents, while the presence of organic molecules such as amines or alcohols favors highly siliceous zeolites such as silicalite or ZSM-5. Various tetra-alkyl ammonium cations favor the formation of certain specific zeolite structures and are known as template ions, although it should not be supposed that the channels and cages form simply by the wrapping of aluminosilicate fragments around suitably shaped cations.

Zeolite catalysts have also found use in the refining industry during the last two decades. Like the silica–alumina catalysts, zeolites also consist of a framework of tetrahedral usually with a silicon atom or an aluminum atom at the center. The geometric characteristics of the zeolites are responsible for their special properties that are particularly attractive to the refining industry (DeCroocq, 1984). Specific zeolite catalysts have shown up to 10,000 times more activity than the so-called conventional catalysts in specific cracking tests. The mordenite-type catalysts are particularly worthy of mention since they have shown up to 200 times greater activity for hexane cracking in the temperature range of 360°C–400°C (680°F–750°F).

Other zeolite catalysts have also shown remarkable adaptability to the refining industry. For example, the resistance to deactivation of the type Y zeolite catalysts containing either noble or nonnoble metals is remarkable, and catalyst life of up to 7 years has been obtained commercially in processing high boiling gas oils in the Unicracking-JHC processes. Operating life depends on the nature of the feedstock, the severity of the operation, and the nature and extent of operational upsets. Gradual catalyst deactivation in commercial use is counteracted by incrementally raising the operating temperature to maintain the required conversion per pass. The more active a catalyst, the lower is the temperature required. When processing for gasoline, lower operating temperatures have the additional advantage that less of the feedstock is converted to isobutane.

Any given zeolite is distinguished from other zeolites by structural differences in its unit cell, which is a tetrahedral structure arranged in various combinations. Oxygen atoms establish the four vertices of each tetrahedron, which are bound to, and enclose, either a silicon (Si) or an aluminum (Al) atom. The vertex oxygen atoms are each shared by two tetrahedrons, so that every silicon atom or aluminum atom within the tetrahedral cage is bound to four neighboring caged atoms through an intervening oxygen. The number of aluminum atoms in a unit cell is always smaller than, or at most equal to, the number of silicon atoms because two aluminum atoms never share the same oxygen.

The aluminum is actually in the ionic form and can readily accommodate electrons donated from three of the bound oxygen atoms. The electron donated by the fourth oxygen imparts a negative, or anionic, charge to the aluminum atom. This negative charge is balanced by a cation from the alkali metal or the alkaline earth groups of the periodic table. Such cations are commonly sodium, potassium, calcium, or magnesium. These cations play a major role in many zeolite functions and help to attract polar molecules, such as water. However, the cations are not part of the zeolite framework and can be exchanged for other cations without any effect on crystal structure.

Zeolites provide the cracking function in many hydrocracking catalysts, as they do in fluid catalytic cracking catalysts. The zeolites are crystalline aluminosilicates, and in many current commercial catalysts, the zeolite used is faujasite. Pentasil zeolites, including silicalite and ZSM-5, are also used in some catalysts for their ability to crack long-chain paraffins selectively. Typical levels are 25–50 wt.% zeolite in the catalysts, with the remainder being the hydrogenation component and a silica (SiO_2) or alumina (Al_2O_3) binder. Exact recipes are guarded as trade secrets.

Crystalline zeolite compounds provide a broad family of solid acid catalysts. The chemistry and structures of these solids are beyond the scope of this book. What is important here is that the zeolites are not acidic as crystallized. They must be converted to acidic forms by ion-exchange processes. In the process of doing this conversion, the chemistry of the crystalline structure is often changed. This complication provides tools for controlling the catalytic properties, and much work has been done on understanding and applying these reactions as a way to make catalysts with higher activities and more desirable selectivity.

As an example, the zeolite faujasite crystallizes with the composition $SiO_2(NaAlO_2)_x(H_2O)_y$. The ratio of silicon to aluminum, expressed here by the subscript x, can be varied in the crystallization from 1 to greater than 10. What does not vary is the total number of silicon and aluminum atoms per unit cell, 192. For reference, zeolites with a ratio of 1–1.5 are called type X; those with ratio greater than 1.5 are type Y.

Both silicon and aluminum in zeolites are found in tetrahedral oxide sites. The four oxides are shared with another silicon or aluminum (except that two aluminum ions are never found in adjacent, linked tetrahedral). Silicon with a plus four charge balances exactly half of the charge of the oxide ions it is linked to; since all of the oxygen atoms are shared, silicon balances all of the charge around it and is electrically neutral. Aluminum, with three positive charges, leaves one charge unsatisfied. Sodium neutralizes this charge.

The sodium, as expected from its chemistry, is not linked to the oxides by covalent bonds as the silicon and aluminum are. The attraction is simply ionic, and sodium can be replaced by other cations by ion-exchange processes. In extensive but rarely published experiments, virtually every metallic and organic cation has been exchanged into zeolites in studies by catalyst designers.

The most important ion exchanged for sodium is the proton. In the hydrogen ion form, faujasite zeolites are very strong acids, with strengths approaching that of oleum. Unfortunately, direct exchange using mineral acids such as hydrochloric acid is not practical. The acid tends to attack the silica–alumina network, in the same way that strong acids attack clays in the activation processes developed by Houdry. The technique adopted to avoid this problem is indirect exchange, beginning with exchange of ammonium ion for the sodium. When heated to a few hundred degrees, the ammonium decomposes, forming gaseous ammonia and leaving behind a proton:

$$R^- \, NH_4^+ \rightarrow R^- \, H^+ + NH_3 \uparrow$$

The step is accompanied by a variety of solid-state reactions that can change the zeolite structure in subtle but important ways. This chemistry and the related structural alterations have been described in many articles (Humphries et al., 1993).

While zeolites provided a breakthrough that allowed catalytic hydrocracking to become commercially important, continued advances in the manufacture of amorphous silica–alumina made these materials competitive in certain kinds of applications. This was important, because patents controlled by Unocal and Exxon dominated the application of zeolites in this area.

Typical catalysts of this type contain 60–80 wt.% of the silica–alumina, with the remainder being the hydrogenation component. The compositions of these catalysts are closely held secrets. Over the years, broad ranges of silica/alumina molar ratios have been used in various cracking applications, but silica is almost always in excess for high acidity and stability. A typical level might be 25 wt.% alumina (Al_2O_3).

Amorphous silica–alumina is made by a variety of precipitation techniques. The whole class of materials traces its beginnings to silica gel technology, in which sodium silicate is acidified to precipitate the hydrous silica–alumina sulfate; sulfuric acid is used as some or all of the acid for this precipitation, and a mixed gel is formed. The properties of this gel, including acidity and porosity, can be varied by changing the recipe—concentrations, order of addition, pH, temperature, aging time, and the like. The gels are isolated by filtration and washed to remove sodium and other ions.

Careful control of the precipitation allows the pore size distributions of amorphous materials to be controlled, but the distributions are still much broader than those in the zeolites. This limits the activity and selectivity. One effect of the reduced activity has been that these materials have been applied only in making middle distillates: diesel and turbine fuels. At higher process severities the poor selectivity results in production of unacceptable amounts of methane (CH_4) to butane (C_4H_{10}) hydrocarbon derivatives.

Hydrocarbon derivatives, especially aromatic hydrocarbon derivatives, can react in the presence of strong acids to form coke. This coke is a complex polynuclear aromatic material that is low in hydrogen. Coke can deposit on the surface of a catalyst, blocking access to the active sites and reducing the activity of the catalyst. Coke poisoning is a major problem in fluid catalytic cracking catalysts, where coked catalysts are circulated to a fluidized bed combustor to be regenerated. In hydrocracking, coke deposition is virtually eliminated by the hydrogenation function of the catalyst.

However, the product referred to as coke is not a single material. The first products deposited are tarry deposits that can, with time and temperature, continue to polymerize. Acid catalyzes these polymerizations. The stable product would be graphite, with very large aromatic sheets and no hydrogen. This product forms only with very high temperature aging, far more severe than that found in a hydrocracker. The graphitic material is both more thermodynamically stable and less kinetically reactive. This kinetic stability results from the lack of easily hydrogenated functional groups.

In a well-designed hydrocracking system, the hydrogenation function adds hydrogen to the tarry deposits. This reduces the concentration of coke precursors on the surface. There is, however, a slow accumulation of coke that reduces activity over 1- to 2-year period. Refiners respond to this slow reduction in activity by raising the average temperature of the catalyst bed to maintain conversions. Eventually, however, an upper limit to the allowable temperature is reached, and the catalyst must be removed and regenerated.

Catalysts carrying coke deposits can be regenerated by burning off the accumulated coke. This is done by service in rotary or similar kilns rather than leaving catalysts in the hydrocracking reactor, where the reactions could damage the metals in the walls. Removing the catalysts also allows inspection and repair of the complex and expensive reactor internals, discussed next. Regeneration of a large catalyst charge can take weeks or months, so refiners may own two catalyst loads: one in the reactor and one regenerated and ready for reload.

The thermal reactions also convert the metal sulfide hydrogenation functions to oxides and may result in agglomeration. Excellent progress has been made since the 1970s in regenerating hydrocracking catalysts; similar regeneration of hydrotreating catalysts is widely practiced.

After combustion to remove the carbonaceous deposits, the catalysts are treated to disperse active metals. Vendor documents claim more than 95% recovery of activity and selectivity in these regenerations. Catalysts can undergo successive cycles of use and regeneration, providing long functional life with these expensive materials.

Hydrocracking allows refiners the potential to balance fuel oil supply and demand by adding VGO cracking capacity. Situations where this is the case include

(1) refineries with no existing VGO cracking capacity, (2) refineries with more VGO available than VGO conversion capacity, (3) refineries where addition of vacuum residuum conversion capacity has resulted in production of additional feedstocks boiling in the VGO range—such as coker gas oil, and (4) refineries that have one of the two types of VGO conversion units but could benefit from adding the second type. In some cases a refiner might add both gas oil cracking and residuum conversion capacity simultaneously.

Those refiners who do choose gas oil cracking as part of their strategy for balancing residual fuel oil supply and demand must decide whether to select a hydrocracker or a fluid catalytic cracking unit. Although the two processes have been compared vigorously over the years, neither process has evolved to be the universal choice for gas oil cracking. Both processes have their advantages and disadvantages, and process selection can be properly made only after careful consideration of many case-specific factors. Among the most important factors are (1) product slate required, (2) amount of flexibility required to vary the product slate, (3) product quality (specifications) required, and (4) the need to integrate the new facilities in a logical and cost-effective way with any existing facilities.

As illustrated earlier for various forms of more conventional hydrocracking, the type of catalyst used can influence the product slate obtained. For example, for a mild hydrocracking operation at constant temperature, the selectivity of the catalyst varies from approximately 65 to approximately 90% by volume. Indeed, several catalytic systems have now been developed with a group of catalysts specifically for mild hydrocracking operations (Sarrazin et al., 2005). Depending on the type of catalyst, they may be run as a single catalyst or in conjunction with a hydrotreating catalyst.

Precious metal catalysts, particularly catalysts incorporating platinum or platinum and palladium, are used in the latter stages of deep desulfurization process. They have excellent performance in hydrogenation of monocyclic aromatic hydrocarbon derivatives and are likely to become more and more important in the years ahead, and refiners seek to hydrocrack higher amounts of the high-boiling feedstocks. Current efforts are seeking to produce such catalysts with increased hydrogenation activity as well as resistance to sulfur and nitrogen poisoning while balancing these characteristics. In addition, work is being done to assign appropriate cracking activity to match specific applications using inorganic (composite) oxides, such as amorphous alumina, silica–alumina, or crystalline silica–alumina, as carriers and optimization of the amount of the precious metals and highly dispersed metal impregnation methods.

The cobalt–molybdenum, nickel–molybdenum, and precious metal catalysts are also available for use in deep desulfurization and aromatics hydrogenation. These catalysts should be helpful in producing diesel fuel with a sulfur content of 50 ppm or less only by substituting catalysts.

Catalysts used in residuum upgrading processes typically use an association of several kinds of catalysts, each of them playing a specific and complementary role (Parkash, 2003; Gary et al., 2007; Speight, 2014, 2017; Hsu and Robinson, 2017). The first major function to be performed is hydrodemetallization (HDM). Therefore the HDM catalyst must desegregate asphaltene constituents and remove as much metals (nickel and vanadium) as possible. One catalyst in particular has been developed by optimizing the support pore structure and acidity, and catalyst allows a uniform distribution of metals deposited, and therefore a high metal retention capacity is reached. A specific HDS catalyst can be placed downstream of the HDM catalyst, and the main function is of such positioning is to desulfurize the already deeply demetallized feedstock as well as to reduce coke precursors (1999). Thus the main function the HDS catalyst is not the same function as that of the HDM catalyst. In addition, for fixed-bed processes, swing guard reactors may be used to improve the protection of downstream catalysts and increases the unit cycle length. For example, the Hyvahl process (*q.v.*) includes two swing guard reactors followed by conventional HDM and HDS reactors (DeCroocq, 1997). The HDM catalyst in the guard reactors may be replaced during unit operation, and the total catalyst amount is replaced at the end of a cycle.

9.4 Process options for heavy feedstocks

The major goal of residuum hydroconversion is cracking of residua with desulfurization, metal removal, denitrogenation, and asphaltene conversion. Residuum hydroconversion process offers the production of kerosene and gas oil and the production of feedstocks for hydrocracking, fluid catalytic cracking, and petrochemical applications. The processes that follow are listed in alphabetical order with no other preference in mind.

9.4.1 Asphaltenic bottom cracking process

The asphaltenic bottom cracking (ABC) process can be used for distillate production (Table 9.1) HDM, asphaltene cracking with moderate HDS as well as sufficient resistance to coke fouling and metal deposition using feedstocks such as vacuum residua, thermally cracked residua, solvent deasphalted bottoms, and bitumen with fixed catalyst beds (Takeuchi, 1982).

The process can be combined with (1) solvent deasphalting for complete or partial conversion of the residuum or (2) HDS to promote the conversion of residue, to treat feedstock with high metals and to increase catalyst life or (3) hydrovisbreaking to attain high conversion of residue with favorable product stability.

In the process the feedstock is pumped up to the reaction pressure and mixed with hydrogen. The mixture is heated to the reaction temperature in the charge heater after a heat exchange and fed to the reactor.

In the reactor, HDM and subsequent asphaltene cracking with moderate HDS take place simultaneously under conditions similar to residuum HDS. The reactor effluent gas is cooled, cleaned up, and recycled to the reactor section, while the separated liquid is distilled into distillate fractions and vacuum residue, which is

Table 9.1 Feedstock and product data for the asphaltenic bottom cracking process.

Feedstock	Arabian Light vacuum residuum	Arabian Heavy vacuum residuum	Cerro Negro vacuum residuum
API	7.0	5.1	1.7
Sulfur (wt.%)	4.0	5.3	4.3
Carbon residue (wt.%)	20.8	23.3	23.6
C7-asphaltenes	7.0	13.1	19,819.8
Nickel (ppm)	223.0	52.0	150.0
Vanadium (ppm)	76.0	150.0	640.0
Products			
Naphtha (vol. %)	6.5	7.7	15.1
API	57.2	57.2	54.7
Distillate (vol. %)	16.0	19.8	21.3
API	34.2[a]	34.2[a]	32.5[a]
Vacuum gas oil (vol.%)	34.3	38.1	32.8
API	24.7	21.6	15.4
Sulfur (wt.%)	0.2	1.7	0.5
Vacuum residuum (vol. %)	46.2	37.9	34.7
API	10.6	7.8	<0.0
Sulfur (wt.%)	0.6	1.7	2.2
Carbon residue (wt.%)	13.6	26.5	13.6
C7-asphaltenes (wt.%)			
Nickel (ppm)	9.0	45.0	117.0
Vanadium (ppm)	11.0	75.0	371.0
Conversion	55.0	60.0	60.0

[a]Estimated.
Source: Modified from Speight, J.G., 2007. The Chemistry and Technology of Petroleum, fourth ed. CRC Press, Taylor & Francis Group, Boca Raton, FL.

further separated by deasphalting (Chapter 19) into deasphalted oil and asphalt using butane or pentane.

In case of the ABC–HDS catalyst combination, the ABC catalyst is placed upstream of the HDS catalyst and can be operated at a higher temperature than the HDS catalyst under conventional residuum HDS conditions. In the VisABC process a soaking drum is provided after heater, when necessary. Hydrovisbroken oil is first stabilized by the ABC catalyst through hydrogenation of coke precursors and then desulfurized by the HDS catalyst.

9.4.2 CANMET hydrocracking process

The CANMET hydrocracking process hydrocracking process for heavy crude oils, extra heavy oil, tar sand bitumen, atmospheric residua, and vacuum residua (Parkash, 2003; Gary et al., 2007; Speight, 2014, 2017; Hsu and Robinson, 2017). Energizing hydrocracking technology, such as that developed by the CANMET process, shows promise for some applications. The scheme is a high conversion, high demetallization, residuum hydrocracking process that, using an additive to inhibit coke formation, achieves conversion of high boiling point hydrocarbon derivatives into lower boiling products. Initially developed to upgrade heavy crude oil, extra heavy oil, and tar sand bitumen as well as residua. The process does not use a catalyst but employs a low-cost additive to inhibit coke formation and allow high conversion of viscous feedstocks (such as heavy crude oil, extra heavy oil, and tar sand bitumen) into lower boiling products using a single reactor (Table 9.2). The process is unaffected by high levels of feed contaminants such as sulfur, nitrogen, and metals. Conversion of over 90% of the 525°C$^+$ (975°F$^+$) fraction into distillates has been attained.

In the process the feedstock and recycle hydrogen gas are heated to reactor temperature in separate heaters. A small portion of the recycle gas stream and the required amount of additive are routed through the oil heater to prevent coking in the heater tubes. The outlet streams from both heaters are fed to the bottom of the reactor.

The vertical reactor vessel is free of internal equipment and operates in a three-phase mode. The solid additive particles are suspended in the primary liquid hydrocarbon phase through which the hydrogen and product gases flow rapidly in bubble form. The reactor exit stream is quenched with cold recycle hydrogen prior to the high-pressure separator. The high-boiling liquids are further reduced in pressure to a hot medium pressure separator and from there to fractionation. The spent additive leaves with the high-boiling fraction and remains in the unconverted vacuum residue.

The vapor stream from the hot high-pressure separator is cooled stepwise to produce middle distillate and naphtha that are sent to fractionation. High-pressure purge of low-boiling hydrocarbon gases is minimized by a sponge oil circulation system. Product naphtha will be hydrotreated and reformed, low-boiling gas oil will be hydrotreated and sent to the distillate pool, the high-boiling gas oil will be processed in the FCC and the pitch will be sold.

The additive, prepared from iron sulfate [$Fe_2(SO_4)_3$], is used to promote hydrogenation and effectively eliminate coke formation. The effectiveness of the dual-role additive permits the use of operating temperatures that give high conversion in a single-stage reactor. The process maximizes the use of reactor volume and provides a thermally stable operation with no possibility of temperature runaway.

Table 9.2 Feedstock and product data for the CANMET process.

Feedstock[a]	
API gravity	4.4
Sulfur (wt.%)	5.1
Nitrogen (wt.%)	0.6
Asphaltenes (wt.%	15.5
Carbon residue (wt.%)	20.6
Metals (ppm)	
Ni	80
V	170
Residuum (>525°C, >975°F) (wt.%)	
Products (wt.%)[b]	
Naphtha (C5—204°C, 400°F)	19.8
Nitrogen (wt.%)	0.1
Sulfur (wt.%)	0.6
Distillate (204°C–343°C, 400°F–650°F)	33.5
Nitrogen (wt.%)	0.4
Sulfur (wt.%)	1.8
Vacuum gas oil (343°C–534°C, 650°F–975°F)	28.5
Nitrogen (wt.%)	0.6
Sulfur (wt.%)	2.3
Residuum (>534°C, >975°F)	4.5
Nitrogen (wt.%)	1.6
Sulfur (wt.%)	3.1

[a]Cold Lake (Canada) heavy oil vacuum residuum.
[b]Residuum: 93.5% by weight.
Source: Modified from Speight, J.G. 2007. The Chemistry and Technology of Petroleum, fourth ed. CRC Press, Taylor & Francis Group, Boca Raton, FL.

The process also offers the attractive option of reducing the coke yield by slurrying the feedstock with less than 10 ppm of catalyst (molybdenum naphthenate) and sending the slurry to a hydroconversion zone to produce low-boiling products (Speight, 2000, 2014, 2017; Parkash, 2003; Ancheyta and Speight, 2007; Gary et al., 2007; Hsu and Robinson, 2017).

9.4.3 H-Oil process

The H-Oil process (Parkash, 2003; Gary et al., 2007; Speight, 2014, 2017; Hsu and Robinson, 2017) is a catalytic process that that uses a single-stage, two-stage, or three-stage ebullated-bed reactor in which, during the reaction, considerable hydrocracking takes place (Fig. 9.4). The process is designed for hydrogenation of residua and other high feedstocks in an ebullated-bed reactor to produce upgraded crude oil products (Parkash, 2003; Gary et al., 2007; Speight, 2014, 2017; Hsu and Robinson, 2017). The process is able to convert all types of feedstocks to either

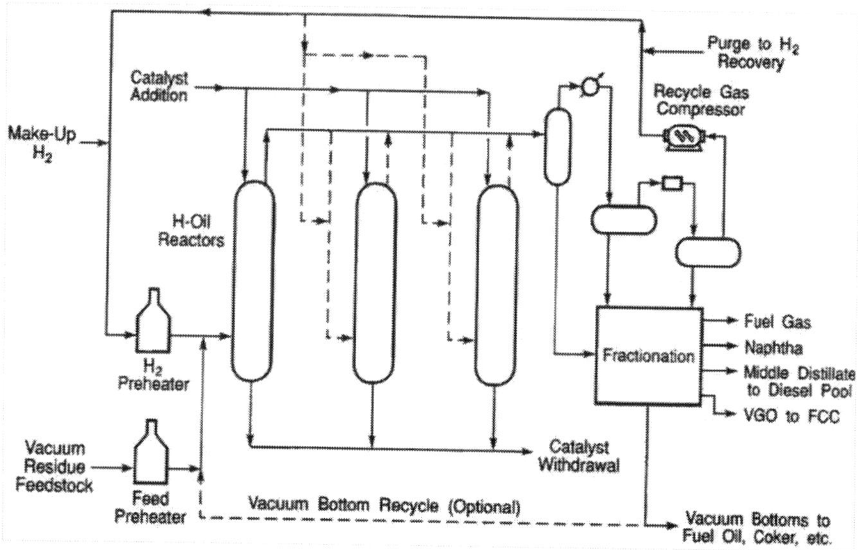

Figure 9.4 H-Oil process.
Source: Modified from Speight, J.G., 2007. The Chemistry and Technology of Petroleum, fourth ed. CRC Press, Taylor & Francis Group, Boca Raton, FL.

distillate products as well as to desulfurize and demetallize residues for feed to coking units or residue fluid catalytic cracking units, for production of low-sulfur fuel oil, or for production to asphalt blending. A modification of the H-Oil process (Hy-C Cracking process) converts high-boiling distillates to middle distillates and kerosene (Table 9.3).

A wide variety of process options can be used with the H-Oil process depending on the specific operation. In all cases a catalytic ebullated-bed reactor system is used to provide an efficient hydroconversion. The system insures uniform distribution of liquid, hydrogen-rich gas, and catalyst across the reactor. The ebullated-bed system operates under essentially isothermal conditions, exhibiting little temperature gradient across the bed (Parkash, 2003; Gary et al., 2007; Speight, 2014, 2017; Hsu and Robinson, 2017). The heat of reaction is used to bring the feed oil and hydrogen up to reactor temperature.

In the process the feedstock (a vacuum residuum) is mixed with recycle vacuum residue from downstream fractionation, hydrogen-rich recycle gas, and fresh hydrogen. This combined stream is fed into the bottom of the reactor whereby the upward flow expands the catalyst bed. The mixed-vapor liquid effluent from the reactor either goes to flash drum for phase separation or the next reactor. A portion of the hydrogen-rich gas is recycled to the reactor. The product oil is cooled and stabilized, and the vacuum residue portion is recycled to increase conversion.

A catalyst of small particle size can be used, giving efficient contact among gas, liquid, and solid with good mass and heat transfer. Part of the reactor effluent is

Table 9.3 Feedstock and product data for the H-Oil process.

Feedstock	Arabian Medium vacuum residuum 65%[a]	Arabian Medium vacuum residuum 90%[a]	Athabasca bitumen
API gravity	4.9	4.9	8.3
Sulfur (wt.%)	5.4	5.4	4.9
Nitrogen (wt.%)			0.5
Carbon residue (wt.%)			
Metals (ppm)	128.0	128.0	
Ni			
V			
Residuum (> 525°C, >975°F) (wt.%)			50.3
Products (wt.%)[b]			
Naphtha (C5—204°C, 400°F)	17.6	23.8	16.0
Sulfur (wt.%)			1.0
Distillate (204°C–343°C, 400°F–650°F)	22.1	36.5	43.0
Sulfur (wt.%)			2.0
Vacuum gas oil (343°C–534°C, 650°F–975°F)	34.0	37.1	26.4
Sulfur (wt.%)			3.5
Residuum (> 534°C, >975°F)	33.2	9.5	16.0
Sulfur (wt.%)			5.7

[a]% Conversion.
[b]% Desulfurization.
Source: Modified from Speight, J.G., 2007. The Chemistry and Technology of Petroleum, fourth ed. CRC Press, Taylor & Francis Group, Boca Raton. FL.

recycled back through the reactors for temperature control and to maintain the requisite liquid velocity. The entire bed is held within a narrow temperature range, which provides essentially an isothermal operation with an exothermic process. Because of the movement of catalyst particles in the liquid–gas medium, deposition of tar and coke is minimized, and fine solids entrained in the feed do not lead to reactor plugging. The catalyst can also be added and withdrawn from the reactor without destroying the continuity of the process. The reactor effluent is cooled by exchange and separates into vapor and liquid. After scrubbing in a lean oil absorber, hydrogen is recycled, and the liquid product is either stored directly or fractionated before storage and blending.

H-Oil and LC-Fining technologies are often practiced commercially at approximately the 85% conversion level.

9.4.4 Hydrovisbreaking process

Briefly, the hydrovisbreaking process (HYCAR process) is a noncatalytic process, is conducted under similar conditions to visbreaking (Chapter 17), and involves treatment with hydrogen under mild conditions. The presence of hydrogen leads to more stable products (lower flocculation threshold) than can be obtained with straight visbreaking, which means that higher conversions can be achieved, producing a lower viscosity product.

The HYCAR process is composed fundamentally of three parts: (1) visbreaking, (2) HDM, and (3) hydrocracking. In the visbreaking section the viscous feedstock (e.g., vacuum residuum or bitumen) is subjected to moderate thermal cracking while no coke formation is induced. The visbroken oil is fed to the demetallization reactor in the presence of catalysts, which provides sufficient pore for diffusion and adsorption of high-molecular-weight constituents. The product from this second stage proceeds to the hydrocracking reactor, where desulfurization and denitrogenation take place along with hydrocracking.

9.4.5 Hyvahl F process

The process is used to hydrotreat atmospheric and vacuum residua to convert the feedstock to naphtha and middle distillates (Peries et al., 1988; Billon et al., 1994; Parkash, 2003; Gary et al., 2007; Speight, 2014, 2017; Hsu and Robinson, 2017).

The main features of this process are its dual-catalyst system and its fixed-bed swing-reactor concept. The first catalyst has a high capacity for metals (to 100% by weight of new catalyst) and is used for both HDM and most of the conversion. This catalyst is resistant to fouling, coking, and plugging by asphaltene constituents (as well as by reacted asphaltene constituents) and shields the second catalyst from the same. Protected from metal poisons and deposition of coke-like products, the highly active second catalyst can carry out its deep HDS and refining functions. Both catalyst systems use fixed beds that are more efficient than moving beds and are not subject to attrition problems.

The swing-reactor design reserves two of the HDM reactor design reserves two of the HDM reactor for use as guard reactors: one of them can be removed from service for catalyst reconditioning and put on standby, while the rest of the unit continues to operate. More than 50% of the metals are removed from the feed in the guard reactors.

In the process the preheated feedstock enters one of the two guard reactors where a large proportion of the nickel and vanadium are adsorbed and hydroconversion of the high-molecular-weight constituents commences. Meanwhile, the second guard reactor catalyst undergoes a reconditioning process and then is put on standby.

From the guard reactors the feedstock flows through a series of HDM reactors that continue the metals removal and the conversion of the high-boiling feedstock.

The next processing stage, HDS, is where most of sulfur, some of the nitrogen, and the residual metals are removed. A limited amount of conversion also takes place. From the final reactor the gas phase is separated, hydrogen is recirculated to the reaction section, and the liquid products are sent to a conventional fractionation section for separation into naphtha, middle distillates, and higher boiling streams.

9.4.6 IFP hydrocracking process

The process features a dual-catalyst system: the first catalyst is a promoted nickel–molybdenum amorphous catalyst. It acts to remove sulfur and nitrogen and hydrogenate aromatic rings. The second catalyst is a zeolite that finishes the hydrogenation and promotes the hydrocracking reaction.

In the two-stage process, feedstock and hydrogen are heated and sent to the first reaction stage where conversion to products occurs (RAROP, 1991, p. 85). The reactor effluent phases are cooled and separated, and the hydrogen-rich gas is compressed and recycled. The liquid leaving the separator is fractionated, the middle distillates and lower boiling streams (Parkash, 2003; Gary et al., 2007; Speight, 2014, 2017; Hsu and Robinson, 2017) are sent to storage, and the high-boiling stream is transferred to the second reactor section and then recycled back to the separator section.

In the single-stage process the first reactor effluent is sent directly to then second reactor, followed by the separation and fractionation steps. The fractionator bottoms are recycled to the second reactor or sold.

9.4.7 Isocracking process

The process has been applied commercially in the full range of process flow schemes: single-stage, once-through liquid; single-stage, partial recycle of the unconverted feedstock; single-stage extinction recycle of oil (100% conversion); and two-stage extinction recycle of oil (Speight, 2000, 2014, 2017; Parkash, 2003; Ancheyta and Speight, 2007; Gary et al., 2007; Hsu and Robinson, 2017). The preferred flow scheme will depend on the feed properties, the processing objectives, and, to some extent, the specified feed rate.

The process uses multibed reactors, and in most applications, a number of catalysts are used in a reactor. The catalysts are dual function being a mixture of hydrous oxides (for cracking) and heavy metal sulfides (for hydrogenation) (Speight, 2000, 2014, 2017; Parkash, 2003; Ancheyta and Speight, 2007; Gary et al., 2007; Hsu and Robinson, 2017). The catalysts are used in a layered system to optimize the processing of the feedstock that undergoes changes in its properties along the reaction pathway (Parkash, 2003; Gary et al., 2007; Speight, 2014, 2017; Hsu and Robinson, 2017). In most commercial Isocracking units, the entire fractionator bottoms fraction is recycled or all of it is drawn as a high-boiling product, depending on whether the low-boiling or high-boiling products are of greater value. If the low-boiling distillate products (naphtha or naphtha/kerosene) are the most valuable products, the higher boiling point distillates (such as diesel) can be recycled to the reactor for conversion rather than drawn as a product (RAROP, 1991, p. 83). Product distribution depends upon the mode of operation. High-boiling feedstocks have been used in the process, and the product yield is very much dependent upon the catalyst and the process parameters (Speight, 2000, 2014, 2017; Parkash, 2003; Ancheyta and Speight, 2007; Gary et al., 2007; Hsu and Robinson, 2017).

9.4.8 LC-Fining process

The LC-Fining process is a hydrocracking process capable of desulfurizing, demetallizing, and upgrading a wide spectrum of high-boiling feedstocks by means of an expanded-bed reactor (RAROP, 1991, p. 61; Parkash, 2003; Gary et al., 2007; Speight, 2014, 2017; Hsu and Robinson, 2017). Operating with the expanded bed allows the processing of high-boiling feedstocks, such as atmospheric residual, vacuum residua, and oil sand bitumen. The catalyst in the reactor behaves like fluid that enables the catalyst to be added to and withdrawn from the reactor during operation. The reactor conditions are near isothermal because the heat of reaction is absorbed by the cold fresh feed immediately owing to through mixing of reactors.

In the process (Fig. 9.5) the feedstock and hydrogen are heated separately and then pass upward in the hydrocracking reactor through an expanded bed of catalyst (Speight, 2007). Reactor products flow to the high-pressure–high-temperature separator. Vapor effluent from the separator is let down in pressure and then goes to the heat exchange and thence to a section for the removal of condensable products and purification (Table 9.4) (Speight, 2000, 2014, 2017; Parkash, 2003; Ancheyta and Speight, 2007; Gary et al., 2007; Hsu and Robinson, 2017).

Liquid is let down in pressure and passes to the recycle stripper. This is a most important part of the high conversion process. The liquid recycle is prepared to the proper boiling range for return to the reactor. In this way the concentration of bottoms in the reactor, and therefore the distribution of products, can be controlled (Table 9.4). After the stripping the recycle liquid is then pumped through the coke precursor removal step where high-molecular-weight constituents are removed. The clean liquid recycle then passes to the suction drum of the feed pump. The product from the top of the recycle stripper goes to fractionation, and any high-boiling oil product is directed from the stripper bottoms pump discharge.

The residence time in the reactor is adjusted to provide the desired conversion levels. Catalyst particles are continuously withdrawn from the reactor, regenerated, and recycled back into the reactor that provides the flexibility to process a wide range of high-boiling feedstock such as atmospheric and vacuum tower bottoms, coal-derived liquids, and bitumen. An internal liquid recycle is provided with a

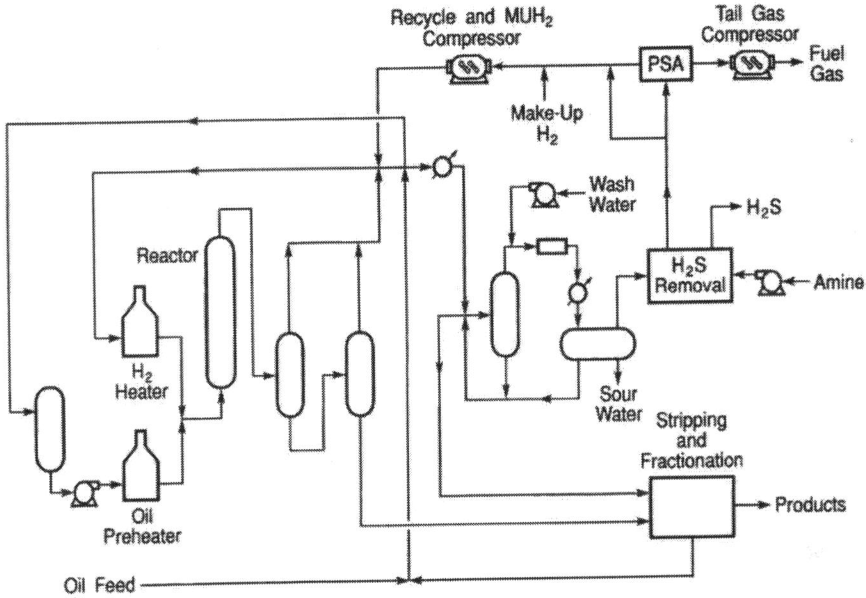

Figure 9.5 LC-Fining process.
Source: Modified from Speight, J.G., 2007. The Chemistry and Technology of Petroleum, fourth ed. CRC Press, Taylor & Francis Group, Boca Raton, FL.

pump to expand the catalyst bed, continuously. As a result of expanded-bed operating mode, small pressure drops and isothermal operating conditions are accomplished. Small-diameter extruded catalyst particles as small as 0.8 mm (1/32 in.) can be used in this reactor.

Although the process may not be the means by which direct conversion of the bitumen to a synthetic crude oil would be achieved it does nevertheless offer an attractive means of bitumen conversion. Indeed, the process would play the part of the primary conversion process from which liquid products would accrue—these products would then pass to a secondary upgrading (hydrotreating) process to yield a synthetic crude oil.

9.4.9 MAKfining process

The process uses a multiple catalyst system in multibed reactors that include quench and redistribution system internals (Parkash, 2003; Gary et al., 2007; Speight, 2014, 2017; Hsu and Robinson, 2017).

In the process (Speight, 2007) the feedstock and recycle gas are preheated and brought into contact with the catalyst in a downflow fixed-bed reactor. The reactor effluent is sent to high- and low-temperature separators. Product recovery is a

Table 9.4 Feedstock and product data for the LC-Fining process.

Feedstock	Kuwait atmospheric residuum	Gach Saran vacuum residuum	Arabian Heavy vacuum residuum	Athabasca bitumen
API gravity	15.0	6.1	7.5	9.1
Sulfur (wt.%)	4.1	3.5	4.9	5.5
Nitrogen (wt.%)				0.4
Products (wt.%)				
Naphtha (C5—205°C, C5—400°F)	2.5	9.7	14.3	11.9
Sulfur (wt.%)				1.1
Nitrogen (wt.%)				
Distillate (205°C–345°C, 400°F–650°F)	22.7	14.1	26.5	37.7
Sulfur (wt.%)				0.7
Nitrogen (wt.%)				
Heavy distillate (345°C–525°C, 650°F–975°F)	34.7	24.1	31.1	30
Sulfur (wt.%)				1.1
Nitrogen (wt.%)				
Residuum (>525°C, >975°F)	35.5	47.5	21.3	12.9
Sulfur (wt.%)				3.4

Source: Modified from Speight, J.G., 2007. The Chemistry and Technology of Petroleum, fourth ed. CRC Press, Taylor & Francis Group, Boca Raton, FL.

stripper/fractionator arrangement. Typical operating conditions in the reactors are 370°C–425°C (700°F–800°F) (single-pass) and 370°C–425°C (700°F–800°F) (recycle) with pressures of 1000–2000 psi (6895–13,790 kPa) (single-pass) and 1500–3000 psi (10,342–20,684 kPa) (recycle). Product yields depend upon the extent of the conversion (Speight, 2000, 2014, 2017; Parkash, 2003; Ancheyta and Speight, 2007; Gary et al., 2007; Hsu and Robinson, 2017).

9.4.10 Microcat-RC process

The Microcat-RC process (also referred to as the M-Coke process) is a catalytic ebullated-bed hydroconversion process that is similar to Residfining (*q.v.*) and operates at relatively moderate pressures and temperatures (Parkash, 2003; Gary et al.,

2007; Speight, 2014, 2017; Hsu and Robinson, 2017). The catalyst particles, containing a metal sulfide in a carbonaceous matrix formed within the process, are uniformly dispersed throughout the feed. Because of their ultrasmall size (10^{-4} in diameter), there are typically several orders of magnitude more of these micro-catalyst particles per cubic centimeter of oil than is possible in other types of hydroconversion reactors using conventional catalyst particles. This results in smaller distances between particles and less time for a reactant molecule or intermediate to find an active catalyst site. Because of their physical structure, micro-catalysts suffer none of the pore-plugging problems that plague conventional catalysts.

In the process, fresh vacuum residuum, micro-catalyst, and hydrogen are fed to the hydroconversion reactor. Effluent is sent to a flash separation zone to recover hydrogen, gases, and liquid products, including naphtha, distillate, and gas oil (Speight, 2000, 2014, 2017; Parkash, 2003; Ancheyta and Speight, 2007; Gary et al., 2007; Hsu and Robinson, 2017). The residuum from the flash step is then fed to a vacuum distillation tower to obtain a 565°C$^-$ (1050°F$^-$) product oil and a 565°C$^+$ (1050°F$^+$) bottoms fraction that contains unconverted feed, micro-catalyst, and essentially all of the feed metals.

Hydrotreating facilities may be integrated with the hydroconversion section or built on a stand-alone basis, depending on product quality objectives and owner preference.

9.4.11 Mild hydrocracking process

The mild hydrocracking process uses operating conditions similar to those of a VGO desulfurizer to convert VGO to significant yields of lower boiling products. Consequently, the flow scheme for a mild hydrocracking unit is virtually identical to that of a VGO desulfurizer.

For example, in a simplified process for VGO desulfurization, the VGO feedstock is mixed with hydrogen makeup gas and preheated against reactor effluent. Further preheating to reaction temperature is accomplished in a fired heater. The hot feed is mixed with recycle gas before entering the reactor. The temperature rises across the reactor due to the exothermic heat of reaction. Catalyst-bed temperatures are usually controlled by using multiple catalyst beds and by introducing recycle gas as an interbed quench medium. Reactor effluent is cooled against incoming feed and air or water before entering the high-pressure separator. Vapors from this separator are scrubbed to remove H_2S before compression back to the reactor as recycle and quench. A small portion of these gases is purged to fuel gas to prevent buildup of light ends. Liquid from the high-pressure separator is flashed into the low-pressure separator. Sour flash vapors are purged from the unit. Liquid is preheated against stripper bottoms and in a feed heater before steam stripping in a stabilizer tower. Water wash facilities are provided upstream of the last reactor effluent cooler to remove ammonium salts produced by denitrogenation of the VGO feedstock.

Variation of this process leads to the hot separator design. The process flow scheme is identical to that described earlier up to the reactor outlet. After initial reactor effluent cooling against incoming VGO feed and makeup hydrogen, a hot separator is installed. Hot liquid is routed directly to the product stabilizer. Hot vapors are further cooled by air and/or water before entering the cold separator. This arrangement reduces the stabilizer feed preheat duty and the effluent cooling duty by routing hot liquid direct to the stripper tower.

The conditions for mild hydrocracking are typical of many low-pressure desulfurization units that for hydrocracking units, in general, are marginal in pressure and hydrogen oil ratio capabilities. For hydrocracking, in order to obtain satisfactory run lengths (approximately 11 months), reduction in feed rate or addition of an extra reactor may be necessary In most cases, since the product slate will be lower boiling than for normal desulfurization service only, changes in the fractionation system may be necessary. When these limitations can be tolerated, the product value from mild hydrocracking versus desulfurization can be greatly enhanced.

In summary, the so-called mild hydrocracking process is a simple form of hydrocracking. The hydrotreaters designed for VGO desulfurization and catalytic cracker feed pretreatment are converted to once-through hydrocracking units and, because existing units are being used, the hydrocracking is often carried out under nonideal hydrocracking conditions.

9.4.12 MRH process

The MRH process is a hydrocracking process designed to upgrade high-boiling feedstocks containing large amount of metals and asphaltene, such as vacuum residua and bitumen, and to produce mainly middle distillates (RAROP, 1991, p. 65; Speight, 2000, 2014, 2017; Parkash, 2003; Ancheyta and Speight, 2007; Gary et al., 2007; Hsu and Robinson, 2017). The reactor is designed to maintain a mixed three-phase slurry of feedstock, fine powder catalyst, and hydrogen and to promote effective contact.

In the process a slurry consisting of high-boiling oil feedstock and fine powder catalyst is preheated in a furnace and fed into the reactor vessel. Hydrogen is introduced from the bottom of the reactor and flows upward through the reaction mixture, maintaining the catalyst suspension in the reaction mixture. Cracking, desulfurization, and demetallization reactions are taken place via thermal and catalytic reactions. In the upper section of the reactor, vapor is disengaged from the slurry, and hydrogen and other gases are removed in a high-pressure separator. The liquid condensed from the overhead vapor is distilled and then flows out to the secondary treatment facilities.

From the lower section of the reactor, bottom slurry oil (SLO) that contains catalyst, uncracked residuum, and a small amount of VGO fraction is withdrawn. VGO is recovered in the slurry separation section, and the remaining catalyst and coke are fed to the regenerator.

Product distribution focuses on middle distillates with the process focused as a residuum processing unit and inserted into a refinery just downstream from the vacuum distillation unit.

9.4.13 RCD Unibon process

The RCD Unibon process (BOC process) is a process to upgrade vacuum residua (RAROP, 1991, p. 67; Speight, 2000, 2014, 2017; Parkash, 2003; Ancheyta and Speight, 2007; Gary et al., 2007; Hsu and Robinson, 2017). There are several possible flow scheme variations involving for the process. It can operate as an independent unit or be used in conjunction with a thermal conversion unit. In this configuration, hydrogen and a vacuum residuum are introduced separately to the heater and mixed at the entrance to the reactor. To avoid thermal reactions and premature coking of the catalyst, temperatures are carefully controlled, and conversion is limited to approximately 70% of the total projected conversion. The removal of sulfur, heptane-insoluble materials, and metals is accomplished in the reactor. The effluent from the reactor is directed to the hot separator. The overhead vapor phase is cooled, condensed, and the separated hydrogen is recycled to the reactor.

Liquid product goes to the thermal conversion heater where the remaining conversion of nonvolatile materials occurs. The heater effluent is flashed and the overhead vapors are cooled, condensed, and routed to the cold flash drum. The bottoms liquid stream then goes to the vacuum column where the gas oils are recovered for further processing, and the residuals are blended into the high-boiling fuel oil pool.

9.4.14 Residfining process

Residfining is a catalytic fixed-bed process for the desulfurization and demetallization of atmospheric and vacuum residua (RAROP, 1991, p. 69; Speight, 2000, 2014, 2017; Parkash, 2003; Ancheyta and Speight, 2007; Gary et al., 2007; Hsu and Robinson, 2017). The process can also be used to pretreat residua to suitably low contaminant levels prior to catalytic cracking.

In the process, liquid feedstock to the unit is filtered, pumped to pressure, preheated, and combined with treat gas prior to entering the reactors. A small guard reactor would typically be employed to prevent plugging/fouling of the main reactors. Provisions are employed to periodically remove the guard while keeping the main reactors online. The temperature rise associated with the exothermic reactions is controlled utilizing either a gas- or liquid-quench. A train of separators is employed to separate the gas and liquid products. The recycle gas is scrubbed to remove ammonia and H_2S. It is then combined with fresh makeup hydrogen before being reheated and recombined with fresh feed. The liquid product is sent to a fractionator where the product is fractionated.

Residfining is an option that can be used to reduce the sulfur, to reduce metals and coke-forming precursors, and/or to accomplish some conversion to lower boiling products as a feed pretreat step ahead of a fluid catalytic cracking unit. There is also a hydrocracking option where substantial conversion of the resid occurs.

9.4.15 Residue hydroconversion process

The residue hydroconversion process is a high-pressure fixed-bed trickle-flow hydrocatalytic process (RAROP, 1991, p. 71). The feedstock can be desalted atmospheric or vacuum residue (Speight, 2000, 2014, 2017; Parkash, 2003; Ancheyta and Speight, 2007; Gary et al., 2007; Hsu and Robinson, 2017). The reactors are of multibed design with interbed cooling, and the multicatalyst system can be tailored according to the nature of the feedstock and the target conversion. For residua with high metals content, a HDM catalyst is used in the front-end reactor(s), which excels in its high metal uptake capacity and good activities for metal removal, asphaltene conversion, and residue cracking. Downstream of the demetallization stage, one or more hydroconversion stages, with optimized combination of catalysts' hydrogenation function and texture, are used to achieve desired catalyst stability and activities for denitrogenation, desulfurization, and craking of high-boiling oils. A guard reactor may be employed to remove contaminants that promote plugging or fouling of the main reactors with periodic removal of the guard reactor while keeping the main reactors online.

9.4.16 Tervahl-H process

In the Tervahl-H process the feedstock and hydrogen-rich stream are heated using heat recovery techniques and fired heater and held in the soak drum as in the Tervahl-T process. The gas and oil from the soaking drum effluent are mixed with recycle hydrogen and separated in the hot separator where the gas is cooled passed through a separator and recycled to the heater and soaking drum effluent. The liquids from the hot and cold separator are sent to the stabilizer section where purge gas and synthetic crude are separated. The gas is used as fuel and the synthetic crude can now be transported or stored.

In the related Tervahl-T process (a thermal process but covered here for convenient comparison with the Tervahl-T process, please see Section 2.7) (LePage et al., 1987), the feedstock is heated to the desired temperature using the coil heater and heat recovered in the stabilization section and held for a specified residence time in the soaking drum. The soaking drum effluent is quenched and sent to a conventional stabilizer or fractionator where the products are separated into the desired streams. The gas produced from the process is used for fuel.

9.4.17 Unicracking process

Unicracking is a fixed-bed catalytic process that employs a high-activity catalyst with a high tolerance for sulfur and nitrogen compounds and can be regenerated (Parkash, 2003, 2014, 2017; Gary et al., 2007; Hsu and Robinson, 2017).

The design is based upon a single-stage or a two-stage system with provisions to recycle to extinction (RAROP, 1991, p. 79).

In the process a two-stage reactor system receives untreated feed, makeup hydrogen, and a recycle gas at the first stage, in which gasoline conversion may be as high as 60% by volume. The reactor effluent is separated to recycle gas, liquid product, and unconverted oil (Speight, 2000, 2014, 2017; Parkash, 2003; Ancheyta and Speight, 2007; Gary et al., 2007; Hsu and Robinson, 2017). The second-stage oil may be either once-through or recycle cracking; feed to the second sage is a mixture of unconverted first-stage oil and second-stage recycle. The process operates satisfactorily for a variety of feedstocks that vary in sulfur content from approximately 1.0 wt.% to approximately 5 wt.%. The rate of desulfurization is dependent on the sulfur content of the feedstock as is catalyst life, product sulfur, and hydrogen consumption (Speight, 2000; Ancheyta and Speight, 2007).

In the process the feedstock and hydrogen-rich recycle gas are preheated, mixed, and introduced into a guard reactor that contains a relatively small quantity of the catalyst. The guard chamber removes particulate matter and residual salt from the feed. The effluent from the guard chamber flows down through the main reactor, where it contacts one or more catalysts designed for removal of metals and sulfur. The catalysts, which induce desulfurization, denitrogenation, and hydrocracking, are based upon both amorphous and molecular-sieve containing supports. The product from the reactor is cooled, separated from hydrogen-rich recycle gas, and either stripped to meet fuel oil flash point specifications or fractionated to produce distillate fuels, upgraded VGO, and upgraded vacuum residuum. Recycle gas, after H_2S removal, is combined with makeup gas and returned to the guard chamber and main reactors.

The most commonly implemented configuration is a single-stage Unicracking design, where the fresh feed and recycle oil are converted in the same reaction stage. This configuration simplifies the overall unit design by reducing the quantity of equipment in high-pressure service and keeping high-pressure equipment in a single train. The two-stage design has a separation system in each reaction stage. However, the optimum flow scheme depends on feedstock capacity and product slate objectives.0

The high efficiency of the process is due to the excellent distribution of the feedstock and hydrogen that occurs in the reactor where a proprietary liquid distribution system is employed. In addition, the process catalyst (also proprietary) was designed for the desulfurization of residua and is not merely an upgraded gas oil hydrotreating catalyst as often occurs in various processes. It is a cobalt–molybdena–alumina catalyst with a controlled pore structure that permits a high degree of desulfurization and, at the same time, minimizes any coking tendencies.

The process uses base metal or noble metal hydrogenation-activity promoters impregnated on combinations of zeolites and amorphous aluminosilicates for cracking activity. The specific metals chosen and the proportions of the metals, zeolite, and nonzeolite aluminosilicates are optimized for the feedstock and desired product balance. This is effective in the production of clean fuels, especially for cases where a partial conversion Unicracking unit and a fluid catalytic cracking unit are integrated.

The Unicracking process converts feedstocks into lower molecular weight products that are more saturated than the feed. Feedstocks include atmospheric gas oil, VGO, fluid catalytic cracking/resid catalytic cracking cycle oil, coker gas oil, deasphalted oil, and naphtha. Hydrocracking catalysts promote sulfur and nitrogen removal, aromatic saturation, and molecular weight reduction. All of these reactions consume hydrogen, and as a result, the volume of recovered liquid product normally exceeds the feedstock. Many units are operated to make naphtha (for petrochemical or motor fuel use) as a primary product.

Unicracking catalysts are designed to function in the presence of H_2S and ammonia (NH_3). This gives rise to an important difference between Unicracking and other hydrocracking processes: the availability of a single-stage design. In a single-stage unit the absence of a stripper between treating and cracking reactors reduces investment costs by making use of a common recycle gas system. Process objectives determine catalyst selection for a specific unit. Product from the reactor section is condensed, separated from hydrogen-rich gas, and fractionated into desired products. Unconverted oil is recycled or used as lube stock, fluid catalytic cracking feedstock, or ethylene plant feedstock.

The advanced partial conversion unicracking (APCU) process is a recent advancement in the area of ultralow-sulfur diesel (ULSD) production and feedstock pretreatment for catalytic cracking units. At low conversions (20%–50%) and moderate pressure, the APCU technology provides an improvement in product quality compared to traditional mild hydrocracking. In the process, high-sulfur feeds such as VGO and high-boiling cycle gas oil are mixed with a heated hydrogen-rich recycle gas stream and passed over consecutive beds of high activity pretreat catalyst and distillate selective unicracking catalyst. This combination of catalysts removes refractory sulfur and nitrogen contaminants, saturates polynuclear aromatic compounds, and converts a portion of the feed to ULSD fuel. The hydrocracked products and desulfurized feedstock for a fluid catalytic cracking unit are separated at reactor pressure in an enhanced hot separator. The overhead products for the separator are immediately hydrogenated in the integrated finishing reactor.

As pretreatment severity is increased, conversion increased in the fluid catalytic cracker and both gasoline and alkylate octane-barrel output per barrel of cat cracker feedstock increase. APCU units can be customized to achieve maximum octane-barrel production in the cat cracker.

Another development in the unicracking family is the HyCycle Unicracking technology that is designed to maximize diesel production for full conversion applications.

9.4.18 Veba Combi-Cracking process

The Veba Combi-Cracking process is a thermal hydrocracking/hydrogenation process for converting residua and other high-boiling feedstocks (RAROP, 1991, p. 81; Parkash, 2003; Gary et al., 2007; Speight, 2014, 2017; Hsu and Robinson, 2017).

The process is based on the Bergius–Pier technology that was used for coal hydrogenation in Germany up to 1945. The high-boiling feedstock is hydrogenated (hydrocracked) using a commercial catalyst and liquid-phase hydrogenation reactor operating at 440°C–485°C (825°F–905°F) and 2175–4350 psi (14,996–29,993 kPa) pressure. The product obtained from the reactor is fed into the hot separator operating at temperatures slightly below the reactor temperature. The liquid and solid materials are fed into a vacuum distillation column, the gaseous products are fed into gas-phase hydrogenation reactor operating at an identical pressure. This high-temperature, high-pressure coupling of the reactor products with further hydrogenation provides a specific process economics.

In the process the residue feed is slurried with a small amount of finely powdered additive and mixed with hydrogen and recycle gas prior to preheating. The feed mixture is routed to the liquid-phase reactors. The reactors are operated in an upflow mode and arranged in series. In a once-through operation, conversion rates of >95% are achieved. Substantial conversion of asphaltene constituents, desulfurization, and denitrogenation takes place at high levels of residue conversion. Temperature is controlled by a recycle gas quench system.

The flow from the liquid-phase hydrogenation reactors is routed to a hot separator, where gases and vaporized products are separated from unconverted material. A vacuum flash recovers distillates in the hot separator bottom product.

The hot separator top product, together with recovered distillates and straight-run distillates, enters the gas-phase hydrogenation reactor. The gas-phase hydrogenation reactor operates at the same pressure as the liquid-phase hydrogenation reactor and contains a fixed bed of commercial hydrotreating catalyst. The operation temperature (340°C–420°C) is controlled by a hydrogen quench. The system operates in a trickle-flow mode, which may not be efficient for some high-boiling feedstocks. The separation of the synthetic crude from associated gases is performed in a cold separator system. The synthetic crude may be sent to stabilization and fractionation unit as required. The gases are sent to a lean oil scrubbing system for contaminant removal and are recycled.

9.5 The future

In a refinery the hydrocracking unit is utilized to upgrade a variety of feeds that range from coker naphtha to various heavy gas oils and viscous feedstocks into lower boiling products. As a result, the hydrocracking process has emerged as the production "workhorse" of middle distillates—diesel, jet fuel, and heating oil—in many refinery configurations. Thus hydrocracking has become an indispensable processing technology to modern crude oil refining and petrochemical industry due to its flexibility to feedstocks and product scheme, and high-quality products. Particularly, high-quality naphtha, jet fuel, diesel, and lube base oil can be produced through this technology. The hydrocracker provides a better balance of gasoline and distillates, improves gasoline yield, octane quality, and can supplement the fluid catalytic cracker to upgrade high-boiling feedstocks. In the hydrocracker, low-boiling fuel oil is converted into lower boiling products under a high hydrogen pressure and over a hot catalyst bed—the main products are naphtha, jet fuel, and diesel oil.

Heavy crude oil, sour crude oil, and tar sand bitumen—which require more energy-intensive processing than conventional crude oil—will contribute a growing fraction of fuels production. As existing reserves of conventional oil are depleted and there is greater worldwide competition for premium (e.g., low-density, low-sulfur) crude oil, refineries will increasingly utilize sour (high-sulfur) crude oil, heavy crude oil, extra heavy oil, and tar sand bitumen to meet demand.

However, like many refinery processes, the hydrocracking process can succumb to the problems encountered in hydrocracking high-boiling feedstocks, which can be directly equated to the amount of complex, higher boiling constituents that may require pretreatment (Speight and Moschopedis, 1979; Speight, 2000, 2014, 2017; Parkash, 2003; Ancheyta and Speight, 2007; Gary et al., 2007; Hsu and Robinson, 2017). Processing these feedstocks is not merely a matter of applying know-how derived from refining conventional crude oils but requires knowledge of composition and properties (Chapter 1: Feedstock Types and Properties). The attempts to modify the process (Section 4.0) have had some measure of success materials that are not only complex in terms of the carbon number and boiling point ranges but also because a large part of this envelope falls into a range of model compounds where very little is known related to the properties.

Hydrocracking adds flexibility and offers the refiner a process that can handle varying feeds and operate under diverse process conditions. Utilizing different types of catalysts can modify the product slate produced. Reactor design and number of processing stages play a role in this flexibility.

Furthermore, it is apparent that the conversion of viscous feedstock requires new lines of thought to develop suitable processing scenarios (Parkash, 2003; Gary et al., 2007; Speight, 2014, 2017; Hsu and Robinson, 2017). Indeed, the use of thermal process (carbon rejection processes) and hydrothermal processes (hydrogen addition processes), which were inherent in the refineries designed to process lower boiling feedstocks, has been a particular cause for concern. This has caused the evolution of processing schemes that accommodate the higher boiling feedstocks (Chapter 10: Nonfossil Fuel Feedstocks) (Khan and Patmore, 1998; Parkash, 2003; Gary et al., 2007; Speight, 2014, 2017; Hsu and Robinson, 2017).

For the high-boiling feedstocks, which will increase in amounts in terms hydrocracking feedstocks, reactor designs will continue to focus on online catalyst addition and withdrawal. Fixed-bed designs have suffered from mechanical inadequacy when used for the higher boiling feedstocks and short catalyst lives—6 months or less—even though large catalyst volumes are used (LHSV typically of 0.5–1.5). Refiners will attempt to overcome these shortcomings by innovative designs, allowing better feedstock flow and catalyst utilization or online catalyst removal. For example, the OCR process, in which a lead, moving-bed reactor is used to demetallized high-boiling feed ahead of the fixed-bed hydrocracking reactors, has seen some success. But whether this will be adequate for continuous hydrocracking high-boiling feedstock remains a question.

Improved catalysts are now available based on a better understanding of asphaltene chemistry, and this will be a focus of catalyst manufacture—to hydrocrack asphaltene constituents without serous deleterious effect on the catalyst. In this respect a reexamination of the CANMET process is warranted since processing Athabasca bitumen using this process gave good result, and the iron-based catalyst also acted as a scavenger for coke formers—other options include the addition of metal oxides as scavengers for coke formers and sulfur. However, the refining industry has remained cautious, and investment in these technologies has been disappointing. In fact, application of hydrogen addition technologies to high-boiling feedstocks only account less than one-thirds of the global residue upgrading capacity.

The use of ebullated-bed technologies was first introduced in the 1960s in an attempt to overcome problems of catalyst ageing and poor distribution in fixed-bed designs. Hydrogen and feed enter at the bottom of the reactor, thereby expanding the catalyst bed. Although catalyst performance can be kept constant because catalyst can be replaced online, the ebullition results in a back mixed reactor, and therefore desulfurization and hydroconversion are less than obtainable in a fixed-bed unit. Currently, in order to limit coking, most commercial ebullating bed units operate in the 70%–85% desulfurization range and 50–70 vol.% resid conversion.

Development work will continue and ebullating bed units will see more use and have a greater impact of resid conversion operations. Improvements are as follows: (1) second-generation catalyst technology, which will allow higher conversion to a stable product, (2) catalyst rejuvenation, which allows spent catalyst to be reused to a greater extent than current operations allow, and (3) new reactor designs raising single train size greater throughput.

Slurry-phase hydrocracking of heavy crude oil and the latest development of dispersed catalysts present strong indications that such technologies will play a role in future refineries. Catalysts for slurry-phase hydrocracking of heavy crude oil have undergone two development phases: (1) heterogeneous solid powder catalysts, which have low catalytic activity and will produce a large number of solid particles in bottom oil making the catalyst difficult to dispose and utilize, and (2) homogeneous dispersed catalysts, which are divided into water-soluble dispersed catalysts and oil-soluble dispersed catalysts. These types of catalysts are highly dispersed and have greater surface-area-to-volume ratio. Therefore they show high catalytic activity and good performance. They are desirable catalysts for slurry-phase hydrocracking of heavy crude oil and will be used more prominently in future hydrocracking operations.

In spite of the numerous process design variations (Section 4.0), process design innovations and hardware innovations will continue. Although conventional (high-pressure) hydrocracking will still be used to address the need to produce more gasoline and diesel, moderate-pressure hydrocracking (where control of the reaction chemistry is more possible), another approach is to introduce a mild hydrocracking unit upstream of the fluid catalytic cracking unit to maintain that unit at full capacity. Alternatively, more refiners will turn to two-stage recycle hydrocracking (TSR hydrocracking) and reverse-staging configurations.

Catalyst improvements will continue to (1) improve hydrocracking activity, (2) improve HDS activity, (3) reduce catalytic deactivation, and (4) increase cycle length. The development in nonprecious metal catalysts, heteropolyanions to improve metal dispersions, beta zeolite, and the acid cracking–based formulations of highly active hydrocracking catalysts has already added (and will continue to add) flexibility in the operations of hydroprocessing units. New formulations that employ amorphous silica–alumina supports and dealuminated Y-zeolites will be readily available and offer high activity with high stability. These designs allow for lower operating pressures, increased run length, and higher gasoline and diesel yields.

Another central focus will be reduction of reducing hydrogen consumption while maintaining product quality. Catalysts that can withstand organic nitrogen contamination are being developed for lower cost, single-stage units. The addition of metal traps upstream of the hydroprocessing unit is a solution to protect highly active catalyst from high metal feeds that will see wider application.

Furthermore, biomass gasification and Fischer–Tropsch synthesis conversion are very likely to be a part of the future refineries as part of the next-generation biofuels developments (Chapter 10: Nonfossil Fuel Feedstocks), refiners will need to monitor closely the latest refinery-related advances as well as future directions in biomass processing. In particular, the response of the refining industry to opportunities for processing heavy crude oil, extra heavy oil, tar sand bitumen, and resids, mandated biofuels usage, and requirements to comply with carbon dioxide, will need to be addressed. Furthermore, gasification with carbon capture and the use of biomass as feedstock should help refiners meet emissions reduction requirements for carbon dioxide.

A major problem for refiners looking to process tight oil in the hydrocracker is the relative lack of gasoil material provided by this feed, potentially leading to existing hydrocracking capacity being underutilized. In order to prevent this underutilization, refiners may look to add additional streams (for instance, cracked stocks) to the hydrocracker feed to ensure that capacity utilization is maintained. Also, the low-density nature of the tight crude oils leads to more light ends (low-boiling gases and liquids) coming from the hydrocracker at the same conversion level while also negatively impacting the cold flow properties of middle distillate products and lowering the octane value of naphtha produced from the unit.

In summary, viscous feeds such as vacuum residue, extra heavy oil, and tar sand bitumen are alternate sources for the production of high-value transportation fuels and abundantly available. These feedstocks are of low quality due to the presence of impurities such as asphaltene constituents, sulfur-containing constituents, nitrogen-containing constituents, and metal-containing constituents. Several process technologies have been developed to upgrade these feeds through fixed-bed, moving-bed, ebullated-bed, slurry-phase reactor, or a combination. Hydrocracking in slurry-phase type reactor is a prominent technology to convert low-value feeds into high-value transportation fuels and petrochemical products. The selection of a suitable process technology is implemented by considering the feedstock properties, product demand, and the economic benefit as well as environmental concerns.

Also, as environmental regulations on transportation fuels continue to tighten, the hydrocracker will be one of the tools available to refiners to meet new product specifications. Hydrocrackers can effectively yield ULSD streams whereas middle distillate range products (light cycle oil, LCO) from FCC units will regularly require additional dearomatization to meet product blending specifications. HCUs can also offer improved flexibility to shift production modes between gasoline and diesel products based on process

Nevertheless, hydrocracking technology needs to increase the flexibility in the light of shifting demand. In the short term, global gasoline and diesel/gasoil demand is still riding high thanks to lower retail prices. Climbing global air traffic, particularly for citizens of China and India, continues to support jet fuel consumption. Diesel is considered a good replacement for high-sulfur bunker fuel. In the longer term, gasoline demand in is expected to be peaking. But diesel is also facing twin threats: a ban of diesel-fueld automobiles in growing numbers of countries and cities triggered by the increasing popularity of zero-NO_x emitting electric vehicles.

References

Ancheyta, J., Speight, J.G., 2007. Hydroprocessing Heavy Oils and Residua. CRC Press, Taylor & Francis Group, Boca Raton, FL.

Billon, A., Morel, F., Morrison, M.E., Peries, J.P., 1994. Converting residues with IPP's Hyvahl and Solvahl processes. Rev. Inst. Français du. Pétrole 49 (5), 495–507.

DeCroocq, D., 1984. Catalytic Cracking of Heavy Petroleum Hydrocarbons. Editions Technip, Paris.

DeCroocq, D., 1997. Major scientific and technical challenges about development of new processes in refining and petrochemistry. Rev. Inst. Français du. Pétrole 52 (5), 469–489.

Dolbear, G.E., 1998. Chapter 7: Hydrocracking: react ions, catalysts, and processes. In: Speight, J.G. (Ed.), Petroleum Chemistry and Refining. Taylor & Francis, Washington, DC.

Gary, J.H., Handwerk, G.E., Kaiser, M.J., 2007. Petroleum Refining: Technology and Economics, fifth ed. CRC Press, Taylor & Francis Group, Boca Raton, FL.

Hsu, C.S., Robinson, P.R. (Eds.), 2017. Handbook of Petroleum Technology. Springer International Publishing AG, Cham, Switzerland.

Khan, M.R., Patmore, D.J., 1998. Chapter 6: Heavy oil upgrading processes. In: Speight, J.G. (Ed.), Petroleum Chemistry and Refining. Taylor & Francis, Washington, DC.

Meyers, R.A. (Ed.), 1997. Handbook of Petroleum Refining Processes, second ed. Marcel Dekker Inc. New York.

Parkash, S., 2003. Refining Processes Handbook. Gulf Professional Publishing, Elsevier, Amsterdam.

Peries, J.P., Quignard, A., Farjon, C., Laborde, M., 1988. Thermal and catalytic ASVAHL processes under hydrogen pressure for converting heavy crudes and conventional residues. Rev. Inst. Français Du. Pétrole 43 (6), 847–853.

RAROP, 1991. In: Kamiya, Y. (Ed.), Heavy Oil Processing Handbook. Research Association for Residual Oil Processing, Agency of Natural Resources and Energy, Ministry of International Trade and Industry, Tokyo.

Sarrazin, P., Bonnardot, J., Wambergue, S., Morel, F., 2005. New mild hydrocracking route produces 10-ppm-sulfur diesel. Hydrocarbon Process. 84, 57–64.

Scherzer, J., Gruia, A.J., 1996. Hydrocracking Science and Technology. Marcel Dekker Inc, New York.

Sherman, J.D., 1998. Synthetic zeolites and other microporous oxide molecular sieves. Proceedings Colloquium on Geology, Mineralogy, and Human Welfare. National Academy of Sciences, Irvine, CA.

Speight, J.G., 2000. The Desulfurization of Heavy Oils and Residua, second ed. Marcel Dekker Inc, New York.

Speight, J.G., 2014. The Chemistry and Technology of Petroleum, fifth ed. CRC Press, Taylor & Francis Group, Boca Raton, FL.

Speight, J.G., 2017. Handbook of Petroleum Refining. CRC Press, Taylor & Francis Group, Boca Raton, FL.

Speight, J.G., Moschopedis, S.E., 1979. The production of low-sulphur liquids and coke from Athabasca bitumen. Fuel Process. Technol. 2, 295.

Swaddle, T.W., 1997. Inorganic Chemistry. Academic Press Inc, New York.

Non–fossil fuel feedstocks

10

10.1 Introduction

The end of the widespread production of liquid fuels and other products within the current refinery infrastructure is considered by some observers to be imminent and could occur during the next several decades but is very dependent upon the remaining reserves and petropolitics (Speight, 2011a; Speight and Islam, 2016). In addition, rightly or wrongly, but without much justification but with much speculation, the combustion of fossil fuels is considered as the largest source of anthropogenic emissions of carbon dioxide (CO_2), which is largely blamed for global warming and climate change (Speight, 2020) although other sources are evident but are often ignored (Speight, 2020). Nevertheless, having been identified as one of the causative agents of climate change, it is necessary to attempt to mitigate the emissions of carbon dioxide from fossil fuel combustion and to offset the depletion of fossil fuels such as the commonly used natural gas, crude oil, and coal (Ragland et al., 1991). Oil shale is of lesser importance having received on-again and off-again popularity as a source of fuels but has never really been recognized as a major source of fuels (Baughman, 1978; Allred, 1982; Scouten, 1990; Lee, 1991, 1996; US DOE, 2004a, b,c; Speight and Islam, 2016; Speight, 2013, 2014, 2019b).

These resources (natural gas, crude oil, and coal) are currently on a depletion curve with estimates of the longevity of these resources varying up to 50 years (Speight and Islam, 2016). However, seeking alternate sources of energy is of critical importance for long-term security and continued economic growth. Supplementing crude oil consumption with renewable biomass resources is a first step toward this goal. The realignment of the chemical industry from one of crude oil refining to a biorefinery concepts is, given time, feasible has become a national goal of many oil-importing countries. However, clearly defined goals are necessary for increasing the use of biomass-derived feedstocks in industrial chemical production and it is important to keep the goal in perspective. In this context the increased use of feedstocks and the production of fuels therefrom should be viewed as one of a range of possible measures for achieving self-sufficiency in energy, rather than a panacea (Demirbaş, 2008, 2009; Speight, 2019b).

As the refining industry evolves even further, and in many cases away from natural gas and crude oil as the major feedstocks, a variety of biomass and waste-derived feedstocks will be used as feedstocks' fuel production and this may be no more evident than the use of biofeedstocks for gasifier units. Moreover, gasification is (1) a well-established technology, (2) has broad flexibility of feedstocks and operation, and (3) is the most environment-friendly route for handling these

The Refinery of the Future. DOI: https://doi.org/10.1016/B978-0-12-816994-0.00010-5

feedstocks for power production. A wide variety of biofeedstocks such as wood pellets, and wood chips, waste wood, plastics, municipal solid waste (MSW), refuse-derived fuel, agricultural and industrial wastes, sewage sludge, switch grass, discarded seed corn, corn stover, and other crop residues will all be used as gasifier feedstocks.

This chapter presents an overview of the production of fuels from alternate (non−fossil fuel) sources in order for the reader to understand the chemical and physical parameters that are involved in the production of alternate fuels. For the purpose of this chapter, three distinct sources of biomass energy are considered: (1) agricultural crops, (2) wood, and (3) industrial and domestic wastes.

10.2 Biomass

Biomass is a term used to describe any material of recent biological origin, including plant materials, such as trees, grasses, and agricultural crops, and even animal manure. Thus biomass (also referred to as biofeedstock) refers to living and recently dead biological materials, which can be used as fuel or for industrial production (Lee, 1996; Wright et al., 2006; Speight, 2008; Lorenzini et al., 2010; Nersesian, 2010).

In terms of the chemical composition, biomass is a mixture of complex organic compounds that contain, for the most part carbon, hydrogen, and oxygen, with small amounts of nitrogen and sulfur, as well as with traces of other elements, including metals. In the most cases the biomass composition comprises approximately carbon 47−53 w/w%, hydrogen 5.9−6.1 w/w%, and oxygen 41−45 w/w%. The presence of a large amount of oxygen in biomass makes a significant difference with fossil hydrocarbons. When used as fuel this is less efficient, but more suited for getting higher value chemicals and bioproducts containing functional entities within the constituent molecules.

The biomass used as industrial feedstock can be supplied by agriculture, forestry, and aquaculture, as well as resulting from various waste materials. The biomass can be classified as follows: (1) agricultural feedstocks, such as sugarcane, sugar beet, and cassava; (2) starch feedstocks, such as wheat, maize, and potatoes; (3) oil feedstocks, such as rapeseed and soy; (4) dedicated energy crops, such as short-rotation coppice (SRC) that includes poplar, willow, and *Eucalyptus*; (5) high-yield perennial grass, such as *Miscanthus* and switchgrass; (6) nonedible oil plants, such as *Jatropha*, *Camellia*, and *Sorghum*; and (7) lignocellulosic waste material, which includes forestry wood, straw, corn stover, bagasse, paper pulp, and algal crops from land farming.

The utilization of biomass through the adoption of the conventional crude oil refinery systems and infrastructure to produce substitutes for fuels and other chemicals currently derived from conventional fuels (coal, oil, and natural gas) is one of the most favored methods to combat fossil fuel depletion as the 21st century matures. In a biorefinery a solid biomass feedstock is converted, through either a thermochemical process (such as gasification, pyrolysis) or a biochemical process (such as hydrolysis, fermentation) into a mixture of organic (such as hydrocarbon, alcohol, and ester derivatives) and inorganic compounds (such as carbon monoxide and hydrogen) that can be upgraded through catalytic reactions to high-value fuels or chemicals (Speight, 2011a, 2014, 2017, 2019c).

In this manner, reducing the national dependence of any country on imported crude oil for long-term security and continued economic growth by supplementing crude oil consumption with renewable biomass resources becomes a first step toward energy self-sufficiency. The realignment of the chemical industry from one of crude oil refining to a biomass refining concepts is, given time, feasible is a worthy goal of many oil-importing countries (Speight and Islam, 2016; Speight and El-Gendy, 2018; Speight, 2019c). However, clearly defined goals are necessary for increasing the use of biomass-derived feedstocks in industrial chemical production and it is important to keep the goal in perspective. In this context the increased use of biofuels should be viewed as one of a range of possible measures for achieving self-sufficiency in energy, rather than the sole panacea (Crocker and Crofcheck, 2006; Langeveld et al., 2010), although there are arguments against the rush to the large-scale production of biofuels (Giampietro and Mayumi, 2009).

Biomass is carbon based and is composed of a mixture of organic molecules containing hydrogen, usually including atoms of oxygen, often nitrogen, and also small quantities of other atoms, including alkali metals, alkaline earth metals, and heavy metals. These metals are often found in functional molecules such as the porphyrin molecules that include chlorophyll that contains magnesium. It is a renewable resource, the utilization of which has received great attention due to environmental considerations and the increasing demands of energy worldwide. Biomass is the oldest form of energy used by humans and traditionally, biomass has been utilized through direct combustion, and this process is still widely used in many parts of the world (Ragland et al., 1991). Since the energy crises of the 1970s, many countries have become interested in biomass as a fuel source to expand the development of domestic and renewable energy sources and reduce the environmental impacts of energy production. Biomass energy (bioenergy) can be an important alternative in a future and more sustainable energy supply. Biomass is material that is derived from plants and there are many types of biomass resources currently used and potentially available. Biomass energy has the potential to be produced and used efficiently and cost competitively, generally in the more convenient forms of gases, liquids, or electricity (Larson and Kartha, 2000).

Biomass feedstocks and fuels exhibit a wide range of physical, chemical, and agricultural/process engineering properties and is subdivided into three different grades (or types) and the feedstock origin determines the so-called biomass generation (Table 10.1). In some cases the third-generation biomass may also include high-yield algal crops that can be fed directly with concentrated carbon dioxide streams resulting from industrial processes, as from coal power plants and from fermentation of sugars—algal cultures can produce various hydrocarbon derivatives as well as various volatile olefin derivatives (Dimian, 2015).

Table 10.1 Different grades of biomass.

Grade	Description
Primary biomass	Produced directly by photosynthesis and includes all terrestrial plants now used for food, feed, fiber, and fuel wood
Secondary biomass	Differs from primary biomass feedstocks in that the secondary feedstocks are a by-product of processing of the primary feedstocks in which there has been a substantial physical or chemical breakdown of the primary biomass and production of by-products; processors may be factories or animals Field processes such as harvesting, bundling, chipping, or pressing do not cause a biomass resource that was produced by photosynthesis (e.g., tree tops and limbs) to be classified as secondary biomass Specific examples of secondary biomass includes sawdust from sawmills, black liquor (which is a by-product of paper making), and cheese whey (which is a by-product of cheese making processes). Manure from concentrated animal feeding operations are collectable secondary biomass resources Vegetable oils used for biodiesel that are derived directly from the processing of oilseeds for various uses are also a secondary biomass resource
Tertiary biomass	Includes postconsumer residues and wastes, such as fats, greases, oils, construction and demolition wood debris, other waste wood from the urban environments, as well as packaging wastes, municipal solid wastes, and landfill gases

Chemically, the forms of biomass include (1) cellulose and related compounds, which can be used for the production of paper and/or bioethanol and (2) long-chain lipid derivatives that can be used in cosmetics or for other specialty chemicals.

Cellulose is an important structural component of the primary cell wall of green plants. Chemically, cellulose is an organic compound with the empirical formula ($C_6H_{10}O_5$) and is a polysaccharide consisting of a linear chain of several hundred to many thousands of linked glucose.

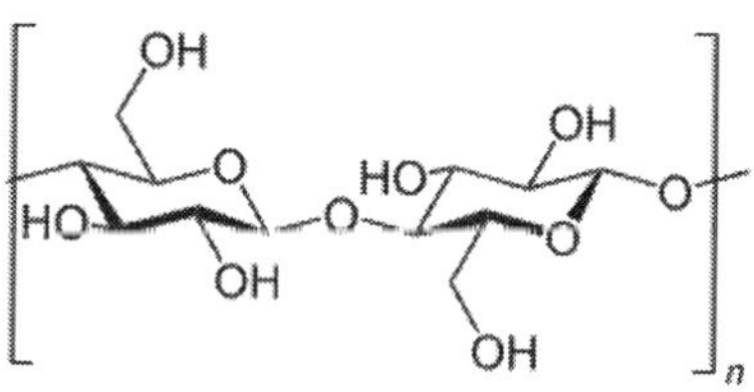

Cellulose.

On the other hand, lipid derivatives are very diverse in both their respective structures and functions. These diverse compounds that make up the lipid family are so grouped because they are insoluble in water (hydrophobic). They are also soluble in other organic solvents such as ether, acetone, and other lipids. Lipids serve a variety of important functions in living organisms and also act as chemical messengers, serve as valuable energy sources, provide insulation, and are the main components of membranes. Major lipid derivatives include the phospholipids derivatives (Fig. 10.1).

Briefly, the phospholipids are a class of lipids that are a major component of all cell membranes. They can form lipid bilayers because of their amphiphilic characteristic. The structure of the phospholipid molecule generally consists of two hydrophobic fatty acid tails and a hydrophilic head consisting of a phosphate group. The two components are usually joined together by a glycerol ($HOCH_2CHOHCH_2OH$) molecule. The free fatty acid derivatives are variable but commonly include the naturally occurring stearic acid, palmitic acid, oleic acid, and linoleic acid (Fig. 10.2).

Other biomass components, which are generally present in minor amounts, include triglycerides, sterols, alkaloids, resins, terpenes, terpenoids, and waxes. This includes everything from primary sources of crops and residues harvested/collected directly from the land to secondary sources such as sawmill residuals, to tertiary sources of postconsumer residuals that often end up in landfills. A fourth source, although not usually categorized as such, includes the gases that result from

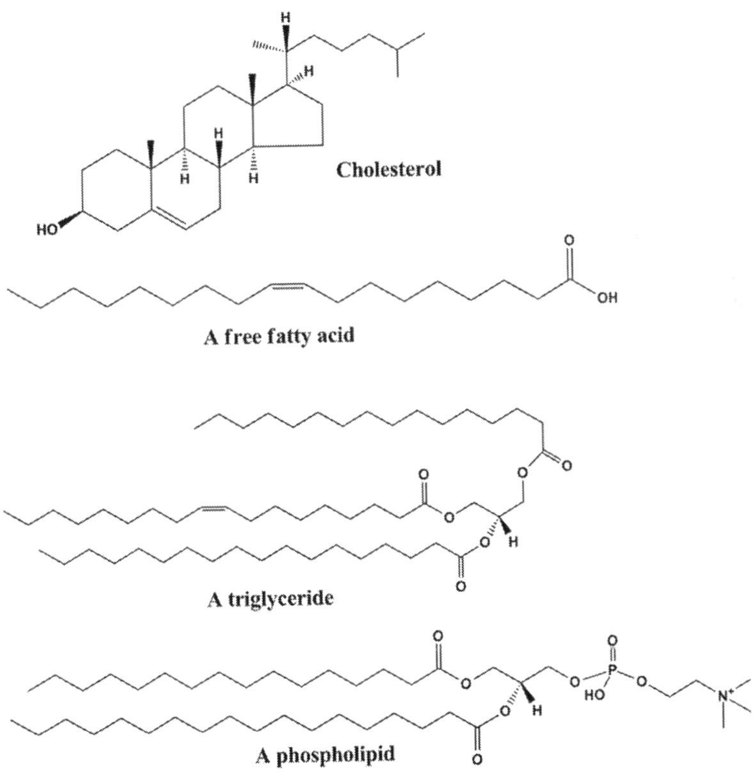

Figure 10.1 Examples of lipid derivatives.

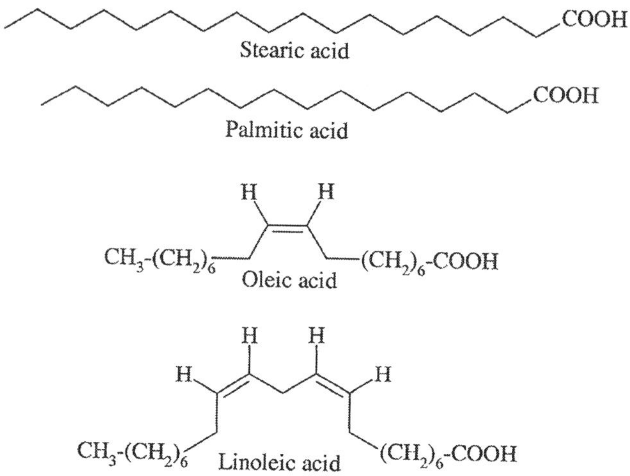

Figure 10.2 Examples of fatty acids.

anaerobic digestion of animal manures or organic materials in landfills (Wright et al., 2006).

Examples of modern biomass use are ethanol production from sugarcane, combined heat and power (often referred to using the acronym CHP) district heating programs, and the cocombustion of biomass in conventional coal-based power plants (Hoogwijk et al., 2005). In industrialized countries the main biomass processes used in the future are expected to be direct combustion of residues and wastes for electricity generation, bioethanol and biodiesel as liquid fuels, and CHP production from energy crops. In fact, biomass can be considered as the best option and has the largest potential, which meets these requirements and could ensure fuel supply in the future (Demirbaş, 2008).

Other biomass components, which are generally present in minor amounts, include diglyceride derivatives, triglyceride derivatives, sterol derivatives, alkaloid derivatives, terpene derivatives, terpenoid derivatives, and waxes. This includes everything from primary sources of crops and residues harvested/collected directly from the land to secondary sources such as sawmill residuals, to tertiary sources of postconsumer residuals that often end up in landfills.

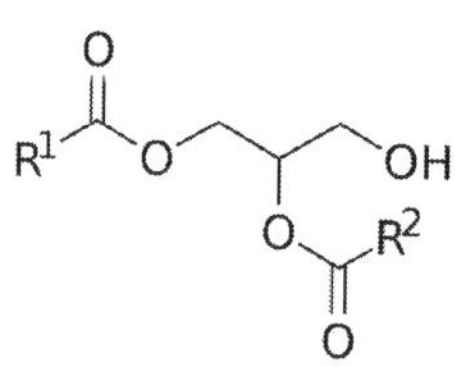

Example of a diglyceride.

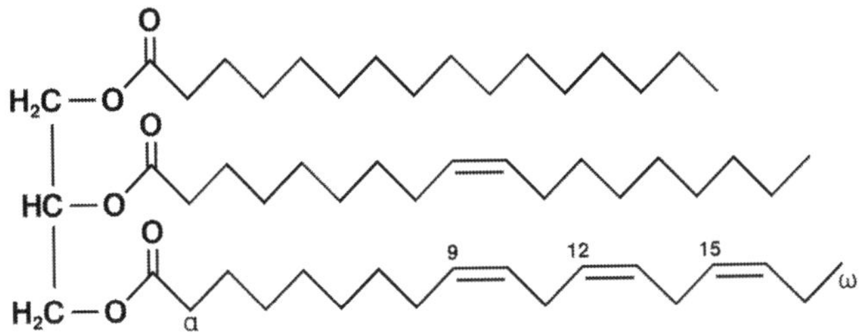

Example of a triglyceride.

10.2.1 Chemical constituents

More generally, biomass feedstocks are recognized or classified by the specific plant content of the feedstock or the manner in which the feedstocks is produced. For example, primary biomass feedstocks are thus primary biomass that is harvested or collected from the field or forest where it is grown. Examples of primary biomass feedstocks currently being used for bioenergy include grains and oilseed crops used for transportation fuel production, plus some crop residues (such as orchard trimmings and nut hulls) and some residues from logging and forest operations that are currently used for heat and power production.

Secondary biomass feedstocks differ from primary biomass feedstocks in that the secondary feedstocks are a by-product of processing of the primary feedstocks. By processing it is meant that there is substantial physical or chemical breakdown of the primary biomass and production of by-products; processors may be factories or animals. Field processes such as harvesting, bundling, chipping, or pressing do not cause a biomass resource that was produced by photosynthesis (e.g., tree tops and limbs) to be classified as secondary biomass. Specific examples of secondary biomass include sawdust from sawmills, black liquor (which is a by-product of paper making), and cheese whey (which is a by-product of cheese making processes). Manure from concentrated animal feeding operations are collectable secondary biomass resources. Vegetable oils used for biodiesel that are derived directly from the processing of oilseeds for various uses are also a secondary biomass resource.

Tertiary biomass feedstock includes postconsumer residues and wastes, such as fats, greases, oils, construction and demolition wood debris, other waste wood from the urban environments, as well as packaging wastes, MSWs, and landfill gases. A category other wood waste from the urban environment includes trimmings from urban trees, which technically fits the definition of primary biomass. However, because this material is normally handled as a waste stream along with other postconsumer wastes from urban environments (and included in those statistics), it makes the most sense to consider it to be part of the tertiary biomass stream. Tertiary biomass often includes fats and greases, which are by-products of the reduction of animal biomass into component parts, since most fats and greases, and some oils, are not available for bioenergy use until after they become a postconsumer waste stream. Vegetable oils derived from processing of plant components and used directly for bioenergy (e.g., soybean oil used in biodiesel) would be a secondary biomass resource, though amounts being used for bioenergy are most likely to be tracked together with fats, greases, and waste oils.

One aspect of designing a refinery for any feedstocks is the composition of the feedstocks. For example, a heavy oil refinery would differ somewhat from a conventional refinery and a refinery for tar sand bitumen would be significantly different to both (Speight, 2008, 2014, 2017). Furthermore, the composition of biomass is variable (Speight, 2008), which reflects the range of heat value (heat content, calorific value) of biomass, which is somewhat lesser than for coal and much lower than the heat value for crude oil, generally falling in the range 6000–8500 Btu/lb (Table 10.2). Moisture content is probably the most important determinant of heating value. Air-dried biomass typically has approximately 15%–20% moisture, whereas the moisture content for oven-dried biomass is around 0%. Moisture content is also an important characteristic of coals, varying in the range of 2%–30%. However, the bulk density (and hence energy density) of most biomass feedstocks is generally low, even after densification, approximately 10% and 40% of the bulk density of most fossil fuels.

Plants (Table 10.3) offer a unique and diverse feedstock for chemicals, and the production of biofuels from biomass requires some knowledge of the chemistry of biomass, the chemistry of the individual constituents of biomass, and the chemical means by which the biomass can be converted to fuel. It is widely recognized that further significant production of plant-based chemicals will only be economically viable in highly integrated and efficient production complexes producing a diverse range of chemical products. This biorefinery concept is analogous to conventional oil refineries and petrochemical complexes that have evolved over many years to maximize process synergies, energy integration, and feedstock utilization to drive down production costs.

Table 10.2 Types of pyrolysis.

Type	Parameters
Slow pyrolysis	
	Several hours to complete Char is the main product
Flash pyrolysis	
	High heating rate of the feedstock High temperature—450°C–1000°C Rapid devolatilization in an inert atmosphere Up to 75 w/w% yield of bio-oil Tendency for oil to be thermally unstable
Fast pyrolysis	
	Takes seconds to complete Up to 60% yield of bio-oil Also yields synthesis gas and char

Table 10.3 Heating value of selected fuels.

Fuel	Btu/lb
Natural gas	23,000
Gasoline	20,000
Crude oil	18,000
Heavy oil	16,000
Coal (anthracite)	14,000
Coal (bituminous	11,000
Wood (farmed trees, dry)	8400
Coal (lignite)	8000
Biomass (herbaceous, dry)	7400
Biomass (corn stover, dry)	7000
Wood (forest residue, dry)	6600
Bagasse (sugarcane)	6500
Wood	6000

In addition, the specific components of plants such as (1) carbohydrates, (2) vegetable oils, (3) plant fibers, and (4) complex organic molecules known as primary and secondary metabolites can be utilized to produce a range of valuable monomers, chemical intermediates, pharmaceuticals, and materials.

10.2.2 Carbohydrates

Plants capture solar energy as fixed carbon during which carbon dioxide is converted to water and sugars $(CH_2O)_x$:

$$CO_2 + H_2O \rightarrow (CH_2O)_x + O_2.$$

The sugars produced are stored in three types of polymeric macromolecules: (1) starch, (2) cellulose, and (3) hemicellulose.

In general, sugar polymers such as cellulose (Fig. 10.3) and starch can be readily broken down to their constituent monomers by hydrolysis, preparatory to conversion to ethanol or other chemicals. In contrast, lignin is an unknown complex structure containing aromatic groups that is totally hypothetical (Fig. 10.4) and is less readily degraded than starch or cellulose.

Although lignocellulose is one of the cheapest and most abundant forms of biomass, it is difficult to convert this relatively unreactive material into sugars. Among

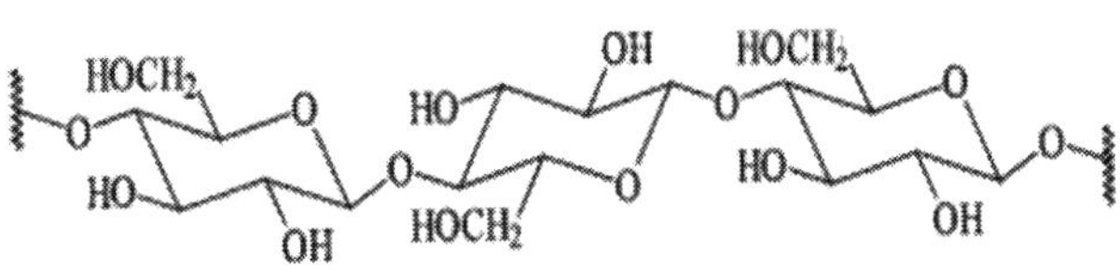

Figure 10.3 Generalized structure of cellulose.

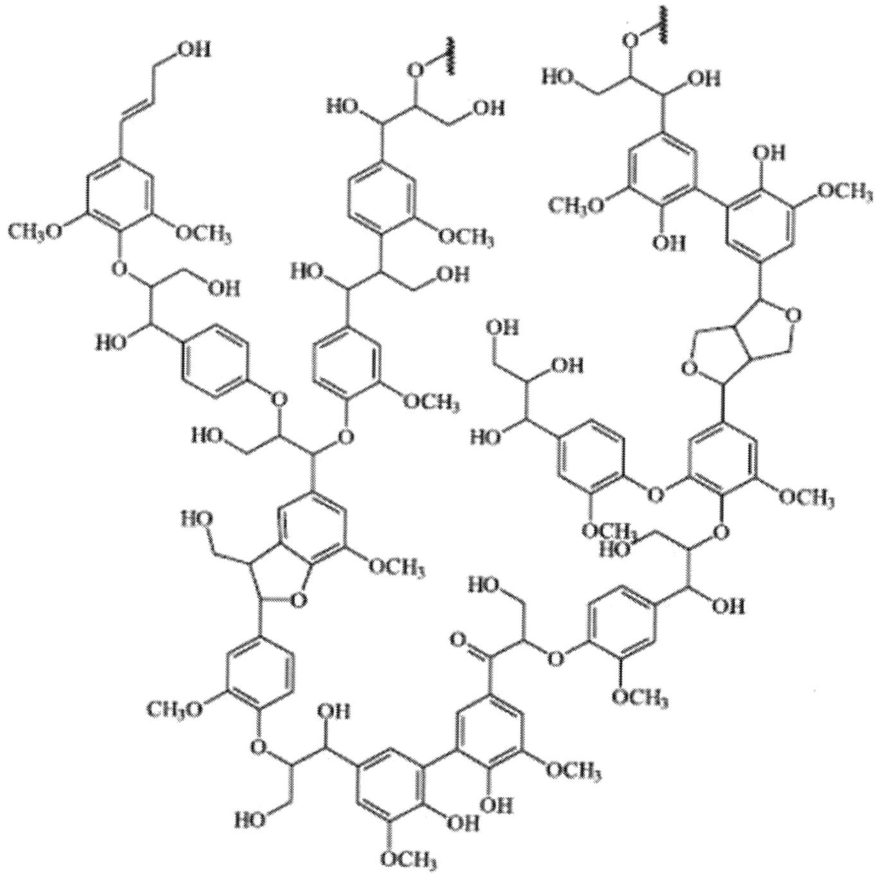

Figure 10.4 Hypothetical structure of lignin to illustrate the complexity of the molecule.

other factors, the walls of lignocellulose are composed of lignin, which must be broken down in order to render the cellulose and hemicellulose accessible to acid hydrolysis. For this reason, many efforts focused on ethanol production from biomass are based almost entirely on the fermentation of sugars derived from the starch in corn grain.

Carbohydrates (starch, cellulose, and sugars): starch is readily obtained from wheat and potato, while cellulose is obtained from wood pulp. The structures of these polysaccharides can be readily manipulated to produce a range of biodegradable polymers with properties similar to those of conventional plastics such as polystyrene foams and polyethylene film. In addition, these polysaccharides can be hydrolyzed, catalytically or enzymatically, to produce sugars, a valuable fermentation feedstock for the production of ethanol, citric acid, lactic acid, and dibasic acids such as succinic acid.

10.2.3 Vegetable oils

Vegetable oils (sometimes referred to as vegetable fats) are oils extracted from seeds, or less often, from other parts of fruits. Like animal fats, vegetable fats are mixtures of triglycerides. Soybean and rapeseed oils are examples of fats from seeds, while olive and palm oils are examples of fats from other parts of fruits. In common usage, vegetable oils may refer exclusively to vegetable fats that are liquid at room temperature. Vegetable oils are usually edible; nonedible oils derived mainly from petroleum are termed mineral oils. The predominant source of vegetable oils in many countries is rapeseed oil. Vegetable oils are a major feedstock for the oleo-chemicals industry (surfactants, dispersants, and personal care products) and are now successfully entering new markets such as diesel fuel, lubricants, polyurethane monomers, and functional polymer additives and solvents.

However, most of natural oils have only a limited application in their original form, as a consequence of their specific chemical composition. They therefore often undergo a chemical or physical modification. Due to the continuous technological developments, a whole variety of products normally processed by solvent or detergent fractionation can now be obtained with a high degree of selectivity by dry fractionation (Gibon et al., 2009).

Unsaturated vegetable oils can be transformed through partial or complete hydrogenation into oils of higher melting point. The hydrogenation process involves sparging the oil at high temperature and pressure with hydrogen in the presence of a catalyst, typically a nickel compound. As each carbon–carbon double bond is chemically reduced to a single bond, two hydrogen atoms each form single bonds with the two carbon atoms. An oil may be hydrogenated to increase resistance to oxidation (which can turn the oil rancid) or to change its physical characteristics. As the degree of saturation increases, the viscosity and the melting point of the oil increase.

Vegetable oils are used as an ingredient or component in many manufactured products. They are used to make soaps, skin products, candles, perfumes, and other personal care and cosmetic products. Some oils are particularly suitable as drying oils and are used in making paints and other wood treatment products. Vegetable oils are increasingly being used in the electrical industry as insulators since vegetable oils are bridgeable if spilled and are not toxic to the environment—they also have high flash and fire points. However, vegetable oils are less stable chemically, so they are generally used in systems where they are not exposed to oxygen and they are more expensive than crude oil distillate. More important in the present context, vegetable oils are also used as the starting material for biodiesel, which can be used like conventional diesel. Some vegetable oils are used in unmodified vehicles, but straight vegetable oil, also known as pure plant oil, needs specially prepared vehicles that have a method of heating the oil to reduce the viscosity.

10.2.4 Plant fibers

Lignocellulosic fibers extracted from plants such as hemp and flax can replace cotton and polyester fibers in textile materials and glass fibers in insulation products. Lignin is a complex chemical compound that is most commonly derived from wood and is an integral part of the cell walls of plants, especially in tracheids, xylem fibers, and sclereids. The chemical structure of lignin is unknown and, at best, can only be represented by hypothetical formulas.

Lignin (Latin: lignum—wood) is one of most abundant organic compounds on the Earth after cellulose and chitin. By way of clarification, chitin $(C_8H_{13}O_5N)_n$ is a long-chain polymeric polysaccharide of beta-glucose that forms a hard, semitransparent material found throughout the natural world. Chitin is the main component of the cell walls of fungi and is also a major component of the exoskeletons of arthropods, such as the crustaceans (e.g., crab, lobster, and shrimp), and the insects (e.g., ants, beetles, and butterflies), and of the beaks of cephalopods (e.g., squids and octopuses).

Lignin makes up approximately one-quarter to one-third of the dry mass of wood and is generally considered to be a large, cross-linked hydrophobic, aromatic macromolecule with molecular mass that is estimated to be in excess of 10,000. Degradation studies indicate that the molecule consists of various types of substructures, which appear to repeat in random manner.

Lignin fills the spaces in the cell wall between cellulose, hemicellulose, and pectin components and is covalently linked to hemicellulose. Lignin also forms covalent bonds to polysaccharides, which enables cross-linking to different plant polysaccharides. Lignin confers mechanical strength to the cell wall (stabilizing the mature cell wall) and therefore the entire plant.

10.2.5 Waste

As already stated, waste (refuse) comes in a variety of forms and must be dealt with on a day-to-day basis. Industry produces huge amounts of industrial waste and domestic waste makes a large contribution to the general waste problem. In spite of the recognition, many insidious waste products escape (inadvertently or deliberately) into the surrounding environment. Thus there are numerous pollution incidents. On the other hand, there are technologies available for the treatment of most of the wastes we produce. The level of treatment is largely a matter of cost, but conversion of waste to new products is a concept that has long been ready to hatch.

Domestic waste (also known as rubbish, garbage, trash, or junk) is unwanted or undesired material (although the old adage one man's waste is another man's treasure sometimes applies). Waste is the general term; though the other terms are used loosely as synonyms, they have more specific meanings. Thus (1) rubbish or trash are mixed household waste, including paper and packaging; (2) food waste or garbage (North America) is kitchen and table waste, and (3) junk or scrap is metallic or industrial material. There are other categories of waste as well: sewage, ash, manure, and plant materials from garden operations, including grass cuttings, fallen leaves, and pruned branches. On the other hand, industrial waste is waste produced by industrial operations such as factories, mills, and mines and has existed since the onset of the Industrial Revolution. Chemical waste and toxic waste are two additional (but specific) designations of industrial waste.

MSW is a waste type that includes predominantly household waste (domestic waste) with sometimes the addition of commercial wastes collected by a municipality within a given area. They are in either solid or semisolid form and generally exclude industrial hazardous waste. The term residual waste relates to waste left from household sources containing materials that have not been separated out or sent for reprocessing.

10.2.6 Energy crops

Biomass currently provides varying proportions (depending upon the country) of the primary energy supply, but, as a word of caution, the production of energy crops may compete with traditional agricultural and forestry uses of land. It is essential to create integrated biomass production systems that landowners can use to help meet the growing energy demands of any nation. For example, the production of fast growing short-rotation woody crops on agricultural lands is one such approach that shows considerable promise. In addition to using woody biomass for energy for power and heat generation by means of cofiring and gasification, woody crops with their high hemicellulose and cellulose content are well suited for biorefining to yield liquid fuels such as methanol, ethanol, and distillable oil (sometimes referred to as pyrolysis oil or bio-oil), as well as other products, such as specialty chemicals (Tables 10.4 and 10.5) (Speight, 2019c).

Crops are the annual or seasonal yield of any plant that is grown to be harvested as food, as livestock fodder, fuel, or for any other economic purpose in significant quantities. This category includes crop species as well as agricultural techniques related to cropping. The products from crops are not only as primary source of human foods and animal feed, but also as source of timber, fibers, and biomass energy. In addition, crops have also an essential function to maintain ecological

Table 10.4 Typical plants used as a source of energy.

Type of biomass	Plant species	Predominant use
Wood	*Butea monosperma, Casuarina equisetifolia, Eucalyptus globulus, Leucaena leucocephala, Melia azadirachta, Tamarix dioica*	Firewood
Starch	Cereals, millets, root, and tuber crops, e.g., potato	Bioethanol
Sugar	Sugarcane, sugar beet	Bioethanol
Hydrocarbons	*Euphorbia lathyris, Asclepias speciosa, Copaifera multijuga*, algae	Biodiesel
Wastes	Crop residues, animal/human refuge, sewage	Biogas

Table 10.5 Biomass liquefaction by pyrolysis.

Biomass	Pyrolysis, 550°C, no air			
		Gases	H_2, CO, CO2, C_nH_{2n+2}	
			Liquids	C_nH_{2n+2}
		Char	C	
	Combustion			
		Gases	CO_2, H_2O	

systems and natural environment. Most of crop production is used as foods, but in the last century (the 20th century) crops were also cultivated for nonfood use—examples are pharmaceutical and nutritional products, chemical derivative products such as adhesive, paints, polymer, plastics, and industrial oils in forms of biodiesel, transmission fluids, and lubricants.

Thus, by definition, energy crops are plants grown specifically for use as a fuel. Although growing crops for fuel dates from mediaeval times, in their modern form energy crops are the most recent and innovative renewable energy option. Energy crops are important as a renewable energy technology because their use will produce a variety of economic, environmental, and energy benefits. Commercial energy crops are typically densely planted, high-yielding crop species where the energy crops will be burnt to generate power. Woody crops such as willow and poplar are widely utilized, as well as tropical grasses such as *Miscanthus* and elephant grass (*Pennisetum purpureum*).

Grasses are usually herbaceous plants with narrow leaves growing from the base. They include the true grasses of the Poaceae (or Gramineae) family, as well as the sedges (Cyperaceae) and the rushes (Juncaceae). The true grasses include cereals, bamboo, and the grasses of lawns (turf) and grassland. Sedges include many wild marsh and grassland plants, and some cultivated ones such as water chestnut (*Eleocharis dulcis*) and papyrus sedge (*Cyperus papyrus*). Most of the interest in grass biomass tends to focus on economics, but there is a list of traits that should be considered and valued when evaluating a potential solid biomass energy source. These traits are beneficial to society in general or impact the suitability of biomass for a farm operation.

If the carbohydrate content is desired for the production of biogas, whole crops, such as maize, Sudan grass, millet, white sweet clover, and many others, can be made into silage after which they can converted into biogas. On the other hand, crop residues are the residues remaining after crops have been harvested. Crop residues typically contain 40 w/w% of the nitrogen (N), 80 w/w% of the potassium (K), and 10 w/w% of the phosphorus (P) applied to the soil in the form of fertilizer. If these residues are subjected to direct combustion for energy, only a small percentage of the nutrients are left in the ash.

The valuable portion of sugar and starch crops (in terms of biofuel production) is the stalks and leaves, which are composed mainly of cellulose. The individual six-carbon sugar units in cellulose are linked together in extremely long chains by a stronger chemical bond that exists in starch. In starch crops, most of the six-carbon sugar units are linked together in long, branched chains (starch). Yeast cannot use these chains to produce ethanol. The starch chains must be broken down into individual six-carbon units or groups of two units. The starch conversion process is relatively simple because the bonds in the starch chain can be broken in an inexpensive manner by the use of heat and enzymes, or by a mild acid solution.

Sugar crops include a variety of plants such as fodder beets, fruit crops, Jerusalem artichokes, sugar beets, sugarcane, and sweet *Sorghum*. Interest in ethanol production from such agricultural crops has prompted the development of sugar crops that have not been cultivated on a widespread commercial basis in many countries. Preparation is basically a crushing and extraction of the sugars, which the yeast can immediately use. But sugar crops must be dealt with fairly quickly before their high sugar and water content causes spoilage. Because of the danger of such spoilage, the storage of sugar crops is not practical.

As with starch, cellulose must be broken down into sugar units before it can be used by yeast to make ethanol. However, the breaking of the cellulose bonds is much more complex and costly than the breaking of the starch bonds. Breaking the cellulose into individual sugar units is complicated by the presence of lignin, a complex compound surrounding cellulose, which is even more resistant than cellulose to enzymatic or acidic pretreatment (Hwang and Obst, 2003). Because of the high cost of converting liquefied cellulose into fermentable sugars, agricultural residues (as well as other crops having a high percentage of cellulose) are not yet a practical feedstock source for small ethanol plants.

Crop residues (cobs, stems, leaves, in particularly straw and other plant matter) left in agricultural fields after harvest could potentially be used for solid biofuel production. Due to high energy content, straw is one of the best crop residues for solid biofuels. However, straw has several disadvantages—it has a higher ash content, which results in lower calorific value. In order to improve its bulk density, the straw is generally baled before transportation. Straw burning requires a specific technology. There are four basic types of straw burners: those that accept shredded, loose straw; burners that use densified straw products such as pellets, briquettes, or cubes and straw logs; small, square bale burners and round bale burners. To be suitable for heat and electricity production, straw should not have a large content of moisture, preferably not more than 20% as the moisture reduces the boiler efficiency. Also straw color as well as straw chemistry should be considered before burning as it indicates the quality of the straw. Most crop residues are returned to the soil, and the humus resulting from their decomposition helps maintain soil nutrients, soil porosity, water infiltration and storage, as well as reducing soil erosion.

Wood is the oldest fuel known to man. Burning wood rather than fossil fuels can reduce the carbon dioxide emissions responsible for global climate change. Wood fuel is carbon dioxide (CO_2) neutral. It gives off only as much carbon dioxide when burned as it stores during its lifetime. In addition, wood fuel has very low levels of sulfur, a chemical that contributes to acid rain.

Regularly coppiced plantations will actually absorb more carbon dioxide than mature trees—since carbon dioxide absorption slows once a tree has grown. Growing crops for fuel, particularly wood coppice, offers very promising developments for the future. Short-rotation arable coppicing, using fast growing willows, is currently seen as an important source of fuel for electricity generation. The overall process involves several stages—growing over 2 or 3 years, cutting and converting to wood chip, storage and drying, and transport to a power plant for combustion. And the combustion process can be very efficient, given the development of advanced cogeneration techniques.

Energy crop fuel contains almost no sulfur and has significantly less nitrogen than fossil fuels; therefore reductions in pollutants causing acid rain (SO_2) and smog (NO_x) may be realized. For example, the use of energy crops will greatly reduce greenhouse gas emissions. Burning fossil fuels removes carbon that is stored underground and transfers it to the atmosphere. Burning energy crops, on the other hand, releases carbon dioxide but as their growth requires carbon dioxide, there is no net release of carbon into the atmosphere, that is, it creates a closed carbon cycle. Furthermore, where energy crops are gasified, there is a net reduction of carbon dioxide. In addition, substantial quantities of carbon can be captured in the soil through energy crop root structures, creating a net carbon sink.

An additional environmental benefit is in water quality, as energy crop fuel contains less mercury than coal. Also, energy crop farms using environmentally proactive designs will create water quality filtration zones as well as the uptake and sequestering pollutants such as phosphorus from soils that leach into water bodies. Also, growing energy crops on agricultural land that might otherwise be converted to residential or industrial use will reduce erosion/chemical runoff and enhance wildlife habitat. This will give energy producers and consumers will have available a renewable energy option with uniquely desirable characteristics. For example, energy crops differ from other sources of renewable energy in virtue of the fact that they can be grown to meet the needs of the market, whereas other renewable resources (e.g., wind and wave power) must be harnessed where and when they occur.

Energy crops are low-cost and low-maintenance crops grown solely for energy production (not for food). The crops are processed into solid, liquid, or gaseous fuels, such as pellets, bioethanol, or biogas. The fuels are burned to generate power or heat. Furthermore, the term energy crops can be used both for biomass crops that simply provide high output of biomass per hectare for low inputs, and for those that provide specific products that can be converted into other biofuels such as sugar or starch for bioethanol by fermentation, or into vegetable oil for biodiesel by transesterification. Energy crops such as grasses, *Miscanthus* oilseed crops, short-rotation woody crops, residual herbaceous biomass, starch crops, sugar crops, and switchgrass can be converted to liquid biofuels by biochemical and thermochemical conversion process (Table 10.6).

In the biochemical process, bacteria, yeasts, and enzymes also break down carbohydrates. For example, the fermentation process used to make wine, changes biomass liquids into alcohol, a combustible fuel. A similar process is used to turn corn into ethanol, which is mixed with gasoline to make gasohol. Also, when bacteria break down biomass, methane and carbon dioxide are produced. This methane can be captured, in sewage treatment plants and landfills, for example, and burned for

Table 10.6 Summary of the methods for the conversion of biomass to fuels.

Biomass			
	Extraction	Transesterification	Biodiesel
	Hydrolysis	Fermentation	Biogas Ethanol
	Gasification	Synthesis gas	Biogas Hydrogen Methanol Ethanol
	Pyrolysis		Hydrogen Bio-oil
	Hydrotreating		Diesel

heat and power. Also, biomass oils, such as soybean and canola oils, can be chemically converted into a liquid fuel similar to diesel fuel, and into gasoline additives. Used cooking oil has been used as a source to make biodiesel.

In the thermochemical process the plant matter is broken down into gaseous products, liquid products, and a carbonaceous solid (commonly referred to as char). These products can then be processed further and refined into useful fuels such as methane and alcohol. Another approach is to take these fuels and run them through fuel cells, converting the hydrogen-rich fuels into electricity and water, with few or no emissions. However, the direct conversion thermal processes, such as combustion, may encounter the same problems as those encountered when coal is the feedstock (Speight, 2013). The conversion of biomass into other useful forms such as gaseous fuels or liquid fuels is considered as an alternative way to make use of biomass energy. Perennial crops that regenerate annually from buds at the base of the plant offer the greatest potential for energy-efficient production. These include (1) cordgrass and switchgrass, (2) Jerusalem artichoke, (3) *Miscanthus*, (4) reed plants, (5) residual herbaceous biomass, (6) SRC, and (7) *Sorghum*.

10.2.6.1 Cordgrass and switchgrass

Cordgrass (genus *Spartina*), also called marsh grass or salt grass, is a genus of 16 species of perennial grasses in the family Poaceae, which is found on marshes and tidal mud flats of North American, Europe, and Africa and often forms dense colonies. Some species are planted as soil binders to prevent erosion and a few are considered invasive species in areas outside their native range. Prairie cordgrass (*Spartina pectinata*) and gulf cordgrass (*Spartina spartinae*) are the most widely distributed North American species.

Cordgrasses are erect, tough, long-leaved plants that range from 1 to 10 ft in height. Most species grow in clumps, with short flower spikes alternating along and often adherent to the upper portion of the stems. Many spread though rhizomes (underground stems) that send up new plants. One of the variations of cordgrass—referred to as smooth cordgrass—has smooth, blade-like leaves that taper to a point. The leaves grow 12–20 in. in length and 0.5 in. in wide and has round, hollow stems and a strong, interconnected root system. Smooth cordgrass grows in two forms: a short form that grows to 2 ft tall, and a tall form that can reach 7 ft.

Switchgrass (*Panicum virgatum*) is a perennial sod-forming grass with thick strong stems. It is a perennial warm-season bunchgrass that is native to North America, where it occurs naturally from Canada southwards into the United States and Mexico. Switchgrass is one of the dominant species of the central North American tallgrass prairie and can be found in prairie remnants (grassland areas in the Western and Midwestern United States and Canada that remain, to some extent, undisturbed), in native grass pastures, and naturalized along roadsides.

The advantages of switchgrass as an energy crop are that it is fast growing, remarkably adaptable, and high yielding. Further advantages of switchgrass are that it can be harvested, using conventional equipment, either annually or semiannually for 10 years or more before replanting is needed and that it is able to reach deep into the soil for water and use water very efficiently.

Further advantages of switchgrass are that it can be harvested, using conventional equipment, either annually or semiannually for 10 years or more before replanting is needed and that it is able to reach deep into the soil for water and use water very efficiently. Besides showing great promise as an energy crop for energy production, switchgrass also restores vital organic nutrients to farmed-out soils and with its extensive network of stems and roots (the plants extend nearly as far below ground as above), it is also a valuable soil stabilization plant.

Switchgrass has the potential to be a versatile bioenergy feedstock since the energy content is comparable to that of wood with significantly lower initial moisture content. Switchgrass is very suitable substrate and produces high ethanol yield using current simultaneous saccharification and fermentation technology. Extensive analysis of ash and alkali content of switchgrass indicates that it typically has relatively low alkali content and should have low slagging potential in coal-fired combustion systems. As an agro-fiber source for pulping, switchgrass has a relatively high cellulose content, low ash content, and good fiber-length-to-width ratios. Switchgrass reaches its full yield potential after the third year planted, producing approximately 6–8 ton/ac; that is, 500 gal of ethanol/ac.

The utilization of energy crops such as switchgrass (*Panicum virgatum*, L., Poaceae) is a concept with great relevance to current ecological and economic issues on a global scale. Development of a significant national capacity to utilize perennial forage crops, such as switchgrass as biofuels could provide an important new source of energy from perennial cropping systems, which are compatible with conventional farming practices, would help reduce degradation of agricultural soils, lower national dependence on foreign oil.

10.2.6.2 Jerusalem artichoke

The Jerusalem artichoke (*Helianthus tuberosus*, also called sunroot, sunchoke, or earth apple) is a species of sunflower (that is native to central North America), which is a herbaceous perennial plant that grows up to 5–10 ft tall with opposite leaves on the upper part of the stem but alternate below. The leaves have a rough, hairy texture. Larger leaves on the lower stem are broad-ovoid-acute and can be up to 12 in. long, while the leaves higher on the stem are smaller and narrower.

The tubers are often elongated and uneven, typically 3–4 in. long and 1–2.2 in. thick with a crisp and crunchy texture when raw. They vary in color from pale brown to white, red, or purple.

The Jerusalem artichoke has shown excellent potential as an alternative sugar crop. A member of the sunflower family, this crop is native to North America and well adapted to northern climates. Like the sugar beet, the Jerusalem artichoke produces sugar in the top growth and stores it in the roots and tuber. It can grow in a variety of soils, and it is not demanding of soil fertility. The Jerusalem artichoke is a perennial; small tubers left in the field will produce the next season's crop, so no ploughing or seeding is necessary. The high-fructose syrups that can be derived from the tubers produced by the Jerusalem artichoke may be used for the production of ethanol and other industrial raw materials. Jerusalem Artichokes also produce a large amount of top growth, which may also prove to be a useful source of biomass for energy purposes.

10.2.6.3 Miscanthus

Miscanthus (also called silvergrass) is a hardy perennial grass that produces a crop of bamboo-like cane up to 15-ft tall. *Miscanthus* is high in lignin and lignocellulose fiber. Lignocellulose is the term used to describe the three-dimensional polymeric composites formed by plants as structural material. It consists of variable amounts of cellulose, hemicellulose, and lignin.

Briefly, lignocellulosic feedstocks are composed primarily of carbohydrate polymers (cellulose and hemicellulose) and phenolic polymers (lignin). Lower concentrations of various other compounds, such as proteins, acids, salts, and minerals, are also present. Cellulose and hemicellulose, which typically make up two-thirds of cell wall dry matter (dry matter: the portion of biomass that is not water), are polysaccharides that can be hydrolyzed to sugars and then fermented to ethanol. Process performance, in this case ethanol yield from biomass, is directly related to cellulose, hemicellulose, and individual sugar concentration in the feedstock. Lignin cannot be used in fermentation processes; however, it may be useful for other purposes.

Miscanthus can be grown in a cool climate and on many types of arable land. *Miscanthus* does not require a big input of fertilizers due to its capability to recycle large amounts of nutrients. *Miscanthus* has a similar calorific value per unit weight as wood and therefore could possibly be used in the same power plant or those designed for agricultural residues.

Miscanthus is well equipped for high productivity under relatively cool temperatures and may require substantial amounts of water for maximal growth (its growth could therefore also have valuable environmental benefits by acting as absorbing disposal areas for wastewater and some industrial effluents). Furthermore, *Miscanthus* seems to grow well in most soil conditions (thin droughty soils) but appears to thrive within areas, which are currently best suited to maize production. The advantages of *Miscanthus* as an energy crop are that it multiplies very rapidly, has a high yield which is relatively dry, and can be harvested annually (from its second season onwards) compared with every 2–4 years for SRC. Further advantages are that *Miscanthus* can be grown and harvested with existing farm machinery, it requires little or no pesticide/fertilizer input after establishment and the harvest can use the same infrastructure for storage and transport as SRC. Finally, *Miscanthus* has a similar calorific value per unit weight as wood and therefore could possibly be used in the same power plant or those designed for agricultural residues.

10.2.6.4 Reed plants

Reed is a common name for several tall, grass-like plants that are commonly found in wetlands. Reed plants are a potentially prolific producer of biomass, capable of yielding 20–25 ton/ha (2.47 ac) of dry matter annually for a number of years. They can grow up to 6 m, are spread by means of stout rhizomes (continuously growing horizontal underground stems, which puts out lateral shoots and adventitious roots at intervals) and stolons (also known as runners, which are horizontal connections between organisms), and are commonly found in swampy ground and shallow water throughout temperate and subtropical areas.

Reed canary grass is a robust perennial grass, widely distributed across temperate regions of Europe, Asia, and also North America. It occurs in wet places, along the margins of rivers, streams, lakes, and pools. The species spreads naturally by creeping rhizomes, but plants can also be raised from seed. The advantages of Reed Canary grass as an energy crop are its good adaptation to cool temperate climates and poor wet soil conditions and, conversely, its ability to withstand drought. Crucially, for the purposes of biomass production, reed canary grass is also able to attain high dry matter content earlier than *Miscanthus*. The crop responds well to nitrogen and phosphate and it may be used in a bed system to remove nutrients from wastewater, as well as to stabilize areas at risk of soil erosion.

10.2.6.5 Residual herbaceous biomass

Residual herbaceous biomass (straw) is the main residual herbaceous material for energy application. As it is a residual product, its availability for energy purposes is driven by the cereal markets and does not have autonomous market behavior. In addition, farms consume significant quantities of straw internally—as bed material for livestock, grain drying, etc. Some straw is also chaffed and returned back to the field as soil ameliorator. The net straw yield per hectare for energy application also depends on the crop, the grain yield per hectare, climate and cultivation conditions, etc. Nevertheless, one can roughly estimate that the average straw yield per hectare is approximately 50%–65% of the grain yield per hectare from cereals and oilseeds.

Similar to herbaceous crops, straw usually has lower moisture content than woody biomass. Conversely, it has a lower calorific value, bulk density, ash melting point and higher content of ash, and problematic inorganic component such as chlorine, potassium, and sulfur, which cause corrosion and pollution. The last two drawbacks can be relatively easily overcome by leaving straw on the field for a while. In such a way rainfall provides a natural leaching process and separates a large part of the potassium and the chlorine. Alternatively, fresh straw can be directly shipped to the gasification plant, where it is washed by dedicated facilities at moderate temperatures (50°C–60°C; 120°F–140°F). Due to washing, the initially low moisture content of straw becomes higher in both cases; hence, a mandatory drying is applied afterward. In both cases also the content of corrosive components is reduced, but not completely taken out. In order to decrease handling costs, straw and dedicated herbaceous energy crops are usually baled before being shipped to the gasification plant. The weight and the size of bales depend on the baling equipment and on the requirements of the gasification plant (Luque and Speight, 2015).

The simplest form of agricultural biomass energy use involves direct combustion of cellulosic crops or residues, such as hay, straw, or corn fodder, to heat space or produce steam. Such fuels are useful for heating farm buildings and small commercial buildings in rural areas and for drying crops. Ideally, energy crops should be produced on land not needed for food production. This use should not increase the erosion hazard or cause other environmental damage. On the other hand, a variety of crops can be grown specifically to provide sources of energy and, once established, a stand of perennial biomass/energy crop is expected to remain productive for a period of 6 years or more.

10.2.6.6 Short-rotation coppice

Short-rotation woody crops (SRC) refer to fast growing deciduous trees, which are grown as energy crops, such as willow and poplar trees. The species of SRC that are most suitable, and therefore most popular, for use as energy crops are poplar and willow (and possibly also birch) because they both require deep, moisture-retentive soils for proper growth. Willow, in particular, is able to endure periods of water logging and is therefore better suited to wetter soils.

The species of SRC that are most suitable, and therefore most popular, for use as energy crops are poplar and willow (and possibly also birch) because they both require deep, moisture-retentive soils for proper growth. Willow, in particular, is able to endure periods of water logging and is therefore better suited to wetter soils. SRC is harvested during winter when the dry matter percentage of the coppice is at its highest and it is then bundled or immediately chipped. It may then be stored for a few weeks in order to reduce its moisture content to a satisfactory level for use in energy production. Dry SRC can then be burnt under controlled conditions to produce other fuels, gas, or liquid, which are then used for electricity generation.

10.2.6.7 Sorghum

Sorghum is an annual tropical grass with large genetic variation that is a crop with the potential for energy production. It is a genus of flowering plants in the grass family Poaceae. Seventeen of the 25 species are native to Australia with the range of some extending to Africa, Asia, and Central America as well as to islands in the Indian Ocean and the Pacific Ocean. One species is grown for grain, while many others are used as fodder plants, either cultivated in warm climates throughout the world or naturalized, in pasture lands.

Sweet *Sorghum* has been selected for its sugar content and is normally grown for molasses production. Forage *Sorghum* has been selected for high yields of reasonably good-quality animal feed. *Sorghum* varieties producing tall plants with large stems make the best candidates for biomass production. Both sweet and forage *Sorghum* have a high potential for lodging. Lodging can result in harvest problems with ensuing loss of yield from both initial and ratoon crops.

Sweet *Sorghum* is a name given to varieties of a species of *Sorghum*. This crop has been cultivated on a small scale in the past for the production of table syrup, but other varieties can be grown for the production of sugar. The most common types of *Sorghum* species are those used for the production of grain. Sweet *Sorghum* can be considered as an energy crop, because it can be grown in all continents, in tropical, subtropical, temperate regions as well as in poor-quality soils. Sweet *Sorghum* is a warm-season crop that matures earlier under high temperatures and short days. Sweet *Sorghum* is an extraordinarily promising multifunctional crop not only for its high economic value but also for its capacity to provide a very wide range of renewable energy products, industrial commodities, and food and animal feed products. Sweet *Sorghum* biomass is rich in readily fermentable sugars and thus it can be considered as an excellent raw material for fermentative hydrogen production—hydrogen is an important commodity for the refining industry and new sources are continually sought (Parkash, 2003; Gary et al., 2007; Speight, 2014, 2017; Hsu and Robinson, 2017). Sweet *Sorghum* crops produce sugar syrups that could form the basis of fermentation processes for methane or ethanol production, and some of the forage types of the plant may be suitable for biomass production.

10.2.7 Wood

Wood is a porous and fibrous structural tissue found in the stems and roots of trees and other woody plants. It is a natural composite of cellulose fibers that are strong in tension and embedded in a matrix of lignin that resists compression. Wood is sometimes defined as only the secondary xylem in the stems of trees, or it is defined more broadly to include the same type of tissue elsewhere such as in the roots of trees or shrubs. In a living tree, it performs a support function, enabling woody plants to grow large or to stand up by themselves. Wood also conveys water and nutrients between the leaves, other growing tissues, and the roots. As a result of this structure, wood has lent itself to a variety of uses throughout recorded history.

10.2.7.1 History

The use of wood by humans from biblical times to modern computer applications are reviewed here. In peace and in war, in the Old and the New World, man first turned to wood for his basic needs and later learned to use advanced science to employ wood as his most sophisticated raw material, being infinitely versatile and an easily renewable source.

The manner in which wood was used by early cultures is difficult to determine, as wood artifacts have largely disappeared. Certainly the use of wood for fire is one of the first and most significant contributions of this resource to the development of society. No doubt man built early pole structures from the small trees growing along the rivers and later he would build more solid structures from planks, turf, mud, and adobe. The Scandinavians developed the basic principles of timber framing, which were probably known in Europe in the Bronze Age and framing eventually became the preeminent method of wood building in the Western World, reflecting developments in structural engineering that had been worked out with wood mostly through trial and error.

One of the first uses of wood for water transport was probably a hollowed-out log. Around 4000 BC, the Egyptians were making ships from bundles of reeds and their earliest wooden boats copied the hull frame of the reed boats. For larger vessels the Egyptians imported cedar from Lebanon. One reason for the northward expansion of Egypt's influence was to ensure its cedar supply. Records show that the Egyptian shipbuilder could use wood on a grand scale. Queen Hatshepsut's barge, built in 1500 BC to transport granite obelisks from Aswan to Thebes, had a displacement of some 7500 ton, and 30 oar-powered tugs were needed to tow it. According to Theophrastus, a pupil of Aristotle, the ship-building woods in Ancient Greece were are silver fir, fir, and cedar. Silver fir is used for lightness; for merchant ships, fir is used because of its resistance to rot. In Syria and Phoenicia, cedar is used because of the lack of fir.

Technological improvement in land transport was slower than that of water transport. From 7000 BC onward, wood sledges were used for heavy loads such as stones, and archeologists reason that the massive stones in the great monument at Stonehenge on Salisbury Plain, England, must have been moved on sledges placed on rollers, which may have inspired the discovery of the wheel. Thus another significant contribution of wood to the ancient world was for war devices. Examples include the catapult, which enabled a man to attack his enemy from a safe distance, the battering ram and scaling ladder, the tortoise, and the siege tower. Although the choice of materials for these purposes was quite limited, the properties of wood made it eminently suitable. High strength and low weight were highly valued characteristics of wood then, just as they are currently. These siege engines were integral to the expansion of both Greek and Roman civilizations and of the science, technology, and philosophy that developed under the tutelage of the great thinkers and teachers of the times.

Ancient man was using wood to conquer his world as well as build it and explore it. Then some unknown woodman in Ancient Greece invented a primitive wooden lathe, and man found himself on the threshold of the age of machines. When he entered that age, he would find ways to make wood work for him to unprecedented degrees. From the basic concept of the lathe and the ability to shape wood to circular symmetry developed new concepts of both materials use and machine development.

In Europe the water-and-wood phase reached a high plateau around the 16th century with the work of Leonardo da Vinci and his talented contemporaries. At this time the availability of timber diminished, particularly in many European countries. The scarcity was caused by the expansion of agriculture, the increasing use of wood as a structural material and fuel, and from growing demands of the smelting furnaces. To smelt one cannon took several tons of wood. By the 17th century Europeans were turning to coal for the domestic hearth, and when the secret of smelting metal with coal was discovered, coal became the unique basis for industrial technology until late in the 19th century.

In the early 19th-century America, a seemingly inexhaustible supply of timber existed. The technology here was geared to exploiting the use of all natural resources to make up for the scarcity in capital and labor. But the technological advances of the 19th century, along with the increasing population, would have a major impact on American forests. Railroads, telegraph lines, charcoal-fueled steel mills, and other industries were consuming immense quantities of wood. The Civil War made a heavy demand, too. One gun factory alone used 28,000 walnut trees for gunstocks. During the latter half of the 19th century, the volume of lumber produced each year rose from 4000 million board feet to approximately 35,000 million.

Up to the latter part of the 19th century, no appreciable systematic research on wood occurred—no research of the type we now call wood science. Wood had been used by early experimenters to make instruments and other research equipment, and early engineers had used it as a construction material and a material with which to work out engineering problems and designs. Methods for pulping wood to make paper had been worked out by the paper industry, too. Further, both cotton and wood had been used by chemists as a source of cellulose for man-made fibers. This led to work on cellulose acetate reactions with solvents that led to the ability to produce that compound as both film and fiber. These advances provided a base for the subsequent technology of nylon and established the principles by which countless numbers and kinds of linear high polymers can be synthesized. The carriage business provided an early milestone for a new era of wood research. In 1889 the Carriage Builders Association has concerns related to the scarcity of northern oak, a species long preferred for their craft.

During the World War II, wood research covered the whole gamut of possible wartime uses of wood but after the war the importance of timber products declined, on a relative scale, as the importance of minerals increased, due in part to abundant low-cost energy in the form of coal and then crude oil. It is worth noting, however, that tonnage of timber products produced in the United States then exceeded that of all metals and plastics combined, just as it does currently. So, while timber declined in relative importance and public awareness, it remained the major product of American manufacture.

10.2.7.2 Types of wood

The amount and types of wood fuel used vary considerably between regions, mainly due to different local situations and conditions. However, some of the use of wood is derived from the use of black liquor from the pulp and paper industries.

Composition of black liquor

Element	**w/w%**
Carbon	35.7
Hydrogen	3.7
Nitrogen	≥0.1
Oxygen	35.8
Sulfur	4.4
Chlorine	0.3
Potassium	1.1
Sodium	19.0

Source: Kavalov, B., Peteves, S.D., 2005. Status and Perspectives of Biomass-to-Liquid Fuels in the European Union. European Commission, Directorate General Joint Research Centre (DG JRC), Institute for Energy, Petten, The Netherlands (Kavalov and Peteves, 2005).

The quality of wood fuels is determined for fuel types by choosing for each delivery batch the limit values for the energy density, moisture content, and particle size of the fuel as received from the quality. The energy density is dependent on net calorific value (Table 10.7), moisture content, bulk density, and particle size of the fuel concerned. When choosing limit values for energy density and moisture, the interdependence of different characteristics should be considered by using characteristic values typical of different wood fuels. In deliveries of different wood fuel blends or mixtures (e.g., bark/sawdust, cutter chips/grinding dust/other wood residues), the parties should agree upon the application of quality classification and the quality determination of fuel considering the use and safety issues. Quality limits for other characteristics of the fuel can be specified case by case for mechanical properties of the fuel (i.e., oversize particles) and other properties. When preparing delivery agreements, differences in properties due to possible seasonal variations should also be considered and agreed by parties separately.

Table 10.7 The gross calorific value of wood as one of a variety of fuels.[a]

Fuel	Gross calorific value, Btu/lb[a]
Alcohol, 96%	
Anthracite	14,000–14,500
Bituminous coal	7300–10,000
Butane	20,900
Charcoal	12,800
Coal	8000–14,000
Diesel	19,300
Ethanol	12,800
Gasoline	20,400
Gasoline	7000
Kerosene	154,000
Wood (dry)	5500–8500

1 kJ/kg = 1 J/g = 0.4299 Btu/lb_m = 0.23884 kcal/kg; 1 Btu/lb = 2.326 kJ/kg = 0.55 kcal/kg; 1 kcal/kg = 4.1868 kJ/kg = 1.8 Btu/lb_m.
[a]The gross calorific value assumes that the water of combustion is entirely condensed and that the heat contained in the water vapor is recovered.

The components of wood include cellulose, hemicellulose, lignin, extractives, lipids, proteins, simple sugars, starches, water, hydrocarbon derivatives, ash, and other compounds. The proportion of these wood constituents varies between species, and there are distinct differences between hardwoods and softwoods.

10.2.7.2.1 Hardwood

As the name suggests, the wood from these trees is generally harder than that of softwoods, but there are significant exceptions. In both groups there is an enormous variation in actual wood hardness, with the range in density in hardwoods completely, including that of softwoods; some hardwoods (such as balsa) are softer than most softwoods, while yew is an example of a hard softwood. Trees grown in tropical climates are generally hardwood. Hardwood grows faster than softwood but have shorter fibers compared to softwood. Hardwoods or deciduous woods have a higher proportion of cellulose, hemicelluloses, and extractives than softwoods, but softwoods have a higher proportion of lignin (Table 10.8). Generally, hardwoods that provide long-burning fires contain the greatest total heating value per unit of volume.

Hardwoods have a more complex structure than softwoods and are often much slower growing as a result. The dominant feature separating "hardwoods" from softwoods is the presence of pores, or vessels. The vessels may show considerable variation in size, shape of perforation plates (simple, scalariform, reticulate, foraminate), and structure of cell wall, such as spiral thickenings.

Hardwoods are employed in a large range of applications, including fuel, tools, construction, and the manufacture of charcoal. Solid hardwood joinery tends to be expensive compared to softwood. In the past, tropical hardwoods were easily available, but the supply of some species, such as teak and mahogany, is now becoming scarce due to overexploitation. Hardwoods may be used in a variety of objects but

Table 10.8 Composition of different biomass types (w/w%, dry basis).

Type	Cellulose	Hemicellulose	Lignin	Others	Ash
Soft wood	41	24	28	2	0.4
Hard wood	39	35	20	3	0.3
Pine bark	34	16	34	14	2
Straw (wheat)	40	28	17	11	7
Rice husks	30	25	12	18	16
Peat	10	32	44	11	6

are most frequently seen in furniture or musical instruments because of their density, which adds to durability, appearance, and performance. Different species of hardwood lend themselves to different end uses or construction processes due to the variety of characteristics apparent in different timbers, including density, grain, pore size, growth and fiber pattern, flexibility, and ability to be steam bent.

10.2.7.2.2 Softwood

Softwood is usually wood from gymnosperm trees such as pine and spruce, which often reproduce using cones and occasionally nuts. The trees classified as softwoods have needle- or scale-like leaves that, with a few exceptions, remain on the tree all through the year. Hence, softwood trees are sometimes called evergreens. Botanically, they are known as gymnosperms and instead of bearing seeds from flowers, gymnosperms have exposed seeds in cones.

Within the softwood and hardwood groups, there is a considerable variation in actual wood hardness, the range of density in hardwoods completely, including that of softwoods. Some hardwoods (such as balsa) are softer than most softwoods, while the hardest hardwoods are much harder than any softwood. In short, the terms softwood and hardwood are archaic with questionable meaning and often belie the properties of the wood.

Softwoods are generally most used by the construction industry and are also used to produce paper pulp, and card products. In many of these applications, there is a constant need for density and thickness monitoring, and gamma-ray sensors have shown good performance in this case. Certain species of softwood are more resistant to insect attack from woodworm, as certain insects prefer damp hardwood. Softwoods that give a fast burning, cracking blaze are less dense and contain less total heating value per unit of volume.

10.2.8 Composition and properties

Wood is the hard, fibrous substance found beneath bark in the stems and branches of trees and shrubs. Practically all commercial wood, however, comes from trees. It is plentiful and replaceable. Since a new tree can be grown where one has been cut, wood has been called the world's only renewable natural resource. Wood consists of cellulose ($C_6H_{19}O_5$), resins, lignin, various inorganic//salts, and water, which is reflected in the ultimate analysis of wood (Table 10.9). The quantity of water present has great effect on the heating value and ranges from 25 to 50 w/w% in green wood, and from 10 to 20 w/w% in air-dried wood (Table 10.10).

Wood cut in the spring and summer contains more water than that cut in the early part of the winter. A cord (8 ft long by 4 ft wide by 4 ft high) of hard wood, such as ash or maple, is approximately equal in heating value to 1 ton of bituminous coal; soft woods, such as pine and poplar, have less than half this amount. Wood burns with a long flame and is kindled, the fire quickly reaches its maximum intensity, and a relatively small quantity of ash is formed. Wood is too expensive for industrial use, except in a few special cases, where freedom from dirt and smoke is necessary. Of other cellulose materials, shavings, sawdust, and straw are used for fuel in some places. They are bulky and difficult to handle, while their heat value, which depends on the amount of moisture they contain, is seldom more than from one-third to one-half that of good coal. Such waste matter as spent tanbark and bagasse (crushed sugarcane), and the pulp from sugar beets is sometimes used for fuel for evaporation of the steam, but owing to the large amount of moisture they contain, the heat value is very low.

All woods dried to the same moisture content contain approximately the same heat value per pound—from 6200 to 7500 Btu for fully dried wood and 5500 to 8500 Btu for air-dried seasoned wood (Table 10.7). However, the heat content of any fire depends on wood density, resin, ash, and moisture. A general rule for estimating heat value of firewood is one cord of well-seasoned hardwood (weighing approximately 2 ton) burned in an airtight, draft-controlled wood stove with a 55%–65% efficiency is equivalent to approximately 175 gal of no. 2 fuel oil or 225 thm of natural gas consumed in normal furnaces having 65%–75% efficiencies. There are also differences in types of wood. Softwoods usually contain a lot of resin that has high energy content so the total energy content of a softwood is usually higher than the energy content of a hardwood (often by approximately 5%). Also, softwood tends to burn up faster than hardwood and has other characteristics that reduce their attractiveness as fuel since the typical density of softwood is usually lower than the density of hardwood can be equated to less weight in a cord of softwood and the extra 5% of volatile fuel will not make up for loss in weight.

Briefly, a cord of wood is a stack of wood comprising 128 ft^3 (3.62 m^3)—the standard dimensions of the cord are 4 ft by 4 ft by 8 ft, including air space and bark and one cord contains approximately 1.2 US tons (oven-dry), which is equivalent to 2400 lb (1089 kg).

Table 10.9 Ultimate analysis of wood (w/w%, dry-ash-free).

Element	Average of 11 hardwoods	Average of 9 softwoods	Oak bark	Pine bark
C	50.2	52.7	52.6	54.9
H	6.2	6.3	5.7	5.8
O	43.5	40.8	41.5	39.0
N	0.1	0.2	0.1	0.2
S	–	0.0	0.1	0.1

Table 10.10 Properties of various woody feedstocks compared to coal and natural gas.

	Bituminous coal	Natural gas	Wood	Bark	Willow	Forest residues
Ash, w/w%	8.5–10.9	0	0.4–0.5	3.5–8	1.1–4.0	1–3
Moisture, w/w%	5–10	0	5–60	45–65	50–60	50–60
Volatile matter, w/w%	25–40	100	>70	70–77	>70	>70
Ash melting point, °C	1100–1400		1400–1700	1300–1700	n.a.	n.a.
C, w/w%	76–87	75	46–52	46–52	47–51	48–52
H, w/w%	3.5–5	24	6.2–604	4.6–6.8	5.8–6.7	6.0–6.2
N, w/w%	0.8–1.5	0.9	0.1–0.5	0.3–0.8	0.2–0.8	0.3–0.5
O w/w%	2.8–11.3	0.9	36–42	24.3–42.4	40–46	40–44
S, w/w%	0.5–3.1	0	<0.05	<0.05	0.02–0.10	<0.05
Cl, w/w%	<0.1		0.01–0.03	0.01–0.03	0.02–0.05	0.01–0.04
K, w/w%	0.003	–	0.02–0.05	0.1–0.4	0.2–0.5	0.1–004
Ca, w/w%	4–12		0.1–1.5	0.02–0.08	0.2–0.7	0.2–0.9

Two most important properties of any papermaking cellulosic raw material are (1) the amount of cellulose fiber and (2) the length of the fibers. The amount of cellulose fiber in wood determines the pulp yield, ease of pulping, and cost of pulp produced.

10.2.8.1 Chemical composition

The chemical composition of wood varies from species to species but is approximately 50 w/w% carbon, 42 w/w% oxygen, 6 w/w% hydrogen, 1 w/w% nitrogen, and 1 w/w% other elements (mainly calcium, potassium, sodium, magnesium, iron, and manganese).

10.2.8.1.1 Cellulose

Cellulose, the major chemical constituent of wood, is in many respects the most important. It is also the most easily defined and described. Wood cellulose is chemically defined as $(C_6H_{10}O_5)_n$. Cellulose is a high-molecular-weight, stereoregular, and linear polymer of repeating beta-D-glucopyranose units. Simply speaking, it is the chief structural element and major constituent of the cell wall of trees and plants. The empirical formula for cellulose is $(C_6H_{10}O_5)_n$ where n is the degree of polymerization (DP).

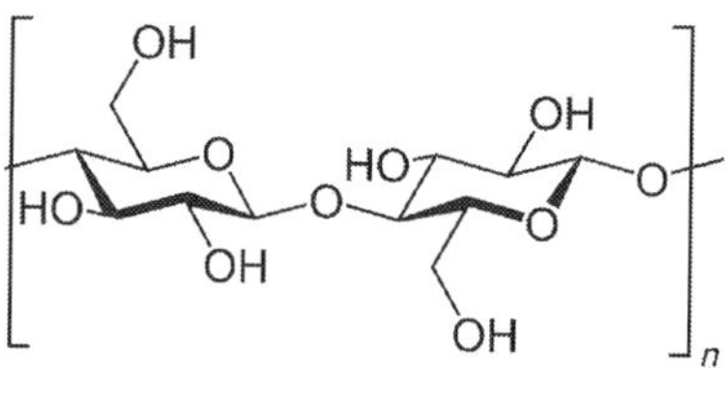

Generalized structure of cellulose.

10.2.8.1.2 Hemicellulose

Hemicellulose derivatives, which make up 20%–35% of the dry weight of wood, are the second important constituent of wood, are also sugar polymers; and unlike cellulose, which is made only from glucose, hemicelluloses consist of glucose and several other water-soluble sugar derivatives that are produced during photosynthesis. In the hemicellulose family the DP is lower than in cellulose and are composed of shorter molecular chains that are found in cellulose. There are many varieties of hemicelluloses, and they markedly differ in composition in softwoods and hardwoods—generally, hemicellulose derivatives are in a relatively greater proportion in hardwoods than in softwoods.

Hemicellulose (hemi-cellulose) is a constituent of woods that is, like cellulose, a polysaccharide, but less complex and easily hydrolysable. Hemicellulose derivatives are polysaccharides that are often associated with cellulose but have a very different compositions.

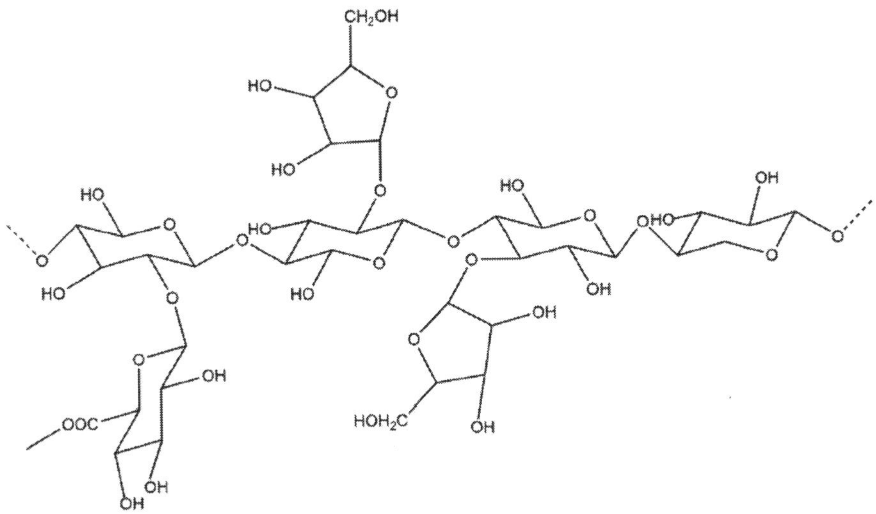

Generalized structure of hemicellulose.

Unlike cellulose, hemicellulose consists of 50–3000 sugar units as opposed to 7000–15,000 glucose molecules per polymer in cellulose. Hemicelluloses are classified according to the main sugar residue in the backbone as xylan derivatives, mannan derivatives, and glucan derivatives. Depending on the plant species, developmental stage, and tissue type, various subclasses of hemicellulose may exist, which may be grouped into two general categories based on the hydration of the fibers. Low-hydration polysaccharide derivatives function primarily to stabilize the cell wall through hydrogen-bonding interactions with cellulose and covalent interaction with lignin. They are water soluble due to their branched structure. The second type is hemicellulose derivatives composed mainly of hydrocolloids (often called gums, are hydrophilic polymers, of vegetable, animal, microbial or synthetic origin, that generally contain many hydroxyl groups and may be polyelectrolytes), which function primarily as an extracellular energy and raw material storage system and as a water-retention mechanism in seeds.

Hemicellulose derivatives contain many different sugar monomers, while cellulose only contains anhydrous glucose. For example, in addition to glucose, the sugar monomers in hemicellulose derivatives can include the five-carbon sugars xylose and arabinose, the six-carbon sugars mannose and galactose, and the six-carbon deoxy-sugar rhamnose.

Xylose is, in most cases, the sugar monomer present in the largest amount, although in softwoods mannose can be the most abundant sugar, leading to the production of ethanol (Keller, 1996; Galbe and Zacchi, 2002). Not only regular sugars can be found in hemicellulose, but also their acidified form, for instance, glucuronic acid and galacturonic acid can be present.

10.2.8.1.3 Lignin

Lignin is a complex constituent of the wood that cements the cellulose fibers together. Lignin is brown in color. Lignin is largely responsible for the strength and rigidity of plants. Lignin is a class of complex organic polymers that form key structural materials in the support tissues of vascular plants and some algae. Lignin is particularly important in the formation of cell walls, especially in wood and bark, because it does not rot easily. Chemically, lignins are cross-linked phenolic polymers.

The composition of lignin varies from species to species—as example of composition from an aspen sample has carbon 63.4 w/w%, hydrogen 5.9%, and oxygen 30% (by difference), mineral as 0.7% ash, which corresponding to the approximate formula $(C_{31}H_{34}O_{11})_n$. As a biopolymer, lignin is unusual because of the heterogeneity and lack of a defined primary structure. Its most commonly noted function is the support through strengthening of wood.

Structurally, lignin is a cross-linked polymer with a molecular mass in excess of 10,000. It is relatively hydrophobic and rich in aromatic subunits. The actual DP is difficult to measure, since the material is heterogeneous. Thus

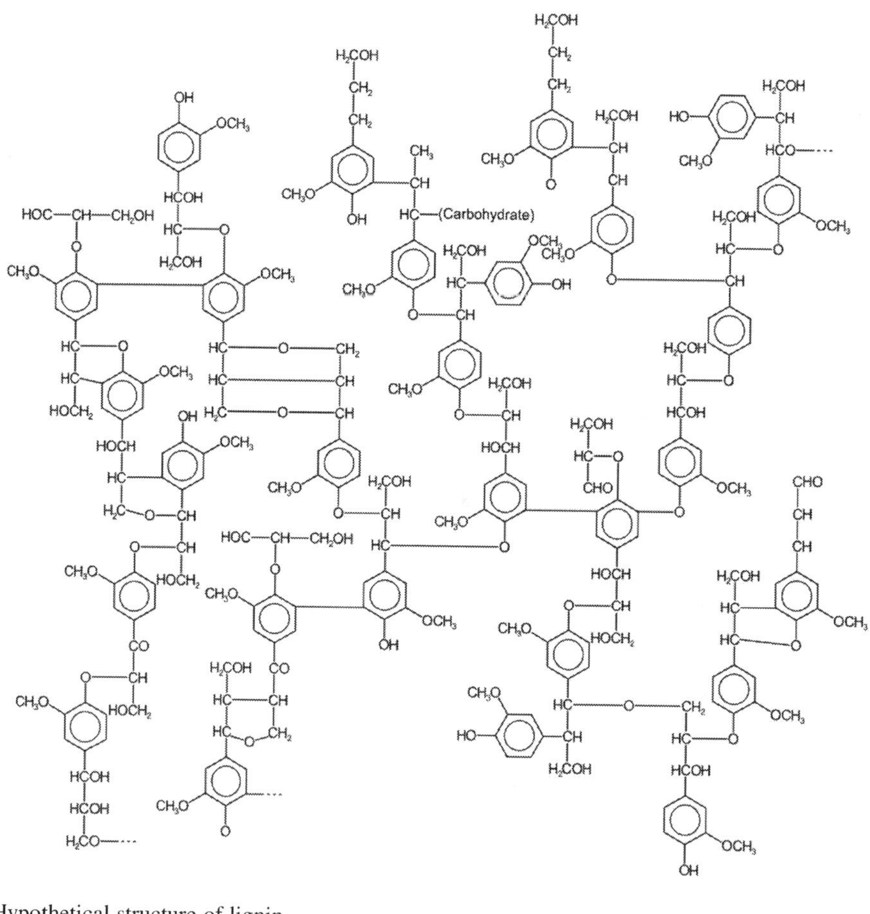

Hypothetical structure of lignin.

The lignol derivatives that cross-link are of three main types, all derived from phenylpropane and are 4-hydroxy-3-methoxy phenylpropane; 3,5-dimethoxy-4-hydroxy phenylpropane; and 4-hydroxy phenylpropane. Thus different types of lignin have been described depending on the means of isolation. The three common monolignols presented next are (1) *trans*-coniferyl alcohol, (2) *trans*-sinapyl alcohol, and (3) *trans-p*-coumaryl alcohol.

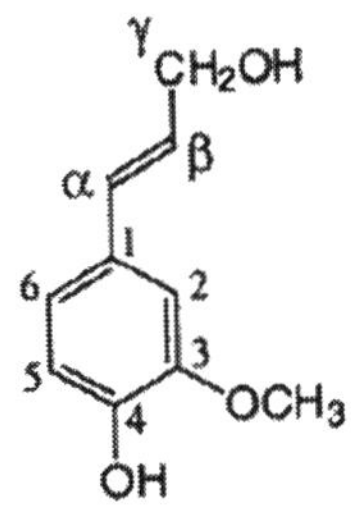

trans-Coniferyl alcohol.

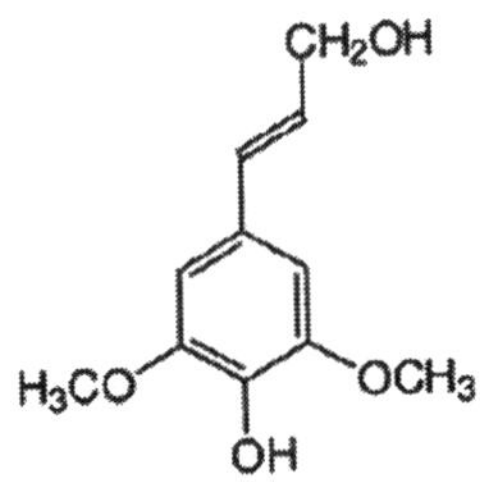

trans-Sinapyl alcohol.

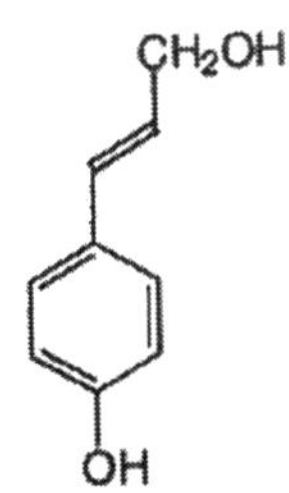

trans-p-Coumaryl alcohol.

Thus lignin can be defined as a polyphenolic material arising primarily from enzymic dehydrogenative polymerization of three phenylpropanoid units (*p*-hydroxy-cinnamyl alcohols). The proportions of the precursors in lignins vary with their botanical origin. The typical structural elements of softwood lignins are derived principally from *trans*-coniferyl alcohol (90%) with the remainder of the structure consisting predominantly of *trans-p*-coumaryl alcohol. In contrast, the lignin derivatives in hardwood are composed predominantly of *trans*-coniferyl alcohol and *trans*-sinapyl alcohol in varying ratios (approximately 50% for each alcohol).

Structural and other chemical issues aside, lignin fills the spaces in the cell wall between cellulose, hemicellulose, and pectin components (structural acidic heteropolysaccharide derivatives contained in the primary cell walls of terrestrial plants), especially in vascular and support tissues. It is covalently linked to hemicellulose and therefore cross-links different plant polysaccharides, conferring mechanical strength to the cell wall and, by inference, to the whole plant.

By way of explanation, pectin is a structural acidic heteropolysaccharide contained in the primary cell walls of terrestrial plants. The main component is galacturonic, a sugar acid derived from galactose.

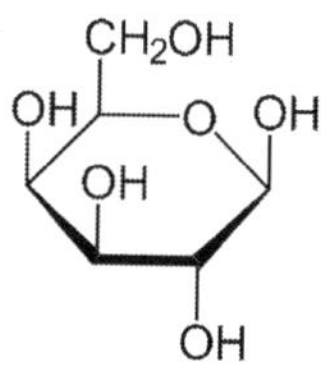

Galactose.

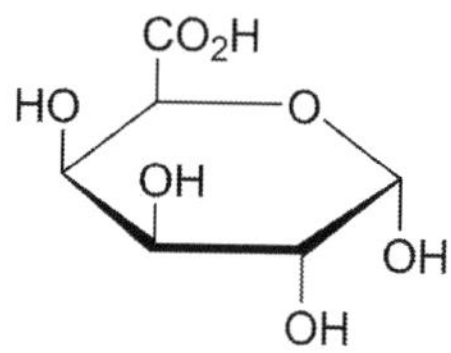

Galacturonic acid.

Lignin plays a crucial part in conducting water in plant stems—the polysaccharide constituents of the plant cell are hydrophilic and thus permeable to water, whereas lignin is more hydrophobic and less permeable to water. The cross-linking of polysaccharides by lignin is an obstacle for water absorption to the cell wall and, thus, the presence of lignin makes it possible for the vascular tissue of the plant to conduct water efficiently.

10.2.8.1.4 Solvent-extractable materials

The structure created by hydrogen bonds results in the typical material properties of the chemical constituents of wood confers insolubility in most solvents. For isolation of, for example, cellulose from wood, a direct nitration of wood yields undegraded cellulose trinitrate, which is soluble in organic solvents. On the other hand, the glycosidic linkages are easily cleaved by strong mineral acids and therefore cellulose can be hydrolyzed to simple sugars. However, for a complete hydrolysis of cellulose, concentrated acid solutions must be used in order to achieve the necessary swelling and at least a partial destroying of the ordered regions. Furthermore, although native lignin derivatives behave as an insoluble and three-dimensional network, the isolated lignin derivatives exhibit maximum solubility in a variety of solvents, including dioxane, acetone, methyl cellosolve, tetrahydrofuran, dimethylformamide, and dimethyl sulfoxide.

More generally, the soluble materials or extractives in wood consist of those components that are soluble in neutral organic solvents. The dichloromethane-extractable content of wood is a measure of such substances such as waxes, fats, resins, photo-sterols, and nonvolatile hydrocarbon derivatives. The amount of extractives is highly dependent on seasoning or drying of wood. The ethanol–benzene-extractable content of the wood consists of certain other dichloromethane-insoluble components such as low-molecular-weight carbohydrates, salts, and other water-soluble substances. Most water-soluble and volatile compounds are removed during pulping. The extractives reduce pulp yield, increase pulping and bleaching chemical consumption, and create problems such as foaming during papermaking if not removed.

For isolation of the solvent-extractable constituents from wood, the different methods can be used. Volatile extractives are represented by high-volatile compounds that can be separated by water distillation. They are mainly composed of monoterpene derivatives and other volatile terpene derivatives, terpenoid derivatives as well as of many different low-molecular compounds. Resin is the collective name for the lipophilic extractives (with the exception of phenolic substances). Resin extractives can be extracted with organic solvents. Water-soluble compounds consist of various phenol derivatives, carbohydrates, glycoside derivatives, and soluble salts, which can be extracted by cold or hot water.

Most plant resins are composed of terpenes. Specific components include alpha-pinene, beta-pinene as well as the monocyclic terpenes limonene and terpinolene, with and smaller amounts of the tricyclic sesquiterpene derivatives.

Limonene.

Some wood resins also contain a high proportion of resin acid derivative. On the other hand, rosins are less volatile and consist, inter alia, of diterpane derivatives.

In addition, wood resins are divided into free acids, for example, resin acid and fatty acid, and neutral compounds (such as fats and waxes). The resin fraction is soluble in organic solvents but insoluble in water, and therefore it can be extracted with organic solvents, such as hexane, dichloromethane, diethyl ether, acetone, or ethanol. Different nonpolar and polar solvents can be selected for isolation of the different types of the soluble-extractable constituents of wood.

10.2.9 Chemistry and uses

The utilization of biomass to produce valuable products by thermal processes is an important aspect of biomass technology (Speight, 2008, 2011b). Biomass pyrolysis gives usually rise to three phases: (1) gases, (2) condensable liquids, and (3) char/coke. However, there are various types of related kinetic pathways ranging from very simple paths to more complex paths and all usually include several elementary processes occurring in series or in competition. As anticipated, the kinetic paths are different for cellulose, lignin, and hemicelluloses (main basic components of biomass) and also for usual biomasses according to their origin, composition, and inorganic contents.

The main biomass constituents—hemicellulose, cellulose, and lignin—can be selectively devolatilized into value-added chemicals. This thermal breakdown is guided by the order of thermochemical stability of the biomass constituents that ranges from hemicellulose (as the least stable constituent) to the more stable—lignin exhibits an intermediate thermal degradation behavior. Thus wood constituents are decomposed in the order of hemicellulose–cellulose–lignin, with a restricted decomposition of the lignin at relatively low temperatures. With prolonged heating, condensation of the lignin takes place, whereby thermally largely stable macromolecules develop. Whereas both hemicellulose and cellulose exhibit a relatively high devolatilization rate over a relatively narrow temperature range, thermal degradation of lignin is a slow-rate process that commences at a lower temperature when compared to cellulose.

Thus biomass, unlike natural gas or crude oil, offers a wide variety of compositions. In addition, the specific components of plants such as carbohydrates, vegetable oils, plant fiber, and complex organic molecules known as primary and secondary metabolites can be utilized to produce a range of valuable monomers, chemical intermediates, pharmaceuticals, and materials: (1) carbohydrates, (2) vegetable oils, (3) plant fibers, and (4) specialty chemicals.

Carbohydrates (starch, cellulose, and sugars): starch is readily obtained from wheat and potato, while cellulose is obtained from wood pulp. Polysaccharides can be hydrolyzed, catalytically or enzymatically, to produce sugars, a valuable fermentation feedstock for the production of ethanol, citric acid, lactic acid, and dibasic acids such as succinic acid. Vegetable oils: vegetable oils are obtained from seed oil plants such as palm, sunflower, and soya. The predominant source of vegetable oils in many countries is rapeseed oil. Plant fibers, such as lignocellulosic, can be extracted from plants such as hemp and flax can replace cotton and polyester fibers in textile materials and glass fibers in insulation products. Plants can synthesize highly complex bioactive molecules (specialty chemicals) that are often beyond the power of laboratories, and a wide range of chemicals is currently extracted from plants for a wide range of markets from crude herbal remedies through to very high–value pharmaceutical intermediates.

More generally, biomass feedstocks are recognized by the specific chemical content of the feedstock or the manner in which the feedstocks are produced. However, the chemical composition of biomass varies considerably. Predictably, the chemical and molecular composition of biomass impacts its subsequent decomposition. The rate of decomposition is an early facet of the dynamic process for converting biomass and is dependent on the quality of biomass (chemical and molecular composition) as well as other factors, such as process parameters.

Since the thermal stabilities of the main biomass constituents partially overlap and the thermal treatment is not specific, a careful selection of temperatures, heating rates, and gas and solid residence times is required to make a discrete degasification possible when applying a step-wise increase in temperature. Depending on these process conditions and parameters such as composition of the biomass and the presence of catalytically active materials, the product mixture is expected to contain degradation products from hemicellulose, cellulose, or lignin.

The main biofeedstock constituents hemicellulose, cellulose, and lignin can be selectively devolatilized into value-added chemicals—this thermal breakdown is guided by the order of thermochemical stability of the biomass constituents that ranges from hemicellulose (fast degassing/decomposition from 200°C to 300°C, 390°F to 570°F) as the least stable natural product to the more stable cellulose (fast degassing/decomposition from 300°C to 400°C, 570°F to 750°F). Lignin exhibits an intermediate thermal degradation behavior (gradual degassing/decomposition from 250°C to 500°C, 480°F to 930°F).

The chemical components of wood are decomposed in the order of hemicellulose–cellulose–lignin, with a restricted decomposition of the lignin at relatively low temperatures. In the further course of heating a recondensation of the lignin takes place, whereby thermally largely stable macromolecules develop. Whereas both hemicellulose and cellulose exhibit a relatively high devolatilization rate over a relatively narrow temperature range, thermal degradation of lignin is a slow-rate process that commences at a lower temperature when compared to cellulose (Bajus, 2010).

The past abundance of biomass, particularly word, and the dispersion of the industry have worked against advances in technology for the efficient production, conversion, and use of wood products. Fortunately, and despite its relatively recent origin as a recognized field of study, wood science has had an appreciable effect on wood technology as well as science in general. The study of wood chemistry has contributed to our understanding of the principal components of wood—cellulose and lignin—and their reactions. Early research on hydrolysis of cellulose was prompted by fuel needs in the World War I but contributed much to our knowledge of this form of chemical reaction.

The Madison Process as it was described in the 1940s was developed to hydrolyze softwood species, which is particularly valuable or the production of fuels such as ethanol (Keller, 1996; Galbe and Zacchi, 2002). The hemi-cellulose sugars were recovered in the form of furfural, and only one stage was required. If hardwood species are used, a two-stage process is more desirable to maximize the recovery of both the hemicellulose and cellulose. High yields of hemi-cellulose products can be obtained at the milder prehydrolysis conditions compared to the higher temperature required to maximize glucose yields from the cellulose in the second stage.

The several acid hydrolysis processes now being promoted to produce ethanol from wood, but they do it in different types of equipment at slightly different acid–temperature–time conditions. The stake process uses a horizontal screw reactor. The Iotech process uses a high-pressure, short-time hydrolysis followed by rapid release of pressure. The twin-screw extruder process developed by New York University uses a high-pressure, special reactor design. The plug-flow reactor under study at Dartmouth and another developed by American can use different methods to pump the wood and acid into the reactor. The New Zealand process under license to Ultra Systems is probably a modern version of Madison Process.

After the hydrolysis of the cellulose, the processes could be identical if the same products were to be recovered. Some of the two-stage processes differ in that the residue from the first stage is delignified with a lignin solvent. This dissolves the lignin to leave only cellulose to be hydrolyzed in the second stage. The lignin is then recovered by distilling off the solvent. This is in contrast to hydrolyzing the first-stage residue to solubilize the cellulose and leave the lignin.

Wood is bulky, has less than half the heat of combustion of fuel oil, and in its green state is heavy to ship. Furthermore the cost of a wood-burning system may be three to four times that of a gas-burning installation because of fuel storage, handling, and air quality control systems. These drawbacks have kindled interest in production of liquid and gaseous fuels from wood. Much research is devoted to improving existing technology and devising new approaches, but such fuels are still expensive compared with crude oil–based fuels.

Finally, closely related to the conversion of wood to liquid or gaseous fuel is the use of the chemical storehouse that is wood to produce a wide range of silvichemicals. Many processes of these types already form the basis of chemical production on a commercial scale. But the potential to use wood as a chemical feedstock is much greater than has so far been realized. Whole wood can be gasified, liquefied, or pyrolyzed in ways comparable with those used for coal to yield a wide variety of chemicals. Cellulose, as a glucose polymer, can be hydrolyzed to the glucose monomer by acid or enzymes, and the glucose then fermented to ethanol. The ethanol can be used as a fuel or as a source of other important chemicals such as ethylene or butadiene.

As an alternative, use of glucose as substrate for fermentation would make possible production of antibiotics, vitamins, and enzymes. Hemicelluloses can easily be converted to simple sugars, which can be used to produce ethanol or furfural, a potential raw material for nylon or other synthetics.

Lignin can be pyrolyzed, hydrogenated, and hydrolyzed to yield phenols, which can be further processed to benzene. Once the technology and economics are feasible, future plants will manufacture a variety of these very significant chemicals from wood, now derived from crude oil or other resources.

Electricity generation from woody biomass grew from 59.5 to 79.6 TWh between 1990 and 2001, yielding a 2.7% average annual growth. As the second largest renewable electricity source after hydropower, solid biomass accounted for 5.6% of renewable electricity generation in 2001. This share is up from 4.6% in 1990. Of the electricity generated from solid biomass, 52.3% (41.6 TWh) is accounted for by the United States, where it makes up 14.6% of the country's renewable electricity production. The second largest producer of electricity from solid biomass is Finland (8.2 TWh), where it represents 37.8% of renewable electricity supply. Other big producers are Japan and Canada; in fact, electricity is produced from solid biomass in most OECD member countries. This mainly involves the combustion of commercial woody fuels in modern devices. For example, wood chip–fired cogeneration plants for heat and power. Other applications are industrial heat supply and cofiring with coal for large-scale power generation.

Charcoal continues to be used as an important industrial source of energy. For example, in Brazil, some 6 million tons of charcoal is produced every year for use in heavy industry, such as steel and alloy production. The industrial demand for charcoal in the last few years has led to new, more efficient, and large-scale technologies, mainly aimed at improving charcoal yield and quality. Furthermore, although fuel wood is mainly a local source of energy, there are signs of an international trade in wood fuel developing between European and North American countries.

The dynamics of wood fuel flow are complex and very site-specific. The development of sustainable wood energy systems remains one of the most critical issues to be addressed by policy makers and community planners. With society giving increasing attention to sustainability issues, in the case of wood energy in both developing and developed countries, economic, environmental, and social issues deserve particular attention.

Most of the uses of wood are accounted for by combustion in intermediate- or large-scale units outside the forest industries (e.g., in schools, hospitals, barracks, or district heating plants), with minor volumes going to the production of charcoal. Very small volumes were used in a few European countries to generate electricity or to manufacture solid fuels (e.g., briquettes). No wood is used at present in the region to make synthetic liquid or gaseous fuels. Use of energy wood by the forest industries and users has grown faster than use by households.

10.3 Waste

Waste is the result of human activities or the by-product from a process (or processes) for which no use is planned or foreseen. Thus waste in the context of this book is actually is a by-product of the human chemical, physical, and economic system. The words domestic and industrial are qualifiers of the source of the waste and, to some extent, are also descriptive of the contents of the waste. Once a material has been designated as waste, it remains waste until it has been fully recovered and no longer poses a potential threat to the environment or a use is found for the waste. However, in nature (which is typically a balanced system unless disturbed by human activities), there is no waste. Since the Industrial Revolution, human society has developed economies that are largely unrelated from nature and the natural order of events. They have created public and private wealth, material well-being, a plentiful food supply, comfortable living conditions, personal freedoms, and enhanced culture, but at a price.

Relevant to the disposal of waste, it can be used as feedstocks for various processes such as (1) gasification, (2) pyrolysis, or (3) incineration are streams such as chemical waste, medical, waste, paper waste, plastic waste, and textile waste, as well as many other types of other carbonaceous waste, which may form part of the industrial waste and domestic waste. However, although waste is a very general category, it must be remembered that as is the case for the feedstocks for any process, these raw materials require different processes for optimal operation.

Thus domestic waste (also known as rubbish, garbage, trash, or junk) is unwanted or undesired material (although the old adage one man's waste is another man's treasure sometimes applies) (Table 10.11). Waste is the general term; though the other terms are used loosely as synonyms, they have more specific meanings. Thus (1) rubbish or trash are mixed household waste, including paper and packaging; (2) food waste or garbage (North America) is kitchen and table waste; and (3) junk or scrap is metallic or industrial material. There are other categories of waste as well: sewage, ash, manure, and plant materials from garden operations, including grass cuttings, fallen leaves, and pruned branches. On the other hand, industrial waste is waste produced by industrial operations such as factories, mills, and mines and has existed since the onset of the Industrial Revolution. Chemical waste and toxic waste are two additional (but specific) designations of industrial waste (Table 10.11).

MSW is a type of waste that includes predominantly household waste (domestic waste) with, on occasion, the addition of commercial wastes collected by a municipality within a given area. They are in either solid or semisolid form and generally exclude industrial hazardous waste. The term residual waste relates to waste left from household sources containing materials that have not been separated out or sent for reprocessing.

There are five broad categories of MSW: (1) biodegradable waste, such as food and kitchen waste and green waste; (2) recyclable material such as paper, glass, cans metals, and certain types of plastic; (3) inert waste such as construction and demolition waste, dirt, rocks, and debris; (4) composite waste that includes waste clothing and waste plastics; and (5) domestic hazardous waste (also called household hazardous waste) and toxic waste such as discarded medications, paints, chemicals, light bulbs, fluorescent tubes, spray cans, fertilizer containers, pesticide containers, batteries, and shoe polish.

Some components of waste—such as plastic bottles, metals, glass, or paper—can be recycled once recovered from the waste stream. The biodegradable components of wastes (such as paper and food waste), including agricultural waste (Tables 10.11 and 10.12), can be composted or anaerobically digested to produce soil improvers and alternate (renewable) fuels though a variety of reactions (Table 10.13). Biomass is carbon based and is composed of a mixture of organic molecules containing hydrogen, usually including atoms of oxygen, often nitrogen, and also small quantities of other atoms, including alkali metals, alkaline earth metals, and heavy metals. These metals are often found in functional molecules such as the porphyrin molecules that include chlorophyll that contains magnesium. Also, if biomass not used in a sustainable manner, biodegradable waste can contribute to greenhouse gas emissions and, by implication, climate change.

On the other hand, there is also electronic waste that is a waste consisting of any broken or unwanted electrical or electronic appliance. While there is no generally

Table 10.11 Major sources of waste.

Dredging and irrigation	Waste consists of soil and sediments removed from waterways, harbours, estuaries, and irrigation canals. The quantities may be considerable and contain hazardous materials discharged from industrial and farming activities. It is not suited for energy production.
Farming, livestock rearing, dairy activities	Waste composed largely of spoilt food, manure, crop waste, and waste from chemical or pesticide use. Not all of the agricultural waste generated can be put to energy production; some of this waste is returned to the land as part of good agricultural practice.
Industrial	Some of the major industrial sources of waste are construction and demolition, fabrication, light and heavy manufacturing, refineries, chemical plants and non–nuclear power plants. Some large industrial facilities have their own recycling initiatives and operate their own landfills; it is difficult to determine the amount of waste discarded.
Mining and quarrying	Waste consists of mine tailings (silts, fine sands, or other aggregate materials) and may pose problems to the environment due to large quantities of waste produced and, in some cases, its hazardous nature. Not suited for energy production.
Nuclear power and nuclear defense	The civilian nuclear power industry, defense facilities, and nuclear research projects generate dangerous radioactive waste. The toxicity due to exposure and concerns over long range health and environmental effects make nuclear waste extraordinarily difficult to dispose of safely.
Residential, commercial, and institutional	Homes, commercial or institutional businesses, construction and demolition activities, municipal services, and treatment plants (inclusive of waste incinerators) are major contributors to waste streams.

accepted definition of electronic waste, in most cases electronic waste consists of electronic products that were used for data processing, telecommunications, or entertainment in private households and businesses that are now considered obsolete, broken, irreparable, or of no further use due to planned obsolescence. Despite its common classification as a waste, disposed electronics are a considerable category of secondary resource due to their significant suitability for direct reuse (e.g., many fully functional computers and components are discarded during upgrades), refurbishing, and material recycling of its constituents. It is a point of concern considering that many components of such equipment are considered toxic and are not biodegradable but they are not precursors to fuels and other than recognition though the previous paragraph will not be considered in the context of the present text.

10.3.1 Domestic and industrial waste

Much of what human society discards contains usable material, much of it in the form of recoverable energy. Paper, wood, cloth, food waste, and plastics are the main potential energy sources in waste. The remainder of the waste consists of glass, metals, and miscellaneous rubble. Domestic waste is typically disposed of by tipping it into large holes in the ground—landfill sites. Sometimes the waste is incinerated first and only the remaining ash and noncombustible material is sent to a landfill. Increasingly, a proportion of the waste is being separated for recycling at some stage along the way.

Domestic waste could also provide feedstock for a number of other conversion systems, all of which could recover useful energy while reducing the requirement for landfill sites. However, whatever the energy technology, domestic waste is a low-grade fuel. Its consistency is variable and not well suited to mechanical handling systems; the proportions of the various constituents will vary from load to load; the moisture content and heating value will vary; and the proportion of noncombustible material will keep the heating value low. All of this can lead to inefficient combustion if the process is not well controlled, making it more difficult to control toxic emissions from plastics and other materials. There is also a potential conflict between the recycling of materials and the recovery of energy from those materials. The main benefit of domestic waste as a fuel source is that, as with most other waste streams, energy technology can reduce the waste disposal problem.

10.3.2 Effects of waste

Wastes generated from the domestic and industrial sources increase continuously with rising population. In general, the lack of facilities for disposal of waste caused overall landfill sites resulting in hazards for the environment and for public health. These effects include (1) air pollution, (2) pollution of surface waters, (3) changes in soil fertility, and (4) changes in the landscape and visual discomfort. The historic approach to solid waste has been to bury it in landfill. This is becoming increasingly problematic because the diminishing availability of suitable landfill sites and the increasingly stringent conditions being applied to landfill, meaning that charges

Table 10.12 Amounts (w/w%) of cellulose, hemicellulose, and lignin common agricultural residues and wastes.

Agricultural residue	Cellulose	Hemicellulose	Lignin
Bamboo	41−49	24−28	24−26
Coastal Bermuda grass	25	35.7	6.4
Corn cobs	45	35	15
Corn stover	35	28	16−21
Cotton seed hairs	80−90	5−20	0
Grasses	25−40	35−50	10−30
Hardwood stem	40−50	24−40	18−25
Leaves	15−20	80−85	0
Newspaper	40−55	25−40	18−30
Nut shells	25−30	25−30	30−40
Paper	85−99	0	0−15
Primary wastewater solids	8−15	NA	24−29
Rice straw	40	18	5.5
Softwood stem	45−50	25−35	25−35
Solid cattle manure	1.6−4.7	1.4−3.3	2.7−5.7
Sorted refuse	50−60	10−20	15−20
Sugarcane bagasse	32−48	19−24	23−32
Sweet *Sorghum*	27	25	11
Swine waste	6.0	28	−
Switch grass	30−51	10−50	5−20
Waste papers from chemical pulps	60−70	10−20	5−10
Wheat straw	33−40	20−25	15−20

Table 10.13 Common reactions in anaerobic digestion process.

Substrate	Reactions
Alcohol derivatives	$4CH_3OH \rightarrow 3CH_4 + CO_2 + 2H_2O$
	$CH_3OH + H_2 \rightarrow CH_4 + H_2O$
	$4HCOO^- + 2H^+ \rightarrow CH_4 + 2H_2O$
Monosaccharide derivatives	$C_6H_{12}O_6 + 2H_2O \rightarrow 2H_2 +$ butyrate $+ 2HCO_3^- + 3H^+$
	$C_6H_{12}O_6 + 4H_2O \rightarrow 4H_2 + 2$ acetate $+ 2HCO_3^- + 4H^+$
Organic acid derivatives	Butyrate $+ 2H_2O \rightarrow 2H_2 + 2$ acetate $+ H^+$
	Propionate $+ 3H_2O \rightarrow 3H_2 +$ acetate $+ HCO_3^- + H^+$
	$HCOO^- + 3H_2 + H^+ \rightarrow CH_4 + 2H_2O$
	$CH_3COO^- + H_2O \rightarrow CH_4 + HCO_3^-$
	$4CH_2NH_2 + 2H_2O + 4H^+ \rightarrow 3CH_4 + CO_2 + 4NH_4^+$
Methanogenic substances	$2(CH_3)_2NH + 2H_2O + 2H^+ \rightarrow 3CH_4 + CO_2 + 2NH_44$
	$(CH_3)_3N + 6H_2O + 4H^+ \rightarrow 9CH_4 + 3CO_2 + 4NH_42CH_3CH_2$-$N(CH_3)_2 + 2H_2O \rightarrow 3CH_4 + CO_2 + 2CH_3CH_2NH_2$
	$4H_2 + HCO_3^- + H^+ \rightarrow CH_4 + 3H_2O$
	$4H_2 + 2HCO_3^- + H^+ \rightarrow$ acetate $+ 4H_2O$
Sulfate derivatives	$4H_2 + SO_4^{2-} \rightarrow HS^- + 3H_2O + OH^-$

have increased and will continue to increase. Major problems associated with landfills are the leachate containing toxic heavy metals and the methane gas that is produced.

Indeed, the challenge for waste disposal arises from the joint storage of hazardous materials (including toxic sludge, oil products, dyeing residues, and metallurgical slag) and solid domestic waste. This situation is likely to generate inflammable, explosive, or corrosive mixtures and combinations thereof. On the other hand, the presence of easily degradable household may facilitate the decomposition of complex hazardous components and thus diminish environmental pollution.

Another negative aspect is the fact that several recyclable and useful materials are stored in the same place as materials that cannot be recycled; consequently, these materials blend together and become chemically and biologically contaminated, which renders their retrieval rather difficult.

Thus the problems faced by waste management activities may be summarized as follows: (1) storage in open grounds is the most used method to remove waste ultimately; (2) existing landfills may be located in sensitive places, which are those places in the close proximity to lodgings, surface or ground water, leisure areas); (3) existing waste landfills may be improperly designed from an environmental protection point of view, thus allowing for water and soil pollution in those areas; and (4) currently waste landfills may require a review of waste-handling practices insofar as waste layers are not compacted and there is no strict control of the quality and quantity of waste that is dumped on the landfill leading to the potential for fire and/or the emanation of unpleasant odors.

All of the abovementioned facts lead to the conclusion that specific measures need to be taken with regard to waste management, which would be adequate in each phase of the waste dumping process. Environmental monitoring activities should comprise the observance of these measures. However, one answer to these issues is to convert the waste to usable products either through (1) the production of gaseous fuels, or (2) the production of liquid fuels, or (3) the production of solid fuels. Such efforts may not only solve the depletion of fuels from fossil sources but also assist in the disposal of waste materials and the ensuing environmental issues. However, before entering upon the process descriptions for waste conversion, it is necessary to understand the composition of domestic and industrial waste.

Generating waste at current levels is incompatible with a sustainable future. While the problem of waste is serious, a variety of initiatives are being taken to address the various threats. These include moves toward waste minimization, waste segregation and recycling, cleaner production with regular waste audits, green chemistry, renewable energy and energy efficiency, and developing the concept of industrial ecosystems. The issues involved are much more than technical problem.

This is a major issue because of the variability of industrial solid, liquid, and gaseous wastes, and the capacity of modern processing industries to produce huge quantities of waste. Fortunately, regulatory processes are now such that the numerous disasters caused previously should not be repeated. However, there are remaining problem sites that constitute long-term hazards.

Thus, in order to reduce the amount of landfill, the amount of waste must be reduced. This involves either (1) cutting back on the use of many materials or (2) use of the waste by conversion to useful products. Either option would reduce the amount of waste sent to landfill sites. The first option certainly reduces the amount of landfill material but is often more difficult to achieve. However, the second option offers the attractive proposition of the production of fuel products. Thus waste conversion becomes an attractive option to landfill disposal and the result is the generation of a usable product in the form of a gaseous, liquid, or solid fuel. However, ancillary of the second option is the heterogeneity of waste material. In fact, it is obvious that many waste streams are not subject to direct processing and will require special measures in the form of specific pretreatment of the waste prior to processing.

One form of pretreatment is separation and recycling of waste components thereby removing a portion of the waste stream for recycling and other uses. The result of this separation at the source is the remaining residual waste (i.e., the waste stream from which recyclable materials have been removed) that is not destined for any use other than landfill is sent to the conversion reactor.

References

Allred, V.D. (Ed.), 1982. Oil Shale Processing Technology. Center for Professional Advancement, East Brunswick, NJ.

Bajus, M., 2010. Pyrolysis technologies for biomass and waste treatment to fuels and chemical production. Pet. Coal 52 (1), 1–10.

Baughman, G.L., 1978. Synthetic Fuels Data Handbook, second ed. Cameron Engineers, Inc., Denver, CO.

Crocker, M., Crofcheck, C., 2006. Reducing national dependence on imported oil, Energeia, vol. 17. Center for Applied Energy Research, University of Kentucky, Lexington, KY, no. 6.

Demirbaş, A., 2008. Biofuels sources, biofuel policy, biofuel economy and global biofuel projections. Energy Convers. Manage. 49, 2106–2116.

Demirbaş, A., 2009. Biofuels: Securing the Planets Future Energy Needs. Springer Verlag, London, United Kingdom.

Dimian, A.C., 2015. Biorefinery: the future of the chemical process industries. Bull. Rom. Chem. Soc. 22 (1), 15–44.

Galbe, M., Zacchi, G., 2002. A review of the production of ethanol from softwood. Appl. Microbiol. Biotechnol. 59 (6), 618–628.

Gary, J.H., Handwerk, G.E., Kaiser, M.J., 2007. Petroleum Refining: Technology and Economics, Fifth ed. CRC Press, Taylor & Francis Group, Boca Raton, FL.

Giampietro, M., Mayumi, K., 2009. The Biofuel Delusion: The Fallacy of Large-Scale Agro-Biofuel Production. Earthscan, Washington, DC.

Gibon, V., Ayala, J.V., Dijckmans, P., Maes, J., De Greyt, W., 2009. OCL—oilseeds and fats. Crop. Lipids 16 (4), 193–200.

Hoogwijk, M., Faaij, A., Eickhout, B., de Vries, B., Turkenburg, W., 2005. Potential of biomass energy out to 2100, for four IPCC SRES land-use scenarios. Biomass Bioenergy 29, 225–257.

Hsu, C.S., Robinson, P.R. (Eds.), 2017. Handbook of Petroleum Technology. Springer International Publishing AG, Chaim, Switzerland.

Hwang, B., Obst, J.R., 2003. Basic studies on the pyrolysis of lignin compounds. In: Proceedings of the IAWPS 2003 International Conference on Forest Products: Better Utilization of Wood for Human, Earth and Future, vol. 2. The Korean Society of Wood Science and Technology, International Association of Wood Products Societies, Daejeon, Korea, April 21–24, pp. 1165–1170.

Kavalov, B., Peteves, S.D., 2005. Status and Perspectives of Biomass-to-Liquid Fuels in the European Union. European Commission, Directorate General Joint Research Centre (DG JRC), Institute for Energy, Petten, The Netherlands.

Keller, F.A., 1996. Integrated bioprocess development for bioethanol production. In: Wyman, C.E. (Ed.), Handbook on Bioethanol: Production and Utilization. Taylor & Francis, Washington, DC, pp. 351–379.

Langeveld, H., Sanders, J., Meeusen, M. (Eds.), 2010. The Biobased Economy. Earthscan, Washington, DC.

Larson, E.D., Kartha, S., 2000. Expanding roles for modernized biomass energy. Energy Sustain. Dev. 4, 15–25.

Lee, S., 1991. Oil Shale Technology. CRC Press, Taylor & Francis Group, Boca Raton, FL.

Lee, S., 1996. Alternative Fuels. CRC Press, Taylor & Francis Group, Boca Raton, FL.

Lorenzini, G., Biserni, C., Flacco, G., 2010. Solar, Thermal, and Biomass Energy. WIT Press, Boston, MA.

Luque, R., Speight, J.G. (Eds.), 2015. Gasification for Synthetic Fuel Production: Fundamentals, Processes, and Applications. Woodhead Publishing, Elsevier, Cambridge, United Kingdom.

Nersesian, R.L., 2010. Energy for the 21st Century: A Comprehensive Guide to Conventional and Alternative Energy Sources. Earthscan, Washington, DC.

Parkash, S., 2003. Refining Processes Handbook. Gulf Professional Publishing, Elsevier, Amsterdam, The Netherlands.

Ragland, K.W., Aerts, D.J., Baker, A.J., 1991. Properties of wood for combustion analysis. Bioresour. Technol. 37, 161–168.

Scouten, C.S., 1990. Oil shale. Fuel Science and Technology Handbook. Marcel Dekker Inc, New York, pp. 795–1053, Chapters 25 to 31.

Speight, J.G., 2008. Handbook of Synthetic Fuels: Properties, Processes, and Performance. McGraw-Hill, New York.

Speight, J.G., 2011a. An Introduction to Petroleum Technology, Economics, and Politics. Scrivener Publishing, Beverly, MA.

Speight, J.G. (Ed.), 2011b. The Biofuels Handbook. The Royal Society of Chemistry, London, United Kingdom.

Speight, J.G., 2013. The Chemistry and Technology of Coal, third ed. CRC-Taylor and Francis Group, Boca Raton, FL.

Speight, J.G., 2014. The Chemistry and Technology of Petroleum, fifth ed. CRC-Taylor and Francis Group, Boca Raton, FL.

Speight, J.G., 2017. Handbook of Petroleum Refining. CRC Press, Taylor & Francis Group, Boca Raton, FL.

Speight, J.G., 2019b. Synthetic Fuels Handbook: Properties, Processes, and Performance, second ed. McGraw-Hill, New York.

Speight, J.G., 2019c. Handbook of Petrochemical Processes. CRC Press, Taylor & Francis Group, Boca Raton, FL.

Speight, J.G., 2020. Global Climate Change Demystified. Scrivener Publishing, Beverly, MA.

Speight, J.G., Islam, M.R., 2016. Peak Energy – Myth or Reality. Scrivener Publishing, Beverly, MA.

Speight, J.G., El-Gendy, N.S., 2018. Introduction to Petroleum Biotechnology. Gulf Professional Publishing Company, Elsevier, Cambridge, MA.

US DOE, 2004a. Strategic Significance of America's Oil Shale Reserves, I. Assessment of Strategic Issues. <http://www.fe.doe.gov/programs/reserves/publications>.

US DOE, 2004b. Strategic Significance of America's Oil Shale Reserves, II. Oil Shale Resources, Technology, and Economics. <http://www.fe.doe.gov/programs/reserves/publications>.

US DOE, 2004c. America's Oil Shale: A Roadmap for Federal Decision Making. USDOE Office of US Naval Petroleum and Oil Shale Reserves. <http://www.fe.doe.gov/programs/reserves/publications>.

Wright, L., Boundy, R., Perlack, R., Davis, S., Saulsbury, B., 2006. Biomass Energy Data Book, first ed. Oak Ridge National Laboratory, Oak Ridge, TN, Office of Planning, Budget and Analysis, Energy Efficiency and Renewable Energy, United States Department of Energy. Contract No. DE-AC05-00OR22725.

Production of fuels from nonfossil fuel feedstocks

11

11.1 Introduction

Nonfossil fuel feedstocks can be converted into liquid or gaseous forms for the production of electric power, heat, chemicals, or gaseous and liquid fuels. The predominant conversion processes are direct liquefaction, indirect liquefaction, physical extraction, thermochemical conversion, biochemical conversion, and electrochemical conversion (Luque and Speight, 2015). More generally, the production of biofuels from lignocellulosic feedstocks can be achieved through two very different processing routes: (1) the thermochemical platform and (2) the bioconversion platform (Chapter 14: A Biorefinery). While each platform is adequate to the task, depending upon the feedstock, there is no clear candidate for best pathway between the various thermochemical technologies and the biochemical technologies.

The thermochemical platform typically uses a combination of pyrolysis, gasification, and catalysis to transform the feedstock into gaseous products—one of which is synthesis gas and then into fuels or chemicals (Luque and Speight, 2015; Speight, 2019). The production of synthesis gas (also referred to as syngas) through pyrolysis is accompanied by the generation of char, which can then be gasified to provide process heat and energy for the thermochemical platform. A variety of commercial-scale processes exist to transform fossil fuels such as coal or natural gas into liquid fuels, including Fischer–Tropsch fuels. However, the use of a nonfossil fuel feedstock instead of a fossil fuel feedstock changes the composition of synthesis gas, creating a more heterogeneous intermediate product and increasing the difficulty in downstream catalysis. A range of technical problems must be overcome before the nonfossil fuel feedstocks become commercially viable substitutes for fossil feedstocks in the production of second-generation biofuels. However, elements of the thermochemical platform are highly suitable for bioenergy production.

This platform combines the process elements of pretreatment, pyrolysis, gasification, cleanup, and conditioning to generate a mixture of hydrogen, carbon monoxide, carbon dioxide, and other gases. The products of this platform may be viewed as intermediate products, which can then be assembled into chemical building blocks and eventually end products. In the platform the pretreatment is often required and may involve drying, grinding, and screening of the feedstock in order to create a substrate that can easily be fed into the reaction chamber. The technology required for this stage is already available on a commercial basis and is often associated with primary or secondary wood processing, or agricultural residue collection and distribution. The thermochemical platform provides an opportunity for a number of additional coproducts, as well as energy in the form of heat or electricity

The Refinery of the Future. DOI: https://doi.org/10.1016/B978-0-12-816994-0.00011-7

and biofuels. Each component (such as carbon monoxide, CO; carbon dioxide, CO_2; methane, CH_4; and hydrogen, H_2) of the product gas stream may be recovered, separated, and utilized.

On the other hand, the bioconversion platform typically uses a combination of physical or chemical pretreatment and enzymatic hydrolysis to convert lignocellulose into its component monomers. This platform (examples are anaerobic digestion and fermentation) uses biological agents to carry out a structured deconstruction of lignocellulose components and combines process elements of pretreatment with enzymatic hydrolysis to release carbohydrates and lignin from the wood. The advantage of the bioconversion platform is that it provides a range of intermediate products, including glucose, galactose, mannose, xylose, and arabinose, which can be relatively easily processed into value-added bioproducts. The bioconversion platform also generates a quantity of lignin or lignin components, depending upon the pretreatment, lignin components may be found in the hydrolyzate after enzymatic hydrolysis, or in the wash from the pretreatment stage.

Once hydrolyzed, six-carbon sugars can be fermented to ethanol using age-old yeasts and processes. Five-carbon sugars, however, are more difficult to ferment; new yeast strains are being developed that can process these sugars, but issues remain with process efficiency and the length of fermentation. Other types of fermentation, including bacterial fermentation under aerobic and anaerobic conditions, can produce a variety of other products from the sugar stream, including lactic acid. Bioconversion proceeds at lower temperatures and lower reaction rates and can offer high selectivity for products. Ethanol production is a biochemical conversion technology used to produce energy from nonfossil fuel feedstocks, depending upon the type and properties of the feedstock (Tables 11.1–11.3) (Prakash and Karunanithi, 2008). For ethanol production, biochemical conversion researchers have focused on a process model of dilute acid hydrolysis of hemicelluloses followed by enzymatic hydrolysis of cellulose. Biodiesel production is a biochemical conversion technology used to produce energy from oilseed crops.

Cellulosic materials can be used to produce ethanol that represents an important, renewable liquid fuel for motor vehicles. The production of ethanol from nonfossil fuel feedstocks is one way to reduce both the consumption of crude oil and environmental pollution. In order to produce ethanol from cellulosic materials a

Table 11.1 The biomass resource base.

Resource	Products
Cereals, grains, sugar crops	Can be fermented to produce ethanol
Cellulosic materials	Can be converted into alcohol or synthesis gas
Oilseed crops	Can be converted into methyl esters
Organic waste material	Can be converted into biodiesel
	Mature and organic household waste might be converted into biogas

Table 11.2 Elemental composition, ash yield, and calorific heating value (dry basis) of various biomass feedstocks.

Fuel	C (wt. %)	H (wt. %)	O (wt. %)	N (wt. %)	S (wt. %)	Ash (wt.%)	Heating value (kJ/g)
Birch wood	48.8	6.0	44.2	0.5	0.01	0.5	20.0
Pine wood	49.3	6.0	44.2	0.5	0.01	0.5	20.1
Bark	47.2	5.6	46.9	0.3	0.07	3.9	20.9
Wheat straw	49.6	6.2	43.6	0.6	n.a.	4.7	18.6
Miscanthus	49.5	6.2	43.7	0.6	n.a.	3.3	18.5
Sugar cane	49.5	6.2	43.8	0.5	n.a.	3.7	18.5
Reed grass	49.4	6.3	42.7	1.6	n.a.	8.8	18.8
Peat	53.1	5.5	38.1	1.3	0.2	5.6	20.5
Coal	80.4	5.0	6.7	1.3	0.53	7.0	30.4

Table 11.3 Composition of different biomass types (% w/w, dry basis).

Type	Cellulose	Hemicellulose	Lignin	Others	Ash
Softwood	41	24	28	2	0.4
Hardwood	39	35	20	3	0.3
Pine bark	34	16	34	14	2
Straw (wheat)	40	28	17	11	7
Rice husks	30	25	12	18	16
Peat	10	32	44	11	6

pretreatment process is used to reduce the sample size, break down the hemicelluloses to sugars, and open up the structure of the cellulose component. The cellulose portion is hydrolyzed by acids or enzymes into glucose sugar that is fermented to ethanol. The sugar derivatives from the hemicellulose feedstocks are also fermented to ethanol.

The fermentation process requires pretreatment of the feedstock by chemical, physical, or biological means to reduce the complex carbohydrates to simple sugars. This type of pretreatment is often referred to as hydrolysis. The resulting sugars can then be fermented by the yeast and bacteria employed in the process. Furthermore, feedstocks that have a high content of starch and sugar are most easily hydrolyzed. Cellulosic feedstocks, including the major fraction of organics in municipal solid waste (MSW), are more difficult to hydrolyze, requiring more extensive pretreatment. Fermentation is generally used industrially to convert substrates such as glucose to ethanol for use in beverage, fuel, and chemical applications and to other chemicals (e.g., lactic acid used in producing renewable plastics) and products (e.g., enzymes for detergents). Strictly speaking, fermentation is an enzymatically controlled anaerobic process although the term is sometimes more loosely applied to include aerobic processing as well.

The bioconversion platform is an industrial option that might be used in a biorefinery (Chapter 12: Synthesis Gas and the Fischer–Tropsch Process) for producing fuels from nonfossil fuel feedstocks using biochemical reactions and/or biochemical agents. For example, fermentation or anaerobic digestion to produce fuels and chemicals from organic sources is a bioconversion platform. The bioconversion platform therefore has the ability to serve as the basis for wood-based biorefining operations, generating value-added bioproducts as well as fuel and energy for the forest sector.

This chapter focuses on the production of fuels from three biocarbonaceous sources: (1) crops, (2) wood, and (3) waste, especially when the constituents of the waste are carbonaceous in nature and can be traced to an origin from biomaterials.

11.2 Types of fuels

Biofuels are primarily used to fuel vehicles, but can also fuel engines or fuel cells for electricity generation. A brief introduction is warranted here with a more complete description available elsewhere.

The options for the production of biofuels to replace natural gas and crude oil are in active development (Table 12.2). The focus is on the use of cheap organic matter (usually cellulose, agricultural, and sewage waste) in the efficient production of liquid and gas biofuels which yield a high net energy gain. The carbon in biofuels was recently extracted from atmospheric carbon dioxide by growing plants, so burning it does not result in a net increase of carbon dioxide in the atmosphere of the Earth. As a result, biofuels are seen by many as a way to reduce the amount of carbon dioxide released into the atmosphere by using them to replace nonrenewable sources of energy.

11.2.1 Gaseous fuels

Most nonfossil fuel feedstocks are easier to gasify than coal because they are more reactive with higher ignition stability. This characteristic also makes them easier to process thermochemically into higher value fuels such as methanol or hydrogen. Ash content is typically lower than for most coals, and sulfur content is much lower than for many fossil fuels. Unlike coal ash, which may contain toxic metals and other trace contaminants, the ash may be used as a soil amendment to help replenish nutrients removed by harvest. Several of the nonfossil fuel feedstocks stand out for their peculiar properties—such as high silicon or alkali metal contents—which may require special precautions for harvesting, processing, and combustion equipment. Note also that mineral content of the feedstock can vary as a function of soil type and the timing of the harvesting of the feedstock. In addition, when the nonfossil fuel feedstock is heated with no oxygen or only approximately one-third the oxygen needed for efficient combustion (amount of oxygen and other conditions) determine if the feedstock is actually gasified or is thermally decomposed (by pyrolysis), a mixture of carbon monoxide and hydrogen (synthesis gas) is produced.

Combustion is a function of the mixture of oxygen with the hydrocarbon fuel. Gaseous fuels mix with oxygen more easily than liquid fuels, which in turn mix more easily than solid fuels. Synthesis gas therefore inherently burns more efficiently and cleanly than the solid feedstock from which it was produced. Thus gasification can improve the efficiency of large-scale power facilities, such as those for forest industry residues and specialized facilities such as black liquor recovery boilers of the pulp and paper industry, both major sources of power. Like natural gas, synthesis gas can also be burned in gas turbines, a more efficient electrical generation technology than steam boilers to which solid biomass and fossil fuels are limited.

11.2.2 Liquid fuels

Ethanol is the predominant fuel produced from crops and has been used as fuel in many countries including the United States since 1908.

Currently, the production of ethanol by fermentation of corn-derived carbohydrates is the main technology used to produce liquid fuels from nonfossil fuel resources. Furthermore, amongst different biofuels, suitable for application in transport, bioethanol and biodiesel seem to be the most feasible ones at present. The key advantage of bioethanol and biodiesel is that they can be mixed with conventional petrol and diesel respectively, which allows using the same handling and distribution infrastructure. Another important strong point of bioethanol and biodiesel is that when they are mixed at low concentrations ($\leq$10% bioethanol in petrol and $\leq$20% biodiesel in diesel), no engine modifications are necessary.

Alternatively, nonfossil fuel feedstocks can be converted into fuels and chemicals indirectly (by gasification to synthesis gas followed by catalytic conversion to liquid fuels) or directly to a liquid product by thermochemical means. Direct thermochemical conversion processes include pyrolysis, liquefaction, and solvolysis. Biologically produced alcohols, most commonly ethanol and methanol, and less commonly propanol and butanol are produced by the action of microbes and enzymes through fermentation.

Methanol is a colorless, odorless, and nearly tasteless alcohol and is also produced from crops and is also used as a fuel. Methanol, such as ethanol, burns more completely but releases as much or more carbon dioxide than its gasoline counterpart. Propanol and butanol are considerably less toxic and less volatile than methanol. In particular, butanol has a high flashpoint of 35°C (95°F), which is a benefit for fire safety, but may be a difficult for starting engines in cold weather.

Biodiesel is a diesel-equivalent fuel derived from biological sources (such as vegetable oils) which can be used in unmodified diesel-engine vehicles. It is thus distinguished from the straight (unused) vegetable oil or waste vegetable oil that is used as fuel in some diesel vehicles. In the current context, biodiesel refers to the alkyl ester products produced by the transesterification of vegetable oil or animal fat. Biodiesel fuel is a fuel made from the oil of certain oilseed crops such as soybean, canola, palm kernel, coconut, sunflower, safflower, corn, and hundreds of other oil-producing crops. The oil is extracted by the use of a press and then mixed in specific proportions with other agents, which causes a chemical reaction. The results of this reaction are two products, biodiesel and soap. After a final filtration the biodiesel is ready for use. After curing the glycerin soap that is produced as a byproduct can be used as is or can have scented oils added before use. In general, biodiesel compares well to crude oil−based diesel. Pure biodiesel fuel (100% esters of fatty acids) is called B100. When blended with diesel fuel the designation indicates the amount of B100 in the blend, for example, B20 is 20% B100 and 80% diesel, and B5 used in Europe contains 5% B100 in diesel.

Hydrocarbon derivatives are products of various plant species belonging to different families, which convert a substantial amount of photosynthetic products into latex. The latex of such plants contains liquid hydrocarbon derivatives of high molecular weight (10,000). These hydrocarbon derivatives can be converted into high-grade transportation fuel (i.e., crude oil). Therefore hydrocarbon producing plants are called crude oil plants or petroplants and their crop as petrocrop. Natural gas is also one of the products obtained from hydrocarbon derivatives. Thus crude oil plants can be an alternative source for obtaining crude oil to be used in diesel engines. Normally, some of the latex-producing plants of families such as Euphorbiaceae, Apocynaceae, Asclepiadaceae, Sapotaceae, Moraceae, and Dipterocarpaceae are petroplants. Similarly, sunflower (family Composiae), *Hardwickia pinnata* (family Leguminosae) are also petroplants. Some algae also produce hydrocarbon derivatives.

However, hydrocarbon derivatives, as such, are not usually produced from crops, there being an insufficient amount of the hydrocarbon derivatives present in the plant tissue to make the process economical. However, biodiesel is produced from crops thereby offering an excellent renewable fuel for diesel engines.

Biooil is a product that is produced by a totally different process than that used for biodiesel production. The process (fast pyrolysis, flash pyrolysis) occurs when solid fuels are heated at temperatures between 350°C and 500°C (570°F and 930°F) for a very short period of time (<2 seconds). The biooils currently produced are suitable for use in boilers for electricity generation. In another process the feedstock is fed into a fluidized bed (at 450°C−500°C, 840°F−930°F) and the feedstock flashes and vaporizes. The resulting vapors pass into a cyclone where solid particles, char, are extracted. The gas from the cyclone enters a quench tower where they are quickly cooled by heat transfer using biooil already made in the process. The biooil condenses into a product receiver, and any noncondensable gases are returned to the reactor to maintain process heating. The entire reaction from injection to quenching takes only 2 seconds.

11.2.3 Solid fuels

Examples of solid fuels from nonfossil fuel (i.e., biomass) feedstocks include wood and wood-derived charcoal as well as dried dung, particularly cow dung. One widespread use of such fuels is in home cooking and heating. The biofuel may be burned on an open fireplace or in a suitably constructed stove. The efficiency of this process may vary widely, from 10% for a well-made fire (even less if the fire is not made carefully) up to 40% for a custom-designed charcoal stove. Inefficient use of fuel is a cause of deforestation (though this is negligible compared to deliberate destruction to clear land for agricultural use), but more importantly, it means that more work has to be put into gathering fuel; thus the quality of cooking stoves has a direct influence on the viability of biofuels.

11.3 Fuel production

In the context of this book, crops are by definition, plants that can be grown and harvested extensively for profit or subsistence and may refer either to the harvested parts or to the harvest in a more refined state. Most crops are cultivated in agriculture or aquaculture. A crop may include macroscopic fungus (such as mushrooms), or algae (alga-culture). The issue is related to the use of crops for energy production and crops for human consumption. Most crops are harvested as food for humans or fodder for livestock and, therefore, crop management is essential.

The amount of hemicellulose and cellulose in wood and the chemical products desired determine the general type of process that might be used to hydrolyze wood. Hardwoods yield more five-carbon sugars than softwoods. Since, at this time, only the six-carbon sugars from cellulose are readily fermentable, softwoods are desired for ethanol production, but they are not as widely available as hardwoods. Hardwoods are more widely available now, so considerable effort has been expended to develop processes to utilize their unique constituents.

The main components of wood cells are cellulose (an insoluble substance which is the main constituent of plant cell walls and of vegetable fibers such as cotton. It is a polysaccharide consisting of chains of glucose monomers.), hemicellulose (a class of substances which occur as constituents of the cell walls of plants and are polysaccharides of simpler structure than cellulose), and lignin (a complex organic polymer deposited in the cell walls of many plants, making them rigid and woody), forming some 99% w/w of the wood material. Cellulose and hemicellulose are formed by long chains of carbohydrates, whereas lignin is a complicated component of polymeric phenolic derivatives. Lignin is rich in carbon and hydrogen, which are the main heat-producing elements. Thus the calorific value (CV) of lignin is higher than that of cellulose and hemicellulose (carbohydrates). Wood and bark also contain so-called extractives, such as terpene derivatives and phenol derivatives. The amount of wood extractives is relatively small compared to the amount of extractives from bark and foliage.

The nitrogen (N) content of wood is approximately 0.75% w/w but does vary from one tree species to another. For example, nitrogen-fixing alder (*Alnus* sp.) contains twice as much nitrogen as most coniferous trees. Wood has practically no sulfur (S) at all, as its share in wood is 0.05% at the highest. Compared to many other fuels, the wood has a relatively low carbon content (some 50% of dry weight) and high oxygen content (some 40%), which leads to relatively low heating value per dry weight.

The collection of solid wastes is usually organized on a communal basis; in developing countries though, it may be organized (to a greater or lesser extent) on an informal basis. The treatment and disposal of solid wastes are definitely connected. Treatment is applied to recover useful substances or energy, to reduce waste volume, or to stabilize waste remains to be dumped or disposed of in landfills. Wastes may be treated before disposal to reduce the volume or to alter the characteristics of the waste which can be achieved by various physical, chemical and biological processes while incineration can be used to destroy some toxic organic chemicals. Where a method of waste disposal is not specified, the choice of disposal route will typically depend on (1) the availability of facilities, (2) volume of waste material, and (3) hydro-geological characteristics; the influence of industrial and environmental lobby groups must also be taken into account.

Currently, the acceptable methods of waste disposal are (1) incineration, (2) gasification, (3) pyrolysis as well as a variety of other processes that are site and waste specific.

There are several basic types of processes used to generate energy from crops, which are presented alphabetically as (1) anaerobic digestion; (2) combustion; (3) fermentation; (4) gasification, (5) incineration, landfilling; and (6) pyrolysis as well as several processes that are listed under other processes. Each process has its own particular aspects and process application is dependent upon the type of feedstock and the desired product(s).

11.3.1 Anaerobic digestion

Anaerobic digestion is the decomposition of biological wastes by microorganisms, usually under wet conditions, in the absence of air (oxygen), to produce a gas comprising mostly methane and carbon dioxide. Most organic substances can become subject to anaerobic digestion so long as there are warm, wet, and airless conditions. For example, marsh gas is a product of the anaerobic digestion of vegetation at the bottom of ponds, this gas rises to the surface and bubbles, and the gas is also combustible. With the aid of human intervention there are two products of this process, biogas and landfill gas. The chemical processes behind the production of these gases are very complex.

A digester system (the anaerobic digester) is a device that promotes the decomposition of manure or digestion of the organics in manure to simple organics and gaseous biogas products. The anaerobic digestion process is a multistage biological waste treatment process whereby bacteria, in the absence of oxygen, decompose organic matter to carbon dioxide, methane, and water. In this way the waste sludge is stabilized and the obnoxious odor is removed. The process can however be described adequately and simply as occurring in two stages, involving two different types of bacteria. The process occurs in the absence of air, the decomposition in this case is caused not by heat but by bacterial action. In the first stage, the organic material present in the feed sludge is converted into organic acids [also called volatile fatty acids (VFAs)] by acid-forming bacteria. In the second stage, these organic acids serve as the substrate (food) for the strictly anaerobic methane-producing bacteria, which converts the acids into methane and carbon dioxide. The end result of the process is a well-established sludge in which 40%–60% of the volatile solids are destroyed. Finally, a combustible gas is produced consisting of 60%–75% methane and the remainder largely being carbon dioxide.

Thus in a typical anaerobic digestion process, the feedstock is allowed to decompose in an anaerobic (oxygen-deficient) environment. A number of factors affect the entire process. They include (1) the temperature of the substrate, (2) the loading rate, (3) the pH, (4) the residence time, (5) the concentration of nutrients, and (6) the presence of any toxic substances.

In terms of temperature the optimal temperature (where digestion and gasification proceed at the highest rate) is 35°C (95°F). Below 15°C (59°F), the rate is so slow that very little gas is produced. Temperature is also dependent on the bacterial populations: mesophilic or thermophilic. Mesophilic bacteria prefer temperatures ranging from 30°C to 40°C. Thermophilic bacteria prefer temperatures ranging from 50°C to 60°C (140°F).

The loading rate is the amount of fermentable matter that is fed into the digester per cubic meter of digester capacity. Any change in loading rate affects the balance inside the digester, so it should be kept constant. For a given capacity, if the loading rate is increased the fermentation period is correspondingly increased. The common range of solid concentration is 7%–9% and digression from this range can cause fermentation to be retarded. In terms of the alkalinity/acidity (pH), the optimal gas formation occurs at pH of 7–8. If the pH becomes too acidic, gas production could stop altogether.

The retention time is the amount of time fermentable material resides inside the digester. It has been observed that maximum gas production takes place within the first 4 weeks and then gradually tapers off. Detention time can be significantly reduced if the temperature is raised or the contents of the digester are agitated, or the supply of nutrients in the digester is augmented. Usually, some type of nutrient source is needed to help stimulate gas production. Bacteria use nitrogen, phosphorus, and potassium for their nutrients. Once these are available, fermentation proceeds very quickly. However, caution is advised that, although rare, toxic substances such as copper can inhibit gas production if found in large quantities.

Overall, the anaerobic digestion process is a complex process that requires strict anaerobic conditions (oxidation–reduction potential < -200 mV) to proceed and depends on the coordinated activity of a complex microbial association to transform organic material into mostly carbon dioxide and methane (Table 11.2). Despite the successive steps, hydrolysis is generally considered as rate limiting. The hydrolysis step degrades both insoluble organic material and high molecular weight compounds (lipids, polysaccharides, proteins, and nucleic acids) into soluble organic substances (amino acids and fatty acids). The components formed during hydrolysis are further split during acidogenesis, the second step. VFAs are produced by acidogenic bacteria along with ammonia (NH_3), carbon dioxide (CO_2), hydrogen sulfide (H_2S), and other by-products. The third stage in anaerobic digestion is acetogenesis, where the higher organic acids and alcohols produced by acidogenesis are further digested by acetogens to produce mainly acetic acid and as well as carbon dioxide and hydrogen. The final stage of methanogenesis produces methane by two groups of methanogenic bacteria: (1) the first group splits acetate into methane and carbon dioxide and (2) the second group uses hydrogen as an electron donor and carbon dioxide as the acceptor to produce methane. Thus:

$$CH_3CO_2H \rightarrow CH_4 + H_2$$

$$CO_2 + H_2 \rightarrow CH_4 + 2H_2O$$

The digestion process is continuous—fresh feedstock must be added continuously or at predetermined frequent intervals. The gas formed during digestion is removed continuously. In high-rate digestion, stabilized sludge is displaced from the digester during feeding. In low-rate digestion, sludge "supernatant" is normally removed as the feed sludge is added, stabilized sludge is removed at less frequent intervals.

It is essential that any organic acid derivatives formed in the first stage of the waste treatment process are converted to methane at the same rate at which they are formed. If not, the acidic species accumulate and ultimately lower the pH, leading to inhibition of the second stage of the digestion process and digester failure. The temperature must be maintained within certain ranges—heating increases the activity of the anaerobic bacteria reducing the required digestion time—and a pH of 7.0–7.5 is recommended to encourage the production of methane. Also, the gas leaving the digester is almost saturated with water vapor which condenses as the gas stream cools there causing problems. To mitigate this potential problem, it is essential to remove as much of the moisture as possible before the gas comes into contact with the gas system devices. This can be accomplished by the inclusion of water traps which should be located as close to the digester as possible and at a low point in the gas line so that all piping should be sloped towards the water trap.

Several types of digesters have been developed including the floating drum, the fixed dome, the bag, the plastic tube, the plug flow and the upflow anaerobic sludge blanket digesters. A digester is an airtight vessel or enclosure in which bacteria decomposes the nonfossil fuel feedstock in water to produce gaseous products, often referred to as biogas. Both batch digesters and continuous digesters are commercially available but while batch systems may be suitable for some applications, they are labor intensive and in developed countries are probably not appropriate to the steady production of gas on any significant scale.

Mixing may be used in both batch and continuous digesters to enhance contact between the bacteria and substrate. It has, however, been argued that while mixing may increase conversion rates, it is not energy efficient. In two-stage systems, only the first stage is mixed. Multistage systems provide optimum conditions for the successive digestive steps, and further development may prove them to be economical for some nonfossil fuel feedstocks.

In a process of manure and straw mixture digestion, for the first 72 hours the yield of methane was minimal (essentially 0%) and the yield of canon dioxide is virtually quantitative (c. 100%)—in this period, digestion occurred as aerobic fermentation to carbon dioxide. The yields of methane and carbon dioxide gases were approximately equal after 250 hours, and after 500 hours the digestion reached the stationary phase. The methane content of the biogas was in the range of 73%–79% for the runs, the remainder being principally carbon dioxide. During a 720-hour digestion period, approximately 80%–85% v/v of the biogas was produced in the first 360–430 hours which is an indication that the digester retention time can be designed to a shorter period (360–430 hours) instead of the longer period (720 hours).

The final stage in the process is the disposal of any waste. Essentially, all the water entering the system will be present in the effluent. Direct recycle of the water after separation from the sludge cannot be practiced, as this will result in a buildup of toxic substances and eventual failure of the system. It is possible that some fraction of the water may be safely recycled, although data on this aspect have not been seen. Treatment of this large volume of water to remove the toxic constituents is costly and is likely to be uneconomical other than in large-scale operations.

Typically, it can be assumed that the water, which is considered to have some nutritional value, can be used for the irrigation of growing crops. However, serious consideration must be given to the effect of trace metals and other toxic materials present in the water and the effects of these constituents on the crops. In addition, subject to the presence of nontoxic constituents (or the lack of toxic constituents) in the sludge, it also may be used for land application and will generally have higher nitrogen content than the original material. On an absolute (feedstock-to-sludge) basis, there may be less nitrogen retention due to losses of the nitrogen as ammonia (NH_3), and there may be additional losses on land application id the nitrogen in the sludge is in a more volatile form than the nitrogen in the original feedstock. Nevertheless, the sludge resulting from the digestion of animal manure is generally considered (subject to a critical analysis of the constituents of the sludge) to have improved fertilizer value over the original feedstock. During digestion, the VFA concentration is lower and the pH higher (Demirbaş and Ozturk, 2004). Nearly all digesters plants have ancillary processes to treat and manage all of the by-products. Before storage and use, the gas stream is dried and sweetened by the removal of sulfur compounds (Chapter 9: Hydrocracking). The sludge liquor mixture has to be separated by one of a variety of ways, the most common of which is filtration. Excess water is also sometimes treated in sequencing batch reactors for discharge into irrigation systems.

Essentially, all organic materials can be digested except for the stable woody materials since the anaerobic microorganisms are unable to degrade lignin. The biogas which is formed has a high CV and is considered as a renewable energy source. The main disadvantages of this process are (1) the possible presence of volatile siloxanes in the biogas that can cause serious damage in the energy users engine and boiler due to the formation of microcrystalline silica; and (2) the increased concentration of heavy metals and various industrial organic chemicals in the residual sludge due to the significant reduction of the organic fraction during digestion, leaving the mineral and nondegradable fraction untouched.

Anaerobic digestion produces a clean and environment-friendly fuel, although it contains carbon dioxide, water vapor, hydrogen, hydrogen sulfide, and siloxane derivatives (i.e., compounds having a molecular structure based on a chain of alternate silicon and oxygen atoms, –O–Si–O–Si–, especially—as in silicone derivatives—with organic groups attached to the silicon atoms).

As a note of caution, a mixture of biogas and air can be explosive. Methane in concentrations of between 5% and 15% v/v in air is explosive and air should not be allowed to enter the digester or gasholder. All piping and equipment must be sealed properly to prevent gas from escaping to the outside. In addition, all electrical installations must be of the explosion-proof type, as the smallest spark could ignite escaped gases.

11.3.2 Combustion

Direct combustion involves burning the energy crop and then using the resulting hot combustion gases to raise steam. The steam is, in turn, used to drive a steam turbine which drives a generator to produce electricity. The conversion efficiency from energy crop to energy is fairly low, especially for small systems, but this is balanced by the relatively low capital cost of direct combustion systems and the fact that the technology is tried and tested. Furthermore, combustion facilities can burn many types of biomass fuel, including wood, agricultural residues, wood pulping liquor, MSW and refuse-derived fuel which convert the feedstock into several forms of useful energy for commercial or industrial uses: hot air, hot water, steam, and electricity.

A furnace is the simplest combustion technology—in a furnace the fuel is combusted in the combustion chamber in which the feedstock is converted into heat energy. As the feedstock combusts, hot gases are released and these hot gases contain approximately 85% of the potential energy of the fuel. Thus a nonfossil fuel feedstock-fired boiler is a more adaptable direct combustion technology because a boiler transfers the heat of combustion into steam. Steam can be used for electricity, mechanical energy, or heat. Nonfossil fuel feedstock boilers supply energy at low cost for many industrial and commercial uses.

In a stationary or traveling grate combustor an automatic feeder distributes the fuel onto a grate, where the fuel burns. Combustion air enters from below the grate. In the stationary grate design, ashes fall into a pit for collection. In contrast, a traveling grate system has a moving grate that drops the ash into a hopper. In another configuration, a pile burner consists of separate cells and each cell consists of an upper and a lower combustion chamber for use in a step-wise burner configuration. In this type of equipment the feedstock is combusted burns on a grate in the lower chamber, thereby releasing volatile gases which are then combusted in the upper (secondary) combustion chamber. However, pile burners must be shut down periodically to remove the mineral ash.

Fluidized bed combustors burn the fuel in a hot bed of granular material, such as sand. Injection of air into the bed creates turbulence resembling a boiling liquid. The turbulence distributes and suspends the fuel. This design increases heat transfer and allows for operating temperatures up to 970°C (1700°F), reducing nitrogen oxide (NO_x) emissions. Fluidized bed combustors can handle high mineral mater (high ash-producing) feedstocks.

However, conventional combustion equipment is not designed for burning agricultural residues. Straw and grass contain alkali (potassium and sodium) compounds, which are also present in all annual crops and crop residues and in the annual growth of trees and plants. During combustion, alkali combines with silica, which is also present in agricultural residues. This reaction causes slagging and fouling problems in conventional combustion equipment designed for burning wood at higher temperatures. The volatile alkali constituents lower the fusion temperature of ash and in conventional combustion equipment having furnace gas exit temperatures above 790°C (1450°F), combustion of agricultural residue causes the formation of a troublesome slag that can deposit on heat transfer surfaces. Specially designed boilers with lower furnace exit temperatures could reduce slagging and fouling from combustion of these fuels. Low-temperature gasification is an option for using these fuels for efficient energy production while avoiding the slagging and fouling problems encountered in direct combustion.

Two cogeneration arrangements, or cycles, are possible for combining electric power generation with industrial steam production. Steam can be used in an industrial process first and then routed through a turbine to generate electricity. This arrangement is frequently referred to as called a bottoming cycle. In the alternate arrangement, steam from the boiler passes first through a turbine to produce electric power. The steam exhaust from the turbine is then used for industrial processes or for space and water heating. This arrangement is referred to as a topping cycle.

The direct-fired gas turbine is another combustion technology for converting nonfossil fuel feedstock to electricity. In this technology, fuel pretreatment reduces the feedstock to a particle size of less than 2 mm and a moisture content of less than 25%. Then the fuel is burned with compressed air. Cleanup of the combustion gas reduces particulate matter before the gas expands through the turbine stage. The turbine drives a generator to produce electricity.

Cofiring a nonfossil fuel feedstock as a secondary fuel in a coal-burning power plant using high-sulfur coal could help reduce the overall sulfur content of the feedstocks thereby offering a partial mitigation of the emissions of sulfur dioxide and nitrogen oxide. Also, cofiring decreases net carbon dioxide emissions from the power plant—if the fuel comes from a sustainable source, remembering that this is a paper exercise and may not occur in real time. Also, cofiring of such feedstocks may require recombustion of the wood fuel or boiler modifications to maintain boiler efficiency.

11.3.3 Fermentation

Fermentation is the conversion of organic carbon−containing compounds (and not mineral carbonates) by microorganisms for the production of fuels and chemicals such as alcohols, acids, or energy-rich gases. Cellulose hydrolysis and fermentation can be achieved by two different process schemes, depending upon where the fermentation is carried out: (1) separate hydrolysis and fermentation (SHF) or (2) simultaneous saccharification and fermentation (SSF).

In SHF process, the hydrolysis is carried out in one vessel and the hydrolyzate is then fermented in a second reactor. Approximately 73% of the cellulose is converted to ethanol in 48 hours, while the remainder of the cellulose, hemicelluloses, and lignin is burned. The most important parameters are the hydrolysis section yield, product quality, and the required enzyme loading which are all interrelated. Yields are typically higher in more dilute systems where inhibition of enzymes by glucose and cellobiose is minimized. Increasing the amount of enzyme loading can help to overcome inhibition and increase yield and concentration. Increased reaction times also make higher yields and concentrations. Cellulase enzymes from different organisms can result in markedly different performances.

In SSF process, both the hydrolysis and fermentation are carried out in the same vessel. In this process, yeast ferments the glucose to ethanol as soon as the glucose is produced, preventing the sugars from accumulating and the end-product inhibition. Using the yeast, *Candida brassicae* and the Genencor enzyme (by Genencor International), the yield increases to 79% and the ethanol concentration produced is 3.7%.

11.3.4 Gasification

Gasification is a high-temperature process that produces gas which can then be used in an internal combustion engine or fuel cell. Briefly, gasification was used as long ago as the early 1800s. The process was rather crude and the fuel was most often coal. The gas product was used for heating and lighting. In the current context, the gasification process is amenable to a variety of nonfossil fuel feedstocks such as waste rice hulls, wood waste, grass, and the dedicated energy crops. Gasification is a clean process with few air emissions and, when crops are used as the feedstock, little or no production of mineral ash. The process produces a fuel gas from crops by heating them under carefully controlled temperature, pressure, and atmospheric conditions. A key to gasification is using less air or oxygen than is usually found in combustor. The product (biogas or fuel gas), such as natural gas, can be burned in high-efficiency gas turbines.

The gasifier is the heart of the gasification process, no matter what the feedstock is (Speight, 2013, 2014a,b; Luque and Speight, 2015). Gasifiers are designed to process the feedstock in a variety of ways consistent with the type of fuel, the end use of the gas, the size of the process, and the source of oxygen. The oxygen may be introduced as a pure gas or may come from air or steam. Some gasifiers operate under pressure; others do not.

The simplest type of gasifier is the fixed bed countercurrent gasifier. The feedstock is fed into the top of the reactor and moves downwards as a result of the conversion of the feedstock and the removal of the mineral ash. The air intake is at the bottom and the gas leaves at the top. The feedstock moves in a flow that is countercurrent to the gas flow and passes through the drying, the distillation, reduction, and the oxidation zones. The major advantages of this type of gasifier are (1) the simplicity of the unit, (2) the high charcoal burn-out, and (3) the internal heat exchange leading to low gas exit temperatures and high gasification efficiency. In this way, feedstocks with a high moisture content (up to 50% w/w) can be used. However, major disadvantages of the process are (1) the high yield of tar and (2) the high yields of pyrolysis products because the pyrolysis gas is not taken into the oxidation zone. This may be of minor importance if the gas is used for direct heat applications, in which the tar is burned. In case the gas is used for engines, gas cleaning is required, resulting in problems of tar-containing condensates.

In the conventional downdraft gasifier (sometimes called the coflow gasifier), feedstock is fed at the top of the reactor and air intake is at the top or from the sides. The gas leaves at the bottom of the reactor, so the fuel and the gas move in the same direction. The pyrolysis gases are lead through the oxidation zone (with high temperatures) and/or more or less burnt or cracked. Therefore the producer gas has low tar content and is suitable for engine applications. In practice however, a tar-free gas is seldom if ever achieved over the whole operating range of the equipment. Because of the lower level of organic components in the condensate, downdraft gasifiers suffer less from environmental objections than updraft gasifiers.

Successful operation of a downdraft gasifier requires drying the feedstock to a moisture content that is less than 20% w/w of the feedstock. The advantage of the downdraft design is the very low tar content of the producer gas but there are disadvantages which are (1) the high amounts of ash and dust particles in the gas, (2) the inability to operate on a number of unprocessed fuels, often pelletization or briquetting of the feedstock is necessary, (3) the outlet gas has a high temperature leading to a lower gasification efficiency, and (4) it is preferable that the moisture content of the feedstock is less than 20% w/w.

A more recent development is the open core gasifier design for gasification of small-sized feedstock with a high content of mineral matter (high ash-forming propensity). The producer gas is not tar-free; it contains approx. 0.05 kg tar per kg gas. In the open core gasifier the air is sucked over the whole cross section from the top of the bed. This facilitates better oxygen distribution since the oxygen will be consumed over the whole cross section, so that the solid bed temperature will not reach the local extremes (hot spots) observed in the oxidation zone of conventional gasifiers due to poor heat transfer. Moreover, the air nozzles in conventional gasifiers generate caves and create obstacles that may obstruct solid flow, especially for solids of low bulk like rice husk. On the other hand, the entry of air through the top of the bed creates a downward flow of the pyrolysis gases and transports the tars products to the combustion zone. Thus flow problems due to the caking of rice husk caused by back mixing of tar are avoided.

However, before a nonfossil fuel feedstock can be gasified, it must be pretreated to meet the processing constraints of the gasifier. This typically involves size reduction, and drying to keep moisture contents below specific levels. Thereafter, the gasification process involves heating the feedstock in the presence of low levels of oxygen (i.e., less than required for complete combustion to carbon dioxide and water). Above certain temperatures the feedstock will break down into a gas stream and a solid residue. The composition of the gas stream is influenced by the operating parameters of the gasifier, with some gasification processes more suited than others to producing a gas for methanol production. In particular, simple gasification with air creates a synthesis gas stream that is diluted with large quantities of nitrogen. This nitrogen is detrimental to subsequent processing to methanol and so techniques using indirect gasification or an oxygen feed are preferred. For large-scale gasification, pressurized systems are considered to be more economic than atmospheric systems. Once the economic optimum synthesis gas is available the methanol synthesis takes place. This typically uses a copper–zinc catalyst at a temperature of 200°C–280°C and a pressure of 750–1500 psi.

However, most crop or agricultural residues are not found throughout the year but are available only at the time of harvest. This makes collection easy but creates storage problems if the residues have to be saved for use during other months of the year, especially due to its low bulk density. The amount available depends upon the harvesting time, storage-related characteristics, and the storage facility.

In the gasifier the feedstock is converted into a gaseous mixture of hydrogen, carbon monoxide, carbon dioxide, and other compounds by applying heat under pressure in the presence of steam and a controlled amount of oxygen. The feedstock produces synthesis gas (Chapter 8: Desulfurization, Denitrogenation, and Demetalization).

$$C_6H_{12}O_6 + O_2 + H_2O \rightarrow CO + CO_2 + H_2 + \text{other products}$$

The above reaction uses glucose as a surrogate for cellulose. Biomass feedstocks have a highly variable composition and complexity, with cellulose as one major component.

Willow crops have been shown to be a good fuel for farm-based power production using advanced gasification technology. The fuel gas can be used for generating electricity, using micro-turbines modified to operate on low-Btu (British thermal unit) gas, or for other farm energy needs.

Agricultural residues can be divided into two groups: crop residues and agro-industrial residues. Crop residues are plant materials left behind in the farm after removal of the main crop produce. The remaining materials could be of different sizes, shapes, forms, and densities like straw, stalks, sticks, leaves, haulms, fibrous materials, roots, branches, and twigs. The agro-industrial residues are by-products of the postharvest processes of crops such as cleaning, threshing, sieving, and crushing. These could be in the form of husk, dust, and straw. Furthermore, the quantity of agricultural residues produced differs from crop to crop and is affected by soil type and irrigation conditions. Production of agricultural residues is directly related to the corresponding crop production and ratio between the main crop produce and the residues, which varies from crop to crop and, at times, with the variety of the seeds in one crop itself. Thus for known amounts of crop production, it may be possible to estimate the amounts of agricultural residues produced using the residue to crop ratio.

The gasification process could play a significant role in meeting the goal of greenhouse gas mitigation insofar as it is possible that both in the transition phase to a hydrogen economy and in the steady state, a significant fraction of hydrogen might be derived from domestically abundant crops. In addition, the cofiring applications of crops and other feedstocks—for example, a biomass feedstock with coal —can provide up to 15% of the total energy input of the fuel mixture. Such concepts address greenhouse gas mitigation by cofiring biomass and coal to offset the losses of carbon dioxide to the atmosphere that are inherent in coal combustion processes (even with the best-engineered capture and storage of carbon). Since the growth of biomass fixes atmospheric carbon, its combustion leads to no net addition of atmospheric carbon dioxide even if vented. Thus cofiring of crops or crop residues (or other biomass with coal) in an efficient coal gasification process, affording the opportunity for capture and storage of carbon dioxide, could lead to a net reduction of atmospheric carbon dioxide. Cheaper, though less plentiful, biomass residue could supplant crops as gasifier feedstock leading to a less significant impact on the environment than would farming bioenergy crops.

A cautionary note is that most crop or agricultural residues are not found throughout the year but are available only at the time of harvest. This makes collection easy but creates storage problems if the residues have to be saved for use during other months of the year, especially due to its low bulk density. The amount available depends upon the harvesting time, storage-related characteristics, and the storage facility.

11.3.5 Incineration

The concept of the incineration (combustion) of domestic waste and industrial waste to energy (heat) and/or gaseous fuels is one of the first suggestions for the disposal of waste and the generation of energy. With this in mind, waste incineration has been and continues to be practice in many parts of the word. In fact, cocombustion of waste in (large and efficient) coal-fired power plants is an especially attractive option because of the high conversion efficiency of these plants. However, there are issues that need attention and while it is not intended here to use the following comments to be a critique of waste incineration, here are lessons to be learned and changes to be made in this manner of waste removal.

Although not in keeping with the main context of this text, for example, the production of fuels, the incineration of wastes is worthy of mention insofar as waste combustion (although the main goal of incineration is volume reduction with the sterilization of the waste as a significant side-effect) can be a source of energy, but generally. Thus the incineration process may also be used to produce steam and electricity. However, as the moisture content of the waste increases, a self-sustaining combustion process is not possible. Nevertheless, domestic waste and industrial waste are feedstocks for incinerators.

Nevertheless, incineration is recognized as a waste treatment technology for the disposal of the organic constituents of the waste. Incineration and other high-temperature waste treatment systems convert the waste into flue gas, particulate matter, ash, and heat, which can in turn be used to generate electricity. The flue gases are cleaned for pollutants before it is dispersed in the atmosphere.

Furthermore, incineration with energy recovery is one of the several waste-to-energy technologies such as gasification and anaerobic digestion that can be applied to the production of alternate fuels.

Modern incinerators reduce the volume of the original waste by 95%–96%, depending upon composition and degree of recovery of materials such as metals from the ash for recycling. While incineration does not replace landfilling, it does reduce the necessary volume significantly. In fact, incineration has particularly strong benefits for the treatment of certain types of waste in niche areas such as clinical waste and hazardous waste where pathogens and toxins can be destroyed by high temperatures.

Incineration is a controlled combustion process used for reducing solid wastes to carbon dioxide, water, flue gas, and ash. The ash residue may require disposal at a mono-fill or sanitary landfill given its toxicity, leaching potential, and contaminants. In addition, atmospheric emissions must be controlled to maintain good air quality. Not all solid waste can be incinerated; wastes such as white goods, industrial process waste, street and drain cleanings, and scrap steel cannot be fed to an incinerator and require alternative disposal methods. One of the most difficult aspects of waste as feedstock, whether for incineration or gasification is its heterogeneous nature. Waste composition is influenced by numerous local conditions, such as population structure and local waste separation and recycling regulations. In addition, the composition of local waste is subject to major seasonal or even daily variations. The variability in chemical composition, water and ash content, heating value, and the presence of a number of substances, such as sulfur, chlorides or metals, affects the performance of thermal conversion processes.

Modern incineration plants reflect significant advances in addressing the technical and practical difficulties of material handling, combustion control, flue gas clean up, and conformance with increasingly stringent air pollution control regulations. Energy recovery and production offer a direct economic benefit and lessen demands on processing and disposal requirements. The types, quantities, and distribution of solid waste with fuel potential, particularly urban solid wastes, have received increased attention over the past decade due to rapidly escalating energy costs.

11.3.5.1 Types of incinerators

The incinerator is, simply, a furnace for burning refuse and modern incinerators include pollution mitigation equipment such as flue-gas cleaning. There are various types of incinerator plant design: (1) simple incinerator, (2) fixed or moving grate incinerator, (3) rotary kiln incinerator, and (4) the fluidized bed incinerator.

The simple incinerator is an older and simpler kind of incinerator that is, essentially, a brick-lined cell with a metal grate over a lower ash pit, with one opening in the top or side for loading and another opening in the side for removing incombustible solids (clinker). Many small incinerators formerly found in apartment houses have now been replaced by waste compactors.

The fixed or moving grate incinerator is a large fixed hearth incinerator with a moving grate. The moving grate enables the movement of waste through the combustion chamber to be optimized to allow a more efficient and complete combustion. These incinerators are typically used for combustion of municipal waste and are thus referred to as MSW incinerators.

In the fixed or moving grate incinerator the waste is introduced through the "throat" at one end of the grate, from where it moves down over the descending grate to the ash pit in the other end. Here, the ash is removed through a water lock. Part of the combustion air (primary combustion air) is supplied through the grate from below. This airflow also has the purpose of cooling the grate itself. Cooling is important for the mechanical strength of the grate, and many moving grates are also water cooled internally. Secondary combustion air is supplied into the boiler at high speed through nozzles over the grate. It facilitates complete combustion of the flue gases by introducing turbulence for better mixing and by ensuring a surplus of oxygen.

The incinerator must be designed to ensure that the flue gases reach a temperature of at least 850°C (1560°F) in order to ensure proper breakdown of organic toxins. This includes backup auxiliary burners (often fueled by oil), which are fired into the boiler in case the heating value of the waste becomes too low to reach this temperature alone. The flue gas is then cooled by heat transfer and heat the steam to typically 400°C (750°F) at a pressure of 550–600 psi for the electricity generation in the turbine. At this point, the flue gas has a temperature of around 200°C (390°F) and is passed to the flue gas cleaning system.

The rotary kiln incinerator has a primary chamber and secondary chamber. The primary chamber consists of an inclined refractory lined cylindrical tube. Movement of the cylinder on its axis facilitates the movement of waste. In the primary chamber, there is conversion of solid fraction to gases, through volatilization, destructive distillation, and partial combustion reactions. The secondary chamber is necessary to complete gas phase combustion reactions.

The clinker spills out at the end of the cylinder. A tall flue gas stack, fan, or steam jet supplies the needed draft. Ash drops through the grate, but many particles are carried along with the hot gases. The particles and any combustible gases may be combusted in an afterburner. To control air pollution the combustion product gases are further treated with acid gas scrubbers to remove sulfuric acid and emissions of nitric acid and then routed through baghouses to remove particulate matter before the gas is released into the atmosphere.

The fluidized bed incinerator uses a strong airflow through a sand bed until a point is reached where the sand particles separate to let the air through and mixing and churning occurs, thus a fluidized bed is created and fuel and waste can now be introduced. The sand with the pretreated waste and/or fuel is kept suspended on pumped air currents and takes on a fluid-like character. The bed is thereby thoroughly mixed and agitated keeping small inert particles and air in a fluid-like state. This allows all of the mass of waste, fuel, and sand to be fully circulated through the furnace.

The heat produced by an incinerator can be used to generate steam that may then be used to drive a turbine in order to produce electricity. However, the amount of energy produced is very much dependent upon the type and composition of the waste used as the feedstock. Much of the current thinking involves coincineration of a specified amount of the waste (calculated on the basis of the carbon content of the waste) with, for example, coal to produce a consistent amount of energy. However, the incineration process has a number of outputs such as (1) the mineral ash and (2) the emission to the atmosphere of flue gas. Before the flue gas cleaning, the flue gases may contain significant amounts of particulate matter, heavy metals, dioxin derivatives, furan derivatives, sulfur dioxide, hydrochloric acid, and polynuclear aromatic hydrocarbon derivatives (carcinogens).

The most publicized concerns related to the incineration of MSWs involve the fear that it produces significant amounts of emissions of dioxin and furan derivatives which are considered by many to be serious health hazards. Older generation incinerators that were not equipped with modern gas cleaning technologies were indeed significant sources of dioxin emissions. However, modern incinerators (due to advances in emission control designs and stringent regulations) must limit and even mitigate such emissions.

The quantity of pollutants in the flue gas from incineration plants is reduced by several processes. For example, taking an example from the gas processing industry (Chapter 9: Hydrocracking), particulate matter is collected by particle filtration most often electrostatic precipitators and/or baghouse filters (Mokhatab et al., 2006; Speight, 2019). Hydrogen chloride and sulfur dioxide are removed in scrubbers or as in a dry desulfurization process by injection of a limestone ($CaCO_3$) as a slurry into the flue gas stream before the particle filtration step. Wastewater from scrubbers must subsequently pass through a wastewater treatment plant. Nitrogen oxide (NO_x) emissions are either reduced by catalytic reduction with ammonia in a catalytic converted (a selective catalytic reduction process) or by a high-temperature reaction with ammonia in the furnace (a selective noncatalytic reduction process). Heavy metals are often adsorbed on injected active carbon powder (such as activate charcoal), which is collected by the particle filtration. A as is the case when coal is combusted, incineration of a biomass-based feedstock or organic waste also produces fly ash and bottom ash.

By the way of explanation, heavy metals are generally defined as metals with relatively high density, atomic weight, or atomic number. Fly ash is a combustion product that is composed of the particulates (fine particles of burned fuel) that are driven out of coal-fired boilers together with the flue gases. Fly ash is, thus, that portion of the ash that escapes up the chimney or stack. On the other hand, bottom ash is part of the noncombustible residue of combustion in a power plant, boiler, furnace, or incinerator. In an industrial context, it typically comprises traces of combustibles embedded in forming clinkers and sticking to hot side walls of a furnace during its operation. The clinker materials fall by themselves into the bottom hopper of a coal-burning furnace and are cooled. The above portion of the ash is also referred to as bottom ash.

The total amount of ash produced by MSW incineration ranges from 15% to 20% w/w of the waste and the fly ash amounts to 0%–10% w/w of the total ash. The fly ash, by far, constitutes more of a potential health hazard than does the bottom ash because the fly ash often contains high concentrations of heavy metals such as lead, cadmium, copper, and zinc as well as small amounts of dioxin derivatives and furan derivatives. The bottom ash seldom contains significant levels of heavy metals.

While fly ash is always regarded as hazardous waste, bottom ash is generally considered safe for regular landfills after a certain level of testing defined by the local legislation. Ash, which is considered hazardous, may generally only be disposed of in landfills which are carefully designed to prevent pollutants in the ash from leaching into underground aquifers—or after chemical treatment to reduce the leaching characteristics of the ash.

In spite of the promise shown by the use of incinerators for waste disposal and, in some cases, energy production (in the form of fuel gases), the use of incinerators for waste incineration is often controversial. However, the concerns over the health effects of the emissions of furan derivatives and dioxin derivatives have been significantly lessened by advances in emission control designs and very stringent new governmental regulations that have resulted in large reductions in the amount of dioxin derivatives and furan derivatives emissions emitted into the environment. However, there may be concerns related to the health effects of furan and dioxin emissions into the atmosphere from older generation incinerators.

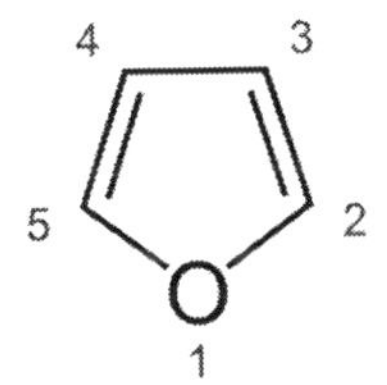

Furan

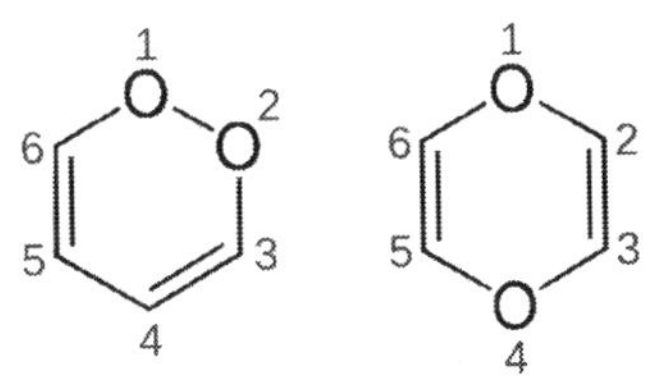

1,2-Dioxin 1,4-Dioxin

Also, incinerators emit varying levels of heavy metals such as (alphabetically) arsenic, cadmium, chromium, lead manganese, mercury, nickel, and vanadium, which can be toxic at very minute levels. Other advanced alternative technologies are available such as biological treatment combined with anaerobic digestion and gasification.

11.3.5.2 Incineration plants

Modern incineration plants generate electricity and heat that can be sold to the regional electric grid and can sell steam to district heating systems or industrial customers. The bottom ash residue remaining after combustion has been shown to be a nonhazardous solid waste that can be safely landfilled or possibly reused. In densely populated areas, finding space for additional landfills is becoming increasingly difficult.

Incineration of MSW avoids the release of carbon dioxide and methane. Every ton of MSW incinerated prevents approximately 1 ton of carbon dioxide equivalents from being released to the atmosphere. Incineration of medical waste and sewage sludge produces an end-product ash that is sterile and nonhazardous. On the other hand, the end-product ash must still be safely sent for disposal. Many countries deny the use of incineration as an option for the treatment of their MSWs because of the potential of environmental pollution. In addition, incinerators produce soot particles ($PM_{2.5}$, nanoparticles, i.e., particles with a size less than 2.5 μm, $<1 \times 10^{-6}$ m) in the furnace. Even with modern particle filtering of the flue gases —such as baghouse filters—some of these small particles are emitted to the atmosphere.

However, incineration is not a complete method of disposal, its main advantage is that it produces a residue that is substantially reduced in volume and may be relatively inert. In addition, the efficiency of an incinerator is, on occasion, due to excessive throughput or choice of operating parameters that are less than desired. For example, the operating temperature inside the plant should be at the specified temperature (usually on the order of 850°C, 1560°F) to reduce the possible effects of the most dangerous air pollutants and the production of dioxins. A lower than specified combustion temperature implies incomplete incineration, leaving part of the waste in the ashes and other combustion residues. Furthermore, the exhaust gases may be unnecessarily polluting because of the low combustion temperature and the by-products of such combustion may be sent to an unprotected landfill for disposal.

Incineration plants are usually classified as mass-burn systems or refuse-derived fuel (RDF) systems. In the mass-burn system, all waste is incinerated; they are usually large facilities, over 200 tons (4000 lb) per day. Most mass-burn systems burn waste in a single combustion chamber with excess air. The waste is burned on a sloping moving grate which helps agitate the MSW and mixes the waste with combustion air; many different proprietary grate systems exist. Grate design is critical because it must agitate the waste sufficiently to provide complete combustion and minimize ash carryover in the flue gas which contributes to emissions.

In the boiler/furnace portion of the incinerator the waste characteristics of interest are the CV, moisture content, proportion of noncombustibles, and other components such as heavy metals, chlorine, and sulfur whose presence during combustion will result in the need for flue gas clean up. The capacity of the incinerator's furnace/boiler is roughly proportional to the CV of the waste. Mass-burn produces by-products that have damaging consequences on the environment. Pollution control systems can be used to mitigate such risks. The technologies used to recover energy from MSW include (1) waterwall incineration and (2) modular incineration.

Waterwall incineration involves the combustion of unprocessed MSW (mass burning) or processed MSW in a furnace with integral boiler tubes. Waterwall units have been built as small as 75–100 tons per day capacity. However, the cost per ton of rated capacity of such units is relatively high. A more common unit size is 250–300 tons per day, while waterwall mass fired units have been built as large as 750–1200 tons per day capacity.

The advantages of waterwall incineration are as follows: (1) preprocessing of waste is not required, except for large bulky items; (2) the ash produced could be used for block manufacture and (3) auxiliary fuel is not required during normal steady-state operation. On the other hand, the challenges of waterwall incineration are as follows: (1) capital and operating costs are high; (2) complexity of operation, control, and maintenance; and (3) reserve furnace capacity is not available during downtime. Modular incineration is used in smaller plants and is an assembly of prefabricated major components assembled in the field to form a total operation. Modular incinerators have been built in individual unit sizes up to 100 tons per day, combined into plants of just over 400 tons per day.

Modular systems are similar to mass-burn systems in that these systems not only combust unprocessed MSW but also feature two combustion chambers and the MSW is charged into the system by a hydraulic ram and combustion takes place on a series of stationary hearths. The MSW is pushed from one hearth to the next by hydraulic rams. Two types of modular systems have been built and operated: starved air (slightly oxygen deficient) and excess air (greater than stoichiometric).

The advantages of modular incineration are (1) low cost and (2) shorter field construction time given their modular construction. The disadvantages of modular incineration are as follows: (1) waste burnout is not always complete (increased ash quantities and reduced energy recovery); (2) combustion control is less effective (possibility of discharge of trace organics); (3) lower quality of equipment; and (4) less redundancy that larger mass-fired waterwall and refuse-derived fuel waste to energy plants.

Feedstock preparation plays an important role in any thermal waste conversion approach. RDF is produced when MSW is processed by mechanical means (picking), separation of inert materials (glass, minerals, and ferrous and nonferrous metals), flailing, screening, shredding, and conditioning to produce a more homogenous material prior to incineration. RDF systems utilize pretreated MSW. Several types of refuse-derived fuel are generated—coarse, fluff, powder, densified. They differ in complexity, power requirements, particle size, and whether the material is compacted into pellets or briquettes. RDF can be used as the primary fuel source in a dedicated boiler or cofired with a conventional fossil fuel as natural gas or landfill gas. The heating value of nearly all solid waste fuels is a function of carbon content. Ash content is generally low, but the amount of moisture is highly variable and depends upon moisture generation plus the effects of processing, handling, and storage.

Scrap tires can be processed into a waste-derived fuel and mixed with other fuels without major alteration to the existing combustion facility. Tire-derived fuels (TDFs) are commonly used as a supplemental fuel in cement kilns which are technologically and environmentally superior for selective waste disposal due to favorable internal kiln conditions as flame temperature (1850°C, 3360°F), residence time, alkaline environment, high gas turbulence and mixing which ensure complete combustion, and the added benefit of having no bottom ash disposal problem since the residual materials (e.g., steel from steel-belted tires) are incorporated into the clinker. The rationale behind the use of TDF is as follows: (1) excellent combustion characteristics and high heating value; (2) lower cost and good availability; (3) relatively low sulfur content, especially on a Btu basis; and (4) a reduction in the environmental burden and health effects of tire stockpiles.

Typically, the sulfur content of scrap tires and TDFs is on the order of 1.5%–2.0% w/w. Considering the emission potential of sulfur dioxide per the generated energy value of the fuel, the result is closer to that for low sulfur-containing coals.

11.3.6 Landfilling

With respect to municipal and industrial solid wastes and provided that there is no shortage of land with suitable geological formations, landfill remains the principal final disposal route for the majority of wastes, even in highly industrialized countries. Where there is treatment, it is usually designed to reduce the volume of waste to be landfilled and includes compaction, shredding, baling, and incineration. Most solid wastes therefore will directly be disposed of in sanitary landfills.

The prefix sanitary is mainly to be understood as providing some protection for the population against airborne dust and litter, stench, rodents, and insects. Most of these nuisances can be prevented by the prompt covering of freshly dumped waste with soil. However, to be able to call a waste tip or landfill sanitary from an environmental viewpoint requires more measures to be taken. The main environmental problem associated with landfilling is pollution of groundwater. Rainwater percolating through solid waste tends to carry large amounts of pollutants to groundwater aquifers if the underlying strata are pervious or fissured. Thus wells drawing from the aquifers will be extracting groundwater contaminated by the leachate; such a situation is often difficult to remedy. Studies have shown that the leachate from solid wastes may have a pollution load up to 15–20 times higher than domestic wastewater.

Landfill tends to predominate as a waste disposal mode because it is regarded as an effective but low-cost method of disposal, also for hazardous waste. Even where other methods are more suitable for environmental reasons, the higher capital and (short-term) running costs mean that they cannot compete without government intervention. However, such cost calculations take no account of the longer term. In the long run, landfill of hazardous materials may impose a larger financial burden than other methods because of the high cost of ensuring that the site remains secure for the time it takes for the waste to be rendered harmless.

Disposal to landfill of untreated wastes other than inert material is becoming less acceptable, partly in response to a few well-documented instances where poorly designed and operated landfills are giving rise to pollution problems, and partly as a result of greater public awareness of the issues involved.

11.3.7 Pyrolysis

Pyrolysis is the thermal decomposition of (in the current context) organic materials at elevated temperatures in an inert atmosphere. The process involves a change of chemical composition of the material to be pyrolyzed and is irreversible. In general, pyrolysis of organic substances produces volatile products and leaves a solid residue enriched in carbon (char)—as an example, pyrolysis is one of the processes involved in the reduction of char from wood. As another example, the process is used frequently in the chemical industry, for example, to produce ethylene, many forms of carbon, and other chemicals from crude oil, coal, and wood, to produce coke from coal. In the current context, other applications of pyrolysis would convert biomass into synthesis gas and waste plastic materials into usable oil and char.

Generally, pyrolysis is a medium temperature method that produces gas, oil, and char from crops which can then be further processed into useful fuels or feedstock. Pyrolysis is often considered to be the gasification of, say, biomass in the absence of oxygen but the chemistry of each process may differ significantly. In general, biomass does not gasify as easily as coal, and it produces other hydrocarbon compounds in the gas mixture exiting the gasifier; this is especially true when no oxygen is used. The products of biomass pyrolysis include gases such as methane, hydrogen, carbon monoxide, carbon dioxide, biooil, and char. As a result, typically an extra step must be taken to reform these products with a catalyst to yield a clean synthesis gas mixture of hydrogen, carbon monoxide, and carbon dioxide. Thus when organic matter is heated at increasing temperatures in open containers, a variety of processes generally occur, in successive or overlapping stages (Table 11.4).

Depending upon the operating conditions (temperature, heating rate, particle size, and solid residence time), pyrolysis can be divided into three subclasses: conventional, thermal or catalytic, and fast or flash. For example, to maximize biooil production, fast/flash pyrolysis is used, heating the biomass to approximately 500°C (930°F) for less than 10 seconds. In plastics pyrolysis, the macromolecular structures are broken down into smaller molecules (oligomers and monomers); further degradation depends on temperature, residence time, presence of catalysts, and other process conditions. Pyrolysis aims to convert waste into other valuable chemicals, which can be used as feedstock for downstream industrial processes or transportation fuels.

Fast pyrolysis is a thermal decomposition process that occurs at moderate temperatures with a high heat transfer rate to the feedstock particles and a short hot vapor residence time in the reaction zone. Several reactor configurations have been shown to assure this condition and to achieve yields of liquid product as high as 75% based on the starting dry weight of the feedstock. These configurations include bubbling fluid beds, circulating and transported beds, cyclonic reactors, and ablative reactors.

Table 11.4 Examples of the stages that occur during pyrolysis.

Temperature (°C)	Event
<100	Volatiles, including some water, evaporate. Heat-sensitive substances may partially decompose.
100	Water remaining absorbed in the material is evolved. Water trapped in crystal structure of hydrates may be evolved at somewhat higher temperatures. Some solid substances such as fats, sugars, and waxes may melt.
100–500	Organic molecules break down. Most sugar derivatives start decomposing at 160°C–180°C. Cellulose decomposes at approximately 350°C. Lignin starts decomposing at approximately 350°C and continues releasing volatile products up to 500°C. The decomposition products usually include water, carbon monoxide, and/or carbon dioxide as well as a large number of organic compounds. The nonvolatile residues typically become richer in carbon and form large disordered molecules, with colors ranging between brown and black.
200–300	When oxygen has not been excluded, the carbonaceous residue may start to burn in a highly exothermic reaction releasing carbon dioxide and/or monoxide. At this stage, some of the nitrogen still remaining in the residue may be oxidized into nitrogen oxides (such as NO_2 and N_2O). Sulfur and other elements (such as chlorine and arsenic) may be oxidized and volatilized.

Briefly, ablative pyrolysis is conceptually very different compared to other fast pyrolysis methods. Most pyrolysis processes depend on having a high heat-transfer rate, which determines the rate of the reaction, and, therefore, these systems employ small feedstock particles. Ablative pyrolysis depends on heat transfer from the hot reactor wall to melt the wood particles contacting the wall under pressure. Unlike other pyrolysis processes, because the reaction rates are not limited by heat transfer through the feedstock particle, large feedstock particles up to lumber size may also be processed by an ablative pyrolysis reactor.

Fast pyrolysis of biomass produces a liquid product (referred to as pyrolysis oil or biooil) that can be readily stored and transported. Pyrolysis oil is a renewable liquid fuel and can also be used for the production of chemicals. Fast pyrolysis has now achieved a commercial success for the production of chemicals and is being actively developed for producing liquid fuels. Pyrolysis oil has been successfully tested in engines, turbines, and boilers and been upgraded to high-quality hydrocarbon fuels. In the 1990s several fast pyrolysis technologies reached near-commercial status and the yields and properties of the generated liquid product depend on the feedstock, the process type and conditions, and the product collection efficiency.

Molten salts such as lithium chloride–potassium chloride with 10% w/w cuprous chloride (LiCl–KCl–10% CuCl) at 520°C (970°F) are used to pyrolyze polymeric waste to yield a fuel oil. At a process temperature of 420°C (790°F) the gaseous fraction is minimized; this enables a higher liquid and solid fraction to be recovered. The recovered fractions consist of low-boiling oils, aromatic derivatives, paraffin waxes, and monomers.

Hydrocracking of polymer waste involves reaction with hydrogen over a catalyst in a stirred batch autoclave at moderate temperatures and pressures. Typical plastic waste feedstock such as polyethylene (PE), PE terephthalate, polystyrene, polyvinyl chloride, and mixed polymers are reacted in the presence of a catalyst (platinum, nickel, molybdenum, and iron on acid solids such as alumina, zeolites, and sulfonated zirconia) at 150°C–400°C (300°F–750°F) and hydrogen at 3–10 MPa to produce a high-quality gasoline.

Catalytic cracking involves a suitable catalyst to carry out the cracking reaction. The presence of catalyst lowers the reaction temperature and time. Catalytic degradation yields a narrow product distribution with a peak at lower molecular weight hydrocarbon derivatives and occurs at lower temperatures. Using solid acid catalysts, HZSM-5, HY, rare earth metal-exchanged Y-type zeolite, and silica-alumina for polymeric waste yields 30% gasoline and 20% high-boiling oil. The differences amongst catalysts arise in the production of coke and gas fractions. Formation of coke is not desirable as it results in catalytic deactivation by fouling and plugging the pore paths. Catalytic cracking, owing to its lower operating temperature, appears to be more technologically sound than thermal pyrolysis, however it is more costly. Addition of catalyst enhances the conversion and fuel quality, as compared to pure thermal pyrolysis.

There are numerous proposed mechanisms for catalytic degradation. These include the carbonium ion mechanism and the free radical mechanism. In the carbonium ion (R^+) mechanism, there are four steps: initiation, depropagation, isomerization, and aromatization. In the initiation step the carbonium ion is generated through proton addition to an olefinic linkage, beta-scission, or through the abstraction of a hydride ion by low molecular weight carbonium ions.

In the depropagation step the molecular weight of the main polymer chain can be successively reduced through beta-scission. In the isomerization step the carbonium ion intermediates can undergo hydrogen atom shift or carbon atom shift, leading to rearrangement of the molecules. Some of the carbonium ion intermediates can also undergo cyclization reactions, which leads to the formation of aromatics (aromatization step).

In the free radical mechanism, there are three known radical mechanism steps: initiation, propagation, and termination. Normally, random bond breaking generates a radical in the initiation step after which the radical is propagated via a number of reactions such as beta-scission, radical abstraction, and radical decomposition. Finally, recombination of two radicals leads to the termination step. Thus:

Initiation:

$$R_1R_2 \rightarrow R_1^{\cdot} + R_2^{\cdot}$$

Example: $RCH_2CH = CHCH_2R + HX \rightarrow RCH_2CH_2{}^+CHCH_2R + X$

Propagation:

$$R_1^{\cdot} + R_3 \rightarrow R_3$$

$$R_1^{\cdot} \rightarrow C_x + R_4$$

Example: $RCH_2CH_2{}^+CHCH_2CH_2CH_2R \rightarrow RCH_2CH_2CH = CH_2 + {}^+CH_2CH_2R$

$$RCH_2CH_2CH_2R + R^+ \rightarrow RCH_2{}^+CHCH_2R + RH$$

Termination:

$$R_4^{\cdot} + R_5^{\cdot} \rightarrow R_4R_5$$

Thermal cracking (pyrolysis) involves the degradation of polymeric materials by heating in the absence of oxygen and is usually conducted between 350°C and 900°C (650°F and 1650°F). The products formed are a carbonized char and a volatile fraction that may be separated into condensable hydrocarbon oil consisting of paraffins, iso-paraffins, olefins, naphthenes, aromatics and a noncondensable high CV gas.

The thermal pyrolysis of polymeric waste at temperatures higher than 480°C (895°F) yields a gas–liquid hydrocarbon. At 650°C–760°C (1200°F–1400°F) more of the gaseous product is generated; conversely, at lower temperatures, up to 85% of the product is liquid hydrocarbon derivatives. The portion of each fraction and their precise composition depends primarily on the properties of the waste plastic material(s) and the process parameters (temperature, pressure, residence time of the feedstock in the thermal zone). Both the gaseous and the liquid forms of the polymeric waste can be utilized as a feedstock to petrochemical facilities (Speight, 2019). Generally, thermal cracking results in liquids with low octane value and high residue content at moderate temperatures, this is an inefficient process for producing gasoline ranges. The gaseous products obtained by thermal pyrolysis are not suitable for use as fuel products and require further refining to be upgraded to useable fuel products. A few researchers have sought to improve thermal pyrolysis of waste plastics without employing the use of catalysts. However, these changes either yielded insignificant improvements or added another level of complexity and costs to the system.

Typically, discarded tires can be reduced in size by grinding, chipping, pelletizing and passed through a classifier to remove the steel belting after which the chips are pyrolyzed for 1 hour at a temperature of 300°C–500°C (570°F–930°F) and then heated for 2 hours in a closed retort. The products are then distilled (purification and analysis); this yields oil and gas and char. However, discarded tires can also be debeaded, shredded to 25 mm and ground to 24 mesh as a feed preparation step for occidental flash pyrolysis which involves flash pyrolysis and product collection. The pyrolytic reaction occurs without the introduction of hydrogen or using a catalyst. This yields a gaseous stream that is passed to a quench tower from which fuel oil and gas (recycled to char fluidized and pyrolysis reactor as a supplemental fuel) and carbon black (35% w/w) are produced.

Hydrothermal liquefaction and pyrolysis are often confused since both are thermochemical processes where organic compounds are converted to liquid products. However, in hydrothermal liquefaction, macromolecules are decomposed into unstable, reactive low molecular weight products in the presence of a catalyst. These fragments react further to produce high molecular weight compounds that may be constituents of the oil product.

The hydrothermal liquefaction process involves solvolysis, decarboxylation, hydrogenolysis, and hydrogenation, leading to challenging mechanistic transformation studies. The process involves converting biomass to an oily liquid by contacting the feedstock with water at elevated temperatures (300°C–350°C, 570°F–660°F) with sufficient pressure to maintain the water primarily in the liquid phase residence times up to 30 minutes. During solvolysis the solvent reacts with the molecule, helping to begin the degradation of the feedstock. Depolymerization of hemicelluloses, cellulose, and lignin further degrades the feedstock to lower molecular weight products. In decarboxylation, loss of carbon dioxide can lead to rearrangement of the molecule. Molecules can also be further transformed via hydrogenolysis (water molecules split into H^+ and OH^- and react with molecules) and hydrogenation (H_2 is added to the molecule).

Alkali may be added to promote organic conversion. The primary product is an organic liquid with reduced oxygen content (approximately 10%) and the primary byproduct is water containing soluble organic compounds.

11.3.8 Other processes

There are several other processes by which organic waste material can be converted to potentially useful products. The most common composting techniques in the current context (alphabetically and not in order of preference) are (1) acid hydrolysis, (2) briquetting, (3) enzymatic hydrolysis, and (4) transesterification.

11.3.8.1 Acid hydrolysis

Hydrolysis is any chemical reaction in which a molecule of water ruptures one or more chemical bonds. The term is used broadly for substitution, elimination, and fragmentation reactions in which water is the nucleophile:

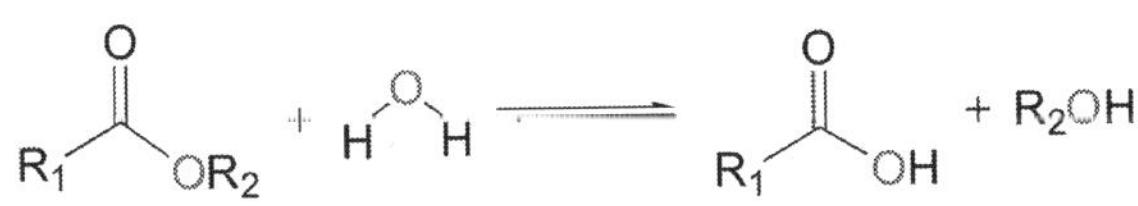

Acid hydrolysis is the means by which cellulosic material can be converted to lower molecular weight products and thence to fuels. Enormous amounts of inexpensive and renewable lignocellulose waste are generated each year from forestry and agricultural production. These include fruit processing residues, dairy industry wastes, corn and sugar by-products, and paper industry wastes. Lignocellulosic feedstock can be converted to biofuels such as ethanol and hydrogen via simple sugars. Lignocellulose is a composite of cellulose fibers embedded in a cross-linked lignin–hemicellulose matrix. It gives structure and strength to plants. Although abundant, lignocellulosic waste is difficult to convert to fermentable sugars because of its complex chemical structure: its three major components [cellulose (crystalline and amorphous) hemicellulose and lignin] must be processed separately (Lee et al., 2007).

Cellulose consists of linear, highly ordered chains of glucose and is one of the most abundant biopolymers on the planet. When produced by plants, it has both highly amorphous regions containing large voids and other irregularities as well as tightly packed crystalline regions. It also accumulates in the environment because it is resistant to most forms of degradation. Hemicellulose is a complex polymer of a variety of sugars. This mix of sugars mainly consists of six-carbon and five-carbon sugars. Lignin is a polyphenolic polymer that fills the spaces in the cell wall between cellulose, hemicellulose, and pectin components. It is covalently linked to hemicellulose and crosslinks different plant polysaccharides, conferring mechanical strength to the cell wall, and by extension, the plant as a whole. The content of cellulose, hemicellulose, and lignin in common agricultural residues depends upon the source and origin of the feedstock (Table 11.1). However, because lignocellulosic feedstock is so complex, it is difficult to use as a feedstock for biofuel. A variety of physical, chemical, and enzymatic processes have been developed to fractionate lignocellulose into the major plant components of hemicellulose, cellulose, and lignin.

Ethanol is produced from lignocellulose via pretreatment, hydrolysis, fermentation, and distillation. The goal of pretreatment is to increase the surface area of lignocellulosic material, making the polysaccharides more susceptible to hydrolysis, by separating the xylose and lignin from the crystalline cellulose. Pretreatment must also avoid the degradation or loss of precious carbohydrates and avoid the formation of by-products that can inhibit subsequent steps. However, pretreatment often produces biological inhibitors, which affect fermentation. A large variety of pretreatment processes have been developed. Common pretreatments are steam explosion, acid treatment, biological methods, and comminution. These methods can be used singly or in combination.

Steam explosion involves saturation of the pores of plant materials with steam followed by rapid decompression. The explosive expansion of steam reduces the plant material to separated fibers, increasing the accessibility of polysaccharides to subsequent hydrolysis. Ammonia fiber explosion is similar to the steam explosion except that liquid ammonia is used. It is very effective on agricultural residues but has not been successful in pretreating woody feedstock. Biological pretreatments employ microorganisms that produce lignin-degrading enzymes (ligninase). Comminution is an integral part of pretreatment and uses a hammer mill to produce particle sizes that can pass through 3 mm screen openings.

The mechanisms by which pretreatments improve the digestibility of lignocellulose are not well understood. Pretreatment effectiveness has been correlated with the removal of hemicellulose and lignin (lignin solubilization is beneficial for subsequent hydrolysis but may also produce derivatives that inhibit enzyme activity). Some pretreatments reduce the crystallinity of cellulose, which improves reactivity, but this does not appear to be the key for many successful pretreatments.

Hemicellulose is readily hydrolyzed to pentoses (five-carbon sugars) but pentoses are difficult to ferment. The cellulose hydrolyzes to hexoses (six-carbon sugars). Crystalline cellulose is difficult to hydrolyze but the resulting hexose derivatives are readily fermented. The three basic methods for hydrolyzing structural polysaccharides to fermentable sugars (such as glucose, xylose, and arabinose) are concentrated acid hydrolysis or dilute acid hydrolysis or enzymatic hydrolysis.

Acid treatment is the use of acid to hydrolyze cellulosic materials. Two acid processes hydrolyze both hemicellulose and cellulose with very little pretreatment beyond comminution of the lignocellulosic material to particles of approximately 1 mm in size. These are concentrated and dilute acid hydrolysis. Concentrated acid hydrolysis dissolves carbohydrates in a woody feedstock to form a homogenous gelatine with acid where cellulose is extremely susceptible to hydrolysis. Typically, this is achieved with 70%–90% v/v H_2SO_4 at room temperature, leaving lignin. Before fermentation, the solution of oligosaccharides is diluted to 4% v/v H_2SO_4 and boiled for 4 hours or processed using an autoclave at 120°C (248°F) for 1 hour to yield monosaccharides. Following neutralization with limestone, the sugar solution is fermented. The procedure takes 10–12 hours. Concentrated acid hydrolysis is attractive because it is relatively simple, and high sugar yields (approach 100% of theoretical hexose yields in pure samples and 90% in mixed samples which replicate municipal waste mixtures) are achieved. However, corrosion-resistant equipment is necessary and acid recovery is expensive.

Dilute acid hydrolysis (1% acid w/w) greatly reduces the amount of acid required to hydrolyze lignocellulose. The process is accelerated at elevated temperatures: 100°C–160°C (212°F–320°F) for hemicellulose and 180°C–220°C (356°F–428°F) for cellulose. The high temperatures cause oligosaccharides released from the lignocellulose to decompose, greatly reducing yields of simple sugars to only 55%–60% w/w of the theoretical yield. The decomposition products include toxins (acetic acid, furfural) that inhibit fermentation.

11.3.8.2 Briquetting

A briquette (also spelled briquet) is a compressed block of coal dust (Speight, 2013) or other combustible material (such as charcoal, sawdust, wood chips, peat, or paper) used for fuel as well as for kindling to start a fire.

Historically, briquettes (especially coal and coke briquettes) have been used for fuel for approximately 100 years. Traditionally, briquetting technology was established for developing countries to produce briquettes of local residues, for use in household cooking stoves and restaurants. Later, as the capacities of the machines increased, briquettes were used in industrial boilers to create heat, steam, and power for industry and power plants. Within the past three decades, briquetting has also found its way to households in industrialized countries as consumer logs for wood-burning stoves and fireplaces. However, with the advent of modern fuel systems, the use of briquettes use has declined and use for these products is found most often in the operation of barbecue units. While still a marketable product for such use, briquettes as fuel (such as smokeless fuel) are not used often for domestic and industrial heating. In more recent years, as the focus on renewable energy has grown, the applications for briquettes have grown concurrently, as have different technologies and new applications.

Briquette manufacture (briquetting) involves the collection of combustible materials that are not usable as such because of their low density and compressing them into a solid fuel product of any convenient shape that can be burned Using waste materials as a means to produce briquettes are investigated in this section. Briquettes from waste materials are commonly made from a combustible material and binder. Combustible materials include char, low-grade biomass, and bagasse. Low-grade biomass includes grasses, weeds, and thinning branches (i.e., forest waste resulting from logging), agricultural waste, sawdust, wood shavings, and leaves.

A binding agent is usually necessary to increase the cohesion of the combustible materials. If the combustible material is not well-bound, the briquette will crumble when removed from the mold. The binding agents can be sourced based upon cost, local sources, and combustion properties. They can include animal manure, treated and dewatered sewage sludge, starch, wax, clay, molasses, cement, wood pitch glue, and local plant resin or synthetic resin. A binder must not cause smoke or gummy deposits, and the creation of excess dust must also be avoided. For this reason, the use of noncombustible binding agents such as clay, cement, and other adhesive minerals is kept to a minimum. Typically, starch is used because it is relatively cheap and easily available. Since bagasse briquetting is often done on site or close to a sugar factory, molasses is usually the binding agent used for bagasse briquettes.

In the briquetting step the mixture is tightly compacted through a manually operated or automatic press or extruder. The press or extruder for briquetting must be well-designed, strongly built and capable of agglomerating the mixture sufficiently for it to be handled through the drying process. The extruder forms a roll of charcoal while the press favors the production of large pieces (chunks) of charcoal. As explained earlier, sawdust briquettes are formed under sufficiently high pressure to produce cohesion between wood particles. Briquettes often need to be further dried after the briquetting step. Briquettes are dried in the sunlight approximately three days before use. Rolls of charcoal formed from extruders will break into chunks during the drying process.

11.3.8.3 *Enzymatic hydrolysis*

Enzymatic hydrolysis is a process in which enzymes facilitate the cleavage of bonds in molecules with the addition of the elements of water and plays an important role in the human system for the digestion of food. In the current context the process involves the addition of cellulases (a group of enzymes which, when acting together, hydrolyze cellulose) to hydrolyze pretreated a lignocellulosic feedstock into fermentable sugars. The process involves several key steps: (1) transfer of enzymes from the bulk aqueous phase to the surface of the cellulose, (2) adsorption of the enzymes and formation of enzyme-substrate complexes, (3) hydrolysis of the cellulose, (4) transfer of the hydrolysis products from the surface of the cellulosic particles to the bulk aqueous phase, and (5) hydrolysis of the sugar-type products to glucose in the aqueous phase. The overall rate of the process is influenced by the structural features of the lignocellulosic feedstock and the composition and source of the cellulases.

Thus the enzymatic hydrolysis process was developed to better utilize both cellulose and hemicellulose from lignocellulosic materials. The enzymatic process must be preceded by extensive pretreatment to separate cellulose, hemicellulose, and lignin. Because milder conditions are used compared to acid hydrolysis of cellulose, subsequent enzymatic hydrolysis of the cellulose does not degrade pentoses during prehydrolysis. Cellulose is a homopolysaccharide of glucose linked by 1,4-glycoside-type bonds (bonds in which a sugar is bound to another functional group, also known as glycosidic bond). Enzymatic hydrolysis of cellulose proceeds in several steps to break glycoside-type bonds by the action of a system of enzymes known as cellulase. The system of enzymes also contains hemicellulase to hydrolyze any hemicellulose not solubilized by prehydrolysis. Unfortunately, enzymatic technologies are not likely to be suitable for unsorted MSW due to its high heterogeneity, and further research is necessary to convert wastes that are very high in hemicellulose.

In a fermentation process, glucose is fermented to produce a dilute ethanol–water solution (dilute beer). Xylose is difficult to convert to ethanol but does occur. Lignin, which is not susceptible to biological transformation, can be chemically upgraded or, more frequently, simply burned as boiler fuel.

SSF has been developed for fermenting sugars released from lignocellulose. The process combines hydrolysis and fermentation to overcome end-product inhibition. By combining hydrolysis and fermentation, glucose is rapidly removed before it can inhibit further hydrolysis. In the process the feedstock is milled and then prehydrolyzed to yield a mixture of pentoses (xylose and arabinose) and fiber. The mixture is neutralized with limestone and mixed with cellulose and hemicellulose enzymes. The cellulose and any remaining hemicellulose are solubilized to hexose (glucose) and pentoses (xylose and arabinose), which are immediately fermented to ethanol. The optimum temperature for the hydrolysis/fermentation reaction is between the optimum temperature for cellulase activity and the best temperature for the yeast. Lignin, as usual, is separated from the mixture and used as boiler fuel or converted to high-value octane enhancers that can be blended with gasoline.

11.3.8.4 Transesterification

In the 1970s it was found that the viscosity of fresh vegetable oils (biooils) could be reduced through a simple process, referred to as transesterification.

Transesterification is the process by which triglycerides (vegetable oils, waste cooking oils, and fats) are reacted with methanol or ethanol to produce methyl or ethyl esters respectively along with the coproduct glycerol. This process allowed biooils from natural sources to be converted to biodiesel.

Biodiesel is diesel fuel produced from biomass or some alternative source as opposed to petroleum-derived biodiesel. Biodiesel is advantageous compared to conventional biodiesel in that it is renewable, biodegradable, nontoxic, and has low emissions. Recent research has suggested that biodiesel can be produced from waste oils such as used cooking vegetable oil and from inedible animal tallow collected from slaughterhouses and other meat processing units.

Vegetable oil–based biodiesel has been shown to be comparable to conventional diesel. It has low viscosity, low flash points, high vapor pressure, and is easy to process. Animal fats are preferred to vegetable oils because they are much cheaper and there is a much larger supply. Animal fats cannot be used directly because their physical properties cause poor atomization, poor vapor–air mixing, low pressure, incomplete combustion, and engine deposits. Transesterification is a means to improve all of these disadvantages. Biodiesel from animal fats have higher CVs and cetane numbers but are less stable to oxidation and has a higher cold filter plugging point as compared to biodiesel from vegetable oils.

In the conventional synthesis of biodiesel the transesterification occurs in the presence of methanol and base (usually sodium hydroxide). In this reaction the biooil (a triester or triglyceride) reacts with methanol and a catalyst to produce glycerol and three new methyl-substituted esters. Mechanistically, the triglycerides are converted to diglycerides, which are in turn converted to monoglycerides. Finally, the monoglyceride is converted to glycerol. A high ratio of methanol to oil is required for transesterification. The high ratio of methanol also helps the glycerol that is formed phase separate. Once the glycerol phase separates from the biodiesel, the biodiesel is purified via water washing, vacuum drying, and filtration. The energy-intensive purification of the biodiesel is one of the many reasons its processing is not energy efficient.

Production of biodiesel is dependent upon a number of variables. These include (1) the feedstock, (2) the catalyst, (3) the solvent, (4) the temperature and the pressure, and (5) the ratio of alcohol to the oil.

The feedstock usually includes animal tallow (animal fat) and waste cooking vegetable oil. Most studies on animal tallow focus on freshly collected animal tallow, which is low in free fatty acid value. If the animal tallow is not immediately converted, the acid value increases due to hydrolysis caused by the presence of water and degradation. Free fatty acids can react with basic catalysts, thus using up the catalyst, and the so-called soap products are difficult to separate from the desired biodiesel. Waste cooking oils also have higher amounts of free fatty acids and higher levels of moisture.

The high levels of moisture and free fatty acids are usually dealt with through some sort of acidic pretreatment that causes acid-catalyzed transesterification of the free fatty acids, followed by alkaline transesterification of the biooils. Numerous studies, in fact, have introduced an acidic pretreatment or simply used an acidic catalyst when the feedstock has a large amount of free fatty acids.

Conventionally, alkaline bases are used as the catalyst for transesterification; however, acidic catalysts are also beginning to be used. Common basic catalysts used include sodium hydroxide, potassium hydroxide, or their alkoxides or carbonates. Acidic catalysts commonly used are sulfuric acid and hydrochloric acid. Novel enzymatic catalysts are being tested and scientific literature indicates that they show excellent activity and reduction in undesired coproducts; however, they have significantly higher production costs.

Typically, homogenous strong bases are used as the catalyst for biodiesel production; however, there have been numerous studies on heterogeneous catalysts. Heterogeneous catalysts are preferred in large-scale productions because of their ease of separation. Obviously, any new catalyst must be inexpensive since the final product (biodiesel) must be inexpensive. Thus eggshells have been proposed as potential catalysts for biodiesel production. When 3% w/w calcined eggshells (1000°C, 1830°F, 2 hours) were added to a 9:1 molar ratio mixture of methanol and oil, the biodiesel yield exceeded 95% after 3 hours. The reaction conditions and yield were similar to the homogenous catalytic reaction; however, you now have the benefit of a low-cost, heterogeneous catalyst.

The most common alcohols used are methanol and ethanol; however, methanol is used more frequently because it is lower in cost. When methanol is used as the solvent, the products are methyl esters. Ethanol is being further investigated because the viscosities of ethanol–ethyl ester mixtures are lower, and ethanol can be produced from renewable or waste resources. Furthermore, ethanol separates from glycerol more effectively.

Transesterification can occur at room temperature if the reaction is left long enough. However, usually, the temperature is kept close to the boiling point of the alcohol used at atmospheric pressure. It has been noted, though, that optimal reaction temperature and pressures depend upon the type and quality of the waste cooking oil being processed. Alcohol in excess of the stoichiometric proportion necessary is commonly used. For acid-catalyzed conversion of oils with high amounts of free fatty acids, much higher ratios are used compared to base-catalyzed transesterification. Specifically, ratios can range from 6:1 alcohol to oil for base-catalyzed esterification all to way to 250:1 for acid-catalyzed reactions.

Using wastes as sources for biodiesel has several major advantages: (1) avoid the fuel versus food debate since the feedstock is nonedible, (2) reduce the final cost of the fuel since nonedible feedstock is typically cheaper, (3) recycle waste, and (4) the produced fuel is biodegradable. However, the production of biodiesel also has its disadvantages. Transesterification is normally achieved using alkaline catalysts. These catalysts are corrosive and form unwanted by-products that are expensive to separate. Homogenous acid catalysts can also be used, but these are difficult to recycle, operate at very high temperatures, and have significant corrosion and environmental issues.

References

Demirbaş, A., Ozturk, T., 2004. Anaerobic digestion of agricultural solid residues. Int. J. Green Energy 1, 483–494.

Luque, R., Speight, J.G. (Eds.), 2015. Gasification for Synthetic Fuel Production: Fundamentals, Processes, and Applications. Woodhead Publishing, Elsevier, Cambridge.

Mokhatab, S., Poe, W.A., Speight, J.G., 2006. Handbook of Natural Gas Transmission and Processing. Elsevier, Amsterdam.

Prakash, N., Karunanithi, T., 2008. Kinetic modeling in biomass pyrolysis – a review. J. Appl. Sci. Res. 4, 1627–1636.

Speight, J.G., 2013. The Chemistry and Technology of Coal, third ed. CRC Press, Taylor & Francis Group, Boca Raton, FL.

Speight, J.G., 2014a. The Chemistry and Technology of Petroleum, fifth ed. CRC Press, Taylor & Francis Group, Boca Raton, FL.

Speight, J.G., 2014b. Gasification of Unconventional Feedstocks. Gulf Professional Publishing Company, Elsevier, Oxford.

Speight, J.G., 2019. Natural Gas: A Basic Handbook, second ed. Gulf Publishing Company, Elsevier, Cambridge, MA.

Synthesis gas and the Fischer–Tropsch process

12

12.1 Introduction

Synthesis gas (also known by the shortened name, syngas) is the name given to a gas mixture that contains varying amounts of carbon monoxide and hydrogen generated by the gasification of a carbon containing fuel to a gaseous product with a heating value. Examples include (1) steam reforming of natural gas (or higher molecular weight) liquid hydrocarbon derivaitves to produce hydrogen, (2) the gasification of crude oil resids, (3) the gasification of coal, and (4) the gasification of waste in some types of waste-to-energy facilities. The name comes from their use as intermediates in creating synthetic natural gas (SNG) and for producing ammonia or methanol (Table 8.1). Synthesis gas is also used as an intermediate in producing synthetic crude oil for use as a fuel or as a lubricant by way of the Fischer–Tropsch process.

Gasification to produce synthesis gas can proceed from a wide range of carbonaceous organic feedstocks, including biomass and waste plastic (Gay et al., 1980; Bhattacharaya et al., 1999; Baxter, 2005; Brigwater, 2003; EPCI, 2004; Vélez et al., 2009; Arena, 2012; Brar et al., 2012; Basu, 2013). The resulting synthesis gas burns cleanly into carbon dioxide and water vapor:

$$CO + H_2 \rightarrow CO_2 + H_2O$$

Alternatively, synthesis gas may be converted efficiently to methane via the Sabatier reaction at 400°C under pressure, or to a diesel-like synthetic fuel via the Fischer–Tropsch process.

$$CO_2 + 4H_2 \rightarrow CH_4 + 2H_2O$$

The inorganic components of the feedstock, such as metals and minerals, are trapped in an inert char (the carbonaceous residue of the reaction), which may have use as a fertilizer.

When used as an intermediate in the large-scale, industrial synthesis of hydrogen and ammonia, it is also produced from natural gas (via the steam reforming reaction) as follows:

$$CH_4 + H_2O \rightarrow CO + 3H_2$$

The manufacture of gas mixtures of carbon monoxide and hydrogen has been an important part of chemical technology for more than 100 years. Originally, such

The Refinery of the Future. DOI: https://doi.org/10.1016/B978-0-12-816994-0.00012-9

mixtures were obtained by the reaction of steam with incandescent coke and were known as water gas. Used first as a fuel, water gas soon attracted attention as a source of hydrogen and carbon monoxide for the production of chemicals, at which time it gradually became known as synthesis gas. Eventually, steam-reforming processes, in which steam is reacted with natural gas (methane) or crude oil naphtha over a nickel catalyst, found wide application for the production of synthesis gas.

A modified version of steam reforming known as autothermal reforming, which is a combination of partial oxidation (POX) near the reactor inlet with conventional steam reforming further along the reactor, improves the overall reactor efficiency and increases the flexibility of the process. POX processes using oxygen instead of steam also found wide application for synthesis gas manufacture, with the special feature that they could utilize low-value feedstocks such as high-boiling crude oil residua. In recent years, catalytic POX employing very short reaction times (milliseconds) at high temperatures (850°C–1000°C) is providing still another approach to synthesis gas manufacture. Nearly complete conversion of methane, with close to 100% selectivity to H_2 and CO, can be obtained with a rhenium monolith under well-controlled conditions. Experiments on the catalytic POX of *n*-hexane conducted with added steam give much higher yields of H_2 than can be obtained in experiments without steam, a result of much interest in obtaining hydrogen-rich streams for fuel cell applications.

There uses of synthesis gas include use as a chemical feedstock and in gas-to-liquid processes (Mangone, 2002), which use Fisher–Tropsch chemistry to produce liquid fuels as feedstock for chemical synthesis, as well as being used in the production of fuel additives, including diethyl ether and methyl *t*-butyl ether, acetic acid, and its anhydride, synthesis gas could also make an important contribution to chemical synthesis through conversion to methanol. There is also the option in which stranded natural gas is converted to synthesis gas production followed by conversion to liquid fuels.

12.2 Gasification of coal

Gasification is a process that converts carbonaceous materials, such as coal, crude oil, or biomass, into carbon monoxide and hydrogen by reacting the raw material at high temperatures with a controlled amount of oxygen (Luque and Speight, 2015). For the purposes of this text the process has been used for centuries with coal as the feedstock and, therefore, coal can be used in this introduction to the gasification process. I fact, with coal being denounced by many observers as a nondesirable fossil fuel, coal is no longer as popular as it was several decades ago. However, the attack on coal may be unjustified—the observers who lobby against coal do not mention that coal processing plants have made significant progress in the effort to reduce the emissions form such plants. In fact, gasification of any carbonaceous feedstock is a very efficient method for extracting energy from many different types of organic materials and also has applications as a clean waste disposal technique. In the process the feedstock or the feedstock char is converted to gaseous products by reaction with steam, oxygen, air, hydrogen, carbon dioxide, or a mixture of these. In fact, the resulting gas mixture from the coal gasification plants goes by various names, but the most desirable mixture is synthesis gas (syngas—a mixture of carbon monoxide and hydrogen) which is itself a fuel.

The advantage of gasification is that using the synthesis gas is more efficient than direct combustion of the original fuel; more of the energy contained in the fuel is extracted. Synthesis gas may be burned directly in internal combustion engines, used to produce methanol and hydrogen, or converted via the Fischer–Tropsch process into synthetic fuel. Gasification can also begin with materials that are not otherwise useful fuels, such as biomass or organic waste. In addition, the high-temperature combustion refines out corrosive ash elements, such as chloride and potassium, allowing clean gas production from otherwise problematic fuels.

Gasification of coal has been, and continues to be, widely used on industrial scales to generate electricity. However, almost any type of organic carbonaceous material can be used as the raw material for gasification, such as wood, biomass, or even plastic waste. Thus gasification may be an important technology for renewable energy. In particular, biomass gasification is carbon neutral. Furthermore, because of the historical use of coal for gasification purposes (Ishi, 1982; Ricketts et al., 2002; Hotchkiss, 2003; Speight, 2013), it is the feedstock against which the suitability of all other feedstocks is measured. Therefore inclusion of coal among the gasification feedstocks in this section is warranted.

Another advantage of gasification-based energy systems is that when oxygen is used in the gasifier in place of air, the carbon dioxide produced by the process is in a concentrated gas stream, making it easier and less expensive to separate and capture. Once the carbon dioxide is captured, it can be sequestered and prevented from escaping to the atmosphere, where it could otherwise potentially contribute to the greenhouse effect.

12.2.1 Chemistry

Gasification chemistry is reasonably simple and straightforward and, hence, gasification processes are reasonably efficient. For many years such processes were used to manufacture illuminating gas (coal gas) for gas lighting, before electric lighting became widely available. The simplest method, and the first used, was to heat coal in a retort in the absence of air, partially converting coal to gas with a residue of coke; the Scottish engineer William Murdock used this technique in pioneering the commercial gasification of coal in 1792. Murdock licensed his process to the Gas Light and Coke Company in 1813, and in 1816, the Baltimore Gas Company, the first coal gasification company in the United States, was established. The process of heating coal to produce coke and gas is still used in the metallurgical industry.

Gasification relies on chemical processes at elevated temperatures >700°C (>1290°F), which distinguishes it from biological processes such as anaerobic digestion that produce biogas. In a gasifier the carbonaceous material undergoes several different processes: (1) pyrolysis of the carbonaceous feedstock, (2) combustion of the pyrolysis products, and (3) gasification of the remaining char.

Pyrolysis (devolatilization) is the thermal degradation of an organic substance in the absence of air to produce char, pyrolysis oil, and synthesis gas, for example, the conversion of wood to charcoal. The pyrolysis process occurs as the carbonaceous feedstock heats up. Volatiles are released and char is produced, resulting in up to 70% weight loss for coal. The process is very dependent on the properties of the carbonaceous material and determines the structure and composition of the char, which will then undergo gasification reactions.

The combustion process occurs as the volatile products and some of the char reacts with oxygen to form carbon dioxide and carbon monoxide, which provides heat for the subsequent gasification reactions:

$$2C + O_2 \rightarrow CO_2$$

Gasification is the decomposition of hydrocarbon derivatives into a synthesis gas by carefully controlling the amount of oxygen present, such as the conversion of coal into town gas. The gasification process occurs as the char reacts with carbon dioxide and steam to produce carbon monoxide and hydrogen:

$$C + H_2O \rightarrow H_2 + CO_2$$

In addition, the gas phase water gas shift reaction reaches equilibrium quickly at the temperatures in a gasifier. This removes carbon dioxide from the reactor and provides water for the gasification reaction:

$$CO + H_2O \leftrightarrow CO_2 + H_2$$

In essence, a limited amount of oxygen or air is introduced into the reactor to allow some of the organic material to be burned to produce carbon monoxide and energy, which drives a second reaction that converts further organic material to hydrogen and additional carbon monoxide.

Currently, hydrogen is produced from coal by gasification and the subsequent processing of the resulting synthesis gas. In its simplest form, coal gasification works by first reacting coal with oxygen and steam under high pressures and temperatures to form a synthesis gas consisting primarily of carbon monoxide and hydrogen. This synthesis gas is cleaned of virtually all of its impurities and shifted to produce additional hydrogen. The clean gas is sent to a separation system to recover hydrogen.

The most complete conversion of coal or coke to gas that is feasible was achieved by reacting coal continuously in a vertical retort with air and steam. The gas (producer gas) obtained in this manner has a relatively low heat content per unit volume of gas (100–150 Btu/ft.^{3}). The development of a cyclic steam–air process in 1873 made possible the production of a gas of higher thermal content (300–350 Btu/ft.^{3}), composed chiefly of carbon monoxide and hydrogen, and known as water gas. By adding oil to the reactor, the thermal content of gas was increased to 500–550 Btu/ft.^{3}; this became the standard for gas distributed to residences and industry. Since 1940, processes have been developed to produce continuously a gas equivalent to water gas; this involves the use of steam and essentially pure oxygen as a reactant. A more recently developed process reacts coal with pure oxygen and steam at an elevated pressure (450 psi) to produce a gas that may be converted to synthetic natural gas.

12.2.2 Processes

The most common modern coal gasification process uses lump coal in a vertical retort. In the process, coal is fed at the top with air, and steam is introduced at the bottom and the gas (air and steam) rising up the retort heat the coal in its downward flow and react with the coal to convert it to gas. Ash is removed at the bottom of the retort. Using air and steam as reacting gases results in a producer gas; using oxygen and steam results in a water gas. Increasing operating pressure increases the productivity.

Two other processes currently in commercial use react finely powdered coal with steam and oxygen. One of these, the Winkler process, uses a fluidized bed in which the powdered coal is agitated with the reactant gases. The Koppers–Totzek process operates at a much higher temperature, and the powdered coal is reacted while it is entrained in the gases passing through the reactor. The ash is removed as a molten slag at the bottom of the reactor.

As crude oil supplies decrease, the desirability of producing gas from coal may increase, especially in those areas where natural gas is in short supply. It is also anticipated that costs of natural gas will increase, allowing coal gasification to compete as an economically viable process. Research in progress on a laboratory and pilot-plant scale should lead to the invention of new process technology by the end of the century, thus accelerating the industrial use of coal gasification.

Thus the products of coal gasification consist of carbon monoxide, carbon dioxide, hydrogen, methane, and some other gases in proportions dependent upon the specific reactants and conditions (temperatures and pressures) employed within the reactors, and the treatment steps which the gases undergo subsequent to leaving the gasifier. Similar chemistry can also be applied to the gasification of coke derived from crude oil and other sources. The reaction of coal or coal char with air or oxygen to produce heat and carbon dioxide could be called gasification, but it is more properly classified as combustion. The principal purposes of such conversion are the production of synthetic natural gas as a substitute gaseous fuel and synthesis gases for the production of chemicals and plastics.

In all cases of commercial interest, gasification with steam, which is endothermic, is an important chemical reaction. The necessary heat input is typically supplied to the gasifier by combusting a portion of the coal with oxygen added along with the steam. From the industrial viewpoint the final product is either synthesis of gas, medium-Btu gas, or substitute natural gas. Each of the gas types has potential industrial applications.

In the chemical industry, synthesis gas from coal is a potential alternative source of hydrogen and carbon monoxide. This mixture is obtained primarily from the steam reforming of natural gas, natural gas liquids, or other crude oil liquids. Fuel users in the industrial sector have studied the feasibility of using medium-Btu gas instead of natural gas or oil for fuel applications. Finally, the natural gas industry is interested in substitute natural gas, which can be distributed in existing pipeline networks.

The conversion of the gaseous products of coal gasification processes to synthesis gas, a mixture of hydrogen (H_2) and carbon monoxide (CO), in a ratio appropriate to the application, needs additional steps, after purification. The product gases —carbon monoxide, carbon dioxide, hydrogen, methane, and nitrogen—can be used as fuels or as raw materials for chemical or fertilizer manufacture.

12.2.3 Gasifiers

The focal point of any gasification-based system is the gasifier. A gasifier converts hydrocarbon feedstock into gaseous components by applying heat under pressure in the presence of steam.

A gasifier differs from a combustor in that the amount of air or oxygen available inside the gasifier is carefully controlled so that only a relatively small portion of the fuel burns completely. The POX process provides the heat and rather than combustion, most of the carbon-containing feedstock is chemically broken apart by the heat and pressure applied in the gasifier, resulting in the chemical reactions that produce synthesis gas. However, the composition of the synthesis gas will vary because of dependence upon the conditions in the gasifier and the type of feedstock.

Minerals in the fuel (i.e., the rocks, dirt and other impurities which do not gasify) separate and leave the bottom of the gasifier either as an inert glass-like slag or other marketable solid products.

Sulfur impurities in the feedstock are converted to hydrogen sulfide (H_2S) and carbonyl sulfide (COS), from which sulfur can be extracted, typically as elemental sulfur. Nitrogen oxides (NO_x), other potential pollutants, are not formed in the oxygen-deficient (reducing) environment of the gasifier. Instead, ammonia (NH_3) is created by nitrogen–hydrogen reactions and can be washed out of the gas stream.

Four types of gasifier are currently available for commercial use: (1) the countercurrent fixed bed gasifier, (2) the cocurrent fixed bed gasifier, (3) the fluidized gasifier, and (4) the entrained flow—sometime referred to as the entrained bed gasifier (Speight, 2013).

The countercurrent fixed bed (up draft) gasifier consists of a fixed bed of carbonaceous fuel (e.g., coal or biomass) through which the "gasification agent" (steam, oxygen and/or air) flows in countercurrent configuration. The ash is either removed dry or as a slag. The slagging gasifiers require a higher ratio of steam and oxygen to carbon in order to reach temperatures higher than the ash fusion temperature. The nature of the gasifier means that the fuel must have high mechanical strength and must be noncaking so that it will form a permeable bed, although recent developments have reduced these restrictions to some extent. The throughput for this type of gasifier is relatively low. Thermal efficiency is high as the gas exit temperatures are relatively low. However, this means that tar and methane production is significant at typical operation temperatures, so product gas must be extensively cleaned before use or recycled to the reactor.

The cocurrent fixed bed (down draft) gasifier is similar to the countercurrent type, but the gasification agent gas flows in cocurrent configuration with the fuel (downwards, hence the name down draft gasifier). Heat needs to be added to the upper part of the bed, either by combusting small amounts of the fuel or from external heat sources. The produced gas leaves the gasifier at a high temperature, and most of this heat is often transferred to the gasification agent added in the top of the bed, resulting in energy efficiency on level with the countercurrent type. Since all tars must pass through a hot bed of char in this configuration, tar levels are much lower than the countercurrent type.

In the fluidized bed gasifier the fuel is fluidized by oxygen (or air) and steam. The ash is removed dry or as high-density agglomerates. The temperatures are relatively low in dry ash gasifiers, so the fuel must be highly reactive; low-grade coals are particularly suitable. The agglomerating gasifiers have slightly higher temperatures and are suitable for higher rank coals. Fuel throughput is higher than for the fixed bed, but not as high as for the entrained flow gasifier. The conversion efficiency is rather low, so recycle or subsequent combustion of solids is necessary to increase conversion. Fluidized bed gasifiers are most useful for fuels that form highly corrosive ash that would damage the walls of slagging gasifiers. Also, a disadvantage of biomass feedstocks is that they generally contain high levels of corrosive ash.

In the entrained flow gasifier a dry pulverized solid, an atomized liquid fuel, or a fuel slurry is gasified with oxygen (much less frequent: air) in cocurrent flow. The gasification reactions take place in a dense cloud of very fine particles. Most coal is suitable for this type of gasifier because of the high operating temperatures and because the coal particles are well separated from one another. The high temperatures and pressures also mean that a higher throughput can be achieved but thermal efficiency is somewhat lower as the gas must be cooled before it can be cleaned with existing technology. The high temperatures also mean that tar and methane are not present in the product gas; however, the oxygen requirement is higher than for the other types of gasifiers. All entrained flow gasifiers remove the major part of the ash as a slag as the operating temperature is well above the ash fusion temperature. A smaller fraction of the ash is produced either as a very fine dry fly ash or as black fly ash slurry. Some fuels, in particular, certain types of biomasses, can form slag that is corrosive for ceramic inner walls that serve to protect the gasifier outer wall. However, some entrained bed type of gasifiers do not possess a ceramic inner wall but have an inner water or steam cooled wall covered with partially solidified slag. These types of gasifiers do not suffer from corrosive slag. Some fuels have ashes with very high ash fusion temperatures. In this case mostly limestone is mixed with the fuel prior to gasification. Addition of limestone will usually suffice for the lowering the fusion temperatures. The fuel particles must be much smaller than for other types of gasifiers. This means the fuel must be pulverized, which requires somewhat more energy than for the other types of gasifiers. By far the most energy consumption related to entrained bed gasification is not the milling of the fuel but the production of oxygen used for the gasification.

In integrated gasification combined-cycle (IGCC) systems the synthesis gas is cleaned of its hydrogen sulfide, ammonia, and particulate matter and is burned as fuel in a combustion turbine (much like natural gas is burned in a turbine). The combustion turbine drives an electric generator. And hot air from the combustion turbine can be channeled back to the gasifier or the air separation unit, while exhaust heat from the combustion turbine is recovered and used to boil water, creating steam for a steam turbine-generator.

The use of these two types of turbines—a combustion turbine and a steam turbine—in combination, known as a combined cycle, is a reason why gasification-based power systems can achieve unprecedented power-generation efficiencies. Currently, commercially available gasification-based systems can operate at around 42% efficiencies; in the future, these systems may be able to achieve efficiencies approaching 60%. A conventional coal-based boiler plant, by contrast, employs only a steam turbine generator and is typically limited to 33%–40% efficiency.

Higher efficiency means that less fuel is required to generate the rated power, resulting in better economics (which can mean lower costs to the consumer) and the formation of fewer greenhouse gases—a 60%-efficient gasification power plant can cut the formation of carbon dioxide by 40% compared to a typical coal combustion plant.

12.3 Gasification of crude oil fractions

Because of the ever-increasing need for crude oil products and the increasing scarcity of conventional crude oil, high conversion refineries are moving to gasification of feedstocks for the development of alternative fuels and to enhance equipment usage (Speight, 2011, 2014a). A major trend in the refining industry market demand for refined products will be in synthesizing fuels from simple basic reactants (e.g., synthesis gas) when it becomes uneconomical to produce super clean transportation fuels through conventional refining processes. Fischer–Tropsch plants together with IGCC systems will be integrated with or even into refineries, which will offer the advantage of high-quality products.

In addition, the concept of using other feedstocks to blend with crude oil feedstocks bring to the fore the concept of a gasification refinery, which would use gasification technology as is the case of the Sasol refinery in South Africa (Chadeesingh, 2011). The refinery would produce synthesis gas (from the carbonaceous feedstock) from which liquid fuels would be manufactured using the Fischer–Tropsch synthesis technology. Synthesis gas is used as a source of hydrogen or as an intermediate in producing hydrocarbon derivatives via the Fischer–Tropsch synthesis.

As crude oil supplies decrease, the desirability of producing gas from other carbonaceous feedstocks will increase, especially in those areas where natural gas is in short supply. It is also anticipated that costs of natural gas will increase, allowing coal gasification to compete as an economically viable process. Research in progress on a laboratory and pilot-plant scale should lead to the invention of new process technology by the end of the century, thus accelerating the industrial use of coal gasification.

12.3.1 Feedstocks

The most common, and perhaps the best, feedstocks for steam reforming are low boiling saturated hydrocarbon derivatives that have a low sulfur content, including natural gas, refinery gas, liquefied petroleum gas (LPG), and low-boiling naphtha.

Natural gas is the most common feedstock for hydrogen production since it meets all the requirements for reformer feedstock. Natural gas typically contains more than 90% methane and ethane with only a few percent of propane and higher boiling hydrocarbon derivatives. Natural gas may (or most likely will) contain traces of carbon dioxide with some nitrogen and other impurities. Purification of natural gas, before reforming, is usually relatively straight forward. Traces of sulfur must be removed to avoid poisoning the reformer catalyst; zinc oxide treatment in combination with hydrogenation is usually adequate.

Low molecular weight refinery gas, containing a substantial amount of hydrogen, can be an attractive steam reformer feedstock since it is produced as a by-product. Processing of refinery gas will depend on its composition, particularly the levels of olefins and of propane and heavier hydrocarbon derivatives. Olefins, that can cause problems by forming coke in the reformer, are converted to saturated compounds in the hydrogenation unit. Higher boiling hydrocarbon derivatives in refinery gas can also form coke, either on the primary reformer catalyst or in the preheater. If there is more than a few percent of C_3 and higher compounds, a promoted reformer catalyst should be considered, in order to avoid carbon deposits.

Refinery gas from different sources varies in suitability as hydrogen plant feed. Catalytic reformer off-gas, for example, is saturated, very low in sulfur, and often has high hydrogen content. The process gases from a coking unit or from a fluid catalytic cracking unit are much less desirable because of the content of unsaturated constituents. In addition to olefins, these gases contain substantial amounts of sulfur that must be removed before the gas is used as feedstock. These gases are also generally unsuitable for direct hydrogen recovery, since the hydrogen content is usually too low. Hydrotreater off-gas lies in the middle of the range. It is saturated, so it is readily used as hydrogen plant feed. Content of hydrogen and heavier hydrocarbon derivatives depends to a large extent on the upstream pressure. Sulfur removal will generally be required.

As hydrogen use has become more widespread in refineries, hydrogen production has moved from the status of a high-tech specialty operation to an integral feature of most refineries. This has been made necessary by the increase in hydrotreating and hydrocracking, including the treatment of progressively heavier feedstocks (Speight, 2007). The continued increase in hydrogen demand over the last several decades is a result of the conversion of crude oil to match changes in product slate and the supply of viscous, high-sulfur oil, and in order to make lower boiling, cleaner, and more salable products. There are also many reasons other than product quality for using hydrogen in processes adding to the need to add hydrogen at relevant stages of the refining process and, most important according to the availability of hydrogen.

Hydrogen has historically been produced during catalytic reforming processes as a by-product of the production of the aromatic compounds used in gasoline and in solvents. As reforming processes changed from fixed-bed to cyclic to continuous regeneration, process pressures have dropped and hydrogen production per barrel of reformate has tended to increase. However, hydrogen production as a by-product is not always adequate to the needs of the refinery and other processes are necessary. Thus hydrogen production by steam reforming or by POX of residua has also been used, particularly where heavy crude oil, extra heavy crude oil, and tar sand bitumen are available. Steam reforming is the dominant method for hydrogen production and is usually combined with pressure-swing adsorption (PSA) to purify the hydrogen to greater than 99% by volume.

The gasification of residua and coke to produce hydrogen and/or power may become an attractive option for refiners. The premise that the gasification section of a refinery will be the garbage can for deasphalter residues, high-sulfur coke, as well as other refinery wastes is worthy of consideration.

Of the processes that are available for the production of the hydrogen, many can be considered dual processes insofar as they also produce carbon monoxide and, therefore, are considered as producers of synthesis gas. For example, most of the external hydrogen is manufactured by steam–methane reforming or by oxidation processes. Other processes such as ammonia dissociation, steam–methanol interaction, or electrolysis are also available for hydrogen production, but economic factors and feedstock availability assist in the choice between processing alternatives.

The processes described in this section are those gasification processes by which hydrogen is produced for use in other parts of the refinery.

12.3.2 Chemistry

In steam reforming, low-boiling hydrocarbon derivatives such as methane are reacted with steam to form hydrogen:

$$CH_4 + H_2O \rightarrow 3H_2 + CO \quad \Delta H_{298K} = +97,400\ \text{Btu/lb}$$

H is the heat of reaction. A more general form of the equation that shows the chemical balance for higher boiling hydrocarbon derivatives is

$$C_nH_m + nH_2O \rightarrow (n + m/2)H_2 + nCO$$

The reaction is typically carried out at approximately 815°C (1500°F) over a nickel catalyst packed into the tubes of a reforming furnace. The high temperature also causes the hydrocarbon feedstock to undergo a series of cracking reactions, plus the reaction of carbon with steam:

$$CH_4 \rightarrow 2H_2 + C$$

$$C + H_2O \rightarrow CO + H_2$$

Carbon is produced on the catalyst at the same time that hydrocarbon is reformed to hydrogen and carbon monoxide. With natural gas or similar feedstock, reforming predominates and the carbon can be removed by reaction with steam as fast as it is formed. When higher boiling feedstocks are used, the carbon is not removed fast enough and builds up, thereby requiring catalyst regeneration or replacement. Carbon buildup on the catalyst (when high-boiling feedstocks are employed) can be avoided by addition of alkali compounds, such as potash, to the catalyst thereby encourage or promoting the carbon–steam reaction.

However, even with an alkali-promoted catalyst, feedstock cracking limits the process to hydrocarbon derivatives with a boiling point less than 180°C (350°F). Natural gas, propane, butane, and low-boiling naphtha are most suitable. Prereforming, a process that uses an adiabatic catalyst bed operating at a lower temperature, can be used as a pretreatment to allow heavier feedstocks to be used with lower potential for carbon deposition (coke formation) on the catalyst.

After reforming the carbon monoxide in the gas is reacted with steam to form additional hydrogen (the water–gas shift reaction):

$$CO + H_2O \rightarrow CO_2 + H_2 \quad \Delta H_{298\ K} = -16,500\ \text{Btu/lb}$$

This leaves a mixture consisting primarily of hydrogen and carbon monoxide that is removed by conversion to methane:

$$CO + 3H_2O \rightarrow CH_4 + H_2O$$

$$CO_2 + 4H_2 \rightarrow CH_4 + 2H_2O$$

The critical variables for steam reforming processes are (1) temperature, (2) pressure, and (3) the steam/hydrocarbon ratio. Steam reforming is an equilibrium reaction, and conversion of the hydrocarbon feedstock is favored by high temperature, which in turn requires higher fuel use. Because of the volume increase in the reaction, conversion is also favored by low pressure, which conflicts with the need to supply the hydrogen at high pressure. In practice, materials of construction limit temperature and pressure.

On the other hand, and in contrast to reforming, shift conversion is favored by low temperature. The gas from the reformer is reacted over iron oxide catalyst at 315°C–370°C (600°F–700°F) with the lower limit being dictated activity of the catalyst at low temperature.

Hydrogen can also be produced by POX of hydrocarbon derivatives in which the hydrocarbon is oxidized in a limited or controlled supply of oxygen:

$$2CH_4 + O_2 \rightarrow CO + 4H_2 \quad \Delta H_{298K} = -10,195\ \text{Btu/lb}$$

The shift reaction also occurs and a mixture of carbon monoxide and carbon dioxide is produced in addition to hydrogen. The catalyst tube materials do not limit the reaction temperatures in POX processes and higher temperatures may be used that enhance the conversion of methane to hydrogen. Indeed, much of the design and operation of hydrogen plants involves protecting the reforming catalyst and the catalyst tubes because of the extreme temperatures and the sensitivity of the catalyst. In fact, minor variations in feedstock composition or operating conditions can have significant effects on the life of the catalyst or the reformer itself. This is particularly true of changes in molecular weight of the feed gas, or poor distribution of heat to the catalyst tubes.

Since the high temperature takes the place of a catalyst, POX is not limited to the lower boiling feedstocks that are required for steam reforming. POX processes were first considered for hydrogen production because of expected shortages of lower boiling feedstocks and the need to have available a disposal method for higher boiling, high-sulfur streams such as asphalt or crude oil coke.

Catalytic POX, also known as auto-thermal reforming, reacts oxygen with a low-boiling feedstock and by passing the resulting hot mixture over a reforming catalyst. The use of a catalyst allows the use of lower temperatures than in noncatalytic POX and which causes a reduction in oxygen demand.

The feedstock requirements for catalytic POX processes are similar to the feedstock requirements for steam reforming and low-boiling hydrocarbon derivatives from refinery gas to naphtha are preferred. The oxygen substitutes for much of the steam in preventing coking and a lower steam/carbon ratio is required. In addition, because a large excess of steam is not required, catalytic POX produces more carbon monoxide and less hydrogen than steam reforming. Thus the process is more suited to situations where carbon monoxide is the more desirable product such as, for example, as synthesis gas for chemical feedstocks.

12.3.3 Commercial processes

In spite of the use of low-quality hydrogen (that contain up to 40% by volume hydrocarbon gases), a high-purity hydrogen stream (95%–99% by volume hydrogen) is required for hydrodesulfurization, hydrogenation, hydrocracking, and petrochemical processes. Hydrogen, produced as a by-product of refinery processes (principally hydrogen recovery from catalytic reformer product gases), often is not enough to meet the total refinery requirements, necessitating the manufacturing of additional hydrogen or obtaining supply from external sources.

12.3.3.1 Heavy residue gasification

High-boiling residua are gasified and the produced gas is purified to clean fuel gas. As an example, solvent deasphalter residuum is gasified by POX method under pressure of approximately 570 psi and at a temperature on the order of 1300°C–1500°C (2370°F–2730°F). The high temperature generated gas flows into the specially designed waste heat boiler, in which the hot gas is cooled and high pressure saturated steam is generated. The gas from the waste heat boiler is then heat exchanged with the fuel gas and flows to the carbon scrubber, where unreacted carbon particles are removed from the generated gas by water scrubbing.

The gas from the carbon scrubber is further cooled by the fuel gas and boiler feed water and led into the sulfur compound removal section, where hydrogen sulfide (H_2S) and carbonyl sulfide (COS) are removed from the gas to obtain clean fuel gas. This clean fuel gas is heated with the hot gas generated in the gasifier and finally supplied to the gas turbine at a temperature of 250°C–300°C (480°F–570°F).

The exhaust gas from the gas turbine having a temperature of approximately 550°C–600°C (1020°F–1110°F) flows into the heat recovery steam generator consisting of five heat exchange elements. The first element is a superheater in which the combined stream of the high pressure saturated steam generated in the waste heat boiler and in the second element (high pressure steam evaporator) is superheated. The third element is an economizer, the fourth element is a low pressure steam evaporator, and the final or the fifth element is a deaerator heater. The off gas from heat recovery steam generator having a temperature of approximately 130°C (265°F) is emitted into the air via stack.

In order to decrease the nitrogen oxide (NO_x) content in the flue gas, two methods can be applied. The first method is the injection of water into the gas turbine combustor. The second method is to selectively reduce the nitrogen oxide content by injecting ammonia gas in the presence of de-NO_x catalyst that is packed in a proper position of the heat recovery steam generator. The latter is more effective that the former to lower the nitrogen oxide emissions to the air.

12.3.3.2 Hybrid gasification process

In the Hybrid gasification process a slurry of coal and residual oil is injected into the gasifier where it is pyrolyzed in the upper part of the reactor to produce gas and chars. The chars produced are then partially oxidized to ash. The ash is removed continuously from the bottom of the reactor.

In this process, coal and vacuum residue are mixed together into slurry to produce clean fuel gas. The slurry fed into the pressurized gasifier is thermally cracked at a temperature of 850°C–950°C (1560°F–1740°F) and is converted into gas, tar and char. The mixture oxygen and steam in the lower zone of the gasifier gasify the char. The gas leaving the gasifier is quenched to a temperature of 450°C (840°F) in the fluidized bed heat exchanger, and is then scrubbed to remove tar, dust, and steam at around 200°C (390°F).

The coal and residual oil slurry is gasified in the fluidized-bed gasifier. The charged slurry is converted to gas and char by thermal cracking reactions in the upper zone of the fluidized bed. The produced char is further gasified with steam and oxygen that enter the gasifier just below the fluidizing gas distributor. Ash is discharged from the gasifier and indirectly cooled with steam and then discharged into the ash hopper. It is burned with an incinerator to produce process steam. Coke deposited on the silica sand is removed in the incinerator.

12.3.3.3 Hydrocarbon gasification

The gasification of hydrocarbon derivatives to produce hydrogen is a continuous, noncatalytic process that involves POX of the hydrocarbon. Air or oxygen (with steam or carbon dioxide) is used as the oxidant at 1095°C–1480°C (2000°F–2700°F). Any carbon produced (2%–3% by weight of the feedstock) during the process is removed as a slurry in a carbon separator and pelletized for use either as a fuel or as raw material for carbon-based products.

12.3.3.4 Hypro process

The hypro process is a continuous catalytic method for hydrogen manufacture from natural gas or from refinery effluent gases. The process is designed to convert natural gas:

$$CH_4 \rightarrow C + 2H_2$$

Hydrogen is recovered by phase separation to yield hydrogen of approximately 93% purity; the principal contaminant is methane.

12.3.3.5 Pyrolysis processes

There has been recent interest in the use of pyrolysis processes to produce hydrogen. Specifically the interest has focused on the pyrolysis of methane (natural gas) and hydrogen sulfide.

Natural gas is readily available and offers relatively rich stream of methane with lower amounts of ethane, propane and, butane also present. The thermocatalytic decomposition of natural gas hydrocarbon derivatives offers an alternate method for the production of hydrogen:

$$C_nH_m \rightarrow nC + (m/2)H_2$$

If a hydrocarbon fuel such as natural gas (methane) is to be used for hydrogen production by direct decomposition, the process that is optimized to yield hydrogen production may not be suitable for the production of high-quality carbon black by-product intended for the industrial rubber market. Moreover, it appears that the carbon produced from high-temperature (850°C–950°C; 1560°F–1740°F) direct thermal decomposition of methane is soot-like material with high tendency for the catalyst deactivation. Thus if the object of methane decomposition is hydrogen production, the carbon by-product may not be marketable as high-quality carbon black for rubber and tire applications.

The production of hydrogen by direct decomposition of hydrogen sulfide has been studied extensively and a process proposed. Hydrogen sulfide decomposition is a highly endothermic process and equilibrium yields are poor (Clark and Deswarte, 2015). At temperatures less than 1500°C (2730°F) the thermodynamic equilibrium is unfavorable toward hydrogen formation. However, in the presence of catalysts such as platinum–cobalt (at 1000°C; 1830°F), disulfides of molybdenum or tungsten Mo or W at 800°C (1470°F), or other transition metal sulfides supported on alumina (at 500°C–800°C; 930°F– 1470°F), decomposition of hydrogen sulfide proceeds rapidly. In the temperature range of approximately 800°C–1500°C (1470°F–2730°F), thermolysis of hydrogen sulfide can be treated simply:

$$H_2S \rightarrow H_2 + 1/xS_x \quad \Delta H_{298\ K} = +34,300\ Btu/lb$$

where $x = 2$. Outside this temperature range, multiple equilibria may be present depending on temperature, pressure, and relative abundance of hydrogen and sulfur.

Above approximately 1000°C (1830°F), there is a limited advantage to using catalysts since the thermal reaction proceeds to equilibrium very rapidly. The hydrogen yield can be doubled by preferential removal of either H_2 or sulfur from the reaction environment, thereby shifting the equilibrium. The reaction products must be quenched quickly after leaving the reactor to prevent reversible reactions.

12.3.3.6 Shell gasification process

The shell gasification (POX) process is a flexible process for generating synthesis gas, principally hydrogen and carbon monoxide, for the ultimate production of high-purity high pressure hydrogen, ammonia, methanol, fuel gas, town gas, or reducing gas by reaction of gaseous or liquid hydrocarbon derivatives with oxygen, air, or oxygen-enriched air.

The most important step in converting high-boiling residue to industrial gas is the POX of the oil using oxygen with the addition of steam. The gasification process takes place in an empty, refractory-lined reactor at temperatures of approximately 1400°C (2550°F) and pressures between 29 and 1140 psi (196 and 845 kPa). The chemical reactions in the gasification reactor proceed without catalyst to produce gas containing carbon amounting to some 0.5%–2% by weight, based on the feedstock. The carbon is removed from the gas with water, extracted in most cases with feed oil from the water and returned to the feed oil. The high reformed gas temperature is utilized in a waste heat boiler for generating steam. The steam is generated at 850–1565 psi (5884–10,787 kPa). Some of this steam is used as process steam and for oxygen and oil preheating. The surplus steam is used for energy production and heating purposes.

12.3.3.7 Steam–methane reforming

Steam–methane reforming is a catalytic process that involves a reaction between natural gas or other low-boiling hydrocarbon derivatives and steam. Steam–methane reforming is the benchmark process that has been employed over a period of several decades for hydrogen production. The process involves reforming natural gas in a continuous catalytic process in which the major reaction is the formation of carbon monoxide and hydrogen from methane and steam.

Thus the first reforming step catalytically makes methane react (the chief chemical constituent of natural gas) to form hydrogen and carbon monoxide in an endothermic (heat-absorbing) reaction:

$$CH_4 + H_2O = CO + 3H_2 \quad \Delta H_{298\ K} = +97,400\ Btu/lb$$

Higher molecular weight feedstocks can also be reformed to hydrogen:

$$C_3H_8 + 3H_2O \rightarrow 3CO + 7H_2$$

That is,

$$C_nH_m + nH_2O \rightarrow nCO + (0.5m + n)H_2$$

In the actual process the feedstock is first desulfurized by passage through activated carbon, which may be preceded by caustic and water washes. The desulfurized material is then mixed with steam and passed over a nickel-based catalyst (730°C–845°C, 1350°F–1550°F and 400 psi, 2758 kPa). Effluent gases are cooled by the addition of steam or condensate to approximately 370°C (700°F), at which point carbon monoxide reacts with steam in the presence of iron oxide in a shift converter to produce carbon dioxide and hydrogen in which the carbon monoxide is then "shifted" with steam to form additional hydrogen and carbon dioxide in an exothermic (heat-releasing) reaction:

$$CO + H_2O = CO_2 + H_2 \quad \Delta H_{298\ K} = -41.16\ kJ/mol\ (3)$$

The carbon dioxide (usually by amine washing) leaving hydrogen separated for its commercial use, the hydrogen is usually a high-purity ($>99\%$) material.

Since the presence of any carbon monoxide or carbon dioxide in the hydrogen stream can interfere with the chemistry of the catalytic application, a third stage is used to convert of these gases to methane:

$$CO + 3H_2 \rightarrow CH_4 + H_2O$$

$$CO_2 + 4H_2 \rightarrow CH_4 + 2H_2O$$

For many refiners, sulfur-free natural gas (CH_4) is not always available to produce hydrogen by this process. In that case, higher boiling hydrocarbon derivatives (such as propane, butane, or naphtha) may be used as the feedstock to generate hydrogen (*q.v.*).

The net chemical process for steam methane reforming is then given by

$$CH_4 + 2H_2O \rightarrow CO_2 + 4H_2 \quad \Delta H_{298\ K} = +165.2\ kJ/mol\ (4)$$

Indirect heating provides the required overall endothermic heat of reaction for the steam–methane reforming.

One way of overcoming the thermodynamic limitation of steam reforming is to remove either hydrogen or carbon dioxide as it is produced, hence shifting the thermodynamic equilibrium toward the product side. The concept for sorption-enhanced methane steam reforming is based on in situ removal of carbon dioxide by a sorbent such as calcium oxide (CaO).

$$CaO + CO_2 \rightarrow CaCO_3$$

Sorption enhancement enables lower reaction temperatures, which may reduce catalyst coking and sintering, while enabling the use of less expensive reactor wall materials. In addition, heat release by the exothermic carbonation reaction supplies most of the heat required by the endothermic reforming reactions. However, energy is required to regenerate the sorbent to its oxide form by the energy-intensive calcination reaction, that is,

$$CaCO_3 \rightarrow CaO + CO_2$$

Use of a sorbent requires either that there be parallel reactors operated alternatively and out of phase in reforming and sorbent regeneration modes, or that sorbent be continuously transferred between the reformer/carbonator and regenerator/calciner.

In autothermal (or secondary) reformers the oxidation of methane supplies the necessary energy and carried out either simultaneously or in advance of the reforming reaction. The equilibrium of the methane steam reaction and the water–gas shift reaction determines the conditions for optimum hydrogen yields. The optimum conditions for hydrogen production require high temperature at the exit of the reforming reactor (800°C–900°C; 1470°F–1650°F), high excess of steam (molar steam-to-carbon ratio of 2.5 to 3), and relatively low pressures (below 450 psi). Most commercial plants employ supported nickel catalysts for the process.

The steam–methane reforming process described briefly above would be an ideal hydrogen production process if it was not for the fact that large quantities of natural gas, a valuable resource, are required as both feed gas and combustion fuel. For each mole of methane reformed, more than 1 mol of carbon dioxide is coproduced and must be disposed. This can be a major issue as it results in the same amount of greenhouse gas emission as would be expected from direct combustion of natural gas or methane. In fact, the production of hydrogen as a clean burning fuel by way of steam reforming of methane and other fossil-based hydrocarbon fuels is not in environmental balance if in the process, carbon dioxide and carbon monoxide are generated and released into the atmosphere, although alternate scenarios are available. Moreover, as the reforming process is not totally efficient, some of the energy value of the hydrocarbon fuel is lost by conversion to hydrogen but with no tangible environmental benefit, such as a reduction in emission of greenhouse gases. Despite these apparent shortcomings, the process has the following advantages: (1) produces 4 mol of hydrogen for each mole of methane consumed, (2) feedstocks for the process (methane and water are readily available), (3) the process is adaptable to a wide range of hydrocarbon feedstocks, (4) operates at low pressures, less than 450 psi, (5) requires a low steam/carbon ratio on the order of 2.5 to 3, (6) good utilization of input energy—up to 93%, (7) can use catalysts that are stable and resist poisoning, and (8) good process kinetics.

Liquid feedstocks, either LPG or naphtha, can also provide backup feed, if there is a risk of natural gas curtailments. The feed handling system needs to include a surge drum, feed pump, vaporizer (usually steam-heated) followed by further heating before desulfurization. The sulfur in liquid feedstocks occurs as mercaptans, thiophene derivatives, or higher boiling compounds. These compounds are stable and will not be removed by zinc oxide; therefore, a hydrogenation unit will be required. In addition, as with refinery gas, olefins must also be hydrogenated if they are present.

The reformer will generally use a potash-promoted catalyst to avoid coke buildup from cracking of the heavier feedstock. If LPG is to be used only occasionally, it is often possible to use a methane-type catalyst at a higher steam/carbon ratio to avoid coking. Naphtha will require a promoted catalyst unless a preformer is used.

12.3.3.8 Steam-naphtha reforming

Steam-naphtha reforming is a continuous process for the production of hydrogen from liquid hydrocarbon derivatives and is, in fact, similar to steam–methane reforming that is one of several possible processes for the production of hydrogen from low boiling hydrocarbon derivatives other than ethane. A variety of naphtha-types in the gasoline boiling range may be employed, including feeds containing up to 35% aromatics. Thus following pretreatment to remove sulfur compounds, the feedstock is mixed with steam and taken to the reforming furnace (675°C–815°C, 1250°F–1500°F, 300 psi, 2068 kPa), where hydrogen is produced.

12.3.3.9 Synthesis gas generation

The synthesis gas generation process is a noncatalytic process for producing synthesis gas (principally hydrogen and carbon monoxide) for the ultimate production of high-purity hydrogen from gaseous or liquid hydrocarbon derivatives.

In this process a controlled mixture of preheated feedstock and oxygen is fed to the top of the generator where carbon monoxide and hydrogen emerge as the products. Soot, produced in this part of the operation, is removed in a water scrubber from the product gas stream and is then extracted from the resulting carbon–water slurry with naphtha and transferred to a fuel oil fraction. The oil–soot mixture is burned in a boiler or recycled to the generator to extinction to eliminate carbon production as part of the process.

The soot-free synthesis gas is then charged to a shift converter where the carbon monoxide reacts with steam to form additional hydrogen and carbon dioxide at the stoichiometric rate of 1 mol of hydrogen for every mole of carbon monoxide charged to the converter.

The reactor temperatures vary from 1095°C to 1490°C (2000°F to 2700°F), while pressures can vary from approximately atmospheric pressure to approximately 2000 psi (13,790 kPa). The process has the capability of producing high-purity hydrogen although the extent of the purification procedure depends upon the use to which the hydrogen is to be put. For example, carbon dioxide can be removed by scrubbing with various alkaline reagents, while carbon monoxide can be removed by washing with liquid nitrogen or, if nitrogen is undesirable in the product, the carbon monoxide should be removed by washing with copper–amine solutions.

This particular POX technique has also been applied to a whole range of liquid feedstocks for hydrogen production. There is now serious consideration being given to hydrogen production by the POX of solid feedstocks such as crude oil coke (from both delayed and fluid-bed reactors), lignite, and coal, as well as crude oil residua.

The chemistry of the process, using naphthalene as an example, may be simply represented as the selective removal of carbon from the hydrocarbon feedstock and further conversion of a portion of this carbon to hydrogen:

$$C_{10}H_8 + 5O_2 \rightarrow 10CO + 4H_2$$

$$10CO + 10H_2O \rightarrow 10CO_2 + 10H_2$$

Although these reactions may be represented very simply using equations of this type, the reactions can be complex and result in carbon deposition on parts of the equipment, thereby requiring careful inspection of the reactor.

12.3.3.10 Texaco gasification process

The Texaco gasification process (Texaco POX process) is a gasification process for generating synthetic gas, principally hydrogen and carbon monoxide. The characteristic of Texaco gasification process is to inject feedstock together with carbon dioxide, steam, or water into the gasifier. Therefore solvent deasphalted residua, or crude oil coke rejected from any cooking method, can be used as feedstock for this gasification process. The produced gas from this gasification process can be used for the production of high-purity high-pressurized hydrogen, ammonia, and methanol. The heat recovered from the high temperature gas is used for the generation of steam in the waste heat boiler. Alternatively the less expensive quench type configuration is preferred when high pressure steam is not needed or when a high degree of shift is needed in the downstream CO converter.

In the process the feedstock, together with the feedstock carbon slurry recovered in the carbon recovery section, is pressurized to a given pressure, mixed with high-pressure steam, and then blown into the gas generator through the burner together with oxygen.

The gasification reaction is a POX of hydrocarbon derivatives to carbon monoxide and hydrogen:

$$C_xH_{2y} + x/2O_2 \rightarrow xCO + yH_2$$

$$C_xH_{2y} + xH_2O \rightarrow xCO + (x+y)H_2$$

The gasification reaction is instantly completed, thus producing gas mainly consisting of H_2 and CO ($H_2 + CO = > 90\%$). The high-temperature gas leaving the reaction chamber of the gas generator enters the quenching chamber linked to the bottom of the gas generator and is quenched to 200°C–260°C (390°F–500°F) with water.

12.4 Gasification of other feedstocks

Gasification offers more scope for recovering products from waste than incineration. When waste is burnt in a modern incinerator the only practical product is energy, whereas the gases, oils, and solid char from pyrolysis and gasification can not only be used as a fuel but also purified and used as a feedstock for petrochemicals and other applications (Table 12.1 and Fig. 12.1). Many processes also produce a stable granulate instead of an ash that can be more easily and safely utilized. In addition, some processes are targeted at producing specific recyclables such as metal alloys and carbon black. From waste gasification, in particular, it is feasible to produce hydrogen, which many see as an increasingly valuable resource.

The process can be used in conjunction with gas engines (and potentially gas turbines) to obtain higher conversion efficiency than conventional fossil-fuel energy generation. By displacing fossil-fuels, waste pyrolysis and gasification can help meet renewable energy targets, address concerns related to global climate change,

Table 12.1 Illustration of the role of synthesis gas in the production of various and products.

Feedstock	Gasification	Synthesis gas	Gas cleaning	Fischer–Tropsch process		
					Naphtha[a]	
						Gasoline
					Kerosene	
						Diesel fuel
						Jet fuel
					Wax	
						Naphtha[a]
						Diesel fuel
						Jet fuel
					Methanol	
						Naphtha[a]

[a]Gasoline blend stock.

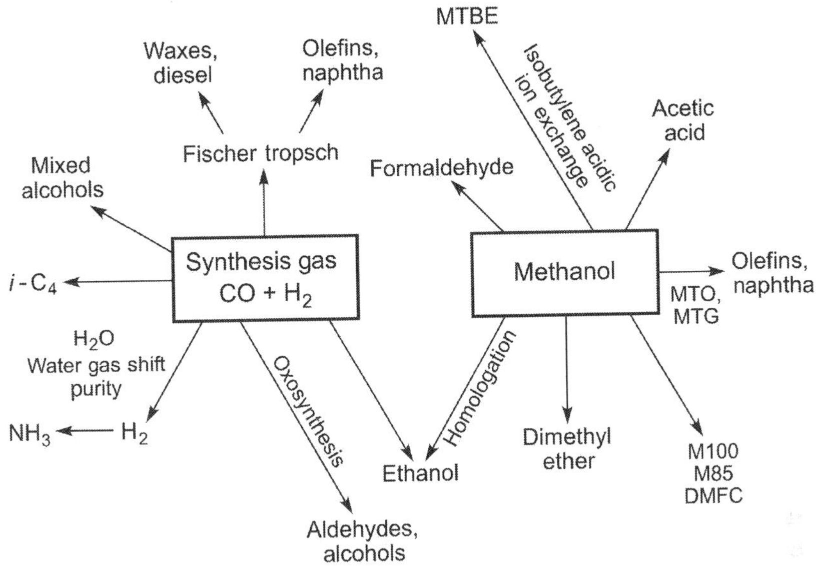

Figure 12.1 Routes to chemicals from synthesis gas and methanol.

and contribute to achieving Kyoto Protocol commitments. Conventional incineration, used in conjunction with steam-cycle boilers and turbine generators, achieves lower efficiency.

In addition to the feedstock derived above, the gasification process can accept a variety of feedstocks, but the reactor must be selected on the basis of the feedstock properties and behavior in the process, especially when coal and biomass are considered as gasification feedstocks (Speight, 2014b). Furthermore, because of the historical use of coal for gasification purposes (Speight, 2013), it is the feedstock against which the suitability of all other feedstocks is measured. Therefore inclusion of coal among the gasification feedstocks in this section is warranted.

Many of the processes fit well into a modern integrated approach to waste management. They can be designed to handle the waste residues and are fully compatible with an active program of composting for the waste fraction that is subject to decay and putrefaction. This, by analogy with coal, the high-temperature conversion of waste (Chapter 10: Nonfossil Fuel Feedstocks) is a downdraft gasification process that gasifies the feed material within a controlled and limited oxygen supply. Combustion of the feed material is prevented by the limited oxygen supply. The temperature within the reactor reaches 2700°C (4890°F), at which point molecular dissociation takes place. The pollutants that were contained within the feedstock such as dioxin derivatives, and furan derivatives as well as pathogens are completely cracked into harmless compounds.

All metal components in the waste stream are converted into a castable iron alloy/pig iron for use in steel foundries. The mineral fraction is reduced to a nonleaching vitrified glass that can be used for road construction and/or further processed into a mineral wool for insulation. All of the organic material is fully converted to a fuel quality synthesis gas that can be used to produce electrical energy, heat, methanol, or used in the production of various other chemical compounds. The resultant synthesis gas, with a hydrogen/carbon monoxide ratio approximately equal to one, is also capable of being used for the production of Fischer–Tropsch fuels. Under certain conditions, heat from the reactor could be used for district heating, industrial steam production, or water desalination plants.

A wide range of materials can be handled by gasification technologies, and specific processes have been optimized to handle particular feedstock (e.g., tire pyrolysis and sewage sludge gasification), while others have been designed to process mixed wastes. For example, recovering energy from agricultural and forestry residues, household and commercial waste, materials recycling (auto-shredder residue, electrical and electronic scrap, tires, mixed plastic waste, and packaging residues) are feasible processes.

12.5 The Fischer–Tropsch process

Although the focus of this section is the production of hydrocarbon derivatives from synthesis gas, it is worthy of note that all or part of the clean synthesis gas can also be used (1) as chemical building blocks to produce a broad range of chemicals using processes well established in the chemical and petrochemical industry, (2) as a fuel producer for highly efficient fuel cells, which run off the hydrogen made in a gasifier, or perhaps in the future, hydrogen turbines and fuel cell-turbine hybrid systems, and (3) as a source of hydrogen that can be separated from the gas stream and used as a fuel or as a feedstock for refineries, which use the hydrogen to upgrade crude oil products. However, the decreasing availability and increased price of crude oil has been renewed the world-wide interest in the production of liquid hydrocarbon derivatives from carbon monoxide and hydrogen using metal catalysts, also known as Fischer–Tropsch process or the Fischer–Tropsch synthesis (Chapter 10: Nonfossil Fuel Feedstocks) (Speight, 2013, 2014a).

The synthesis of hydrocarbon derivatives from the hydrogenation of carbon monoxide over transition metal catalysts was discovered in 1902 when Sabatier and Sanderens produced methane from hydrogen and carbon monoxide mixtures passed over nickel, iron, and cobalt catalysts. In 1923 Fischer and Tropsch reported the use of alkalized iron catalysts to produce liquid hydrocarbon derivatives rich in oxygenated compounds.

The Fischer–Tropsch process is a catalyzed chemical reaction in which carbon monoxide and hydrogen are converted into liquid hydrocarbon derivatives of various forms. Typical catalysts used are based on iron and cobalt. The principal purpose of this process is to produce a synthetic crude oil substitute for use as synthetic lubrication oil or as synthetic fuel. The hydrocarbon derivatives synthesized in the process are made primarily of liquid alkane derivatives (Table 12.2); by-products are olefin derivatives, alcohol derivatives, and solid paraffin derivatives (waxes).

Table 12.2 Carbon chain groups in the range of Fischer–Tropsch products.

Carbon number	Group name
C1–C2	SNG
C3–C4	LPG
C5–C7	Low boiling naphtha
C8–C10	High boiling naphtha
C11–C16	Middle distillate
C17–C30[a]	Low melting wax
C31–C60	High melting wax

LPG, Liquefied petroleum gas; *SNG*, synthetic natural gas.
[a]The C17 *n*-alkane (*n*-heptadecane) is the first member of the series that is not fully liquid under ambient conditions (melting point: 21°C (70°F)).

The required gas mixture of carbon monoxide and hydrogen (synthesis gas) is created through a reaction of coke or coal with water steam and oxygen, at temperatures over 900°C. In the past, town gas and gas for lamps were a carbon monoxide–hydrogen mixture, made by gasifying coke in gas works. In the 1970s, it was replaced with imported natural gas (methane). Coal gasification and Fischer–Tropsch hydrocarbon synthesis together cause a two-stage sequence of reactions which allows the production of liquid fuels like diesel and petrol out of the solid combustible coal.

The Fischer–Tropsch synthesis took its first serious place in industry in 1935 at Ruhrchemie in Oberhausen. By the beginning of the 1940s, some 600,000 tonnes of liquid hydrocarbon derivatives were produced per year in German facilities, made from coal using Fischer–Tropsch synthesis. Licensed by Ruhrchemie, four facilities in Japan, as well as a plant in France and in Manchuria, were in service. After World War II, competition from crude oil made petrol production from coal unprofitable. The only new production facilities were in South Africa, for political reasons, built starting in 1950 in Sasolburg.

The Fischer Tropsch synthesis is, in principle, a carbon chain–building process, where methylene groups are attached to the carbon chain. The actual reactions that occur have been, and remain, a matter of controversy, as it has been the last century since the 1930s.

$$(2n+1)H_2 + nCO \rightarrow C_nH_{(2n+2)} + nH_2O$$

Even though the overall Fischer–Tropsch process is described by the following chemical equation

$$(2n+1)H_2 + nCO \rightarrow C_nH_{(2n+2)} + nH_2O$$

the initial reactants in the above reaction (i.e., CO and H_2) can be produced by other reactions such as the partial combustion of a hydrocarbon:

$$C_nH_{(2n+2)} + \frac{1}{2}nO_2 \rightarrow (n+1)H_2 + nCO$$

For example (when $n = 1$), methane [in the case of the gas-to-liquids (GTL) applications]:

$$2CH_2 + O_2 \rightarrow 4H_2 + 2CO$$

Or by the gasification of any carbonaceous source, such as biomass:

$$C + H_2O \rightarrow H_2 + CO$$

The energy needed for this endothermic reaction is usually provided by the exothermic combustion reaction with air or oxygen:

$$2C + O_2 \rightarrow 2CO$$

The Fischer–Tropsch reaction is dependent of a catalyst, mostly an iron or cobalt catalyst where the reaction takes place. There is either a low temperature Fischer–Tropsch (LTFT) process or a high temperature Fischer–Tropsch (HTFT) process, with temperatures ranging between 200°C–240°C (390°F–465°F) for the low temperature Fischer–Tropsch process and 300°C–350°C (570°F–650°F) for the HTFT process. The low temperature Fischer–Tropsch process uses either an iron or a cobalt catalyst, while the HTFT process uses an iron catalyst.

The different catalysts include also nickel-based and ruthenium-based catalysts, which also have enough activity for commercial use in the process. But the availability of ruthenium is limited and the nickel-based catalyst has high activity but produces too much methane, and, additionally, the performance at high pressure is poor, due to the production of volatile carbonyls. This leaves only cobalt and iron as practical catalysts, and this study will only consider these two. Iron is cheap, but cobalt has the advantage of higher activity and longer life, though it is on a metal basis 1000 times more expensive than iron catalyst.

For large-scale commercial Fischer–Tropsch reactors, heat removal and temperature control are the most important design features to obtain optimum product selectivity and long catalyst lifetimes. Over the years, basically four Fischer–Tropsch reactor designs have been used commercially (Chadeesingh, 2011). These are the (1) the multitubular fixed bed reactor, (2) the slurry reactor, or (3) the fluidized bed reactor, and (4) the circulating bed reactor.

The fixed bed reactor consists of thousands of small tubes with the catalyst as surface-active agent in the tubes. Water surrounds the tubes and regulates the temperature by settling the pressure of evaporation. The slurry reactor is widely used and consists of fluid and solid elements, where the catalyst has no particular position, but moves around the reactor around as small pieces of catalyst together with the reaction components. The slurry and fixed bed reactor are used in the low temperature Fischer–Tropsch process. The fluidized bed reactors are diverse but characterized by the fluid behavior of the catalyst (Chadeesingh, 2011). The fluidized bed reactor is used in the HTFT process.

The Fischer–Tropsch process is an established technology and already applied on a large scale, although its popularity is hampered by high capital costs, high operation and maintenance costs, and the uncertain and volatile price of crude oil, although this concept has been offset somewhat by the development of tight oil (i.e., crude oil in tight formations—low permeability—formations in the United States).

Under most circumstances the production of synthesis gas by reforming natural gas will be more economical than from coal gasification, but site specific factors need to be considered. In fact, any technological advance in this field (such as better energy integration or the oxygen transfer ceramic membrane reformer concept) will speed up the rate at which the synfuels technology will become common practice.

There are large coal reserves that may increasingly be used as a fuel source during oil depletion. Since there are large coal reserves in the world, this technology could be used as an interim transportation fuel if conventional oil were to become more expensive. Furthermore, combination of biomass gasification and Fischer–Tropsch synthesis (Chapter 7: Deasphalting and Dewaxing) is a very promising route to produce transportation fuels from renewable or green resources.

Often a higher concentration of some types of hydrocarbon derivatives is wanted, which might be achieved by changed reaction conditions. Nevertheless, the product range is wide and infected with uncertainties, due to the lack of knowledge of the details of the process and of the kinetics of the reaction. Since the different products have quite different characteristics such as boiling point, physical state at ambient temperature and thereby different use and ways of distribution, often only a few of the carbon chains is wanted. As an example, the low temperature Fischer–Tropsch process is used when longer carbon chains are desired, because lower temperature increases the portion of longer chains. But too low temperature is not wanted, because of reduced activity. When the desired products are shorter carbon chains, such as crude oil, the longer ones might be cracked into shorter chains.

The yield of diesel is, therefore, highly dependent on the chain growth probability, which again is dependent on pressure, temperature, feed gas composition, catalyst type, catalyst composition, and reactor design. The desire to increase the selectivity of some favorable products leads to a need of understanding the relation between reaction conditions and chain growth probability, which, in turn, request a mathematical expression for the growth probability in order to make a suitable model of the process. The different attempts to model the growth probability, α, have resulted in some models that are regarded in literature as appropriate to describe the product distribution. Two will be presented here to show the influence of temperature and partial pressure.

12.6 Fuels and petrochemicals

The fuels petrochemical industry, as the name implies, is based upon the production of chemicals from, initially, crude oil. However, there is more to the industry than just crude oil products. The petrochemical industry also deals with chemicals manufactured from the by-products of crude oil refining, such as natural gas, natural gas liquids, and (in the context of this book) other feedstocks such as coal, oil shale, and biomass. The structure of the industry is extremely complex, involving thousands of chemicals and processes and there are many interrelationships within the industry with products of one process being the feedstocks of many others. For most chemicals the production route from feedstock to final products is not unique but includes many possible alternatives. As complicated as it may seem, however, this structure is comprehensible, at least in general form.

At the beginning of the production chain are the raw feedstocks: crude oil, natural gas, and alternate carbonaceous feedstocks tar. From these are produced a relatively small number of important building blocks that include primarily, but not exclusively, the lower boiling olefins and aromatic derivatives, such as ethylene, propylene, butylene isomers, butadiene, benzene, toluene, and the xylene isomers. These building blocks are then converted into a complex array of thousands of intermediate chemicals. Some of these intermediates have commercial value in and of themselves, and others are purely intermediate compounds in the production chains. The final products of the petrochemical industry are generally not consumed directly by the public but are used by other industries to manufacture consumer goods.

Thus on a scientific basis, as might be expected, the petrochemical industry is concerned with the production and trade of petrochemicals that have a wide influence on lifestyles though the production of commodity chemicals and specialty chemicals that have a marked influence on lifestyles:

Crude oil/natural gas → bulk chemicals (commodity chemicals) →
specialty chemicals

The basis of the petrochemical industry and, therefore, petrochemicals production consists of two steps: (1) feedstock production from primary energy sources to feedstocks and (2) and petrochemicals production from feedstocks.

Crude oil/natural gas → feedstock production → petrochemical products

This simplified equation encompasses the multitude of production routes available for most chemicals. In the actual industry, many chemicals are products of more than one method, depending upon local conditions, corporate polices, and desired by-products. There are also additional methods available, which have either become obsolete and are no longer used, or which have never been used commercially but could become important as technology, supplies, and other factors change. Such versatility, adaptability, and dynamic nature are three of the important features of the modern petrochemical industry.

Thus the fuels and petrochemical industry began as suitable by-products became available through improvements in the refining processes. As the decades of the 1920s and 1930s closed, the industry had developed in parallel with the crude oil industry and has continued to expand rapidly since the 1940s as the crude oil refining industry was able to provide relatively cheap and plentiful raw materials (Speight, 2002; Hsu and Robinson, 2017; Gary et al., 2007; Lee et al., 2007; Speight, 2011). The supply-demand scenario as well as the introduction of many innovations has resulted in basic chemicals and plastics, becoming the key building blocks for manufacture of a wide variety of durable and nondurable consumer goods. Chemicals and plastic materials provide the fundamental building blocks that enable the manufacture of the vast majority of consumer goods. Moreover, the demand for chemicals and plastics is driven by global economic conditions, which are directly linked to demand for consumer goods.

At the start of the production chain is the selection and preparation of the feedstock from which the petrochemicals will be produced. Typically, the feedstock is a primary energy source (such as crude oil, natural gas, coal and biomass) are extracted and then converted into feedstocks (such as naphtha, gas oil, and/or methanol). In the production of petrochemicals the feedstocks are converted into basic petrochemicals, such as ethylene ($CH_2 = CH_2$) and aromatic derivatives, which are then separated from each other. Thus petrochemicals or products derived from these feedstocks, along with other raw materials, are converted to a wide range of products (Fig. 12.1).

Therefore the history of the industry has always been strongly influenced by the supply of primary energy sources and feedstocks. Thus the petrochemical industry directly interfaces with the crude oil industry and the natural gas industry, which provides the feedstocks (Speight, 2014a, 2017, 2019), and especially the downstream sector, as well as the potential for the introduction and use of nonconventional feedstocks (Chapter 2: Introduction to Refining Processes). A major part of the petrochemical industry is made up of the polymer (plastics) industry (Speight, 2014a, 2019). The petrochemical industry is currently the biggest of the industrial chemicals sectors and petrochemicals represent the majority of all chemicals shipped between the continents of the world (EPCA, 2016). Petrochemicals have a history that began in the 19th century that has experience many changes.

In the petrochemical industry the organic chemicals produced in the largest volumes are methanol (methyl alcohol, CH_3OH), ethylene ($CH_2 = CH_2$), propylene ($CH_3CH = CH_2$), butadiene ($CH_2 = CHCH = CH_2$), benzene (C_6H_6), toluene ($C_6H_5CH_3$), and the xylene isomers ($H_3CC_6H_4CH_3$). Ethylene, propylene, and butadiene, along with butylene isomers, are collectively called olefins, which belong to a class of unsaturated aliphatic hydrocarbon derivatives having the general formula C_nH_{2n}. Olefin derivatives contain one or more double bonds ($>C = C<$), which make them chemically reactive and, hence, starting materials for many products. Benzene, toluene, and xylenes, commonly referred to as aromatics, are unsaturated cyclic hydrocarbon derivatives containing one or more rings.

As stated above, some of the chemicals and compounds produced in a refinery are destined for further processing and as raw material feedstocks for the fast growing petrochemical industry. Such nonfuel uses of crude oil products are sometimes referred to as its nonenergy uses. Crude oil products and natural gas provide two of the basic starting points for this industry; methane, naphtha, including benzene, toluene, and the xylene isomers and refinery gases that contain olefin derivatives such as ethylene, propylene, and, potentially, all of the butylene isomers (Parkash, 2003; Gary et al., 2007; Speight, 2014a; Hsu and Robinson, 2017; Speight, 2017, 2019).

Petrochemical intermediates are generally produced by chemical conversion of primary petrochemicals to form more complicated derivative products. Petrochemical derivative products can be made in a variety of ways: directly from primary petrochemicals; through intermediate products which still contain only carbon and hydrogen; and through intermediates which incorporate chlorine, nitrogen, or oxygen in the finished derivative. In some cases, they are finished products; in others, more steps are needed to arrive at the desired composition.

Of all the processes used, one of the most important is polymerization (Speight, 2014a, 2019). It is used in the production of products such as plastics, fibers, and synthetic rubber, the main finished petrochemical derivatives. Some typical petrochemical intermediates are (1) vinyl acetate, $CH_2=CHOCOCH_3$, that is used in the manufacture of paint, paper, and textile coatings; (2) vinyl chloride, $CH_2=CHCl$, that is used in the manufacture of polyvinyl chloride polymer that is generally known as polyvinyl chloride (PVC), resin manufacture, ethylene glycol, CH_2OHCH_2OH, for polyester textile fibers, styrene which is important in rubber, and the manufacture of plastics. The end products number in the thousands, some going on as inputs into the chemical industry for further processing. The more common products made from petrochemicals include adhesives, plastics, soaps, detergents, solvents, paints, drugs, fertilizer, pesticides, insecticides, explosives, synthetic fibers, synthetic rubber, and flooring and insulating materials.

Petrochemical products include such items as plastics, soaps and detergents, solvents, drugs, fertilizers, pesticides, explosives, synthetic fibers and rubbers, paints, epoxy resins, and flooring and insulating materials. Petrochemicals are found in products as diverse as aspirin, luggage, boats, automobiles, aircraft, polyester clothes, and recording discs and tapes.

The petrochemical industry has grown with the crude oil industry (Goldstein, 1949; Steiner, 1961; Hahn, 1970) and is considered by some to be a mature industry. However, as is the case with the latest trends in changing crude oil types, it must also evolve to meet changing technological needs (Speight, 2011). The manufacture of chemicals or chemical intermediates from a variety of raw materials is well established (Wittcoff and Reuben, 1996). And the use of natural gas and crude oil is an excellent example of the conversion of such raw materials to more valuable products. The individual chemicals made from natural gas and crude oil are numerous and include industrial chemicals, household chemicals, fertilizers, and paints, as well as intermediates for the manufacture of products, such as synthetic rubber and plastics.

Petrochemicals are generally considered chemical compounds derived from crude oil either by direct manufacture or indirect manufacture as by-products from the variety of processes that are used during the refining of crude oil. Gasoline, kerosene, fuel oil, lubricating oil, wax, asphalt, and the like are excluded from the definition of petrochemicals, since they are not, in the true sense, chemical compounds but are in fact intimate mixtures of hydrocarbon derivatives.

The classification of materials as petrochemicals is used to indicate the source of the chemical compounds, but it should be remembered that many common petrochemicals can be made from other sources, and the terminology is, therefore, a matter of source identification.

The starting materials for the petrochemical industry are obtained from crude oil in one of two general ways. They may be present in the raw crude oil and, as such, are isolated by physical methods, such as distillation or solvent extraction (Parkash, 2003; Gary et al., 2007; Speight, 2014a; Hsu and Robinson, 2017; Speight, 2017, 2019). On the other hand, they may be present, if at all, in trace amounts and are synthesized during the refining operations. In fact, unsaturated (olefin) hydrocarbon derivatives, which are not usually present in crude oil, are nearly always manufactured as intermediates during the various refining sequences (Parkash, 2003; Gary et al., 2007; Speight, 2014a; Hsu and Robinson, 2017; Speight, 2017, 2019).

The manufacture of chemicals from crude oil is based on the ready response of the various compound types to basic chemical reactions, such as oxidation, halogenation, nitration, dehydrogenation addition, polymerization, and alkylation. The low-molecular-weight paraffins and olefins, as found in natural gas and refinery gases, and the simple aromatic hydrocarbon derivatives have so far been of the most interest because it is individual species that can be readily be isolated and dealt with. A wide range of compounds is possible, many are being manufactured and we are now progressing to the stage in which a sizable group of products is being prepared from the heavier fractions of crude oil. For example, the various reactions of asphaltene constituents (Speight, 1994, 2014a) indicate that these materials may be regarded as containing chemical functions and are, therefore, different and are able to participate in numerous chemical or physical conversions to, perhaps, more useful materials. The overall effect of these modifications is the production of materials that either afford good-grade aromatic cokes comparatively easily or the formation of products bearing functional groups that may be employed as a nonfuel material.

For example, the sulfonated and sulfomethylated materials and their derivatives have satisfactorily undergone tests as drilling mud thinners, and the results are comparable to those obtained with commercial mud thinners. In addition, these compounds may also find use as emulsifiers for the in situ recovery of heavy oils. These are also indications that these materials and other similar derivatives of the asphaltene constituents, especially those containing such functions as carboxylic or hydroxyl, readily exchange cations and could well compete with synthetic zeolites. Other uses of the hydroxyl derivatives and/or the chloro-asphaltenes include high-temperature packing or heat transfer media.

Reactions incorporating nitrogen and phosphorus into the asphaltene constituents are particularly significant at a time when the effects on the environment of many materials containing these elements are receiving considerable attention. Here we have potential slow-release soil conditioners that only release the nitrogen or phosphorus after considerable weathering or bacteriological action. One may proceed a step further and suggest that the carbonaceous residue remaining after release of the hetero-elements may be a benefit to humus-depleted soils, such as the gray-wooded and solonetzic soils. It is also feasible that coating a conventional quick-release inorganic fertilizer with a water-soluble or water-dispersible derivative will provide a slower release fertilizer and an organic humus-like residue. In fact, variations on this theme are multiple.

Nevertheless, the main objective in producing chemicals from crude oil is the formation of a variety of well-defined chemical compounds that are the basis of the petrochemical industry. It must be remembered, however, that the ease of separation of a particular compound from crude oil does not guarantee its use as a petrochemical building block. Other parameters, particularly the economics of the reaction sequences, including the costs of the reactant equipment, must be taken into consideration.

Petrochemicals are made, or recovered from, the entire range of crude oil fractions, but the bulk of petrochemical products are formed from the lighter (C_1-C_4) hydrocarbon gases as raw materials. These materials generally occur in natural gas, but they are also recovered from the gas streams produced during refinery, especially cracking, operations. Refinery gases are also particularly valuable because they contain substantial amounts of olefins that, because of the double bonds, are much more reactive than the saturated (paraffin) hydrocarbon derivatives. Also important as raw materials are the aromatic hydrocarbon derivatives (benzene, toluene, and xylene), that are obtained in rare cases from crude oil and, more likely, from the various product streams. By means of the catalytic reforming process, non-aromatic hydrocarbon derivatives can be converted to aromatics by dehydrogenation and cyclization (Parkash, 2003; Gary et al., 2007; Speight, 2014a; Hsu and Robinson, 2017; Speight, 2017, 2019).

A highly significant proportion of these basic petrochemicals is converted into plastics, synthetic rubbers, and synthetic fibers. Together these materials are known as polymers, because their molecules are high-molecular-weight compounds made up of repeated structural units that have combined chemically. The major products are polyethylene, PVC, and polystyrene, all derived from ethylene and polypropylene, derived from monomer propylene. Major raw materials for synthetic rubbers include butadiene, ethylene, benzene, and propylene. Among synthetic fibers, the polyesters, which are a combination of ethylene glycol and terephthalic acid (made from xylene), are the most widely used. They account for about one-half of all synthetic fibers. The second major synthetic fiber is nylon, its most important raw material being benzene. Acrylic fibers, in which the major raw material is the propylene derivative acrylonitrile, make up most of the remainder of the synthetic fibers.

12.6.1 Gaseous fuels and chemicals

Synthesis gas is a very important chemical intermediate for many relevant processes including the production of methanol and Fischer–Tropsch synthesis of synthetic fuels (Chapter 10: Nonfossil Fuel Feedstocks) and, thus, is an essential industrial feedstock in several catalytic processes.

12.6.1.1 Ammonia

In the process, natural gas is first natural gas (methane) or LPGs (such as propane) or crude oil naphtha into hydrogen by steam reforming (Chapter 8: Desulfurization,

Denitrogenation, and Demetallization). The hydrogen is then combined with nitrogen to produce ammonia by the Haber–Bosch process:

Steam reforming process:

$$CH_4 + H_2O \rightarrow CO + 3H_2$$

$$C_3H_8 + 3H_2O \rightarrow 3CO + 7H_2$$

$$C_nH_m + nH_2O \rightarrow nCO + (0.5m + n)H_2$$

The Haber–Bosch process, also called the Haber process which was developed it in the first decade of the 20th century, is a nitrogen fixation process and is the main industrial procedure for the production of ammonia. The process converts atmospheric nitrogen (N_2) to ammonia (NH_3) by a reaction with hydrogen (H_2) using a metal catalyst under high temperatures and pressures:

$$N_2 + 3H_2 \rightarrow 2NH_3$$

Starting with the feedstock (natural gas), the first process leading to hydrogen production is the removal of sulfur compounds from the feedstock because sulfur—sulfur compounds deactivate the catalysts used in subsequent steps. Sulfur removal requires catalytic hydrogenation to convert sulfur compounds in the feedstock to hydrogen sulfide:

$$H_2 + RSH \rightarrow RH + H_2S \ \text{(gas)}$$

The gaseous hydrogen sulfide is then adsorbed (Chapter 9: Hydrocracking) and removed by passing it through the beds of zinc oxide where it is converted to solid zinc sulfide:

$$H_2S + ZnO \rightarrow ZnS + H_2O$$

Catalytic steam reforming of the sulfur-free feedstock is then used to form hydrogen plus carbon monoxide (synthesis gas):

$$CH_4 + H_2O \rightarrow CO + 3H_2$$

The next step is the shift conversion to convert the carbon monoxide to carbon dioxide and more hydrogen:

$$CO + H_2O \rightarrow CO_2 + H_2$$

The carbon dioxide is then removed either by absorption in aqueous ethanolamine solution or by adsorption in a PSA unit using a solid adsorbent (Chapter 9: Hydrocracking). The final step in producing the hydrogen is to use catalytic methanation to remove any small residual amounts of carbon monoxide or carbon dioxide from the hydrogen:

$$CO + 3H_2 \rightarrow CH_4 + H_2O$$

$$CO_2 + 4H_2 \rightarrow CH_4 + 2H_2O$$

To produce the desired end-product ammonia the hydrogen is then catalytically reacted with nitrogen (derived from process air) to form anhydrous liquid ammonia —the Haber–Bosch process (also known as the ammonia synthesis loop):

$$3H_2 + N_2 \rightarrow 2NH_3$$

Due to the nature of the (typically multipromoted magnetite) catalyst used in the ammonia synthesis reaction, only very low levels of oxygen-containing compounds (especially carbon monoxide, carbon dioxide, and water) can be tolerated in the synthesis gas. Relatively pure nitrogen can be obtained by an air separation step, but additional oxygen removal may be required.

The steam reforming, shift conversion, carbon dioxide removal and methanation steps each operate at absolute pressures of approximately 375–525 psi, and the ammonia synthesis loop operates at absolute pressures ranging from 900 to 2700 psi depending upon the process design.

12.6.1.2 Hydrogen

Hydrogen can be produced from the gasification product gas through the steam–methane reforming and water gas shift reaction.

Steam–methane reforming reaction:

$$CH_4 + H_2O\ CO + 3H_2$$

Water gas shift reaction:

$$CO + H_2O \rightleftharpoons CO_2 + H_2$$

Using a dual fluidized bed gasification system with carbon dioxide adsorption along with suitable catalysts, it is possible to achieve a hydrogen yield up to 70 vol. % direct in the gasifier (Soukup et al., 2009). Another sources of hydrogen is glycerol, a byproduct derived from the production of biodiesel. One approach to alleviate this problem is to transform glycerol into valuable chemicals such as hydrogen and syngas (Lin, 2013).

12.6.1.3 Synthetic natural gas

Natural gas that is produced from coal or biomass is known as synthetic natural gas or substitute natural gas (SNG). The typical catalyst for methanation is nickel, and the main reaction is represented by the equation.

$$CO + 3H_2 \rightarrow CH_4 + H_2O$$

Nickel-based catalysts are also active in water–gas shift and hydration of higher hydrocarbon derivatives, such as olefin derivatives.

Typically a hydrogen-to-carbon monoxide ratio of 3 is necessary, which is obtained by a water–gas shift reactor before methanation. In some types of reactors, such as fluidized bed reactors, the water–gas shift can be carried out also in parallel with the methanation, so no external adjustment of the hydrogen-to-carbon monoxide ratio is necessary. Methanation can be done at atmospheric pressure, although according to the thermodynamics higher pressure is preferred.

Since nickel-based catalysts are sensitive to poisoning by sulfur-containing compounds, gas treatment (Chapter 9: Hydrocracking) is quite important before the methanation, and sulfur compounds have to be removed to below 0.1 ppm.

12.6.2 *Liquid fuels and chemicals*

Liquid transportation hydrocarbon fuels and various other chemical products can be produced from syngas via the well-known and established catalytic Fischer–Tropsch process (Chapter 10: Nonfossil Fuel Feedstocks). Depending on the source of the synthesis gas, the technology is often referred to as coal-to-liquid technology and/or GTL technology.

12.6.2.1 *Fischer–Tropsch liquids*

Fischer–Tropsch liquids are well known for their being virtually sulfur, nitrogen, and heteroatom free, and they are said to carry a product premium in excess of conventional crude oil derived liquid products such as diesels and gas oils.

Conventional refinery processes can be used for upgrading Fischer–Tropsch liquids and Fischer–Tropsch waxes. The LTFT process is employed in the production of waxes, which are converted into naphtha or kerosene (diesel oil) after a hydroprocessing step. The HTFT process is employed in the production of alpha-olefins (i.e., $RCH = CH_2$) (Chapter 10: Nonfossil Fuel Feedstocks). Fuels produced with Fischer – Tropsch synthesis are of a high quality due to a very low content of aromatic constituents and zero content of sulfur constituents. The product stream consists of various fuel types: LPG, gasoline, diesel fuel, and jet fuel.

12.6.2.2 *Methanol*

Methanol (CH_3OH, MeOH, methyl alcohol, wood alcohol, or wood spirits) can be produced from fossil or renewable resources and can be used either directly as a transportation fuel or can be converted further to hydrocarbon derivatives (Olsbye et al., 2012; Tian et al., 2015).

Methanol is produced from synthesis gas by the hydrogenation of carbon oxides over a suitable (copper oxide, zinc oxide, or chromium oxide-based) catalyst:

$$CO + 2H_2 \leftrightarrow CH_3OH$$

$$CO_2 + 3H_2 \leftrightarrow CH_3OH + H_2O$$

The first reaction is the primary methanol synthesis reaction and a small amount of carbon dioxide (2%–10% v/v) in the feedstock acts as a promoter of this primary reaction and helps maintain catalyst activity. The reactions are exothermic and give a net decrease in molar volume. Therefore the equilibrium is favored by high pressure and low temperature. During production, heat is released and has to be removed to keep optimum catalyst life and reaction rate; 0.3% of the produced methanol reacts further form side products such as dimethyl ether, formaldehyde, or higher molecular weight alcohols.

The catalyst deactivates primarily because of loss of active copper due to physical blockage of the active sites by large by-product molecules; poisoning by halogens or sulfur in the synthesis gas, which irreversibly form inactive copper salts; and sintering of the copper crystallites into larger crystals, which then have a lower surface-to-volume ratio. Conventionally, methanol is produced in two-phase systems in which the reactants and products form the gas phase and the catalyst being the solid phase.

12.6.2.3 Dimethyl ether

Dimethyl ether (DME and CH_3OCH_3) is generally produced by dehydration of methanol. The methanol production and dehydration processes are combined in one reactor, such that dimethyl ether is produced directly from synthesis gas slightly more efficient than methanol. In addition, the direct synthesis of dimethyl ether allows a hydrogen–carbon monoxide ratio of approximately 1, which is an advantage for oxygen blown gasifiers.

Like methanol, dimethyl ether has promising features as fuel candidate with both the Otto and the diesel engine. With adaptations to the engine and the fuel system, dimethyl ether can be used in diesel engines, leading to higher fuel efficiency and lower emissions. In Otto engines, dimethyl ether can be used with LPG. Also, since dimethyl ether is as easily reformed as methanol, it has a big potential as fuel for fuel cell vehicles. Dimethyl ether has similar physical properties as LPG and can be handled as a liquid, using the same infrastructure as LPG.

12.6.2.4 Methanol-to gasoline and olefins

In the 1970s, at the time of the oil embargos, Mobil developed and commercialized a methanol-to-gasoline process. The methanol-to-olefins (MtO) synthesis is a commercially attractive process because of the high demand of ethylene ($CH_2=CH_2$) and propylene ($CH_3CH=CH_2$).

Currently, these compounds are produced mainly through noncatalytic steam cracking of fossil fuels (naphtha). The MtO process uses zeolite-based catalysts that efficiently convert methanol into ethylene and propylene.

12.6.2.5 Other processes

The methanol-to-diesel process first converts methanol into propylene, which is followed by olefin oligomerization (conversion to distillates), then product separation-plus-hydrogenation. The process would yield mostly kerosene and diesel, along with a small yield of naphtha (or gasoline) and low-boiling hydrocarbon derivatives. The near-zero sulfur/polyaromatics diesel fuel resulting from this process would differ from the more conventional Fischer–Tropsch diesel product only in cetane number (>52 via methanol-to-synfuel compared to >70 cetane for Fischer–Tropsch diesel). The incidental gasoline stream not only would be near zero sulfur but also have commercially acceptable octane ratings (92 RON, 80 MON) and maximally 11% v/v aromatic derivatives.

The catalytic conversion of synthesis gas to mixed alcohols is of great importance because mixed alcohols are valuable additives to gasoline to increase the octane number and reduce the environmental pollution. Furthermore, a benefit of the mixed alcohol synthesis is the high resistance of the catalysts against sulfur poisoning and the fact that the gas cleaning facilities can be simpler as in other syntheses. Mixed alcohols can also be converted to high-quality fuels via dehydration and oligomerization. Typically, the alkali-doped oxides (zinc and chromium oxides) and alkali-doped sulfides (molybdenum sulfides) are used as catalysts for mixed alcohols synthesis.

Depending on process conditions and catalysts, the main primary products are generally methanol and water. The reaction mechanism of the mixed alcohol synthesis is represented simply as:

$$nCO + 2nH_2 \rightarrow C_nH_{2n+1}OH + (n-1)H_2O$$

Owing to reaction stoichiometry the proposed carbon monoxide ratio is 2, but the optimal ratio is in practice closer to 1 because of water–gas shift reaction that occurs in parallel with the alcohol formation.

12.7 The future

The petrochemical industry is concerned with the production and trade of petrochemicals and has a direct relationship with the crude oil industry, especially the downstream sector of the industry. The petrochemical industries are specialized in the production of petrochemicals that have various industrial applications. The petrochemical industry can be considered to be a subsector of the *crude oil industry* since without the crude oil industry the petrochemical industry cannot exist. Thus crude oil is the major prerequisite raw material for the production of petrochemicals either in qualities or quantities. In addition, the petrochemical industry is subject to the geopolitics of the crude oil industry, with each industry being reliant upon the other for sustained survival.

In the 1970s, as a result of various oil embargos, coal-liquefaction processes seemed on the point of commercialization and would have provided new sources of coal liquids for chemical use, as well as fulfilling the principal intended function of producing alternate fuels. Because of the varying price of crude oil, this prospect is unlikely to come to fruition in the immediate future, due to the question of economic viability rather than not technical feasibility. The combination of these and other factors has contributed to sharpening the focus on the use of coal for the production of heat and power, and lessening or eclipsing its possible use as a starting point for other processes.

The growth and development of petrochemical industries depends on a number of factors and also varies from one country to another either based on technical know-how, marketability, and applicability of these petrochemicals for the manufacture of petrochemical products through petrochemical processes which are made feasible by knowledge and application of petrochemistry. Moreover, *petrochemistry* is a branch of chemistry (chemistry being a branch of natural science concerned with the study of the composition and constitution of substances and the changes such substances undergo because of changes in the molecules that make up such substances) that deals with crude oil, natural gas, and their derivatives.

However, not all of the petrochemical or commodity chemical materials produced by the chemical industry are made in one single location but groups of related materials are often made in adjacent manufacturing plants to induce industrial symbiosis as well as material and utility efficiency and other economies of scale (integrated manufacturing). Specialty and fine chemical companies are sometimes found in similar manufacturing locations as petrochemicals but, in most cases, they do not need the same level of large-scale infrastructure (e.g., pipelines, storage, ports, and power) and, therefore, can be found in multisector business parks. This will continue as long as the refining industry continues to exist in its present form (Favennec, 2001; Speight, 2011).

Currently, natural gas and crude oil are the main sources of the raw materials for the production of petrochemicals because they (natural gas and crude oil) are the least expensive, most readily available, and can be processed most easily into the primary petrochemicals. However, as the current century progresses and, the changes to crude oil supply that might be anticipated during the next five decades (Speight, 2011), there is a continuing need to assess the potential of other sources of petrochemicals.

For example, coal could well see a revitalization of use, understanding that there is the need to adhere to the various environmental regulations that apply to the use of any fossil fuel. Coal carbonization was the earliest and most important method to produce chemicals. For many years, chemicals that have been used for the manufacture of such diverse materials as nylon, styrene, fertilizers, activated carbon, drugs, and medicine, as well as many others have been made from coal. These products will expand in the future as natural gas and crude oil resources become strained to supply petrochemical feedstocks and coal becomes a predominant chemical feedstock once more. The ways in which coal may be converted to chemicals include carbonization, hydrogenation, oxidation, solvent extraction, hydrolysis, halogenation, and gasification (followed by conversion of the synthesis gas to chemical products) (Speight, 2013, 2014a). In some cases, such processing does not produce chemicals in the sense that the products are relatively pure and can be marketed as even industrial grade chemicals. Therefore the concept of using coal as a major source of chemicals can be very real indeed.

Compared to crude oil crude, shale oils obtained by retorting of world's oil shales in their multitude and dissimilarity; are characterized by wide boiling range and by large concentrations of heteroelements; and also by high content of oxygen-, nitrogen-, or sulfur-containing compounds. The chemical potential of oil shales as retort fuel to produce shale oil and from that liquid fuel and specialty chemicals has been used so far to a relatively small extent. While the majority of countries are discovering the real practical value of shale oil, in Estonia retorting of its national resource kukersite oil obtained for production of a variety of products is in use for 75 years already. Using stepwise cracking motor fuels have been produced and even exported before World War II. At the same time, shale oils possess molecular structures of interest to the specialty chemicals industry and also a number of non-fuel specialty products have been marketed based on functional group, broad range concentrate, or even pure compound values.

Based on large quantity of oxygen-containing compounds in heavy fraction, asphalt blending material, road asphalt and road oils, construction mastics, anticorrosion oils, and rubber softeners are produced. Also, the pyrolysis of the lower boiling fractions of shale oil leads to the production of benzene (C_6H_6) and toluene ($C_6H_5CH_3$) for the production of benzoic acid ($C_6H_5CO_2H$) as well as for solvent mixtures. Middle shale oil fractions having antiseptic properties are used to produce effective oil for the impregnation of wood as a major shale oil–derived specialty product. Water soluble phenols are selectively extracted from shale oil, fractionated and crystallized for the production of pure 5-methyl resorcinol and other alkyl resorcinol derivatives and high-value intermediates to produce tanning agents, epoxy resins and adhesives, diphenyl ketone and phenol-formaldehyde adhesive resins, rubber modifiers, chemicals, and pesticides. Some conventional products such as coke and distillate boiler fuels are produced from shale oil as by-products. New market opportunities for shale oil and its fractions may be found improving the oil conversion and separation techniques. In the petrochemical industry the organic chemicals produced in the largest volumes are methanol, ethylene, propylene, butadiene, benzene, toluene, and xylene isomers.

Basic chemicals and plastics are the key building blocks for manufacturing of a wide variety of durable and nondurable consumer goods. The demand for chemicals and plastics is driven by global economic conditions, which are directly linked to demand for consumer goods. The petrochemical industry continues to be impacted by the globalization and integration of the world economy. In the future, manufacturing processes introduced in recent years will continue to result in the adaptation of the industry to new feedstocks which will chase shifts in the ratio of products produced. This, in turn, will lead to the potential for a supply/demand imbalance, particularly for smaller downstream petrochemical derivatives. In addition, growing environmental concerns and the variability of crude oil prices (usually upward) will expedite the development and commercialization of chemical products from source other than natural gas and crude oil. As a result, feedstocks and technologies previously considered economically impractical will rise to meet the increasing demand.

There is, however, the ever-present political uncertainty arises from the occurrence of natural gas and crude oil resources in countries (provider countries) other than user countries. This has serious global implications for the supply and demand of petrochemicals and raw materials. In addition, the overall expansion of the population and an increase in individual purchasing power has resulted in an increase in demand for finished goods and greater consumption of energy in China, India, and Latin America.

However, the continued development of shale gas (tight gas) resources as well as crude oil from tight formation as well as the various technological advances to recover these resources (such as the combination of vertical hydraulic fracturing and horizontal drilling) will lead to a considerable rise in natural gas production and crude oil production. This new potential will cause many countries to reexamine their natural gas reserves and crude oil reserves to pursue development of their own nationally occurring gas plays and crude oil plays.

The production of chemicals from biomass is becoming an attractive area of investment for industries in the framework of a more sustainable economy. From a technical point of view a large fraction of industrial chemicals and materials from fossil resources can be replaced by their bio-based counterparts. Nevertheless, fossil-based chemistry is still dominant because of optimized production processes and lower costs. The best approach to maximize the valorization of biomass is the processing of biological feedstocks in integrated biorefineries where both bio-based chemicals and energy carriers can be produced, similar to a traditional crude oil refinery. The challenge is to prove, together with the technical and economic feasibility, an environmental feasibility, in terms of lower impact over the entire production chain.

Biomass is essentially a rich mixture of chemicals and materials and, as such, has a tremendous potential as feedstock for making a wide range of chemicals and materials with applications in industries from pharmaceuticals to furniture. Various types of available biomass feedstocks, including waste, and the different pretreatment and processing technologies being developed to turn these feedstocks into platform chemicals, polymers, materials, and energy.

There are several viable biological and chemical transformation pathways from sugars to building blocks. A large quantity of sugar to building block transformations can be done by aerobic fermentation employing fungi, yeast, or bacteria. Chemical and enzymatic transformations are also important process options. It should be noted, however, that pathways with more challenges and barriers are less likely be considered viable industrial processes. In addition to gasification followed by Fischer–Tropsch chemistry of the gaseous product (synthesis gas), chemical reduction, oxidation, dehydration, bond cleavage, and direct polymerization predominated. Enzymatic biotransformations comprise the largest group of biological conversions and some biological conversions can be accomplished without the need for an intermediate building block. 1,3-Propanediol ($HOCH_2CH_2CH_2OH$) is an example where a set of successive biological processes convert sugar directly to an end product. Each pathway has its own set of advantages and disadvantages. Biological conversions, of course, can be tailored to result in a specific molecular structure but the operating conditions must be relatively mild. Chemical transformations can operate at high throughput but, unfortunately, less conversion specificity is achieved.

Bio-based feedstocks may present a sustainable alternative to petrochemical sources to satisfy the ever-increasing demand for chemicals. However, the conversion processes needed for these future biorefineries will likely differ from those currently used in the petrochemical industry. Biotechnology and chemo-catalysis offer routes for converting biomass into a variety of chemicals that can serve as starting-point chemicals. While a host of technologies can be leveraged for biomass upgrading, the outcome can be significant because there is the potential to upgrade the bioderived feedstocks while minimizing the loss of carbon and the generation of by-products.

In fact, biomass offers a source of carbon from the biosphere as an alternative to fossilized carbon laid down tens of millions of years ago. Anything that grows and is available in nonfossilized form can be classified as biomass, including arable crops, trees, bushes, animal by-products, human and animal waste, waste food, and any other waste stream that rots quickly and which can be replenished on a rolling timeframe of years or decades. One of the attractions of biomass is its versatility: under the right circumstances, it can be used to provide a sustainable supply of electricity, heat, transport fuels, or chemical feedstocks in addition to its many other uses. One of the drawbacks of biomass, especially in the face of so many potential end uses, is its limited availability, even though the precise limitation is the subject of debate. Compared with the level of attention given to biomass as a source of electricity or heat, relatively little attention has been paid to biomass as a chemical feedstock. However, in a world in which conventional feedstocks are becoming constrained and countries are endeavoring to meet targets for reducing carbon dioxide emissions, there is a question as to whether biomass is too good to burn.

Developments in homogeneous and heterogeneous catalysis have led the way to effective approaches to utilizing renewable sources; however, further advances are needed to realize technologies that are competitive with established petrochemical processes. Catalysis will play a key role, with new reactions, processes, and concepts that leverage both traditional and emerging chemo- and biocatalytic technologies.

Thus new knowledge and better technologies are needed in dealing with chemical transformations that involve milder oxidation conditions, selective reduction, and dehydration, better control of bond cleavage, and improvements to direct polymerization of multifunctional monomers. For biological transformations, better understanding of metabolic pathways and cell biology, lower downstream recovery costs, increased utility of mixed sugar streams, and improved molecular thermal stability are necessary. While it is possible to prepare a very large number of molecular structures from the top building blocks, there is a scarcity of information about these behaviors of the molecular products and industrial processing properties. A comprehensive database on biomolecular performance characteristics would prove extremely useful to both the public and private sectors. Nevertheless, here is a significant market opportunity for the development of bio-based products from the four-carbon building blocks. In order to competitive with petrochemical-derived products, there is a significant technical challenge and should be undertaken with a long-term perspective.

The petrochemical industry, which is based on natural gas and crude oil, competes with the energy providing industry for the same fossil raw material. Dwindling oil and gas reserves, concern regarding the greenhouse effect (carbon dioxide emissions) and worldwide rising energy demand raise the question of the future availability of fossil raw materials. Biotechnological, chemical, and engineering solutions are needed for the utilization of this second generation biorenewable based supply chain. One approach consists of the concept of a biorefinery. Also gasification followed by Fischer–Tropsch chemistry is a promising pathway. Short and medium term a feedstock mix with natural gas and crude oil dominating can most likely be expected. In the long term, due to the final limited availability of oil and gas, biomass will prevail. Prior to this change to occur great research and developments, efforts must be carried out to have the necessary technology available when needed.

In summary the petrochemical industry gives a series of added-value products to the natural gas and crude oil industry but, like any other business, suffers from issues relating to maturity. The reasons relating to the maturity of the industry are (1) expired patents, (2) varying demand, (3) matching demand with capacity, and intense competition. Actions to combat the aches and pains of maturity are to restructure capacity achieving mega sizes, downstream, and restructuring business practices. Strategies followed by some companies to combat maturity include exit, focus on core business, and exploit a competitive advantage.

Nevertheless, the petrochemical industry is and will remain a necessary industry for the support of modern and emerging lifestyles. In order to maintain an established petrochemical industry, strategic planning is the dominating practice to maintain the industry (replace imports, export, new products, alternate feedstocks such as the return to the chemicals-from-coal concept and the acceptance of feedstocks, such as oil shale and biomass) including developing criteria for selecting products/projects. After the oil crises of the 1970s (even though it is now four decades since these crises), it is necessary to cope with the new environment of product demand through the response to new growth markets and security of feedstock supply. Mergers, alliances, and acquisitions could well be the dominating practice to combat industry maturity and increased market demand as one of the major activities. Other strategies are the focus on core business (the production of chemicals) and, last but not least, the emergence (or, in the case of coal, the reemergence) of alternate feedstocks to ensure industry survival.

In summary, synthesis gas offers many routes to industrial chemicals. They can be classified in a direct and indirect path. The direct path involves methanation, Fischer–Tropsch chemistry, and the synthesis of oxygenates. The direct conversion deals with the straight hydrogenation of carbon monoxide to paraffins, olefins, and oxygen-containing products. Best known in the direct hydrogenation of carbon monoxide is the Fischer–Tropsch synthesis, yielding the mixtures of mainly linear alkanes and/or alkenes. Mechanistically it can be described as a reductive oligomerization of carbon monoxide following a geometric progression (Schulz–Flory distribution). With α-values close to one broad product distributions are obtained, whereas small α-values predominantly yield methane. The indirect path embraces carbonylation, methanol, and methyl formate chemistry (Keim, 1986).

The great desire for the synthesis of ethylene glycol based on carbon monoxide chemistry is also underlined by the research efforts being reported by various companies. Besides the direct hydrogenation also routes via methanol or formaldehyde are possible. Also, the reductive carbonylation of formaldehyde yields glycol dialdehyde, which can be hydrogenated to ethylene glycol (Keim, 1986).

References

Arena, U., 2012. Process and technological aspects of municipal solid waste gasification. A review. Waste Manage. 32, 625–639.

Baxter, L., 2005. Biomass-coal co-combustion: opportunity for affordable renewable energy. Fuel 84 (10), 1295–1302.

Basu, P., 2013. Biomass Gasification, Pyrolysis and Torrefaction, Practical Design and Theory, second ed. Academic Press, Inc, New York.

Bhattacharaya, S., Mizanur Rahman Siddique Md., A.H., Pham, H.-L., 1999. A study in wood gasification on low tar production. Energy 24, 285–296.

Brar, J.S., Singh, K., Wang, J., Kumar, S., 2012. Cogasification of coal and biomass: a review. Int. J. For. Res. 2012 (2012), 1–10.

Brigwater, A.V. (Ed.), 2003. Pyrolysis and Gasification of Biomass and Waste. CPL Press, Newbury, Berkshire.

Chadeesingh, R., 2011. The Fischer–Tropsch Process. In: Speight, J.G. (Ed.), The Biofuels Handbook. The Royal Society of Chemistry, London, Part 3, Chapter 5.

Clark, J.H., Deswarte, F. (Eds.), 2015. Introduction to Chemicals from Biomass. second ed. John Wiley & Sons Inc, Hoboken, NJ.

EPCI, 2004. The Gasification of Residual Plastics Derived from Municipal Recycling Facilities. The Environment and Industry Council, Canadian Plastics Industry Association, Ottawa, ON.

EPCA, 2016. 50 Years of Chemistry for You. European Petrochemical Association, Brussels, <https://epca.eu/>.

Favennec, J.-P. (Ed.), 2001. Petroleum Refining: Refinery Operation and Management. Editions Technip, Paris.

Gary, J.H., Handwerk, G.E., Kaiser, M.J., 2007. Petroleum Refining: Technology and Economics, fifth ed. CRC Press, Taylor & Francis Group, Boca Raton, FL.

Gay, R.L., Barclay, K.M., Grantham, L.F., Yosim, S.J., 1980. Chapter 17: Fuel production from solid waste. In: Symposium on Thermal Conversion of Solid Waste and Biomass. Symposium Series No. 130, American Chemical Society, Washington, DC., pp. 227–236.

Goldstein, R.F., 1949. The Petrochemical Industry. E. & F. N. Spon, London.

Hahn, A.V., 1970. The Petrochemical Industry: Market and Economics. McGraw-Hill, New York.

Hotchkiss, R., 2003. Coal gasification technologies. Proc. Inst. Mech. Eng., A 217 (1), 27–33.

Hsu, C.S., Robinson, P.R. (Eds.), 2017. Handbook of Petroleum Technology. Springer International Publishing AG, Cham.

Ishi, S., 1982. Coal gasification technology. Energy 15 (7), 40–48.

Keim, W., 1986. C1 chemistry: potential and developments. Pure Appl. Chem. 58 (6), 825–832.

Lee, S., Speight, J.G., Loyalka, S., 2007. Handbook of Alternative Fuel Technologies. CRC Press, Taylor & Francis Group, Boca Raton, FL.

Lin, Y.-C., 2013. Catalytic valorization of glycerol to hydrogen and syngas. Int. J. Hydrogen Energy 38 (6), 2678–2700.

Luque, R., Speight, J.G. (Eds.), 2015. Gasification for synthetic fuel production: fundamentals, processes, and applications. Woodhead Publishing, Elsevier, Cambridge.

Mangone, C., 2002. Gas to Liquids—Conversions Produce Extremely Pure Base Oils. Machinery Lubrication Magazine, Independent Lubricant Manufacturers Association (ILMA).

Olsbye, U., Svelle, S., Bjorgen, M., Beato, P., Janssens, T.V.W., Joensen, F., et al., 2012. Conversion of methanol to hydrocarbons: how zeolite cavity and pore size controls product selectivity. Angew. Chem. Int. Ed. 51 (24), 5810–5831.

Parkash, S., 2003. Refining Processes Handbook. Gulf Professional Publishing, Elsevier, Amsterdam.

Ricketts, B., Hotchkiss, R., Livingston, W., Hall, M., 2002. Technology status review of waste/biomass co-gasification with coal. In: Proceedings. Inst. Chem. Eng. Fifth European Gasification Conference. Noordwijk, April 8–10.

Soukup, G., Pfeifer, C., Kreuzeder, A., Hofbauer, H., 2009. In situ carbon dioxide capture in a dual fluidized bed biomass steam gasifier—bed material and fuel variation. Chem. Eng. Technol. 32, 348–354.

Speight, J.G., 1994. Chapter 2: Chemical and physical studies of petroleum asphaltene constituents. In: Yen, T.F., Chilingarian, G.V. (Eds.), Asphaltene constituents and Asphalts. I. Developments in Petroleum Science, vol. 40. Elsevier, Amsterdam.

Speight, J.G., 2002. Chemical Process and Design Handbook. McGraw-Hill Publishers, New York.

Speight, J.G., 2011. The Refinery of the Future. Gulf Professional Publishing Company, Elsevier, Oxford.

Speight, J.G., 2013. The Chemistry and Technology of Coal, third ed. CRC Press, Taylor & Francis Group, Boca Raton, FL.

Speight, J.G., 2014a. The Chemistry and Technology of Petroleum, fifth ed. CRC Press, Taylor & Francis Group, Boca Raton, FL.

Speight, 2014b. Gasification of Unconventional Feedstocks. Gulf Professional Publishing Company, Elsevier, Oxford.

Speight, J.G., 2017. Handbook of Petroleum Refining. CRC Press, Taylor & Francis Group, Boca Raton, FL.

Speight, J.G., 2019. Handbook of Petrochemical Processes. CRC Press, Taylor & Francis Group, Boca Raton, FL.

Steiner, H., 1961. Introduction to Petroleum Chemicals. Pergamon Press, New York.

Tian, P., Wei, Y., Ye, M., Liu, Z., 2015. Methanol to olefins (MTO): from fundamentals to commercialization. ACS Catal. 5 (3), 1922–1938.

Vélez, F.F., Chejne, F., Valdés, C.F., Emery, E.J., Londoño, C.A., 2009. Cogasification of colombian coal and biomass in a fluidized bed: an experimental study. Fuel 88 (3), 424–430.

Wittcoff, H.A., Reuben, B.G., 1996. Industrial Organic Chemicals. John Wiley & Sons Inc, New York.

Types and properties of fuels from nonfossil fuel sources 13

13.1 Introduction

Currently, in fact since the oil crises of the 1970s and crude oil from tight formations notwithstanding, there has been and still is a growing concern approximately in the future of energy supply as well as the depletion of natural gas (NG) and crude oil reserves that can lead to future shortages of these resources. For these reasons, alternative fuels are being used and developed to replace traditional fuels. One application of these fuels is in a spark-ignited gaseous-fueled internal combustion engines (ICE). These alternative gaseous fuels are derived from a variety of feedstock such as biomass, waste products, and coal. Different feedstocks introduce variation in fuel composition, properties, and performance, the effects of which have not been fully investigated. One key property yet to be explored for these fuels is knock tendency.

Alternate fuels, also known as nonconventional fuels and advanced fuels, are any materials or substances that can be used as fuels, other than conventional fuels, such as the fossil fuels NG, crude oil, coal, extra heavy oil, tar sand bitumen, and oil shale. Some well-known alternate fuels include biodiesel, bioalcohol (methanol, ethanol, propanol, and butanol), refuse-derived fuel, biohydrogen, vegetable oil, and other biomass sources.

Finally, a note of explanation. The prefix "bio" as often used when referencing alternate fuels from biological material such as nonfossil fuel feedstocks (Chapter 10: Nonfossil Fuel Feedstocks) should not be allowed to confuse the situation. Methanol and biomethanol are the same in terms of properties and behavior—it is the source that is the difference. There is no difference between the properties of methanol (from say nonbiological sources) and biomethanol (from biological sources). In fact, to allay any possible confusion and erroneous thoughts of property differences, there may be a strong case for omitting the prefix bio in such cases. However, in some cases—such as biodiesel and crude oil-based diesel—there are vast chemical differences between the constituents, the properties, and the behavior of these two diesel fuels. Biodiesel is produced from nonhydrocarbon starting materials such as animal fats or vegetable oils, renewable resources that come from plants such as jatropha, soybean, sunflowers, corn, olive, peanut, palm, coconut, safflower, canola, sesame, and cottonseed by the process known as transesterification (Chapter 11: Production of Fuels From Nonfossil Fuel Feedstocks).

The Refinery of the Future. DOI: https://doi.org/10.1016/B978-0-12-816994-0.00013-0

13.2 Gaseous fuels

By definition, gaseous fuels are those fuels that exist in the gaseous phase under ambient conditions (room temperature and pressure). Gaseous fuels—which are produced by a variety of processes (Table 13.1) (Speight, 2013, 2014a)—are the most convenient fuels requiring the least amount of handling and simplest and most maintenance free burner systems. Gas is delivered on tap via a distribution network and so is suited to a high population or industrial density. However, large consumers do have gas holders and some produce their own gas.

13.2.1 Synthetic/gases

In the context of this book a synthetic gas (Table 13.2) is a gas that is not naturally occurring and which has been produced by a process. Examples of the types of

Table 13.1 Gaseous fuels from different sources.

1. Fuels naturally found in nature:
   a. Natural gas: (1) From crude oil reservoirs and (2) from natural gas reservoirs; most volatile fraction from the distillation of oil, petroleum gas (mainly butane and propane)
   b. Methane from coal mines (firedamp)
   c. Methane the digestion of biomass by bacteria (landfill gas and digester gas)
2. Fuel gases produced from solid fuel
   a. Gases derived from coal
      i. Pyrolysis and carbonization: the product of the heating of a solid or liquid to devolatilize the substance to form a char, coke, or charcoal
      ii. Coke-oven gas: liberated in the high temperature carbonization (coking) of coal.
      iii. Gas from charcoal production
      iv. Gasification—the product of complete gasification of a solid or liquid feedstock—that is, including the char. Only residue is ash
         A. Producer gas: low calorific value (low heat content) gas from partial combustion of coal in air
         B. Water gas: medium calorific value (medium heat content) gas from gasification of coal with steam
         C. Carbureted water gas: medium-to-high calorific value (medium-to-high heat content) gas.
         D. Blue gas or water gas: a higher calorific value gas produced by intermittently blasting the incandescent bed with air and steam such that the overall heat balance is maintained; often used as a synthesis gas in the chemical industry
   b. Gases derived from waste and biomass
3. Gases derived from crude oil
   a. Oil gas: also called refinery gas; formed by the thermal cracking of crude oil.
   b. LPG
   c. Refinery gas: formed by the cracking higher boiling fractions
   d. Gas from heavy oil gasification of from resid gasification
4. Gas from some fermentation process

LPG, Liquefied petroleum gas.

Table 13.2 Types of gases produced by gasification of carbonaceous feedstocks.[a]

1. *Producer gas*
Simplest method of gasification; available for well over 100 years
Air is blown through an incandescent bed of the feedstock
Enough air to maintain temperature; insufficient air to complete the combustion reaction
Composition: a mixture on the order of 50% v/v nitrogen (from air), 29% v/v CO (incomplete combustion), and 4% v/v CO_2
Very low calorific value; not suitable for distribution but can be used on site as fuel gas
Steam is added to modify the products. Endothermic reactions (water gas) result in the formation of hydrogen
Chemistry:
Initially, oxidation of feedstock in the presence of available oxygen:
$2C + O_2 \rightarrow 2CO$ (exothermic)
Then competition for oxygen between the carbon and the water. The water–gas reactions.
$C + H_2O \rightarrow CO + H_2$ (endothermic)
$C + 2H_2O \rightarrow CO_2 + 2H_2$
$CO + H_2O \rightarrow CO_2 + H_2$
The ratio of steam and air may be used to modify the composition of the gas produced
2. *Blue water gas*
The feedstock bed is simultaneously blasted with air followed by steam
Air reaction is exothermic; steam reaction is endothermic
Mixture of gas produced is higher in quality than produced gas
Composition: a mixture on the order of 5% v/v CO_2, 40% v/v CO, 49% H_2, 1% v/v methane, <5% v/v nitrogen
If oxygen is used instead of air the process can be continuous
3. *Carbureted water gas*
Low-quality gas is enriched by adding a carburetor
Fuel oil is sprayed into a brick lined chamber during the blow period of the blue water gas plant with air. This heats the bricks to around 1000°C (1830°F)
During the make period the air is turned off and the oil is cracked into smaller hydrocarbon derivatives in the now heated chamber
Composition: mainly methane

[a]Using coal as the example of the feedstock.

synthetic gases produced from various feedstocks are (alphabetically) (1) biogas, (2) blue water gas (BWG), (3) carbureted water gas, (4) coal gas and coke-oven gas, (5) producer gas, (6) refuse gas, (7) water gas, and (8) wood gas.

13.2.1.1 Biogas

Gasification technology is an attractive route for the production of fuel gases from various types of biomass (van Doorn, 1995; Faaij et al., 1997). By gasification, solid biomass is converted into a combustible gas mixture normally called producer gas consisting primarily of hydrogen (H_2) and carbon monoxide (CO), with lesser amounts of carbon dioxide (CO_2), water (H_2O), methane (CH_4), and higher boiling (higher molecular weight) hydrocarbon derivatives (C_xH_y), as well as nitrogen (N_2) and particulates.

A variety of biomass gasifiers has been developed and can be grouped into four major classes: (1) fixed-bed updraft or counter-current gasifier, (2) fixed-bed downdraft or cocurrent gasifier, (3) bubbling fluidized bed gasifier, and (4) circulating fluidized bed gasifier. Differentiation is based on the means of supporting the biomass in the reactor vessel, the direction of flow of both the biomass and oxidant, and the way heat is supplied to the reactor. The processes occurring in any gasifier include drying, pyrolysis, reduction, and oxidation. The unique feature of the updraft gasifier is the sequential occurrence of these processes: they are separated spatially and, therefore, temporally.

As an illustration, using the updraft gasifier, biomass and air are fed in an opposite direction. In the highest zone, biomass is heated up and releases its moisture. In the pyrolysis zone, biomass undergoes a further increase in temperature and decomposes into hydrocarbon derivatives, gas products, and char in the temperature range of 150°C–500°C (300°F–930°F). The major reactions are

$$\text{Biomass} \rightarrow C_xH_y + C_xH_yO_z + H_2O + CO_2 + CO + H_2$$

The hydrocarbon fraction consists of methane to high-boiling tar. The composition of this fraction can be influenced by many parameters, such as particle size of the biomass, temperature, pressure, heating rate, residence time, and catalysts. The operative reactions are

$$C + CO_2 \rightarrow 2CO$$

$$C + H_2O \rightarrow H_2 + CO$$

$$C + 2H_2O \rightarrow H_2 + CO_2$$

$$C + 2H_2 \rightarrow CH_4$$

$$CO + H_2O \rightarrow CO_2 + H_2$$

$$CO + 3H_2 \rightarrow CH_4 + H_2O$$

The composition of the producer gases varies widely with the properties of the biomass, the gasifying agent, and the process conditions. Depending on the nature of the raw solid feedstock and the process conditions, the char formed from pyrolysis contains 20%–60% of the energy input. Therefore the gasification of char is an important step for the complete conversion of the solid biomass into gaseous products and for an efficient utilization of the energy in the biomass.

The producer gases from the reduction zone rise beyond the reduction zone. When they come into contact with the cooler biomass, the temperature drops down and the aforementioned reactions are frozen. The unreacted char further undergoes the oxidation with air in the lowest zone:

$$C + O_2 \rightarrow CO_2$$

As a result, ash is left at the bottom of the reactor. The produced carbon dioxide flows upward and is involved in the reactions in the reduction zone. The heat released in the oxidation zone drives both the reduction and pyrolysis processes.

13.2.1.2 Blue water gas

BWG (sometimes referred to as blue gas) is produced in a similar manner to produce gas but allows the production of a higher heat content gas by the intermittent blasting the incandescent bed with air and steam such that the overall heat balance is maintained. The products of the air blast contain the nitrogen that reduces heat content of the product gas.

The process follows the similar principles to the production of producer gas with the exception that the problem of nitrogen dilution is overcome. The feedstock bed is simultaneously blasted with air followed by steam—the air reaction is exothermic causing the bed to increase in temperature and this is balanced by the endothermic reaction of the steam. The typical composition of the product gas is in the order of carbon dioxide 5% v/v, carbon monoxide 41% v/v, hydrogen 49% v/v, methane 1% v/v, and nitrogen 4% v/v. If oxygen is used instead of air, the process can be continuous and coke is preferred to coal because coal can continue to devolatilize in the blow period, thereby reducing process efficiency—there are also issues related to process efficiency that arise from the use of caking coal as the feedstock.

The BWG may be enriched by adding a carburetor in which heavy (high-density and high-boiling) fuel oil is sprayed into a brick lined chamber during the blow period of the BWG plant with air. During the make period the air is turned off and the oil is cracked into smaller hydrocarbon derivatives in the now heated chamber. This produces a mixture of hydrocarbon derivatives (predominantly methane) which increases the heat content and enriches the gas product.

The problem of nitrogen dilution in manufactured gas streams was overcome by the BWG process in which the incandescent fuel bed would be alternately blasted with air followed by steam. The air reactions during the blow cycle are exothermic, heating up the bed, while the steam reactions during the make cycle are endothermic and cool down the bed. The products from the air cycle contain noncalorific nitrogen and are exhausted out the stack, while the products of the steam cycle are kept as BWG. This gas is composed almost entirely of carbon monoxide and hydrogen and burns with a pale blue flame similar to NG.

13.2.1.3 Carbureted water gas

Carbureted water gas is the result of combining the water gas and oil gas methods. Briefly, oil gas is the gas formed by the thermal cracking of crude oil. If oil is sprayed onto heated checker work (refractory) it cracks to form lower gaseous hydrocarbon derivatives. The types of hydrocarbon derivatives depend entirely on the type of feedstock but do cause an increase in the heat content of the product gas.

In the oil gas method an oil product (typically heavy fuel oil, as described previously) is sprayed into the hot water gas chamber to produce a better quality (higher heat content) gas. The amount of the heavy fuel ratio is a determinant of the quality of the gas. In the 19th century and early 20th century, this was the method used to produce the town gas but has largely been superseded by NG in countries with an abundant supply. As supplies of NG diminish and various carbonaceous feedstocks are used for gas production, the process may once again rise in importance.

13.2.1.4 Coal gas and coke-oven gas

Coal gas is a flammable gaseous fuel produced from coal and is obtained by in the carbonization of coal especially at high temperatures for the production of coke. In the process the coal feedstock is heated strongly in the absence of air. Town gas is a more general term referring to manufactured gaseous fuels produced for sale to consumers and municipalities. The coal gas contains a variety of gases, including hydrogen, carbon monoxide, methane, and other volatile hydrocarbon derivatives, including gasolines olefin derivatives; small quantities of noncalorific gases such as carbon dioxide may also be present.

These gas product is cleaned and used as fuel or as petrochemical feedstocks (Speight, 2013, 2019a,b). If coke is not required and another carbonaceous feedstock is used, steam injection at the end of the cycle reacts with the coke to form BWG. This reduces the heat content of the product gas produced, but the thermal efficiency of conversion rises.

Coke-oven gas is also a fuel gas having a medium calorific value that is produced during the manufacture of metallurgical coke by heating bituminous coal to temperatures of 900°C–1000°C (1650°F–1830°F) in a chamber from which air is excluded. The main constituents are approximately 50% v/v hydrogen, 30% v/v methane, 3% v/v higher molecular weight hydrocarbon derivatives, 7% v/v carbon monoxide, 3% v/v carbon dioxide, and 7% v/v nitrogen.

Typically, coke-oven gas is obtained from a battery comprising a number of narrow, vertical chambers, or ovens (approximately 18 in. wide, 16 ft. high, and 40 ft. long) built of silica brick that are separated by heating ducts, such that heat is transmitted to the coal through both sides of the chamber walls. The ovens are slightly tapered so that one end is wider than the other to facilitate the horizontal discharge of the coke.

In the process, crushed coal is charged from overhead bunkers into the ovens, which are sealed at each end by refractory-lined sheet doors and heated for about 24 hours. The hot coke is then discharged—approximately 12% w/w of the coal feedstock is converted into gas. The hot gases evolved from the coal pass through a gas space at the top of the oven and into a collecting main prior to quenching and treatment to remove dust, tar, oil, and gaseous impurities such as ammonia and hydrogen sulfide

13.2.1.5 Producer gas

The gas is produced by blowing air and sometimes steam through an incandescent fuel bed (the process is self-heating). The reaction with air is exothermic, but insufficient air is added, hence carbon monoxide is produced. Steam addition results in the formation of hydrogen by the water gas reaction. This is endothermic and hence balances out the exothermic air reaction. Producer gas is low calorific value and is, hence, typically used as an on-site fuel gas.

In the process—using coal as the example, of the carbonaceous feedstock—air is blown through an incandescent bed of coal. The amount of air is sufficient to maintain the process temperature but insufficient to complete the combustion reaction. The gas products is a mixture containing approximately 50% v/v nitrogen, form the air, 29% v/v carbon monoxide, due to incomplete combustion of the feedstock, and 4% v/v carbon dioxide, as well as sundry other products, including hydrocarbon derivatives. The heat content of the gas is low and, although not suitable for distribution to consumers, can be used on site as a fuel gas. In order to modify the product distribution, steam can be used in the process, and endothermic reactions (such as the water gas reaction) result in the formation of hydrogen.

Initially, the oxidation of the feedstock (coal) in the presence of the available oxygen occurs after which there is competition for oxygen between the carbon and the water gas reactions occur:

$$2C_{feedstock} + O_2 \rightarrow 2CO \text{ (exothermic)}$$

$$C_{feedstock} + H_2O \rightarrow CO + H_2 \text{ (endothermic)}$$

$$C_{feedstock} + 2H_2O \rightarrow CO_2 + 2H_2\text{(endothermic)}$$

$$CO + H_2O \rightarrow CO2 + H_2 \text{ (exothermic)}$$

The ratio of steam and air may be used to modify the composition of the gas that is produced.

13.2.1.6 Refuse gas

There are two ways of producing gaseous fuel from refuse or waste: (1) gasification of the refuse at high temperatures by partial oxidation and then conversion of materials containing carbon into synthesis gas (mainly hydrogen and carbon monoxide) and (2) production of biogas, mainly methane, by the anaerobic digestion (AD) of the refuse in a landfill (Chapter 6: Catalytic Cracking, and Chapter 7: Deasphalting and Dewaxing).

The principle behind waste gasification and the production of gaseous fuels is that waste contains carbon, and it is this carbon that is converted to gaseous products via the usual gasification chemistry. This, when fuel is fed to a gasifier, water and volatile matter are released fast and a char residue is left to react further. The char gasification is what mainly controls the conversion achieved in the process. From solid carbon, product gas is formed according to the following main reactions:

$$C + H_2O \rightarrow CO + H_2$$

$$C + CO_2 \rightarrow 2CO$$
$$C + 2H_2 \rightarrow CH_4$$

In addition to the reactions of solid carbon the most important reaction is the water–gas shift reaction, which takes place in the gas phase:

$$CO + H_2O \leftrightarrow H_2 + CO_2$$

The product gas generally contains large amounts of hydrogen and carbon monoxide and a small amount of methane, as well as carbon dioxide and steam, and in air gasification nitrogen. In addition, a significant amount of other organic components in the gas, known as tar, is formed. Tar formation is a well-known phenomenon in the thermal reactions of coal (Speight, 2013) and can also be anticipated to form in the gasification or pyrolysis of any complex carbonaceous material.

13.2.1.7 Water gas

Water gas is a flammable gaseous mixture that consists predominantly of carbon monoxide and hydrogen with small amounts of methane, carbon dioxide, and nitrogen. It is typically produced by blowing air and then steam over red-hot coke or coal and is used as a fuel or after carbureting as an illuminant.

Historically, the reaction of coke and water was of great importance in the generation of a combustible gas. Coke can be produced by heating coal in an airtight container. One of the by-products produced was coal gas, which could be burned. An even better method of generating a flammable gas is

$$C(s) + H_2O \leftrightarrow CO(g) + H_2(g)$$

The gaseous product, a mixture of hydrogen and carbon, is known as water gas, or synthesis gas is combustible. Some of the carbon monoxide can be removed using the water gas shift reaction:

$$H_2O(g) + CO(g) \leftrightarrow H_2(g) + CO_2(g)$$

While water gas is no longer used for household purposes, these reactions are still important in the generation of hydrogen.

13.2.1.8 Wood gas

Wood can be used to make both liquid and gaseous fuels. When wood is heated in the absence of air, or with a reduced air supply, it is possible to produce a liquid fuel which can be used in a similar way to conventional oil fuels. It can be used to run ICE in vehicles or generators. The gas produced from wood is a mixture of hydrogen and carbon monoxide which is similar to the coal gas that was made before the arrival of NG from the North Sea. This wood gas can be used in ICE or in gas turbines that can be used to power generators.

Gasification of wood has the potential for providing all fuel needs for small oil- or gas-fired boilers and supplemental fuel for large boilers. The gasification is carried out at elevated temperatures (500°C–1500°C, 930°F–2730°F) and at atmospheric or elevated pressures. The process involves conversion of biomass, which is carried out in the absence of air or with less air than the stoichiometric requirement of air for complete combustion. Partial combustion produces carbon monoxide as well as hydrogen which are both combustible gases. The producer gas can serve in different ways: it can be burned directly to produce heat or used as a fuel for gas engines and gas turbines to generate electricity; in addition, it can also be used as a feedstock (synthesis gas) in the production of chemicals, for example, methanol.

13.2.2 Composition and properties

Prior to the development of NG resources, virtually all fuel and lighting gas was manufactured from coal and the history of NG cleaning has its roots in town gas cleaning (Mokhatab et al., 2006; Speight, 2013, 2014a, 2018, 2019a). The by-product coal tar produced during the manufacture of gas from coal was an important feedstock for the chemical industry. The development of manufactured gas paralleled that of the industrial revolution and urbanization.

The naturally occurring gas is also found in coal fields (generally referred *t* as firedamp, coalbed methane—coal gas is gas that is manufactured from coal). The properties of the gas are used to describe the various gases (Table 13.3) (Mokhatab et al., 2006; Speight, 2013, 2014a, 2018, 2019a). Gas from coal seams (coalbed methane) is of equal quality to gas from crude oil fields but is much more difficult to extract.

13.2.2.1 Composition

The gas products produced from nonfossil fuel feedstocks are combustible mixtures of a variety of hydrocarbon constituents and nonhydrocarbon constituents. The nonhydrocarbon gas portion of the may be composed of nitrogen (N_2), carbon dioxide (CO_2), helium (He), hydrogen sulfide (H_2S), water vapor (H_2O), and other sulfur compounds [such as carbonyl sulfide (COS) and mercaptans (e.g., CH_3SH)] and trace amounts of other gases. Carbon dioxide and hydrogen sulfide are commonly referred to as acid gases since they form corrosive compounds in the presence of water (Speight, 2014b, 2019a). Nitrogen, helium, and carbon dioxide are often referred to as diluents since none of these burn, and thus they have no heating value.

The composition of the gas steams vary depending on the feedstock from which the gas was produced as well as the process by which the gas was produced. The different constituents of a gas stream can be separated using their different physical

Table 13.3 Nomenclature of relevant properties of the various gas products.[a]

Gas	Relevant property
Biogas	Gas produced from a biological source; also referred to as biogas
Blue water gas	Higher heat content (calorific value) gas produced from coal
Casinghead gas	Gas extracted from an oil well by separation at the surface
Carbureted water gas	Higher heat content (calorific value) gas produced from coal
Coalbed methane	Natural gas from coal seams
Dry gas	High methane content
High-Btu gas	High energy gas; also called high heat-content gas
Landfill gas	Gas produced by the decay of organic matter in a landfill
Lean gas	High methane content
Low-Btu gas	Low energy gas; also called low heat-content gas
Medium-Btu gas	Medium energy gas; also called medium heat-content gas
Producer gas	Very low heat content (calorific value) gas produced from coal
Residue gas	Gas remaining after the condensing process; predominantly methane
Sour gas	High concentration of hydrogen sulfide (H_2S)
Sweet gas	Low concentration of hydrogen sulfide or no hydrogen sulfide
Synthesis gas	A mixture of carbon monoxide and hydrogen
Wet gas	High concentration of higher boiling hydrocarbon derivatives (C_5-C_{10})
Wood gas	Gas produced from wood

[a]Listed alphabetically.

properties as weight, boiling point, or vapor pressure (Mokhatab et al., 2006; Speight, 2013, 2014a, 2018, 2019a).

13.2.2.2 Properties

The properties of unrefined gas streams are variable because the composition of gas stream is never constant are best understood by investigating the properties and behavior of the constituents (Speight, 2019a).

13.2.2.2.1 Density

Density is a physical property of matter is a measure of the relative heaviness of hydrocarbon derivatives and other chemicals at a constant volume, and each constituents of NG has a unique density associated with it. Another term, specific gravity is commonly used is relation to the properties of hydrocarbon derivatives. The specific gravity of a substance is a comparison of its density to that of water. For most chemical compounds (i.e., those that are solid or liquid) the density is measured relative to water (1.00). For gases the density is more likely to be compared to the density of air (also given the number 1.00 but this is arbitrary and bears no relationship to the density of water). As a comparison, the density of liquefied NG is approximately 0.41–0.5 kg/L, depending on temperature, pressure, and composition; in comparison, the density of water is 1.0 kg/L. Density values [including those of the hydrocarbon derivatives in the gas stream (Table 2.3)] are given at room temperature unless otherwise indicated by the superscript figure. For example, 2.487^{15} indicates a density of 2.487 g/cm at 15°C (59°F). Thus the density of any gas compared to the density of air is the vapor density and is a very important characteristic of the constituents of a gas stream (Speight, 2019a).

13.2.2.2.2 Heat of combustion

All of the hydrocarbon constituents of a gas stream are combustible, but nonflammable nonhydrocarbon components (carbon dioxide, hydrogen sulfide, nitrogen, and helium) detract slightly from the heating value of the gas stream. The heat of combustion (energy content) of a gas stream is the amount of energy that is obtained from the burning of a volume of the gas is measured in British thermal units (Btu). The value of a gas stream is calculated by its Btu content. One Btu is the quantity of heat required to raise the temperature of 1 lb of water of 1°F at atmospheric pressure.

Thus the energy content of a gas stream is variable because the gas has variations in the amount and types of energy gases (methane, ethane, propane, and butane) it contains: the more noncombustible gases in the gas stream, the lower the energy (Btu). In addition, the volume mass of energy gases which are present in a gas stream accumulation also influences the Btu value of the gas. The more carbon atoms in a hydrocarbon gas, the higher the Btu value (heat content) of the gas.

13.2.2.2.3 Volatility, flammability, and explosive properties

The boiling point (boiling temperature) of a substance is the temperature at which the vapor pressure of the substance is equal to atmospheric pressure.

At the boiling point a substance changes its state from liquid to gas. A stricter definition of boiling point is the temperature at which the liquid and vapor (gas) phases of a substance can exist in equilibrium. When heat is applied to a liquid, the temperature of the liquid rises until the vapor pressure of the liquid equals the pressure of the surrounding atmosphere (gases). At this point there is no further rise in temperature, and the additional heat energy supplied is absorbed as latent heat of vaporization to transform the liquid into gas. This transformation occurs not only at the surface of the liquid (as in the case of evaporation) but also throughout the volume of the liquid, where bubbles of gas are formed. The boiling point of a liquid is lowered if the pressure of the surrounding atmosphere (gases) is decreased. On the other hand, if the pressure of the surrounding atmosphere (gases) is increased, the boiling point is raised. For this reason, it is customary when the boiling point of a substance is given to include the pressure at which it is observed, if that pressure is other than standard, that is, 760 mm of mercury or 1 atm (STP, standard temperature and pressure).

The boiling points of petroleum fractions are rarely, if ever, distinct temperatures. It is, in fact, more correct to refer to the boiling ranges of the various fractions; the same is true of a gas stream. To determine these ranges the material in question is tested in various methods of distillation, either at atmospheric pressure or at reduced pressure. Thus the boiling points of the hydrocarbon constituents of a gas stream increase with molecular weight and the initial boiling point of a gas stream corresponds to the boiling point of the most volatile constituent (i.e., methane) (Speight, 2019a).

The flash point of petroleum or a petroleum product, including gas, is the temperature to which the product must be heated under specified conditions to give of sufficient vapor to form a mixture with air that can be ignited momentarily by a specified flame (ASTM D56, D92, and D93). As with other properties, the flash point is dependent on the composition of the gas and the presence of other hydrocarbon constituents (Speight, 2019a).

The fire point is the temperature to which the gas must be heated under the prescribed conditions of the method to burn continuously when the mixture of vapor and air is ignited by a specified flame (ASTM D92). In addition to the fire point, and also from the viewpoint of safety, information related to the flash point is of most significance at or slightly above the maximum temperatures (30°C–60°C, 86°F–140°F) that may be encountered in storage, transportation, and use of liquid petroleum products, in either closed or open containers. In this temperature range the relative fire and explosion hazard can be estimated from the flash point. For products with flash point below 40°C (104°F), special precautions are necessary for safe handling. Flash points above 60°C (140°F) gradually lose their safety significance until they become indirect measures of some other quality.

The flash point of a petroleum product is also used to detect contamination. A substantially lower flash point than expected for a product is a reliable indicator that a product has become contaminated with a more volatile product, such as gasoline. The flash point is also an aid in establishing the identity of a particular petroleum product.

A further aspect of volatility that receives considerable attention is the vapor pressure of petroleum and its constituent fractions. The vapor pressure is the force exerted on the walls of a closed container by the vaporized portion of a liquid. Conversely, it is the force that must be exerted on the liquid to prevent it from vaporizing further (ASTM D323). The vapor pressure increases with temperature for any given gasoline, liquefied petroleum gas, or other product. The temperature at which the vapor pressure of a liquid, either a pure compound of a mixture of many compounds, equals 1 atm (14.7 psi, absolute) is designated as the boiling point of the liquid.

The flammable range is expressed by the lower explosive limit (LEL) and the upper explosive limit (UEL). The LEL is the concentration of a gas stream in the air below which the propagation of a flame will not occur on contact with an ignition source. The UEL is the concentration of NG in the air above which the propagation of a flame will not occur on contact with an ignition source.

13.3 Liquid fuels

By definition, liquid fuels are combustible or energy-generating molecules that can be harnessed to create mechanical energy, usually producing kinetic energy; they

Table 13.4 Examples of properties that are used in the specification for liquid fuels.

Property	Definition
Density	The mass per unit volume of a fuel
Cloud point	The temperature at which a sample has visible cloudiness
Pour point	The temperature at which the fuel will no longer pour
Freezing point	The temperature at which a liquid becomes solid
Flash point	The temperature above which the vapor above a fuel in air becomes flammable
Viscosity	A measure of the resistance of a fluid to flow
Gum content	The quantity of remaining solid residue after a sample has been heated and evaporated for a prescribed period of time in air
Ash content	A measure of the mineral matter in a fuel
Sulfur content	A measure of the amount of sulfur-containing constituents in a fuel
Trace metal content	A measure of the amount of Va, Ni, Cu, and Fe in a fuel
Aromatic content	The percentage of aromatic hydrocarbon derivatives in the fuel
Thermal stability	The resistance of the fuel to thermal degradation

also must take the shape of their container. It is the fumes of liquid fuels that are flammable instead of the fluid. Most liquid fuels in widespread use are derived from fossil fuels. However, there are several types, such as hydrogen fuel (for automotive uses), ethanol, and biodiesel, which are also categorized as a liquid fuel. Many liquid fuels play a primary role in transportation and the economy.

The main advantage of liquid fuels over their gaseous alternatives is the extremely high thermal energy/volume ratio. Hence, they have become the conventional gas for transport applications. Liquid fuels (light and heavy fuel oils) are also used in furnaces and boilers. Not surprisingly the most heavily specified of all fuels are the transport fuels that have a variety of physical parameter that must be met and are described as specifications (Table 13.4) (Speight, 2014a, 2015a).

13.3.1 Methanol

Methanol (Table 13.5) is another possible replacement for conventional motor fuels. In fact, it has been considered as a possible large-volume motor fuel substitute at various times during gasoline shortages. In fact, it was often used in the early 20th century to power automobiles before the introduction of inexpensive gasoline. Later, synthetically produced methanol was widely used as a motor fuel in Germany during the Second World War. Again, during the oil crisis of 1970s, methanol blending with motor fuel received attention due to its availability and low cost.

Methanol can be produced from biomass essentially any primary energy source. Thus the choice of fuel in the transportation sector is to some extent determined by the availability of biomass. As regards the difference between hydrogen and methanol production costs, conversion of NG, biomass, and coal into hydrogen is generally more energy efficient and less expensive than conversion into methanol.

Table 13.5 General properties of methanol.

Chemical formula	CH_3OH or CH_4O
Molecular weight	32.04 g/mol
Appearance	Colorless liquid
Density	0.792 g/cm^3
Melting point	−97.6°C (−143.7°F; 175.6K)
Boiling point	64.7°C (148.5°F; 337.8K)
Solubility in water	Miscible

Table 13.6 General properties of ethanol.

Chemical formula	C_2H_6O
Molecular weight	46.069 g/mol
Appearance	Colorless liquid
Density	0.7893 g/cm^3 (at 20°C, 68°F)
Melting point	−114.14°C ± 0.03°C (−173.45°F ± 0.05°F)
Boiling point	78.24°C ± 0.09°C (172.83°F ± 0.16°F)
Solubility in water	Miscible

Methanol is poisonous and burns with an invisible flame. Similar to ethanol, methanol has a high octane rating and hence is suitable for Otto engine. Methanol is a colorless, odorless and nearly tasteless alcohol and is also produced from crops and is also used as a fuel. Methanol, like ethanol, burns more completely but releases as much or more carbon dioxide than its gasoline counterpart. The balance is often seen as the various biprocesses that draw carbon dioxide from the atmosphere so there is no net modern release, as there is for fossil fuels.

Methanol is known as wood alcohol. Generally, methanol is easier to find than ethanol. Methanol is produced from synthetic gas or biogas and evaluated as a fuel for ICE. The production of methanol is a cost intensive chemical process. It is commonly used in biodiesel production for its reactivity with vegetable oils. Methanol can be used as one possible replacement for conventional motor fuels. Many tests have shown promising results using 85%−100% v/v methanol as a transportation fuel in automobiles, trucks, and buses.

13.3.2 Ethanol

Ethanol (Table 13.6) is produced from organic feedstocks, such as biomass, by a fermentation process. However, recent efforts have focused on the production of ethanol from waste materials using a gasification process. The overall process consists of a thermochemical conversion of synthesis gas which is then converted to higher molecular weight alcohols by a catalytic process after which high-purity ethanol is separated by distillation. The key differentiating factors are the catalysts and their operating parameters.

Most current research efforts have centered on bacteria, enzyme, and acid/solvent hydrolysis fermentation based ethanol production. However, the action of bacteria and enzymes is often very dependent on the feedstock. In addition, these processes generate a small quantity of ethanol over a period of days and the dilute aqueous ethanol product must be distilled to recover the ethanol. In the gasification-synthesis process, various carbonaceous feedstocks can be used, with appropriate modifications in the synthesis gas production step.

13.3.3 Propanol and butanol

Propanol (Tables 13.7 and 13.8) and butanol (Tables 13.9–13.11) exist in various isomeric forms.

Table 13.7 General properties of 1-propanol.

Chemical formula	C_3H_8O
Molecular weight	60.096 g/mol
Appearance	Colorless liquid
Odor	Mild, alcohol-like
Density	0.803 g/mL
Melting point	−126°C (−195°F)
Boiling point	97°C–98°C (206°F–208°F)
Solubility in water	Miscible

Table 13.8 General properties of 2-propanol.

Chemical formula	C_3H_8O
Molecular weight	60.096 g/mol
Appearance	Colorless liquid
Density	0.786 g/cm^3 (20°C, 68°F)
Melting point	−89°C (−128°F)
Boiling point	82.6°C (180.7°F)
Solubility in water	Miscible
Solubility	Miscible with benzene, chloroform, ethanol, ether, glycerin, acetone

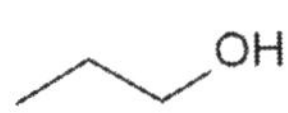

n-Propanol

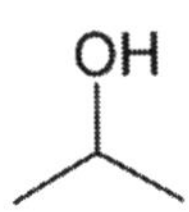

Iso-propanol

Table 13.9 General properties of *n*-butanol.

Chemical formula	$C_4H_{10}O$
Molecular weight	74.123 g/mol
Appearance	Colorless
Odor	Harsh, alcoholic, and sweet
Density	0.81 g/cm^3
Melting point	−89.8°C; −129.6°F
Boiling point	117.7°C; 243.9°F
Solubility in water	73 g/L at 25°C
Solubility	Soluble in acetone; miscible with ethanol, ethyl ether

Table 13.10 General properties of 2-butanol.

Chemical formula	$C_4H_{10}O$
Molecular weight	74.123 g/mol
Density	0.808 g/cm
Melting point	−115°C; −175°F
Boiling point	98°C−100°C; 208°F−212°F
Solubility in water	290 g/L

Table 13.11 General properties of *tert*-butanol.

Chemical formula	$C_4H_{10}O$
Molecular weight	74.123 g/mol
Appearance	Colorless solid
Odor	Camphor-like
Density	0.775 g/mL
Melting point	25°C−26°C; 77°F−79°F
Boiling point	82°C−83°C; 179°F−181°F
Solubility in water	Miscible

n-Butanol

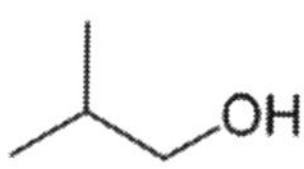

Iso-butanol

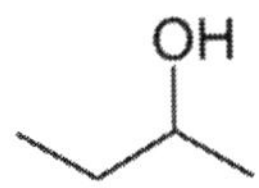

sec-Butanol

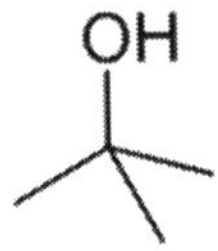

tert-Butanol

The unmodified term butanol usually refers to the straight-chain isomer with the alcohol functional groups at the terminal carbon, which is also known as *n*-butanol or 1-butanol. The straight-chain isomer with the alcohol at an internal carbon is *sec*-butanol or 2-butanol. The branched isomer with the alcohol at a terminal carbon is iso-butanol (isobutanol) or 2-methyl-1-propanol, and the branched isomer with the alcohol at the internal carbon is *tert*-butanol or 2-methyl-2-propanol. These two alcohols are considerably less toxic and less volatile than methanol. In particular, butanol has a high flash point of 35°C (95°F), which is a benefit for fire safety but may be a difficulty for starting engines in cold weather.

The butanol isomers have different melting and boiling points. *n*-Butanol and isobutanol have limited solubility, *sec*-butanol has substantially greater solubility, while *tert*-butanol is miscible with water above *tert*-butanol's melting point. The hydroxyl group makes the molecule polar, promoting solubility in water, while the longer hydrocarbon chain mitigates the polarity and reduces solubility.

13.3.4 Hydrocarbon fuels

Hydrocarbon derivatives, as such, are not usually produced from crops, there being insufficient amount of the hydrocarbon derivatives present in the plant tissue to make the process economical. However, biodiesel is produced from crops, thereby offering an excellent renewable fuel for diesel engines.

The distillable oil obtained from the fast pyrolysis of biomass has high oxygen content. Ketone derivatives and aldehyde derivatives, carboxylic acid derivatives, ester derivatives, aliphatic and aromatic alcohol derivatives, and ether derivatives have been detected in significant quantities. Because of the reactivity of oxygenated groups, the main problems of the oil are instability. The hydrodeoxygenation (HDO) of bio-oil in the presence of a cobalt molybdate catalyst was studied, which is a most important route for upgrading the oil.

$$-(CH_2O)- + H_2 \rightarrow -(CH_2)- + H_2O$$

The reaction has strong analogies with typical refinery processes, such as the hydrodesulfurization and hydrodenitrogenation processes (Parkash, 2003; Gary et al., 2007; Speight, 2011a, 2014a; Hsu and Robinson, 2017; Speight, 2017). In general, most of the HDO studies have been performed using existing hydrodesulfurization catalysts (NiMo and CoMo on suitable carriers). However, these types of catalyst need activation using a suitable sulfur source, and this is a major drawback when using low-to-no sulfur resources.

13.3.4.1 Naphtha and gasoline

The naphtha-gasoline fraction is a crude oil-derived liquid mixture (Table 13.12), primarily used as fuel in ICE, specifically in spark-ignition engines. However, gasoline is not typically produced directly from crude oil but is a blend of various fraction—the basic fraction is the low-boiling naphtha fraction.

The amount of naphtha that can be distilled from oil crude depends on where the crude is obtained. For example, heavy crude oil, such as Venezuelan heavy (high-density) crude oil, yields very little naphtha (approximately 5% v/v), whereas West Texas crude oil or Arabian light (low-density) crude oil yields approximately 30% v/v naphtha. The density of gasoline is on the order of 0.71−0.77 kg/L. An important characteristic of the finished gasoline is the octane number or octane rating, which is a measure of how resistant the gasoline is to the abnormal combustion phenomenon known as predetonation (also known as knocking, pinging, spark knock, and other names). The octane number is measured relative to a mixture of 2,2,4-trimethylpentane and *n*-heptane. Octane number a measure of the gasoline quality for the prevention of early ignition that leads to cylinder knocks. Higher octane numbers are preferred in ICE. For gasoline production, aromatic derivatives, naphthene (cycloalkane) derivatives, and iso-alkane derivative derivatives are highly desirable, whereas olefin derivatives and *n*-paraffin derivatives are less desirable.

The typical composition of gasoline hydrocarbon derivatives (% volume) is as follows: 4%−8% v/v alkane derivatives; 2%−5% v/v alkene derivatives; 25%−40% iso-alkane derivatives; 3%−7% v/v cycloalkane derivatives; 1%−4% cycloalkene derivatives, and 20%−50% aromatic derivatives (including 0.5%−2.5% v/v benzene) (Table 4.2). Additives and blending agents are added to the hydrocarbon mixture to improve the performance and stability of gasoline. These compounds include antiknock agents, antioxidants, metal deactivators, lead scavengers, antirust agents, antiicing agents, upper cylinder lubricants, detergents, and dyes.

Methanol and ethanol have a long history as alternative transportation fuels. The properties allow for a higher compression ratio and shorter burn time, which lead to theoretical efficiency advantages over gasoline in an internal combustion engine. Both methanol and ethanol fuel are technically and economically suitable for ICE (Table 13.13).

The hydrothermal liquefaction (HTL) process or direct liquefaction is a promising technology to treat waste streams from various sources and produce valuable bioproducts such as biocrude oil. A major problem with commercializing the HTL

Table 13.12 Typical constituents of gasoline.

Component	Composition (% w/w)[a]
n-Alkane derivatives	
C5	3
C6	12
C7	2
C9	1
C10–C13	1
Total of *n*-alkane derivatives	19
Branched alkane derivatives	
C4	2
C5	15
C6	8
C7	2
C8	2
C9	2
C10–C13	1
Total of branched alkane derivatives	32
Cycloalkane derivatives	
C6	3
C7	1
C8	1
Total of cycloalkane derivatives	5
Olefin derivatives	
C6	2
Total of olefin derivatives	2
Aromatic derivatives	
Benzene	3
Toluene	5
Xylene isomers	7
Ethylbenzene	1

(Continued)

Table 13.12 (Continued)

Component	Composition (% w/w)[a]
C3-benzene derivatives	4
C4-benzene derivative	8
Others	3
Total aromatic derivatives	31
Other constituents	11

[a]Approximate; varies with different crude oils.

Table 13.13 Physical and chemical properties of methanol, ethanol, and gasoline.

Property	Methanol	Ethanol	Gasoline
	CH_3OH	C_2H_5OH	C_4-C_{12}
Molecular weight (g/mol)	32	46	~114
Specific gravity	0.789 (298K)	0.788 (298K)	0.739 (288.5K)
Vapor density rel. to air	1.10	1.59	3.0–4.0
Liquid density (g/cm at 298K)	0.79	0.79	0.74
Boiling point (K)	338	351	300–518
Melting point (K)	175	129	
Vapor pressure @311K (psia)	4.6	2.5	8–10
Heat of evaporation (Btu/lb)	472	410	135
Heating value (kBtu/gal)			
Lower	58	74	111
Upper	65	85	122
Tank design pressure (psig)	15	15	15
Viscosity (cp)	0.54	1.20	0.56
Flash point (K)	284	287	228
Flammability/explosion limits			
(%) Lower (LFL)	6.7	3.3	1.3
(%) Upper (UFL)	36	19	7.6
Autoignition temperature (K)	733	636	523–733
Solubility in H_2O (%)	Miscible (100%)	Miscible (100%)	Negl. (~0.01)
Azeotrope with H_2O	None	95% EtOH	Immiscible

LFL, Lower flammability limit; *UFL*, upper flammability limit.

processes for biomass conversion is that it remains uneconomical when compared to the costs of diesel or gasoline production. High-transportation costs of large quantities of biomass increase production costs, and poor conversion efficiency coupled with a lack of understanding complex reaction mechanisms inhibits the growth of the process commercially.

In the HTL process, biomass is reacted in liquid water at elevated temperature and pressure. The phase equilibria in the HTL process are very complicated due to the presence of water, supercritical carbon dioxide, alcohols, as well as the so-called biocrude oil that is a mixture with a wide molecular weight distribution and consists of various kinds of molecules. Biocrude oil contains 10%–13% w/w oxygen and is upgraded by catalytic HDO in a central facility. Preliminary process studies on the conversion of various biomass types into liquid fuels have indicated that HTL is more attractive than pyrolysis or gasification.

In the HTL process the biomass, typically as a 25% w/w slurry in water, is treated at temperatures of 300°C–430°C (570°F–805°F) at varying pressures in the presence of liquid water for 5–20 minutes to yield a mixture of gas (mainly carbon dioxide) and synthetic crude oil (sometimes referred to as biocrude oil). Subsequent processing is necessary of the synthetic crude oil to useable fuel.

Biomass, such as wood, with a lower energy density is converted to biocrude oil with a higher energy density, organic compounds including mainly alcohols and acids, gases mainly including CO_2. Water is also a by-product. In the products, CO_2, the main component of the gas product, can be used to represent all gas produced, and methanol and ethanol represent organic compounds. The weight fraction of each component is assigned on the basis of the data of the vacuum flash of biocrude oil.

13.3.4.2 Kerosene and diesel

Gasification followed by Fischer–Tropsch (FT) synthesis (FTS) is currently the most promising method for upgrading low-value biomass to high-value liquid fuels and chemicals (Speight, 2019a, 2019b) (Speight, 2013, 2014a, 2019a). To produce synthesis gas from a biomass fuel the following procedures are necessary: (1) gasification of the biomass, (2) cleaning of the product gas, and (3) use of the synthesis gas to produce liquid fuel via FTS.

Synthesis gas (a mixture of carbon monoxide and hydrogen) produced by gasification of fossil fuels or biomass can be converted into a large number of organic compounds that are useful as chemical feedstocks, fuels, and solvents. Many of the conversion technologies were developed for coal gasification but process economics have resulted in a shift to natural gas–derived synthesis gas. These conversion technologies successively apply similarly to biomass-derived synthesis gas.

The biofuels to be considered as relevant technologies by both developing and industrialized countries due to include energy security reasons, environmental concerns, foreign exchange savings, and socioeconomic issues related to the rural sector. Catalytic conversion will be a primary tool for industry to produce valuable fuels, chemicals, and materials from biomass platform chemicals. Catalytic conversion of biomass is best developed for synthesis gas or synthesis gas. Economic considerations dictate that the current production of liquid fuels from synthesis gas translates into the use of NG as the hydrocarbon source. Biomass is the only renewable material that can meet our demand for carbon-based liquid fuels and chemicals. Biofuels as well as green motor fuels produced from biomass by FTS are the most modern biomass-based transportation fuels. Green motor fuels are the renewable replacement to crude oil-based diesel. Biomass energy conversion facilities are important for obtaining bio-oil by pyrolysis. The aim of FT process is the synthesis of long-chain hydrocarbon derivatives from carbon monoxide and hydrogen gas mixture.

The products from the FT process are mainly aliphatic straight-chain hydrocarbon derivatives (C_xH_y). Besides the C_xH_y also branched hydrocarbon derivatives, unsaturated hydrocarbon derivatives, and primary alcohols are formed in minor quantities. The FT process is capable of producing liquid hydrocarbon fuels from synthesis gas. The large hydrocarbon derivatives can be hydrocracked to form mainly diesel of excellent quality. The process for producing liquid fuels from biomass, which integrates biomass gasification with the FT process, converts a renewable feedstock into a clean fuel.

FT is a process for producing mainly straight-chain hydrocarbon derivatives from a synthesis gas rich in CO and H_2. Catalysts are usually employed. Typical operating conditions for FTS are temperatures of 200°C–350°C and very high pressure depending on the desired products. The product range includes low-boiling hydrocarbon derivatives such as methane (CH_4) and ethane (C_2H_6), propane (C_3H_8) and butane (C_4H_{10}), gasoline (C_5–C_{12}), diesel (C_{13}–C_{22}), and waxes (C_{23}–C_{33}). The distribution of the products depends on the catalyst and the process conditions (temperature, pressure, and residence time). The synthesis gas must have very low tar and particulate matter content.

The FT process was established in 1923 by German scientists Franz Fischer and Hans Tropsch. The main aim of the FT process is the synthesis of long-chain hydrocarbon derivatives from the carbon monoxide and hydrogen gas mixture (synthesis gas). The process chemistry is described by simple equations (although it is much more complex than the simple equations indicate):

$$n\mathrm{CO} + (n + m/2)\mathrm{H_2} \rightarrow \mathrm{C}_n\mathrm{H}_m + n\mathrm{H_2O}$$

where n is the average length of the hydrocarbon chain and m is the number of hydrogen atoms per carbon. All reactions are exothermic and the product is a mixture of different hydrocarbon derivatives in that paraffin derivatives and olefin derivatives are the main constituents.

Typical operation conditions for the FT process are a temperature range of 475K–625K and pressures of 15–40 bar, depending on the process. The kind and quantity of liquid product obtained is determined by the reaction temperature, pressure and residence time, the type of reactor, and the catalyst used. Iron catalysts have a higher tolerance for sulfur, are cheaper, and produce more olefin products and alcohols. However, the lifetime of the Fe catalysts is short and in commercial installations generally limited to 8 weeks. Bulk iron (Fe) catalysts are the catalysts of choice for converting low H_2/CO ratio synthesis gas produced by gasification of biomass or coal to fuels via the FT process. These relatively low-cost catalysts have low methane selectivity and high water–gas shift activity. However, development of a bulk Fe catalyst that combines high FT activity, low methane selectivity, high attrition resistance, and long-term stability is still elusive and presents a widely recognized barrier to the commercial deployment of the FT process for biomass conversion. The critical property determining the activity and deactivation of Fe catalysts for the FT process appears not to be Fe in the metallic state but the carburized Fe surface.

The design of a biomass gasifier integrated with a FT process reactor must be aimed at achieving a high yield of liquid hydrocarbon derivatives. For the gasifier, it is important to avoid methane formation as much as possible and convert all carbon in the biomass to mainly carbon monoxide and carbon dioxide. The gas cleaning is an important process before the FT process. Gas cleaning is even more important for the integration of a biomass gasifier and a catalytic reactor. To avoid poisoning the process catalyst, tar, hydrogen sulfide, carbonyl sulfide, ammonia, hydrogen cyanide, alkali, and dust particles must be removed thoroughly.

Synthetic diesel fuels from the FT process can have excellent autoignition characteristics. The diesel fuel is composed of only straight-chain hydrocarbon derivatives and has no aromatics or sulfur. Reaction parameters are temperature, pressure, and H_2/CO ratio. The product composition from the FT process strongly is influenced by catalyst composition—for example, the product from cobalt catalyst has a higher content of paraffin derivatives and product from the use of an iron catalyst has a higher contents of olefin derivatives and oxygenate derivatives. The design of a biomass gasifier integrated with a FT process reactor must be aimed at achieving a high yield of liquid hydrocarbon derivatives. For the gasifier, it is important to avoid methane formation as much as possible and convert all carbon in the biomass to mainly carbon monoxide and carbon dioxide.

13.3.5 Biodiesel

Biodiesel fuels are attracting increasing attention worldwide as blending components or direct replacements for diesel fuel in vehicle engines (Table 13.14). Biodiesel refers to a diesel-equivalent, processed fuel derived from biological sources. Biodiesel fuel typically comprises lower alkyl fatty acid (chain length C14–C22), esters of short-chain alcohols, primarily, methanol or ethanol. Various methods have been reported for the production of biodiesel from vegetable oil, such as direct use and blending, microemulsification, pyrolysis, and transesterification. Among these, transesterification is an attractive and widely accepted technique. The purpose of the transesterification process is to lower the viscosity of the oil. The most important variables affecting methyl ester yielded during the transesterification reaction are the molar ratio of alcohol to vegetable oil and the reaction temperature. Methanol is the commonly used alcohol in this process, due in part to its low cost.

Table 13.14 Examples of specifications for crude oil-derived diesel and biodiesel fuel.

Property	Diesel	Biodiesel
Standard test method	ASTM D975	ASTM D6751
Composition	HC[a] (C10–C21)	FAME[b] (C12–C22)
Kin. viscosity (mm^2/s) at 313K	1.9–4.1	1.9–6.0
Specific gravity (g/mL)	0.85	0.88
Flash point (K)	333–353	373–443
Cloud point (K)	258–278	270–285
Cetane number	238–258	258–289
	87	77

[a]Hydrocarbon derivatives.
[b]Fatty acid methyl esters.

Methyl esters of vegetable oils (biodiesel) have several outstanding advantages over other new-renewable and clean engine fuel alternatives. Biodiesel fuel is a renewable substitute fuel for crude oil diesel or petrodiesel fuel made from vegetable or animal fats; it can be used in any mixture with petrodiesel fuel, as it has very similar characteristics, but it has lower exhaust emissions. Biodiesel fuels are attracting increasing attention worldwide as a blending component or a direct replacement for diesel fuel in vehicle engines. Biodiesel, as an alternative fuel for ICE, is defined as a mixture of mono-alkyl esters of long-chain fatty acids (fatty acid methyl esters) derived from a renewable lipid feedstock, such as vegetable oil or animal fat.

Biodiesel is a liquid that varies in color—between golden and dark brown—depending on the production feedstock. It is practically immiscible with water, has a high boiling point, and low vapor pressure. Typical methyl ester biodiesel has a flash point on the order of 150°C (300°F), making it rather nonflammable. Biodiesel has a density of 0.88 g/cm^3, less than that of water. Biodiesel uncontaminated with starting material can be regarded as nontoxic. Biodiesel has a viscosity similar to diesel derived from crude oil (often referred to as petrodiesel) and can be used as an additive in formulations of diesel to increase the lubricity of ultra-low sulfur diesel (ULSD) fuel, which is advantageous because it has virtually no sulfur content. Much of the world uses a system known as the "B" factor to state the amount of biodiesel in any fuel mix, in contrast to the or "E" system used for fuels containing ethanol. For example, pure biodiesel is referred to as B100, while fuel containing 20% biodiesel is labeled B20.

Biodiesel is a renewable fuel that is safe, biodegradable, and reduces air pollutants, such as particulates, carbon monoxide, and hydrocarbon derivatives. Blends of 20% biodiesel with 80% crude oil diesel (B20) can generally be used in unmodified diesel engines. Biodiesel can also be used in its pure form (B100) but may require certain engine modifications to avoid maintenance and performance problems. Biodiesel has approximately 5%–8% less energy density, but better lubricity and more complete combustion can make the energy output of a diesel engine only 2% less per volume when compared to petrodiesel.

A series of standards applied to biodiesel ensure that the following important factors in the fuel production process are satisfied: (1) complete reaction, (2) removal of glycerin, (3) removal of catalyst, (4) removal of alcohol, (5) absence of free fatty acids, (6) low-to-no sulfur. Pure, nonblended biodiesel can be poured straight into the tank of any diesel vehicle. As with normal diesel, low-temperature biodiesel is sold during winter months to prevent viscosity problems. Some older diesel engines still have natural rubber parts that will be affected by biodiesel, but in practice, these rubber parts should have been replaced long ago.

The temperature at which pure (B100) biodiesel starts to gel varies significantly and depends upon the mix of esters and, therefore, the feedstock oil used to produce the biodiesel. For example, biodiesel produced from varieties of canola seed starts to gel at approximately $-10°C$ (50°F). Biodiesel produced from tallow tends to gel at around 16°C (61°F). Some users modify the vehicles to permit the use of biodiesel without mixing and without the possibility of gelling at low temperatures, which is a similar practice to the one used for running straight vegetable oil. A second fuel tank is installed, which is insulated and a heating coil—a temperature sensor is installed to notify the driver when the fuel is warm enough to burn, the driver then switches which tank the engine is drawing from. Biodiesel may contain small but problematic quantities of water. Although biodiesel is hydrophobic (nonmiscible with water), it is hygroscopic to the point of attracting water from the atmosphere. In addition, there may be water that is residual to processing or resulting from storage tank condensation.

Biodiesel fuel has several advantages over fossil fuel. The most celebrated advantage is that biodiesel fuel is less costly to make than gasoline. As an added bonus, biodiesel is environmentally friendly fuel. The use of biodiesel in an unmodified diesel engine will substantially reduce the emissions of harmful unburned hydrocarbon derivatives, sulfates, polycyclic aromatic hydrocarbon derivatives, and carbon monoxide.

Biodiesel fuel can be used in almost any diesel engine found in cars, trucks, tractors, planes, trains, boats, buses, and almost all agricultural equipment and most diesel engines can use biodiesel without having to be modified in any way. Biodiesel fuel reduces carbon dioxide emissions by up to 80%, thereby reducing the formation of acid rain because it produces no sulfur dioxide, which is the key element in creating acid rain. Furthermore, biodiesel burns much cleaner and also takes considerably less time to degrade in the event of a spill, usually on the order of 28 days, which is approximately four times faster than gasoline or regular diesel fuel. In fact, the extra lubrication provided by biodiesel fuel helps improve the longevity of the engine, as well as boosting engine performance, also helping eliminate engine knocks and noise. In addition, biodiesel fuel can be stored in any type of tank and has a much higher flash point at 300°C (570°F) than regular diesel, at 150°C (300°F).

Biodiesel is a light to dark yellow liquid that is immiscible with water and has a high boiling point and low vapor pressure. Typical methyl ester biodiesel has a flash point on the order of 150°C (300°F), making it nonflammable. Biodiesel has a density of 0.88 g/cm^3. Biodiesel has a viscosity similar to diesel fuel derived from crude oil (the current term to differentiate the different forms of diesel) and can be used as an additive in formulations of diesel to increase the lubricity of ULSD fuel. Much of the world uses a system known as the "B" factor to state the amount of biodiesel in any fuel mix, in contrast to the "BA" or "E" system used for ethanol mixtures. For example, fuel containing 20% v/v biodiesel is labeled B20, whereas pure biodiesel (100% v/v) is referred to as B100.

Biodiesel is safe, biodegradable, and reduces air pollutants, such as particulates, carbon monoxide, and hydrocarbon derivatives. Blends of 20% biodiesel with 80% petroleum diesel (B20) can generally be used in unmodified diesel engines. Biodiesel can also be used in its pure form (B100) but may require certain engine modifications to avoid the maintenance and performance problems.

Another source of liquid fuels is waste cooking oil; usually biodiesel is the desired product. In fact, waste cooking oil is a valuable asset for conversion to liquid fuels and the quantity of biodiesel, which could be produced from waste cooking oil, can be quantified in a manner similar to that of biodiesel produced direct from agriculture. The only difference is that rather than calculating crop yields, it is necessary to quantify the amount of waste oil available.

13.3.6 Other fuels

The bio-oil obtained from the fast pyrolysis of biomass has high oxygen content, especially when produced from wood (Table 13.15). Because of the reactivity of oxygenated groups, the main problem of the oil is related to instability and incompatibility with hydrocarbon-type products and there is the need for HDO of the bio-oil, which can be accomplished using a cobalt molybdate catalyst.

Table 13.15 Properties of pyrolysis oil from various wood feedstocks.

Property	Birch	Pine	Poplar	Various
Solids (% w/w)	0.06	0.03	0.045	0.01–1
pH	2.5	2.4	2.8	2.0–3.7
Water (% w/w)	18.9	17.0	16.8	15–30
Density	1.25	1.24	1.20	1.2–1.3
Viscosity cSt/50°C	28	28	13.5	13–80
Ash (% w/w)	0.004	0.03	0.007	0.004–0.3
Carbon residue (% w/w)	20	16	–	14–23
C (% w/w)	44.0	45.7	48.1	32–49
H (% w/w)	6.9	7.0	5.3	6.9–8.6
N (% w/w)	<0.1	<0.1	0.14	0.0–0.2
S (% w/w)	0.00	0.02	0.04	0.0–0.05
O (% w/w)	49.0	47.0	46.1	44–60
Na + K (ppm)	29	22	2	5–500
Ca (ppm)	50	23	1	4–600
Flash point (°C)	62	95	64	50–100
Pour point (°C)	−24	−19	–	−36 to −9

The main HDO reaction is represented by the following simple equation:

$$-(CH_2O)- + H_2 \rightarrow -(CH_2)- + H_2O$$

This is the most important route of chemical upgrading. The reaction has strong analogies with typical refinery hydrogenations such as hydrodesulfurization and hydrodenitrogenation. In general, most of the HDO studies have been performed using existing hydrodesulfurization catalysts (NiMo and CoMo on suitable carriers). Such catalyst need activation using a suitable sulfur source, and this is a major drawback when using nearly sulfur-free resources such as bio-oil.

The primary objective catalytic partial HDO is to increase the energetic value of the oil by removing bound oxygen in the form of water. HDO of bio-oils involves treating the oils at moderate temperatures with high-pressure hydrogen in the presence of heterogeneous catalysts. The process was carried out in two distinct stages, a first stage at relatively low temperatures (250°C–300°C, 480°F–570°F), aimed to stabilize the bio-oil and a second stage at higher temperatures (300°C–400°C, 570°F–750°F) to deoxygenate the intermediate product. Different types of catalysts were screened, ranging from conventional sulfide-promoted catalysts used in the hydrodesulfurization process (i.e., $NiMo/Al_2O_3$, $CoMo/Al_2O_3$) to novel nonsulfide-promoted catalyst based on noble metal catalysts (i.e., Ru/Al_2O_3). Operating conditions were optimized to obtain the highest yield of a hydrocarbon such as liquid product.

It is possible to produce low-boiling alkanes by aqueous-phase reforming of biomass-derived oxygenates such as sorbitol, which can be obtained from glucose by hydrogenation. The production of alkanes from aqueous carbohydrate solutions would be advantageous because of the easy separation of the alkanes from water. Much hydrogen is needed to reduce biomass-derived oxygenates to alkanes as shown in the following equation:

$$C_6H_{14}O_6 + 6H_2 \rightarrow C_6H_{14} + 6H_2O \tag{13.1}$$

It was shown in earlier work that hydrogen can be produced in a catalytic process at 500K in water from biomass-derived polyols having a C/O stoichiometry of 1:1 such as sorbitol.

$$C_6H_{14}O_6 + 6H_2O \rightarrow 13H_2 + 6CO_2$$

The overall reaction is an exothermic process and converts approximately 1.5 mol of sorbitol into 1 mol of hexane:

$$19C_6O_6H_{14} \rightarrow 13C_6H_{14} + 36CO_2 + 42H_2O$$

Approximately 95% of the heating value but only 30% of the mass of the biomass-derived reactant is retained in the product. The remaining 70% of the biomass is found as CO_2 and water. A maximum yield of 61% of hydrogen with respect to sorbitol was obtained over Pt/Al_2O_3 under optimized reaction conditions.

Destructive reaction of cellulose is characterized by a decreasing polymerization degree and, thus, thermal degradation of cellulose proceeds through a gradual degradation, decomposition and charring on heating at lower temperatures, and a rapid volatilization accompanied by the formation of levoglucosan on pyrolysis at higher temperatures. The glucose chains in cellulose are first cleaved to glucose and, in a second stage, glucosan is formed by the splitting off of one molecule of water. Initial degradation reactions include depolymerization, hydrolysis, oxidation, dehydration, and decarboxylation.

13.4 Solid fuels

In the present context the term solid fuel is the term applied to wood or any other nongaseous or nonliquid fuel produced from wood.

Burning wood as a fuel is the largest current use of biomass-derived energy and, as such, wood can be used in many forms as a solid fuel for cooking or hearting, occasionally for steam engines (locomotives) and for steam turbines for the generation of electricity. The particular form of wood fuel used depends upon (among other things) its source, quantity, and quality. Available forms include logs, bolts, blocks, firewood, stove–wood (often from split blocks), charcoal, chips, sheets, pellets, and sawdust. The wastes from sawmills and the construction industry by-products also include various forms of lumber tailings.

The main components of wood cells are cellulose, hemicellulose, and lignin, forming approximately 99% w/w of the wood material. Cellulose and hemicellulose are composed of long chains of carbohydrate moieties, whereas lignin is a complicated component of polymeric phenol constituents. Lignin is rich in carbon and hydrogen, which are the main heat-producing elements. Thus the calorific value of lignin is higher than that of cellulose and hemicellulose (carbohydrates). Wood and bark also contain so-called extractives, such as terpenes, fats, and phenols. The amount of wood extractives is relatively small compared to the amount of extractives from bark and foliage. The nitrogen (N) content of wood is approximately 0.75% w/w but does vary from one tree species to another. The occurrence of sulfur-containing species in wood is minimal and usually not more than 0.05% w/w.

Wood fuel is a general term applied to wood and wood products (including coppices, scrubs, branches, and forest waste) as well as industrial waste that is burned primarily for heating, cooking, or lighting.

The use of wood as a fuel for fires to produce heat and light is the current largest use of wood as an energy source. The particular form of wood fuel used depends upon (among other things) its source, quantity, and quality (Table 13.12). Available forms include logs, bolts, blocks, firewood, stove–wood (often from split blocks), charcoal, chips, sheets, pellets, and sawdust. Sawmill waste and construction industry by-products also include various forms of lumber tailings.

Fuelwood (i.e., wood itself) is the most common solid fuel and continues to be widely used as a major source of energy for households, especially in developing countries. Charcoal is also increasingly used in many African countries by urban dwellers, as a result of a relentless process of migration of people from rural areas toward urban centers.

Firewood can be sold by weight, but freshly cut (green) wood can contain from 40% to 60% moisture by weight, whereas properly seasoned wood contains only 15%–20%. Select the driest wood when buying by weight. Green wood will shrink approximately 8% in volume (i.e., approximately cubic feet per cord) when properly seasoned.

Wood should be dried as much as possible before burning. Properly seasoned wood has approximately 7700 Btu/lb versus only approximately 5000 Btu/lb available from green wood. For best results, season or air-dry wood for at least six to 8 months after cutting, which should reduce the moisture content down to 15%–20% w/w.

The combustion of wood is a chemical reaction. If the burning process is complete, the carbon and hydrogen in the wood are converted to carbon dioxide and water:

$$C + O_2 \rightarrow CO_2$$

$$2H_2 + O_2 \rightarrow 2H_2O$$

In practice the burning process is never complete. The combustion always produces small amounts of unburned hydrocarbon derivatives (CO, C_xH_y).

Depending on the combustion efficiency, the ash may contain combustibles. The ash deformation temperature is the temperature at which first signs of rounding due to melting of the tip or edges of the test piece occur. The ash flow temperature is the temperature at which the ash is spread out over the supporting tile in a layer, the height of which is one-third of the height of the test piece at the ash hemisphere temperature. The ash fusibility temperature is the temperature at which the ash melts and is characteristic physical state of the ash obtained by heating under specific conditions.

13.4.1 Fuelwood

The most common form of wood fuel at the moment is logs. These will usually come from local sources and can be brought from a variety of outlets—for example, coal merchants, farmers, tree surgeons. It is important that the logs are dry and well-seasoned. Burning wet or unseasoned wood is less efficient and can cause harmful buildup of deposits in the chimney over a very short time. Thick coatings of creosote or resinous material can cause chimney fires or prevent the chimney functioning properly. This can allow harmful fumes to escape into the dwelling. Efficient burning is achieved by setting the appliance to burn fast after stoking. This will ensure that all the gases are fully burned. Only set to slow burn when all wood has been reduced to charcoal and ash. Newly added wood set to burn slowly creates smoke and creosote in the chimney. The stove should not be banked up with logs for overnight burning. A bright fire which has turned the wood into charcoal should be left with the day's ash, no secondary air and minimal primary air.

13.4.2 Logs and wood chips

Log-fired heating conjures up images of open fires and log stoves for many people, but there are also sophisticated, controllable, log-boilers available which can provide central heating and hot water. Logs are readily available in many areas where there are existing markets for domestic fuelwood. Logs should be well seasoned before burning to ensure the most efficient combustion and should ideally be stored under cover in a well-ventilated log store for a year or more.

Logs are ideal for providing heat for domestic buildings and are suitable for heating loads of up to around 50 kW. Log heating systems do require manual stoking once a day, which makes logs less convenient than more automated heating systems such as woodchips and pellets. Log-fired systems are available with outputs greater than 50 kW to heat larger building such as village halls; however, these larger systems will require more frequent fueling.

Wood chips are made from whole trees, branch wood, or coppice products, which have been mechanically shredded by a chipping machine. For some types of boilers, the wood needs to have been air-dried before chipping, or the chips dried before burning. Wood chips are a bulky fuel and sufficient storage and delivery access needs to be considered when designing a heating system. Transport costs can be high, but if wood chip is sourced within 20 mi, it can be a very cost-effective fuel.

The potential for woodchip heating in many countries is high. The use of a timber resource for local woodchip heating would provide a valuable economic return and stimulate the rural economy. Woodchip systems can provide automated, clean, and convenient heating for larger domestic properties with outputs of 20–30 kW, up to large-scale systems for hospitals, factories, schools, and district heating schemes with heating loads in the megawatt range. Woodchips can also be used to fuel combine heat and power plants in which the heat produced during electricity generation is used to provide hot water and is not lost as in conventional power stations.

Logs that have not been seasoned should be stored under cover but open to allow free air circulation for at least a year. Some logs may take 3–4 years to fully season. Bring the fuel into the house a few days before you want to use it to get it as dry as possible. Wood from different trees has different heat values. Wood fuel has typically less than half the calorific value of coal and smokeless fuel, so you must be prepared to use a greater volume of wood to heat your home or room, unless you use both wood and mineral solid fuel.

Wood chips are generally a by-product of the arboriculture industry or created from waste timber. The fuel is used in screw-fed boilers that have the capacity to fuel local micro-chip systems and larger buildings such as leisure center, where commercial sized boilers are needed. It is important to ensure stored that wood chips do not contain excessive quantities of moisture.

These products are manufactured from sawdust and bound or pressed together into a log shape. They incorporate an inflammable substance that makes the product easy to light. Some are suitable for use on open fires. They are relatively expensive but will give a fire lasting for 2 or 3 hours. They are clean and handy to use and are ideal for a quick fire on a cold night. The product can be purchased from supermarkets, garages, other shops as well as from some coal merchants.

13.4.3 Pellets and briquettes

Pellets made of compressed sawdust or wood shavings have been available in many parts of the world for more than several decades. Because they are compressed, pellets offer a more concentrated form of fuel than wood chips. Consequently they need less storage space and are easier to handle. The manufacture of wood pellets requires more energy than woodchips and the capital cost for production plant is high; pellets are, therefore, more expensive than chips.

Wood pellets are usually made of compressed sawdust that is a waste product from other industries (e.g., sawmills). The pellets are held together by the natural lignin in the wood—typically no binder glue is needed. Lignin makes up approximately one quarter to a third of dry wood. It strengthens the wood as well as having water proofing properties.

Wood pellets have a very low moisture content, typically 6%–10% which means that wood pellets are a very good source of energy because very little of the energy in the pellets is used to evaporate moisture. The general properties of wood pellets are

Moisture: 3.5%–5.5%
Ash: 0.2%–0.5%
Calorific value: 8800 Btu/lb
Bulk density: 600–750 kg/m

The pellets are extremely dense and the low humidity content (below 10%) allows them to be burned with very high-combustion efficiency. The regular geometry of the pellets and small size allow automatic feeding with very fine calibration. Their high density also permits compact storage and rational transport over long distance.

Wood chips are a solid fuel made from woody that are often used as a raw material for technical wood processing. In industry, processing of bark chips is often separated after peeling the logs due to different chemical properties. Wood chips can also be used as a wood fuel.

Pellets can be used to fuel a varicty of appliances and heating systems. The smallest are pellet stoves with outputs of up to around 9 kW which are suitable for heating individual rooms. These stoves are electronically controlled and can deliver regulated heat output and only need fueling once every few days. Pellet boilers are available in a wide range of outputs from small domestic scale to large industrial scale to heat schools or hospitals. For the domestic user, pellets offer the most user-friendly form of wood heating. In Scandinavia, wood pellets are delivered by tanker and are pumped into storage silos, which feed, automatically into the boiler.

Wood pellets are used on specially designed stoves. Pellets are typically made from timber waste from sawmills. The wood goes through a fairly lengthy process of transformation before it is finally extruded as hard pellets. In spite of the processing the fuel is still carbon neutral and so is less harmful to the environment than other fossil fuels. For commercial undertakings, as with any wood fuel, wood pellets do not attract climate change levy tax. There are at present a limited number of pellet producing plants in the United Kingdom although there is some encouragement to produce from local sources. Some pellets are being imported in the meantime. The pellets for domestic room heaters and boilers are being supplied in prepacked bags, usually 22 lb (10 kg) in weight, which makes them a handy size to store and load into the hopper. For the larger boiler models, pellets can be delivered in bulk, but an appropriate silo must be available, connected to the boiler. At the moment, most of the pellets will be supplied by the stove shop that markets the boilers, but see our links page below for other suppliers.

Fuel briquettes (sometime called Fuel briquettes)—especially briquettes manufactured from coal (Speight, 2013) emerged as a significant business enterprise in the 20th century. In the 1950s, several economic methods were developed to make briquettes without a binder. A multitude of factories throughout the world produced literally tens of millions of tons of usable and economic material that met the household and industrial energy needs. During the two World Wars, households in many European countries made their own briquettes from soaked waste paper and other combustible domestic waste using simple lever-operated presses (Lardinois and Klundert, 1993). Modern industrial briquetting machines, although much larger and more complex, operate on the same principle although the marketed briquettes are now sold at a premium for occasional backyard barbeques rather than for everyday use.

For over 100 years informal waste collectors in Cairo have separated and dried organic waste products for sale as solid fuel for domestic use. This process faded somewhat when fossil fuel sources became available. Switching to conventional fuels may prove advantageous to those who can afford them, but given the economic and energy conditions in many cities, urban and agricultural wastes remain a viable alternative fuel.

Briquetting is undergoing resurgence, principally due to the convergence of three critical factors. First, the recent developments in briquette processing and binding have dramatically changed the economics of using fuel briquettes as an energy resource. Second, a shortage of fuelwood has become increasingly severe in most of the developing countries. Finally, there has been a steady increase by environmental concerns to address the problem of domestic and urban waste disposal, a dilemma that briquetting can help remedy.

13.4.3.1 Briquette manufacture

Generally, briquette manufacture (briquetting) involves the collection of combustible materials that are not usable as such because of their low density, and compressing them into a solid fuel product of any convenient shape that can be burned like wood or charcoal. Thus the material is compressed to form a product of higher bulk density, lower moisture content, and uniform size, shape, and material properties. Briquettes are easier to package and store, cheaper to transport, more convenient to use, and their burning characteristics are better than those of the original organic waste material.

The raw material of a briquette must bind during compression; otherwise, when the briquette is removed from the mold, it will crumble. Improved cohesion can be obtained with a binder but also without, since under high temperature and pressure, some materials such as wood bind naturally. A binder must not cause smoke or gummy deposits, while the creation of excess dust must also be avoided. Two different sorts of binders may be employed. Combustible binders are prepared from natural or synthetic resins, animal manure or treated, dewatered sewage sludge. Noncombustible binders include clay, cement, and other adhesive minerals. Although combustible binders are preferable, noncombustible binders may be suitable if used in sufficiently low concentrations. For example, if organic waste is mixed with too much clay, the briquettes will not easily ignite or burn uniformly. Suitable binders include starch (5%–10% w/w) or molasses (15%–25% w/w) although their use can prove expensive. It is important to identify additional, inexpensive materials to serve as briquette binders in Kenya and their optimum concentrations. The exact method of preparation depends upon the material being briquetted as illustrated in the following three cases of compressing sugar bagasse, sawdust, and urban waste into cooking briquettes.

13.4.3.2 Bagasse briquettes

Surplus bagasse presents a disposal problem for many sugar factories. Briquetting technology remains simple and involves the following steps: (1) size reduction in which the bagasse is chopped, rolling or hammered, (2) drying in which moisture is removed by open air drying or by using forced, heated air in a large rotating drum, (3) carbonization in which the bagasse is combusted in a limited supply of oxygen in a buried pit or trench until it carbonizes into charcoal, (4) feedstock preparation in which the carbonized bagasse is mixed with a binder such a clay or molasses, (5) compaction and extrusion in which the material is passed through a machine-operated or manually operated extruder to form rolls of charcoal, (6) drying in which the rolls are air-dried for 1–3 days, causing them to break into chunks, and (7) packaging in which the briquettes in are made ready for sales.

13.4.3.3 Sawdust briquettes

Sawdust is waste material from all types of primary and secondary wood processing. Between 10% and 13% w/w of a log is reduced to sawdust in milling operations. Sawdust is bulky and is, therefore, expensive to store and transport. Also, the calorific value of sawdust is quite low, so that briquetting is an ideal way to reduce the bulk, to increase the density, and thus to increase the calorific value. The equipment required for producing sawdust briquettes consist of a drier, a press, and an extruder with a tapered screw and a large revolving disk.

Sawdust briquettes are formed under sufficiently high pressure to produce cohesion between wood particles. In the process the lignin softens and binds the briquette, so no additional binder is required. The use of sawdust briquettes has several advantages, including (1) price which is usually approximately the same as fuelwood but is much more convenient to use as they do not require further cutting and chopping, (2) good burning characteristics in any kind of solid fuel stove and boiler, (3) quick ignition followed by clean burning and only leaving 1%–6% w/w mineral ash, (4) sulfur-free and burn without producing an odor, and (5) a heat content that is almost equivalent to the heat content of medium quality coal.

13.4.3.4 Urban waste briquettes

Solid waste (refuse) disposal is one of the most serious urban environmental problems in developing countries. Many municipal authorities collect and adequately dispose (in places other than landfill sites) less than half of these wastes. This failure is attributed to (1) an inadequate number of landfill sites, (2) a variety of environmental regulations, (3) the absence of sufficient capacity for waste processing and recycling, and last, but not least, (4) the planned obsolescence of packaging and many of the items that form the basis of the waste.

Open or crude dumping is the most common method used by municipal authorities. Waste poses a health hazard when it lies scattered in the streets and at the dumping sites. It is now an accepted environmental philosophy that wastes have value and should be utilized based on principles of reduce, reuse, recover, and recycle. Through recycling, urban wastes can be transformed into useful products. Waste paper and leaves, in particular, provide a potentially important, alternative source of cooking fuel.

Recycling organic combustible materials into fuel briquettes contributes to solving urban needs such as income-generation, insufficient land for waste disposal and maintaining environmental quality. Since the earth's resources are finite, greater resource recovery and utilization are essential to achieve an acceptable level of organic waste management. Enhancing the recovery of organic waste can restore various natural cycles, thus preventing the loss of raw materials, energy, and nutrients.

On the other hand, the demand for energy in many countries is expected to add to the emission of greenhouse gas through burning of fossil fuels. There is urgent need to promote climate-friendly technologies and fuel briquetting appears to be one such technology that addresses the multiple needs of society and the environment.

Current needs are focused on finding better binders for bagasse briquettes, improved calorific values, and combustion by producing higher density briquettes, introducing more efficient extrusion methods, and reducing production costs.

13.4.4 Charcoal

The most commonly known solid fuel produced from wood is charcoal, but there are other sources such as coconut shells and crop residues. Charcoal is produced by heating wood at conditions that severely restrict the amount of oxygen. When wood is heated slowly to approximately 280°C (535°F), an exothermic reaction occurs. The usual carbonization procedure is prolonged, heating to a final temperature of 400°C–500°C (750°F–930°F) in the absence of air. The yield of charcoal is approximately 25% w/w on a dry basis.

Thus charcoal is produced in kilns by pyrolysis, that is, breaking down the chemical structure of wood under high temperature in the absence of air. During the process, first the water is driven from the wood (drying), and then the pyrolysis starts when the temperature in the kiln is high enough. When the pyrolysis is complete, the kiln gradually cools down, after which the charcoal can be removed from the kiln. Because some of the wood is burned to drive off the water, dry wood produces better charcoal at a higher efficiency. Typically, approximately two-third of the energy is lost in the process, but charcoal has advantages over fuelwood like stoves with higher efficiency, higher convenience and easier distribution.

Commercial charcoal is found either as (1) lump charcoal, (2) briquetted charcoal, or as (3) extruded charcoal. Lump charcoal is made directly from hardwood material and usually produces far less ash than briquettes. Briquettes are made by compressing charcoal, typically made from sawdust and other wood by-products, with a binder (usually starch). Extruded charcoal is made by extruding either raw ground wood or carbonized wood into logs without the use of a binder—the heat and pressure of the extruding process hold the charcoal together. If the extrusion is made from raw wood material, the extruded logs must be subsequently carbonized.

Charcoal quality can be specified and measured in various ways which are usually dictated by the end-use. Efficiency in use normally means transferring the maximum amount of the heat content of the charcoal to the object to be heated and depends on the use of properly designed charcoal burning equipment.

Charcoal burns by reacting with oxygen of the air at a glowing red heat to form colorless carbon monoxide gas, which then burns with a blue flame with more oxygen from the air to produce carbon dioxide gas. Due to the heat liberated by both of these reactions, the charcoal reaches a glowing red and radiates heat energy and the hot carbon dioxide gas leaves the combustion zone. The gas temperature falls as it transfers heat and it passes off into the room. Carbon monoxide gas can be given off by burning charcoal. It is very poisonous and ventilation of rooms where charcoal is burning is essential.

Char is one of the thermal degradation products of wood. These products consist of moisture, volatiles, char, and ash. Volatiles are further subdivided into gases and liquid (including tarry products). Some properties vary with species, location within the tree, and growth conditions. Other properties depend on the combustion environment.

Char is typically a fine-grained charcoal high in organic carbon and largely resistant to decomposition. It is also produced by the pyrolysis of other (nonwoody) plant and waste feedstocks. During the pyrolysis, several changes take place in wood as the temperature is increased and the different chemical components undergo the thermal degradation that affects the performance of wood. The extent of the changes depends on the temperature level and length of time under exposure conditions.

For example, permanent changes in the properties of the wood start occurring at temperatures above 65°C (150°F), with the amount depending on the temperature, pH of wood, moisture content, heating medium, exposure period, and species of wood. This is likely to be due to depolymerization reactions involving no significant carbohydrate weight loss. Chemical bonds begin to break at temperatures on excess of 100°C (212°F) as carbohydrate and lignin weight losses of various types that increase with the temperature (wood pyrolysis). Between 100°C (212°F) and 200°C (390°F), wood dehydrates as it loses water vapor and other noncombustible gases and liquids (including carbon dioxide). With prolonged exposure to higher temperatures, wood starts to form char. From 200°C to 300°C (390°F to 570°F), wood components begin to undergo significant pyrolysis, and significant amounts of carbon monoxide and high-boiling-point tar are evolved—hemicellulose and lignin constituents are pyrolyzed in the range 200°C–450°C (390°F–840°F).

Dehydration reactions are primarily responsible for pyrolysis of hemicellulose and lignin and result in a high char yield for wood. Although cellulose remains mostly unpyrolyzed, its thermal degradation can be accelerated in the presence of water, acids, and oxygen. As the temperature increases, the degree of polymerization of cellulose decreases further, and any tar formed undergoes cracking to lighter gases and repolymerization to char while streaming through the hot char residue. All wood components end their volatile emissions at approximately 450°C (840°F), and at higher temperatures the remaining wood residue is char, which undergoes further degradation (if oxygen is absent) to a higher carbon residue or (if oxygen is present) to carbon dioxide, carbon monoxide, and water.

Upon completion of pyrolysis the char has a porosity on the order of 0.8–0.9 (for wood heated under combustion conditions) and the initial char density is 10%–20% that of the dry wood and as the char continues to react the density decreases further.

The oldest and probably still the most widely used method for charcoal production is the earth kiln. Two varieties exist, the earth pit kiln and the earth mound kiln. An earth pit kiln is constructed by first digging a small pit in the ground. Then the wood is placed in the pit and lit from the bottom, after which the pit is first covered with green leaves or metal sheets and then with earth to prevent complete burning of the wood. The earth mound kiln is built by covering a mound or pile of wood on the ground with earth. The mound is preferred over the pit where the soil is rocky, hard, or shallow, or the water table is close to the surface. Mounds can also be built over a long period, by stacking gathered wood in position and allowing it to dry before covering and burning.

Earth kilns can be made at minimal cost, and are often used near wood resources, since they can be made entirely from local materials. Earth kilns can be made in any size, with the duration of the process ranging from 3 days to 2 months.

Gross variations in the quality of the charcoal can occur, because in one batch some of the wood is burned and some of the wood is only partly carbonized. Efficiencies are generally low, around 10%–20% by weight and 20%–40% in energy terms. The efficiency and the quality varies depending on the construction of the kiln (e.g., walls can be lined with rocks or bricks and external chimneys can be used), and the monitoring of the carbonization process.

Several other types of charcoal kilns have been developed, which generally have higher efficiencies but also require higher investments than the earth kiln. Two often-used types are fixed kilns made of mud, clay, bricks, and portable steel kilns. Fixed kilns usually have a beehive shape. Smaller beehives are usually made of mud and are not very durable. Larger beehives are made of bricks and have external chimneys. Beehive kilns have an opening for loading the wood and unloading the charcoal, which is closed after loading.

Portable steel kilns can be made from oil drums and can be used both in horizontal and vertical position. They generally have a short lifetime. When used in the horizontal position, an opening is made in the side, through which the wood is loaded. For the vertical kiln the top is cut out and used as a lid.

The production of briquettes (briquetting or densification) is used to improve characteristics of materials for transport and use as energy source. Raw materials include sawdust, loose crop residues, and charcoal fines. The material is compacted under pressure, and depending on the material, the pressure, and the speed of densification, additional binders may be needed to bind the material. The two main briquetting technologies are the piston press and the screw press. In the piston press the material is punched into a die by a ram with a high pressure. In the screw press the material is compacted continuously by a screw. With the screw press, generally briquettes of higher quality can be produced.

Charcoal is produced by heating wood at conditions that severely restrict the amount of oxygen. When wood is heated slowly to approximately 280°C (535°F), an exothermic reaction occurs. The usual carbonization procedure is prolonged, heating to a final temperature of 400°C–500°C (750°F–930°F) in the absence of air. The term charcoal also includes charcoal made from bark.

Charcoal is produced commercially in either kilns or continuous furnaces from primary wood. Processing residues and from low-quality round wood. A kiln is used if the raw material is in the form of round wood, sawmill slabs, or edgings. The most common kilns in the United States are constructed of poured concrete. They have a capacity of 40–100 cd of wood and operate on approximately a 7- to 12-day cycle. If the raw material is in the form of sawdust, shavings or milled wood and bark, it is converted to charcoal in a continuous multiple-hearth furnace commonly referred to as a Herreshoff furnace. The capacity is usually at least 1 t of charcoal per hour. The yield of charcoal is approximately 25% by weight on a dry basis.

The proximate analysis of charcoal is approximately 20%–25% volatile matter, 70%–75% fixed carbon, and 5% ash. The higher heating value (HHV) is approximately 12,000 Btu/lb (28,000 J/kg). Charcoal briquettes have a HHV of only approximately 10,000–11,000 Btu/lb (23,000–25,000 J/kg) due to added ingredients. To alleviate the air pollution problem associated with charcoal kilns and furnaces, the gases from the kilns and furnaces are burned. The gases can be burned with additional fossil fuel to recover heat and steam, or they can be burned in afterburners to nearly eliminate visible air pollution and odors.

Charcoal was an important industrial raw material in the United States for iron ore reduction until it was replaced by coal in the early 1880s. Charcoal production did continue to increase, however, due to the demand for the by-products—acetic acid, methanol, and acetone.

The increase in the production of charcoal since the 1940s reflects the use of charcoal briquettes for home recreational cooking. The charcoal presently produced is nearly all consumed for cooking as charcoal briquettes. Some charcoal is used in certain metallurgical and filtration processes and horticultural uses. In Brazil, charcoal is produced in beehive-type kilns from naturally and plantation-grown trees for use as an iron ore reducing agent because Brazil does not have abundant supplies of coking coal. In many developing countries, charcoal is the preferred fuel for domestic cooking. Often it is made in pit-type kilns, but beehive and portable sheet metal kilns are also used where the construction materials are available.

13.4.5 Coke

The carbonization of coal is essentially a process for the production of a coke (the carbonaceous residue) by thermal decomposition (with simultaneous removal of distillate) of organic substances (Speight, 2013):

$$C_{organic} \rightarrow C_{coke/char/carbon} + \text{liquids} + \text{gases}$$

The process may also be referred to as destructive distillation and has been applied to a whole range of organic materials, but more particularly to natural products such as wood, sugar, and vegetable matter to produce charcoal. Coal usually yields "coke" that is physically dissimilar from charcoal and has the more familiar honeycomb-type structure.

The thermal decompositions of coal is a strong function of coal type or rank. Low rank coals, such as lignite and subbituminous coal, produce relatively high levels of light gases and very little tar. Bituminous coal produces significantly more tar than the low rank coals and moderate amounts of light gases. Higher rank coal produces relatively low levels of both low-boiling gases and tar. In fact, only a limited range of coals produces acceptable metallurgical cokes. These coals are in the bituminous rank range but not all bituminous coals are caking coals. Prime coking coals are expensive and not always available nationally. It is quite possible that remaining indigenous coals available for coke making are poorer in coking quality.

The thermal decomposition of coal is a complex sequence of events which can be described in terms of several important physicochemical changes, such as the tendency of the coal to soften and flow when heated (plastic properties or the relationship) to carbon-type in the coal. In fact, some coals become quite fluid at temperatures of the order of 400°C–500°C (750°F–930°F) and there is a considerable variation in the degree of maximum plasticity, the temperature of maximum plasticity, as well as the plasticity temperature range for various coals. Indeed, significant changes also occur in the structure of the char during the various stages of devolatilization.

At temperatures of just over 100°C (212°F) the coal residual moisture evolves. The evolution of low-boiling gases begins at temperatures of 200°C–500°C (390°F–930°F); these early gases consist mainly of carbon monoxide (CO) and carbon dioxide (CO_2) and low-boiling hydrocarbons. Tar formation was seen in low heating rate experiments to begin at around 330°C (625°F) and increases to temperatures above 530°C (985°F). Cross-linking reactions are thought to occur at different temperatures—depending on the coal and heating rates—due to different kinetics of the competing processes of bond breaking, vaporization, and bond formation. Pyrolysis experiments conducted with a number of coals indicate that early cross-linking begins in the range of 400°C–500°C (750°F–930°F), later cross-linking continues as temperatures increase. The exact temperatures at which these pyrolysis steps occur are dependent on many factors and it is likely that heating rate and coal type have the largest effect on the temperatures at which these steps occur.

On the other hand, coke is also produced by the destructive distillation (thermal decomposition with simultaneous removal of distillate) of crude oil residua (Parkash, 2003; Gary et al., 2007; Speight, 2011a, 2014a; Hsu and Robinson, 2017; Speight, 2017). That formed in catalytic cracking operations is usually nonrecoverable, as it is often employed as fuel for the process. The composition of the coke varies with the source of the crude oil, but in general, large amounts of high-molecular-weight complex hydrocarbon derivatives (rich in carbon but correspondingly poor in hydrogen) make up a high proportion. The solubility of petroleum coke in carbon disulfide has been reported to be as high as 50%–80%, but this is in fact a misnomer, since the coke is the insoluble, honeycomb material that is the end product of thermal processes. Three physical structures of coke can be produced by delayed coking: (1) shot coke, (2) sponge coke, or (3) needle coke.

Shot coke is an abnormal type of coke resembling small balls. Due to mechanisms not well understood, the coke from some coker feedstocks forms into small, tight, nonattached clusters that look like pellets, marbles, or ball bearings. It usually is a very hard coke, that is, low Hardgrove grindability index (Speight, 2013). Such coke is less desirable to the end users because of difficulties in handling and grinding. It is believed that feedstocks high in asphaltene constituents and low American Petroleum Institute (API) favor shot coke formation. Blending aromatic materials with the feedstock and/or increasing the recycle ratio reduces the yield of shot coke. Fluidization in the coke drums may cause the formation of shot coke.

Occasionally, the smaller shot coke may agglomerate into ostrich egg-sized pieces. While shot coke may look like it is entirely made up of shot, most shot coke is not 100% shot. Interestingly, even sponge coke may have some measurement of embedded shot coke. The test for such a measurement is not precise, but a low shot coke percentage is sometimes specified for anode grades of petroleum coke. In the case of fuel coke production, there is a general belief that shot structure of coke means that the coker is being run at the most economic conditions.

Sponge coke is the common type of coke produced by delayed coking units. It is in a form that resembles a sponge and has been called honeycombed. Sponge coke, mostly used for anode-grade is dull and black, having porous, amorphous structure.

Needle coke (acicular coke) is a special quality coke produced from aromatic feedstocks is silver-gray, having crystalline broken needle structure, and is believed to be chemically produced through cross-linking of condensed aromatic hydrocarbon derivatives during coking reactions. It has a crystalline structure with more unidirectional pores and is used in the production of electrodes for the steel and aluminum industries and is particularly valuable because the electrodes must be replaced regularly.

Petroleum coke is employed for a number of purposes, but its chief use is in the manufacture of carbon electrodes for aluminum refining, which requires a high-purity carbon—low in ash and sulfur free; the volatile matter must be removed by calcining. In addition to its use as a metallurgical reducing agent, petroleum coke is employed in the manufacture of carbon brushes, silicon carbide abrasives, and structural carbon (e.g., pipes and Raschig rings), as well as calcium carbide manufacture from which acetylene is produced:

$$\text{Coke} \rightarrow CaC_2$$

$$CaC_2 + H_2O \rightarrow HC \equiv CH$$

13.5 Fuel quality

General aspects of fuel quality (whatever the source of the fuel) are assessed by the measurement of physical properties such as relative density (specific gravity), refractive index or viscosity, or by empirical tests such as pour point or oxidation stability that are intended to relate to behavior in service. In some cases the evaluation may include tests in mechanical rigs and engines either in the laboratory or under actual operating conditions.

Measurements of bulk properties are generally easy to perform and, therefore, quick and economical. Several properties may correlate well with certain compositional characteristics and are widely used as a quick and inexpensive means to determine those. The most important properties of a whole crude oil are its boiling-point distribution, its density (or API gravity), and its viscosity. Density and viscosity are measured for secondary reasons. The former helps to estimate the paraffinic character of the fuel, and the latter permits the assessment of its undesirable residual material that cause resistance to flow. Sulfur content, another crucial and primary property of a crude oil, is also readily determined. Certain composite characterization values, calculated from density and mid-boiling-point correlate better with molecular composition than density alone.

However, as a preliminary step, the value of any product is judged by the characteristics of the sample as determined by laboratory tests. The sample used for the test(s) must be representative of the bulk material or data will be produced that are not representative of the material and will, to be blunt, incorrect no matter how accurate or precise the test method. In addition, the type and cleanliness of sample containers are important if the container is contaminated or is made of the material that either reacts with the product or is a catalyst, the test results may be wrong.

Thus the importance of the correct sampling of any fuel destined for analysis should always be over emphasized. Incorrect sampling protocols can lead to erroneous analytical data from which performance of the product in service cannot be accurately deduced. In addition, adequate records of the circumstances and conditions during sampling have to be made, for example, in sampling from storage tanks, the temperatures and pressures of the separation plant, and the atmospheric temperature would be noted.

Gaseous fuels derived from the gasification and pyrolysis of municipal solid waste, crop and wood waste, biomass, scrap tires (pyrolysis gas and synthesis gas), biogas from AD, and landfill gas were compared to NG.

Liquid waste derived fuels such as used lubricant engine oils, diesel and gasoline-like fuel, straight and biodiesel blends, biocrude oil, and pyrolysis-derived liquids were compared to regular petroleum-derived fuels (diesel fuel, fuel oil, gasoline, etc.) using common fuel characteristics as cetane number, octane number, carbon residue, cloud, flash, pour point, acid number (mg KOH/g), viscosity, ash, iodine value (IV) (oxidative stability), and heating value as an indication of its suitability for application in combustion ignition and spark ignition engines.

The octane number is a measure of the gasoline quality for the prevention of early ignition that leads to cylinder knocks. Higher octane numbers are preferred in ICE. For gasoline production, aromatics, naphthenes, and iso-alkane derivatives are highly desirable, whereas olefin derivatives and *n*-paraffin derivatives are less desired. On the other hand, the cetane number is a dimensionless descriptor related to the ignition delay time a fuel experiences upon injection into the combustion chamber of a diesel engine. The higher the cetane number, the shorter the ignition delay time and vice versa. A cetane scale using hydrocarbon derivatives has been established and is commonly used with hexadecane being the high cetane number reference compound on this scale (CN = 100).

The viscosity is the thickness of the fluid (sometimes referred to as the heaviness of the fluid) that is the resistance the fluid offers to motion which is being deformed by stress (either shear stress or tensile stress). Thus viscosity is a measure of the internal resistance of a fluid to flow and may be thought of as a measure of fluid friction. The less viscous the fluid is, the greater its ease of movement. There are actually two quantities that are called viscosity. The quantity defined previously is sometimes referred to as the dynamic viscosity, the absolute viscosity, or the simple viscosity to distinguish it from the kinematic viscosity which is the ratio of the viscosity of a fluid to its density. Thus the kinematic viscosity is a measure of the resistive flow of a fluid under the influence of gravity.

The flash point of a volatile liquid is the temperature at which the oil will vaporize and form an ignitable mixture in air. The flash requires an ignition source and the vapor may cease to burn when the source of ignition is removed. The flash point should not be confused with the autoignition temperature, which does not require an ignition source. As an example, gasoline has a flash point of on the order of approximately −40°C (−40°F) and is more flammable than ethylene glycol (antifreeze) which has a flash point of 111°C (232°F). A closely related and less common term is the fire point which is typically several degrees above the flash point and is the temperature at which the flame becomes self-sustained so as to continue burning the liquid (at the flash point, the flame does not need to be sustained).

The cloud point is the temperature at which the first solids appear in a fuel but the fuel can still flow, these solids can lead to fuel filter plugging. Several other methods exist for determining the low-temperature properties of biodiesel. These are the cold filter plugging point and low-temperature flow test. The low-temperature properties of biodiesel are again influenced by the properties of the individual components. On the other hand, the pour point is the temperature at which the oil stops flowing like a liquid. The pour point is usually a few degrees below the cloud point and is the temperature at which the fuel can no longer be freely poured. Any reduction in the pour point depends on both the composition and properties of the oil, as well as the type of pour point depressant used. Alkyl aromatic derivatives and aliphatic polymers are the two types of pour point depressants that are commercially available. Most commercially available pour point depressants are organic polymers but nonpolymeric substances such as phenyl tri-stearyl oxy silane and pentaerythritol tetrastearate may also be effective.

Mixtures such as biodiesel do not possess melting points, but rather melting ranges, this is reflected in the specifications used in biodiesel standards (cold flow). The melting points of fatty esters, which are the main constituents of biodiesel, generally increase with chain length (although chains with odd numbers of carbons have slightly lower melting points that the preceding even-number chain) and increasing saturation.

The IV is a measure of the total unsaturation of a sample and has often been used in connection with issues such as oxidative stability, with the implication that materials with a high IV are less stable to oxidation than those with a low IV. The IV is also molecular weight dependent; for example, ethyl esters have a lower IV than methyl esters although the kind of ester is highly unlikely to affect oxidative stability of the double bonds in a fatty acid chain. Therefore the iodide value should not be considered in the choice of optimized biodiesel components.

The heating value (heat content, calorific value, Btu value) of a fuel may be expressed as HHV or lower heating value (LHV). The LHV is the total quantity of sensible heat released during combustion of a specified quantity of fuel. The HHV is the LHV plus the enthalpy of evaporation of water vapor resulting from combustion (can be substantial for hydrogen rich fuels, typical for renewable energy

Table 13.16 Summary of the data retrieved from the ultimate analysis and proximate analysis of solid fuels.

Ultimate analysis	% w/w			
	Carbon Hydrogen Nitrogen Oxygen Sulfur			
Proximate analysis	% w/w	As received[a]	Dry[b]	Dry-ash free[c]
	Moisture	√	0	0
	Ash	√	√	0
	Volatile matter	√	√	√
	Fixed carbon	√	√	√

[a]Often designated as "AR".
[b]Often designated as "MF" (moisture free).
[c]Often designated as "MMF" (mineral moisture free of DAF—dry, ash free).

applications). The HHV is usually favored because its value allows easier checking of energy calculations and when examining technologies for implementation.

Biomass pyrolysis liquids were of significant interest because of its potential as biofuels for combustion systems (engines, boilers, and turbines). The quality of pyrolysis liquids was determined by several physical and chemical properties, namely, phase separation, solid and water content of the liquid product, and HHVs of the organic phase and liquid product. Phase separation was one of the main problems of pyrolysis liquids in view of its application and commercialization. Low solids content was required because the solids can contain metals with catalytic activity that causes the liquid to become unstable during storage. Solids were capable of causing mechanical problems in turbines and engines, because of its erosion power. Moderate water content (<27% w/w) appeared beneficial from the point of view of viscosity and atomization, although higher water contents can lead to a low heating value and to a phase-separated pyrolysis liquid. The heating value of the liquid product was necessary for evaluating the possibility of using pyrolysis liquids as fuels and for checking the energy balance of the process.

Finally, waste derived solid fuels (briquettes, coal supplements, granulated fuels, and pyrolysis solids) were compared using proximate analysis and heating value. The concept of proximate analysis was developed for coal as a convenient and effective means for determining the distribution of products obtained by heating coal under a set of standard conditions (Table 13.16) (Speight, 2013, 2015b). This particular group that fall within the proximate analysis group of tests has been used widely as the basis for coal characterization in connection with coal utilization. The proximate analysis (ASTM D3172) of coal may also be considered as the determination of the general properties of coal and is, in reality, the determination of moisture content, volatile matter content, ash yield, and (by difference) fixed carbon yield in contrast to the ultimate analysis of coal that provides the elemental composition (Fig. 8.5). Similarly, these test methods can be adapted to determining the properties of solid fuels.

The moisture, volatile matter, and ash results are typically among the primary parameters used for assessing the quality of the fuel. The moisture result is utilized for calculating the dry basis results of other analytical results. The ash result is utilized in the ultimate analysis calculation of oxygen by difference (ASTM D3176) and for calculating material balance and ash load purposes in industrial boiler systems. The volatile matter result indicates the coke/char yield from thermal processes and provides additional information relating to the combustion characteristics of the solid fuel. Fixed carbon is a calculated value of the difference between 100 and the sum of the moisture, ash, and volatile matter where all values are of the same moisture reference base.

Thus the objective of the proximate analysis is to determine the amount moisture, volatile matter yield, ash yield, and fixed carbon from the fuel sample. Mineral matter is not directly measured but may be obtained by one of a number of empirical formula either from the yield of mineral ash or from data derived from the ultimate analysis.

The variables are measured in percent by weight (% w/w) and are calculated in several different bases. The AR (as-received) basis is the most widely used basis in industrial applications and puts all variables into consideration and uses the total weight as the basis of measurement. The AD (air-dried) basis neglect the presence of moistures other than inherent moisture. The DB (dry basis) leaves out all moistures, including surface moisture, inherent moisture, and other moistures. The DAF (dry, ash free) basis neglect all moisture and ash constituent in the fuel. The DMMF (dry, mineral-matter-free) basis leaves out the presence of moisture and mineral matter in coal, such as quartz, pyrite, calcite, and clay. Examples of these basis are given in the following table (Speight, 2013, 2015b).

Proximate analysis	AR	AD	DB	DAF
Moisture (% w/w)	3.3	2.7		
Ash (% w/w)	22.1	22.2	22.8	
Volatile matter (% w/w)	27.3	27.5	28.3	36.6
Fixed carbon (% w/w)	47.3	47.6	48.9	63.4

AD, Air-dried; *AR*, as-received; *DAF*, dry, ash free; *DB*, dry-basis.

The proximate analysis of the solid fuel covers the determination of moisture, volatile matter, ash content, and fixed carbon [initial sample mass − (volatile matter + ash content)]. Volatile matter and fixed carbon were the dominant components to produce heat. Density, compressive strength, and ignition temperature were important for storage, transportation, and combustion safety.

Finally, the moisture content is probably the most important determinant of heating value of a feedstock or fuel. Air-dried biomass typically has approximately 15%–20% w/w moisture, whereas the moisture content for oven-dried biomass is typically at 0% w/w. However, the bulk density (and, hence, the energy density) of most biomass feedstocks is generally low, even after densification, and is typically on the order of 10%–40% of the bulk density of most fossil fuels. Liquid biofuels have comparable bulk densities to fossil fuels.

References

Faaij, A., van Doorn, J., Curvers, A., Waldheim, L., Olsson, E., van Wijk, A., et al., 1997. Characterization and availability of biomass waste and residues in the Netherlands for gasification. Biomass Bioenergy 12 (4), 225–240.

Gary, J.H., Handwerk, G.E., Kaiser, M.J., 2007. Petroleum Refining: Technology and Economics, fifth ed. CRC Press, Taylor & Francis Group, Boca Raton, FL.

Hsu, C.S., Robinson, P.R. (Eds.), 2017. Handbook of Petroleum Technology. Springer International Publishing AG, Cham.

Mokhatab, S., Poe, W.A., Speight, J.G., 2006. Handbook of Natural Gas Transmission and Processing. Elsevier, Amsterdam.

Parkash, S., 2003. Refining Processes Handbook. Gulf Professional Publishing, Elsevier, Amsterdam.

Speight, J.G., 2013. The Chemistry and Technology of Coal, third ed. CRC Press, Taylor & Francis Group, Boca Raton, FL.

Speight, J.G., 2014a. The Chemistry and Technology of Petroleum, fifth ed. CRC Press, Taylor & Francis Group, Boca Raton, FL.

Speight, J.G., 2014b. Oil and Gas Corrosion Prevention. Gulf Professional Publishing Company, Elsevier, Oxford.

Speight, J.G., 2015a. Handbook of Petroleum Product Analysis, second ed. John Wiley & Sons Inc, Hoboken, NJ.

Speight, J.G., 2015b. Handbook of Coal Analysis, second ed. John Wiley & Sons Inc, Hoboken, NJ.

Speight, J.G., 2017. Handbook of Petroleum Refining. CRC Press, Taylor & Francis Group, Boca Raton, FL.

Speight, J.G., 2018. Handbook of Natural Gas Analysis. John Wiley & Sons Inc, Hoboken, NJ.

Speight, J.G., 2019a. Natural Gas: A Basic Handbook, second ed. Gulf Publishing Company, Elsevier, Cambridge, Massachusetts.

Speight, J.G., 2019b. Handbook of Petrochemical Processes. CRC Press, Taylor & Francis Group, Boca Raton, FL.

van Doorn, J. 1995. Characterization of energy crops and biomass and waste streams, Netherlands Energy Research Foundation. Report ECN-C-95-047, Petten.

A biorefinery

14

14.1 Introduction

The previous chapters have focused on the production of fuels by a variety of processes from fossil fuel feedstocks and also from nonfossil fuel feedstocks. This leads to the concept by which fuels might be produced from nonfossil fuel feedstocks in a central location such as in tandem (on-site) with the production of crude oil–based products or, as may be the case in the future (Speight, 2011a), in a refinery solely dedicated to the conversion of, say, biomass-type feedstocks only. This has already commenced insofar as the application of biotechnology to refinery processes had been advocated (Chapter 8: Desulfurization, Denitrogenation, and Demetallization) (El-Gendy and Speight, 2015). The thinking behind the concept of the biorefinery is, in essence, the creation of a system for a variety of fuels that are not based on crude oil which, under this all-encompassing umbrella, would lead to the production of a gaseous, liquid, and solid fuels (Taylor, 2008; El-Gendy and Speight, 2015).

The biorefinery, given time, and the depletion of the supplies of the fossil fuel resources could well be integrated with the conventional crude oil refinery or, in some cases, even replace the crude oil refiners at a future date. Gaseous, liquid, and solid biofuels may be made from a variety of biomass sources, including waste organic materials, through a number of conversion routes. Several types of product suitable for road and other forms of transportation fuel can be produced, including, in the future, aviation fuel, perhaps using microalgae for feedstock supply (Di Visconte et al., 2019). Current products include bioethanol, biodiesel, and other bio-based products such as biobutanol. Feedstocks are varied and could include oil-based crops such as oil-seed rape, used for the production of biodiesel, and sugar- and starch-based crops such as sugar beet and sugar cane, which are used for the production of bioethanol. Such crops that have a food use are often referred to as "first-generation" bioenergy crops. All lignocellulosic biomass (biomass that is mostly plant cell walls that have a high carbon content) can be converted to liquid fuels through biological processes (esterification and fermentation) or through thermochemical routes such as pyrolysis. Such dedicated bioenergy crops with no food use are often referred to as second-generation bioenergy crops. Hydrogen may also be produced from biomass, but this is still at the research stage.

The supply of crude oil, the basic feedstock for refineries and for the petrochemicals industry, is finite and its dominant position will become unsustainable as supply/demand issues erode its economic advantage over other alternative

The Refinery of the Future. DOI: https://doi.org/10.1016/B978-0-12-816994-0.00014-2

feedstocks. This situation will be mitigated to some extent by the exploitation of more technically challenging fossil resources and the introduction of new technologies for fuels and chemicals production from natural gas, crude oil, and coal (Speight, 2008, 2013a,b, 2014, 2017).

However, the use of fossil resources at current rates will have serious and irreversible consequences for the global climate. Consequently, there is a renewed interest in the utilization of plant-based matter as a raw material feedstock for the chemicals industry. Plants accumulate carbon from the atmosphere via photosynthesis, and the widespread utilization of these materials as basic inputs into the generation of power, fuels, and chemicals is a viable route to reduce greenhouse gas (GHG) emissions.

Thus the crude oil and petrochemicals industries are coming under increasing pressure not only to compete effectively with global competitors utilizing more advantaged hydrocarbon feedstocks but also to ensure that its processes and products comply with increasingly stringent environmental legislation.

Reducing dependence of any country on imported crude oil is of critical importance for long-term security and continued economic growth. Supplementing crude oil consumption with renewable biomass resources is the first step toward this goal. The realignment of the chemical industry from one of petrochemical refining to a biorefinery concept is, given time, feasible has become a national goal of many oil-importing countries. However, clearly defined goals are necessary for construction of a biorefinery and increasing the use of biomass-derived feedstocks in industrial chemical production, and it is important to keep the goal in perspective. In this context the increased use of biofuels should be viewed as one of a range of possible measures for achieving self-sufficiency in energy, rather than a panacea (Crocker and Crofcheck, 2006).

A biorefinery is the means by which biomass can be converted to other products—in the current context, the other products are biofuels that have the potential to replace certain crude oil–derived fuels. Biomass refers to (1) energy crops grown specifically to be used as fuel, such as fast-growing trees or switch grass; (2) agricultural residues and by-products, such as straw, sugarcane fiber, and rice hulls; and (3) residues from forestry, construction, and other wood-processing industries (NREL, 2003; Wright et al., 2006). Biomass is a renewable energy source, unlike the fossil fuel resources (natural gas, crude oil, and coal) and, like the fossil fuels, biomass is a form of stored solar energy but produced in lesser time than the fossil fuels (Speight, 2008). The energy of the Sun is captured through the process of photosynthesis in growing plants. One advantage of biofuel in comparison to most other fuel types is it is biodegradable and thus relatively harmless to the environment if spilled.

Thus a biorefinery is a facility that integrates biomass conversion processes and equipment to produce fuels, power, and chemicals from biomass. The biorefinery concept is the modern crude oil refineries, which produces multiple fuels and products from crude oil (Parkash, 2003; Gary et al., 2007; Speight, 2014, 2017, 2019a; Hsu and Robinson, 2017). Industrial biorefineries have been identified as the most promising route to the creation of a new domestic bio-based industry.

By producing multiple products, a biorefinery can take advantage of the differences in biomass components and intermediates and maximize the value derived from the biomass feedstock. A biorefinery might, for example, produce one or several low-volume, but high-value, chemical products and a low-value, but high-volume, liquid transportation fuel while generating electricity and process heat for its own use and perhaps enough for sale of electricity. The high-value products enhance profitability, the high-volume fuel helps meet national energy needs, and the power production reduces costs and avoids GHG emissions.

Thus a key to accessing the integrated production of chemicals, materials, and fuels is the development of biorefineries. The biorefinery concept is analogous to that of an oil refinery. A biorefinery would integrate a variety of conversion processes to produce multiple product streams such as motor fuels, heat, electricity, and chemicals from biomass. Industrial-scale biorefineries have been identified as the most promising route to the creation of a sustainable bio-based economy. Plant biomass consists of carbohydrates, lignin, proteins, and fats, with a variety of minor products such as vitamins, dyes, and flavors. Biorefineries combine the essential technologies to transform biological raw materials into a range of industrially useful intermediates. By producing multiple products, a biorefinery maximizes the value derived from a biomass feedstock. A biorefinery could produce one or more low-volume, high-value chemical products together with a low-value, high-volume liquid transportation fuel while generating electricity and process heat for its own use and/or export.

In theory, a biorefinery can use all kinds of biomass, including wood and dedicated agricultural crops, plant- and animal-derived waste, municipal waste, and aquatic biomass (algae, seaweeds). A biorefinery produces a spectrum of marketable products and energy, including intermediate and final products: food, feed, materials, chemicals, fuels, power, and/or heat. However, the differences in the various biomass feedstocks may dictate that a biorefinery be constructed and operated on the basis of the chemical composition of the feedstock and the mean by which the feedstock is to be processed.

This chapter presents a brief review of the different biomass feedstocks that might be used in a biorefinery and offers descriptions of the various conversion options that can be applied to the production of fuels and other products from biomass.

14.2 The biorefinery

Biorefining offers a method to accessing the integrated production of chemicals, materials, and fuels. However, contrary to many opinions, the concept of biorefining is not new when activities such as production of vegetable oils, beer, and wine requiring pretreatment are considered. Many of these activities are known to have been in practice for millennia. However, the differences in the various biomass feedstocks require a divergence in the methods used to convert the feedstocks to fuels and chemicals (Ruth, 2004; Speight, 2014, 2017, 2019a). In addition, a modern biorefinery, such as a crude oil refinery, will be a facility that integrates biomass conversion processes and the relevant equipment to produce fuels, power, and chemicals from the biofeedstocks. The various conversion processes would enable the production of multiple product streams—such as motor fuels and other chemicals—from a variety of biofeedstocks such as the inclusion of gasification and fermentation processes to name only two possible processes options (Speight, 2011b).

In short, a biorefinery should combine the essential technologies to transform biological raw materials into a range of industrially useful intermediates. However, it cannot be overemphasized that the type of biorefinery would have to be amenable to the character of the feedstock. For example, the crop biorefinery would use raw materials such as cereals or maize, and the lignocellulose biorefinery would use raw material with high cellulose content, such as straw, wood, and paper waste—the operation both of these biorefineries would be substantially different.

However, the different compositional nature of the biomass feedstock, compared to crude oil, will require the application of a wider variety of processing tools in the biorefinery.

The bioconversion platform typically uses a combination of physical or chemical pretreatment and enzymatic hydrolysis to convert lignocellulose into its component monomers. Once liberated, the carbohydrate components of wood may be processed into a number of chemical and fuel products such as the production and blending as an oxygenate additive in liquid fuels. Other chemical products can be used to create consumer products, such as bioplastics, or as platform chemicals in a number of industrial applications. The development of better ways to separate lignin from the lignocellulose matrix during bioconversion has created the possibility of developing value-added lignin-based products as well. The bioconversion platform therefore has the ability to serve as the basis for full-fledged wood-based biorefining operations, generating value-added bioproducts as well as fuel and energy for the forest sector.

Using components of these platforms, biomass can provide a sustainable, renewable source of bioenergy. This report illustrates how evolutions in technology may be combined to create truly revolutionary processes that can transform the energy sector. It is also shown how each technological platform might be used to generate other valuable chemical products or energy thus creating a biorefinery. An important conclusion is that the development of the biorefinery should take precedence over specific biofuel and bioenergy projects.

Although a number of new bioprocesses have been commercialized, it is clear that economic and technical barriers still exist before the full potential of this area can be realized. The biorefinery concept could significantly reduce production costs of plant-based chemicals and facilitate their substitution into existing markets. This concept is analogous to that of a modern oil refinery in that the biorefinery is a highly integrated complex that will efficiently separate biomass raw materials into individual components and convert these into marketable products such as energy, fuels, and chemicals. By analogy with crude oil, every element of the plant feedstock will be utilized, including the low-value lignin components.

A key requirement for the biorefinery is the ability of the refinery to develop process technology that can economically access and convert the five- and six-membered ring sugars present in the cellulose and hemicellulose fractions of the lignocellulosic feedstock. Although engineering technology exists to effectively separate the sugar-containing fractions from the lignocellulose, the enzyme technology to economically convert the five ring sugars to useful products requires further development.

Plants are very effective chemical mini-factories or refineries insofar as they produce chemicals by specific pathways. The chemicals they produce are usually essential manufacture (called metabolites), including sugars and amino acids that are essential for the growth of the plant, as well as more complex compounds. Unlike crude oil derived in petrochemicals where most chemicals are built from the bottom up, biofeedstocks already have some valuable products to skim off the top before being broken down and used to build new molecules.

As a feedstock, biomass can be converted by thermal or biological routes to a wide range of useful forms of energy, including process heat, steam, electricity, as well as liquid fuels, chemicals, and synthesis gas (Chapter 12: Synthesis Gas and the Fischer–Tropsch Process). As a raw material, biomass is a nearly universal feedstock due to its versatility, domestic availability, and renewable character. At the same time, it also has its limitations. For example, the energy density of biomass is low compared to that of coal, liquid crude oil, or crude oil–derived fuels. The heat content of biomass, on a dry basis (7000–9000 Btu/lb), is at best comparable with that of a low-rank coal or lignite, and substantially (50%–100%) lower than that of anthracite, most bituminous coals, and crude oil. Most biomass, as received, has a high burden of physically adsorbed moisture, up to 50% by weight. Thus without substantial drying, the energy content of a biomass feed per unit mass is even less.

These inherent characteristics and limitations of biomass feedstocks have focused on the development of efficient methods of chemically transforming and upgrading biomass feedstocks in a refinery. The refinery would be based on two platforms to promote different product slates: (1) the sugar platform and (2) the thermochemical platform (Speight, 2011b). The sugar-base involves breakdown of biomass into raw component sugars using chemical and biological means. The raw fuels may then be upgraded to produce fuels and chemicals that are interchangeable with existing commodities such as transportation fuels, oils, and hydrogen.

Although a number of new bioprocesses have been commercialized, it is clear that economic and technical barriers still exist before the full potential of this area can be realized. One concept gaining considerable momentum is the biorefinery that could significantly reduce production costs of plant-based chemicals and facilitate their substitution into existing markets. This concept is analogous to that of a modern oil refinery in that the biorefinery is a highly integrated complex that will efficiently separate biomass raw materials into individual components and convert these into marketable products such as energy, fuels, and chemicals.

By analogy with crude oil, every element of the plant feedstock will be utilized, including the low-value lignin components. However, the different compositional nature of the biomass feedstock, compared to crude oil, will require the application of a wider variety of processing tools in the biorefinery. Processing of the individual components will utilize conventional thermochemical operations and state-of-the-art bioprocessing techniques. The production of biofuels in the biorefinery complex will service existing high-volume markets, providing economy-of-scale benefits and large volumes of by-product streams at minimal cost for upgrading to valuable chemicals. A pertinent example of this is the production of glycerol (glycerin) as a by-product in biodiesel plants (Figs. 14.1 and 14.2).

Glycerol has high functionality and is a potential platform chemical for conversion into a range of higher value chemicals. The high-volume product streams in a biorefinery need not necessarily be a fuel but could also be a large-volume chemical intermediate such as ethylene or lactic acid.

In addition, a variety of method techniques can be employed to obtain different product portfolios of bulk chemicals, fuels, and materials. Biotechnology-based conversion processes can be used to ferment the biomass carbohydrate content into sugars that can then be further processed. As one example, the fermentation path to lactic acid shows promise as a route to biodegradable plastics. An alternative is to employ thermochemical conversion processes that use pyrolysis or gasification of biomass to produce synthesis gas that can be used in a wide range of chemical processes (Chapter 12: Synthesis Gas and the Fischer–Tropsch Process) (Speight, 2019a).

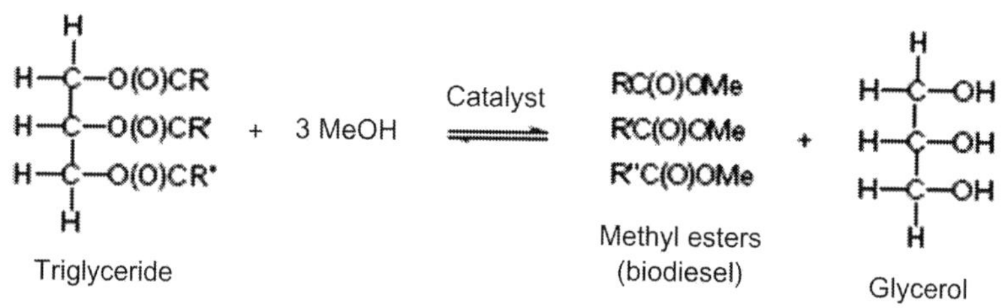

Figure 14.1 The chemistry of biodiesel production.

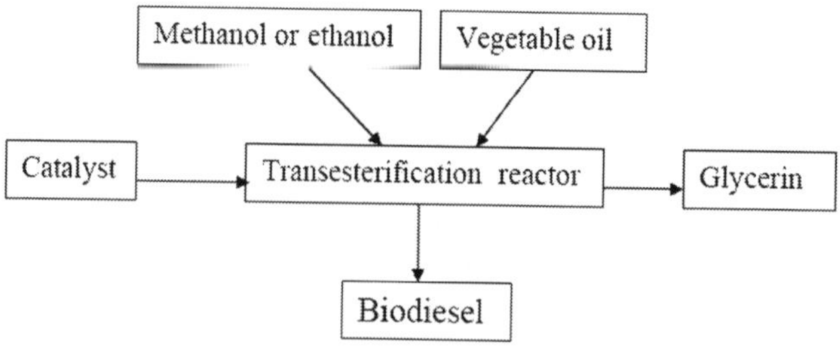

Figure 14.2 Catalytic production of biodiesel.

If the biorefinery is truly analogous to an oil refinery in which crude oil is separated into a series of products, such as gasoline, heating oil, jet fuel, and petrochemicals, the biorefinery can take advantage of the differences in biomass components and intermediates and maximize the value derived from the biomass feedstock. A biorefinery might, for example, produce one or several low-volume, but high-value, chemical products and a low-value, but high-volume, liquid transportation fuel while generating electricity and process heat for its own use and perhaps enough for sale of electricity. The high-value products enhance profitability, the high-volume fuel helps meet national energy needs, and the power production reduces costs and avoids GHG emissions.

14.3 Process options

A crude oil refinery is a series of integrated unit processes by which crude oil can be converted to a slate of useful (salable) products and, as currently configured, is unsuitable for processing raw, or even partially processed, biomass. A typical refinery might be suitable for processing products such as gases, liquid, or solid products from biomass processing (Table 14.1). These products from biomass might be acceptable as a single feedstock to a specific unit or, more likely, as a feedstock to be blended with refinery streams to be coprocessed in various refinery units.

Thus a biorefinery might, in the early stages of development, be a series of unit processes that covert biomass to a primary product that requires further processing to become the final saleable product. The analogy is in the processing of bitumen afro tar sand that is fist processed to a synthetic crude oil (primary processing) and then sent to a refinery for conversion to saleable fuel products (Speight, 2008, 2014, 2017).

Biomass can be converted into commercial fuels, suitable to substitute for fossil fuels. These can be used for transportation, heating, electricity generation, or anything else fossil fuels are used for. The conversion is accomplished through the use of several distinct processes that include both biochemical and thermal conversions to produce gaseous, liquid, and solid fuels that have high energy contents, are easily transportable, and are therefore suitable for use as commercial fuels.

The basic types of processes used to generate energy from biomass as might be incorporated into a biorefinery are (1) anaerobic digestion, (2) combustion, (3) fermentation and hydrolysis, (4) gasification, (5) pyrolysis, and (6) transesterification.

14.3.1 Anaerobic digestion

Anaerobic digestion is a natural process and is the microbiological conversion of organic matter to methane in the absence of oxygen. The biochemical conversion of biomass is completed through alcoholic fermentation to produce liquid fuels and anaerobic digestion or fermentation, resulting in biogas (hydrogen, carbon dioxide, ammonia, and methane) usually through four steps (hydrolysis, acidogenesis,

Table 14.1 Simplified representation of the potential products from a biorefinery.

Feedstock[a]				
	Pretreatment[b]			
	Gasification/ pyrolysis			
		Gas conditioning		
		Gas separation		
			Hydrogen	
			Methane	Chemicals
			Ethane	Chemicals
			Propane	Chemicals
			Butane	Chemicals
			Pentane plus	
				Gasoline
			Synthesis gas	
				Methanol
				Ethanol
				Propanol
				Butanol
				Hydrocarbon fuels
	Enzymatic hydrolysis			
		Sugar products		
			Fermentation	
				Ethanol
		Lignin intermediates		
			Various products	
				Chemicals

[a]Biomass, waste.
[b]Whenever necessary.

acetogenesis, and methanogenesis) (Table 14.2). The decomposition is caused by natural bacterial action in various stages and occurs in a variety of natural anaerobic environments, including water sediment, water-logged soils, natural hot springs, ocean thermal vents, and the stomach of various animals (e.g., cows). The digested organic matter resulting from the anaerobic digestion process is usually called digestate.

Symbiotic groups of bacteria perform different functions at different stages of the digestion process. There are four basic types of microorganisms involved. Hydrolytic bacteria break down complex organic wastes into sugars and amino acids. Fermentative bacteria then convert those products into organic acids. Acidogenic microorganisms convert the acids into hydrogen, carbon dioxide, and

Table 14.2 The various stages of anaerobic digestion.

Stage	Chemical changes
1. Hydrolysis	Carbohydrates→sugars Fats→fatty acids Proteins→amino acids
2. Acidogenesis	Sugars→carbon acids + alcohols + hydrogen + carbon dioxide + ammonia Fatty acids→carbon acids + alcohols + hydrogen, carbon dioxide + ammonia Amino acids→carbon acids + alcohols + hydrogen, carbon dioxide + ammonia
3. Acetogenesis	Carbon acids + alcohols→acetic acid + carbon dioxide + hydrogen
4. Methanogenesis	Acetic acid→methane + carbon dioxide

acetate. Finally, the methanogenic bacteria produce biogas from acetic acid, hydrogen, and carbon dioxide.

The process of anaerobic digestion occurs in a sequence of stages involving distinct types of bacteria. Hydrolytic and fermentative bacteria first break down the carbohydrates, proteins, and fats present in biomass feedstock into fatty acids, alcohol, carbon dioxide, hydrogen, ammonia, and sulfides. This stage is called hydrolysis (or liquefaction). In the next stages, acetogenic (acid-forming) bacteria further digest the products of hydrolysis into acetic acid, hydrogen, and carbon dioxide. Methanogenic (methane-forming) bacteria then convert these products into biogas.

The combustion of digester gas can supply useful energy in the form of hot air, hot water, or steam. After filtering and drying, digester gas is suitable as fuel for an internal combustion engine, which, combined with a generator, can produce electricity. Future applications of digester gas may include electric power production from gas turbines or fuel cells. Digester gas can substitute for natural gas (methane) or propane (and butane) in space heaters, refrigeration equipment, cooking stoves, or other equipment. Furthermore, compressed digester gas can be used as an alternative transportation fuel. Also, there are three principal by-products of anaerobic digestion: (1) biogas, (2) acidogenic digestate, and (3) methanogenic digestate.

Since the gas is not released directly into the atmosphere and the carbon dioxide that comes from an organic source with a short carbon cycle biogas does not contribute to increasing atmospheric carbon dioxide concentrations; because of this, it is considered to be an environment-friendly energy source. The production of biogas is not a steady stream; it is highest during the middle of the reaction. In the early stages of the reaction, little gas is produced because the number of bacteria is still small. Toward the end of the reaction, only the hardest to digest materials remain, leading to a decrease in the amount of biogas produced.

The second by-product (acidogenic digestate) is a stable organic material that not only comprises largely lignin and chitin but also a variety of mineral components in a matrix of dead bacterial cells; some plastic may be present. This resembles domestic compost and can be used as compost or to make low-grade building products such as fiberboard.

The third by-product is a liquid (methanogenic digestate) that is rich in nutrients and can be an excellent fertilizer dependent on the quality of the material being digested. If the digested materials include low levels of toxic heavy metals or synthetic organic materials such as pesticides or polychlorobiphenyls, the effect of digestion is to significantly concentrate such materials in the digester liquor. In such cases, further treatment will be required in order to dispose of this liquid properly. In extreme cases the disposal costs and the environmental risks posed by such materials can offset any environmental gains provided by the use of biogas. This is a significant risk when treating sewage from industrialized catchments.

Nearly all digestion plants have ancillary processes to treat and manage all of the by-products. The gas stream is dried and sometimes sweetened before storage and use. The sludge liquor mixture has to be separated by one of a variety of ways, the most common of which is filtration. Excess water is also sometimes treated in sequencing batch reactors for discharge into sewers or for irrigation.

In recent years, increasing awareness that anaerobic digesters can help control the disposal and odor of animal waste has stimulated renewed interest in the technology. New digesters now are being built because they effectively eliminate the environmental hazards of dairy farms and other animal feedlots.

Controlled anaerobic digestion requires an airtight chamber, called a digester. To promote bacterial activity, the digester must maintain a temperature of at least 68°F. Using higher temperatures, up to 150°F, shortens processing time and reduces the required volume of the tank by 25%–40%. However, there are more species of anaerobic bacteria that thrive in the temperature range of a standard design (mesophilic bacteria) than there are species that thrive at higher temperatures (thermophilic bacteria). High-temperature digesters also are more prone to upset because of temperature fluctuations, and their successful operation requires close monitoring and diligent maintenance.

The gas stream produced in a digester (digester gas) is actually a mixture of gases, with methane and carbon dioxide making up more than 90% of the total. Biogas typically contains smaller amounts of hydrogen sulfide, nitrogen, hydrogen, methyl mercaptans, and oxygen. Methane is a combustible gas. The energy content of digester gas depends on the amount of methane it contains. Methane content varies from approximately 55%–80% v/v and a typical digester gas, with a methane concentration of 65% v/v contains approximately 600 Btu of energy per cubic foot.

There are three basic digester designs and all of them can trap methane and reduce fecal coliform bacteria, but they differ in cost, climate suitability, and the concentration of manure solids they can digest. A covered lagoon digester, as the name suggests, consists of a manure storage lagoon with a cover. The cover traps gas produced during decomposition of the manure. This type of digester is the least expensive of the three. A complete mix digester converts organic waste to biogas in a heated tank above or below ground. A mechanical or gas mixer keeps the solids in suspension. Complete mix digesters are expensive to construct and cost more than plug-flow digesters to operate and maintain.

Complete mix digesters are suitable for larger volumes of feedstock having solids concentration of 3%–10% w/w. Typically, the reactor is a circular steel or poured concrete container and during the digestion process, the manure slurry is continuously mixed to keep the solids in suspension. The gas product accumulates at the top of the digester.

Plug-flow digesters are suitable for ruminant animal manure that has a solids concentration of 11%–13%. A typical design for a plug-flow system includes a manure collection system, a mixing pit, and the digester itself. In the mixing pit the addition of water adjusts the proportion of solids in the manure slurry to the optimal consistency. The digester is a long, rectangular container, usually built below-grade, with an airtight, expandable cover. A plug-flow digester requires minimal maintenance. Waste heat from the engine-generator can be used to heat the digester. Inside the digester, suspended heating pipes allow hot water to circulate. The hot water heats the digester to keep the slurry at 25°C–40°C (77°F–104°F), a temperature range suitable for methane-producing bacteria. The hot water can come from recovered waste heat from an engine generator fueled with digester gas or from burning digester gas directly in a boiler.

14.3.2 Combustion

The simplest, cheapest, and most common method of obtaining energy from biomass is direct combustion. Any organic material, which has a water content low enough to allow for sustained combustion, can be burned to produce energy. The heat of combustion can be used to provide space or process heat, water heating, or, through the use of a steam turbine, electricity. In the developing world, many types of biomass such as dung and agricultural wastes are burned for cooking and heating. In fact, biofuels such as wood or dried dung have been used since man learned to control fire—there were no attempts to refine these feedstocks other than drying to reduce the water content.

Combustion facilities can burn many types of biomass fuel, including wood, agricultural residues, wood pulping liquor, municipal solid waste, and refuse-derived fuel. Combustion technologies convert biomass fuels into several forms of useful energy for commercial or industrial uses: hot air, hot water, steam, and electricity. The process involves burning the energy crop and then using the resulting hot combustion gases to raise steam. The steam is, in turn, used to drive a steam turbine that drives a generator to produce electricity. The conversion efficiency from energy crop to energy is fairly low, especially for small systems, but this is balanced by the relatively low capital cost of direct combustion systems and the fact that the technology is tried and tested. Furthermore, using the waste heat produces much better efficiencies and economics.

A furnace is the simplest combustion technology. In a furnace, biomass fuel burns in a combustion chamber, converting biomass into heat energy. As the biomass burns, hot gases are released. These hot gases contain approximately 85% of the fuel's potential energy. Commercial and industrial facilities use furnaces for heat either directly or indirectly through a heat exchanger in the form of hot air or water. In fact, a boiler fired by biomass is a more adaptable direct combustion technology because a boiler transfers the heat of combustion into steam. Steam can be used for electricity, mechanical energy, or heat. Biomass boilers supply energy at low cost for many industrial and commercial uses.

Pile burners consist of cells, each having an upper and a lower combustion chamber. Biomass fuel burns on a grate in the lower chamber, releasing volatile gases. The gases burn in the upper (secondary) combustion chamber and pile burners must be shut down periodically to remove ash. Although capable of handling high-moisture fuels and fuels mixed with dirt, pile burners have become obsolete with the development of more efficient combustion designs with automated ash-removal systems.

In a stationary or traveling grate combustor, an automatic feeder distributes the fuel onto a grate, where the fuel burns. Combustion air enters from below the grate. In the stationary grate design, ashes fall into a pit for collection. In contrast a traveling grate system has a moving grate that drops the ash into a hopper. On the other hand, fluidized-bed combustors burn biomass fuel in a hot bed of granular material, such as sand. Injection of air into the bed creates turbulence resembling a boiling liquid. The turbulence distributes and suspends the fuel. This design increases heat transfer and allows for operating temperatures below 970°C (1700°F), reducing nitrogen oxide (NO_x) emissions. Fluidized-bed combustors can handle high-ash fuels and agricultural biomass residue.

Conventional combustion equipment is not designed for burning agricultural residues. Straw and grass contain alkali (potassium and sodium) compounds that are also present in all annual crops and crop residues and in the annual growth of trees and plants. During combustion, alkali combines with silica, which is also present in agricultural residues. This reaction causes slagging and fouling problems in conventional combustion equipment designed for burning wood at higher temperatures.

Volatile alkali lowers the fusion temperature of ash. In conventional combustion equipment having furnace gas exit temperatures above 1450°F, combustion of agricultural residue causes slagging and deposits on heat transfer surfaces. Specially designed boilers with lower furnace exit temperatures could reduce slagging and fouling from combustion of these fuels. Low-temperature gasification may be another method of using these fuels for efficient energy production while avoiding the slagging and fouling problems encountered in direct combustion.

Two cogeneration arrangements, or cycles, are possible for combining electric power generation with industrial steam production. Steam can be used in an industrial process first and then routed through a turbine to generate electricity. This arrangement is called a bottoming cycle. In the alternate arrangement, steam from the boiler passes first through a turbine to produce electric power. The steam exhaust from the turbine is then used for industrial processes or for space and water heating. This arrangement is called a topping cycle.

The direct-fired gas turbine is another combustion technology for converting biomass to electricity. In this technology, fuel pretreatment reduces biomass to a particle size of less than 2 mm and a moisture content of less than 25%. Then the fuel is burned with compressed air. Cleanup of the combustion gas reduces particulate matter before the gas expands through the turbine stage. The turbine drives a generator to produce electricity.

14.3.3 Fermentation and hydrolysis

A number of processes allow biomass to be transformed into gaseous fuels such as methane or hydrogen (Sørensen et al., 2006). One pathway uses algae and bacteria that have been genetically modified to produce hydrogen directly instead of the conventional biological energy carriers. A second pathway uses plant material such as agricultural residues in a fermentation process leading to biogas from which the desired fuels can be isolated. This technology is established and in widespread use for waste treatment, but often with the energy produced only for on-site use, which often implies less than maximum energy yields. Finally, high-temperature gasification supplies a crude gas, which may be transformed into hydrogen by a second reaction step. In addition to biogas, there is also the possibility of using the solid by-product as a biofuel.

Traditional fermentation plants producing biogas are in routine use, ranging from farms to large municipal plants. As feedstock they use manure, agricultural residues, urban sewage, and waste from households, and the output gas is typically 64% methane. The biomass conversion process is accomplished by a large number of different agents, from the microbes decomposing and hydrolyzing plant material, over the acidophilic bacteria dissolving the biomass in aquatic solution, and to the strictly anaerobic methane bacteria responsible for the gas formation. Operating a biogas plant for a period of some months usually makes the bacterial composition stabilize in a way suitable for obtaining high conversion efficiency (typically above 60%, the theoretical limit being near 100%), and it is found important not to vary the feedstock compositions abruptly, if optimal operation is to be maintained. Operating temperatures for the bacterial processes are only slightly above ambient temperatures, for example, in the mesophilic region around 30°C.

The production of ethanol from corn is a mature technology that holds much potential (Nichols et al., 2006). Substantial cost reductions may be possible, however, if cellulose-based feedstocks are used instead of corn. The feed for all ethanol fermentations is sugar—traditionally a hexose (a six-carbon or "C6" sugar) such as those present naturally in sugar cane, sugar beet, and molasses. Sugar for fermentation can also be recovered from starch, which is actually a polymer of hexose sugars (polysaccharide).

Biomass, in the form of wood and agricultural residues such as wheat straw, is viewed as a low-cost alternative feed to sugar and starch. It is also potentially available in far greater quantities than sugar and starch feeds. As such it receives significant attention as a feed material for ethanol production. Like starch, wood and agricultural residues contain polysaccharides. However, unlike starch, while the cellulose fraction of biomass is principally a polymer of easily fermented C6 sugars, the hemicellulose fraction is principally a polymer of C5 sugars; with quite different characteristics for recovery and fermentation, the cellulose and hemicellulose in biomass are bound together in a complex framework of crystalline organic material known as lignin.

There are several different methods of hydrolysis: (1) concentrated sulfuric acid, (2) dilute sulfuric acid, (3) nitric acid, and acid pretreatment followed by enzymatic hydrolysis. The greatest potential for ethanol production from biomass, however, lies in enzymatic hydrolysis of cellulose. The enzyme cellulase, now used in the textile industry to stone wash denim and in detergents, simply replaces the sulfuric acid in the hydrolysis step. The cellulase can be used at lower temperatures, 30°C–50°C, which reduces the degradation of the sugar. In addition, process improvements now allow simultaneous saccharification and fermentation (SSF). In the saccharification and fermentation process, cellulase and fermenting yeast are combined, so that as sugars are produced, the fermentative organisms convert them to ethanol in the same step.

Once the hydrolysis of the cellulose is achieved, the resulting sugars must be fermented to produce ethanol. In addition to glucose, hydrolysis produces other six-carbon sugars from cellulose and five-carbon sugars from hemicellulose that are not readily fermented to ethanol by naturally occurring organisms. They can be converted to ethanol by genetically engineered yeasts that are currently available, but the ethanol yields are not sufficient to make the process economically attractive. It also remains to be seen whether the yeasts can be made hardy enough for production of ethanol on a commercial scale.

The fermentation processes to produce propanol and butanol from cellulose are fairly tricky to execute, and the *Clostridium acetobutylicum* currently used to perform these conversions produces an extremely unpleasant smell, and this must be taken into consideration when designing and locating a fermentation plant. This organism also dies when the butanol content of whatever it is fermenting rises to 7%. For comparison, yeast dies when the ethanol content of its feedstock hits 14%. Specialized strains can tolerate even greater ethanol concentrations—so-called turbo yeast can withstand up to 16% ethanol. However, if ordinary *Saccharomyces* yeast can be modified to improve its ethanol resistance, scientists may yet one day produce a strain of the Weizmann organism with a butanol resistance higher than the natural boundary of 7%. This would be useful because butanol has a higher energy density than ethanol, and because waste fiber left over from sugar crops used to make ethanol could be made into butanol, raising the alcohol yield of fuel crops without there being a need for more crops to be planted.

Wet milling and dry milling are the means by which grain and straw fractions are processed into a variety of end products. The processes encompass fermentation and distilling of grains (wheat, rye, or maize). Wet milling starts with water-soaking the grain adding sulfur dioxide to soften the kernels and loosen the hulls, after which it is ground. It uses well-known technologies and allows separation of starch, cellulose, oil, and proteins. Dry milling grinds whole grains (including germ and bran). After grinding the flour is mixed with water to be treated with liquefying enzymes and, further, cooking the mash to breakdown the starch. This hydrolysis step can be eliminated by adding saccharification enzymes and fermenting yeast to the fermenter (SSF).

After fermentation the mash (called beer) is sent through a multicolumn distillation system, followed by concentration, purification, and dehydration of the alcohol. The residue mash (stillage) is separated into a solid (wet grains) and liquid (syrup) phase that can be combined and dried to produce distiller's dried grains with solubles, to be used as cattle feed. Its nutritional characteristics and high vegetable fiber content make distiller's dried grains with solubles unsuitable for other animal species.

14.3.4 Gasification

Gasification is a high-temperature process in which the thermal degradation of the feedstock occurs in the presence of an externally supplied oxidizing (oxygen containing) agent, for example, air, steam, and oxygen. Various gasification concepts have been developed over the years, mainly for the purposes of power generation. However, efficient biomass-to-liquids (BTL) production imposes completely different requirements for the composition of the gas. In power generation the gas is used as a fuel, while in BTL processing the gas is used as a chemical feedstock to obtain other products (Table 14.3). This difference has implications with respect to the purity and composition of the gas.

14.3.4.1 Gasification chemistry

In the gasifier, biomass is converted into a gaseous mixture of hydrogen, carbon monoxide, carbon dioxide, and other compounds by applying heat under pressure in the presence of steam and a controlled amount of oxygen. The biomass produces synthesis gas. Thus using glucose as a surrogate for cellulose and remembering that biomass has highly variable composition and complexity, with cellulose as one major component:

$$C_6H_{12}O_6 + O_2 + H_2O \rightarrow CO + CO_2 + H_2 + \text{other products}$$

Table 14.3 Options for fuels production via gasification.

Option	Use
Fischer–Tropsch liquids	Blend with crude oil–derived liquids[a] Use as substitute for crude oil–derived liquids
Methanol	Convert to naphtha/gasoline Use directly as fuel Convert to dimethyl ether
Dimethyl ether	Use directly as fuel

[a]Subject to no issues arising from incompatibility of the liquids.

Gasification occurs through the thermal decomposition of biomass with the help of an oxidant such as pure oxygen or oxygen-enriched air to yield a combustible gas such as synthesis gas rich in carbon monoxide and hydrogen. The synthesis gas is posttreated, by steam reforming or partial oxidation, to convert the hydrocarbon derivatives produced by gasification into hydrogen and carbon monoxide. The carbon monoxide is then put through the shift process to obtain a higher fraction of hydrogen, by carbon dioxide removal and methanation or by pressure swing adsorption (Mokhatab et al., 2006; Speight, 2019b).

The basic gasification of biomass (such as wood scrap) is a well-known process, taking place in pyrolysis (oxygen supply far below what is required for complete combustion, the fraction called "equivalence ratio") or fluidized-bed type of reactors. Conditions such as operating temperature determine whether hydrogen is consumed or produced in the process. Hydrogen evolution is largest for near-zero equivalence ratios, but the energy conversion efficiency is highest at an equivalence ratio around 0.25. The hydrogen fraction (in this case typically some 30%) must be separated for most fuel-cell applications, as well as for long-distance pipeline-transmission. In the pyrolysis-type application, gas production is low and most energy is in the oily substances that must be subsequently reformed in order to produce significant amounts of hydrogen. Typical operating temperatures are around 850°C. An overall energy conversion efficiency of around 50% is attainable, with considerable variations. Alternative concepts use membranes to separate the gases produced, and many reactor types uses catalysts to help the processes to proceed in the desired direction, notably at a lower temperature (down to some 500°C, 930°F).

Environmental concerns include disposal of associated tars and ashes, particularly for the fluidized bed reactors, where these substances must be separated from the flue gas stream (in contrast to the pyrolysis plants, where most tar and ash deposits at the bottom of the reactor). Concerns over biomass transportation are similar to those mentioned previously for fermentation, and a positive fertilizer effect can also in many cases be derived from the gasification residues.

Methanol production from natural gas feed consists of three steps: (1) synthesis gas generation—in the case of natural gas feed, synthesis gas production consists of converting methane (CH_4) into carbon monoxide (CO) and hydrogen (H_2) via steam reforming; (2) synthesis gas upgrading—primarily removal of CO_2, plus any contaminants such as sulfur; and (3) methanol synthesis and purification—reacting the carbon monoxide, hydrogen, and steam over a catalyst in the presence of a small amount of CO_2 and at elevated temperature and pressure. The methanol synthesis is an equilibrium reaction, and excess reactants must be recycled to optimize yields.

Methods proposed for the production of methanol from biomass involve the conversion of the biomass to a suitable synthesis gas, after which processing steps are very similar to those developed for methanol from natural gas. However, the gasification techniques proposed are still at a relatively early stage of development using biomass feed, and the methods are based on similar techniques used widely already with natural gas as feed.

14.3.4.2 Gasifiers

The gasifier is the heart of the gasification process and each gasifier is designed to process the fuel in a variety of ways consistent with the type of fuel, the end use of the gas, the size of the process, and the source of oxygen. The oxygen may be introduced as a pure gas or may come from air or steam.

The gasification process is amenable to a variety of biomass feedstocks such as waste rice hulls, wood waste, grass, and the dedicated energy crops. Gasification is a clean process with few air emissions and, when crops are used as the feedstock, little or no ash. However, before biomass can be gasified, it must be pretreated to meet the processing constraints of the gasifier, which typically involves size reduction, and drying to keep moisture contents below specific levels. Thereafter, biomass gasification involves heating biomass in the presence of low levels of oxygen (i.e., less than required for complete combustion to carbon dioxide and water).

The composition of the gas stream is influenced by the operating conditions for the gasifier, with some gasification processes more suited than others to producing a gas for methanol production. In particular, simple gasification with air creates a synthesis gas stream that is diluted with large quantities of nitrogen. This nitrogen is detrimental to subsequent processing to methanol and so techniques using indirect gasification or an oxygen feed are preferred. For large-scale gasification, pressurized systems are considered to be more economic than atmospheric systems.

The simplest type of gasifier is the fixed bed counter current gasifier. The biomass is fed at the top of the reactor and moves downward as a result of the conversion of the biomass and the removal of ashes. The air intake is at the bottom and the gas leaves at the top. The biomass moves in counter current to the gas flow and passes through the drying zone, the distillation zone, reduction zone, and the oxidation zone. The major advantages of this type of gasifier are its simplicity, high charcoal burn-out, and internal heat exchange leading to low gas exit temperatures and high gasification efficiency. In this way, also fuels with high moisture content (up to 50% w/w) can be used. Major drawbacks are the high amounts of tar and pyrolysis products, because the pyrolysis gas is not lead through the oxidation zone. This is of minor importance if the gas is used for direct heat applications, in which the tar is burnt. In case the gas is used for engines, gas cleaning is required, resulting in problems of tar-containing condensates.

In the conventional downdraft gasifier (sometime called the coflow gasifier), biomass is fed at the top of the reactor, and air intake is at the top or from the sides. The gas leaves at the bottom of the reactor, so the fuel and the gas move in the same direction. The pyrolysis gasses are led through the oxidation zone (with high temperatures) and/or more-or-less burnt or cracked. Therefore the producer gas has low tar content and is suitable for engine applications. In practice, however, a tar-free gas is seldom if ever achieved over the whole operating range of the equipment. Because of the lower level of organic components in the condensate, downdraft gasifiers suffer less from environmental objections than updraft gasifiers.

Successful operation of a downdraft gasifier requires drying the biomass fuel to a moisture content of less than 20%. The advantage of the downdraft design is the very low tar content of the producer gas. However, disadvantages of the downdraft gasifier are (1) the high amounts of ash and dust particles in the gas; (2) the inability to operate on a number of unprocessed fuels, often pelletization or briquetting of the biomass is necessary; (3) the outlet gas has a high temperature leading to a lower gasification efficiency; and (4) the moisture content of the biomass must be less than 25% w/w.

A more recent development is the open-core gasifier design for gasification of small-sized biomass with high-ash content and the gas is not always tar-free. In the open-core gasifier, the air is sucked over the whole cross section from the top of the bed. This facilitates better oxygen distribution, since the oxygen will be consumed over the whole cross section, so that the solid bed temperature will not reach the local extremes (hot spots) observed in the oxidation zone of conventional gasifiers due to poor heat transfer. Moreover, the air nozzles in conventional gasifiers generate caves and create obstacles that may obstruct solid flow especially for solids of low bulk like rice husk. On the other hand, the entry of air through the top of the bed creates a downward flow of the pyrolysis gases and transports the tar products to the combustion zone. Thus flow problems due to the caking of rice husk caused by back mixing of tar are avoided.

The gasification process could play a significant role in meeting the goal of GHG mitigation. It is likely that both in the transition phase to a hydrogen economy and in the steady state, a significant fraction of hydrogen might be derived from domestically abundant crops. In addition, the cofiring applications of crops (and other biomass) biomass with coal, biomass can provide up to 15% of the total energy input of the fuel mixture. Such concepts address GHG mitigation by cofiring biomass and coal to offset the losses of carbon dioxide to the atmosphere that are inherent in coal combustion processes (even with the best engineered capture and storage of carbon). Since growth of biomass fixes atmospheric carbon, its combustion leads to no net addition of atmospheric carbon dioxide even if vented. Thus cofiring of crops or crop residues (or other biomass with coal) in an efficient coal gasification process, affording the opportunity for capture and storage of carbon dioxide, could lead to a net reduction of atmospheric carbon dioxide. Cheaper, though less plentiful, biomass residue could supplant crops as gasifier feedstock leading to a less significant impact on the environment than would farming bioenergy crops.

Along similar lines, willow biomass crops have been shown to be a good fuel for farm-based power production using advanced gasification technology (Pian et al., 2006). The fuel gas can be used for generating electricity, using microturbines modified to operate on low-Btu gas, or for other farm energy needs. Willow biomass was found to make a fuel for ash-rejection gasifiers with a predicted net gasification efficiency of approximately 85%. Analysis showed that developing a method to cogasify willow with various amounts of low-cost wastes, such as dairy farm animal waste, can be an excellent way to reduce the fuel cost, to increase the overall fuel availability and help work around problems resulting from seasonal availability of bioenergy crops. Cogasification of dairy farm wastes along with willow offers an economical way to dispose of the wastes and manage nutrient flows on a dairy farm.

Agricultural residues can be divided into two groups: (1) crop residues and (2) agroindustrial residues. Crop residues are plant materials left behind in the farm after removal of the main crop produce. The remaining materials could be of different sizes, shapes, forms, and densities such as straw, stalks, sticks, leaves, haulms, fibrous materials, roots, branches, and twigs.

Due to high energy content, straw is one of the best crop residues for solid biofuels. However, straw has several disadvantages—it has a higher ash content, which results in lower calorific value. In order to improve its bulk density, the straw is generally baled before transportation. Straw burning requires a specific technology. There are four basic types of straw burners: those that accept shredded, loose straw; burners that use densified straw products such as pellets, briquettes or cubes, and straw logs; small, square bale burners and round bale burners. To be suitable for heat and electricity production, straw should not have a large content of moisture, preferably not more than 20% as the moisture reduces the boiler efficiency. Also straw color as well as straw chemistry should be considered before burning as it indicates the quality of the straw.

The agroindustrial residues are by-products of the postharvest processes of crops such as cleaning, threshing, sieving, and crushing. These could be in the form of husk, dust, and straw. Furthermore, the quantity of agricultural residues produced differs from crop to crop and is affected by soil type and irrigation conditions. Production of agricultural residues is directly related to the corresponding crop production and ratio between the main crop produce and the residues, which varies from crop to crop and, at times, with the variety of the seeds in one crop itself. Thus for known amounts of crop production, it may be possible to estimate the amounts of agricultural residues produced using the residue to crop ratio.

Most crop or agricultural residues are not found throughout the year but are available only at the time of harvest. The amount available depends upon the harvesting time, storage-related characteristics, and the storage facility.

14.3.4.3 Synthesis gas

Synthesis gas (also called syngas) is the name given to a gas mixture that contains varying amounts of carbon monoxide and hydrogen generated by the gasification of a carbon-containing fuel to a gaseous product with a heating value. Examples include steam reforming of natural gas or liquid hydrocarbon derivatives to produce hydrogen, the gasification of coal and in some types of waste-to-energy gasification facilities. The name comes from their use as intermediates in creating synthetic natural gas and for producing ammonia or methanol. Synthesis gas is also used in the Fischer–Tropsch synthesis for the production of synthetic crude oil for use as a fuel or lubricant (Chapter 12: Synthesis Gas and the Fischer–Tropsch Process).

Synthesis gas, a product of biomass gasification, is composed primarily of hydrogen and carbon monoxide, also called biosyngas (Cobb, 2007). The production of

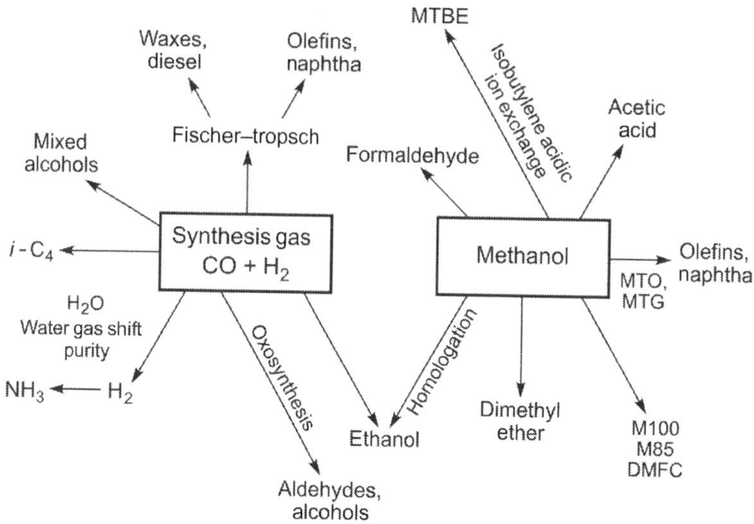

Figure 14.3 Potential products from heavy feedstock gasification.

high-quality synthesis gas from biomass is the feedstock for BTL production (Fig. 14.3). Also, the synthesis gas generated from biomass is not the same as biogas. Biogas is a clean and renewable form of energy generated from biomass that could very well substitute for conventional sources of energy. The gas is generally composed of methane (55%–65% v/v), carbon dioxide (35%–45% v/v), nitrogen (0%–3% v/v), hydrogen (0%–1% v/v), and hydrogen sulfide (0%–1% v/v).

The production of synthesis gas through pyrolysis is accompanied by the generation of char, which can then be gasified to provide process heat and energy for the thermochemical platform. A variety of commercial-scale processes exist to transform fossil fuels such as natural gas, crude oil, and coal into liquid fuels, including Fischer–Tropsch fuels. However, the use of biomass instead of fossil fuels changes the composition of synthesis gas, creating a more heterogeneous intermediate product and increasing the difficulty in downstream catalysis.

The calorific value (heat content) of the gas is the prime factor for power generation—the higher the value, the better. Hence, the availability in the gas of any compounds that increase calorific value is generally welcomed—product gas that contains carbon monoxide (CO), hydrogen (H_2), various hydrocarbon derivatives [methane (CH_4), ethylene (C_2H_4), ethane (C_2H_6), tars, and chars]. The presence of inert components [water (H_2O), carbon dioxide (CO_2), and nitrogen (N_2)] is also acceptable, provided it is kept within certain limits.

The production of synthesis gas from biomass is indeed the novel component in the gas-to-liquids concept—obtaining synthesis gas from fossil raw materials (natural gas, crude oil, and coal) is a relatively mature technology. For example, the Fischer–Tropsch process can be used to convert synthesis gas to liquid fuels needed for transportation (Chapter 12: Synthesis Gas and the Fischer–Tropsch

Process) and a variety of other catalytic processes can turn synthesis gas into a myriad of chemicals or other potential fuels or products (Speight, 2019a).

14.3.5 Pyrolysis

Pyrolysis is a medium temperature method that produces gas, oil, and char from crops which can then be further processed into useful fuels or feedstock. Wood and many other similar types of biomass that contain lignin and cellulose (agricultural wastes, cotton gin waste, wood wastes, peanut hulls, etc.) can be converted through a thermochemical process, such as pyrolysis, into solid, liquid, or gaseous fuels. Pyrolysis, used to produce charcoal since the dawn of civilization, is still the most common thermochemical conversion of biomass to commercial fuel.

During pyrolysis, biomass is heated in the absence of air and breaks down into a complex mixture of liquids, gases, and a residual char. If wood is used as the feedstock, the residual char is what is commonly known as charcoal. With more modern technologies, pyrolysis can be carried out under a variety of conditions to capture all the components and to maximize the output of the desired product be it char, liquid, or gas.

Pyrolysis is often considered to be the gasification of biomass in the absence of oxygen. However, the chemistry of each process may differ significantly. In general, biomass does not gasify as easily as coal, and it produces other hydrocarbon compounds in the gas mixture exiting the gasifier; this is especially true when no oxygen is used. As a result, typically an extra step must be taken to reform these hydrocarbon derivatives with a catalyst to yield a clean synthesis gas mixture of hydrogen, carbon monoxide, and carbon dioxide.

Fast pyrolysis is a thermal decomposition process that occurs at moderate temperatures with a high heat-transfer rate to the biomass particles and a short hot vapor residence time in the reaction zone. Several reactor configurations have been shown to assure this condition and to achieve yields of liquid product as high as 75% based on the starting dry biomass weight. They include bubbling fluid beds, circulating and transported beds, cyclonic reactors, and ablative reactors. Fast pyrolysis of biomass produces a liquid product, pyrolysis oil, or biooil that can be readily stored and transported. Pyrolysis oil is a renewable liquid fuel and can also be used for production of chemicals.

Direct hydrothermal liquefaction involves converting biomass to an oily liquid by contacting the biomass with water at elevated temperatures (300°C–350°C) with sufficient pressure to maintain the water primarily in the liquid phase (12–20 MPa) for residence times up to 30 minutes. Alkali may be added to promote organic conversion. The primary product is an organic liquid with reduced oxygen content (approximately 10%), and the primary by-product is water-containing soluble organic compounds (http://www1.eere.energy.gov/biomass/pyrolysis.html).

The importance of the provisions for the supply of feedstocks as crops and other biomass is often underestimated since it is assumed that the supplies are inexhaustible. While this may be true over the long term, short-term supply of feedstocks can be as much as risk as any venture.

14.3.6 Transesterification

The transesterification process is a means of biodiesel production (Marchetti et al., 2005) in which glycerin is separated from the fat or vegetable oil. The process leaves behind two products: (1) methyl esters that are the components of biodiesel and (2) glycerin that is a valuable by-product usually sold to be used in soaps and other products.

Biodiesel (fatty acid methyl esters; FAME) is a notable alternative to the widely used crude oil–derived diesel fuel, since it can be generated by domestic natural sources such as soybeans, rapeseeds, coconuts, and even recycled cooking oil and thus reduces dependence on diminishing crude oil fuel from foreign sources. In addition, because biodiesel is largely made from vegetable oils, it reduces lifecycle GHG emissions by as much as 78% (Ban-Weiss et al., 2007).

Vegetable oils and animal fats belong to an extensive family of chemicals called lipids. Lipids are bioproducts from the metabolism of living creatures. As a result, they can be found widely distributed in nature. Their biofunctions are diverse, but they are most known for their energy storage capacity. Most lipids can easily dissolve in common organic solvents, meaning that they are hydrophobic. If a lipid is solid at 25°C, it is classified as fat; otherwise, it is oil. Typically, fats are produced by animals and oils by plants, but both are mainly made of triglyceride (TG) molecules, which are tri-esters of glycerol (a triol) and free fatty acids (long alkyl chain carboxylic acids). Other glyceride species, such as diglycerides and monoglycerides, are obtained from TGs by the substitution of one and two fatty acid moieties, respectively, with hydroxyl groups (Lotero et al., 2006).

14.3.6.1 Feedstocks

Soybeans are widely used as a source of biodiesel. However, a variety of oils can be used to produce biodiesel. These include (1) virgin oil, (2) waste vegetable oil, and (3) animal fats.

In terms of a virgin oil feedstock, rapeseed oil and soybean oil are most commonly used, soybean oil alone accounting for approximately 90% of all fuel stocks; it also can be obtained from field pennycress and *Jatropha*, while other crops such as mustard, flax, sunflower, canola, pal oil, hemp, and even algae have been used. Waste vegetable oil is any oil produced from the source that has been used and cannot be used again for the original purpose. Animal fats include tallow, lard, yellow grease, chicken fat, and the by-products of the production of omega-3 fatty acids from fish oil. Many advocates suggest that waste vegetable oil is the best source of oil to produce biodiesel. However, the available supply is drastically less than the amount of crude oil–based fuel that is burned for transportation and home heating in the world.

14.3.6.2 Transesterification

Transesterification (alcoholysis) is the conversion of triacylglycerol lipids by alcohols to alkyl esters without first isolating the free fatty acids (May, 2004). Transesterification of a vegetable oil was conducted as early as 1853, many years before the first diesel engine became functional. The prime model of Rudolf Diesel, a single 10 ft (3 m) iron cylinder with a flywheel at its base, ran on its own power for the first time in Augsburg, Germany, on August 10, 1893. Diesel later demonstrated his engine and received the Grand Prix (highest prize) at the World Fair in Paris, France in 1900. This engine stood as an example of the vision of Rudolf Diesel because it was powered by peanut oil, which is a biofuel, though not biodiesel, since it was not transesterified. He believed that the utilization of biomass fuel was the real future of his engine and that the use of vegetable oils for engine fuels would become as important as crude oil and the coal-tar products of the present time.

The purpose of transesterification of vegetable oils to their methyl esters (biodiesel) process is to lower the viscosity of the oil. The transesterification reaction is affected by alcohol type, molar ratio of glycerides to alcohol, type and amount of catalyst, reaction temperature, reaction time and free fatty acids, and water content of vegetable oils or animal fats. The transesterification reaction proceeds with or without a catalyst by using primary or secondary monohydric aliphatic alcohols having 1–8 carbon atoms as follows (Demirbaş, 2006a,b):

$$\text{Triglycerides} + \text{monohydric alcohol} \leftrightarrow \text{glycerin} + \text{mono-alkyl esters}$$

Generally, the reaction temperature near the boiling point of the alcohol is recommended (Çanakçı and Özsezen, 2005). The reactions take place at low temperatures (approximately 65°C) and at modest pressures (2 atm, 1 atm = 14.7 psi = 101.325 kPa). Biodiesel is further purified by washing and evaporation to remove any remaining methanol. The oil (87%), alcohol (9%), and catalyst (1%) are the inputs in the production of biodiesel (86%), the main output (Lucia et al., 2006). Pretreatment is not required if the reaction is carried out under high pressure (9000 kPa) and high temperature (240°C), where simultaneous esterification and transesterification take place with maximum yield obtained at temperatures ranging from 60°C to 80°C at a molar ratio of 6:1 (Barnwal and Sharma, 2005). The alcohols employed in the transesterification are generally short-chain alcohols such as methanol, ethanol, propanol, and butanol. It was reported that when transesterification of soybean oil using methanol, ethanol, and butanol was performed, 96%–98% of ester could be obtained after 1 hour (Dmytryshyn et al., 2004).

14.3.6.3 Catalytic transesterification

Transesterification reactions can be catalyzed by alkalis (Korytkowska et al., 2001; Dmytryshyn et al., 2004; Stavarache et al., 2005; Varghaa and Truterb, 2005;

Meher et al., 2006a), acids (Lee et al., 2000; Goff et al., 2004; Lopez et al., 2005; Liu et al., 2006), or enzymes (Watanabe et al., 2000, 2002; Ghanem, 2003; Reyes-Duarte et al., 2005; Royon et al., 2007; Shah and Gupta, 2007; Bernardes et al., 2007). The catalytic transesterification of vegetable oils with methanol is an important industrial method used in biodiesel synthesis. Also known as methanolysis, this reaction is well studied and established using acids or alkalis, such as sulfuric acid or sodium hydroxide as catalysts. However, these catalytic systems are less active or completely inactive for long-chain alcohols. Usually, industries use sodium or potassium hydroxide or sodium or potassium methoxide as catalyst, since they are relatively cheap and quite active for this reaction (Macedo et al., 2006). Enzyme-catalyzed procedures, using lipase as catalyst, do not produce side reactions, but the lipases are very expensive for industrial scale production, and a three-step process was required to achieve a 95% conversion. The acid-catalyzed process is useful when a high amount of free acids are present in the vegetable oil, but the reaction time is very long (48–96 hours), even at the boiling point of the alcohol, and a high molar ratio of alcohol was needed (20:1 wt/wt to the oil) (Stavarache et al., 2005).

The transesterification process is catalyzed by alkaline metal alkoxides and hydroxides, as well as sodium or potassium carbonates. Alkali-catalyzed transesterification with short-chain alcohols, for example, generates high yields of methyl esters in short reaction times (Jeong and Park, 1996). The alkaline catalysts show high performance for obtaining vegetable oils with high quality, but a question often arises, that is, the oils contain significant amounts of free fatty acids which cannot be converted into biodiesels but to a lot of soap (Furuta et al., 2004). These free fatty acids react with the alkaline catalyst to produce soaps that inhibit the separation of the biodiesel, glycerin, and wash water (Çanakçı and Van Gerpen, 2003). TGs are readily transesterified in a batch operation in the presence of alkaline catalyst at atmospheric pressure and at a temperature of approximately 60°C–70°C with an excess of methanol (Srivastava and Prasad, 2000). It often takes at least several hours to ensure that the alkali (NaOH or KOH) catalytic transesterification reaction is complete. Moreover, removal of these catalysts is technically difficult and brings extra cost to the final product (Demirbaş, 2002, 2003). Nevertheless, they are a good alternative since they can give the same high conversions of vegetable oils just by increasing the catalyst concentration to 1 or 2 mol.%. Alkaline metal alkoxides (as CH_3ONa for the methanolysis) are the most active catalysts, since they give very high yields (>98%) in short reaction times (30 minutes) even if they are applied at low molar concentrations (0.5 mol.%) (Schuchardta et al., 1998).

The transesterification process is catalyzed by sulfuric, hydrochloric, and organic sulfonic acids. In general, acid-catalyzed reactions are performed at high alcohol-to-oil molar ratios, low-to-moderate temperatures and pressures, and high acid catalyst concentrations (Lotero et al., 2006). These catalysts give very high yields in alkyl esters but these reactions are slow, requiring typically temperature above 100°C and more than 3 hours to complete the conversion (Meher et al., 2006b). Studies of the acid-catalyzed system have been very limited in number.

No commercial biodiesel plants to date have been reported to use the acid-catalyzed process. Despite its relatively slow reaction rate, the acid-catalyzed process offers benefits with respect to its independence from free fatty acid content and the consequent absence of a pretreatment step. These advantages favor the use of the acid-catalyzed process when using waste cooking oil as the raw material (Zhang et al., 2003).

Enzyme (such as lipase)-catalyzed reactions have advantages over traditional chemical-catalyzed reactions: the generation of no by-products, easy product recovery, mild reaction conditions, and catalyst recycling. Also, enzymatic reactions are insensitive to free fatty acids and water content in waste cooking oil (Kulkarni and Dalai, 2006). As for the enzyme-catalyzed system, it requires a much longer reaction time than the other two systems (Zhang et al., 2003). The enzyme reactions are highly specific and chemically clean. Because the alcohol can be inhibitory to the enzyme, a typical strategy is to feed the alcohol into the reactor in three steps of 1:1 mole ratio each. The reactions are very slow, with a three-step sequence requiring from 4 to 40 hours, or more. The reaction conditions are modest, from 35°C to 45°C (Van Gerpen et al., 2004). The main problem of the enzyme-catalyzed process is the high cost of the lipases used as catalyst (Royon et al., 2007).

Synthesis of biodiesel using enzymes such as *Candida antarctica*, *Candida rugosa*, *Pseudomonas cepacia*, immobilized lipase (Lipozyme RMIM), *Pseudomonas* sp., and *Rhizomucor miehei* is well reported in the literature. The best yield (98% w/w) was obtained by using *P. cepacia* lipase immobilized on celite at 50°C in the presence of 4%–5% (w/w) water in 8 hours (Shah and Gupta, 2007).

14.3.6.4 Noncatalytic supercritical methanol transesterification

The transesterification of TGs by supercritical methanol (SCM), ethanol, propanol, and butanol has proved to be the most promising process. Recently, a catalysts-free method was developed for biodiesel production by employing SCM (Saka and Kusdiana, 2001). The supercritical treatment at 350°C, 43 MPa, and 240 seconds with a molar ratio of 42 in methanol is the optimum condition for transesterification of rapeseed oil to biodiesel fuel (Kusdiana and Saka, 2004a).

To achieve more moderate reaction conditions, further effort was made through the two-step preparation (Kusdiana and Saka, 2004b). In this method, oils/fats are, first, treated in subcritical water for hydrolysis reaction to produce fatty acids. After hydrolysis the reaction mixture is separated into oil phase and water phase by decantation. The oil phase (upper portion) is mainly fatty acids, while the water phase (lower portion) contains glycerol in water. The separated oil phase is then mixed with methanol and treated at supercritical condition to produce FAME thorough methyl esterification. After removing unreacted methanol and water produced in reaction, FAME can be obtained as biodiesel. Therefore in this process, methyl esterification is the main reaction for FAME formation, while in the one-step method, transesterification is the major one (Pinto et al., 2005; Saka and Minami, 2006).

Reaction by SCM has some advantages: (1) glycerides and free fatty acids are reacted with equivalent rates, (2) the homogeneous phase eliminates diffusive problems, (3) the process tolerates great percentages of water in the feedstock catalytic process that requires the periodical removal of water in the feedstock or in the intermediate stage to prevent catalyst deactivation, (4) the catalyst removal step is eliminated, and (5) if high methanol/oil ratios are used, total conversion of the oil can be achieved in a few minutes (Vera et al., 2005).

Some disadvantages of the one-stage supercritical method are clear: (1) it operates at very high pressures, (2) the high temperatures bring along proportionally high heating and cooling costs, (3) high methanol/oil ratios—usually on the order of 42—involve high costs for the evaporation of the unreacted methanol, and (4) the process as posed to date does not explain how to reduce free glycerol to less than 0.02% as established in the ASTM D6584 or other equivalent international standards.

14.3.6.5 Process parameters

The main factors affecting transesterification are the molar ratio of glycerides to alcohol, catalyst, reaction temperature and pressure, reaction time and the contents of free fatty acids, and water in oils. The free fatty acids and moisture content are key parameters for determining the viability of the vegetable oil transesterification process (Meher et al., 2006b). In the transesterification, free fatty acids and water always produce negative effects, since the presence of free fatty acids and water causes soap formation, consumes catalyst, and reduces catalyst effectiveness, all of which result in a low conversion (Demirbaş and Karslıoğlu, 2007). These free fatty acids react with the alkaline catalyst to produce soaps that inhibit the separation of the biodiesel, glycerin, and wash water (Çanakçı and Van Gerpen, 2003; Pinto et al., 2005). To carry the base-catalyzed reaction to completion, a free fatty acid value lower than 3% is needed (Meher et al., 2006b). The presence of water has a greater negative effect on transesterification than that of the free fatty acids. In the transesterification of beef tallow catalyzed by sodium hydroxide (NaOH) in the presence of free fatty acids and water, the water and free fatty acid contents must be maintained at specified levels (Ma et al., 1998; Kusdiana and Saka, 2004a; Bala, 2005).

Transesterification can occur in different temperatures depending on the type of oil employed (Ma and Hanna, 1999). A few works reported the reaction at room temperature (Marinetti, 1962, 1966; Graboski and McCormick, 1998; Encinar et al., 2002). Generally, the conversion rate increases with reaction time. The transesterification of rice bran oil with methanol was studied at molar ratios of 4:1, 5:1, and 6:1 (Gupta et al., 2007). At molar ratios of 4:1 and 5:1, there was significant increase in yield when the reaction time was increased from 4 to 6 hours. Among the three molar ratios studied, ratio 6:1 gave the best results.

One of the most important factors that affect the yield of ester is the molar ratio of alcohol to TG. Although the stoichiometric molar ratio of methanol to TG for transesterification is 3:1, higher molar ratios are used to enhance the solubility and to increase the contact between the TG and alcohol molecules (Noureddini et al., 1998). In addition, investigation of the effect of molar ratio on the transesterification of sunflower oil with methanol showed that the molar ratio varied from 6:1 to 1:1 and concluded that 98% conversion to ester was obtained at a molar ratio of 6:1 (Freedman et al., 1986). Another important variable affecting the yield of methyl ester is the type of alcohol to TG. In general, short-chain alcohols such as methanol, ethanol, propanol, and butanol can be used in the transesterification reaction to obtain high methyl ester yields.

Catalysts used for the transesterification of TGs are classified as alkali, acid, and enzyme. Alkali-catalyzed transesterification is much faster than acid-catalyzed transesterification and is most often used commercially (Ma and Hanna, 1999) and, quite often, for the base-catalyzed transesterification the best yields were obtained when the catalyst was used in small concentration, that is, 0.5% wt/wt of oil (Stavarache et al., 2005). On the other hand, data show that during the production of free and bound ethyl ester (FAEE) from castor oil, hydrochloric acid is much more effective than sodium hydroxide at higher reaction temperatures (Meneghetti et al., 2006).

14.4 Benefits

The projections for the continued use of fossil fuels indicate that there will be at least another five decades of fossil fuel use (especially in the cases of coal and crude oil) before biomass and other forms of alternate energy take hold and are able to provide sufficient energy not only to subsidize the production of energy from fossil fuels but also to replace the fossil fuels. Furthermore, estimations that the era of fossil fuels (natural gas, crude oil, coal, and biomass, and carbonaceous wastes) will be almost over when the cumulative production of the fossil resources reaches 85% of their initial total reserves may or may not have some merit. In fact, the relative scarcity (compared to a few decades ago) of crude oil was real but it seems that the remaining reserves make it likely that there will be an adequate supply of energy for several decades. The environmental issues are very real and require serious and continuous attention.

Technologies that ameliorate the effects of fossil fuels combustion on acid rain deposition, urban air pollution, and global warming must be pursued vigorously (Vallero, 2008). There is a challenge that must not be ignored and the effects of acid rain in soil and water leave no doubt in relation to the need to control its causes. Indeed, recognition of the need to address these issues is the driving force behind recent energy strategies as well as a variety of research and development programs.

While regulations on the GHG carbon dioxide (CO_2) would be an immediate hurdle to deployment of coal plants, gasification plants are in the best position compared to other coal-based alternatives to capture carbon dioxide. However, with the continued uncertainty of carbon dioxide regulation, there is industry reluctance to make large investments in projects with high emissions of carbon dioxide since a cost-effective solution for reducing such emissions is not yet available. Nevertheless, the reduction in GHG emissions can be an enhancing factor for gasification in the long run because the carbon dioxide from a gasification plant is more amenable to capture.

As new technology is developed, emissions may be reduced by repowering in which aging equipment is replaced by more advanced and efficient substitutes. Such repowering might, for example, involve an exchange in which an aging unit is exchanged for a newer combustion chamber, such as the atmospheric fluidized-bed combustor or the pressurized fluidized-bed combustor.

Indeed, recognition of the production of these atmospheric pollutants in considerable quantities every year has led to the institution of national emission standards for many pollutants. Using sulfur dioxide as the example, the various standards are not only very specific but will also become more stringent with the passage of time. Atmospheric pollution is being taken very seriously and there is also the threat, or promise, of heavy fines and/or jail terms for any pollution-minded miscreants who seek to flaunt the laws (Vallero, 2008). Nevertheless, a trend to the increased use of fossil fuels will require more stringent approaches to environmental protection issues than we have ever known at any time in the past. The need to protect the environment is strong.

The careless combustion of fossil fuels can account for the large majority of the sulfur oxides and nitrogen oxides released to the atmosphere. Whichever technologies succeed in reducing the amounts of these gases in the atmosphere should also succeed in reducing the amounts of urban smog, those notorious brown and gray clouds that are easily recognizable at some considerable distances from urban areas, not only by their appearance but also by their odor.

$$SO_2 + H_2O \rightarrow H_2SO_4(\text{sulfurous acid})$$

$$2SO_2 + O_2 \rightarrow 2SO_3$$

$$SO_3 + H_2O \rightarrow H_2SO_4(\text{sulfuric acid})$$

$$2NO + H_2O \rightarrow HNO_2 + HNO_3(\text{nitrous acid} + \text{nitric acid})$$

$$2NO + O_2 \rightarrow 2NO_2$$

$$NO_2 + H_2O \rightarrow HNO_3(\text{nitric acid})$$

Current awareness of these issues by a variety of levels of government has resulted, in the United States, of the institution of the Clean Fossil fuels Program to facilitate the development of pollution abatement technologies. And it has led to successful partnerships between government and industry (United States Department of Energy, 1993). In addition, there is the potential that new laws, such as the passage in 1990 of the Clean Air Act Amendments in the United States

(United States Congress, 1990; Stensvaag, 1991), will be a positive factor and supportive of the controlled clean use of fossil fuels. However, there will be a cost but industry is supportive of the measure and confident that the goals can be met.

Besides fuel and product flexibility, gasification-based systems offer significant environmental advantages over competing technologies, particularly coal-to-electricity combustion systems. Gasification plants can readily capture carbon dioxide, the leading GHG, much more easily and efficiently than coal-fired power plants. In many instances, this carbon dioxide can be sold, creating additional value from the gasification process.

Carbon dioxide captured during the gasification process can be used to help recover oil from otherwise depleted oil fields. The Dakota Gasification plant in Beulah, North Dakota, captures its carbon dioxide while making substitute natural gas and sells it for enhanced oil recovery. Since 2000 this plant has captured and sent the carbon dioxide via pipeline to EnCana's Weyburn oil fields in Saskatchewan, Canada, where it is used for enhanced oil recovery. More than five million tons of carbon dioxide have been sequestered.

It is very likely that a biorefinery, by producing multiple products, can take advantage of the differences in biomass components and intermediates and maximize the value derived from the biomass feedstock. A biorefinery might, for example, produce one or several low-volume, but high-value, chemical products and a low-value, but high-volume, liquid transportation fuel while generating electricity and process heat for its own use and perhaps enough for sale of electricity. The high-value products enhance profitability, the high-volume fuel helps meet national energy needs, and the power production reduces costs and avoids GHG emissions.

In fact, a biorefinery can be considered to be an integral unit that can accept various biological nonfood feedstocks and convert them into a range of useful products, including chemicals, energy, and materials. Bio-based products are a great curiosity question of whether it gets to replace fossil-based products. Biomass, a renewable source of carbon, is guaranteed a place in the new energy portfolio for the foreseeable future.

Limitations in optimal plant size are caused by feedstock transportation needs: larger plants demanding larger distances to fulfill feedstock requirements year round. Long transportation distances are especially harmful for feedstocks with high concentrations of water (transport of which is expensive but not effective), minerals, or organic components (required to maintain local soil quality). In contrast to fossil feedstocks that can generally be recovered following the exact timing of its demand most biomass types (wood being the exception) are harvested only during a short period of the year. Year round biomass availability requires expensive storage facilities, while crops with high water concentrations cannot be stored over long periods.

In many traditional biorefineries, the very large volumes that are processed often result in the duplication of equipment because larger equipment cannot be built because of physical limitations. The choice may be to use unit operations that show only small economy of scale benefits such as the usage of membrane processes, instead of evaporation using heat for concentration purposes. Another strategy could be to convert the desired components in intermediates that can be conveniently recovered or even to leave the component in the process water and subsequently convert these components to biogas that can be used on-site or fed to the grid.

Until sufficient quantities of biofeedstocks are guaranteed to operate a large biorefinery, constructing a modular biorefinery based on two (or more) different conversion platforms to accommodate different feedstocks and to promote different product slates and the association of such a facility with a crude oil-based refining operation may be the most economical near-term reality.

In summary the biorefinery is emerging form the conceptual stage and there are strong indications that the concept should be formally adopted as the main avenue for biofuels as well as the added value by-products. The use of the by-products of a biorefinery is expected to rise in the use of chemicals and materials industries.

References

Bala, B.K., 2005. Studies on biodiesels from transformation of vegetable oils for diesel engines. Energy Educ. Sci. Technol. 15, 1–43.

Ban-Weiss, G.A., Chen, J.Y., Buchholz, B.A., Dibble, R.W., 2007. A numerical investigation into the anomalous slight NO_x increase when burning biodiesel—a new (old) theory. Fuel Process. Technol. 88, 659–667.

Barnwal, B.K., Sharma, M.P., 2005. Prospects of biodiesel production from vegetable oils in India. Renewable Sustainable Energy Rev. 9, 363–378.

Bernardes, O.L., Bevilaqua, J.V., Leal, M.C.M.R., Freire, D.M.G., Langone, M.A.P., 2007. Biodiesel fuel production by the transesterification reaction of soybean oil using immobilized lipase. Appl. Biochem. Biotechnol. 137–140, 105–114.

Çanakçı, M., Özsezen, A.N., 2005. Evaluating waste cooking oils as alternative diesel fuel. Gazi Univ. J. Sci. 18, 81–91.

Çanakçı, M., Van Gerpen, J., 2003. A pilot plant to produce biodiesel from high free fatty acid feedstocks. Trans. ASAE 46, 945–955.

Cobb, J.T. Jr., 2007. Production of synthesis gas by biomass gasification. In: Proceedings. Spring National Meeting AIChE. 22–26 April 2007, Houston, TX.

Crocker, M., Crofcheck, C., 2006. Reducing national dependence on imported oil, Energeia, Vol. 17. Center for Applied Energy Research, University of Kentucky, Lexington, KY, No. 6.

Demirbaş, A., 2002. Biodiesel from vegetable oils via transesterification in supercritical methanol. Energy Convers. Manage. 43, 2349–2356.

Demirbaş, A., 2003. Biodiesel fuels from vegetable oils via catalytic and non-catalytic supercritical alcohol transesterification and other methods: a survey. Energy Convers. Manage. 44, 2093–2109.

Demirbaş, A., 2006a. Biodiesel from sunflower oil in supercritical methanol with calcium oxide. Energy Convers. Manage. 48, 937–941.

Demirbaş, A., 2006b. Biodiesel production via non-catalytic SCF method and biodiesel fuel characteristics. Energy Convers. Manage. 47, 2271–2282.

Demirbaş, A., Karslıoğlu, S., 2007. Biodiesel production facilities from vegetable oils and animal fats. Energy Sources, A 29, 133–141.

Di Visconte, G.S., Spicer, A., Chuck, C.J., Allen, M.J., 2019. The microalgae biorefinery: a perspective on the current status and future opportunities using genetic modification. Appl. Sci. 9, 4793.

Dmytryshyn, S.L., Dalai, A.K., Chaudhari, S.T., Mishra, H.K., Reaney, M.J., 2004. Synthesis and characterization of vegetable oil derived esters: evaluation for their diesel additive properties. Bioresour. Technol. 92, 55–64.

El-Gendy, N.S., Speight, J.G., 2015. Handbook of Refinery Desulfurization. CRC Press, Taylor & Francis Group, Boca Raton, FL.

Encinar, J.M., Gonzalez, J.F., Rodriguez, J.J., Tejedor, A., 2002. Biodiesel fuels from vegetable oils: transesterification of *Cynara cardunculus* L. oils with ethanol. Energy Fuel 16, 443–450.

Freedman, B., Butterfield, R.O., Pryde, E.H., 1986. Transesterification kinetics of soybean oil. J. Am. Oil Chem. Soc. 63, 1375–1380.

Furuta, S., Matsuhasbi, H., Arata, K., 2004. Biodiesel fuel production with solid superacid catalysis in fixed bed reactor under atmospheric pressure. Catal. Commun. 5, 721–723.

Gary, J.H., Handwerk, G.E., Kaiser, M.J., 2007. Petroleum Refining: Technology and Economics, Fifth ed. CRC Press, Taylor & Francis Group, Boca Raton, FL.

Ghanem, A., 2003. The utility of cyclodextrins in lipase-catalyzed transesterification in organic solvents: enhanced reaction rate and enantioselectivity. Org. Biomol. Chem. 1, 1282–1291.

Goff, M.J., Bauer, N.S., Lopes, S., Sutterlin, W.R., Suppes, G.J., 2004. Acid-catalyzed alcoholysis of soybean oil. J. Am. Oil Chem. Soc. 81, 415–420.

Graboski, M.S., McCormick, R.L., 1998. Combustion of fat and vegetable oil derived fuels in diesel engines. Prog. Energy Combust. Sci. 24, 125–164.

Gupta, P.K., Kumar, R., Panesar, P.S., Thapar, V.K., 2007. Parametric studies on bio-diesel prepared from rice bran oil. Agric. Eng. IX, April.

Hsu, C.S., Robinson, P.R. (Eds.), 2017. Handbook of Petroleum Technology. Springer International Publishing AG, Cham.

Jeong, G.T., Park, D.H., 1996. Batch (one- and two-stage) production of biodiesel fuel from rapeseed oil. Appl. Biochem. Biotechnol. 131, 668–679.

Korytkowska, A., Barszczewska-Rybarek, I., Gibas, M., 2001. Side-reactions in the transesterification of oligoethylene glycols by methacrylates. Des. Monomers Polym. 4, 27–37.

Kulkarni, M.G., Dalai, A.K., 2006. Waste cooking oils—an economical source for biodiesel: a review. Ind. Eng. Chem. Res. 45, 2901–2913.

Kusdiana, D., Saka, S., 2004a. Effects of water on biodiesel fuel production by supercritical methanol treatment. Bioresour. Technol. 91, 289–295.

Kusdiana, D., Saka, S., 2004b. Two-step preparation for catalyst-free biodiesel fuel production: hydrolysis and methyl esterification. Appl. Biochem. Biotechnol. 115, 781–791.

Lee, Y., Park, S.H., Lim, I.T., Han, K., Lee, S.Y., 2000. Preparation of alkyl (R)-(-)-3-hydroxybutyrate by acidic alcoholysis of poly-(R)-(-)-3-hydroxybutyrate. Enzyme Microb. Technol. 27, 33–36.

Liu, Y., Lotero, E., Goodwin Jr., J.G., 2006. Effect of water on the sulfuric acid catalyzed esterification. J. Mol. Catal. A Chem. 245, 132–140.

Lopez, D.E., Goodwin Jr, J.G., Bruce, D.A., Lotero, E., 2005. Transesterification of triacetin with methanol on solid acid and base catalysts. Appl. Catal. A Gen. 295, 97–105.

Lotero, E., Goodwin Jr., J.G., Bruce, D.A., Suwannakarn, K., Liu, Y., Lopez, D.E., 2006. The catalysis of biodiesel synthesis. Catalysis 19, 41–83.

Lucia, L.A., Argyropoulos, D.S., Adamopoulos, L., Gaspar, A.R., 2006. Chemicals and energy from biomass. Can. J. Chem. 8, 960–970.

Ma, F., Hanna, M.A., 1999. Biodiesel production: a review. Bioresour. Technol. 70, 1–15.

Ma, F., Clements, L.D., Hanna, M.A., 1998. The effect of catalyst, free fatty acids, and water on the transesterification of beef tallow. Trans. ASAE 41, 1261–1264.

Macedo, C.C.S., Abreu, F.R., Tavares, A.P., Alves, M.P., Zara, L.F., Rubim, J.C., et al., 2006. New heterogeneous metal-oxides based catalyst for vegetable oil transesterification. J. Braz. Chem. Soc. 17, 1291–1296.

Marchetti, J.M., Miguel, V.U., Errazu, A.F., 2005. Possible methods for biodiesel production. Renewable Sustainable Energy Rev. 11, 1300–1311.

Marinetti, G.V., 1962. Hydrolysis of lecithin with sodium methoxide. Biochemistry 1, 350–353.

Marinetti, G.V., 1966. Low temperature partial alcoholysis of triglycerides. J. Lipid Res. 7, 786–788.

Meher, L.C., Kulkarni, M.G., Dalai, A.K., Naik, S.N., 2006a. Transesterification of Karanja (*Pongamia pinnata*) oil by solid basic catalysts. Eur. J. Lipid Sci. Technol. 108, 389–397.

Meher, L.C., Sagar, D.V., Naik, S.N., 2006b. Technical aspects of biodiesel production by transesterification – a review. Renewable Sustainable Energy Rev. 10, 248–268.

Meneghetti, P.S.M., Meneghetti, M.R., Wolf, C.R., Silva, E.C., Lima, G.E.S., Coimbra, D. A., et al., 2006. Ethanolysis of castor and cottonseed oil: a systematic study using classical catalysts. JAOCS 83, 819–822.

Mokhatab, S., Poe, W.A., Speight, J.G., 2006. Handbook of Natural Gas Transmission and Processing. Elsevier, Amsterdam.

Nichols, N.N., Dien, B.S., Bothast, R.J., Cotta, M.A., 2006. The corn ethanol industry. In: Minteer, S. (Ed.), Alcoholic Fuels. CRC Press, Taylor & Francis Group, Boca Raton, FL, Chapter 4.

Noureddini, H., Harkey, D., Medikonduru, V., 1998. A continuous process for the conversion of vegetable oils into methyl esters of fatty acids. JAOCS 75, 1775–1783.

NREL, 2003. Dollars From Sense. National Renewable Energy Laboratory, Golden, CO. Available from: http://www.nrel.gov/docs/legosti/fy97/20505.pdf.

Parkash, S., 2003. Refining Processes Handbook. Gulf Professional Publishing, Elsevier, Amsterdam.

Pian, C.C.P., Volk, T.A., Abrahamson, L.P., White, E.H., Jarnefeld, J., 2006. Biomass gasification for farm-based power generation applications, Transactions on Ecology and the Environment, 92. Wessex Institute of Technology, Press, p. 267.

Pinto, A.C., Guarieiro, L.N.N., Rezende, M.J.C., Ribeiro, N.M., Torres, E.A., Lopes, W.A., et al., 2005. Biodiesel: an overview. J. Braz. Chem. Soc. 16, 1313–1330.

Reyes-Duarte, D., Lopez-Cortes, N., Ferrer, M., Plou, F., Ballesteros, A., 2005. Parameters affecting productivity in the lipase-catalysed synthesis of sucrose palmitate. Biocatal. Biotransform. 23, 19–27.

Royon, D., Daz, M., Ellenrieder, G., Locatelli, S., 2007. Enzymatic production of biodiesel from cotton seed oil using t-butanol as a solvent. Bioresour. Technol. 98, 648–653.

Ruth, M., 2004. Development of a biorefinery optimization model. Renewable Energy Modeling Series Forecasting the Growth of Wind and Biomass. National Bioenergy Centre, National Renewable Energy Laboratory, Golden, CO, <http://www.epa.gov/cleanenergy/pdf/ruth2_apr20.pdf>.

Saka, S., Kusdiana, D., 2001. Biodiesel fuel from rapeseed oil as prepared in supercritical methanol. Fuel 80, 225–231.

Saka, S., Minami, E., 2006. A novel non-catalytic biodiesel production process by supercritical methanol as NEDO "high efficiency bioenergy conversion project". In: The 2nd Joint International Conference on "Sustainable Energy and Environment (SEE 2006)". 21–23 November, Bangkok.

Schuchardta, U., Serchelia, R., Vargas, R.M., 1998. Transesterification of vegetable oils: a review. J. Braz. Chem. Soc. 9, 199–210.

Shah, S., Gupta, M.N., 2007. Lipase catalyzed preparation of biodiesel from Jatropha oil in a solvent free system. Process. Biochem. 42, 409–414.

Sørensen, B.E., Njakou, S., Blumberga, D., 2006. Gaseous fuels biomass. In: Proceedings. World Renewable Energy Congress IX. WREN, London.

Speight, J.G., 2008. Synthetic Fuels Handbook: Properties, Processes, and Performance. McGraw-Hill, New York.

Speight, J.G., 2011a. The Refinery of the Future. Gulf Professional Publishing, Elsevier, Oxford.

Speight, J.G. (Ed.), 2011b. The Biofuels Handbook. Royal Society of Chemistry, London.

Speight, J.G., 2013a. The Chemistry and Technology of Coal, Third ed. CRC Press, Taylor & Francis Group, Boca Raton, FL.

Speight, J.G., 2013b. Coal-Fired Power Generation Handbook. Scrivener Publishing, Salem, MA.

Speight, J.G., 2014. The Chemistry and Technology of Petroleum, Fifth ed. CRC Press, Taylor & Francis Group, Boca Raton, FL.

Speight, J.G., 2017. Handbook of Petroleum Refining. CRC Press, Taylor & Francis Group, Boca Raton, FL.

Speight, J.G., 2019a. Handbook of Petrochemical Processes. CRC Press, Taylor & Francis Group, Boca Raton, FL.

Speight, J.G., 2019b. Natural Gas: A Basic Handbook, Second ed. Gulf Publishing Company, Elsevier, Cambridge, MA.

Srivastava, A., Prasad, R., 2000. Triglycerides-based diesel fuels. Renewable Sustainable Energy Rev. 4, 111–133.

Stavarache, C., Vinatoru, M., Nishimura, R., Maed, Y., 2005. Fatty acid methyl esters from vegetable oil by means of ultrasonic energy. Ultrason. Sonochem. 12, 367–372.

Stensvaag, J.-M., 1991. Clean Air Act Amendments: Law and Practice. John Wiley and Sons Inc., New York.

Taylor, G., 2008. Biofuels and the refinery concept. Energy Policy 36, 4406–4409.

United States Congress, 1990. Public Law 101–549. An Act to Amend the Clean Air Act to Provide for Attainment and Maintenance of Health protective National Ambient Air Quality Standards, and for Other Purposes. 15 November.

United States Department of Energy, 1993. Clean Fossil Fuels Technology Demonstration Program. DOE/FE-0272. United States Department of Energy, Washington, DC, February.

Vallero, D., 2008. Fundamentals of Air Pollution, Fourth ed. Elsevier, London.

Van Gerpen, J., Shanks, B., Pruszko, R., Clements, D., Knothe, G., 2004. Biodiesel Analytical Methods: August 2002–January 2004. National Renewable Energy Laboratory, NREL/SR-510-36240, July.

Varghaa, V., Truterb, P., 2005. Biodegradable polymers by reactive blending transesterification of thermoplastic starch with poly(vinyl acetate) and poly(vinyl acetate-*co*-butyl acrylate. Eur. Polym. J. 41, 715–726.

Vera, C.R., D'Ippolito, S.A., Pieck, C.L., Parera, J.M., 2005. Production of biodiesel by a two-step supercritical reaction process with adsorption refining. In: 2nd Mercosur

Congress on Chemical Engineering and 4th Mercosur Congress on Process Systems Engineering (ENPROMER-2005). 14–18 August, Rio de Janeiro.

Watanabe, Y., Shimada, Y., Sugihara, A., 2000. Continuous production of biodiesel fuel from vegetable oil using immobilized candida antarctica lipase. J. Am. Oil Chem. Soc. 77, 355–360.

Watanabe, Y., Shimada, Y., Sugihara, A., Tominaga, T., 2002. Conversion of degummed soybean oil to biodiesel fuel with immobilized *Candida antarctica* Lipase. J. Mol. Catal. B Enzym. 17, 151–155.

Wright, L., Boundy, R., Perlack, R., Davis, S., Saulsbury, B., 2006. Biomass Energy Data Book: Edition 1. Office of Planning, Budget and Analysis, Energy Efficiency and Renewable Energy, United States Department of Energy. Contract No. DE-AC05-00OR22725. Oak Ridge National Laboratory, Oak Ridge, TN.

Zhang, Y., Dube, M.A., McLean, D.D., Kates, M., 2003. Biodiesel production from waste cooking oil: 1. Process design and technological assessment. Bioresour. Technol. 89, 1–16.

The refinery of the future and technology integration

15

15.1 Introduction

With the entry into, and the evolution of, the 21st century, crude oil refining technology continues to experience a period of necessary innovation that is driven by the increasing supply of viscous feedstocks heavy oils with decreasing quality and the fast increases in the demand for clean and ultraclean vehicle fuels and petrochemical raw materials. In fact, upgrading processes of petroleum residues and other viscous feedstocks (because the availability of these feedstocks in the market with reasonable prices) raises several issues related to refining these refinability of these feedstocks (Table 15.1).

This chapter presents suggestions and opinions of the means by which conventional refinery processes and the refinery configuration (Fig. 15.1) will evolve during the next three-to-five decades. While profitability is the key component of any business and is the subject of several works on the refinery of the future, there is more to the issues. For example, a major issue that is very often ignored in treatises on profitability, there is the major issue of feedstock (crude oil) depletion leading to two major question: (1) what is the nature and processability of the feedstock that will be the replacement for the current refinery feedstocks? and (2) in what manner will digitization assist the refinery, if at all, when the current types of feedstocks

Table 15.1 Examples of issues related to the refinability of viscous feedstocks.

Issue	Comment
Coke formation	Causes catalyst deactivation
	Can be reduced by dispersion the asphaltene constituents
Metal deposition	Causes catalyst deactivation
Catalytic cracking	Costly; there is a need for a low-cost effective process
Catalyst regenerability	Important; directly related to the process operating cost
Catalyst reuse	Important; directly related to the process operating cost
Nano-sized catalyst	Reduces deactivation of catalysts
	Enhances mass transfer conditions
	Efficient methods are needed to regenerate the catalyst
Supercritical solvents	Promising alternatives for hydrogen in hydroprocessing
Process control	Impacts product yield and quality
Catalytic partial oxidation[a]	Indirect production liquid fuels

[a]Also known as catalytic gasification.

The Refinery of the Future. DOI: https://doi.org/10.1016/B978-0-12-816994-0.00015-4

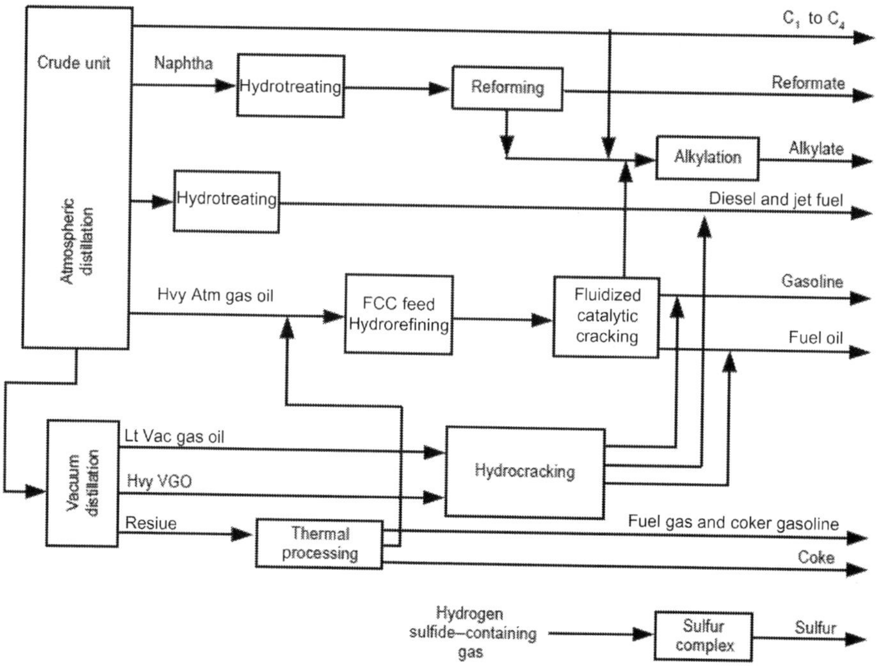

Figure 15.1 Typical layout of a conventional crude oil refinery.

are in short supply or are no longer available? (Speight, 2014, 2017; Alawad and Al-Zubaidi, 2019; Countryman, 2019).

Thus material relevant to (1) comparisons of current feedstocks with heavy oil and biofeedstocks, (2) evolution of refineries since the 1950s, (3) properties and refinability of heavy oil and biofeedstocks, (4) thermal processes versus hydroprocesses, and (5) evolution of products to match the environmental market, with more than passing mention of the effects of feedstocks from coal and oil shale.

The chapter also presents the changes that will be necessary to accommodate the increasingly popular concept of production fuels and other useful products from biomass. There will also be a major trend in the refining industry to produce products from simple basic reactants (e.g., synthesis gas) when it becomes uneconomical to produce clean transportation fuels through conventional refining processes. Fischer–Tropsch plants will be integrated with or even into refineries, which will offer the advantage of high-quality products (Stanislaus et al., 2000; Speight, 2014, 2017).

15.2 Refinery configurations

Hydrocarbon-based energy is important and a crude oil refinery is an industrial processing plant that is the collection of integrated process units (Parkash, 2003;

Gary et al., 2007; Speight, 2014, 2017; Hsu and Robinson, 2017). The crude oil feedstock is typically a blend of two or more crude oils, often with heavy oil or even tar sand bitumen blended into a maximum amount. With the depletion of known crude oil reserves, refining companies have to seek crude oil in places other than the usual sources of supply.

In this section the various types of refinery are described (alphabetically without any preference to one particular type of refinery) with a strong leaning to the future of fuel production. These types are (1) the conventional refinery, (2) the biomass refinery, (3) the coal liquids refinery, (4) the gasification refinery, and (5) the shale oil refinery. But first, the conventional crude oil refinery leads off this section with a discussion of the changes and the future priorities.

15.2.1 The conventional crude oil refinery

The definition of crude oil is confusing and variable (Chapter 1: Feedstock Types and Properties) and has been made even confusing by the introduction of other terms that add little, if anything, to crude oil definitions and terminology (Zittel and Schindler, 2007; Speight, 2008, 2014, 2017). Furthermore, there are different classification schemes based on (1) economic and/or (2) geological criteria. For example, the economic definition of conventional oil is "oil that can be produced with current technology under present economic conditions." The problem with this definition is that it is not very precise and changes whenever the economic or technological aspects of oil recovery change. In addition, there are other classifications based on the American Petroleum Institute (API) gravity of the oil such as "conventional oil is crude oil having a viscosity above 17°API." However, these definitions do not change the definition stated elsewhere (Chapter 1: Feedstock Types and Properties) that has been used through this book.

In recent years the average quality of crude oil has become, has deteriorated, and continues to do so as more heavy oil, extra heavy oil, and tar sand bitumen are being sent to refineries (Speight, 2008, 2014, 2017). This has caused the nature of crude oil refining to change considerably. Indeed, the declining reserves of lighter crude oil have resulted in an increasing need to develop options to desulfurize and upgrade the heavy feedstocks, specifically heavy oil and bitumen. This has resulted in a variety of process options that specialize in sulfur removal during refining.

In addition, the general trend throughout refining has been to produce more products from each barrel of crude oil and to process those products in different ways to meet the product specifications for use in modern engines. Overall, the demand for gasoline has rapidly expanded and the same has also developed for gas oils and fuels for domestic central heating, and fuel oil for power generation, as well as for light distillates and other inputs, derived from crude oil, for the petrochemical industries.

As the need for the lower boiling products developed, crude oil yielding the desired quantities of the lower boiling products became less available and refineries had to introduce conversion processes to produce greater quantities of lighter products from the higher boiling fractions. The means by which a refinery operates in terms of producing the relevant products depends not only on the nature of the crude oil feedstock but also on its configuration (i.e., the number of types of the processes that are employed to produce the desired product slate) and the refinery configuration is, therefore, influenced by the specific demands of a market. Therefore refineries need to be constantly adapted and upgraded to remain viable and responsive to ever-changing patterns of crude supply and product market demands. As a result, refineries have been introducing increasingly complex and expensive processes to gain higher yields of lower boiling products from the higher boiling fractions and residua.

Finally, the yields and quality of refined crude oil products produced by any given oil refinery depends on the mixture of crude oil used as feedstock and the configuration of the refinery facilities. Light/sweet crude oil is generally more expensive and has inherent great yields of higher value low-boiling products such naphtha, gasoline, jet fuel, kerosene, and diesel fuel. Heavy sour crude oil is generally less expensive and produces greater yields of lower value higher boiling products that must be converted into lower boiling products. Thus the conventional crude oil refinery has been adapted to accept (1) heavy crude oil; (2) high-acid crude oil; (3) opportunity crude oil; (4) foamy oil; (5) tight oil, that is, crude oil from tight or low-to-no permeability; (6) extra heavy oil; and (7) tar sand bitumen. Generally, this has required to refinery to be composed of units that are capable of processing the heavier (high-viscosity, high-sulfur) feedstocks.

However, the configuration of refineries may vary from refinery to refinery. Some refineries may be more oriented toward the production of gasoline (large reforming and/or catalytic cracking), whereas the configuration of other refineries may be more oriented toward the production of middle distillates such as jet fuel and gas oil. Changes in the characteristics of conventional crude oil can be specified and will trigger changes in refinery configurations and corresponding investments. The future crude slate is expected to consist of larger fractions of both heavier, sourer crudes and extra-light inputs, such as natural gas liquids. There will also be a shift toward bitumen, such as Canadian oil sands and Venezuelan heavy oil. These changes will require investment in upgrading, either at field level to process tar sand bitumen and oil shale into synthetic crude oil shale either at a field site or at the refinery level (Hedrick et al., 2006). Where this upgrading capacity will be built is likely to be strongly influenced by greenhouse gas policy. In fact, the crude oil and petrochemical industries are coming under increasing pressure not only to compete effectively with global competitors utilizing more advantaged hydrocarbon feedstocks but also to ensure that its processes and products comply with increasingly stringent environmental legislation.

Crude oil refining has grown increasingly complex in the last 20 years. Lower quality crude oil, crude oil price volatility, and environmental regulations that require cleaner manufacturing processes and higher performance products present new challenges to the refining industry. Improving processes and increasing the efficiency of energy use are key to meeting the challenges and maintaining the viability of the crude oil refining industry.

There is also the need for a refinery to be able to accommodate opportunity crude oils and/or high-acid crude oils as part of the refinery slate (Chapter 1: Feedstock Types and Properties). Typically, opportunity crude oils are often dirty and need cleaning before refining by removal of undesirable constituents such as high-sulfur, high-nitrogen, and high-aromatic (such as polynuclear aromatic) components. A controlled visbreaking treatment would clean up such crude oils by removing these undesirable constituents (which, if not removed, would cause problems further down the refinery sequence) as coke or sediment.

On the other hand, high-acid crude oils cause corrosion in the atmospheric and vacuum distillation units. In addition, overhead corrosion is caused by the mineral salts, magnesium, calcium, and sodium chloride, which are hydrolyzed to produce volatile hydrochloric acid, causing a highly corrosive condition in the overhead exchangers. Therefore these salts present a significant contamination in opportunity crude oils. Other contaminants in opportunity crude oils, which are shown to accelerate the hydrolysis reactions, are inorganic clays and organic acids.

In addition to taking preventative measure for the refinery to process these high-margin crude oils without serious deleterious effects on the equipment, refiners will need to develop programs for detailed and immediate feedstock evaluation so that they can understand the qualities of a crude oil very quickly and it can be valued appropriately. There is also the need to assess the potential impact of contaminants, such as metals or acidity, in crudes so that the feedstock can be correctly valued and management of the crude processing can be planned.

15.2.2 The biorefinery

A biorefinery is a facility that integrates biomass conversion processes and equipment to produce fuels, power, and chemicals from biomass. The biorefinery concept is analogous to the crude oil refinery, which produces multiple fuels and products from crude oil (Chapter 2: Introduction to Refining Processes) (Speight, 2011b, 2014, 2017).

The use of fossil resources at current rates may have serious and irreversible consequences for the global climate, although other factors are also involved that are often ignored by the proponents of climate change because of fossil fuel use (Speight, 2020). Whatever the rationale and however the numbers are manipulated, the supply of crude oil, the basic feedstock for refineries and for the petrochemical industry, is finite and its dominant position will become unsustainable as supply/demand issues erode its economic advantage over other alternative feedstocks (Speight and Islam, 2016). This situation will be mitigated to some extent by the exploitation of more technically challenging fossil resources and the introduction of new technologies for fuel and chemical production from natural gas and coal.

Consequently, there is a renewed interest in the utilization of plant-based matter (biomass) as a raw material feedstock for the chemical industry (Marcilly, 2003; Lynd et al., 2005; Huber and Corma, 2007; Lynd et al., 2009). Plants accumulate carbon from the atmosphere via photosynthesis and the widespread utilization of these materials as basic inputs into the generation of power, fuels, and chemicals is a viable route to reduce greenhouse gas emissions.

Biomass is a renewable energy source, unlike the fossil fuel resources (natural gas, crude oil, coal, and oil shale). One advantage of fuel from biomass (biofuel), in comparison to most other fuel types, is it is biodegradability, and thus relatively harmless to the environment if spilled. In addition, a biofuel is any fuel that is derived from biomass, that is, recently living organisms or their metabolic by-products. Biofuel has also been defined as any fuel with an 80% minimum content by volume of materials derived from living organisms harvested within 10 years preceding its manufacture.

The production of fuels and chemicals from renewable plant-based feedstocks utilizing the state-of-the-art conversion technologies presents an opportunity to maintain competitive advantage and contribute to the attainment of national environmental targets. Bioprocessing routes have a number of compelling advantages over conventional petrochemical production. However, it is only in the last decade that rapid progress in biotechnology has facilitated the commercialization of a number of plant-based chemical processes.

Plants offer a unique and diverse feedstock for chemicals. Plant biomass can be gasified to produce synthesis gas, a basic chemical feedstock and also a source of hydrogen for a future hydrogen economy. In addition, the specific components of plants such as carbohydrates, vegetable oils, plant fiber, and complex organic molecules known as primary and secondary metabolites can be utilized to produce a range of valuable monomers, chemical intermediates, pharmaceuticals, and materials. More generally, biomass feedstocks are recognized by the specific plant content of the feedstock or the manner in which the feedstock is produced (Speight, 2008, 2011b).

For example, primary biomass feedstocks that are currently being used for bioenergy include grains and oilseed crops used for transportation fuel production, plus some crop residues (such as orchard trimmings and nut hulls) and some residues from logging and forest operations that are currently used for heat and power production. In the future, it is anticipated that a larger proportion of the residues inherently generated from food crop harvesting, as well as a larger proportion of the residues generated from ongoing logging and forest operations, will be used for bioenergy.

Secondary biomass feedstocks differ from primary biomass feedstocks in that the secondary feedstocks are a by-product of processing of the primary feedstocks. Specific examples of secondary biomass include sawdust from sawmills, black liquor (which is a by-product of paper making), and cheese whey (which is a by-product of cheese-making processes). Vegetable oils used for biodiesel that are derived directly from the processing of oilseeds for various uses are also a secondary biomass resource. Tertiary biomass feedstock includes fats, greases, oils, construction and demolition wood debris, other waste wood from the urban environments, as well as packaging wastes, municipal solid wastes, and landfill gases. A category other than wood waste from the urban environment includes trimmings from urban trees.

The simplest, cheapest, and most common method of obtaining energy from biomass is direct combustion. Any organic material, with a water content low enough to allow for sustained combustion, can be burned to produce energy. The heat of combustion can be used to provide space or process heat, water heating or, through the use of a steam turbine, electricity. In the developing world, many types of biomass such as dung and agricultural wastes are burned for cooking and heating.

In fact, result in some form of organic residue after their primary use has been fulfilled. These organic residues can be used for energy production through direct combustion or biochemical conversion. Most crop residues are returned to the soil, and the humus resulting from their decomposition helps maintain soil nutrients, soil porosity, water infiltration and storage, as well as reducing soil erosion. Crop residues typically contain 40% of the nitrogen (N), 80% of the potassium (K), and 10% of the phosphorous (P) applied to the soil in the form of fertilizer. If these residues are subjected to direct combustion for energy, only a small percentage of the nutrients are left in the ash. Similarly, soil erosion will increase.

In addition to applying biomethods to crude oil itself such as the (1) the denitrogenation of fuels, (2) removal of heavy metals, and (3) transformation of heavy crudes into light crudes (Le Borgne and Quintero, 2003; Bhatia and Sharma, 2006), biorefining offers a key method to accessing the integrated production of chemicals, materials, and fuels. While the biorefinery concept is analogous to that of an oil refinery, there are significant differences—particularly in the character and properties of the respective feedstocks (Speight, 2008, 2014, 2017).

While the primary function of a biorefinery is to make biofuels, refiners are looking at the various types of biomass and the by-products of biofuel production, and asking what they can do with them. In the place of intermediate hydrocarbons, the building blocks at their disposal are a plant's sugars, starches, fats, and proteins. Some chemicals will be synthesized using enzymes or genetically engineered microorganisms, and some will be produced using the inorganic catalysts used in traditional chemical processes.

The biomass could be supplied by anything from corn, sugarcane, grasses, wood, and soybeans to algae. The relatively modest biomass required to meet this demand means that corn or other food crops could be used without creating the competition for food that has arisen over corn-based ethanol fuel in the United States, Shanks says, "You can actually talk about bringing biomass-derived chemicals online without perturbing food supplies."

In a manner similar to the crude oil refinery, a biorefinery would integrate a variety of conversion processes to produce multiple product streams such as motor fuels and other chemicals from biomass. In short, a biorefinery would combine the essential technologies to transform biological raw materials into a range of industrially useful intermediates. However, the type of biorefinery would have to be differentiated by the character of the feedstock. For example, the crop biorefinery would use raw materials such as cereals or maize, and the lignocellulose biorefinery would use raw material with high cellulose content, such as straw, wood, and paper waste.

In addition, a variety of methods can be employed to obtain different product portfolios of bulk chemicals, fuels, and materials (Speight, 2008). Biotechnology-based conversion processes can be used to ferment the biomass carbohydrate content into sugars that can then be further processed. An alternative is to employ thermochemical conversion processes that use pyrolysis or gasification of biomass to produce a hydrogen-rich synthesis gas that can be used in a wide range of chemical processes (Speight, 2008; Shires et al., 2010). For example, a biorefinery using lignin as a feedstock would produce a range of valuable organic chemicals and liquid fuels that, at the present time, could supplement or even replace equivalent or identical products currently obtained from crude oil, coal, or gas.

By producing multiple products a biorefinery can take advantage of the differences in biomass components and intermediates and maximize the value derived from the biomass feedstock. A biorefinery might, for example, produce one or several low-volume, but high-value, chemical products and a low-value, but high-volume, liquid transportation fuel, while generating electricity and process heat for its own use and perhaps enough for sale of electricity. The high-value products enhance profitability, the high-volume fuel helps meet national energy needs, and the power production reduces costs and avoids greenhouse gas emissions.

As a feedstock, and in general terms, the various types of biomass feedstocks can be converted by thermal or biological routes to a wide range of useful forms of energy, including process heat, steam, electricity, as well as liquid fuels, chemicals, and synthesis gas. As a raw material, biomass is a nearly universal feedstock due to its versatility, domestic availability, and renewable character but there are limitations. For example, the energy density of biomass is low compared to that of crude oil, or crude oil–derived fuels. The heat content of biomass, on a dry basis (7000–9000 Btu/lb) is at best comparable with that of a low-rank coal or lignite, and substantially (50%–100%) lower than that of anthracite, most bituminous coals, and crude oil. Most biomass, as received, has a high burden of physically adsorbed moisture, up to 50% by weight and, without substantial drying, the energy content of a biomass feed per unit mass is even less.

These inherent characteristics and limitations of biomass feedstocks have focused on the development of efficient methods of chemically transforming and upgrading biomass feedstocks in a refinery. The refinery would be based on two "platforms" to promote different product slates (Speight, 2011b). Thus, by analogy with crude oil, every element of the plant feedstock will be utilized, including the low-value lignin components. However, the different compositional nature of the biomass feedstock, compared to crude oil, will require the application of a wider variety of processing tools in the biorefinery. Processing of the individual components will utilize conventional thermochemical operations and state-of-the-art bioprocessing techniques. The production of biofuels in the biorefinery complex will service existing high-volume markets, providing economy-of-scale benefits and large volumes of by-product streams at minimal cost for upgrading to valuable chemicals.

The construction of both large biofuel and renewable chemical production facilities coupled with the pace at which bioscience is being both developed and applied demonstrates that the utilization of nonfood crops will become more significant. The biorefinery concept provides a means to significantly reduce production costs such that a substantial substitution of petrochemicals by renewable chemicals becomes possible. However, significant technical challenges remain before the biorefinery concept can be realized.

In order to accommodate liquids from biomass, consideration should be given to the following modes of integration: (1) the stand-alone concept in which the fuels from the biomass-derived fuels remain self-contained, (2) the refinery-integrated concept in which the biomass-derived fuels are produced in a facility that is integrated with an existing crude oil refinery, (3) the coprocessing concept in which the biomass-derived fuels are coprocessed with crude oil–derived fuels in existing refinery process units, and (4) the refinery conversion concept in which an existing refinery unit is adapted to process the biomass-derived fuel. If the biomass-derived fuel is to be processed with coal- or oil shale–derived fuels, additional considerations are needed that are dictated by the properties of the blended feedstock(s). Whichever option is chosen, there must be recognition of the potential for incompatibility of the components of the blended feedstock as well as the potential insatiably of the blended feedstock that can deteriorate considerably with time and/or temperature (Mushrush and Speight, 1995, 1998; Speight, 2013, 2014, 2017).

Finally, and in relation to the previous paragraph, the concept of the biorefinery must include consideration of the layout of the refinery, such as (1) a complete one-site full-service refinery along the lines of the crude oil refinery in which all units are at a single site or (2) the location of each of the various individual (satellite) units with each unit near to (or at) the feedstock site from which with each unit sends the product to a central site for further and/or final processing.

15.2.3 The coal liquids refinery

Refinery feedstocks from coal (coal liquids) have not been dealt with elsewhere in this text but descriptions are available from other sources (Speight, 1998, 2013). While coal has been labeled (perhaps unjustifiability) the "bad boy" of fossil fuel use, there could well be a time when coal comes back into favor as a source of liquid fuels. As always there are two sides to every argument and it need to be stated that environmental technology has advanced to the stage where cleanup of emission from fossil fuel such as coal is efficient but not always recognized by the proponents of climate change because of fossil fuel use.

In terms of coal-to-liquids technology, the Bergius process was one of the early processes for the production of liquid fuels from coal. In the process, lignite or subbituminous coal is finely ground and mixed with heavy oil recycled from the process. Catalyst is typically added to the mixture and the mixture is pumped into a reactor. The reaction occurs at between 400°C and 500°C (750°F and 930°F) and hydrogen pressure. The reaction produces gas, naphtha, kerosene, middle distillate, and heavy oil (Speight, 2008, 2013):

$$nC_{coal} + (n+1)H_2 \rightarrow C_nH_{2n+2}$$

A number of catalysts have been developed over the years, including catalysts containing tungsten, molybdenum, tin, or nickel.

The different fractions can be sent to a conventional refinery for further processing to yield of synthetic fuel or a fuel blending stock of the desired quality. It has been reported that as much as 97 w/w% of the coal carbon can be converted to synthetic fuel but this very much depends on the coal type, the reactor configuration, and the process parameters. However, liquid products from coal are generally different from those produced by crude oil refining, particularly as they can contain substantial amounts of phenol derivatives. Therefore there will always be some question about the place of coal liquids in refining operations. For this reason, there have been some investigations of the characterization and next-step processing of coal liquids.

The composition of coal liquids produced from coal depends very much on the character of the coal and on the process conditions and, particularly, on the degree of hydrogen addition to the coal. In fact, current concepts for refining the products of coal liquefaction processes have relied, for the most part, on already existing crude oil refineries, although it must be recognized that the acidity (i.e., phenol content) of the coal liquids and their potential incompatibility with conventional crude oil (including heavy oil) may pose new issues within the refinery system (Speight, 2007; Speight, 2008, 2013, 2014, 2017).

The other category of coal liquefaction processes invokes the concept of the indirect liquefaction of coal. In these processes the coal is not converted directly into liquid products but involves a two-stage conversion operation in which coal is first converted (by reaction with steam and oxygen) to produce a gaseous mixture that is composed primarily of carbon monoxide and hydrogen (syngas; synthesis gas). The gas stream is subsequently purified (to remove sulfur, nitrogen, and any particulate matter) after which it is catalytically converted to a mixture of liquid hydrocarbon products.

The synthesis of hydrocarbons from carbon monoxide and hydrogen (synthesis gas) (the Fischer–Tropsch synthesis) is a procedure for the indirect liquefaction of coal (Speight, 2008, 2013). Thus coal is converted to gaseous products at temperatures in excess of 800°C (1470°F), and at moderate pressures, to produce synthesis gas:

$$[C]_{coal} + H_2O \rightarrow CO + H_2$$

The gasification may be attained by means of any one of several processes or even by gasification of coal in place (underground, or in situ, gasification of coal) (Speight, 2008, 2013).

In practice, the Fischer–Tropsch reaction is carried out at temperatures of 200°C–350°C (390°F–660°F) and at pressures of 75–4000 psi. The hydrogen/carbon monoxide ratio is usually 2.2:1 or 2.5:1. Since up to three volumes of hydrogen may be required to achieve the next stage of the liquid production, the synthesis gas must then be converted by means of the water–gas shift reaction to the desired level of hydrogen:

$$CO + H_2O \rightarrow CO_2 + H_2$$

After this the gaseous mix is purified and converted to a wide variety of hydrocarbons:

$$nCO + (2n + 1)H_2 \rightarrow C_nH_{2n+2} + nH_2O$$

These reactions result primarily in low- and medium-boiling aliphatic compounds suitable for gasoline and diesel fuel. Synthesis gas can also be converted to methanol, which can be used as a fuel, fuel additive, or further processed into gasoline via the Mobil M-gas process. However, in terms of liquids from coal that can be integrated into a refinery, this represents the most attractive option and does not threaten to bring on incompatibility problems as can occur when phenol derivatives are present in the coal liquids (Mushrush and Speight, 1995, 1998; Speight, 2013, 2014).

A major challenge for refining coal is related to air pollution issues and mining hazards, but conversion of coal to liquids is not likely to be abandoned because of (1) abundance; (2) low, relatively nonvolatile prices; (3) gasification can be the key to environmental acceptance; and (4) synthetic fuels from coal via gasification can be cleaner than crude oil–derived hydrocarbon fuels derived from crude oil. While such a scheme is not meant to replace other fuel-production systems, it would certainly be a fit into a conventional refinery—gasification is used in many refineries to produce hydrogen and a gasification unit is part of the flexicoking process (Chapter 2: Introduction to Refining Processes).

15.2.4 The gasification refinery

As crude oil supplies are depleted, the desirability of producing gas from other carbonaceous feedstocks will increase, especially in those areas where natural gas is in short supply. It is also anticipated that the future of any process depends very much on the effect of gasification processes on the surrounding environment. It is these environmental effects and issues that will direct the success of gasification. In fact, there is the distinct possibility that within the foreseeable future the gasification process will increase in popularity in crude oil refineries—some refineries may even be known as gasification refineries.

A gasification refinery would have, as the center piece, gasification technology as is the case of the Sasol refinery in South Africa (Couvaras, 1997). The refinery would produce synthesis gas (from the carbonaceous feedstock) (Shires et al., 2010; Speight, 2013, 2014, 2017) from which liquid fuels would be manufactured using the Fischer–Tropsch synthesis technology. The refinery would produce synthesis gas (from the carbonaceous feedstock) from which liquid fuels would be manufactured using the Fischer–Tropsch synthesis technology. In fact, gasification to produce synthesis gas (Chapter 12: Synthesis Gas and the Fischer–Tropsch Process) can proceed from any carbonaceous material, including biomass. Inorganic components of the feedstock, such as metals and minerals, are trapped in an inert and environmentally safe form as char, which may have used as a fertilizer. Biomass gasification is therefore one of the most technically and economically convincing energy possibilities for a potentially carbon neutral economy.

The manufacture of gas mixtures of carbon monoxide and hydrogen has been an important part of chemical technology for about a century. Originally, such mixtures were obtained by the reaction of steam with incandescent coke and were known as *water gas*. Eventually, steam reforming processes, in which steam is reacted with natural gas (methane) or crude oil naphtha over a nickel catalyst, found wide application for the production of synthesis gas.

Synthesis gas (syngas) is the name given to a gas mixture that contains varying amounts of carbon monoxide and hydrogen generated by the gasification of a carbon-containing fuel to a gaseous product with a heating value. Examples include the gasification of coal or crude oil residua (Speight, 2008). Synthesis gas is used as a source of hydrogen or as an intermediate in producing hydrocarbons via the Fischer–Tropsch synthesis.

In fact, gasification to produce synthesis gas can proceed from any carbonaceous material, including biomass. Inorganic components of the feedstock, such as metals and minerals, are trapped in an inert and environmentally safe form as char, which may have used as a fertilizer. Biomass gasification is therefore one of the most technically and economically convincing energy possibilities for a potentially carbon neutral economy.

The manufacture of gas mixtures of carbon monoxide and hydrogen has been an important part of chemical technology for about a century. Originally, such mixtures were obtained by the reaction of steam with incandescent coke and were known as water gas. Eventually, steam reforming processes, in which steam is reacted with natural gas (methane) or crude oil naphtha over a nickel catalyst, found wide application for the production of synthesis gas.

A modified version of steam reforming known as autothermal reforming, which is a combination of partial oxidation near the reactor inlet with conventional steam reforming further along the reactor, improves the overall reactor efficiency and increases the flexibility of the process. Partial oxidation processes using oxygen instead of steam also found wide application for synthesis gas manufacture, with the special feature that they could utilize low-value feedstocks such as crude oil residues, heavy crude oil, extra heavy crude oil, and tar sand bitumen. In recent years, catalytic partial oxidation employing very short reaction times (milliseconds) at high temperatures (850°C–1000°C) is providing still another approach to synthesis gas manufacture (Hickman and Schmidt, 1993).

In a gasifier the carbonaceous material undergoes several different processes: (1) pyrolysis of carbonaceous fuels, (2) combustion, and (3) gasification of the remaining char. The process is very dependent on the properties of the carbonaceous material and determines the structure and composition of the char, which will then undergo gasification reactions.

The conversion of the gaseous products of gasification processes to synthesis gas, a mixture of hydrogen (H_2) and carbon monoxide (CO), in a ratio appropriate to the application, needs additional steps, after purification. The product gases—carbon monoxide, carbon dioxide, hydrogen, methane, and nitrogen—can be used as fuels or as raw materials for chemical or fertilizer manufacture.

15.2.4.1 Gasifiers

A gasifier differs from a combustor in that the amount of air or oxygen available inside the gasifier is carefully controlled so that only a relatively small portion of the fuel burns completely. The partial oxidation process provides the heat and rather than combustion, most of the carbon-containing feedstock is chemically broken apart by the heat and pressure applied in the gasifier resulting in the chemical reactions that produce synthesis gas. However, the composition of the synthesis gas will vary because of dependence upon the conditions in the gasifier and the type of feedstock. Minerals in the fuel (i.e., the rocks, dirt, and other impurities that do not gasify) separate and leave the bottom of the gasifier either as an inert glass-like slag or other marketable solid products.

In terms of the gasification process, there are four types of gasification units that are proven and, therefore, currently available for commercial use: (1) the countercurrent fixed bed, (2) cocurrent fixed bed, (3) the fluidized bed, and (4) the entrained flow (Speight, 2008).

The countercurrent fixed bed (up draft) gasifier consists of a fixed bed of carbonaceous fuel (e.g., coal or biomass) through which the gasification agent (steam, oxygen, and/or air) flows in countercurrent configuration. The ash is either removed dry or as a slag. The nature of the gasifier means that the fuel must have high mechanical strength and must be noncaking so that it will form a permeable bed, although recent developments have reduced these restrictions to some extent. The throughput for this type of gasifier is relatively low. Thermal efficiency is high as the gas exit temperatures are relatively low and, as a result, tar and methane production is significant at typical operation temperatures, so product gas must be extensively cleaned before use or recycled to the reactor.

The cocurrent fixed bed (downdraft) gasifier is similar to the countercurrent type, but the gasification agent gas flows in cocurrent configuration with the fuel (downward, hence the name downdraft gasifier). Heat needs to be added to the upper part of the bed, either by combusting small amounts of the fuel or from external heat sources. The produced gas leaves the gasifier at a high temperature, and most of this heat is often transferred to the gasification agent added in the top of the bed. Since all tars must pass through a hot bed of char in this configuration, tar levels are much lower than the countercurrent type.

In the fluidized bed gasifier the fuel is fluidized in oxygen (or air) and steam. The temperatures are relatively low in dry ash gasifiers, so the fuel must be highly reactive; low-grade coals are particularly suitable. The agglomerating gasifiers have slightly higher temperatures and are suitable for higher rank coals. Fuel throughput is higher than for the fixed bed, but not as high as for the entrained flow gasifier. The conversion efficiency is typically low, so recycle or subsequent combustion of solids is necessary to increase conversion. Fluidized bed gasifiers are most useful for fuels that form highly corrosive ash that would damage the walls of slagging gasifiers. The ash is removed dry or as heavy agglomerates—a disadvantage of biomass feedstocks is that they generally contain high levels of corrosive ash.

In the entrained flow gasifier a dry pulverized solid, an atomized liquid fuel, or a fuel slurry is gasified with oxygen (much less frequent: air) in cocurrent flow. The high temperatures and pressures also mean that a higher throughput can be achieved but thermal efficiency is somewhat lower as the gas must be cooled before it can be sent to a gas processing facility. All entrained flow gasifiers remove the major part of the ash as a slag as the operating temperature is well above the ash fusion temperature. Biomass can form slag that is corrosive for ceramic inner walls that serve to protect the gasifier's outer wall.

In integrated gasification combined cycle (IGCC) systems, the synthesis gas is cleaned of its hydrogen sulfide, ammonia, and particulate matter and is burned as fuel in a combustion turbine (much like natural gas is burned in a turbine). The combustion turbine drives an electric generator. Hot air from the combustion turbine can be channeled back to the gasifier or the air separation unit, while exhaust heat from the combustion turbine is recovered and used to boil water, creating steam for a steam turbine generator. The use of these two types of turbines—a combustion turbine and a steam turbine—in combination, known as a combined cycle, is one reason why gasification-based power systems can achieve unprecedented power generation efficiencies.

Gasification also offers more scope for recovering products from waste than incineration. When waste is burnt in an incinerator, the only practical product is energy, whereas the gases, oils, and solid char from pyrolysis and gasification can not only be used as a fuel but also purified and used as a feedstock for petrochemicals and other applications. Many processes also produce a stable granulate instead of an ash, which can be more easily and safely utilized. In addition, some processes are targeted at producing specific recyclables such as metal alloys and carbon black. From waste gasification, in particular, it is feasible to produce hydrogen, which many see as an increasingly valuable resource.

IGCC is used to raise power from feedstocks such as vacuum residua, cracked residua, and deasphalter pitch. The value of these refinery residuals, including crude oil coke, will need to be considered as part of an overall upgrading project. Historically, many delayed coking projects have been evaluated and sanctioned on the basis of assigning zero value to crude oil coke having high sulfur and high metal content.

While there are many alternate uses for the synthesis gas produced by gasification, and a combination of products/utilities can be produced in addition to power. A major benefit of the IGCC concept is that power can be produced with the lowest sulfur oxide (SO_x) and nitrogen oxide (NO_x) emissions of any liquid/solid feed power generation technology.

15.2.4.2 Fischer–Tropsch synthesis

The synthesis reaction is dependent of a catalyst, mostly an iron or cobalt catalyst where the reaction takes place. There is either a low- or high-temperature Fischer – Tropsch (LTFT or HTFT) process, with temperatures ranging between 200°C and 240°C for LTFT and 300°C and 350°C for HTFT. The HTFT uses an iron catalyst, whereas the LTFT either an iron or a cobalt catalyst. The different catalysts include also nickel- and ruthenium-based catalysts, which also have enough activity for commercial use in the process.

The reactors are the multitubular fixed bed, the slurry, or the fluidized bed (with either fixed or circulating bed). The fixed bed reactor consists of thousands of small tubes with the catalyst as surface-active agent in the tubes. Water surrounds the tubes and regulates the temperature by settling the pressure of evaporation. The slurry reactor is widely used and consists of fluid and solid elements, where the catalyst has no particular position, but flows around as small pieces of catalyst together with the reaction components. The slurry and fixed bed reactors are used in LTFT. The fluidized bed reactors are diverse, but characterized by the fluid behavior of the catalyst.

The HTFT technology uses a fluidized catalyst at 300°C–330°C (570°F–625°F). Originally circulating fluidized bed units were used (Synthol reactors). Since 1989 a commercial-scale classical fluidized bed unit has been implemented and improved upon.

The LTFT technology has originally been used in tubular fixed bed reactors at 200°C–230°C. This produces a more paraffinic and waxy product spectrum than the high-temperature technology. A new type of reactor (the Sasol slurry phase distillate reactor) has been developed and is in commercial operation. This reactor uses a slurry phase system rather than a tubular fixed bed configuration and is currently the favored technology for the commercial production of synfuels.

Under most circumstances the production of synthesis gas by reforming natural gas will be more economical than from coal gasification, but site-specific factors need to be considered. In fact, any technological advance in this field (such as better energy integration or the oxygen transfer ceramic membrane reformer concept) will speed up the rate at which the synfuel technology will become common practice.

There are large coal reserves that may increasingly be used as a fuel source during oil depletion. Since there are large coal reserves in the world, this technology could be used as an interim transportation fuel if conventional oil were to become more expensive. Furthermore, a combination of biomass gasification and Fischer–Tropsch synthesis is a very promising route to produce transportation fuels from renewable or green resources.

Although the focus of this section has been on the production of hydrocarbons from synthesis gas, it is worthy of note that clean synthesis gas can also be used (1) as chemical building blocks to produce a broad range of chemicals using processes well established in the chemical and petrochemical industry, (2) as a fuel producer for highly efficient fuel cells (which run off the hydrogen made in a gasifier) or perhaps in the future, hydrogen turbines and fuel cell–turbine hybrid systems, and (3) as a source of hydrogen that can be separated from the gas stream and used as a fuel or as a feedstock for refineries (which use the hydrogen to upgrade crude oil products).

15.2.5 The shale oil refinery

The almost forgotten fossil energy resource is the oil shale, of which the United States has an abundant supply (Baughman, 1978; Allred, 1982; Scouten, 1990; Lee, 1991, 1996; US DOE, 2004a,b,c; Speight and Islam, 2016; Speight, 2013, 2014, 2019b). Like coal mining and processing to liquids, oil shale also has serious environmental issues that must be considered as part of any development plan. The advancement in environmental technologies to control emission is also advanced and there may be a role that shale oil (i.e., the distillate produced by the thermal decomposition of the kerogen in the oil shale) and oil shale could play in the future.

The processes for producing liquids (shale oil) from oil shale involve heating (retorting) the shale to convert the organic kerogen to a raw shale oil (Speight, 2008, 2012). There are two basic oil shale retorting approaches (1) mining followed by retorting at the surface and (2) in situ retorting, that is, heating the shale in place underground.

Retorting essentially involves destructive distillation (pyrolysis) of oil shale in the absence of oxygen. Pyrolysis (temperatures above 900°F) thermally breaks down (cracks) the kerogen to release the hydrocarbons and then cracks the hydrocarbons into lower weight hydrocarbon molecules. Conventional refining uses a similar thermal cracking process, termed coking, to break down high-molecular-weight constituents of residua, heavy oil, and tar sand bitumen.

Shale oil does contain a large variety of hydrocarbon compounds (Speight, 2008) but also has high-nitrogen content compared to a nitrogen content of 0.2–0.3 wt.% for a typical crude oil. In addition, shale oil also has a high olefin and diolefin content. It is the presence of these olefins and diolefins, in conjunction with high-nitrogen content that gives shale oil the characteristic difficulty in refining and the tendency to form insoluble sediment. Crude shale oil also contains appreciable amounts of arsenic, iron, and nickel that interfere with refining.

Upgrading, or partial refining, to improve the properties of a crude shale oil may be carried out using different options—depending upon the composition and the origin, of the shale oil—and hydrotreating is the option of choice to produce a stable product (Parkash, 2003; Gary and Handwerk, 2001; Speight, 2014, 2017; Hsu and Robinson, 2017). In terms of refining and catalyst activity, the nitrogen content of shale oil is a disadvantage and, if not removed, the arsenic and iron in shale oil will poison and foul the supported catalysts used in hydrotreating. In terms of the use of shale oil residua as a modifier for asphalt, where nitrogen species can enhance binding with the inorganic aggregate, the nitrogen content is beneficial (Speight, 2012).

Naphtha from shale oil usually contains a high percentage of aromatic and naphthenic compounds that are not affected by the various treatment processes. The olefin content, although reduced in most cases by refining processes, will still remain significant. Diolefins and the higher unsaturated constituents will have to be removed from the gasoline product by appropriate treatment processes. The same will apply to the nitrogen- and sulfur-containing constituents.

Catalytic hydrodesulfurization processes are not a good solution for the removal of sulfur constituents from gasoline when high proportions of unsaturated constituents are present. A significant amount of the hydrogen would be used for hydrogenation of the unsaturated components. However, when hydrogenation of the unsaturated hydrocarbons is desirable, catalytic hydrogenation processes would be effective.

Thus shale oil is different to conventional crude oils, and several refining technologies have been developed to deal with this. The primary problems identified in the past were related to the presence of arsenic, nitrogen-containing constituents, and the waxy nature of the crude. Nitrogen and wax problems have been resolved using hydroprocessing approaches, essentially classical hydrocracking and the production high-quality lube stocks, which require that waxy materials be removed or isomerized. However, the arsenic problem remains (Speight, 2012).

In general, oil shale distillates have a much higher concentration of high boiling-point compounds that would favor production of middle distillates (such as diesel and jet fuels) rather than naphtha. Oil shale distillates also had a higher content of olefins, oxygen, and nitrogen than crude oil, as well as higher pour points and viscosities. Aboveground retorting processes tended to yield a lower API-gravity oil than the in situ processes (a 25°API gravity was the highest produced). Additional processing equivalent to hydrocracking would be required to convert oil shale distillates to a lighter range hydrocarbon (gasoline). Removal of sulfur and nitrogen would, however, require hydrotreating.

By comparison, a typical 35°API-gravity crude oil may be composed of up to 50% of gasoline and middle distillate range hydrocarbons. West Texas Intermediate (crude benchmark crude for trade in the commodity futures market) has 0.3% by weight sulfur, and Alaska North Slope crude has 1.1% by weight sulfur. The New York Mercantile Exchange specifications for light sweet (low-density, low-sulfur) crude oils limit sulfur content to 0.42% or less (ASTM D4294) and an API gravity between 37 and 42 degrees (ASTM D287).

A preferred way of treating the shale oil involves using a moving bed reactor followed by a fractionation step to divide the wide-boiling-range crude oil produced from the shale oil into two separate fractions. The lighter fraction is hydrotreated for the removal of residual metals, sulfur, and nitrogen, whereas the heavier fraction is cracked in a second fixed bed reactor normally operated under high-severity conditions.

Arsenic removed from the oil by hydrotreating remains on the catalyst, generating a material that is a carcinogen, an acute poison, and a chronic poison. The catalyst must be removed and replaced when its capacity to hold arsenic is reached.

15.3 The future refinery

Over the past three decades, the refining industry has been challenged by changing feedstocks and product slate. In the 21st century the refining industry will need to become increasingly flexible with improved technologies and improved catalysts to match the varying characteristics of a wide variety of feedstocks and be prepared to address many important challenges. Major forces that lead to change include (1) the variation in the character and properties of the feedstocks accepted by the refinery, (2) continuing concern for the environment, (3) governmental regulation and policy, and (4) higher consumer expectations for fuels and fuel delivery systems, as well as the various aspects of petropolitics—also referred to as geopolitics—which can include national economic and social development agendas, which can figure prominently among energy certainties (Speight, 2011a; Beck, 2019).

In the near term, as the crude oil feedstocks to refineries change, there must be an accompanying change in refinery technology and there is a strong need to revise the processes that are commercially available for heavy crude oil refining. Upgrading of heavy oil, extra heavy oil, and tar sand bitumen by means of catalytic processes requires new generation catalyst along with several modifications of process conditions (Rana et al., 2008). Furthermore, even the production of the light (low-density, low-sulfur) oil from tight (low-permeability) formations does not ease the demand on the refineries for the necessary products. In fact, during the decades leading up to the middle of the 21st century, the evolution future of crude oil refining and the current refinery layout (Fig. 15.1) will be primarily on process modification with some new innovations coming onstream.

The light, sweet crude oils have very different yield patterns for refined products, as well as contaminant levels, impacting refiners' processing units and product slates (Table 15.2). These crude oils typically have higher light and heavy naphtha yields and present increasing challenges to the naphtha complex, typically consisting of naphtha hydrotreating, light naphtha isomerization, and catalyst regeneration platforming units. Also, the lower vacuum gas oil and vacuum residue yields decrease the use of coking units, fluidized-bed catalytic cracking units, and alkylation units, which decrease the contribution of the products of these units to

Table 15.2 Characteristics of crude oil from tight formations.

Composition	Batch-to-batch variability
API gravity	Ranges from 20 to 55 degrees
Sulfur	Low sulfur levels, but H_2S can be an issue
Nitrogen	Low levels
Hydrocarbons	High paraffin content
Metals	Heavy metals (nickel, Ni, and vanadium, V) are low
Alkaline metals	Level of alkaline metals may be high
Other contaminants	Barium and lead may be present
Filterable solids	Greater volume and smaller particle size

the gasoline pool. In fact, each refinery has a unique configuration, set of processing objectives, and equipment limitations. This means a movement from conventional means of refining heavy feedstocks using (typically) coking technologies to more innovative processes (including hydrogen management) that will produce the ultimate amounts liquid fuels from the feedstock and maintain emissions within environmental compliance (Penning, 2001; Davis and Patel, 2004; Speight, 2008). In addition, as with liquid products from coal and oil shale, there is also the issue of compatibility/incompatibility of the low-density, low-sulfur crude oils with the high-density, high-sulfur crude oils as well as the compatibility/incompatibility of the product from the various types of crude oil (Mushrush and Speight, 1995; Speight, 2014, 2017). Thus processes in the future need to be characterized by a high degree of flexibility for handling crudes of variable quality, as well as entirely new feedstocks. Refineries would be tightly controlled to increase performance and efficiency and require less maintenance and laboratory services.

To meet the challenges of the future including (1) the changing trends in the feedstock slate accepted by refineries, (2) the changing product slate, and (3) the increasingly stringent product specifications imposed by environmental legislation, the refinery of the future must be increasingly flexible and produce products with specifications that meet the requirements specified by the users of the products.

Tight oil characteristics can vary greatly from batch to batch, even within the same type of crude oil supply. In addition, the range of API gravity for tight oils can be quite wide, from 20 to 55 degrees, with most at 40-degree gravity and above. Tight oil crudes, in general, have low nitrogen and high paraffin content. Heavy metals, such as nickel and vanadium, are generally low, but alkaline metals (calcium, sodium, and magnesium) may be high. This is highly variable as well. In addition, other contaminants such as barium and lead may be elevated. Filterable solids can be higher than conventional crude oils, with greater volume and smaller particle size.

Select samples of Bakken crude have contained salt concentrations as high as 500 ppm, as well as nonextractable salts. Some samples of Eagle Ford crude have been shown to contain olefins or carbonyls—both fouling precursors that are not typically found in virgin crude oils. As a consequence of these dramatic variations in quality and physical properties, it is increasingly more important for refiners to be able to identify, interpret, and respond quickly to changes in crude feed properties.

In general, the modern refinery must be adapting continually to increasing variability in crude oil quality, which, combined with the blending of tight oils into the standard crude slate, and normal refinery operations can be difficult to maintain. Processing these difficult blends can have a significant negative impact on overall profitability, affecting product quality, unit reliability, and onstream time. Determining how a new crude oil fits into a refinery operation requires a comprehensive understanding of the physical properties and unique characteristics of that crude and how it will interact with the rest of the typical crude slate.

Residuum production is low compared to high volumes of gasoline and distillates. For refineries that are configured for bottom-of-the-barrel upgrading, this can be a negative and limit the amount of tight oil that can be added to the crude blend. In order to balance the mix of products in the crude distillation tower to fit many refinery operations, blending tight oils with heavy asphaltic crude makes sense, as the blend can result in a desirable distillation profile for many refiners. However, this practice can also lead to compatibility issues.

Although asphaltene stability has always played a role in crude blending (Speight, 2014, 2017), the high paraffin content of tight oils greatly increases the potential impact of asphaltene precipitation upon blending, and its negative impact on the refinery process. There are several established and developing test methods that can evaluate an oil, or a blend, for asphaltene stability.

Also, in the near term, the main technological progress will be directed to heavy oil upgrading, cleaner transportation fuel production, and the integration of refining and petrochemical businesses. As outlined elsewhere (Parkash, 2003; Gary and Handwerk, 2001; Speight, 2014, 2017; Hsu and Robinson, 2017), even the tried and true processes will see changes as they evolve. The distillation units will continue to be the mainstay of crude oil refining and the main short-term developments are in improved integration through the use of heat recovery technology and integration of different distillation units (i.e., atmospheric distillation unit and the vacuum distillation unit). In the long term the major developments are the integration of different distillation columns into one reactor (e.g., dividing-wall column) or the development of alternative processing routes allowing for the combination of conversion and distillation (such as reactive distillation). Alternative processes to distillation will also include membranes and technologies such as freeze concentration.

Thermal processes (Chapter 5: Thermal Cracking) will also evolve and become more efficient. While the current processes may not see much change in terms of reactor vessel configuration, there will be changes to the reactor internals and to the nature of the catalysts. For example, the tried and true coking processes will remain the mainstay of refineries coping with an influx of heavy oil and bitumen, but other process options will be used. For example, visbreaking (or even hydrovisbreaking, i.e., visbreaking in an atmosphere of hydrogen or in the presence of a hydrogen donor material) (Chapter 5: Thermal Cracking), the long ignored stepchild of the refining industry, may see a surge in use as a pretreatment process. Management of the process to produce a liquid product that has been freed of the high potential for coke deposition (by taking the process parameters into the region where sediment forms) either in the absence or presence of, for example, a metal oxide scavenger could be valuable ally to catalyst cracking or hydrocracking units.

Further to refining viscous feedstocks, high-conversion refineries will move to gasification of feedstocks for the development of alternative fuels and to enhance equipment usage. In this sense, plasma gasification can be expected to play a greater role than it has done in the past. This process (Rabovitser et al., 2010) is a method and system for producing product gases from a variety of carbonaceous feedstocks in which the feedstock is introduced into a nonthermal plasma reactor at a temperature in the range of about 300°C–700°C (570°F–1290°F) and a pressure in the order of 15–1000 psi. A nonthermal plasma discharge is generated within the nonthermal plasma reactor and the feedstock/oxygen carrier are exposed to the nonthermal plasma discharge, which results in the formation of a product gas composed of hydrocarbon derivatives as well hydrogen and/or carbon monoxide.

In the integration of refining and petrochemical businesses, new technologies based on the traditional fluid catalytic cracking process (Chapter 6: Catalytic Cracking) will be of increased interests to refiners because of their potential to meet the increasing demand for light olefins. Meanwhile, hydrocracking, due to its flexibility, will take the central position in the integration of refining and petrochemical businesses in 21st century. Alternately, operating the catalytic cracking unit solely as a slurry riser cracker (without the presence of the main reactor) followed by separation of coke (sediment) would save the capital outlay required for a new catalytic cracker and might even show high conversion to valuable liquids. The quality (i.e., boiling range) of the distillate would be dependent upon the residence time of the slurry in the pipe.

Scavenger additives such as metal oxides may also see a surge in use. As a simple example, a metal oxide (such as calcium oxide) has the ability to react with sulfur-containing feedstock to produce a hydrocarbon (and calcium sulfide):

$$S_{feedstock} + CaO \rightarrow \text{hydrocarbon product} + CaS + H_2O$$

Propane has been used extensively in deasphalting heavy feedstocks, especially in the preparation of high-quality lubricating oils and feedstocks for catalytic cracking units (Chapter 7: Deasphalting and Dewaxing). The use of propane has necessitated elaborate solvent cooling systems utilizing cold water, which is a relatively expensive cooling agent. In order to circumvent such technology, future units will use solvent systems that will allow operation at elevated temperatures relative to conventional propane deasphalting temperatures, thereby permitting easy heat exchange. In addition, it may be found desirable to integrate dewaxing operations with deasphalting operations by having a common solvent recovery system. This will require changes to the solvent composition and the inclusion of solvents not usually considered to be deasphalting solvents.

Furthermore, as a means of energy reduction for the process, in future deasphalting units the conventional solvent recovery scheme will be retrofitted with supercritical solvent recovery scheme to reap benefits of higher energy efficiency. Other improvements will include variations in the extraction column internals.

For example, the three major properties, which influence the design of extraction column, are interfacial tension, viscosity, and density of phases. Solvent deasphalting extraction column is characterized by low interfacial tension, high viscosities of asphaltene phase, and a density difference between the phases. Extension of these property correlations for solvent deasphalting applications will be suitably validated and corrections made where necessary to improve extraction performance and yields of the products.

Other areas of future process modification will be in the extractor tower internals, studies with higher molecular weight solvent, accurate estimation of physical properties of mix stream, studies in combination with other processes, and firming up design tools for supercritical solvent recovery configuration.

In the long term, new desulfurization technologies (such as the biotechnologies) or the evolution of the older technologies will reduce the need for hydrogen (Chapter 8: Desulfurization, Denitrogenation, and Demetalization). At the same time, refineries are constantly faced with challenges to reduce air pollution and other energy-related issues. Thus traditional end-of-pipe air emission–control technologies will lead to increased energy use and decreasing energy efficiency in the refinery. The crude oil refining industry will face many other challenges—climate change, new developments in automotive technology and biotechnology—which are poised to affect the future structure of refineries.

The increasing focus to reduce the sulfur content of liquid fuels will assure that the role of desulfurization in the refinery increases in importance (Babich and Moulijn, 2003). Currently, the process of choice is the hydrotreater, in which hydrogen is added to the fuel to remove the sulfur from the fuel. Some hydrogen may be lost to reduce the octane number of the fuel, which is undesirable.

Because of the increased attention for fuel desulfurization various new process concepts are being developed with various claims of efficiency and effectiveness. In the major developments in desulfurization three main routes will be (1) advanced hydrotreating (new catalysts, catalytic distillation, processing at mild conditions); (2) reactive adsorption (type of adsorbent used, process design); and (3) oxidative desulfurization (catalyst, process design).

In addition, the most common approaches to upgrading hydrotreaters for clean fuel production will continue to be (1) developing higher activity and more resilient catalysts, (2) replacing reactor internals for increased efficiency, (3) adding reactor capacity to accommodate heavy feedstocks and increase gasoline-diesel production, (4) increasing hydrogen partial pressure, and (5) developing process design and hardware that are more specialized and focusing on process schemes that effectively reduce hydrogen consumption.

However, residuum hydrotreating requires considerably different catalysts and process flows, depending on the specific operation so that efficient hydroconversion through uniform distribution of liquid, hydrogen-rich gas, and catalyst across the reactor is assured. In addition to an increase in guard bed use, the industry will see an increase in automated demetallization of fixed bed systems as well as more units that operate as ebullating-bed hydrocrackers.

For heavy oil upgrading, hydrotreating technology (Chapter 8: Desulfurization, Denitrogenation, and Demetalization) and hydrocracking technology (Chapter 9: Hydrocracking) will be the processes of choice. For cleaner transportation fuel production the main task is the desulfurization of gasoline and diesel. With the advent of various techniques, such as adsorption and biodesulfurization, the future development will be still centralized on hydrodesulfurization techniques.

In fact, hydrocracking (Chapter 9: Hydrocracking) will continue to be an indispensable processing technology to modern crude oil refining and petrochemical industry due to its flexibility to feedstocks and product scheme, and high-quality products. Particularly, high-quality naphtha, jet fuel, diesel, and lube base oil can be produced through this technology. The hydrocracker provides a better balance of gasoline and distillates; improves gasoline yield and octane quality; and can supplement the fluid catalytic cracker to upgrade heavy feedstocks. In the hydrocracker, light fuel oil is converted into lighter products under a high hydrogen pressure and over a hot catalyst bed—the main products are naphtha, jet fuel, and diesel oil.

For the heavy feedstocks (and even for biofeedstocks), which will increase in amounts in terms of hydrocracking feedstocks, reactor designs will continue to focus on online catalyst addition and withdrawal. Fixed bed designs have suffered from (1) mechanical inadequacy when used for the heavier feedstocks and (2) short catalyst lives—6 months or less—even though large catalyst volumes are used (liquid hourly space velocity typically in the order of 0.5–1.5). Refiners will attempt to overcome these shortcomings by innovative designs, allowing better feedstock flow and catalyst utilization or online catalyst removal. For example, the OCR process, in which a lead, moving bed reactor is used to demetallize heavy feed ahead of the fixed bed hydrocracking reactors has seen some success. But whether this will be adequate for continuous hydrocracking heavy feedstock remains a question.

Catalyst development (Chapters 6, 8, and 9) will be key in the modification of processes and the development of new ones to make environmentally acceptable fuels (Rostrup-Nielsen, 2004). Conversion of crude oil is expected to remain the principal source of motor fuels for another 30–50 years, but it is likely that the production of fuel additives in large quantities along with conversion of natural gas will become significant (Sousa-Aguiar et al., 2005). Although crude oil conversion is expected to remain the principal source of fuels and petrochemicals in the future, natural gas reserves are emerging, and will continue to emerge, as a major hydrocarbon resource. This trend has already started to result in a shift toward use of natural gas (methane) as a significant feedstock for chemicals and for fuels as well. As a result, deployment of technology for direct and indirect conversion of methane will probably displace much of the current production of liquefied natural gas.

Innovations have occurred in catalyst materials that have allowed refiners to vastly improve environmental performance, product quality and volume, feedstock flexibility, and energy management without fundamentally changing fixed capital stocks. Advanced design and manufacturing techniques mean that catalysts can now be formulated and manufactured for specific processing units, feedstocks, operating environments, and finished product slates. These efforts will continue and more simper cheaper catalysts will be developed. The precise configuration of the refinery of the future is unknown but it is certain that no two refineries will adapt in exactly the same way.

Over the past century, the refining industry has been innovative and able to develop new processes. This trend will continue innovate and by the period 2030–50, refineries will be more technologically advanced and the products will be more environmentally acceptable.

However, the evolution of the refinery of the future will not be strictly confined to crude oil processes. The major consequence will be a much more environment-friendly product quality. These will be solved in refinery of the future, the refinery beyond 2020 with the development of deep conversion processing, such as residue hydrocracking and the inclusion of processes to accommodate other feedstocks.

The panacea (rather than a Pandora's box) for a variety of feedstocks could well be the gasification refinery (Fig. 15.2). This type of refinery approaches that of a petrochemical complex is capable of not only supplying the traditional refined products (Fig. 15.3), but also meeting much more severe specifications, and petrochemical intermediates such as olefins, aromatics, hydrogen, and methanol.

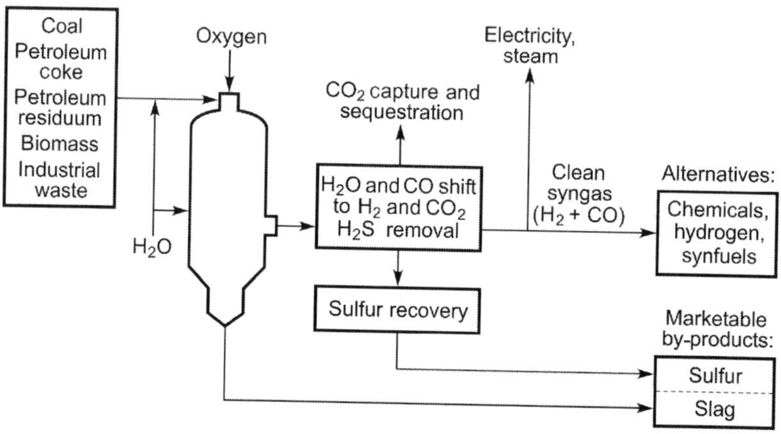

Figure 15.2 The gasification process can accommodate a variety of carbonaceous feedstocks.

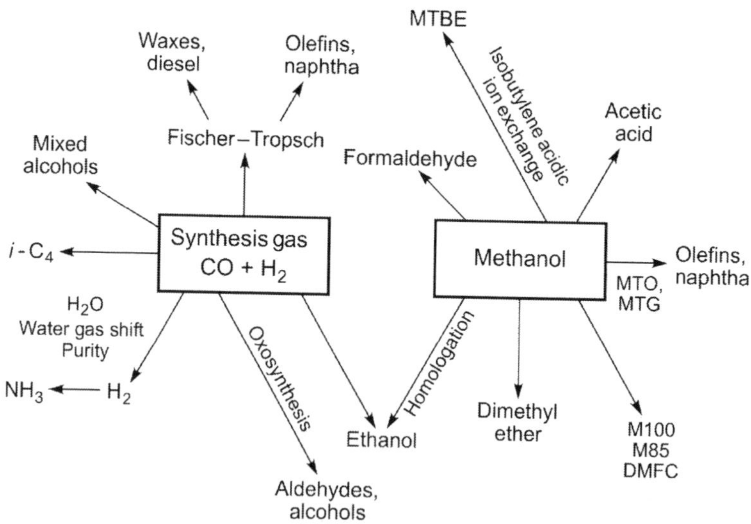

Figure 15.3 Products from a gasification refinery.

Furthermore, as already noted earlier, IGCC can be used to raise power from feedstocks such as vacuum residua and cracked residua and in addition to the production of synthesis gas, a major benefit of the IGCC concept is that power can be produced with the lowest sulfur oxide (SO_x) and nitrogen oxide (NO_x) emissions of any liquid/solid feed power generation technology.

The typical US refinery in the year 2030 will be located at an existing refinery site because economic and environmental considerations will make it difficult and uneconomical to build a new refinery at another site. Many existing refining process may still be in use but they will be more efficient and more technologically advanced and perhaps even rebuilt (reactors having been replaced on a scheduled or as needed basis) rather than retrofitted. However, energy efficiency will still be a primary concern, as refiners seek to combat the inevitable increasing cost of crude oil and refinery operating expenses.

Moreover, the future of the crude oil refining industry will be primarily on processes for the production of improved quality products. In addition to heavy ends deep conversion, there will also be changes in the feedstock into a refinery. Biomass, liquids from coal, and liquids from oil shale will increase in importance. These feedstocks (1) will be sent to refineries or (2) processed at a remote location and then blended with refinery stocks are options for future development and the nature of the feedstocks. Above all, such feedstock must be compatible with refinery feedstocks and not cause fouling in any form.

The refinery of the future will have a gasification section devoted to the conversion of coal and biomass to Fischer–Tropsch hydrocarbons—perhaps even with rich oil shale added to the gasifier feedstock. Many refineries already have gasification capabilities, but the trend will increase to the point (over the next two decades) nearly all refineries feel the need to construct a gasification section to handle residua and other feedstocks.

In addition, with the entry into the 21st century, crude oil refining is experiencing substantial innovation driven by the increasing supply of heavy oils with decreasing quality and the fast increases in the demand for clean and ultraclean vehicle fuels and petrochemical raw materials. As feedstocks to refineries change, there must be an accompanying change in refinery technology. This means a movement from conventional means of refining heavy feedstocks using (typically) coking technologies to more innovative processes (including hydrogen management and microbial transformation) that will produce the ultimate amounts of liquid fuels from the feedstock and maintain emissions within environmental compliance.

Although the advances in petroleum biotechnology have been bioremediation and biodegradation, there are also efforts to produce products from crude oil using biotransformation processes. That would take place through biodesulfurization, biodenitrogenation, biodemetallization, and biodepolymerization of asphaltene constituents. However, the most advanced area is the biodesulfurization, for which pilot plants have been established.

The conventional technologies cannot achieve the target of ultralow sulfur diesel in a cost-effective manner. Thus a combination of biodesulfurization and hydrodesulfurization technologies is suggested to achieve the future goals. In fact, significant progress has been made toward the commercialization of crude oil biodesulfurization. This progress includes the isolation and characterization of crude oil candidates for the biodesulfurization process; improved biocatalyst performance that directly relates to crude oil biodesulfurization (thermotolerant, solvent tolerant, overcome the toxicity of by-products); development of analytical methodology, which led to breakthroughs in the characterization of biodesulfurization-recalcitrant compounds; development of a process concept for crude oil biodesulfurization; and construction and testing of a prototype bench unit. Technical hurdles still need to be overcome to achieve commercialization.

The major obstacles to the economical biodesulfurization of crude oil include biocatalyst specificity, rate, and reusability and stability for a long time under process conditions. Work continues to modify the biocatalyst to increase its effectiveness and to screen other organisms for additional desulfurization capabilities. In addition, mass transfer and separations hurdles must be overcome in crude oils with increased oil viscosity and density.

Thus there is a strong motivation toward crude oil upgrading and desulfurization prior to refining are the economics of higher oil value, lower processing costs, and reduced air pollution that causes smog and acid rain. In addition, the advantages of the application of biotechnology to petroleum industry refining are (1) the potential simplicity of the process, (2) the selectivity of the biocatalysts, (3) low pressure, (4) low temperature, (5) low-to-no hydrogen requirement, (6) low chemical costs, and (7) minimal equipment investment when compared to the conventional refinery operations.

Furthermore, in addition to the whole-cell catalyst option, another plausible option offered by biotechnology is enzyme-based catalysis. The enzyme-based catalysts offer advantages over whole-cell catalysts. For example, an enzyme may function in very low-water-content environments and, in contrast with to whole-cell catalysts, an enzyme may even be less prone to thermal and organic solvent denaturation if the environment is hydrophobic. For the design perspective, it is usually more straightforward to modulate the kinetic and stability properties of a single protein than the properties of an enzymatic cascade, intrinsically constrained by cell metabolism.

Multiple strategies, such as immobilization, solvent engineering, and protein engineering, may be combined to enhance the desired characteristics of the enzyme. Heavy oils require a temperature in the order of 60°C−100°C (140°F−212°F) to be free-flowing fluids, and there is the need to develop microbial species and enzymes that can exist and be active in that temperature range.

Finally, the biotransformation refinery—in which feedstocks are converted into the products using microbes—would allow microbial transformation of the lower sulfur, lower nitrogen, and lower metal portions of the feedstock. This option is coupled with a gasification refinery that would have, as the center piece, gasification technology to produce synthesis gas (from the microbe-resistant portions of the feedstock) from which liquid fuels would be manufactured using the Fischer−Tropsch synthesis technology thereby increasing overall refinery efficiency (conversion).

The production of high-quality fuels will result in a higher demand for related hydrogen and conversion technologies. Furthermore, the trend toward low-sulfur fuels and changes in the product mix of refineries will affect technology choice and needs. For example, the current desulfurization and conversion technologies use relatively large amounts of hydrogen (which is an energy-intensive product) and increased hydrogen consumption will lead to increased energy use and operation expenses, unless more efficient technologies for hydrogen production developed.

The demand for high-value crude oil products will maximize production of transportation fuels at the expense of both residua and light gases. Hydroprocessing of residua will be widespread rather than appearing on selected refineries. At the same time, hydrotreated residua will be the common feedstocks for fluid catalytic cracking units. Additional conversion capacity will be necessary to process increasingly heavier crudes and meet a reduced demand for residua.

Process unit and refinery economics/operations computer models will be optimized, with integration into plant operations via process computer controls. Alternate fuels for power generation will continue to push crude processing toward higher value products, such as transportation fuels and chemicals (Courty and Gruson, 2001). Otherwise heavy crude oils, extra heavy crude oils, and tar sand bitumen that are considered uneconomical to transport to a refinery will be partially refined at their source to facilitate transport; and there will be a new emphasis on partial or full upgrading in situ during the respective recovery operations (Speight, 2014, 2017).

The biomass refinery of the future will not only use multiple feedstocks but also it will be able to shift output from the production of one chemical to another in response to market demands. Given that biomass will be a part of a refinery of the future, refiners may dictate that biomass receives preliminary upgrading at the biomass site before being shipped to the crude oil refinery. In fact, a biorefinery can be considered to be an integral unit that can accept various biological nonfood feedstocks and convert them into a range of useful products, including chemicals, energy, and materials. Products based on biomass may well be the replacements for fossil-based products thereby guaranteeing that biomass, a renewable source of carbon, is guaranteed a place in the new energy portfolio in the not-too-distant future.

Other challenges facing the refining industry include its capital-intensive nature and dealing with the disruptions to business operations that are inherent in industry. It is imperative for refiners to raise their operations to new levels of performance through a thorough study of the behavior of the changing feedstocks and the behavior of these feedstocks in the changing refinery (Table 15.3).

To circumvent these issues, there may be no way out of energy production than to consorting alternative energy sources with crude oil, and not of opposing them. This leads to the concept of alternative energy systems, which is wider ranging and more meaningful than alternative energy sources, because it relates to the actual transformation process of the global energy system (Szklo and Schaeffer, 2005). Alternative energy systems integrate crude oil with other energy sources and pave the way for new systems where refinery flexibility will be a key target, especially when related to the increased use of renewable energy sources.

Table 15.3 Continuing needs in light of the changes in feedstock type and composition.

Need	Comment
Kinetic modeling	Allows the quantitative simulation of commercially important, complex chemical reactions based on molecular reaction paths
Reaction engineering	Integrates the reaction and kinetic phenomena, including catalysis, with transport processes in the areas of reactor design, scale-up, and commercial operation
Automation and control	Enables the optimization of unit operation and economic performance
Compositional modeling	Relates to the nature of the feedstock through explicit reaction pathways on a molecular basis
	Allows quantitative prediction of composition and properties of product streams based on molecular composition

References

Alawad, I., Al-Zubaidi, I., 2019. Advances in upgrading process of petroleum residue: a review. EJERS, Eur. J. Eng. Res. Sci. 4 (6), 104–110. Available from: https://doi.org/10.24018/ejers.2019.4.6.1123.

Allred, V.D. (Ed.), 1982. Oil Shale Processing Technology. Center for Professional Advancement, East Brunswick, NJ.

Babich, I.V., Moulijn, J.A., 2003. Science and technology of novel processes for deep desulfurization of oil refinery streams: a review. Fuel 82 (2003), 607–631.

Baughman, G.L., 1978. Synthetic Fuels Data Handbook., second ed. Cameron Engineers, Inc, Denver, CO.

Beck, R., 2019. A vision for the refinery of 2030. Hydrocarbon Process. 23–24.

Bhatia, S., Sharma, D.K., 2006. Emerging role of biorefining of heavier crude oils and integration of biorefining with petroleum refineries in the future. Pet. Sci. Technol. 24 (10), 1125–1159.

Countryman, T., 2019. Digital strategy – smart moves. Pet. Rev. 22–23.

Courty, P., Gruson, J.F., 2001. Refining clean fuels for the future. Oil & gas science and technology – Revue d'IFP Energies nouvelles. Inst. Fr. Pét. 56 (5), 515–524.

Couvaras, G., 1997. Sasol's slurry phase distillate process and future applications. In: Proceedings. Monetizing Stranded Gas Reserves Conference. Houston, TX, December 1997.

Davis, R.A., Patel, N.M., 2004. Refinery hydrogen management. Petroleum Technology Quarterly. Spring, pp. 29–35.

Gary, J.H., Handwerk, G.E., 2001. Petroleum Refining: Technology and Economics, fourth ed. Marcel Dekker Inc, New York.

Gary, J.G., Handwerk, G.E., and Kaiser, M.J. 2007. Petroleum Refining: Technology and Economics, 5th Edition. CRC Press, Taylor & Francis Group, Boca Raton, Florida. Hedrick, B.W., Seibert, K.D., Crewe, C., 2006. A New Approach to Heavy Oil and Bitumen Upgrading. AM-06-29. UOP 4355B. UOP LLC, Des Plaines, IL.

Hickman, D.A., Schmidt, L.D., 1993. Production of syngas by direct catalytic oxidation of methane. Science 259, 343–346.

Hsu, C.S., Robinson, P.R. (Eds.), 2017. Handbook of Petroleum Technology. Springer International Publishing AG, Cham.

Huber, G.W., Corma, A., 2007. Synergies between bio- and oil refineries for the production of fuels from biomass. Angew. Chem. 46 (38), 7184–7201.

Le Borgne, S., Quintero, R., 2003. Biotechnological processes for the refining of petroleum. Fuel Process. Technol. 81 (2), 155–169.

Lee, S., 1991. Oil Shale Technology. CRC Press, Taylor & Francis Group, Boca Raton, FL.

Lee, S., 1996. Alternative Fuels. CRC Press, Taylor & Francis Group, Boca Raton, FL.

Lynd, L.R., Wyman, C., Laser, M., Johnson, D., Landucci, R., 2005. Strategic biorefinery analysis: review of existing biorefinery examples January 24-July 1, 2002. Subcontract Report NREL/SR-510-34895 October. National Renewable Energy Laboratory, Golden, CO.

Lynd, L.R., Larson, E., Greene, N., Laser, M., Sheehan, J., Dale, B.E., et al., 2009. Biofuels, Bioprod. Biorefin. 3 (2), 113–123.

Marcilly, C., 2003. Present status and future trends in catalysis for refining and petrochemicals. J. Catal. 216 (1–2), 47–62.

Mushrush, G.W., Speight, J.G., 1995. Petroleum Products: Instability and Incompatibility. Taylor & Francis Publishers, Washington, DC.

Mushrush, G.W., Speight, J.G., 1998. Chapter 8: Instability and incompatibility of petroleum products. In: Speight, J.G. (Ed.), Petroleum Chemistry and Refining. Taylor & Francis, Washington, DC.

Parkash, S., 2003. Refining Processes Handbook. Gulf Professional Publishing, Elsevier, Amsterdam.

Penning, R.T., 2001. Petroleum refining: a look at the future. Hydrocarbon Process. 80 (2), 45–46.

Rabovitser, I.K., Nester, S., Bryan, B., 2010. Plasma Assisted Conversion of Carbonaceous Materials into a Gas. United States Patent 7,736,400.

Rana, M.S., Ancheyta, J., Maity, S.K., Marroquin, G., 2008. Comparison between refinery processes for heavy oil upgrading: a future fuel demand. Int. J. Oil Gas. Coal Technol. 1 (3), 250–282.

Rostrup-Nielsen, J.R., 2004. Fuels and energy for the future: the role of catalysis. Catal. Rev. 46 (3&4), 247–270.

Scouten, C.S., 1990. Oil shale. Fuel Science and Technology Handbook. Marcel Dekker Inc, New York, pp. 795–1053, Chapters 25 to 31.

Shires, P., Salazar, N., Ariyapadi, S., 2010. Methods for Producing Synthesis Gas. United States Patent 7,722,690.

Sousa-Aguiar, E.F., Appel, L.G., Mota, C., 2005. Natural gas chemical transformations: the path to refining in the future. Catal. Today 10 (1), 3–7.

Stanislaus, A., Qabazard, H., and Absi-Halabi, M. 2000. Refinery of the future. In: Proceedings. 16th World Petroleum Congress, Calgary, AB, June 11–15.

Speight, J.G., 2008. Handbook of Synthetic Fuels. McGraw-Hill, New York.

Speight, J.G., 2011a. An Introduction to Petroleum Technology, Economics, and Politics. Scrivener Publishing, Beverly, MA.

Speight, J.G., 2011b. A biorefinery. In: Speight, J.G. (Ed.), The Biofuels Handbook. Royal Society of Chemistry, London, Part 1, Chapter 4.

Speight, J.G., 2012. Shale Oil Production Processes. Gulf Professional Publishing Company, Elsevier, Oxford.

Speight, J.G., 2013. The Chemistry and Technology of Coal, third ed. CRC Press, Taylor & Francis Group, Boca Raton, FL.

Speight, J.G., 2014. The Chemistry and Technology of Petroleum, fifth ed. CRC Press, Taylor & Francis Group, Boca Raton, FL.

Speight, J.G., 2017. Handbook of Petroleum Refining. CRC Press, Taylor & Francis Group, Boca Raton FL.

Speight, J.G., 2019b. Synthetic Fuels Handbook: Properties, Processes, and Performance, second ed. McGraw-Hill, New York.

Speight, J.G., 2020. Climate Change Demystified. Scrivener Publishing, Beverly, MA.

Speight, J.G., Islam, M.R., 2016. Peak Energy – Myth or Reality. Scrivener Publishing, Beverly, MA.

Szklo, A., Schaeffer, R., 2005. Alternative energy sources or integrated alternative energy systems? Oil as a modern lance of Peleus for the energy transition. Energy 31, 2513–2522.

US DOE, 2004a. Strategic significance of America's oil shale reserves, I. Assessment of strategic issues. <http://www.fe.doe.gov/programs/reserves/publications>.

US DOE, 2004b. Strategic significance of America's oil shale reserves, II. Oil shale resources, technology, and economics. <http://www.fe.doe.gov/programs/reserves/publications>.

US DOE, 2004c. America's Oil Shale: A Roadmap for Federal Decision Making. USDOE Office of US Naval Petroleum and Oil Shale Reserves. <http://www.fe.doe.gov/programs/reserves/publications>.

Zittel, W., Schindler, J., 2007. Crude Oil: The Supply Outlook. EWG Series No. 3/2007. Energy Watch Group, Berlin.

Conversion factors

1. *General*

   1 acre = 43,560 sq ft
   1 acre foot = 7758.0 bbl
   1 atmosphere = 760 mmHg = 14.696 psia = 29.91 in. Hg
   1 atmosphere = 1.0133 bars = 33.899 ft H_2O
   1 barrel (oil) = 42 gal = 5.6146 *cu* ft
   1 barrel (water) = 350 lb at 60°F
   1 barrel per day = 1.84 cu cm/s
   1 British thermal unit = 778.26 ft lb
   1 centipoise × 2.42 = lb mass/(ft) (h), viscosity
   1 centipoise × 0.000672 = lb mass/(ft) (s), viscosity
   1 cubic foot = 28,317 cu cm = 7.4805 gal
   Density of water at 60°F = 0.999 g/cu cm = 62.367 lb/cu ft = 8.337 lb/gal
   1 gallon = 231 cu in. = 3785.4 cu cm = 0.13368 cu ft
   1 horsepower-hour −0.7457 kWh = 2544.5 Btu
   1 horsepower = 550 ft lb/s = 745.7 W
   1 inch = 2.54 cm
   1 meter = 100 cm = 1000 mm = 10 μm = 10 A
   1 ounce = 28.35 g
   1 pound = 453.59 g = 7000 gr
   1 square mile = 640 ac

2. *Concentration conversions*

   1 part per million (1 ppm) = 1 microgram per liter (1 μg/L)
   1 microgram per liter (1 μg/L) = 1 milligram per kilogram (1 mg/kg)
   1 microgram per liter (μg/L) × 6.243×10^8 = 1 lb per cubic foot (1 lb/ft^3)
   1 microgram per liter (1 μg/L) × 10^{-3} = 1 milligram per liter (1 mg/L)
   1 milligram per liter (1 mg/L) × 6.243×10^5 = 1 pound per cubic foot (1 lb/ft^3)
   1 gram mole per cubic meter (1 g mol/m^3) × 6.243×10^5 = 1 pound per cubic foot (1 lb/ft^3)
   10,000 ppm = 1% w/w
   1 ppm hydrocarbon in soil × 0.002 = 1 lb of hydrocarbons per ton of contaminated soil

3. *Weight conversion*

   1 ounce (1 oz) = 28.3495 grams (28.3495 g)
   1 pound (1 lb) = 0.454 kilogram (0.454 kg)

1 pound (1 lb) = 454 grams (454 g)
1 kilogram (1 kg) = 2.20462 pounds (2.20462 lb)
1 stone (English, 1 st) = 14 pounds (14 lb)
1 ton (United States; 1 short ton) = 2000 pounds (2000 lb)
1 ton (English; 1 long ton) = 2240 pounds (2240 lb)
1 metric ton = 2204.62262 pounds (2204.62262 lb)
1 tonne = 2204.62262 pounds (2204.62262 lb)

4. *Temperature conversions*
$°F = (°C \times 1.8) + 32$
$°C = (°F - 32)/1.8$
$(°F - 32) \times 0.555 = °C$
Absolute zero = −273.15°C
Absolute zero = − 459.67°F

5. *Area*
1 square centimeter (1 cm^2) = 0.1550 square inches
1 square meter (1 m^2) = 1.1960 square yards
1 hectare (1 ha) = 2.4711 acres
1 square kilometer (1 km^2) = 0.3861 square miles
1 square inch (1 $in.^2$) = 6.4516 square centimeters
1 square foot (1 ft^2) = 0.0929 square meters
1 square yard (1 yd^2) = 0.8361 square meters
1 acre (1 ac) = 4046.9 square meters
1 square mile (1 mi^2) = 2.59 square kilometers

6. *Other approximations*
14.7 pounds per square inch (14.7 psi) = 1 atmosphere (1 atm)
1 kiloPascal (kPa) $\times 9.8692 \times 10^{-3}$ = 14.7 pounds per square inch (14.7 psi)
1 yd^3 = 27 ft^3
1 US gallon of water = 8.34 lb
1 imperial gallon of water—10 lb
1 yd^3 = 0.765 m^3
1 acre-inch of liquid = 27,150 gallons = 3.630 ft^3
1 ft depth in 1 ac (in situ) = 1613 × (20%−25% excavation factor) = ~2000 yd^3
1 yd^3 (clayey soils-excavated) = 1.1−1.2 t (United States)
1 yd^3 (sandy soils-excavated) = 1.2−1.3 t (United States)

7. *SI metric conversion factors*

E = exponent; that is, E + 03 = 10^3 and E − 03 = 10^{-3}

Acre-foot × 1.233482	E + 03 = meters cubed
Barrels × 1.589873	E − 01 = meters cubed
Centipoise × 1.000000	E − 03 = Pascal seconds
Darcy × 9.869233	E − 01 = micrometers squared
Feet × 3.048000	E − 01 = meters
Pounds/acre-foot × 3.677332	E − 04 = kilograms/meters cubed
Pounds/square inch × 6.894757	E + 00 = kiloPascals
Dyne per centimeter × 1.000000	E + 00 = milliNewton/meter
Parts per million × 1.000000	E + 00 = milligrams/kilograms

8. *Converting volumes of gas*

Multiply flow of	By	To obtain flow of
Natural gas	0.625	Propane
	0.547	Butane
	0.775	Air
Propane	1.598	Natural gas
	0.874	Butane
	1.237	Air
Butane	1.826	Natural gas
	1.140	Propane
	1.414	Air
Air	1.290	Natural gas
	0.808	Propane
	0.707	Butane

9. *Natural gas conversion table*

ft^3	CCF	Mcf	MMcf	Therm	Dekatherm	Btu	MMBtu	kJ	kWh
0.0009	0.00001	0.000001	0	0.00001	0.0001	0.9482	0.000001	1	0
0.001	0.00001	0.000001	0.000000001	0.00001	0.0001	1	0.000001	1.055	0
1	0.01	0.001	0.000001	0.01	0.001	1000	0.001	1054	0.293
100	1	0.1	0.0001	1	0.1	100000	0.1	105,461.5	29
1000	10	1	0.001	10	1	1,000,000	1	1,054,615	293
1,000,000	10,000	1000	1	10,000	1000	1.00E + 09	1000	1,054,615,000	293,071

Assumed 1 cf = 1000 Btu, for approximate reference only.

Heat content for various fuels

Automotive gasoline	125,000 Btu/gal (gross)	115,400 Btu/gal (net)
Hydrogen	134,200 Btu/kg (gross)	113,400 Btu/kg (net)
Diesel motor fuel	138,700 Btu/gal (gross)	128,700 Btu/gal (net)
Biodiesel	126,206 Btu/gal (gross)	117,093 Btu/gal (net)
Methanol	64,600 Btu/gal (gross)	56,560 Btu/gal (net)
Ethanol	84,600 Btu/gal (gross)	75,670 Btu/gal (net)
Gasohol	120,900 Btu/gal (gross)	112,417 Btu/gal (net)
Aviation gasoline	120,200 Btu/gal (gross)	112,000 Btu/gal (net)
Propane	91,300 Btu/gal (gross)	83,500 Btu/gal (net)
Butane	103,000 Btu/gal (gross)	93,000 Btu/gal (net)
Jet fuel (naphtha)	127,500 Btu/gal (gross)	118,700 Btu/gal (net)
Jet fuel (kerosene)	135,000 Btu/gal (gross)	128,100 Btu/gal (net)
Lubricants	144,400 Btu/gal (gross)	130,900 Btu/gal (net)
Waxes	131,800 Btu/gal (gross)	120,200 Btu/gal (net)
Asphalt and road oil	158,000 Btu/gal (gross)	157,700 Btu/gal (net)
Petroleum coke	143,400 Btu/gal (gross)	
Natural gas		
Wet	1109 Btu/ft^3	
Dry	1027 Btu/ft^3	
Compressed	960 Btu/ft^3	20,551 Btu/pound
Liquid	90,800 Btu/gal (gross)	87,600 Btu/gal (net)
Crude petroleum	138,100 Btu/gal (gross)	131,800 Btu/gal (net)
Fuel oil		
Residual	149,700 Btu/gal (gross) = 138,400 Btu/gal (net)	
Distillate	138,700 Btu/gal (gross) = 131,800 Btu/gal (net)	
Coal		
Anthracite	21.711×10^6 Btu/short ton (2000 lb)	
Bituminous and lignite	21.012×10^6 Btu/short ton (2000 lb)	

Biomass energy conversions

Cord: a stack of wood comprising 128 cubic feet (3.62 m^3); standard dimensions are 4 × 4 × 8 ft, including air space and bark. One cord contains approximately 1.2 US tons (ovendry) = 2400 pounds = 1089 kg

1.0 metric tonne wood = 1.4 cubic meters (solid wood, not stacked)

Energy content of wood fuel (HHV, bone dry) = 18–22 GJ/t (7600–9600 Btu/lb)

Energy content of wood fuel (air dry, 20% moisture) = about 15 GJ/t (6400 Btu/lb)

Energy content of agricultural residues (range due to moisture content) = 10–17 GJ/t (4300–7300 Btu/lb)

Metric tonne charcoal = 30 GJ (=12,800 Btu/lb) (but usually derived from 6 to 12 t air-dry wood, i.e., 90–180 GJ original energy content)

Metric tonne ethanol = 7.94 petroleum barrels = 1262 L

Ethanol energy content = 11,500 Btu/lb = 75,700 Btu/gal = 26.7 GJ/t = 21.1 MJ/L

Ethanol HHV = 84,000 Btu/gal = 89 MJ/gal = 23.4 MJ/L

Ethanol density (average) = 0.79 g/mL (=metric tonnes/m^3)

Metric tonne biodiesel = 37.8 GJ (33.3–35.7 MJ/L)

Biodiesel density (average) = 0.88 g/mL (=metric tonnes/m^3)

Mass conversions

	To				
From	**Pound**	**Kilogram**	**Short ton**	**Long ton**	**Metric ton**
Pound	1	0.4536	5.0×10^{-4}	4.4643×10^{-4}	4.5362×10^{-4}
Kilogram	2.205	1	1.1023×10^{-3}	9.8425×10^{-4}	1.0×10^{-3}
Short ton	2000	907.2	1	0.8929	0.9072
Long ton	2240	1016	1.12	1	1.016
Metric ton	2205	1000	1.102	0.9842	1

Volume and flow rate conversions[a]

1 US gal	=	231 $in.^3$	1 L	=	61.02 $in.^3$
	=	0.1337 ft^3		=	3.531×10^{-2} ft^3
	=	3.785 L		=	0.2624 US gal
	=	0.8321 imperial gal		=	0.2200 imperial gal
	=	0.0238 bbl		=	6.29×10^{-3} bbl
	=	0.003785 m^3		=	0.001 m

Composition of selected feedstocks

Feedstock	Cellulose	Hemicellulose	Lignin
Corn stover	35	28	16–21
Hardwood	45	30	20
Softwood	42	21	26
Bamboo	41–49	24–28	24–26
Switchgrass	44–51	42–50?	13–20
Miscanthus	44	24	17
Arundo donax	31	30	21

Biomass characteristics

Biomass feedstocks and fuels exhibit a wide range of physical, chemical, and agricultural/process engineering properties. Despite their wide range of possible sources, biomass feedstocks are remarkably uniform in many of their fuel properties, compared with competing feedstocks such as coal or petroleum. For example, there are many kinds of coals, gross heating value of which ranges from 20 to 30 GJ/tonne (gigajoules per metric tonne; 8600–12,900 Btu/lb). However, nearly all kinds of biomass feedstocks destined for combustion fall in the range of 15–19 GJ/tonne (6450–8200 Btu/lb). For most agricultural residues the heating values are even more uniform—about 15–17 GJ/tonne (6450–7300 Btu/lb); the values for most woody materials are 18–19 GJ/tonne (7750–8200 Btu/lb). Moisture content is probably the most important determinant of heating value. Air-dried biomass typically has about 15%–20% moisture, whereas the moisture content for ovendried biomass is around 0%. Moisture content is also an important characteristic of coals, varying in the range of 2%–30%. However, the bulk density (and hence energy density) of most biomass feedstocks is generally low, even after densification, about 10% and 40% of the bulk density of most fossil fuels. Liquid biofuels have comparable bulk densities to fossil fuels.

Most biomass materials are easier to gasify than coal because they are more reactive with higher ignition stability. This characteristic also makes them easier to process thermochemically into higher value fuels such as methanol or hydrogen. Ash content is typically lower than for most coals, and sulfur content is much lower than for many fossil fuels. Unlike coal ash, which may contain toxic metals and other trace contaminants, biomass ash may be used as a soil amendment to help replenish nutrients removed by harvest. A few biomass feedstocks stand out for their peculiar properties, such as high silicon or alkali metal contents—these may require special precautions for harvesting, processing, and combustion equipment. Note also that mineral content can vary as a function of soil type and the timing of feedstock harvest. In contrast to their fairly uniform physical properties, biomass fuels are rather heterogeneous with respect to their chemical elemental composition.

Among the liquid biomass fuels, biodiesel (vegetable oil ester) is noteworthy for its similarity to petroleum-derived diesel fuel, apart from its negligible sulfur and ash content. Bioethanol has only about 70% the heating value of petroleum distillates such as gasoline, but its sulfur and ash contents are also very low. Both of these liquid fuels have lower vapor pressure and flammability than their petroleum-based competitors—an advantage in some cases (e.g., use in confined spaces such as mines) but a disadvantage in others (e.g., engine starting at cold temperatures).

Fire hazard properties of flammable liquids, gases, and volatile solids

Substance	Lower flammable limit (v/v%)	Upper flammable limit (v/v%)	Flash point (°C)	Auto ignition temperature (°C)
Acetylene	2	83	–	305
Benzene	1	8	−11	560
Butane	2	9	–	365
Cyclohexane	1	8	−18	260
Ethane	3	12	–	515
Ethanol	3	19	12	363
Ethylene	3	34	–	305
Heptane	1	7	−4	204
Hexane	1	8	−22	223
Hydrogen	4	76	–	–
Methanol	6	37	11	385
Octane	1	7	13	206
Pentane	1	8	−40	285
Propane	3	12	–	515
Propylene	2	11	108	460
Toluene	1	7	4	530
o-Xylene	1	6	17	464
m-Xylene	1	7	25	528
p-Xylene	1	7	25	529

The main parameters that describe the ability of mixtures of substances in air to propagate a flame and/or explode are flammable limits, flash points, and autoignition temperatures. None of these parameters are basic physicochemical properties of a substance or mixture but are defined by standard tests, and for this reason, values are given to the nearest integer. If the test conditions are varied, then the values of the parameters will change. In particular, values will change considerably if the concentration of oxygen in the mixture is changed.

The flammable limits; lower flammable limit and upper flammable limit are the values, by volume, of the concentrations in air, at normal temperature and pressure, between which a mixture can propagate a flame.

The flash point is the lowest temperature, at one atmosphere pressure, at which a liquid gives off flammable vapor in sufficient quantity to ignite when mixed with air at or near the surface of the liquid on application of a flame or spark. Substances with a flash point $<21°C$ are termed "extremely flammable." Substances with flash points between 21°C and 55°C are termed "flammable," while the term "combustible" applies to all other combustible solids and liquids, including those with flash points above 100°C.

The autoignition temperature is the lowest temperature of a glass surface at which droplets of a combustible liquid falling onto the surface will undergo spontaneous combustion. A glass surface is assumed to be chemically inert and in practice, because of catalytic effects, the values for other types of surfaces can be lower.

Aerosols of combustible solids will also propagate a flame in the appropriate concentration range. However, the flammable limits of aerosols are affected by the size of the particles although, in general, the smaller the particle size, the closer the behavior of the aerosol resembles that of a vapor.

Glossary

Abiotic not associated with living organisms; synonymous with abiological.

Abiotic transformation the process in which a substance in the environment is modified by nonbiological mechanisms.

ABN separation a method of fractionation by which petroleum is separated into acidic, basic, and neutral constituents.

Absorber see absorption tower.

Absorption gasoline gasoline extracted from natural gas or refinery gas by contacting the absorbed gas with an oil and subsequently distilling the gasoline from the higher boiling components.

Absorption oil oil used to separate the heavier components from a vapor mixture by absorption of the heavier components during intimate contacting of the oil and vapor; used to recover natural gasoline from wet gas.

Absorption plant a plant for recovering the condensable portion of natural or refinery gas, by absorbing the higher boiling hydrocarbon derivatives in an absorption oil, followed by separation and fractionation of the absorbed material.

Absorption tower a tower or column that promotes contact between a rising gas and a falling liquid so that part of the gas may be dissolved in the liquid.

Acetic acid an acid with the structure of CH_3COOH; acetyl groups are bound through an ester linkage to hemicellulose chains, especially xylan derivatives, in wood and other plants; the natural moisture present in plants hydrolyzes the acetyl groups to acetic acid, particularly at elevated temperatures.

Acetone (CH_3COCH_3) trivial name for propanone, formed by the oxidation of 2-propanol with potassium permanganate.

Acetone–benzol process a dewaxing process in which acetone and benzol (benzene or aromatic naphtha) are used as solvents.

Acid any of a class of substances, the aqueous solutions of which are characterized by a sour taste, the ability to turn blue litmus red, and the ability to react with bases and certain metals to form salts; a substance that yields hydrogen ions when dissolved in water, which can act as a proton (H^+) donor.

Acid anhydride an organic compound that react with water to form an acid.

Acid–base partitioning the tendency for acids to accumulate in basic fluid compartments and bases to accumulate in acidic regions; also called *pH partitioning*.

Acid–base reaction a reaction in which an acidic hydrogen atom is transferred from one molecule to another.

Acid catalyst a catalyst having acidic character; the alumina minerals are examples of such catalysts.

Acid deposition acid rain; a form of pollution depletion in which pollutants, such as nitrogen oxides and sulfur oxides, are transferred from the atmosphere to soil or water; often referred to as atmospheric self-cleaning. The pollutants usually arise from the use of fossil fuels.

Acid detergent fiber (ADF) organic matter that is not solubilized after 1 hour of refluxing in an acid detergent of cetyltrimethylammonium bromide in 1N sulfuric acid; includes cellulose and lignin; this analytical method is commonly used in the feed and fiber industries.

Acid hydrolysis a chemical process in which acid is used to convert cellulose or starch to sugar.

Acidic a solution with a high concentration of H^+ ions.

Acid-insoluble lignin lignin is mostly insoluble in mineral acids and therefore can be analyzed gravimetrically after hydrolyzing the cellulose and hemicellulose fractions of the biomass with sulfuric acid; standard test method ASTM E1721 describes the standard method for determining acid-insoluble lignin in biomass; see American Society for Testing and Materials.

Acidity the capacity of an acid to neutralize a base such as a hydroxyl ion (OH^-).

Acidizing a technique for improving the permeability of a reservoir by injecting acid.

Acid number a measure of the reactivity of petroleum with a caustic solution and given in terms of milligrams of potassium hydroxide that are neutralized by 1 g of petroleum.

Acidophiles metabolically active in highly acidic environments and often have a high heavy metal resistance.

Acid rain the precipitation phenomenon that incorporates anthropogenic acids and other acidic chemicals from the atmosphere to the land and water (see acid deposition).

Acids, bases, and salts many inorganic compounds are available as acids, bases, or salts.

Acid sludge the residue left after treating petroleum oil with sulfuric acid for the removal of impurities; a black, viscous substance containing the spent acid and impurities.

Acid-soluble lignin a small fraction of the lignin in a biomass sample is solubilized during the hydrolysis process of the acid-insoluble lignin method. This lignin fraction is referred to as acid-soluble lignin and may be quantified by ultraviolet spectroscopy; see lignin and acid-insoluble lignin.

Acid treating a process in which unfinished petroleum products, such as gasoline, kerosene, and lubricating oil stocks, are contacted with sulfuric acid to improve their color, odor, and other properties.

Additive a material added to another (usually in small amounts) in order to enhance desirable properties or to suppress undesirable properties.

Adsorbent (sorbent) the solid phase or substrate onto which the sorbate adsorbs.

Adsorption the transfer of a substance from a solution to the surface of a solid resulting in relatively high concentration of the substance at the place of contact.

Adsorption gasoline natural gasoline obtained in the adsorption process from wet gas.

Aerobe an organism that needs oxygen for respiration and hence for growth.

Aerobic in the presence of, or requiring, oxygen; an environment or process that sustains biological life and growth or occurs only when free (molecular) oxygen is present.

Aerobic bacteria any bacteria requiring free oxygen for growth and cell division.

Aerobic conditions conditions for growth or metabolism in which the organism is sufficiently supplied with oxygen.

Aerobic respiration the process whereby microorganisms use oxygen as an electron acceptor.

Aerosol a dispersion of a liquid or solid in a gas.

Agitator a device such as a stirrer that provides complete mixing and uniform dispersion of all components in a mixture; is generally used continuously during the thermal processes and intermittently during fermentation.

Agricultural residue agricultural crop residues are the plant parts, primarily stalks and leaves, not removed from the fields with the primary food or fiber product; examples include corn stover (stalks, leaves, husks, and cobs); wheat straw; and rice straw.

Agricultural waste waste produced at agricultural premises as a result of an agricultural activity.

Air quality maintenance area specific populated area where air quality is a problem for one or more pollutants.

Alcohol the family name of a group of organic chemical compounds composed of carbon, hydrogen, and oxygen. The molecules in the series vary in chain length and are composed of a hydrocarbon plus a hydroxyl group. Alcohol includes methanol and ethanol.

Aldose a sugar derivative in which the carbonyl group of a monosaccharide is an aldehyde.

Algae algae are primitive plants, usually aquatic, capable of synthesizing their own food by photosynthesis; currently being investigated as a possible feedstock for producing biodiesel.

Alicyclic hydrocarbon a compound containing carbon and hydrogen only which has a cyclic structure (e.g., cyclohexane); also collectively called naphthenes.

Aliphatic hydrocarbon a compound containing carbon and hydrogen only which has an open-chain structure (e.g., as ethane, butane, octane, butene) or a cyclic structure (e.g., cyclohexane).

Aliphatic any nonaromatic organic compound having an open-chain structure.

Alkali a soluble mineral salt.

Alkali lignin lignin obtained by acidification of an alkaline extract of wood.

Alkali metal a metal in Group IA on the periodic table; an active metal which may be used to react with an alcohol to produce the corresponding metal alkoxide and hydrogen gas.

Alkanes hydrocarbon derivatives that contain only single carbon–hydrogen bonds. The chemical name indicates the number of carbon atoms and ends with the suffix "-ane."

Alkenes hydrocarbon derivatives that contain carbon–carbon double bonds. The chemical name indicates the number of carbon atoms and ends with the suffix "-ene."

Alkylation a process for manufacturing high-octane blending components used in unleaded petrol or gasoline.

Alkyl groups a group of carbon and hydrogen atoms that branch from the main carbon chain or ring in a hydrocarbon molecule. The simplest alkyl group, a methyl group (CH_3).

Alternative fuel Alumina (Al_2O_3) an inorganic chemical used in separation methods as an adsorbent and in refining as a catalyst.

Ambient air quality the condition of the air in the surrounding environment.

Amine washing a method of gas cleaning whereby acidic impurities such as hydrogen sulfide and carbon dioxide are removed from the gas stream by washing with an amine (usually an alkanolamine).

Anaerobic biological processes that occur in the absence of oxygen.

Anaerobic digestion decomposition of biological wastes by microorganisms, usually under wet conditions, in the absence of air (oxygen), to produce a gas comprising mostly methane and carbon dioxide.

Anhydrous a term meaning without water; transesterification of biodiesel must be an anhydrous process; water in the vegetable oil causes either no reaction or cloudy biodiesel, and water in lye or methanol renders it less useful or even useless, depending on how much water is present.

Amphoteric molecule a molecule that behaves both as an acid and as a base, such as hydroxy pyridine derivative in which the hydroxyl function is acidic and the nitrogen function is basic:

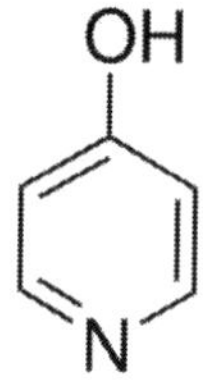

Anaerobe an organism that does not need free-form oxygen for growth. Many anaerobes are even sensitive to free oxygen.

Anaerobic a biologically mediated process or condition not requiring molecular or free oxygen; relating to a process that occurs with little or no oxygen present.

Anaerobic bacteria any bacteria that can grow and divide in the partial or complete absence of oxygen.

Anaerobic respiration the process whereby microorganisms use a chemical other than oxygen as an electron acceptor; common substitutes for oxygen are nitrate, sulfate, and iron.

Antiknock the resistance to detonation or pinging in spark-ignition engines.

Antiknock agent a chemical compound such as tetraethyl lead, which, when added in small amount to the fuel charge of an internal-combustion engine, tends to lessen knocking.

API gravity a measure of the lightness or heaviness of petroleum that is related to density and specific gravity:

$°API = (141.5/sp\ gr@60°F) - 131.5$

Aquatic plants the wide variety of aquatic biomass resources, such as algae, giant kelp, other seaweed, and water hyacinth; certain microalgae can produce hydrogen and oxygen while others manufacture hydrocarbon derivatives and a host of other products; microalgae examples include *Chlorella*, *Dunaliella*, and *Euglena*.

Aquifer a subsurface rock interval that will produce water; often the underlay of a petroleum reservoir.

Arabinan the polymer of arabinose; can be hydrolyzed to arabinose.

Arabinose a five-carbon sugar; a product of hydrolysis of arabinan found in the hemicellulose fraction of biomass.

Aromatic hydrocarbon a hydrocarbon derivative that is characterized by the presence of an aromatic ring or condensed aromatic rings; benzene and substituted benzene, naphthalene and substituted naphthalene, phenanthrene and substituted phenanthrene, as well as the higher condensed ring systems; compounds that are distinct from those of aliphatic compounds or alicyclic compounds.

Aromatics a range of hydrocarbon derivatives that have a distinctive sweet smell and include benzene and toluene occur naturally in petroleum and are also extracted as a petrochemical feedstock, as well as for use as solvents.

Aromatization the conversion of nonaromatic hydrocarbon derivatives to aromatic hydrocarbon derivatives by (1) rearrangement of aliphatic (noncyclic) hydrocarbon derivatives into aromatic ring structures and (2) dehydrogenation of alicyclic hydrocarbon derivatives (naphthenes); see aromatic hydrocarbon.

Asphalt the nonvolatile product obtained by distillation and treatment of an asphaltic crude oil; a manufactured product.

Asphalt cement asphalt that has been especially prepared as to quality and consistency for direct use in the manufacture of bituminous pavements.

Asphalt emulsion an emulsion of asphalt cement in water containing a small amount of emulsifying agent.

Asphaltene fraction (asphaltenes) the brown-to-black powdery material produced by treatment of petroleum, petroleum residua, or bituminous materials with a low-boiling liquid hydrocarbon, for example, pentane or heptane; soluble in benzene (and other aromatic solvents), carbon disulfide, and chloroform (or other chlorinated hydrocarbon solvents).

Asphalt flux an oil used to reduce the consistency or viscosity of hard asphalt to the point required for use.

Asphalt primer a liquid asphaltic material of low viscosity, which, upon application to a nonbituminous surface, to waterproof the surface and prepare it for further construction.

Asphaltene association factor the number of individual asphaltene species that associate in nonpolar solvents as measured by molecular weight methods; the molecular weight of asphaltenes in toluene divided by the molecular weight in a polar nonassociating solvent, such as dichlorobenzene, pyridine, or nitrobenzene.

Asphaltic pyrobitumen see asphaltoid.

Asphaltic road oil a thick, fluid solution of asphalt; usually a residual oil; see also nonasphaltic road oil.

Asphaltite a variety of naturally occurring, dark-brown-to-black, solid, nonvolatile bituminous material that is differentiated from bitumen primarily by a high content of material insoluble in *n*-pentane (asphaltene) or other liquid hydrocarbon derivatives.

Asphaltoid a group of brown-to-black, solid bituminous materials of which the members are differentiated from asphaltites by their infusibility and low solubility in carbon disulfide.

Asphaltum see asphalt.

Associated molecular weight the molecular weight of asphaltene constituents in an associating (nonpolar) solvent, such as toluene.

Atmospheric residuum a residuum obtained by distillation of a crude oil under atmospheric pressure, which boils above 350°C (660°F).

Atmospheric equivalent boiling point (AEBP) a mathematical method of estimating the boiling point at atmospheric pressure of nonvolatile fractions of petroleum.

ASTM International formerly the American Society for Testing and Materials (ASTM): the official organization in the United States for designing standard tests for petroleum and other industrial products.

Atmospheric pressure pressure of the air and atmosphere surrounding us, which changes from day to day; equal to 14.7 psia.

Attainment area a geographic region where the concentration of a specific air pollutant does not exceed federal standards.

Attapulgus clay see fuller's earth.

Autofining a catalytic process for desulfurizing distillates.

Average particle size the weighted average particle diameter of a catalyst.

Aviation gasoline any of the special grades of gasoline suitable for use in certain airplane engines.

Aviation turbine fuel see jet fuel.

Auger a rotating, screw-type device that moves material through a cylinder.

Available production capacity the biodiesel production capacity of refining facilities that are not specifically designed to produce biodiesel.

Average megawatt (MWa or aMW) one megawatt of capacity produced continuously over a period of 1 year; 1 aMW = 1 MW × 8760 hours/year = 8760 MWh = 8,760,000 kWh.

B100 another name for pure (100%) biodiesel.

Background level the average amount of a substance present in the environment. Originally referring to naturally occurring phenomena; used in toxic substance monitoring.

Backup rate a utility charge for providing occasional electricity service to replace on-site generation.

Backup electricity power or services needed occasionally; for example, when on-site generation equipment fails.

Baffle Chamber in incinerator design a chamber designed to settle fly ash and coarse particulate matter by changing the direction and reducing the velocity of the combustion gases.

Bagasse sugarcane waste.

Baghouse a filter system for the removal of particulate matter from gas streams; so called because of the similarity of the filters to coal bags.

Bark the outer protective layer of a tree outside the cambium comprising the inner bark and the outer bark; the inner bark is a layer of living bark that separates the outer bark from the cambium and in a living tree is generally soft and moist; the outer bark is a layer of dead bark that forms the exterior surface of the tree stem; the outer bark is frequently dry and corky.

Barrel (bbl) the unit of measure used by the petroleum industry; equivalent to approximately 42 US gallons or approximately 34 (33.6) Imperial gallons or 159 L; 7.2 bbl are equivalent to 1 tonne of oil (metric).

Barrel of oil equivalent (boe) the amount of energy contained in a barrel of crude oil, that is, approximately 6.1 GJ (5.8 million Btu), equivalent to 1700 kWh.

Base a classification of substances that when combined with an acid will form a salt plus water, usually producing hydroxide ions when dissolved.

Baseload capacity the power output that generating equipment can continuously produce.

Baseload demand the minimum demand experienced by an electric utility, usually 30%–40% of the utility's peak demand.

Base number the quantity of acid, expressed in milligrams of potassium hydroxide per gram of sample that is required to titrate a sample to a specified end point.

Base stock a primary refined petroleum fraction into which other oils and additives are added (blended) to produce the finished product.

Basic nitrogen nitrogen (in petroleum) that occurs in pyridine form.

Basic sediment and water (BS&W, BSW) the material that collects in the bottom of storage tanks, usually composed of oil, water, and foreign matter; also called bottoms, bottom settlings.

Batch distillation a process in which the liquid feed is placed in a single container and the entire volume is heated, in contrast to continuous distillation in which the liquid is fed continuously through the still.

Batch fermentation fermentation conducted from start to finish in a single vessel; see fermentation.

Batch process unit operation where one cycle of feedstock preparation, cooking, fermentation, and distillation is completed before the next cycle is started.

Battery a series of stills or other refinery equipment operated as a unit.

Baumé gravity the specific gravity of liquids expressed as degrees on the Baum (°Bé) scale; for liquids lighter than water:

Sp gr 60°F = 140/(130 + °BJ)

For liquids heavier than water:

Sp gr 60°F = 145/(145 − °BJ)

Bauxite mineral matter used as a treating agent; hydrated aluminum oxide formed by the chemical weathering of igneous rock.

bbl see barrel.

Beer a general term for all fermented malt beverages flavored with hops; a low level (6%−12%) alcohol solution derived from the fermentation of mash by microorganisms.

Beer still the stripping section of a distillation column for concentrating ethanol.

Bentonite montmorillonite (a magnesium−aluminum silicate); used as a treating agent.

Benzene a colorless aromatic liquid hydrocarbon (C_6H_6).

Benzin a refined light naphtha used for extraction purposes.

Benzine an obsolete term for light petroleum distillates covering the gasoline and naphtha range; see ligroine.

Benzol the general term that refers to commercial or technical (not necessarily pure) benzene; also the term used for aromatic naphtha.

Beta-scission the rupture of a carbon−carbon bond, two bonds removed from an aromatic ring.

Billion 1×10^9.

Biobutanol alcohol containing four carbon atoms per molecule, produced from the same feedstocks as ethanol, but with a modified fermentation and distillation process; less water-soluble than ethanol, biobutanol has a higher energy density and can be transported by pipeline more easily.

Bioconversion platform typically uses a combination of physical or chemical pretreatment and enzymatic hydrolysis to convert lignocellulose into its component monomers.

Biochemical conversion process the use of living organisms or their products to convert organic material to fuels, chemicals, or other products.

Biochemical oxygen demand (BOD) a standard means of estimating the degree of water pollution, especially of water bodies that receive contamination from sewage and industrial waste; the amount of oxygen needed by bacteria and other microorganisms to decompose organic matter in water—the greater the BOD, the greater the degree of pollution; biochemical oxygen demand is a process that occurs over a period of time and is commonly measured for a 5-day period, referred to as BOD5.

Biocide any chemical capable of killing bacteria and biorganisms.

Biochemical conversion the use of fermentation or anaerobic digestion to produce fuels and chemicals from organic sources.

Bioconcentration (bioaccumulation) the accumulation of a chemical in tissues of an organism to levels greater than in the environment in which the organism lives.

Biodegradable material of chemical that is capable of decomposing rapidly under natural conditions.

Biodegradable waste waste that is capable of being broken down by plants (including fungi) and animals (including worms and microorganisms); material or chemical.

Biodiesel a fuel derived from biological sources that can be used in diesel engines instead of petroleum-derived diesel; the triglycerides in the biologically derived oils that are produced glycerin, creating a clean-burning, renewable fuel.

Biodiesel blend a blend of biodiesel and diesel fuels—the blend can be with Diesel 1, Diesel 2, or JP8; one standard blend that meets the minimum requirements of the federal EPA Clean Air Act criteria is B20. The number after "B" indicates the percentage of biodiesel included in the blend—in B20, there would be 20% biodiesel and 80% diesel in the fuel blend; a biodiesel blend can come in any mixture percentage, such as B2, B5, B50, and B85.

Biodiesel recipe the most common recipe uses waste vegetable oil (WVO), methanol (wood alcohol), and sodium hydroxide (caustic soda/lye) to produce biodiesel and glycerin; the steps are (1) cleaning/heating waste vegetable oil, (2) titration of the waste vegetable oil sample, (3) combining methanol and sodium hydroxide in exact amounts, (4) mixing at 50°C, (5) settling, (6) separating the biodiesel from the wastes, (7) washing and drying the biodiesel, (8) disposing of wastes.

Bioenergy useful, renewable energy produced from organic matter—the conversion of the complex carbohydrates in organic matter to energy; organic matter may either be used directly as a fuel, processed into liquids and gasses, or be a residual of processing and conversion.

Bioethanol ethanol produced from biomass feedstocks; includes ethanol produced from the fermentation of crops, such as corn, as well as cellulosic ethanol produced from woody plants or grasses.

Biofuel a generic name for a liquid or gaseous fuel that is not derived from petroleum-based fossil fuels or contains a proportion of non–fossil fuel; fuels produced from plants, crops such as sugar beet, rape seed oil or reprocessed vegetable oils or fuels made from gasified biomass; fuels made from renewable biological sources and include ethanol, methanol, and biodiesel; sources include, but are not limited to, corn, soybeans, flaxseed, rapeseed, sugarcane, palm oil, raw sewage, food scraps, animal parts, and rice.

Biogas a combustible gas derived from decomposing biological waste under anaerobic conditions Biogas normally consists of 50%–60% methane. See also landfill gas.

Biogenic material derived from bacterial or vegetation sources.

Biological assessment a specific process required as part of an environmental assessment; an evaluation of potential effects of a proposed project on proposed, endangered, threatened, and sensitive animal and plant species and their habitats.

Biological lipid any biological fluid that is miscible with a nonpolar solvent. These materials include waxes, essential oils, chlorophyll, etc.

Biological oxidation the oxidative consumption of organic matter by bacteria by which the organic matter is converted into gases.

Biomass any organic matter that is available on a renewable or recurring basis, including agricultural crops and trees, wood and wood residues, plants (including aquatic plants), grasses, animal manure, municipal residues, and other residue materials. Biomass is generally produced in a sustainable manner from water and carbon dioxide by photosynthesis. There are three main categories of biomass—primary, secondary, and tertiary.

Biomass fuel liquid, solid, or gaseous fuel produced by conversion of biomass.

Biomass processing residues by-products from processing all forms of biomass that have significant energy potential; the residues are typically collected at the point of processing, they can be convenient and relatively inexpensive sources of biomass for energy.

Bionaphtha (bionaphtha) naphtha produced from biological sources; a term used in some eastern European nations for biodiesel.

Biorefinery a facility that processes and converts biomass into value-added products. These products can range from biomaterials to fuels such as ethanol or important feedstocks for the production of chemicals and other materials.

Biomass to liquid (BTL) the process of converting biomass to liquid fuels.

Biopolymer a high-molecular-weight carbohydrate produced by bacteria.

Biopower the use of biomass feedstock to produce electric power or heat through direct combustion of the feedstock, through gasification and then combustion of the resultant gas, or through other thermal conversion processes. Power is generated with engines, turbines, fuel cells, or other equipment.

Bioreactor a vessel in which a chemical process occurs, which usually involves organisms or biochemically active substances derived from such organisms.

Biorefinery a facility that processes and converts biomass into value-added products. These products can range from biomaterials to fuels such as ethanol or important feedstocks for the production of chemicals and other materials.

Bioremediation a treatment technology that uses biological activity to reduce the concentration or toxicity of contaminants: materials are added to contaminated environments to accelerate natural biodegradation.

Bitumen a naturally occurring material that has little or no mobility under reservoir conditions, which cannot be recovered through a well by conventional oil well production methods, including currently used enhanced recovery techniques; current methods involve mining for bitumen recovery; also, on occasion but incorrectly, referred to as native asphalt; asphalt is a refinery product.

Bituminous containing bitumen or constituting the source of bitumen.

Bituminous rock see bituminous sand.

Bituminous sand a formation in which the bituminous material (see bitumen) is found as a filling in veins and fissures in fractured rock or impregnating relatively shallow sand, sandstone, and limestone strata; a sandstone reservoir that is impregnated with a heavy, viscous black petroleum-like material that cannot be retrieved through a well by conventional production techniques.

Black acid(s) a mixture of the sulfonates found in acid sludge, which are insoluble in naphtha, benzene, and carbon tetrachloride; very soluble in water but insoluble in 30% sulfuric acid; in the dry, oil-free state, the sodium soaps are black powders.

Black liquor solution of lignin residue and the pulping chemicals used to extract lignin during the manufacture of paper.

Black oil any of the dark-colored oils; a term now often applied to heavy oil.

Black soap see black acid.

Black strap the black material (mainly lead sulfide) formed in the treatment of sour light oils with doctor solution and found at the interface between the oil and the solution.

Blown asphalt the asphalt prepared by air blowing a residuum or an asphalt.

Bogging a condition that occurs in a coking reactor when the conversion to coke and light ends is too slow causing the coke particles to agglomerate.

Boiler any device used to burn biomass fuel to heat water for generating steam.

Boiler horsepower a measure of the maximum rate of heat energy output of a steam generator; one boiler horsepower equals 33,480 Btu/h output in steam.

Boiling point a characteristic physical property of a liquid at which the vapor pressure is equal to that of the atmosphere and the liquid is converted to a gas.

Boiling range the range of temperature, usually determined at atmospheric pressure in standard laboratory apparatus, over which the distillation of an oil commences, proceeds, and finishes.

Bone-dry having 0% moisture content. Wood heated in an oven at a constant temperature of 100°C (212°F) or above until its weight stabilizes is considered bone-dry or oven-dry.

Bottled gas usually butane or propane, or butane–propane mixtures, liquefied and stored under pressure for domestic use; see also liquefied petroleum gas.

Bottoming cycle a cogeneration system in which steam is used first for process heat and then for electric power production.

Bottoms the nonvolatile liquid which collects in the bottom of a vessel (tower bottoms, tank bottoms) either during distillation; also the deposit or sediment formed during storage of petroleum or a petroleum product; see also residuum and basic sediment and water.

Brewing generically, the entire beer-making process, but technically only the part of the process during which the beer wort is cooked in a brew kettle and during which time the hops are added; after brewing the beer is fermented.

British thermal unit (Btu) a nonmetric unit of heat, still widely used by engineers; 1 Btu is the heat energy needed to raise the temperature of 1 lb of water from 60°F to 61°F at 1 atm pressure. 1 Btu = 1055 J (1.055 kJ).

Bromine number the number of grams of bromine absorbed by 100 g of oil, which indicates the percentage of double bonds in the material.

Brown acid the oil-soluble petroleum sulfonates found in acid sludge, which can be recovered by extraction with naphtha solvent. Brown-acid sulfonates are somewhat similar to mahogany sulfonates but are more water-soluble. In the dry, oil-free state, the sodium soaps are light-colored powders.

Brown soap see brown acid.

Brønsted acid a chemical species that can act as a source of protons.

Brønsted base a chemical species that can accept protons.

Brown grease waste grease that is the least expensive of the various grades of waste grease.

BS&W see basic sediment and water.

BTEX benzene, toluene, ethylbenzene, and the xylene isomers.

Btu (British thermal unit) the energy required to raise the temperature of 1 lb of water to 1°F.

BTX the collective name given to benzene, toluene, and the xylene isomers (*p*-, *m*-, and *o*-xylene); a group of volatile organic compounds (VOCs) found in petroleum hydrocarbons, such as gasoline, and other common environmental contaminants.

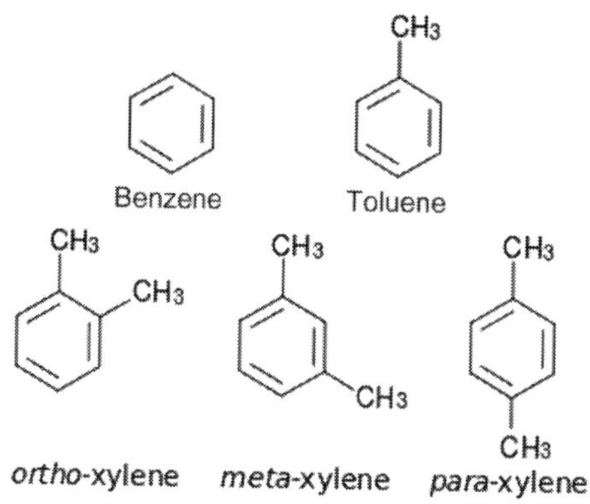

Bubble cap an inverted cup with a notched or slotted periphery to disperse the vapor in small bubbles beneath the surface of the liquid on the bubble plate in a distillation tower.

Bubble cap trays cross-flow trays usually installed in rectifying columns handling liquids free of suspended solids; the bubble caps consist of circular cups inverted over small vapor pipes—the vapor from the tray below passes through the vapor pipes into the caps and curves downward to escape below the rim into the liquid. The rim of each cap is slotted or serrated to break up the escaping vapor into small bubbles, thereby increasing the surface area of the vapor as it passes through the liquid.

Bubble plate a tray in a distillation tower.

Bubble point the temperature at which incipient vaporization of a liquid in a liquid mixture occurs, corresponding with the equilibrium point of 0% vaporization or 100% condensation.

Bubble tower a fractionating tower so constructed that the vapors rising pass up through layers of condensate on a series of plates or trays (see bubble plate); the vapor passes from one plate to the next above by bubbling under one or more caps (see bubble cap) and out through the liquid on the plate where the less volatile portions of vapor condense in bubbling through the liquid on the plate, overflow to the next lower plate, and ultimately back into the reboiler thereby affecting fractionation.

Bubble wash a method of final washing of biodiesel through air agitation. Biodiesel floats above a quantity of water; bubbles from an aquarium air pump and air stone are injected into the water causing the bubbles to rise—at the water/biodiesel interface, the air bubbles carry water up through the biodiesel by surface tension, simple diffusion causes water-soluble impurities in the biodiesel to be extracted into the water, as the bubble reaches the surface and breaks, the water is freed and percolates back down through the biodiesel again.

Butanol though generally produced from fossil fuels, this four-carbon alcohol can also be produced through bacterial fermentation of alcohol.

Bunker a storage tank.

Bunker C oil see No. 6 Fuel oil.

Burner fuel oil any petroleum liquid suitable for combustion.

Burning oil an illuminating oil, such as kerosene (kerosine) suitable for burning in a wick lamp.

Burning point see fire point.

Burning-quality index an empirical numerical indication of the likely burning performance of a furnace or heater oil; derived from the distillation profile and the API gravity, and generally recognizing the factors of paraffin character and volatility.

Butane dehydrogenation a process for removing hydrogen from butane to produce butenes and, on occasion, butadiene.

Butane vapor-phase isomerization a process for isomerizing *n*-butane to iso-butane using aluminum chloride catalyst on a granular alumina support and with hydrogen chloride as a promoter.

Butanol though generally produced from fossil fuels, this four-carbon alcohol can also be produced through bacterial fermentation of alcohol.

By-product a substance, other than the principal product, generated as a consequence of creating a biofuel.

C_1, C_2, C_3, C_4, C_5 fractions a common way of representing fractions containing a preponderance of hydrocarbon derivatives having 1, 2, 3, 4, or 5 carbon atoms, respectively, and without reference to hydrocarbon type.

CAA the Clean Air Act that is the foundation of air regulations in the United States.

Calcining the process of heating a metal oxide or an ore to decompose carbonates, hydrates, or other compounds often in a controlled atmosphere.

Canola a member of the Brassica Family, which includes broccoli, cabbage, cauliflower, mustard, radish, and turnip; it is a variant of the crop rapeseed, with less crucic acid and glucosinolate derivatives than rapeseed; grown for its seed, the seed is crushed for the oil contained within and, after the oil is extracted, the by-product is a protein-rich meal used by the intensive livestock industry.

Capacity the maximum power that a machine or system can produce or carry safely; the maximum instantaneous output of a resource under specified conditions—the capacity of generating equipment is generally expressed in kilowatts or megawatts.

Capacity factor the amount of energy that a power plant actually generates compared to its maximum rated output, expressed as a percentage.

Capillary forces interfacial forces between immiscible fluid phases, resulting in pressure differences between the two phases.

Capillary number, N_c the ratio of viscous forces to capillary forces, and equal to viscosity times velocity divided by interfacial tension.

Carbohydrate a chemical compound made up of carbon, hydrogen, and oxygen; includes sugars, cellulose, and starches.

Carbon chain the atomic structure of hydrocarbon derivatives in which a series of carbon atoms, saturated by hydrogen atoms, form a chain; volatile oils have shorter chains while fats have longer chain lengths, and waxes have extremely long carbon chains.

Carbon dioxide (CO_2) a product of combustion that acts as a greenhouse gas in the atmosphere of the Earth, trapping heat and contributing to climate change.

Carbon-forming propensity see carbon residue.

Carbonization the conversion of an organic compound into char or coke by heat in the substantial absence of air; often used in reference to the destructive distillation (with simultaneous removal of distillate) of coal.

Carbon monoxide (CO) a lethal gas produced by incomplete combustion of carbon-containing fuels in internal-combustion engines. It is colorless, odorless, and tasteless. (As in flavorless, we mean, though it is also been known to tell a bad joke or two.)

Carbon residue the amount of carbonaceous residue remaining after thermal decomposition of petroleum, a petroleum fraction, or a petroleum product in a limited amount of air; also called the coke- or carbon-forming propensity; often prefixed by the terms Conradson or Ramsbottom in reference to the inventor of the respective tests.

Carbon sequestration the absorption and storage of carbon dioxide from the atmosphere; naturally occurring in plants.

Carbon sink a geographical area, the vegetation and/or soil of which soaks up significant carbon dioxide from the atmosphere. Such areas, typically in tropical regions, are increasingly being sacrificed for energy crop production.

Catalyst a substance that accelerates a chemical reaction without itself being affected. In refining, catalysts are used in the cracking process to produce blending components for fuels.

Catalyst selectivity the relative activity of a catalyst with respect to a particular compound in a mixture, or the relative rate in competing reactions of a single reactant.

Catalyst stripping the introduction of steam, at a point where spent catalyst leaves the reactor, in order to strip, that is, remove, deposits retained on the catalyst.

Catalytic activity the ratio of the space velocity of the catalyst under test to the space velocity required for the standard catalyst to give the same conversion as the catalyst being tested; usually multiplied by 100 before being reported.

Catalytic cracking the conversion of high-boiling feedstocks into lower boiling products by means of a catalyst, which may be used in a fixed bed or fluid bed.

Cat cracking see catalytic cracking.

Cellulose fiber contained in leaves, stems, and stalks of plants and trees; most abundant organic compound on earth; it is a polymer of glucose with a repeating unit of $C_6H_{10}O_5$ strung together by ß-glycosidic linkages—the ß-linkages in cellulose form linear chains that are highly stable and resistant to chemical attack because of the high degree of hydrogen bonding that can occur between chains of cellulose; hydrogen bonding between cellulose chains makes the polymers more rigid, inhibiting the flexing of the molecules that must occur in the hydrolytic breaking of the glycosidic linkages—hydrolysis can reduce cellulose to a cellobiose repeating unit, $C_{12}H_{22}O_{11}$, and ultimately to the six-carbon sugar glucose, $C_6H_{12}O_6$.

Ceresin a hard, brittle wax obtained by purifying ozokerite; see microcrystalline wax and ozokerite).

Cetane index an approximation of the cetane number calculated from the density and mid-boiling point temperature; see also diesel index.

Cetane number a number indicating the ignition quality of diesel fuel; a high cetane number represents a short ignition delay time; the ignition quality of diesel fuel can also be estimated from the following formula:

Cetane rating a measure of the combustion quality of diesel fuel.

Characterization factor the UOP characterization factor *K*, which is the ratio of the cube root of the molal average boiling point, T_B, in degrees Rankine (°R = °F + 460), to the specific gravity at 60°F/60°F:

$K = (T_B)^{1/3}/\text{sp gr.}$

The value ranges from 12.5 for paraffin stocks to 10.0 for the highly aromatic stocks; also called the Watson characterization factor.

Chemical waste any solid, liquid, or gaseous material discharged from a process and that may pose substantial hazards to human health and environment.

Chips small fragments of wood chopped or broken by mechanical equipment—total tree chips include wood, bark, and foliage, while pulp chips or clean chips are free of bark and foliage.

Chlorofluorocarbon a family of chemicals composed primarily of carbon, hydrogen, chlorine, and fluorine; used principally as refrigerants and industrial cleansers and have the tendency to destroy the Earth's protective ozone layer.

Clarifier a tank used to remove solids by gravity, to remove colloidal solids by coagulation, and to remove floating oil and scum through skimming.

Clay a silicate mineral that also usually contains aluminum and has particle sizes are less than 0.002 μm; used in separation methods as an adsorbent and in refining as a catalyst.

Clay contact process see contact filtration.

Clay refining a treating process in which vaporized gasoline or other light petroleum product is passed through a bed of granular clay such as fuller's earth.

Clay regeneration a process in which spent coarse-grained adsorbent clays from percolation processes are cleaned for reuse by deoiling them with naphtha, steaming out the excess naphtha, and then roasting in a stream of air to remove carbonaceous matter.

Clay treating see gray clay treating.

Clay wash light (low-boiling, low-density) oil, such as naphtha or kerosene that is used to clean fuller's earth after it has been used in a filter.

Class I area any area designated for the most stringent protection from air quality degradation.

Class II area any area where air is cleaner than required by federal air quality standards and designated for a moderate degree of protection from air quality degradation; moderate increases in new pollution may be permitted in Class II areas.

Clean Air Act (CAA) the national law in the United States that establishes ambient air quality emission standards to be implemented by participating states; originally enacted in 1963, the CAA has been amended several times, most recently in 1990 and includes vehicle emission standards regulating the emission of criteria pollutants (lead, ozone, carbon monoxide, sulfur dioxide, nitrogen oxides and particulate matter); the 1990 amendments added reformulated gasoline (RFG) requirements and oxygenated gasoline provisions.

Clean fuels fuels such as E-10 (unleaded) that burn cleaner and produce fewer harmful emissions compared to ordinary gasoline.

Closed-loop biomass crops grown, in a sustainable manner, for the purpose of optimizing their value for bioenergy and bioproduct uses. This includes annual crops such as maize and wheat, and perennial crops such as trees, shrubs, and grasses such as switchgrass.

Cloud point the temperature at which paraffin wax or other solid substances begin to crystallize or separate from the solution, imparting a cloudy appearance to the oil when the oil is chilled under prescribed conditions.

Coal an organic rock.

Coal tar the specific name for the tar produced from coal.

Coal tar pitch the specific name for the pitch produced from coal.

Coarse materials wood residues suitable for chipping, such as slabs, edgings, and trimmings.

Cogeneration the sequential production of electricity and useful thermal energy from a common fuel source.

Coke a gray-to-black solid carbonaceous material produced from petroleum during thermal processing; characterized by having a high carbon content (95% + by weight) and a honeycomb type of appearance and is insoluble in organic solvents.

Coking a thermal method used in refineries for the conversion of bitumen and residua to volatile products and coke (see delayed coking and fluid coking).

Colloid a stable system of small particles dispersed in another phase; a multiphase system in which one dimension of a dispersed phase is of colloidal size; colloids are the liquid and solid forms of aerosols, foams, emulsions, and suspensions within the colloidal size class.

Colloidal size 0.001–1 μm in any dimension; dispersions where the particle size is in this range are referred to as colloidal aerosols, colloidal emulsions, colloidal foams, or colloidal suspensions.

Color stability the resistance of a petroleum product to color change due to phenomena such as light and aging.

Colza eurasian plant cultivated for its seed and as a forage crop.

Combined heat and power the use of a power station to simultaneously generate both heat and electricity. The steam or hot water generated in the process is utilized either in industrial processes or in community heating.

Combustion (burning) the transformation of biomass fuel into heat, chemicals, and gases through chemical combination of hydrogen and carbon in the fuel with oxygen in the air.

Combustion gases the gases released from a combustion process.

Commercial waste waste arising from premises used wholly or mainly for trade, business, sport, recreation or entertainment, excluding municipal waste and industrial waste.

Composting a resource recovery process where biodegradable waste (such as garden and kitchen waste) is converted, in the presence of oxygen from the air, into a stable granular material which, applied to land, improves soil structure and enriches the nutrient content.

Compound a chemical term denoting a combination of two or more distinct elements.

Compressed natural gas (CNG) natural gas that has been compressed under high pressure (typically 2000–3600 psi).

Compression-ignition engine an engine in which the fuel is ignited by high temperature caused by extreme pressure in the cylinder, rather than by a spark from a spark plug; diesel engine.

Con carbon see carbon residue.

Concentrated acid hydrolysis a method of converting biomass into cellulosic ethanol.

Condensate a mixture of light hydrocarbon liquids obtained by condensation of hydrocarbon vapors: predominately butane, propane, and pentane with some heavier hydrocarbon derivatives and relatively little methane or ethane; see also natural gas liquids.

Conditional use permit a permit, with conditions, allowing an approved use on a site outside the appropriate zoning class.

Conradson carbon residue see carbon residue.

Conservation efficiency of energy use, production, transmission, or distribution that results in a decrease of energy consumption while providing the same level of service.

Construction and demolition waste waste arising from the construction, repair, maintenance and demolition of buildings and structures, including roads. It consists mostly of brick, concrete, hardcore, subsoil, and topsoil; but it can also contain quantities of timber, metal, plastics, and (occasionally) hazardous waste materials.

Contaminant a substance that causes deviation from the normal composition of an environment.

Continuous fermentation a steady-state fermentation system that operates without interruption; each stage of fermentation occurs in a separate section of the fermenter, and flow rates are set to correspond with required residence times.

Continuous flow process a general term for any number of biodiesel production processes that involves the continuous addition of ingredients to produce biodiesel on a continual, round-the-clock basis, as opposed to the batch process.

Controlled waste the UK term for waste that is controlled under the Waste Framework Directive; includes household waste, commercial waste, industrial waste, and agricultural waste.

Conventional biofuels biofuels such as bioethanol and biodiesel, which are typically made from corn, sugarcane and beet, wheat, or oilseed crops such as soy and rapeseed oil.

Conventional crude oil (conventional petroleum) crude oil that is pumped from the ground and recovered using the energy inherent in the reservoir; also recoverable by application of secondary recovery techniques.

Conversion efficiency a comparison of the useful energy output to the potential energy contained in the fuel; the efficiency calculation relates to the form of energy produced and allows a direct comparison of the efficiency of different conversion processes that can be made only when the processes produce the same form of energy output.

Cooker a tank or vessel designed to cook a liquid or extract or digest solids in suspension; the cooker usually contains a source of heat; and is fitted with an agitator.

Cord a stack of wood comprising 128 cubic ft. (3.62 m^3); standard dimensions are 4 ft. by 4 ft. by 8 ft., including air space and bark; one cord contains approximately 1.2 US tons (oven-dry) = 2400 lb = 1089 kg.

Corn stover residue materials from harvesting corn consisting of the cob, leaves, and stalk.

Cp (centipoise) a unit of viscosity.

Cracking a secondary refining process that uses heat and/or a catalyst to break down high-molecular-weight chemical components into lower molecular weight products, which can be used as blending components for fuels.

Cropland Cropland pasture land used for long-term crop rotation. However, some cropland pasture is marginal for crop uses and may remain in pasture indefinitely. This category also includes land that was used for pasture before crops reached maturity and some land used for pasture that could have been cropped without additional improvement.

Crop residue the backbone of sugar and starch crops—the stalks and leaves—is composed mainly of cellulose. The individual six-carbon sugar units in cellulose are linked together in extremely long chains by a stronger chemical bond than exists in starch. As with starch, cellulose must be broken down into sugar units before it can be used by yeast to make ethanol. However, the breaking of the cellulose bonds is much more complex and costly than the breaking of the starch bonds. Breaking the cellulose into individual sugar units is complicated by the presence of lignin, a complex compound surrounding cellulose, which is even more resistant than cellulose to enzymatic or acidic pretreatment. Because of the high cost of converting liquefied cellulose into fermentable sugars, agricultural residues (as well as other crops having a high percentage of cellulose) are not yet a practical feedstock source for small ethanol plants.

Cull tree a live tree, 5.0 in. in diameter at breast height (d.b.h.) or larger that is nonmerchantable for saw logs now or prospectively because of rot, roughness, or species (see definitions for rotten and rough trees.)

Cultivated summer fallow cropland cultivated for one or more seasons to control weeds and accumulate moisture before small grains are planted.

Cut point the boiling-temperature division between distillation fractions of petroleum.

Cutback the term applied to the products from blending heavier feedstocks or products with lighter oils to bring the heavier materials to the desired specifications.

Cutback asphalt asphalt liquefied by the addition of a volatile liquid such as naphtha or kerosene, which, after application and on exposure to the atmosphere, evaporates leaving the asphalt.

Cutting oil an oil to lubricate and cool metal-cutting tools; also called cutting fluid, cutting lubricant.

Cycle stock the product taken from some later stage of a process and recharged (recycled) to the process at some earlier stage.

Cyclic steam injection the alternating injection of steam and production of oil with condensed steam from the same well or wells.

Cyclization the process by which an open-chain hydrocarbon structure is converted to a ring structure, for example, hexane to benzene.

Cyclone a device for extracting dust from industrial waste gases. It is in the form of an inverted cone into which the contaminated gas enters tangential from the top; the gas is propelled down a helical pathway, and the dust particles are deposited by means of centrifugal force onto the wall of the scrubber.

DDGS (dried distillers grain with soluble constituents) a by-product of dry mill ethanol production that is fed to livestock.

Delayed coking a coking process in which the thermal reactions are allowed to proceed to completion to produce gaseous, liquid, and solid (coke) products.

Dealkylation the removal of an alkyl group from aromatic compounds.

Deasphaltened oil the fraction of petroleum after the asphaltene constituents have been removed.

Deasphaltening the removal of a solid powdery asphaltene fraction from petroleum by the addition of the low-boiling liquid hydrocarbon derivatives such as *n*-pentane or *n*-heptane under ambient conditions.

Deasphalting the removal of the asphaltene fraction from petroleum by the addition of a low-boiling hydrocarbon liquid such as *n*-pentane or *n*-heptane; more correctly the removal asphalt (tacky, semisolid) from petroleum (as occurs in a refinery asphalt plant) by the addition of liquid propane or liquid butane under pressure.

Debutanization a distillation process that is to separate butane and lighter components from higher boiling components.

Decant oil the highest boiling product from a catalytic cracker; also referred to as slurry oil, clarified oil, or bottoms.

Decarbonizing a thermal conversion process designed to maximize coker gas-oil production and minimize coke and gasoline yields; operated at essentially lower temperatures and pressures than delayed coking.

Decolorizing the removal of suspended, colloidal, and dissolved impurities from liquid petroleum products by filtering, adsorption, chemical treatment, distillation, bleaching, etc.

Deethanization the distillation process used to separate ethane and lighter components from propane and higher boiling components; also called deethanation.

Degradation the loss of desirable physical properties of EOR fluids, for example, the loss of viscosity of polymer solutions.

Dehydrating agents substances capable of removing water (drying, *q.v.*) or the elements of water from another substance.

Dehydration reaction (condensation reaction) a chemical reaction in which two organic molecules become linked to each other via covalent bonds with the removal of a molecule of water; common in synthesis reactions of organic chemicals.

Dehydrohalogenation removal of hydrogen and halide ions from an alkane resulting in the formation of an alkene.

Dehydrocyclization any process by which both dehydrogenation and cyclization reactions occur.

Dehydrogenation the removal of hydrogen from a chemical compound; for example, the removal of two hydrogen atoms from butane to make butene(s) as well as the removal of additional hydrogen to produce butadiene.

Density the mass (or weight) of a unit volume of any substance at a specified temperature; see also specific gravity.

Desulfurization the removal of sulfur or sulfur compounds from a feedstock.

Diesel 1 and Diesel 2 Diesel 1 is also called kerosene and is not generally used as a fuel oil in diesel vehicles—it has a lower viscosity (it is thinner) than Diesel 2, which is the typical diesel vehicle fuel.

Diesel engine named for the German engineer Rudolph Diesel, this internal-combustion, compression-ignition engine works by heating fuels and causing them to ignite; can use either petroleum- or bio-derived fuel.

Diesel fuel a distillate of fuel oil that has been historically derived from petroleum for use in internal-combustion engines; also derived from plant and animal sources.

Diesel index [aniline point (°F) × API gravity]100.

Diesel, Rudolph German inventor famed for fashioning the diesel engine, which made its debut at the 1900 World's Fair; initially engine to run on vegetable-derived fuels.

Digester an airtight vessel or enclosure in which bacteria decomposes biomass in water to produce biogas.

Direct injection engine a diesel engine in which fuel is injected directly into the cylinder.

Distillate any petroleum product produced by boiling crude oil and collecting the vapors produced as a condensate in a separate vessel, for example, gasoline (light distillate), gas oil (middle distillate), or fuel oil (heavy distillate).

Distillation the primary distillation process which uses high temperature to separate crude oil into vapor and fluids which can then be fed into a distillation or fractionating tower.

Distiller's grains by-product of ethanol production that can be used to feed livestock; alternatively, distillers dried grains with solubles (DDGS).

Dispersion a stable or unstable system of fine particles, larger than colloidal size, evenly distributed in a medium.

Distillate oil any distilled product of crude oil; a volatile petroleum product used for home heating and most machinery.

Distillation the process to separate the components of a liquid mixture by boiling the liquid and then condensing the resulting vapor.

Domestic heating oil see No. 2 Fuel Oil.

Downdraft gasifier a gasifier in which the product gases pass through a combustion zone at the bottom of the gasifier.

Drying the removal of a solvent or water from a chemical substance; also referred to as the removal of solvent from a liquid or suspension.

Dropping point the temperature at which grease passes from a semisolid to a liquid state under prescribed conditions.

Dry gas a gas that does not contain fractions that may easily condense under normal atmospheric conditions.

Dry point the temperature at which the last drop of petroleum fluid evaporates in a distillation test.

Dry mill an ethanol production process in which the entire corn kernel is first ground into flour before processing—in addition to ethanol, dry mills also produce dried distillers grains with solubles (DDGS) which is fed to livestock; and carbon dioxide which is used in food processing and bottling; most new ethanol plants are dry mill facilities.

Dry ton 2000 lb of material dried to a constant weight.

Dutch oven furnace one of the earliest types of furnaces, having a large, rectangular box lined with firebrick (refractory) on the sides and top; commonly used for burning wood.

E10 an alcohol fuel mixture containing 10% ethanol and 90% gasoline by volume.

E85 an alcohol fuel mixture containing 85% ethanol and 15% gasoline by volume, and the current alternative fuel of choice of the US government.

Ebullated bed a process in which the catalyst bed is in a suspended state in the reactor by means of a feedstock recirculation pump that pumps the feedstock upwards at sufficient speed to expand the catalyst bed at approximately 35% above the settled level.

Ecology the study of the interrelationships between organisms and their environment.

E Diesel a blend of ethanol and diesel fuel plus other additives designed to reduce air pollution from heavy equipment, city buses and other vehicles that operate on diesel engines.

Effluent the liquid or gas discharged from a process or chemical reactor, usually containing residues from that process.

Electrostatic precipitator a device used to trap fine dust particles (usually in the size range 30–60 μm) that operate on the principle of imparting an electric charge to particles in an incoming air stream, which are then collected on an oppositely charged plate across a high voltage field.

Elemental analysis the determination of carbon, hydrogen, nitrogen, oxygen, sulfur, chlorine, and ash in a sample as a weight percentage of the whole.

Emission offset a reduction in the air pollution emissions of existing sources to compensate for emissions from new sources.

Emissions substances discharged into the air during combustion; waste substances released into the air or water.

Emulsification to emulsify; to form an emulsion.

Emulsion a suspension of small drops of one liquid in a second with which the first will not mix; can be formed either by mechanical agitation, or by chemical processes; unstable emulsions will separate with time or temperature but stable emulsions will not separate.

Energy balance the difference between the energy produced by a fuel and the energy required to obtain it through agricultural processes, drilling, refining, and transportation.

Energy crops crops grown specifically for their fuel value; include food crops such as corn and sugarcane, and nonfood crops such as poplar trees and switchgrass.

Energy-efficiency ratio a number representing the energy stored in a fuel as compared to the energy required to produce, process, transport, and distribute that fuel.

Enhanced recovery methods that usually involves the application of thermal energy (e.g., steam flooding) to oil recovery from the reservoir.

Entrained bed a bed of solid particles suspended in a fluid (liquid or gas) at such a rate that some of the solid is carried over (entrained) by the fluid.

Environment the external conditions that affect organisms and influence their development and survival.

Environmental assessment (EA) a public document that analyzes a proposed federal action for the possibility of significant environmental impacts—if the environmental impacts will be significant, the federal agency must then prepare an environmental impact statement.

Environmental impact statement (EIS) a statement of the environmental effects of a proposed action and of alternative actions. Section 102 of the National Environmental Policy Act requires an EIS for all major federal actions.

Enzymatic hydrolysis a process by which enzymes (biological catalysts) are used to break down starch or cellulose into sugar.

Enzyme a protein or protein-based molecule that speeds up chemical reactions occurring in living things; enzymes act as catalysts for a single reaction, converting a specific set of reactants into specific products.

Ester any of a large group of organic compounds formed when an acid and alcohol is mixed; methyl acetate (CH_3COOCH_3) is the simplest ester; biodiesel contains methyl stearate.

ETBE see ethyl tertiary butyl ether.

Ethanol (ethyl alcohol, alcohol, or grain spirit) a clear, colorless, and flammable oxygenated hydrocarbon; used as a vehicle fuel by itself (E100 is 100% ethanol by volume), blended with gasoline (E85 is 85% ethanol by volume), or as a gasoline octane enhancer and oxygenate (10% by volume).

Ethers liquid fuel made from a blending an alcohol with isobutylene.

Ethyl tertiary butyl ether (ethyl *t*-butyl ether) ether created from ethanol that can increase octane and reduce the volatility of gasoline, decreasing evaporation and smog formation.

Evaporation the conversion of a liquid to the vapor state by the addition of latent heat or vaporization; a process for concentrating nonvolatile solids in a solution by boiling off the liquid portion of the waste stream.

Expanding clays clay minerals that expand or swell on contact with water, for example, montmorillonite.

Explosive limits the limits of percentage composition of mixtures of gases and air within which an explosion takes place when the mixture is ignited.

Extract the portion of a sample preferentially dissolved by the solvent and recovered by physically separating the solvent.

Extractive distillation the separation of different components of mixtures, which have similar vapor pressures by flowing a relatively high-boiling solvent, which is selective for one of the components in the feed, down a distillation column as the distillation proceeds; the selective solvent scrubs the soluble component from the vapor.

Extractives any number of different compounds in biomass that are not an integral part of the cellular structure—the compounds can be extracted from wood by means of polar and nonpolar solvents, including hot or cold water, ether, benzene, methanol, or other solvents that do not degrade the biomass structure and the types of extractives found in biomass samples are entirely dependent upon the sample itself.

FAAE a term for biodiesel made from any alcohol during its production process.

Fabric filters filters made from fabric materials and used for removing particulate matter from gas streams (see baghouse).

Facies one or more layers of rock that differs from other layers in composition, age or content.

FAME (fatty acid methyl ester) an ester that can be created by a catalyzed reaction between fatty acids and methanol; the constituents in biodiesel are primarily FAMEs, usually obtained from vegetable oils by transesterification.

Fast pyrolysis thermal conversion of biomass by rapid heating to between 450°C and 600°C (842°F and 1112°F) in the absence of oxygen.

Fatty acid a carboxylic acid (an acid with a − COOH group) with long hydrocarbon side chains; feedstocks are first converted to fatty acids and then to biodiesel by transesterification.

Fatty acid alkyl ester see FAAE.

Fatty acid methyl ester see FAME.

FCC fluid catalytic cracking.

FCCU fluid catalytic cracking unit.

Feedstock crude oil as it is fed to the refinery; a refinery product that is used as the raw material for another process; biomass used in the creation of a particular biofuel (e.g., corn or sugarcane for ethanol, soybeans or rapeseed for biodiesel); the term is also generally applied to raw materials used in other industrial processes.

Fermentation conversion of carbon-containing compounds by microorganisms for production of fuels and chemicals such as alcohols, acids, or energy-rich gases.

Fiber products products derived from fibers of herbaceous and woody plant materials; examples include pulp, composition board products, and wood chips for export.

Fine materials wood residues not suitable for chipping, such as planer shavings and sawdust.

Fischer–Tropsch process a process for producing liquid fuels, usually diesel fuel, from natural gas or synthetic gas from gasified coal or biomass.

Fixed bed a stationary bed (of catalyst) to accomplish a process (see fluid bed).

Fixed carbon the carbonaceous residue remaining after heating in a prescribed manner to decompose thermally unstable components and to distill volatiles; part of the proximate analysis group.

Flammability range the range of temperature over which a chemical is flammable.

Flammable a substance that will burn readily.

Flammable liquid a liquid having a flash point below 37.8°C (100°F).

Flammable solid a solid that can ignite from friction or from heat remaining from its manufacture, or which may cause a serious hazard if ignited.

Flashpoint the lowest temperature at which a liquid will produce enough vapor to ignite, if the vapor is flammable.

Flexible-fuel vehicle (flex-fuel vehicle) a vehicle that can run alternately on two or more sources of fuel; includes cars capable of running on gasoline and gasoline/ethanol mixtures, as well as cars that can run on both gasoline and natural gas.

Flue gas gas from the combustion of fuel, the heating value of which has been substantially spent, which is, therefore, discarded to the flue or stack.

Fluid bed use of an agitated bed of inert granular material to accomplish a process in which the agitated bed resembles the motion of a fluid.

Fluid coking a continuous fluidized solids process that cracks feed thermally over heated coke particles in a reactor vessel to gas, liquid products, and coke.

Fluidized-bed boiler a large, refractory-lined vessel with an air distribution member or plate in the bottom, a hot gas outlet in or near the top, and some provisions for introducing fuel; the fluidized bed is formed by blowing air up through a layer of inert particles (such as sand or limestone) at a rate that causes the particles to go into suspension and continuous motion.

Fly ash small ash particles carried in suspension in combustion products; particulate matter produced from mineral matter in coal that is converted during combustion to finely divided inorganic material, which emerges from the combustor in the gases.

Foam a dispersion of a gas in a liquid or solid.

Fodder beets a promising sugar crop that presently is being developed in New Zealand is the fodder beet. The fodder beet is a high-yielding forage crop obtained by crossing two other beet species, sugar beets and mangolds. It is similar in most agronomic respects to sugar beets. The attraction of this crop lies in its higher yield of fermentable sugars per acre relative to sugar beets and its comparatively high resistance to loss of fermentable sugars during storage. Culture of fodder beets is also less demanding than sugar beets.

Forage crops (e.g., forage sorghum, Sudan grass) crops used for ethanol production because, in their early stage of growth, there is very little lignin and the conversion of the cellulose to sugars is more efficient. In addition, the proportion of carbohydrates in the form of cellulose is less than in the mature plant. Since forage crops achieve maximum growth in a relatively short period, they can be harvested as many as four times in one growing season. For this reason, forage crops cut as green chop may have the highest yield of dry material of any storage crop. In addition to cellulose, forage crops contain significant quantities of starch and fermentable sugars which can also be converted to ethanol. The residues from fermentation containing nonfermentable sugars, protein, and other components may be used for livestock feed.

Forest land land at least 10% stocked by forest trees of any size, including land that formerly had such tree cover and that will be naturally or artificially regenerated; includes transition zones, such as areas between heavily forested and nonforested lands that are at least 10% stocked with forest trees and forest areas adjacent to urban and builtup lands; also included are pinyon-juniper and chaparral areas; minimum area for classification of forest land is 1 ac.

Forest residues material not harvested or removed from logging sites in commercial hardwood and softwood stands as well as material resulting from forest management operations such as precommercial thinnings and removal of dead and dying trees.

Forest health a condition of ecosystem sustainability and attainment of management objectives for a given forest area; usually considered to include green trees, snags, resilient stands growing at a moderate rate, and endemic levels of insects and disease.

Fossil fuel solid, liquid, or gaseous fuels formed in the ground after millions of years by chemical and physical changes in plant and animal residues under high temperature and pressure. Oil, natural gas, and coal are examples of fossil fuels.

Fossil fuel resources a gaseous, liquid, or solid fuel material formed in the ground by chemical and physical changes (diagenesis, *q.v.*) in plant and animal residues over geological time; natural gas, petroleum, coal, and oil shale.

Fractional composition the composition of petroleum as determined by fractionation (separation) methods.

Fractional distillation the separation of the components of a liquid mixture by vaporizing and collecting the fractions, or cuts, which condense in different temperature ranges.

Fruit crops (e.g., grapes, apricots, peaches, and pears) another type of feedstock in the sugar crop category. Typically, fruit crops such as grapes are used as the feedstock in wine production. These crops are not likely to be used as feedstocks for production of fuel-grade ethanol because of their high market value for direct human consumption. However, the coproducts of processing fruit crops are likely to be used as feedstocks because fermentation is an economical method for reducing the potential environmental impact of untreated wastes containing fermentable sugars.

Fuel cell a device that converts the energy of a fuel directly to electricity and heat, without combustion.

Fuel cycle the series of steps required to produce electricity. The fuel cycle includes mining or otherwise acquiring the raw fuel source, processing and cleaning the fuel, transport, electricity generation, waste management, and plant decommissioning.

Fuel oil a heavy residue, black in color, used to generate power or heat by burning in furnaces; also called heating oil is a distillate product that covers a wide range of properties; see also No. 1–No. 4 Fuel oils.

Fuel treatment evaluator (FTE) a strategic assessment tool capable of aiding the identification, evaluation, and prioritization of fuel treatment opportunities.

Fuelwood (fuel wood) wood used for conversion to some form of energy, primarily for residential use.

Fuller's earth a clay mineral that has high adsorptive capacity for removing color from oils; Attapulgus clay is a widely used fuller's earth.

Functional group the portion of a molecule that is characteristic of a family of compounds and determines the properties of these compounds.

Furnace an enclosed chamber or container used to burn biomass in a controlled manner to produce heat for space or process heating.

Furnace oil a distillate fuel primarily intended for use in domestic heating equipment.

Galactan the polymer of galactose with a repeating unit of $C_6H_{10}O_5$; found in hemicellulose it can be hydrolyzed to galactose.

Galactose a six-carbon sugar with the formula $C_6H_{12}O_6$; a product of hydrolysis of galactan found in the hemicellulose fraction of biomass.

Gas engine a piston engine that uses natural gas rather than gasoline—fuel and air are mixed before they enter cylinders; ignition occurs with a spark.

Gaseous emissions substances discharged into the air during combustion, typically including carbon dioxide, carbon monoxide, water vapor, and hydrocarbon derivatives.

Gasification a chemical or heat process used to convert carbonaceous material (such as coal, petroleum, and biomass) into gaseous components such as carbon monoxide and hydrogen.

Gasifier a device for converting solid fuel into gaseous fuel; in biomass systems the process is referred to as pyrolitic distillation.

Gasohol a mixture of 10% anhydrous ethanol and 90% gasoline by volume; 7.5% anhydrous ethanol and 92.5% gasoline by volume; or 5.5% anhydrous ethanol and 94.5% gasoline by volume.

Gasoline a volatile, flammable liquid obtained from petroleum that has a boiling range of approximately 30°C–220°C (86°F–428°F) and is used for fuel for spark-ignition internal-combustion engines.

Gas shift process a process in which carbon monoxide and hydrogen react in the presence of a catalyst to form methane and water.

Gas to liquids (GTL) the process of refining natural gas and other hydrocarbon derivatives into longer chain hydrocarbon derivatives, which can be used to convert gaseous waste products into fuels.

Gas turbine (combustion turbine) a turbine that converts the energy of hot compressed gases (produced by burning fuel in compressed air) into mechanical power—often fired by natural gas or fuel oil.

Gel point the point at which a liquid fuel cools to the consistency of petroleum jelly.

Genetically modified organism (GMO) an organism, the genetic material of which has been modified through recombinant DNA technology, altering the phenotype of the organism to meet desired specifications.

Glycerin ($CH_2OH \cdot CHOH \cdot CH_2OH$) a by-product of biodiesel production; each of the hydroxyl (OH) functions is one of the three places where an ester is broken off of the triglyceride molecule (e.g., vegetable oil).

Glycerine (glycerin; glycerol) a liquid by-product of biodiesel production; used in the manufacture of dynamite, cosmetics, liquid soaps, inks, and lubricants.

Grain alcohol see ethanol.

Grassland pasture and range all open land used primarily for pasture and grazing, including shrub and brush land types of pasture; grazing land with sagebrush and scattered mesquite; and all tame and native grasses, legumes, and other forage used for pasture or grazing; because of the diversity in vegetative composition, grassland pasture and range are not always clearly distinguishable from other types of pasture and range; at one extreme, permanent grassland may merge with cropland pasture, or grassland may often be found in transitional areas with forested grazing land.

Grease car a diesel-powered automobile rigged postproduction to run on used vegetable oil.

Greenhouse effect the effect of certain gases in the Earth's atmosphere in trapping heat from the sun.

Greenhouse gases gases that trap the heat of the sun in the Earth's atmosphere, producing the greenhouse effect. The two major greenhouse gases are water vapor and carbon dioxide. Other greenhouse gases include methane, ozone, chlorofluorocarbons, and nitrous oxide.

Grid an electric utility company's system for distributing power.

Gross heating value (GHV) the maximum potential energy in the fuel as-received, considering moisture content (MC).

Growing stock a classification of timber inventory that includes live trees of commercial species meeting specified standards of quality or vigor; cull trees are excluded.

GTL (Gas-to-liquid) a refinery process that converts natural gas into longer chain hydrocarbon derivatives; gas can be converted to liquid fuels via a direct conversion or using a process such as the Fischer–Tropsch process.

Gum an insoluble tacky semisolid material formed as a result of the storage instability and/or the thermal instability of petroleum and petroleum products.

Habitat the area where a plant or animal lives and grows under natural conditions. Habitat includes living and nonliving attributes and provides all requirements for food and shelter.

Hardwoods usually broad-leaved and deciduous trees; one of the botanical groups of dicotyledonous trees that have broad leaves in contrast to the conifers or softwoods—the term has no reference to the actual hardness of the wood—the botanical name for hardwoods is angiosperms; short-rotation, fast growing hardwood trees are being developed as future energy crops, which are uniquely developed for harvest from 5 to 8 years after planting and examples include hybrid poplars (*Populus* sp.), hybrid willows (*Salix* sp.), silver maple (*Acer saccharinum*), and black locust (*Robinia pseudoacacia*).

Hazardous waste a general term for a wide range of waste materials that present different levels of risk. Some present a serious and immediate threat to the population and the environment, for example, those that are toxic and could cause cancer or infectious disease. Others, such as fluorescent tubes or cathode ray tubes in televisions, pose little immediate threat but may cause long-term damage over a period of time.

Heating oil see fuel oil.

Heating value the maximum amount of energy that is available from burning a substance.

Heavy ends the highest boiling portion of a petroleum fraction; see also light ends.

Heavy fuel oil fuel oil having a high density and viscosity; generally residual fuel oil such as No. 5 and No 6 Fuel oils.

Heavy crude oil (heavy oil) oil that is more viscous that conventional crude oil, has a lower mobility in the reservoir, but can be recovered through a well from the reservoir by the application of a secondary or enhanced recovery method; sometimes petroleum having an API gravity of less than 20°.

Hectare common metric unit of area, equal to 2.47 ac. 100 ha = 1 square kilometer.

Hemicellulose consists of short, highly branched chains of sugars in contrast to cellulose, which is a polymer of only glucose, hemicellulose is a polymer of five different sugars; contains five-carbon sugars (usually D-xylose and L-arabinose) and six-carbon sugars (D-galactose, D-glucose, and D-mannose) and uronic acid, which are highly substituted with acetic acid; the branched nature of hemicellulose renders it amorphous and relatively easy to hydrolyze to its constituent sugars compared to cellulose; when hydrolyzed, the hemicellulose from hardwoods releases products high in xylose (a five-carbon sugar); hemicellulose contained in softwoods, by contrast, yields more six-carbon sugars.

Herbaceous nonwoody type of vegetation, usually lacking permanent strong stems, such as grasses, cereals, and canola (rape).

Herbaceous energy crops perennial nonwoody crops that are harvested annually, though they may take 2–3 years to reach full productivity; examples include switchgrass (*Panicum virgatum*), reed canary grass (*Phalaris arundinacea*), miscanthus (*Miscanthus* × *giganteus*), and giant reed (*Arundo donax*).

Herbaceous plants nonwoody species of vegetation, usually of low lignin content such as grasses.

Heteroatom compounds chemical compounds that contain nitrogen and/or oxygen and/or sulfur and/or metals bound within their molecular structure(s).

Heterogeneity the lack of uniformity in reservoir properties such as permeability.

Hexose any of various simple sugars that have six carbon atoms per molecule (e.g., glucose, mannose, and galactose).

Higher heating value (HHV) the potential combustion energy when water vapor from combustion is condensed to recover the latent heat of vaporization. Lower heating value (LHV) is the potential combustion energy when water vapor from combustion is not condensed.

Household waste waste that includes domestic waste from household collection rounds; waste from services such as street sweepings, bulky waste collection, litter collection, hazardous household waste collection, and garden waste collection; waste from civic amenity sites; and wastes separately collected for recycling or composting through bring recycling schemes and curbside recycling schemes. Household waste is a subgroup of municipal solid waste.

Hydraulic fracturing the opening of fractures in a reservoir by high-pressure, high-volume injection of liquids through an injection well.

Hydrocarbon a chemical compound that contains a carbon backbone with hydrogen atoms attached to that backbone.

Hydrocarbonaceous material a material such as bitumen that is composed of carbon and hydrogen with other elements (heteroelements) such as nitrogen, oxygen, sulfur, and metals chemically combined within the structures of the constituents; even though carbon and hydrogen may be the predominant elements, there may be very few true hydrocarbon derivatives.

Hydrocarbon compounds (hydrocarbon derivatives) chemical compounds containing only carbon and hydrogen.

Hydrodesulfurization the removal of sulfur by hydrotreating.

Hydroprocesses refinery processes designed to add hydrogen to various products of refining.

Hydrotreating the removal of heteroatomic (nitrogen, oxygen, and sulfur) species by treatment of a feedstock or product at relatively low temperatures in the presence of hydrogen.

Idle cropland land in which no crops were planted; acreage diverted from crops to soil-conserving uses (if not eligible for and used as cropland pasture) under federal farm programs is included in this component.

Ignitability a characteristic of liquids, the vapors of which are likely to ignite in the presence of ignition source; also characteristic of nonliquids that may catch fire from friction or contact with water and that burn vigorously.

Immiscible two or more fluids that do not have complete mutual solubility and coexist as separate phases.

Incinerator any device used to burn solid or liquid residues or wastes as a method of disposal.

Inclined grate a type of furnace in which fuel enters at the top part of a grate in a continuous ribbon, passes over the upper drying section where moisture is removed, and descends into the lower burning section. Ash is removed at the lower part of the grate.

Incompatibility the immiscibility of petroleum products and also of different crude oils, which is often reflected in the formation of a separate phase after mixing and/or storage.

Indirect injection engine an older model of diesel engine in which fuel is injected into a prechamber, partly combusted, and then sent to the fuel injection chamber.

Indirect liquefaction conversion of biomass to a liquid fuel through a synthesis gas intermediate step.

Industrial wood all commercial round wood products except fuel wood.

Initial boiling point the recorded temperature when the first drop of liquid falls from the end of the condenser.

Iodine value a measure of the number of unsaturated carbon–carbon double bonds in a vegetable oil molecule—double bonds can allow polymerization, leading to the formation of lacquers and possibly blockage and damage to engine or fuel train components; in liquid biofuel applications the iodine value gives a lower cold filter plugging point (CFPP) or cloud point.

Instability the inability of a petroleum product to exist for periods of time without change to the product.

Jatropha a nonedible evergreen shrub found in Asia, Africa, and the West Indies; the seeds contain a high proportion of oil which can be used for making biodiesel.

Jerusalem artichoke a member of the sunflower family, this crop is native to North America and well adapted to northern climates. Like the sugar beet, the Jerusalem artichoke produces sugar in the top growth and stores it in the roots and tuber. It can grow in a variety of soils, and it is not demanding of soil fertility. The Jerusalem artichoke is a perennial; small tubers left in the field will produce the next season's crop, so no plowing or seeding is necessary.

Joule metric unit of energy, equivalent to the work done by a force of 1 N applied over distance of 1 m ($=1\ kg\ m^2/s^2$). 1 J = 0.239 cal (1 cal = 4.187 J).

Kaolinite a clay mineral formed by hydrothermal activity at the time of rock formation or by chemical weathering of rock with high feldspar content; usually associated with intrusive granite rock with high feldspar content.

Kata-condensed aromatic compounds compounds based on linear condensed aromatic hydrocarbon systems, for example, anthracene and naphthacene (tetracene).

Kerosene (also called spelled kerosine in some countries) a light middle distillate that in various forms is used as aviation turbine fuel or for burning in heating boilers or as a solvent, such as white spirit.

Kilowatt Kilowatt hour (kWh) a measure of energy equivalent to the expenditure of 1 kWh for 1 hour. For example, 1 kWh will light a 100-W light bulb for 10 hours. 1 kWh = 3412 Btu.

Klason lignin lignin obtained from wood after the nonlignin components of the wood have been removed with a prescribed sulfuric acid treatment; a specific type of acid-insoluble lignin analysis.

Knock engine sound that results from ignition of the compressed fuel-air mixture prior to the optimal moment.

KOH see potassium hydroxide.

Landfill licensed facilities where waste is permanently deposited for disposal into land. According to the waste hierarchy the final disposal of waste through landfill is the least preferred way of managing waste.

Landfill gas a type of biogas that is generated by decomposition of organic material at landfill disposal sites. Landfill gas is approximately 50% methane. See also biogas.

Life cycle assessment (LCA) the systematic identification and evaluation of all the environmental benefits and disbenefits that result, both directly and indirectly, from a product or function throughout its entire life from extraction of raw materials to its eventual disposal and assimilation into the environment LCA helps to place the assessment of the environmental costs and benefits of these various options, and the development of appropriate and practical waste management policies, on a sound and objective basis.

Light ends the lower boiling components of a mixture of hydrocarbon derivatives; see also heavy ends and light hydrocarbon derivatives.

Light hydrocarbon derivatives hydrocarbon derivatives with molecular weights less than that of heptane (C_7H_{16}).

Light oil the products distilled or processed from crude oil up to, but not including, the first lubricating oil distillate.

Light petroleum petroleum having an API gravity of greater than 20°.

Lignin structural constituent of wood and (to a lesser extent) other plant tissues, which encrusts the walls and cements the cells together; energy-rich material contained in biomass that can be used for boiler fuel.

Lignocellulose plant material made up primarily of lignin, cellulose, and hemicellulose.

Lipid any of a group of organic compounds, including the fats, oils, waxes, sterols, and triglycerides, which are insoluble in water but soluble in nonpolar organic solvents, are oily to the touch, and together with carbohydrates and proteins constitute the principal structural material of living cells.

Liquefied petroleum gas (LPG) propane, butane, or mixtures thereof, gaseous at atmospheric temperature and pressure, held in the liquid state by pressure to facilitate storage, transport, and handling.

Live cull a classification that includes live cull trees; when associated with volume, it is the net volume in live cull trees that are 5.0 in. in diameter and larger.

Logging residues the unused portions of growing-stock and nongrowing-stock trees cut or killed logging and left in the woods.

Lower heating value (LLV, net heat of combustion) the heat produced by combustion of one unit of a substance, at atmospheric pressure under conditions such that all water in the products remains in the form of vapor; the net heat of combustion is calculated from the gross heat of combustion at 20°C (68°F) by subtracting 572 cal/g (1030 Btu/lb) of water derived from one unit mass of sample, including both the water originally present as moisture and that formed by combustion—this subtracted amount is not equal to the latent heat of vaporization of water because the calculation also reduces the data from the gross value at constant volume to the net value at constant pressure and the appropriate factor for this reduction is 572 cal/g.

Lye see sodium hydroxide.

M85 an alcohol fuel mixture containing 85% methanol and 15% gasoline by volume. Methanol is typically made from natural gas but can also be derived from the fermentation of biomass.

Maltenes (maltene fraction) that fraction of petroleum that is soluble in, for example, pentane or heptane; deasphaltened oil; also the term arbitrarily assigned to the pentane-soluble portion of petroleum that is relatively high boiling (>300°C, 760 mm) (see also petrolenes).

Megawatt (MW) a measure of electrical power equal to 1 million watts (1000 kW).

MEK (methyl ethyl ketone) a colorless liquid ($CH_3COCH_2CH_3$) used as a solvent; as a chemical intermediate; and in the manufacture of lacquers, celluloid, and varnish removers.

Membrane technology typically, gas separation processes utilizing membranes that permit different components of a gas to diffuse through the membrane at significantly different rates.

Methanol (wood alcohol) a fuel typically derived from natural gas, but which can be produced from the fermentation of sugars in biomass; a colorless, volatile, inflammable, and poisonous alcohol (CH_3OH) traditionally formed by destructive distillation of wood or, more recently, as a result of synthetic distillation in chemical plants; a fuel typically derived from natural gas, but which can be produced from the fermentation of sugars in biomass.

Methoxide (sodium methoxide, sodium methylate, CH_3O^- Na^+) an organic salt, in pure form a white powder; in biodiesel production, methoxide is a product of mixing methanol and sodium hydroxide, yielding a solution of sodium methoxide in methanol, and a significant amount of heat; making sodium methoxide is the most dangerous step when making biodiesel.

Methyl alcohol see methanol.

Methyl esters see biodiesel.

Methyl t-butyl ether an ether added to gasoline to improve its octane rating and to decrease gaseous emissions; see oxygenate.

Microcarbon residue the carbon residue determined using a thermogravimetric method. See also carbon residue.

Microcrystalline wax wax extracted from certain petroleum residua and having a finer and less apparent crystalline structure than paraffin wax.

Mid-boiling point the temperature at which approximately 50% of a material has distilled under specific conditions.

Middle distillate the distillate boiling between the kerosene and lubricating oil fractions.

Million 1×10^6.

Mill residue wood and bark residues produced in processing logs into lumber, plywood, miscibility: an equilibrium condition, achieved after mixing two or more fluids, which is characterized by the absence of interfaces between the fluids: (1) first-contact miscibility: miscibility in the usual sense, whereby two fluids can be mixed in all proportions without any interfaces forming such as at room temperature and pressure, ethyl alcohol and water are first-contact miscible. (2) Multiple-contact miscibility (dynamic miscibility): miscibility that is developed by repeated enrichment of one fluid phase with components from a second fluid phase with which it comes into contact. (3) Minimum miscibility pressure: the minimum pressure above which two fluids become miscible at a given temperature or can become miscible, by dynamic processes.

Mobility a measure of the ease with which a fluid moves through reservoir rock; the ratio of rock permeability to apparent fluid viscosity.

Modified/unmodified diesel engine traditional diesel engines must be *modified* to heat the oil before it reaches the fuel injectors in order to handle straight vegetable oil. Modified, any diesel engine can run on veggie oil; without modification, the oil must first be converted to biodiesel.

Moisture content (MC) the weight of the water contained in wood, usually expressed as a percentage of weight, either oven-dry or as-received.

Moisture content, dry basis the moisture content expressed as a percentage of the weight of oven wood, that is: [(weight of wet sample − weight of dry sample)/weight of dry sample] × 100.

Moisture content, wet basis moisture content expressed as a percentage of the weight of wood as-received, that is: [(weight of wet sample − weight of dry sample)/weight of wet sample] × 100.

Moisture free basis biomass composition and chemical analysis data is typically reported on a moisture free or dry weight basis—moisture (and some volatile matter) is removed prior to analytical testing by heating the sample at 105°C (221°F) to constant weight; by definition, samples dried in this manner are considered moisture free.

Monosaccharide a simple sugar such as a five-carbon sugar (xylose, arabinose) or six-carbon sugar (glucose, fructose); sucrose, on the other hand is a disaccharide, composed of a combination of two simple sugar units, glucose and fructose.

MTBE methyl tertiary butyl ether is highly refined high-octane light distillate used in the blending of petrol.

Municipal solid waste (MSW) household waste and other wastes collected by a waste collection authority or its contractors, such as municipal parks and gardens waste, beach cleansing waste, and any commercial waste and industrial waste for which the collection authority takes responsibility.

Municipal waste residential, commercial, and institutional postconsumer wastes that contain a significant proportion of plant-derived organic material that constitutes a renewable energy resource; waste paper, cardboard, construction and demolition wood waste, and yard wastes are examples of biomass resources in municipal wastes.

Naphtha a generic term applied to refined, partly refined, or unrefined petroleum products and liquid products of natural gas, the majority of which distills below 240°C (464°F); the volatile fraction of petroleum, which is used as a solvent or as a precursor to gasoline.

Naphthenes cycloparaffin derivatives.

Natural gas the naturally occurring gaseous constituents that are found in many petroleum reservoirs; also there are also those reservoirs in which natural gas may be the sole occupant.

Natural gas liquids (NGL) the hydrocarbon liquids that condense during the processing of hydrocarbon gases that are produced from oil or gas reservoir; see also natural gasoline.

Natural gasoline a mixture of liquid hydrocarbon derivatives extracted from natural gas suitable for blending with refinery gasoline.

Nitrogen fixation the transformation of atmospheric nitrogen into nitrogen compounds that can be used by growing plants.

Nitrogen oxides (NO_x) products of combustion that contribute to the formation of smog and ozone.

Nonforest land land that has never supported forests and lands formerly forested where use of timber management is precluded by development for other uses; if intermingled in forest areas, unimproved roads and nonforest strips must be more than 120-ft wide, and clearings, etc., must be more than 1 ac in area to qualify as nonforest land.

No. 1 Fuel oil a fuel oil that is similar to kerosene and is used in burners where vaporization before burning is usually required and a clean flame is specified.

No. 2 Fuel oil also called domestic heating oil; has properties similar to diesel fuel and heavy jet fuel; used in burners where complete vaporization is not required before burning.

No. 4 Fuel oil a light industrial heating oil and is used where preheating is not required for handling or burning; there are two grades of No. 4 Fuel oil, differing in safety (flash point) and flow (viscosity) properties.

No. 5 Fuel oil a heavy industrial fuel oil that requires preheating before burning.

No. 6 Fuel oil a heavy fuel oil and is more commonly known as Bunker C oil when it is used to fuel ocean-going vessels; preheating is always required for burning this oil.

Nonattainment area any area that does not meet the national primary or secondary ambient air quality standard established (by the Environmental Protection Agency) for designated pollutants, such as carbon monoxide and ozone.

Nonindustrial private an ownership class of private lands where the owner does not operate wood processing plants.

Particulate emissions particles of a solid or liquid suspended in a gas, or the fine particles of carbonaceous soot and other organic molecules discharged into the air during combustion.

Octane number the measure of the resistance of a fuel to self-ignition; the octane number of a fuel is indicated on the pump—the higher the number, the slower the fuel burns; bioethanol typically adds two to three octane numbers when blended with ordinary petroleum—making it a cost-effective octane enhancer; see knock.

Oil from tar sand synthetic crude oil.

Oil mining application of a mining method to the recovery of bitumen.

Oils the portion of the maltene fraction that is not adsorbed by a surface-active material such as clay or alumina.

OOIP (oil originally in place or original oil in place) the quantity of petroleum existing in a reservoir before oil recovery operations begin.

Open-loop biomass biomass that can be used to produce energy and bioproducts even though it was not grown specifically for this purpose; include agricultural livestock waste and residues from forest harvesting operations and crop harvesting.

Overhead (or overhead fraction) the portion of the feedstock, which is vaporized and removed during distillation.

Oxygenate a substance that, when added to gasoline, increases the amount of oxygen in that gasoline blend; includes fuel ethanol, methanol, and methyl tertiary butyl ether (MTBE).

Oxygenated fuels ethanol is an oxygenate, meaning that it adds oxygen to the fuel mixture—more oxygen helps the fuel burn more completely thereby reducing the amount of harmful emissions from the tailpipe; a fuel such as ethanol-blended gasoline that contains a high oxygen content is called oxygenated.

Palm oil a form of vegetable oil obtained from the fruit of the oil palm tree; widely used feedstock for traditional biodiesel production; the palm oil and palm kernel oil are composed of fatty acids, esterified with glycerol just like any ordinary fat.

Paraffin wax the colorless, translucent, and highly crystalline material obtained from the light lubricating fractions of paraffin crude oils (wax distillates).

Particulate a small, discrete mass of solid or liquid matter that remains individually dispersed in gas or liquid emissions.

Particulate emissions solid particles or a liquid droplets suspended in a gas, or the fine particles of carbonaceous soot and other organic molecules discharged into the air during combustion.

Particulate matter (particulates) particles in the atmosphere or on a gas stream that may be organic or inorganic and originate from a wide variety of sources and processes.

Particle size distribution the particle size distribution (of a catalyst sample) expressed as a percent of the whole.

Pay zone thickness the depth of a tar sand deposit from which bitumen (or a product) can be recovered.

Perennial a plant that does not have to be planted every year like traditional row crops.

Permeability the ease of flow of a fluid through the rock.

Petrodiesel petroleum-based diesel fuel, usually referred to simply as diesel.

Petrolenes (petrolenes fraction) the term applied to that part of the pentane- or heptane-soluble material that is low boiling (<300°C, <570°F, 760 mm) and can be distilled without thermal decomposition (see also maltenes).

Petroleum (crude oil) a hydrocarbon-based substance comprising a complex blend of hydrocarbon derivatives derived from crude oil through the process of separation, conversion, upgrading, and finishing, including motor fuel, jet oil, lubricants, petroleum solvents, and used oil.

Petroleum jelly a translucent, yellowish to amber or white, hydrocarbon substance (melting point: 38°C–54°C) having almost no odor or taste, derived from petroleum and used principally in medicine and pharmacy as a protective dressing and as a substitute for fats in ointments and cosmetics; also used in many types of polishes and in lubricating greases, rust preventives, and modeling clay; obtained by dewaxing heavy lubricating oil stocks.

pH a measure of acidity and alkalinity of a solution on a scale with 7 representing neutrality; lower numbers indicate increasing acidity, and higher numbers increasing alkalinity; each unit of change represents a tenfold change in acidity or alkalinity.

Phase a separate fluid that coexists with other fluids; gas, oil, water, and other stable fluids such as microemulsions are all called phases in EOR research.

Phase behavior the tendency of a fluid system to form phases as a result of changing temperature, pressure, or the bulk composition of the fluids or of individual fluid phases.

Phase diagram a graph of phase behavior. In chemical flooding a graph showing the relative volume of oil, brine, and sometimes one or more microemulsion phases. In carbon dioxide flooding, conditions for formation of various liquid, vapor, and solid phases.

Phase properties types of fluids, compositions, densities, viscosities, and relative amounts of oil, microemulsion, or solvent, and water formed when a micellar fluid (surfactant slug) or miscible solvent (e.g., CO_2) is mixed with oil.

Phase separation the formation of a separate phase that is usually the prelude to coke formation during a thermal process; the formation of a separate phase as a result of the instability/incompatibility of petroleum and petroleum products.

Polar aromatics the resin fraction; the constituents of petroleum that are predominantly aromatic in character and contain polar (nitrogen, oxygen, and sulfur) functions in their molecular structure(s).

Pollutant a chemical (or chemicals) introduced into the land water and air systems of that is (are) not indigenous to these systems; also an indigenous chemical (or chemicals) introduced into the land water and air systems in amounts greater than the natural abundance.

Polycyclic aromatic hydrocarbon derivatives (PAHs) polycyclic aromatic hydrocarbon derivatives are a suite of compounds comprising two or more condensed aromatic rings. They are found in many petroleum mixtures, and they are predominantly introduced to the environment through natural and anthropogenic combustion processes.

Photosynthesis process by which chlorophyll-containing cells in green plants concert incident light to chemical energy, capturing carbon dioxide in the form of carbohydrates.

Porosity the percentage of rock volume available to contain water or other fluid.

Porphyrins organometallic constituents of petroleum that contain vanadium or nickel; the degradation products of chlorophyll that became included in the protopetroleum.

Potassium hydroxide (KOH) used as a catalyst in the transesterification reaction to produce biodiesel.

Pour point the lowest temperature at which oil will pour or flow when it is chilled without disturbance under definite conditions.

Primary oil recovery oil recovery utilizing only naturally occurring forces; recovery of crude oil from the reservoir using the inherent reservoir energy.

Primary wood-using mill a mill that converts round wood products into other wood products; common examples are sawmills that convert saw logs into lumber and pulp mills that convert pulpwood round wood into wood pulp.

Process heat heat used in an industrial process rather than for space heating or other housekeeping purposes.

Producer gas fuel gas high in carbon monoxide (CO) and hydrogen (H_2), produced by burning a solid fuel with insufficient air or by passing a mixture of air and steam through a burning bed of solid fuel.

Protein a protein molecule is a chain of up to several hundred amino acids and is folded into a more or less compact structure; in the biologically active state, proteins function as catalysts in metabolism and to some extent as structural elements of cells and tissues; protein content in biomass (in mass percentage) can be estimated by multiplying the mass percentage nitrogen of the sample by 6.25.

Proximate analysis the determination, by prescribed methods, of moisture, volatile matter, fixed carbon (by difference), and ash; the term proximate analysis does not include determinations of chemical elements or determinations other than those named and the group of analyses is defined in ASTM D3172.

Pulpwood round wood, whole-tree chips, or wood residues that are used for the production of wood pulp.

Pyrolysis the thermal decomposition of biomass at high temperatures (greater than 400°F, or 200°C) in the absence of air; the end product of pyrolysis is a mixture of solids (char), liquids (oxygenated oils), and gases (methane, carbon monoxide, and carbon dioxide) with proportions determined by operating temperature, pressure, oxygen content, and other conditions. In contrast to incineration, pyrolysis is the thermal degradation of a substance in the absence of oxygen. This process requires an external heat source to maintain the temperature required. The products produced from pyrolyzing materials are a solid residue and a synthetic gas (syngas). The solid residue (sometimes described as a char) is a combination of noncombustible materials and carbon. The syngas is a mixture of gases (combustible constituents include carbon monoxide, hydrogen, methane, and a broad range of other volatile organic compounds). A proportion of these can be condensed to produce oils, waxes, and tars. If required, the condensable fraction can be collected by cooling the syngas, potentially for use as a liquid fuel.

Pyrolysis oil a bio-oil produced by fast pyrolysis of biomass; typically a dark brown, mobile liquid containing much of the energy content of the original biomass, with a heating value about half that of conventional fuel oil; conversion of raw biomass to pyrolysis oil represents a considerable increase in energy density and it can thus represent a more efficient form in which to transport it.

Quad one quadrillion Btu (10^{15} Btu) = 1.055 EJ, or approximately 172 million barrels of oil equivalent.

Quadrillion 1×10^{15}.

Recovery boiler a pulp mill boiler in which lignin and spent cooking liquor (black liquor) is burned to generate steam.

Refractory lining a lining, usually of ceramic, capable of resisting and maintaining high temperatures.

Rapeseed (*Brassica napus*; rape, oilseed rape or canola) a bright yellow flowering member of the family Brassicaceae (mustard or cabbage family); a traditional feedstock used for biodiesel production; canola is a name taken from Canada oil due to the fact that much of the development of the oil was performed in Canada; see colza.

Rapeseed oil food grade oil produced from rape seed is called canola oil; see colza.

Refuse-derived fuel (RDF) fuel prepared from municipal solid waste; noncombustible materials such as rocks, glass, and metals are removed, and the remaining combustible portion of the solid waste is chopped or shredded.

Renewable Fuel Standard (RFS) legislation enacted by US Congress as part of the Energy Policy Act of 2005, requiring an increasing level of biofuels be used every year, rising to 7.5 billion gallons by 2012.

Residual fuel oil fuel oil obtained by blending the residual product(s) from various refining processes with suitable diluent(s) (usually middle distillates) to obtain the required fuel oil grades.

Residual waste waste remaining to be disposed of after reuse, recycling, composting, and recovery of materials and energy.

Residues bark and woody materials that are generated in primary wood-using mills when round wood products are converted to other products.

Residuum (pl. residua, also known as resid or resids) the nonvolatile portion of petroleum that remains as residue after refinery distillation; hence, atmospheric residuum and vacuum residuum.

Rotation the period of years between establishment of a stand of timber and the time when it is considered ready for final harvest and regeneration.

Round wood products logs and other round timber generated from harvesting trees for industrial or consumer use.

RTFO (Renewable Transport Fuels Obligation) a policy in the United Kingdom (known as the K policy) that places an obligation on fuel suppliers to ensure that a certain percentage of their aggregate sales is made up of biofuels.

Sandstone a sedimentary rock formed by compaction and cementation of sand grains; can be classified according to the mineral composition of the sand and cement.

Saponification the reaction of an ester with a metallic base and water (i.e., the making of soap); occurs when too much lye is used in biodiesel production.

SARA analysis a method of fractionation by which petroleum is separated into saturates, aromatics, resins, and asphaltene fractions.

SARA separation see SARA analysis.

Saturated steam steam at boiling temperature for a given pressure.

Secondary oil recovery the application of energy (e.g., water flooding) to recovery of crude oil from a reservoir after the yield of crude oil from primary recovery diminishes.

Secondary pollutants a pollutant (chemical species) produced by interaction of a primary pollutant with another chemical or by dissociation of a primary pollutant or by other effects within a particular ecosystem.

Second-generation biofuels biofuels produced from biomass or nonedible feedstocks.

Secondary wood processing mills a mill that uses primary wood products in the manufacture of finished wood products, such as cabinets, moldings, and furniture.

Sodium hydroxide (lye, caustic soda, NaOH) strongly alkaline and extremely corrosive; mixing with fluids usually causes heat and can create enough heat to ignite flammables (such as methanol); one of the main reactants for biodiesel production.

Softwood generally, one of the botanical groups of trees that in most cases have needle-like or scale-like leaves; the conifers; also the wood produced by such trees; the term has no reference to the actual softness of the wood; the botanical name for softwoods is gymnosperms.

Soy (soy oil) a vegetable oil pressed from soy beans.

Soybean a bushy, leguminous plant, glycine max, native of Southeast Asia that is grown for the beans, which are used widely in the food industry, for protein in cattle feed and for oil production.

Soy diesel a general term for biodiesel, which accentuates the renewable nature of biodiesel; popular in soy producing regions.

Specific gravity the mass (or weight) of a unit volume of any substance at a specified temperature compared to the mass of an equal volume of pure water at a standard temperature.

Stabilized waste waste that has been treated so that it is chemically stable.

Stand (of trees) a tree community that possesses sufficient uniformity in composition, constitution, age, spatial arrangement, or condition to be distinguishable from adjacent communities.

Starch a molecule composed of long chains of α-glucose molecule linked together (repeating unit $C_{12}H_{16}O_5$); these linkages occur in chains of α-1,4 linkages with branches formed as a result of α-1,6 linkages; widely distributed in the vegetable kingdom and is stored in all grains and tubers (swollen underground plant stems): this polymer is highly amorphous, making it more readily attacked by human and animal enzyme systems and broken down into glucose; gross heat of combustion: Q_v(gross) = 7560 Btu/lb.

Starch crops crops in which most of the six-carbon sugar units are linked together in long, branched chains (starch). Yeast cannot use these chains to produce ethanol. The starch chains must be broken down into individual six-carbon units or groups of two units. The starch conversion process, described in the previous chapter, is relatively simple because the bonds in the starch chain can be broken in an inexpensive manner by the use of heat and enzymes, or by a mild acid solution.

Steam turbine a device for converting energy of high-pressure steam (produced in a boiler) into mechanical power, which can then be used to generate electricity.

Straight vegetable oil (SVO) any vegetable oil that has not been optimized through the process of transesterification; using this type of vegetable oil in a diesel engine requires an engine modification that heats the oil before it reaches the fuel injectors.

Sugar beet a plant, the root of which contains a high concentration of sucrose, which is grown commercially for sugar production; grown in many areas of the United States; must be rotated with nonroot crops (1 beet crop per 4-year period is the general rule). While beet by-products cannot provide fuel for the distillery, the beet pulp and tops are excellent feed in wet or dry form. Or the tops may be left on the field for fertilizer and erosion control.

Sugarcane (simply cane) several species of tall perennial true grasses of the genus *Saccharum*, tribe Andropogoneae, used for sugar production. The plant is 6- to 20–ft. tall with stout, jointed, fibrous stalks that are rich in sucrose; belongs to the grass family Poaceae, an economically important seed plant family that includes maize, wheat, rice, and sorghum, and many forage crops. It is native to the warm temperate to tropical regions of South Asia, Southeast Asia, New Guinea, and South America. Only a minority of states cultivate sugarcane, but there are hybrids (such as *Saccharum spontaneum*), which can be grown further north. High yields per acre of both sugar and crop residue are strong points of sugarcane production. The crop residue, called bagasse, is used in Brazil to provide heat for the distilleries.

Suspension a dispersion of a solid in a gas, liquid, or solid.

Superheated steam steam that is hotter than boiling temperature for a given pressure.

Sustainable an ecosystem condition in which biodiversity, renewability, and resource productivity are maintained over time.

SVO (straight vegetable oil) oil that burns well in many diesel engines but does not start the engine and will coke in the injectors as a hot engine cools; a separate tank of petrodiesel or biodiesel is often used during starting and stopping engine, and an electric valve allows transfer to the straight vegetable oil tank.

Sweet sorghum the name given to varieties of a species of sorghum: *Sorghum bicolor*. This crop has been cultivated on a small scale in the past for production of table syrup, but other varieties can be grown for production of sugar. The most common types of sorghum species are those used for production of grain.

Switchgrass prairie grass native to the United States and known for its hardiness and rapid growth, often cited as a potentially abundant feedstock for ethanol.

Syngas a mixture of carbon monoxide and hydrogen; see synthesis gas.

Synthesis gas a mixture of carbon monoxide (CO) and hydrogen (H_2), which is the product of high-temperature gasification of organic material such as biomass; after cleanup to remove any impurities such as tars, syngas) can be used to synthesize organic molecules such as synthetic natural gas [SNG, methane (CH_4)] or liquid biofuels such as gasoline and diesel fuel via the Fischer–Tropsch process.

Synthetic crude oil (syncrude) a hydrocarbon product produced by the conversion of coal, oil shale, or tar sand bitumen that resembles conventional crude oil; can be refined in a petroleum refinery.

Synthetic ethanol ethanol produced from ethylene, a petroleum by-product.

Tallow produced by meat rendering, which evaporates the moisture and enables the fat, known (the tallow) to be separated from the high-protein solids (the greaves), which are pressed, centrifuged, or subjected to a process of solvent extraction to remove more tallow, before being ground into meat and bone meal (MBM); pure tallow is a creamy-white substance.

Tar the volatile, brown-to-black, oily, viscous product from the destructive distillation of many bituminous or other organic materials, especially coal; a name used for petroleum in ancient texts.

Tar sand (bituminous sand) a formation in which the bituminous material (bitumen) is found as a filling in veins and fissures in fractured rocks or impregnating relatively shallow sand, sandstone, and limestone strata; a sandstone reservoir that is impregnated with a heavy, extremely viscous, black hydrocarbonaceous, petroleum-like material that cannot be retrieved through a well by conventional or enhanced oil recovery techniques; (FE 76-4): the several rock types that contain an extremely viscous hydrocarbon that is not recoverable in its natural state by conventional oil well production methods, including currently used enhanced recovery techniques.

Terpenes naturally occurring molecules that may be classified by the number of isoprene units in the molecule; a prefix in the name indicates the number of terpene units needed to assemble the molecule. Thus:

Hemiterpene derivatives consist of a single isoprene unit. Isoprene itself is considered the only hemiterpene.

Monoterpene derivatives consist of two isoprene units and have the molecular formula $C_{10}H_{16}$.

Sesquiterpene derivatives consist of three isoprene units and have the molecular formula $C_{15}H_{24}$.

Diterpene derivatives are composed of four isoprene units and have the molecular formula $C_{20}H_{32}$.

Triterpene derivatives consist of six isoprene units and have the molecular formula $C_{30}H_{48}$.

Tetraterpene derivatives contain eight isoprene units and have the molecular formula $C_{40}H_{64}$.

Polyterpene derivatives consist of long chains of many isoprene units—natural rubber consists of polyisoprene in which the double bonds are in the *cis*-configuration (the configuration in which two groups are on the same side of an imaginary reference line on the molecule).

Thermal conversion a process that uses heat and pressure to break apart the molecular structure of organic solids.

Thermal treatment the treatment of waste using elevated temperatures as the primary means to change the chemical, physical, or biological character or composition of the waste. Examples of thermal treatment processes are gasification, incineration, and pyrolysis.

Thermochemical conversion the use of heat to chemically change substances from one state to another, such as to produce useful energy products.

Thermochemical platform typically uses a combination of pyrolysis, gasification, and catalysis to transform wood into syngas—the gaseous constituents of wood—and then into fuels or chemicals.

Timberland forest land that is producing or is capable of producing crops of industrial wood, and that is not withdrawn from timber utilization by statute or administrative regulation.

Tipping fee a fee for disposal of waste.

Titration applied to biodiesel, titration is the act of determining the acidity of a sample of waste vegetable oil by the dropwise addition of a known base to the sample while testing with pH paper for the desired neutral reading (pH = 7); the amount of base needed to neutralize an amount of waste vegetable oil determines how much base to add to the entire batch.

Ton (short ton) 2000 lb.

Tonne (imperial ton, long ton, shipping ton) 2240 lb; equivalent to 1000 kg or in crude oil terms approximately 7.5 bbl of oil.

Topping cycle a cogeneration system in which electric power is produced first. The reject heat from power production is then used to produce useful process heat.

Topping and back pressure turbines turbines that operate at exhaust pressure considerably higher than atmospheric (noncondensing turbines); often multistage with relatively high efficiency.

Transesterification the chemical process in which an alcohol reacts with the triglycerides in vegetable oil or animal fats, separating the glycerin and producing biodiesel.

Traveling grate furnace a type of furnace in which assembled links of grates are joined together in a perpetual belt arrangement. Fuel is fed in at one end and ash is discharged at the other.

Trillion 1×10^{12}.

Turbine a machine for converting the heat energy in steam or high-temperature gas into mechanical energy. In a turbine a high-velocity flow of steam or gas passes through successive rows of radial blades fastened to a central shaft.

Turndown ratio the lowest load at which a boiler will operate efficiently as compared to the boiler's maximum design load.

Ultimate analysis the determination of the elemental composition of the organic portion of carbonaceous materials, as well as the total ash and moisture; a description of a fuel's elemental composition as a percentage of the dry fuel weight as determined by prescribed methods; see ASTM International.

Ultralow sulfur diesel (ULSD) ultralow sulfur diesel describes a new EPA standard for the sulfur content in diesel fuel sold in the United States beginning in 2006—the allowable sulfur content (15 ppm) is much lower than the previous US standard (500 ppm), which not only reduces emissions of sulfur compounds (blamed for acid rain) but also allows advanced emission control systems to be fitted that would otherwise be poisoned by these compounds.

Uronic acid a simple sugar, the terminal $-CH_2OH$ group of which has been oxidized to an acid, COOH group; uronic acids occur as branching groups bonded to hemicelluloses such as xylan.

Vacuum distillation a secondary distillation process that uses a partial vacuum to lower the boiling point of residues from primary distillation and extract further blending components.

Viscosity a measure of the ability of a liquid to flow or a measure of its resistance to flow; the force required to move a plane surface of area 1 square meter over another parallel plane surface 1 m away at a rate of 1 m per second when both surfaces are immersed in the fluid; the higher the viscosity, the slower the liquid flows.

VOCs see volatile organic compounds.

Volatile Organic Compounds (VOCs) name given to light organic hydrocarbon derivatives, which escape as vapor from fuel tanks or other sources, and during the filling of tanks. VOCs contribute to smog.

Volatility propensity of a fuel to evaporate.

Waste streams unused solid or liquid by-products of a process.

Waste vegetable oil (WVO) grease from the nearest fryer, which is filtered and used in modified diesel engines or converted to biodiesel through the process of transesterification and used in any diesel-fueled car.

Watershed the drainage basin contributing water, organic matter, dissolved nutrients, and sediments to a stream or lake.

Watt the common base unit of power in the metric system; 1 W equals 1 J per second, or the power developed in a circuit by a current of 1 A flowing through a potential difference of one volt. 1 W = 3.412 Btu/h.

Wet mill an ethanol production facility in which the corn is first soaked in water before processing; in addition to ethanol, wet mills have the ability to produce coproducts such as industrial starch, food starch, high-fructose corn syrup, gluten feed, and corn oils.

Wheeling the process of transferring electrical energy between buyers and sellers by way of an intermediate utility or utilities.

Whole-tree chips wood chips produced by chipping whole trees, usually in the forest, which contain both bark and wood; frequently produced from the low-quality trees or from tops, limbs, and other logging residues.

Whole-tree harvesting a harvesting method in which the whole tree (above the stump) is removed.

Wood a solid lignocellulosic material naturally produced in trees and some shrubs, made of up to 40%–50% cellulose, 20%–30% hemicellulose, and 20%–30% lignin.

Wood alcohol see methanol.

Wort an oatmeal-like substance consisting of water and mash barley in which soluble starch has been turned into fermentable sugar during the mashing process—the liquid remaining from a brewing mash preparation following the filtration of fermentable beer.

Xylan a polymer of xylose with a repeating unit of $C_5H_8O_4$, found in the hemicellulose fraction of biomass, can be hydrolyzed to xylose.

Xylose a five-carbon sugar $C_5H_{10}O_5$; a product of hydrolysis of xylan found in the hemicellulose fraction of biomass.

Yarding the initial movement of logs from the point of felling to a central loading area or landing.

Yeast any of various single-cell fungi capable of fermenting carbohydrates; bioethanol is produced by fermenting sugars with yeast.

Yellow grease a term from the rendering industry which usually means used frying oils from deep fryers and restaurants' grease traps; can also refer to lower quality grades of tallow from rendering plants.

Zeolite a crystalline aluminosilicate mineral that is used as a catalyst and having a particular chemical and physical structure.

Index

Note: Page number followed by "*f*" and "*t*" refer to figures and tables, respectively.

E

F

V

W

X

Z

Made in the USA
Middletown, DE
28 March 2024

52182347R00364